# Advanced Power Generation Systems

*Advanced Power Generation Systems: Thermal Sources* evaluates advances made in heat-to-power technologies for conventional combustion heat and nuclear heat, along with natural sources of geothermal, solar, and waste heat generated from the use of different sources. These advances will render the landscape of power generation significantly different in just a few decades. This book covers the commercial viability of advanced technologies and identifies where more work needs to be done. Since power is the future of energy, these technologies will remain sustainable over a long period of time.

**Key Features**

- Covers power generation and heat engines
- Details photovoltaics, thermo-photovoltaics, and thermoelectricity
- Includes discussion of nuclear and renewable energy as well as waste heat

This book will be useful for advanced students, researchers, and professionals interested in power generation and energy industries.

# Sustainable Energy Strategies

*Series Editor: Yatish T. Shah*

**Other related books by Yatish T. Shah**

For more information on this series, please visit: https://www.routledge.com/ Sustainable-Energy-Strategies/book-series/CRCSES

# Advanced Power Generation Systems

## Thermal Sources

Yatish T. Shah

**CRC Press**
Taylor & Francis Group
Boca Raton  London  New York

CRC Press is an imprint of the
Taylor & Francis Group, an **informa** business

First edition published 2023
by CRC Press
6000 Broken Sound Parkway NW, Suite 300, Boca Raton, FL 33487-2742

and by CRC Press
4 Park Square, Milton Park, Abingdon, Oxon, OX14 4RN

*CRC Press is an imprint of Taylor & Francis Group, LLC*

© 2023 Yatish T. Shah

*Library of Congress Cataloging-in-Publication Data*
Names: Shah, Yatish T., author.
Title: Advanced power generation systems : thermal sources / Yatish T. Shah.
Description: First edition. | Boca Raton : CRC Press, 2023. |
Series: Sustainable energy strategies | Includes bibliographical references and index.
Identifiers: LCCN 2022026714 (print) | LCCN 2022026715 (ebook) |
ISBN 9781032350110 (hbk) | ISBN 9781032356983 (pbk) | ISBN 9781003328087 (ebk)
Subjects: LCSH: Electric power production. | Heat engineering. | Electric power-plants.
Classification: LCC TK1041 .S448 2023 (print) | LCC TK1041 (ebook) |
DDC 621.31/21—dc23/eng/20221017
LC record available at https://lccn.loc.gov/2022026714
LC ebook record available at https://lccn.loc.gov/2022026715

ISBN: 9781032350110 (hbk)
ISBN: 9781032356983 (pbk)
ISBN: 9781003328087 (ebk)

DOI: 10.1201/9781003328087

Typeset in Times
by codeMantra

*This book is dedicated to my wife Mary.*

# Contents

# Sustainable Energy Strategies Series Preface

While fossil fuels (coal, oil, and gas) were the dominant sources of energy during the last century, since the beginning of the twenty-first century an exclusive dependence on fossil fuels is believed to be a non-sustainable strategy due to (a) their environmental impacts, (b) their nonrenewable nature, and (c) their dependence on the local politics of the major providers. The world has also recognized that there are in fact ten sources of energy: coal, oil, gas, biomass, waste, nuclear, solar, geothermal, wind, and water. These can generate our required chemical/biological, mechanical, electrical, and thermal energy needs. A new paradigm has been to explore greater roles of renewable and nuclear energy in the energy mix to make energy supply more sustainable and environmentally friendly. The adopted strategy has been to replace fossil energy with renewable and nuclear energy as rapidly as possible. While fossil energy still remains dominant in the energy mix, by itself, it cannot be a sustainable source of energy for the long future.

Along with exploring all ten sources of energy, sustainable energy strategies must consider five parameters: (a) availability of raw materials and accessibility of product market, (b) safety and environmental protection associated with the energy system, (c) technical viability of the energy system on the commercial scale, (d) affordable economics, and (e) market potential of a given energy option in the changing global environment. There are numerous examples substantiating the importance of each of these parameters for energy sustainability. For example, biomass or waste may not be easily available for a large-scale power system making a very large-scale biomass/waste power system (like a coal or natural gas power plant) unsustainable. Similarly, an electrical grid to transfer power to a remote area or onshore needs from a remote offshore operation may not be possible. Concerns of safety and environmental protection (due to emissions of carbon dioxide) limit the use of nuclear and coal-driven power plants. Many energy systems can be successful at laboratory or pilot scales, but may not be workable at commercial scales. Hydrogen production using a thermochemical cycle is one example. Many energy systems are as yet economically prohibitive. The devices to generate electricity from heat such as thermoelectric and thermophotovoltaic systems are still very expensive for commercial use. Large-scale solar and wind energy systems require huge upfront capital investments, which may not be possible in some parts of the world. Finally, energy systems cannot be viable without the market potential for the product. Gasoline production systems were not viable until the internal combustion engine for the automobile was invented. Power generation from wind or solar energy requires guaranteed markets for electricity. Thus, these five parameters collectively form a framework for sustainable energy strategies.

It should also be noted that the sustainability of a given energy system can change with time. For example, coal-fueled power plants became unsustainable due to their impact on the environment. These power plants are now being replaced by gas-driven power plants. New technology and new market forces can also change the sustainability of the energy system. For example, successful commercial developments of fuel cells and electric cars can make the use of internal combustion engines redundant in the vehicle industry. While an energy system can become unsustainable due to changes in parameters, outlined above, over time, it can regain sustainability by adopting strategies to address the changes in these five parameters. New energy systems must consider long-term sustainability with changing world dynamics and possibilities of new energy options.

Sustainable energy strategies must also consider the location of the energy system. On one hand, fossil and nuclear energy are high-density energies and they are best suited for centralized operations in an urban area, while on the other hand, renewable energies are of low density and they are well-suited for distributed operations in rural and remote areas. Solar energy may be less affordable in locations far away from the equator. Offshore wind energy may not be sustainable if the distance from shore is too great for energy transport. Sustainable strategies for one country may be quite different from another depending on their resource (raw material) availability and local market potential. The current transformation from fossil energy to green energy is often prohibited by the required infrastructure and the total cost of transformation. Local politics and social acceptance also play an important role. Nuclear energy is more acceptable in France than in any other country.

Sustainable energy strategies can also be size-dependent. Biomass and waste can serve local communities well on a smaller scale. As mentioned before, large-scale plants can be unsustainable because of limitations on raw materials. New energy devices that operate well at micro- and nanoscales may not be possible on a large scale. In recent years, nanotechnology has significantly affected the energy industry. New developments in nanotechnology should also be a part of sustainable energy strategies. While larger nuclear plants are considered to be the most cost-effective for power generation in an urban environment, smaller modular nuclear reactors can be the more sustainable choice for distributed cogeneration processes. Recent advances in thermoelectric generators due to advances in nanomaterials are an example of a size-dependent sustainable energy strategy. A modular approach for energy systems is more sustainable at a smaller scale than for a very large scale. In general, a modular approach is not considered as a sustainable strategy for a very large, centralized energy system.

Finally, choosing a sustainable energy system is a game of options. New options are created by either improving the existing system or creating an innovative option through new ideas and their commercial development. For example, a coal-driven power plant can be made more sustainable by using very cost-effective carbon capture technologies. Since sustainability is time, location and size-dependent, sustainable strategies should follow local needs and markets. In short, sustainable energy strategies must consider all ten sources and a framework of five stated parameters under which they can be made workable for local conditions. A revolution in technology (like nuclear fusion) can, however, have global and local impacts on sustainable energy strategies.

The CRC Press Series on Sustainable Energy Strategies will focus on novel ideas that will promote different energy sources sustainable for the long term within the framework of the five parameters outlined above. Strategies can include both improvement in existing technologies and the development of new technologies.

**Yatish T. Shah,**
**Series Editor**

# Preface

We are moving into the electric world. Some will say that future of energy is in electricity and hydrogen. Electricity is at the heart of modern economies and it is providing a rising share of energy services. Recent advances in technology are largely dependent on power and electricity. Demand for electricity is set to increase further as a result of rising household incomes, with the electrification of transport and heat, and growing demand for digitally connected devices and air conditioning. From a long-term sustainability point of view, this makes sense because both electricity and hydrogen are most acceptable for preserving the environment and reducing global warming and their resulting effects on climate change. The usages of electricity and hydrogen are also clean and non-threatening. If these sources can be generated in safe, economical and environmentally acceptable ways, we would have a formidable and sustainable path for future energy needs.

The demand for electricity (or power) is highly distributed. The power needs for industry scale, commercial scale and residential scale are currently supplied by utility scale centralized and large-scale power generation systems. The power need for the transportation system is supplied by small-scale and distributed power generation systems. Finally, at the micro level, power needs of personalized power electronics accessories require power generation at the micro level which sometimes are harvested from natural sources. New computer, medical, electronic and environmental technologies are becoming smaller and faster requiring power from immediate and micro sources. Thus, going forward power needs will be multi-scale and multi-dimensional.

Historically, there has been a limit to the number of things that can run on electricity: many industrial processes and transportation technologies rely on their own combustion engines to run. However, as technology improves, it is increasingly possible for this need for an on-site, fossil-fuel-powered combustion engine to be replaced with electricity by a process known as electrification. The car is a perfect example of this shift. Until very recently, all cars were powered by their own gas-fired pistons: the internal combustion engine (ICE). The ICE is effectively a miniature, on-site, gasoline-fired power plant for cars, that produce energy when the car needs it. With improvements in energy storage technologies, electric vehicles have become viable alternatives to gasoline-powered vehicles and are becoming increasingly popular throughout the country and the world. As more and more people switch from an older ICE car to a newer electric vehicle, electrifying their own transportation, the need for electricity will continue to increase throughout the world.

Another process that's witnessed a similar shift towards electrification with improved technology is home heating. Previously the heating was the domain of natural gas, oil, propane, diesel or even coal. Innovative new and improved technologies, such as *air source heat pumps* and *geothermal heat pumps*, now mean that the energy required to heat your home can come from electricity rather than from burning fuel. Certainly, there are still limits to what can be powered by electricity. For instance, the cost, power requirement and, most importantly, weight of batteries means it is not yet possible to power a commercial airplane exclusively or primarily with electricity. Yet, technology is shifting where those limits lie. Recent technological innovations have the potential to electrify more processes that previously required their own combustion engines. A big component of this could be the new technology pioneered by Heliogen, which, if scalable, promises to run highly energy-intensive processes, like producing steel, with renewable solar energy.

Though the difference between creating energy by burning fossil fuels on-site versus powering things with electricity may seem nuanced, electrification has many ramifications. First, electrification will shift our overall energy consumption habits from a need for fuels to burn locally–gasoline, oil, natural gas and others–to a need for electricity. This means that our nation's demand for fossil fuels will decrease, while our overall demand for electricity will increase. Second, electrification, if done well, can lead to a major decrease in national carbon emissions. There's a limit to how much emissions you can reduce and offset if the only option is "fuel switching" from one form of fossil fuel to another, slightly cleaner one (like from diesel to natural gas). However, if the country switches from burning fossil fuels to running processes on clean electricity generated by renewable energy facilities, we can drive much further carbon reductions across all sectors of the economy. Finally, electrification means individual homes and businesses will have energy freedom and independence. Instead of relying upon fossil fuels that are extracted somewhere in the world, which you have to purchase at volatile, ever-changing rates, by installing solar, you can now produce your own electricity to power your electrified processes.

There is a revolution going on the way we generate power on a large scale. During the last century, the majority of power was generated by fossil fuels and the concepts of internal combustion engine and thermodynamic cycles. Coal, oil and gas were the major sources of fuels. Heat was largely generated by the processes of combustion and nuclear fission. Due to concerns over nuclear safety, the largest source of power was the combustion heat generated from fossil fuels. Over the last several decades there was a realization that while the combustion process was convenient, it was inefficient and detrimental to the environment. The concerns over carbon emission and resulting global warming and its impact on climate change took the center stage. The push was made that future energy needs must be satisfied by protecting the environment. The use of fossil fuel must be reduced and replaced by a greater use of renewable fuels like biomass (and waste), geothermal, solar, wind and hydro energy. If possible, the stringent restrictions on nuclear energy should be relaxed and a new trend has been toward small modular nuclear reactor (SMR). Besides their harmful environmental effect, it also became evident that fossil fuels are non-renewable and have finite lifetime as the source of supply. An indirect conclusion of all of this was that the world should not rely on combustion and internal combustion engine for its power supply. We need other sustainable alternatives for power generation.

While there are ten sources of power generation namely, coal, oil, gas (including hydrogen), nuclear, biomass, waste (mass and heat), geothermal, solar, wind and hydro, they provide power through four fundamental processes; thermal, mechanical (kinetic and potential), electrochemical (and chemical) and hybrid. Combustion, fission, geothermal energy, solar energy along with waste heat are examples of thermal sources, wind and hydro energy are examples of mechanical energy and fuel cell is an example of the electrochemical process to convert chemical energy into electrical energy. Hybrid sources and processes for power generation are numerous and probably the most versatile in their applications on the range, scale and dimensions. The power requirements for personal and self-powered electronics will find the use of hybrid sources most enlightening.

While there are multiple methods for power generation, the present book is focused on heat-to-power technologies. While wind, hydro, fuel cell and hybrid sources of power generation are the newest forms (and they will be the subject of the next book), the conversion of heat to power has been in existence since the beginning of the power industry. The use of the internal combustion engine for power generation has been in existence ever since electricity started taking over our lives. In recent years heat to power industry is going through rapid changes. These changes are two-fold. First, the conventional combustion heat to power industry is making significant efforts to improve its performance by introducing several novel indirect and direct methods for power generation. These include advanced thermodynamic cycles, waste heat conversion and the development of several direct heat to power conversion technologies like photovoltaic, thermophotovoltaic and thermoelectricity technologies. Significant efforts are also being made to reduce carbon emission from combustion heat to power industry. These include the use of different fuels such as biomass, coal/biomass, biomass/gas, etc. and the introduction of various carbon capture and utilization technologies. Alongside, a new generation of nuclear heat to power technologies is being developed to make nuclear power more efficient and safe. Nuclear power industries are also exploring small modular nuclear reactors to make nuclear power more attractive and convenient to developing nations. SMR are also attractive to developing nations because of their flexibility and possible faster installations.

Second, major changes in the power generation from heat that occurred during the past several decades are in the use of two natural sources of heat; geothermal and solar. The developments of heat to power technologies for these two sources are nothing short of mind-boggling. Not only these technologies are environmentally friendly, but they are also becoming more and more economically competitive with the conventional thermal sources of power generation. Both geothermal and solar energy are accessible to most parts of the world and they can be harnessed to generate power both at small and large scales. New advances in technology are making both sources; particularly solar, more attractive globally.

The present book evaluates in detail advances made in heat to power technologies for conventional combustion heat and nuclear heat along with natural sources of geothermal and solar heat and waste heat generated from the use of all sources. For sure, these advances will make the landscape of power generation significantly different a few decades from now than what it is at present. The book goes in detail about the commercial viability of advanced technologies and identifies where more work needs to be done. Since power is the future of energy, these technologies will also remain sustainable over a long period of time.

The book should be useful for all students, researchers and industries interested in the power generation and energy industries in general.

# Author

**Yatish T. Shah** received his BSc in chemical engineering from the University of Michigan, Ann Arbor, USA, and MS and ScD in chemical engineering from the Massachusetts Institute of Technology, Cambridge, USA. He has more than 40 years of academic and industrial experience in energy-related areas. He was chairman of the Department of Chemical and Petroleum Engineering at the University of Pittsburgh, Pennsylvania, USA; dean of the College of Engineering at the University of Tulsa, Oklahoma, USA, and Drexel University, Philadelphia, Pennsylvania, USA; chief research officer at Clemson University, South Carolina, USA; and provost at Missouri University of Science and Technology, Rolla, USA, the University of Central Missouri, Warrensburg, USA, and Norfolk State University, Virginia, USA. He was also a visiting scholar at the University of Cambridge, UK, and a visiting professor at the University of California, Berkley, USA, and Institut für Technische Chemie I der Universität Erlangen, Nürnberg, Germany. He has previously written 13 books related to energy, seven of which are under the *Sustainable Energy Strategies* book series (by Taylor & Francis) of which he is the editor. This book is another addition to this series. He has also published more than 250 refereed reviews, book chapters, and research technical publications in the areas of energy, environment, and reaction engineering. He is an active consultant to numerous industries and government organizations in energy areas.

# 1 Introduction

## 1.1 WHY ELECTRICITY IS THE FUTURE?

We are moving into the electric world. Some will say that future of energy is in electricity and hydrogen. Electricity is at the heart of modern economies, and it is providing a rising share of energy services. Recent advances in technology are largely dependent on power and electricity. Demand for electricity is set to increase further as a result of rising household incomes, with the electrification of transport and heat, and growing demand for digitally connected devices and air conditioning. From a long-term sustainability point of view, this makes sense because both electricity and hydrogen are most acceptable for preserving the environment and reducing global warming and its resulting effects on climate change. The usages of electricity and hydrogen are also clean and non-threatening. If these sources can be generated in safe, economical and environmentally acceptable ways, we would have a formidable and sustainable path for future energy needs.

There are three requirements for the power (electricity) usage; generation, storage and transport. It is the generation part that causes major concerns for the environment. In the past, generation was mainly carried out using fossil fuels and thermodynamic cycles. The processes used were highly inefficient and created major greenhouse gas emissions. When we are moving toward electrical future, it is important that we generate electricity with as little impact on the environment as possible. Rising electricity demand was one of the key reasons why global $CO_2$ emissions from the power sector reached a record high in 2018, yet the commercial availability of a diverse suite of low emissions generation technologies also puts electricity at the vanguard of efforts to combat climate change and pollution. Decarbonized electricity, in addition, could provide a platform for reducing $CO_2$ emissions in other sectors through electricity-based fuels such as hydrogen or synthetic liquid fuels. Renewable energy also has a major role to play in providing access to electricity for all. Growth of fuel cell for power generation provides a win-win solution for power generation and environment protection. If nuclear energy gets more social acceptance, it could play a significant role in providing environment-friendly electricity. The future dictates more use of renewable and nuclear energy. As regard to the fuel, hydrogen would be the most preferred choice because we need to prevent the emission of carbon in the environment. Advanced power generation systems need to be environment friendly, economical and durable.

Historically, many industrial processes and transportation technologies have relied on their own combustion engines to run. However, as power generation technology improves, it is increasingly possible for this need for an on-site, fossil-fuel-powered combustion engine to be replaced with electricity. The car is a perfect example of this shift. Until very recently, all cars were powered by their own internal combustion engine (ICE). The ICE is effectively a miniature, on-site, gasoline-fired power plant for cars, that produce energy when the car needs it. With

DOI: 10.1201/9781003328087-1

advancements in hybrid power trains and improvements in energy storage technologies, electric vehicles have become viable alternatives to gasoline-powered vehicles, and both hybrid and electric vehicles are becoming increasingly popular throughout the country and world. As more and more people switch from an older ICE car to a newer hybrid or electric vehicle, the need for electricity will continue to increase throughout the world. Besides automobiles, the electrification of heavy vehicles is also rapidly expanding. For example, In the UK, train networks and even potentially planes could switch from fossil fuels to electric, which can result in an increasing demand for electricity from the transport sector by 128% between 2015 and 2035. Electric vehicles (EVs) alone are predicted to add 25 terawatt hours (TWh) of electricity demand by 2035, according to a report by Bloomberg New Energy Finance. However, this will be dependent on significant investment in the necessary charging infrastructure to enable a decarbonized transport network across the country.

Another process that has witnessed a similar shift toward electrification with improved technology is home heating. Previously the home heating was the domain of natural gas, oil, propane, diesel or even coal. Innovative new and improved technologies, such as air source heat pumps and geothermal heat pumps, now mean that the energy required to heat your home can come from electricity rather than from burning fuel. For example, currently in the UK about 7% of homes are heated by electricity. With the help of government investment and incentives, this number can significantly grow. An electrified heat network, however, will add a greater strain on the electricity system, particularly in winter months when demand is high. This will require simultaneous investment in the improvement of electrical infrastructure. With the electrification of home heating, the forecast is the increase of 40 GW in demand during peak times – the mornings and the evenings.

The demand for electricity is also increasing by a rapid expansion of self-powered portable electronics and new computer-driven accessories. By 2035, chip manufacturer ARM predicts there will be more than one trillion Internet of Things (IoT) devices globally. This smart technology will be able to turn everything from the morning coffee maker to the bed into intelligent machines, gathering masses of data that can be used to optimize and personalize daily life. Powering all these devices, let alone the vast plains of servers holding all the data they gather, is one of the great challenges for the IoT industry. However, as much as smart devices will demand energy (electricity), they will also help save it.

Currently, there are limits to what can be powered by electricity. For instance, the cost, power requirement and, most importantly, weight of batteries means it is not yet possible to power a commercial airplane exclusively or primarily with electricity. Yet, technology is shifting where those limits lie. With improvements in battery and other energy storage technologies many new things are possible. Recent technological innovations have the potential to electrify more processes that previously required their own combustion engines. A big component of this could be the new technology pioneered by Heliogen [1], which, if scalable, promises to run highly energy-intensive processes, like producing steel and other manufacturing processes with renewable solar energy.

The difference between creating energy by burning fossil fuels on-site versus powering things with electricity has many ramifications. First, electrification will shift our overall energy consumption habits from a need for fuels to burn locally–gasoline, oil, natural gas and others–to a need for electricity. This means that our nation's demand for fossil fuels will decrease, while our overall demand for electricity will increase. Second, electrification, if done well, can lead to a major decrease in national carbon emissions. There's a limit to how much emissions you can reduce and offset if the only option is "fuel switching" from one form of fossil fuel to another, slightly cleaner one (like from diesel to natural gas). However, if the country switches from burning fossil fuels to running processes on clean electricity generated by renewable energy facilities, we can drive much further carbon reductions across all sectors of the economy. This can be even further accelerated if fuel is completely switched to hydrogen and hydrogen-driven fuel cell is more successfully used for both small and large scales commercial applications. Finally, electrification will provide more energy freedom and independence to individual homes and businesses. Instead of relying upon fossil fuels that are extracted somewhere in the world, which you have to purchase at volatile, ever-changing rates; the use of renewable source of electricity and heating will allow one to be in control of one's own need for electricity. Furthermore, one can use self-electricity generation from renewable sources like solar, wind and geothermal sources to generate revenue by selling excess energy to the grid. There are more management and optimization possibilities in the use of electricity than in the use of fuels.

Global electricity consumption continues to increase faster than the world population, leading to an increase in the average amount of electricity consumed per person (per capita electricity consumption), according to the U.S. Energy Information Administration's (EIA) International Energy Statistics. Electricity is used most commonly in buildings for lighting and appliances, in industrial processes for producing goods, and in transportation for powering rail and light-duty vehicles. Nearly all of the increase is attributable to growing electricity consumption in developing countries outside the Organization of Economic Cooperation and Development (OECD). Increases in per capita electricity consumption reflect possible changes in the composition of the economy, such as shifts to more energy-intensive industries, and changes in service demand, such as growing demand for air conditioning and appliances. The increase in consumption from these factors is partially offset by efficiency measures, such as more efficient lighting. Regionally, per capita electricity consumption in a number of countries has been affected by outsourcing energy-intensive industries to other countries. In the United States, total electricity consumption has risen slightly since the early 2000s, but electricity consumption per person decreased by nearly 7% between 2000 and 2017 because of improvements in energy efficiency and changes in the economy that have resulted in less electricity use per unit of economic output (as measured by gross domestic product, or GDP). Growth in global electricity consumption is related to economic growth, but the relationship differs, depending on the country. Per person economic growth can occur independently of growth in per person electricity usage in countries with large, developed economies; largely satisfied residential electricity demand, and relatively smaller portions of economic growth coming from industrial production. Producing a service with greater economic value does not necessarily require any more electricity than a lower-value service. Electricity demand follows two distinct regional paths. In advanced economies, future growth linked to increasing digitalization and electrification is largely offset by energy efficiency improvements. In developing economies, rising incomes, expanding industrial output and a growing services sector push demand firmly up. Developing economies contribute nearly 90% of global electricity demand growth to 2040 in the Stated Policies Scenario, but demand per person in these economies remains 60% lower than in advanced economies [2–4].

In countries with rapidly growing residential electricity consumption and growing energy-intensive activities, electricity use tends to more closely correspond to growth in economic activity. Per capita electricity growth in the economies of less developed non-OECD countries has more than doubled between 2000 and 2017, compared with a nearly flat trend in the economies of more developed OECD countries. At the national level, average per capita electricity consumption values can mask the large variation within a country. For example, the United States population averaged nearly 12,000 (kWh) of electricity consumption per person in 2017, but on a state basis, annual per capita electricity consumption ranged from more than 25,000 kWh in states such as Wyoming and North Dakota to less than 7,000 kWh in states such as Hawaii and California [2–4].

Electricity is an essential part of modern life and important to the U.S. economy. People use electricity for lighting, heating, cooling, and refrigeration and for operating appliances, computers, electronics, machinery, and public transportation systems. Total U.S. electricity consumption in 2020 was about 3.8 trillion kWh and 13 times greater than electricity use in 1950. Total electricity consumption includes retail sales of electricity to consumers and *direct use* of electricity. Direct use of electricity is both produced by and used by the consumer. The industrial sector accounts for the majority of direct use of electricity. In 2020, retail sales of electricity were about 3.66 trillion kWh, equal to 96% of total electricity consumption. Direct use of electricity by all end-use sectors was about 0.14 trillion kWh, or about 4% of total electricity consumption. Total annual U.S. electricity consumption increased in all but 11 years between 1950 and 2020, and 8 of the years with year-over-year decreases occurred after 2007. The highest level of total annual electricity consumption occurred in 2018 at about 4 trillion kWh, when a relatively warm summer and cold winter in most regions of the country contributed to the record-high residential electricity use of nearly 1.5 trillion kWh [2–4].

## 1.2  CHRONOLOGICAL EVOLUTION OF SOURCES FOR POWER GENERATION

The source for power generation has gone through its own cycle. The United States built many coal-fired power plants during the 1970s and 1980s to meet growing electricity demand. National energy policy, responding to concerns about global oil supplies in the 1970s and worries about scarcity of natural gas, favored coal as a domestic and reliable power source. During the two-decade "big buildup" of coal-fired power plants between 1967 and 1987, the United States added 202,416 MW—about two-thirds of the nation's total coal-generating capacity. More recently, new air quality emissions standards and increasing maintenance and replacement costs have led to the retirement of many aging coal-fired power plants. As a result, the nation's capacity to generate electricity from coal is declining. Coal-fired power plant retirements between 2012 and 2030 will reduce capacity by 115,410 MW, more than a third of the former peak capacity to generate electricity from coal (see Figure 1.1) [5].

Advancements in seismic imaging, horizontal drilling, and fracking techniques reversed concerns about scarcity in natural gas markets and vaulted the United States into the largest global producer of natural gas. The increased volume of production

Annual change in electricity generation capacity
Net change from additions and closures of generators

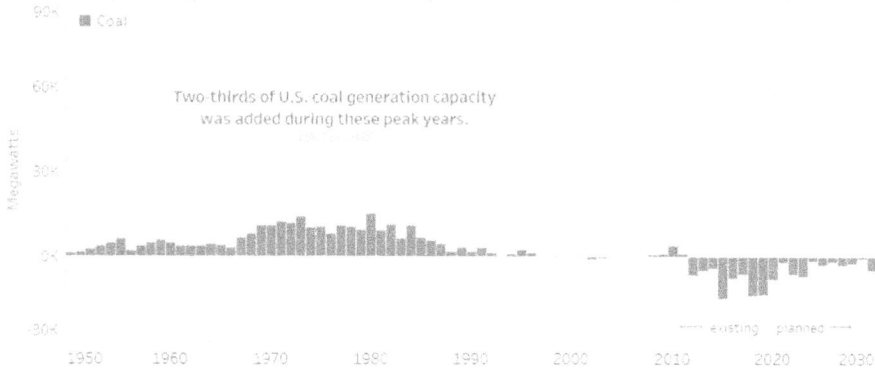

**FIGURE 1.1** Chronological evolution of coal power [5]- data generated under NSF EPSCor project. (Source: U.S.EIA, Univ. of Wyoming, Analysis by Resources and Communities Research Group at Montana State University and Headwaters Economics.)

led to persistently low natural gas prices and a building boom in new high-efficiency combined-cycle natural gas and steam power plants that hastened the retirement of many coal-fired power plants. In just 6 years between 2000 and 2005, 191,745 MW of natural gas capacity were added and natural gas replaced coal as the nation's base-load power that provides enough electricity to meet minimum level of demand (see Figure 1.2) [5]. Renewable energy costs are primarily determined by long-term power purchase agreements, and are less driven by fossil fuel prices (e.g., coal and natural gas). Today, solar and wind energy generators are being built faster than any other energy source. Between 2012 and 2020, natural gas added 35,302 MW of capacity while new wind, solar, and biomass generators totaled 891,383 MW of capacity (see Figure 1.3) [5]. Added renewable electricity capacity generated more electricity than coal for multiple days in April 2020—the first time ever that renewables surpassed coal generation on a daily basis. State regulations and incentives driven by climate policy and declining energy costs are influencing the rise of renewables as a growing component of energy market. The future of renewable sources, particularly, solar, wind and geothermal for power production is very bright.

While the shift from fossil fuels to renewable sources for electricity may appear to be gradual, it is real and it is imbedded in the long-range energy plans of various countries. According to Headwater economics and EIA reports [5], in recent years, the only rising curve for electricity generation is renewable sources. The increase in this slope and decline in all other sources will become more pronounced as the pressure on climate protection increases. It is predicted [1–7] that by 2050, at least half of the power may come from solar and wind sources. In this transformation, the U.S. is not alone. China is expected to install 230 GW of capacity in 2022, including 90 GW of solar, 50 GW of wind, 30 GW of thermal, 20 GW of hydro and 2.3 GW of nuclear, according to a forecast from the China Electricity Council (CEC). China added 180 GW of power capacity in 2021, including 120 GW of wind and solar capacity.

Annual change in electricity generation capacity
Net change from additions and closures of generators

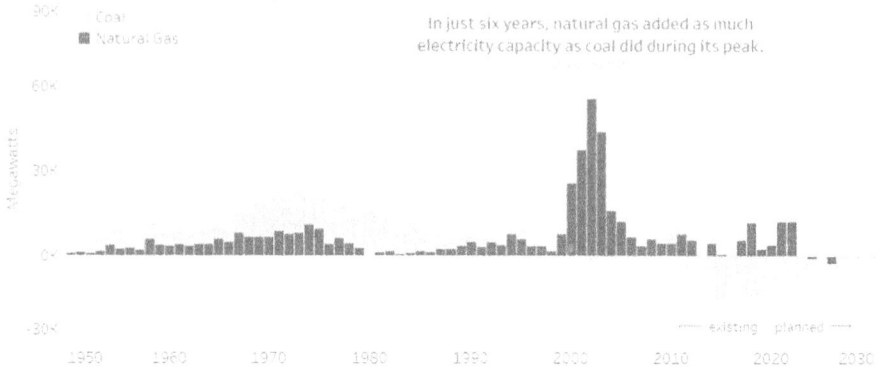

**FIGURE 1.2**  Chronological evolution of gas power [5]-data generated under NSF EPSCor project. (Source: U.S.EIA, Univ. of Wyoming, Analysis by Resources and Communities Research Group at Montana State University and Headwaters Economics.)

Annual change in electricity generation capacity
Net change from additions and closures of generators

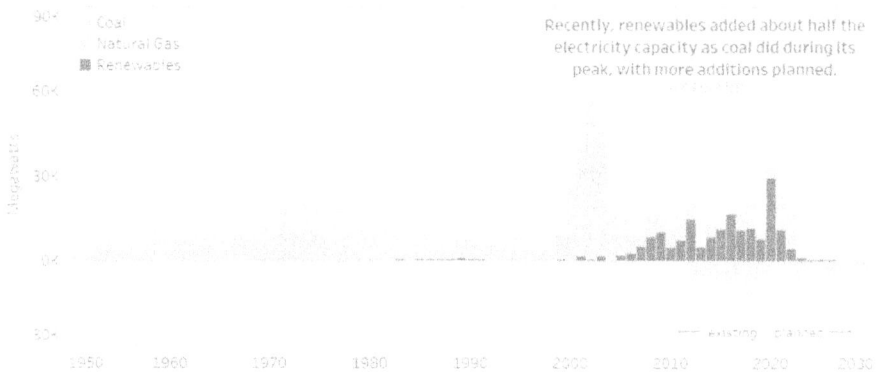

**FIGURE 1.3**  Chronological evolution of renewables power [5]-data generated under NSF EPSCor project. (Source: U.S.EIA, Univ. of Wyoming, Analysis by Resources and Communities Research Group at Montana State University and Headwaters Economics.)

Consequently, the country's non-fossil fuel capacity reached close to 1,120 GW by the end of 2021. The Brazilian Ministry of Mines and Energy has initiated a public consultation on the Ten-Year Energy Expansion Plan 2031 (PTE 2031), which foresees the investment of BRL528bn (US$97bn) in the electricity sector over the 2021–2031 period, including BRL292bn (US$54 bn) in centralized generation, BRL135bn (US$25bn) in distributed generation and BRL101bn (US$19bn) in transmission. The country's installed capacity is expected to increase by 38% from around 200 GW in 2021 to 275 GW in 2031, with 83% of renewables and 17% of non-renewables. The centralized capacity is forecast to increase by 23% to 220 GW in 2031, while the distributed capacity, entirely renewable, should increase five-fold to 37 GW. According to the country's Ministry of Energy and Mineral Resources (ESDM), Indonesia added 1,902 MW of power capacity, including 655 MW of renewables, and 3,821 km of transmission lines. In 2022, the country is expected to commission 2,950 MW of power capacity and to build 4,632 km of transmission lines. The electrification rate should reach 100% in 2022 and the electricity consumption should increase to 1,268 kWh/capita (up from 939 kWh/capita in 2020). Great Britain's last coal power station shut down in 2012, and across the country wind turbines are generating more electricity than fossil fuels and nuclear energy combined, pushing carbon emissions to a new low [1–7].

At its heart, the increasing electrification of our transport, utilities and technology has been driven by a few specific goals, one of which is lowering our reliance on fossil fuels and reducing carbon emissions. It's positive, then, that all projections toward 2035 have us making significant strides toward this vision. Coal will have been completely removed from the electricity system, while gas generation will drop to just 70.8 TWh – down from the 112.2 TWh expected in 2018, according to Bloomberg's forecast. In its place, wind will become the greatest source of electricity producing 138.5 TWh in 2035, up from 52.3 TWh in 2018. The watershed year for wind power will come in 2027 when the wind first overtakes gas to become the biggest contributor to the grid thanks to significant increases in capacity. Nuclear will still play an important role in the energy mix, contributing 45.8 TWh, while solar power market size will more than double during 2017-2030 period. The results of this continued move to lower carbon sources will be significant. Carbon emissions from electricity in 2035 are expected to be 23.81 $gCO_2/MJ$ (grammes of carbon dioxide per megajoule) – less than half what was in 2018. And while there will be many major changes in the electricity system over the next several decades, it is this that is perhaps the most important and optimistic [1–7].

The demand for electricity (or power) is highly distributed. The power needs for industry scale, commercial scale, and residential scale is currently supplied by utility scale centralized and large scale power generation systems. The power needed for the transportation system is supplied by small scale and distributed power generation systems. Finally, at the micro level, power needs of personalized power electronics accessories require power generation at the micro level which sometimes are harvested from natural sources. New computer, medical, electronic and environmental technologies are becoming smaller and faster requiring power from immediate and micro sources. Thus, going forward power needs will be multi-scale and multi-dimensional.

## 1.3   METHODS FOR POWER GENERATION

There are fundamentally two approaches for power generation; indirect and direct. In the past, most power either at large or small scale (as required in transportation sectors) was generated by indirect method where energy conversion (from thermal or mechanical energy etc.) to electrical energy was carried out using thermodynamic cycles. These cycles follow Carnot theorem and are fundamentally inefficient. For example, in a conventional internal combustion engine, only 30% of thermal energy is converted to electrical energy; the remaining is passed onto waste heat. In recent years several direct methods for conversion of thermal, mechanical or chemical energy to electrical energy have become more prominent. The novel indirect methods for power generation include supercritical $CO_2$ cycle and innovative power cycles using liquid among others. The most advanced direct ones are photovoltaics, thermophotovoltaic, thermoelectricity (thermionics) and fuel cell. Others such as piezoelectricity, triboelectricity, and thermo-galvanic electricity are also gaining some momentum; however, these are still too far from being commercially viable.

While there are ten sources of power generation namely, coal, oil, gas (including hydrogen), nuclear, biomass, waste (mass and heat), geothermal, solar, wind and hydro, they provide power through four fundamental processes; thermal, mechanical (kinetic and potential), electrochemical (and chemical) and hybrid. Sources generate these four types of energy which are then converted to electrical energy. Combustion, fission, geothermal energy and solar energy and waste heat are examples of thermal sources; while wind and hydro energy are examples of mechanical energy and fuel cell is an example of electrochemical process to convert chemical energy into electrical energy. Hybrid sources and processes for power generation are numerous and probably the most versatile in their applications on the range, scale and dimensions. The power requirements for personal and self-powered electronics will find the use of nanogenerators and hybrid sources most enlightening. The present book focuses on thermal sources and related hybrid sources for power generation. The next book will examine non-thermal and related hybrid sources for power generation. Within thermal sources, combustion, nuclear heat and waste heat are largely the products of man-made heat. Solar and geothermal heat are natural sources. So far, the most widely used heat source for power generation has been combustion, but in recent years, this dominance has been changing due to environmental concerns. The present book covers recent advances in all of these methods in the next seven chapters.

There is no question that going forward, renewable, nuclear and hybrid sources will become more and more dominant in power generation industries. Environmental concerns will shrink the roles of fossil fuels in power industry. Both economics and environmental protection are becoming more and more favorable for renewable and hybrid sources. This will be amply illustrated in these two books. The role of hydrogen through fuel cells will expand, particularly in transportation industries. Advances described in this and the next book will indicate how rapidly power industry is changing. Environment, efficiency and economics are taking the center stage.

## 1.4   ORGANIZATION OF THE BOOK

The present book on advanced power generation systems-thermal sources is divided into seven additional chapters. Chapters 2–4 are devoted to the subjects of power generation from combustion, nuclear and geothermal heat respectively. The use of combustion heat to generate power goes way back to early days of electricity. As shown earlier, the use of fossil fuels to generate combustion heat which can be transformed into electricity by a variety of heat engines was the dominant method for power generation. The internal combustion engine was used to generate power in both static and mobile situation. Power generation by this method was largely centralized and at a large scale to take advantage of economy of scale. Transportation sector was fully dependent on the workings of the IC engine and the supply of oil.

There were, however, two major problems with this method of power generation. First conversion of heat to power by conventional thermodynamic cycles was highly inefficient. Only 30%–40% of fuel energy was converted to power. Second, the combustion process generated significant pollution including $CO_2$ which, as we know it now, resulted in global warming and climate change. There was also a recognition that fossil fuels are non-renewable, and their limited supply is focused only in some parts of the world and this supply can be very politics dependent. However, massive infrastructure built around fossil fuels and internal combustion engine provided safe and secure power for residential, industrial and transportation sectors. Over the years, the issue of carbon emission by the combustion process became more problematic and required changes in this method of power production.

There were three fundamental issues with power generation by combustion heat. First is the issue of carbon emission and this requires the capture or utilization of $CO_2$ during the process or the use of different types of fuel. Second, the indirect method for power generation by thermodynamic cycles was highly inefficient and this efficiency needed to be improved with innovative process conditions, innovative cycles and advanced industrial gas turbine design. Third, novel methods needed to be implemented to directly convert waste heat to power to improve overall combustion efficiency.

Significant attempts have been made to examine fuels other than coal and oil to reduce carbon emission. Over the last several decades, coal has been replaced by natural gas to reduce carbon emission. Attempts have also been made to reduce carbon emission using coal/biomass, biomass/gas mixtures and more use of biofuels to reduce life cycle carbon emission. Chapter 2 examines all of these options. The efficiency of coal power plants has also been improved using supercritical conditions. The improvement in efficiency, however, is not very large. The issues related to the use of hydrogen and hydrogen-natural gas mixture for large-scale combustion are also being examined. In order to accomplish this advanced industrial turbine also needs to be developed. Hydrogen combustion, if carried out effectively, can create zero carbon emission conditions. In recent years, efforts are also made to improve the efficiency of combined thermodynamic cycles. Power generation by supercritical $CO_2$ cycle has been pursued. The use of novel chemical looping to improve the efficiency of the combined cycle is also being examined. Novel methods such as lean and low-temperature combustion, among others, are also being examined to improve combustion efficiency. While the use of combustion heat to generate power is more likely will be phased out over a long haul, Chapter 2 evaluates all advanced techniques to improve power generation from combustion heat. The subjects of direct power generation from combustion heat using thermophotovoltaics technology and thermoelectricity are covered in Chapters 7 and 8, respectively.

While nuclear power is still resisted in several countries due to concerns over safety and nuclear waste, it is a formidable option for the future because it emits little carbon emission in the environment. Significant advances are being made toward commercialization of several types of Generation III reactors. Significant progress is also made in seven types of generation IV reactors which include; gas-cooled fast reactor, lead cooled fast reactor, molten salt reactor, sodium-cooled fast reactor, supercritical water-cooled reactor, very high-temperature gas reactor and novel Terra power travelling wave reactor. These reactors will significantly improve the efficiency of nuclear power production and reduce the amount of nuclear waste. These topics are discussed in Chapter 3. In recent years, the trend in the nuclear power industry is move toward small modular reactor that can be built off site and can be moved if need arises. Small modular reactors are also cheaper and possibly reduce time for license approval and construction. Their efficiency can also be improved by cogeneration which is more easy to handle than in large-scale reactors. Chapter 3 assesses the progress made in the development of small modular nuclear reactors. The use of supercritical $CO_2$ cycle to improve the efficiency of nuclear power generation is also examined. Finally, Chapter 3 also examines the role of nanotechnology to improve several elements of nuclear energy including better treatment of nuclear waste, better heat transfer with the use of nanofluids, better diagnostic tools and sensors using nanomaterials. Roles of direct conversion of nuclear heat to power by thermophotovoltaics technology and thermoelectricity are examined in detail in Chapters 7 and 8, respectively.

While geothermal energy to power is not as aggressively pursued as solar and wind energy, it carries some advantages. First, unlike solar and wind, it provides dispatchable power that can be relied on a continuous basis over a long period. It also emits very little carbon to the atmosphere. The level of heat recovered from geothermal energy varies depending on the depth of heat recovery and the use of geothermal heat, and the method of power production depends on the level of heat. In the past, most geothermal energy has been recovered from low-temperature conventional hydrothermal resources. While these resources are not the most economical for power production by conventional steam cycles, significant progress has been made to recover power from these sources by organic Rankine cycle and to some extent by Kalina Cycles. Chapter 4 examines recent advances made for these recovery processes. In recent years, significant progresses have also been made to generate power from geothermal energy through four other processes. Enhanced geothermal systems are used to extract heat from deeper and hotter wet heat to convert it into power. Super hot rock geothermal process is used to recover hotter and dry rock deep underground. Advanced closed-loop geothermal systems (Eavor-loop) are used to recover heat from deep sources which do not require any direct and open interactions between working fluids and geothermal reservoir matrix. Finally, there are seven active projects to recover supercritical high enthalpy geothermal power. All these advanced projects will make the recovery of geothermal power more attractive. Supercritical $CO_2$ cycle is also used to improve the efficiency of geothermal heat to power. Finally, the role of nanotechnology, particularly that of nanofluids and nanomaterials to improve geothermal power production is also aggressively pursued. These advanced topics are covered in detail in Chapter 4.

Chapters 5 and 6 cover the different methods for generating power from solar energy. The simplest indirect method to convert heat from solar energy to power is by concentrated solar thermal technology. In order to convert heat from solar energy to power, the solar energy first needs to be concentrated since this allows higher concentration and thereby higher temperature to convert power. The efficiency of power conversion is higher at a higher temperature. There are fundamentally four methods of concentration; two line focus and two point focus. Line focus is carried out by parabolic trough and linear Fresnel and point focus is carried out by parabolic dish and heliostat field-central receiver. The details on choice and their advantages and disadvantages are discussed in detail in Chapter 5. Central tower receiver is very popular in recent years for large-scale power production, and advanced cycles used to generate power from such a system are discussed in detail. These include supercritical steam Rankine cycles, supercritical $CO_2$ cycle, decoupled solar combined cycle and innovative power conversion cycles with liquid. The integration of these cycles with solar tower is also examined in Chapter 5. The temperature of solar concentrator is generally controlled by heat transfer fluid. Besides thermal oil, in recent years other fluids such as direct steam generation, molten salt and compressed gases such as nitrogen, $CO_2$ and air are also investigated. Chapter 5 assesses the advantages and disadvantages of these fluids. Advanced mirror concepts for concentrators and the next generation of receivers including the liquid metal are reviewed. Finally, Chapter 5 examines power generation by various hybrid thermal sources such as CST-coal, CST-gas, CST-biomass, CST-geothermal and CST-PV (or CPV).

Chapter 6 examines advances in photovoltaic technologies which include PV, concentrated PV (CPV), hybrid PV (PVT) and hybrid CPV (CPVT). Solar PV is the fastest moving technology in the world, largely because of its significant cost reduction and the improvement in its efficiency. Significant efforts are being made to test different materials for solar cell. Novel cell architecture and cell configuration based on its application are also aggressively pursued. Numerous industries all over the world are actively engaged in improving solar cell efficiency and power. Innovations are also made in solar cell power maintenance and monitoring. Issues related to large-scale PV power integration with the grid are also pursued. Chapter 6 examines all of these topics in detail. The chapter also evaluates CPV technology and the options for CPV power concentrator design along with the various types of CPV and hybrid CPV power plants. Chapter 6 also examines hybrid PVT and CPVT technologies. PV and CPV can be classified based on the level of concentration (low, medium, high and ultrahigh) and temperature (low, medium, high and ultrahigh) of the working fluids. The chapter examines all of these classifications. The chapter also evaluates the spectral separation of solar energy and its effectiveness in improving the performance of CPV technology. For both PVT and CPVT, the method of cooling is very important. Different methods for passive and active cooling for PVT and CPVT are briefly evaluated in the chapter. Recent developments of innovative cooling methods such as microchannel cooling, jet impingement cooling, earth water heat exchanger cooling, ground coupled central panel cooling systems, use of phase change material, air and water cooling, liquid immersion cooling and heat pipe cooling are examined. The research done on the role of nanofluids for cooling is also briefly reviewed. The applications of CPVT for solar heating and cooling, desalination of water, building and industrial needs, low-cost hydrogen generation and others are briefly evaluated.

Chapter 7 examines Thermophotovoltaics (TPV) technology. TPV system overview along with different types of TPV namely; solar TPV, micron-gap TPV and radioisotope TPV are described. The requirements for effective emitters performance such as spectral control of selective emitters, scalability to large areas, long-term high-temperature stability, ease of integration with TPV system and TPV system efficiency and cost are evaluated. The developments of new materials for TPV are critically assessed. Various TPV applications such as for solar energy, combustion and waste heat, space applications and thermal energy storage are briefly evaluated. Finally, Chapter 7 articulates challenges, recommendations and additional perspectives regarding power generation by TPV technology.

In Chapter 8, the technologies to convert waste to power are evaluated in detail. The chapter demonstrates that waste heat can come from numerous sources and at various temperature levels. It can be accompanied by numerous types of gas pollution. Because composition, temperature level and quantity for waste heat varies significantly, the strategies for its capture and conversion can vary significantly. The largest source of waste heat is at low temperature and this can be converted to power by organic Rankine and Kalina cycles. These cycles are described in the chapter. The chapter also describes the use of the organic Rankine cycle (ORC) for waste heat from biomass combustion, low-temperature geothermal heat, solar heat, mechanical and industrial waste heat sources and waste heat from the aircraft engine. The direct conversion of waste heat to power can be carried by thermoelectricity. The chapter describes this technology and its recent advances in great detail. The applications of thermoelectricity to generate power from the waste heat from human body, industry and homes, transport systems, regenerative computing, solar energy and waste heat from natural gas burning in remote locations are briefly discussed. The commercial potential for thermoelectricity is also evaluated in the chapter. Other evolving methods for waste heat to power such as thermo-ionic device, piezo and tribo-electricity, and thermo-galvanic device are briefly assessed. These technologies are, however, not likely to be commercialized in the near future.

## REFERENCES

1. Shieber, J. Heliogen's new tech could unlock renewable energy for industrial manufacturing. December 19, 2019. A website report put out by MIT Sloan School of Management Career Development Office, Your CDO. https://cdo.mit.edu/blog/2019/12/19/heilogens-new-tech-could-unlock-renewable-energy-for-industrial-manufacturing/
2. IRENA. *Global Energy Transformation: A Roadmap to 2050*. Abu Dhabi: UAE, a report by International Renewable Energy Agency; 2018.
3. IRENA. *Global Renewable Outlook: Energy transformation 2050*. Abu Dhabi: UAE, a report by International renewable Energy Agency; 2020. ISBN 978-92-9260-238-3.
4. International Energy Outlook. *International Energy Outlook 2021*. Washington, DC, a report by Department of Energy; 2021.
5. Haggerty, M. and Pohl, K. The evolution of electric power generation capacity. April, 22, 2020. A website report by Headwaters Economics under Economic Development. https://headwaterseconomics.org/economic-development/evolution-electricity-generation/
6. IRENA. *Future of Solar Photovoltaic Deployment, Investment, Technology, Grid Integration and Socio-Economic Aspects (A Global Energy Transformation Paper)*. Abu Dhabi: UAE, International Renewable Energy Agency; 2019. ISBN 978-92-9260-1.
7. Jones, D. Global electricity review 2022. 2022, March. A website report by https://ember-climate.org/insights/research/global-electricity-review-2022/.

# 2 Advanced Combustion Power Systems

## 2.1 INTRODUCTION

Combustion power is the most basic form of power generation. It has been used since the beginning of the power industry. The use of power in the transportation industry started with the use of an internal combustion engine. Fossil fuels i.e coal, oil and natural gas were the fuels for the internal combustion engine. Unfortunately, the use of internal combustion engine and thermodynamic cycles used to generate power from heat are not very efficient and result in a significant waste of fuel and very negative impacts on environment. Recent awareness of the negative impacts of combustion power generated by fossil fuels has resulted in significant efforts to improve the efficiency of internal combustion engine and minimize its impact on environment. Numerous strategies are used including fuel change, improving the efficiency of turbine technology, changing combustion operating conditions and using other novel thermodynamic cycles. Significant efforts are also made to directly convert heat into electricity by new technologies such as thermoelectricity, photovoltaics, thermophotovoltaics etc. In this chapter, we focus on recent advancements made in some of the basic aspects of combustion power such as the use of different fuels, different operating conditions, different cycles and innovations in turbine technology. All of these mostly relate to indirect conversion of combustion heat to electricity. More advanced direct methods to convert combustion heat to electricity are discussed in the subsequent chapters, particularly in Chapters 6 to 8.

## 2.2 NOVEL METHODS TO IMPROVE PERFORMANCE OF ICE

While it is a forgone conclusion that the future electrical and hydrogen world will most likely exclude the use of internal combustion engine, according to Leach et al. [1] during the transition period ICE still can provide useful service if its performance is improved. Currently, 99.8% of global transport is powered by internal combustion engines (ICEs) and 95% of transport energy comes from liquid fuels made from petroleum [1]. Many alternatives including plug-in hybrid vehicles, battery electric vehicles (BEVs) and other fuels like biofuels and hydrogen are being considered. However, all these alternatives start from a very low base and face very significant barriers to unlimited expansion so that in a near and medium future significant portion of transport energy is still expected to come from conventional liquid fuels powering combustion engines. It is, therefore, important that ICEs are improved in order to reduce the local and global environmental impacts of transport. The study by Leach et al. [1] considers the possible scope for such improvement based on basic principles that govern engine efficiency and the technologies that control exhaust

DOI: 10.1201/9781003328087-2

pollution. The study identifies that significant progress can be made by considering various practical approaches already in the market. For instance, the best in class SI (spark ignition) engines in the U.S. have 14% lower fuel consumption compared to the average. Engine and conventional powertrain developments alone could reduce the fuel consumption by over 30% for light duty vehicles (LDVs). Implementing other technologies such as hybridization and light-weighting could reduce fuel consumption by 50% compared to the current average for LDVs. Current after-treatment technology can ensure that the exhaust pollutant levels meet the most stringent current emissions requirements.

The study considers the implications for transport policy and indicates that all available technologies need to be deployed to mitigate the environmental impact of transport and it would be extremely short-sighted to discourage further development of ICEs by limiting their sales. As electricity generation decarbonizes further, this renewable electricity could be used to manufacture batteries and run BEVs, plug-in hybrids etc. which can result in GHG footprint towards zero on a life cycle basis. The use of renewable electricity will also curtail the use of fossil fuel which is essential for the reduction of carbon emission. Considering the large demand for transportation energy and slow penetration of renewable electricity for BEV and plug-in electric vehicles, in the short and medium terms, combination of electricity and ICE use for transportation industry appears to be the best strategy for the reduction of GHG emission.

According to Leach et al. [1], there is much interest in electro-fuels or e-fuels which can be hydrocarbons—liquid or otherwise - made with $CO_2$ and hydrogen or hydrogen itself. E-fuels will have a very low GHG footprint if they are made using renewable or nuclear energy and biomass instead of fossil fuels. Hydrogen could be made from the electrolysis of water and could be used in fuel cells. However, the production of e-fuels is very energy intensive and the well-to-wheel efficiency of e-fuels is very low. If renewable electricity is available at all for such a purpose, the focus should be on e-fuels for aviation which cannot be realistically powered by batteries [2,3]. However, even replacing only aviation fuel with e-fuels is extremely challenging [1]. In one study, the efficiency of conversion of renewable electricity to e-fuels is estimated to be 44% [3]. The current daily global demand for aviation fuel is 0.049 exajoules, in energy terms. So the global need for renewable electricity to make sufficient e-fuel to replace the current demand for aviation is 0.1114 exajoules or 31 TWh daily. This means the world will need 1,295 GW of continuous carbon-free power generation each day to meet the demand for aviation fuels. This is equivalent to building over 430 nuclear power stations of 3 GW capacity, the size of the Hinkley Point C power station in the U.K. or around 1.1 million wind turbines of 3 MW capacity (assuming a capacity factor of 0.4). It is very unlikely that globally, there is sufficient economic or technical capacity to undertake such a vast expansion of carbon-free electricity in the short to medium term. It is, however, true that as electricity generated from solar and wind energy significantly expands, there may be excess electrical energy from these sources that can be used to generate aviation fuels or generate more hydrogen for fuel cells that can have significant impacts on GHG emission. Leach et al. [1] see approaches such as e-fuels as ways to enable easier expansion of renewable electricity generation rather than a solution to decarbonize transport in the short to medium term.

In the short term, implementing technologies that have been proven commercially viable in conventional, stoichiometric SI engines can lead to an estimated 30% reduction in fuel consumption over a current typical new car. Additional technologies [4–9] not considered in reaching this estimate, including lean-burn SI combustion, water injection, and variable compression ratio are expected to reduce fuel consumption further [1]. In the medium term, lean-burn technologies incorporating some degree of compression ignition are expected to result in a further improvement of ~10%, such that a fuel consumption reduction associated with IC engine improvements alone approaching 40% is feasible for engines used in light duty vehicles, which currently are predominantly SI engines. In dilute gasoline combustion, a flame moves through either premixed or non-premixed (i.e., stratified) mixtures of fuel and air. In this process, the engine dilutes the fuel with either more air than is required to burn it (excess intake air) or recirculated exhaust gases. The Vehicle Technologies Office's (VTO) research focuses on the non-premixed (stratified) version because it offers the highest potential to improve efficiency. These engines can operate on current gasoline and gasoline/ethanol blends and are primarily for automotive and light truck applications. This combustion technology can offer fuel economy improvements of up to 35% relative to a 2009 baseline gasoline vehicle.

In the stratified version of the process, the vehicle injects fuel directly into the cylinder. It times it so that a properly stratified combustible fuel-air mixture occurs near the spark plug at the time of spark. Dilute gasoline combustion results in fuel economy improvements because the engine uses the amount of fuel injected to control the load rather than restricting the intake air flow (throttling) to control it. Most gasoline vehicles on the road have port-fuel-injected (PFI) gasoline engines that use throttling, which is far less efficient. At part load, the combustion products allow the engine to carry out work more efficiently compared to conventional engines. The engine has a lower combustion product temperature at partial loads than a conventional engine would and as a result, loses less heat. VTO is supporting work to address critical challenges which include determining the most efficient fuel-air mixing strategies, which involve issues with port configurations, fuel-spray characteristics, and mixing characteristics, initiating ignition and propagating a flame in stratified mixtures, facing challenges with stochastic misfire and knock (explosive, uncontrolled combustion) and reducing emissions that are different from those that occur with conventional (PFI) engines.

With hybridization, the high efficiency region is greatly extended and with the high efficiency engine concepts that are described here, SI engines may never need to operate with a brake efficiency below 40%. The relatively small expected additional benefit associated with lean-burn technologies is initially surprising, but it must be recalled that the point of comparison is a highly optimized SI engine with low pumping losses and a high compression ratio. One of the key advantages of lean-burn compression ignition technologies is their ability to deliver high efficiency without necessarily requiring a high-octane fuel, potentially providing additional greenhouse gas savings associated with fuel manufacture as well as helping to balance the demand for refinery output streams in terms of their efficiency and design.

Low-temperature combustion (LTC) is a flameless, staged burning of the fuel (gasoline, diesel, or biofuel) in an engine's combustion chamber at temperatures that are lower than what occurs during conventional engine combustion. Research suggests that LTC has the potential for a 20% efficiency improvement over current diesel engines. The lower temperature, flameless combustion results from compression of a fuel-air mixture that has been diluted with either excess air or recirculated exhaust gas. This process raises the density and temperature of the dilute mixture and causes it to autoginite (a process known as compression ignition).

With the LTC process [4–9], the engine compresses a dilute fuel-air mixture, raising its density and temperature. This process, known as compression ignition, causes the fuel-air mixture to autoignite. To dilute the fuel-air mixture so that it has a lower proportion of fuel in it than conventional combustion would, the engine uses either excess intake air or recirculated exhaust gas. Staged burning—the other key element of LTC—is achieved by controlling the timing of the autoignition and rate of heat release. This process works to eliminate excessive combustion rates that can cause engine noise and structural damage, especially at higher loads. DOE is researching a number of forms of LTC, including homogeneous-charge compression ignition (HCCI), premixed-charge compression ignition (PCCI), and reactivity-controlled compression ignition (RCCI). HCCI was one of the early diesel combustion concepts that differed from the conventional diesel process to attract attention. As the name implies, the goal of early HCCI work was to achieve as homogeneous a mixture of air and fuel as possible before ignition—much the same way as in a conventional spark ignition engine. This can be achieved either by injecting fuel into the intake port or directly into the cylinder and allowing sufficient time between injection and ignition to allow complete mixing of air and fuel. The charge then auto-ignites as it is heated by the compressed gases—no spark or other means of forced ignition is used.

In order to address many of the challenges such as limited load range, controllability and knocking posed by HCCI, a number of other concepts have evolved from this homogeneous charge approach and in many cases, charge stratification was introduced. Since the term HCCI may no longer accurately describe many of these systems, the term Low-Temperature Combustion (LTC) can be used as a general term to refer to these and other advanced combustion concepts because the overall goal is to lower combustion temperatures to advantageously alter the chemistry of $NO_x$ and/or soot formation.

One characteristic of HCCI and many other LTC concepts that have evolved from its share is that either all or a significant amount of fuel is premixed with air before ignition occurs. The combustion rate and ignition timing of such premixed LTC concepts are controlled by the chemical kinetics of the mixture. This greatly complicates the control of the combustion process as well as making it sensitive to fuel properties and in-cylinder conditions. Some premixed LTC concepts benefit from low cetane number fuels with volatility characteristics comparable to gasoline. It should be noted that premixing of air and fuel can also be an important factor in "conventional" diesel combustion. While the initial stage in conventional diesel combustion is generally premixed, the combustion of a majority of the fuel occurs after this premixed burn at a rate mainly determined by the rate of mixing of air and unburned/partially burned fuel. The conventional diesel combustion process is thus often referred to as mixing-controlled combustion. This mixing control characteristic greatly simplifies the control of the heat release process [4–9].

LTC offers a number of advantages over workings of today's engines. The fuel/air mixture and combustion product properties enable the engine to be more efficient than conventional combustion engines. Because of the lower combustion temperature, the engine loses less energy through the cylinder walls to the environment. Some of this reduced energy loss allows the cylinder to maintain higher pressure for a longer period of time, enabling the engine to do more work. Some of the energy appears in the form of higher exhaust energy that turbocharging can partly capture. Gasoline-based LTC does not need to throttle intake air to control load, which is a major cause of inefficiency in today's gasoline spark-ignition engines LTC is not restricted by "knock" (explosive, uncontrolled combustion) in the same way gasoline spark-ignition engines are. As a result, LTC allows gasoline engines to have high compression ratios similar to diesels, increasing their fuel economy. LTC may be able to achieve ultra-low exhaust emissions, which could greatly reduce after treatment requirements, cost, and fuel economy penalties.

With a combustion strategy that utilizes LTC, a vehicle fuel economy improvement of 19.4% (over a model year 2015 baseline) was demonstrated in FY19. A number of critical challenges to further develop LTC include expanding the range of engine loads, managing the heat release rate, reducing the lack of transient events, such as changing loads and acceleration, reducing potentially higher hydrocarbon (HC) and carbon monoxide (CO) emissions, understanding if LTC can be more fully effective when combined with fuel that has different specifications than gasoline and diesel and managing difficulty in controlling the start of combustion, because of the lack of a spark or fuel injection.

While much of the work with LTC has focused on premixed LTC concepts, it has been demonstrated that mixing-controlled diesel combustion can also be adopted to produce $NO_x$ emissions in the 0.2 g/kWh range—comparable to those achievable with some premixed LTC concepts [4–9]. Such mixing-controlled approaches could be considered to be the next step in the evolution of conventional diesel combustion beyond the approaches used for example to meet 2004 and 2007 EPA on-road heavy-duty diesel emission standards. They do, however, require advanced "unconventional" hardware to manage PM emissions. These engines require such features as fuel injection systems that provide high injection pressures (as high as 3,000 bar in some prototypes) and air management systems producing levels of boost pressures that require multi-stage turbochargers. Such approaches could be referred to as mixing-controlled LTC concepts. Unlike premixed LTC approaches, it has been shown that mixing-controlled LTC can operate over the entire speed and load range of the engine [9].

New diesel engines show significant (up to 10%) fuel consumption improvement over previous versions, which are already very efficient. Clean diesel combustion can further help. In clean diesel combustion, the burning process takes place in a process fairly similar to conventional diesel combustion. In conventional diesel combustion (also known as diffusion combustion), the rate at which the fuel spray mixes with air inside the cylinder before it reaches the flame determines the rate at which the fuel and air burn in the flame. In clean diesel combustion, more fuel-air mixing occurs prior to the flame. This enables cleaner combustion that produces less soot as well as retains or improves the high efficiency of diesel engines. Adding recirculated exhaust gas to the intake air stream dilutes the fuel-air mixture, resulting in lower combustion temperatures and reducing the formation of $NO_x$. Because fewer emissions form inside the cylinder, clean diesel engines do not have to rely as heavily on after treatment technologies to further reduce emissions.

The Vehicle Technologies Office (VTO) of DOE is supporting research to further improve clean diesel combustion and make it cost-competitive for all passenger and commercial vehicles. This requires pushing the state-of-the-art of technologies such as computer-control, multi-pulse fuel injection, high-pressure fuel injection, use of exhaust gas recirculation, and manipulation of in-cylinder gas flows. Clean diesel combustion suffers from high levels of PM and $NO_x$ in the exhaust and require sophisticated and expensive after-treatment. However, current after-treatment technologies can deliver extremely low levels of $NO_x$ and PM in real-world use—well below the levels required by legislation. VTO's research on clean diesel combustion engines for passenger and commercial vehicles addresses critical challenges which include controlling the amount and temperature of the exhaust gas used for exhaust gas recirculation to minimize emissions, improving the fuel injectors, injection pressure, and control over the fuel spray and spray types in high-pressure and multi-pulse injection and improving lifted-flame combustion, which is when the flame that lifts off of the fuel nozzle stabilizes downstream of the fuel jet. Besides these, clean diesel engines must maintain auto-ignition of the fuel-lean mixture that is immediately upstream of the flame base and improving post-combustion injections for reducing emissions both in-cylinder and through after treatment.

In the medium term, there is scope to make it easier to control these pollutants using gasoline-like fuels rather than diesel, via GCI and RCCI approaches. If a low octane gasoline is used, there is further scope for reducing their GHG footprint through GHG savings in fuels manufacture. Particulate filters are very commonly used in modern diesel engines and are likely to be required even for SI engines. Exhaust particulates are virtually eliminated when particulate filters are used and other sources of PM such as from tire and brake wear become important. If regenerative braking is used as in hybrid electric vehicles, brake wear does not exist. The levels of PM from tire wear will be higher for a BEV because of the higher weight of the vehicle resulting from the weight of the battery when compared to an equivalent HEV using a SI engine. In any case, after-treatment technology has developed to ensure extremely clean exhausts both for SI and diesel engines.

In summary, the fuel consumption and hence the GHG impact of SI engines can be significantly reduced using an existing technology—the best-in-class fuel consumption is already around 14% lower compared to the average. Fuel consumption can be reduced by 30% through engine development alone and with new approaches using compression ignition such as GCI, may exceed diesel-like efficiencies. Using other technologies like hybridization and light-weighting, fuel consumption could be reduced by 50% compared to the current average for light-duty SI engine vehicles. For both SI and CI engines, exhaust pollutants can be controlled to levels well below the requirements of current legislation. Although some countries have ambitions to remove IC engine vehicles (ICEVs), this will take time even in the light-duty sector, as there will be a need for increased electrical power generation capacity (and possibly storage too), changes to the electricity distribution system and the development of smart charging systems and other infrastructure requirements that will be different in different countries. It is particularly difficult, if not impossible, to run heavy-duty road, marine and air transport, which account for more than 50% of global transport energy demand, entirely on electricity because of the very large size and weight of batteries that will be needed. While this change is happening research and development in IC engines should continue because there is still scope for significant improvements in reducing fuel consumption and emissions. Moreover, IC engines burning renewable e-fuels can ease the expansion of renewable electricity generation by providing an alternate storage option when generation exceeds demand.

## 2.3 FLEXIBLE FUELS FOR COMBUSTION POWER USING THERMODYNAMIC CYCLES

A term to describe thermal power generation is, power generation from "Heat Engines." Over 80% of America's electric power generation is from "Heat Engines." It should be noted that nuclear power generation plants are "Heat Engines" as well and also utilize the Rankine steam cycle. As shown later, geothermal power generation plants are also heat engines but utilizing organic Rankine cycles. Over the past several decades, heat engines have used numerous fuels like coal, oil, gas, biomass, hydrogen as well as mixture of fuels. The most commonly used fuels have been coal, oil and gas. Coal and gas are used for static power generation, while oil is used mostly used for transportation power. Power generation by diverse fuels is important for long-term sustainability. In recent years, a combination of coal and biomass, coal and gas, biomass and gas and gas and hydrogen are also investigated. Pure hydrogen is also considered as the ultimate fuel for power generation. Here we briefly examine the use of these fuels. Heat engines are based on thermodynamic cycles and they are indirect methods for power generation.

Two important factors determining the cost of electricity are the efficiency of cycles and cost of fuels. As shown in the subsequent sections, the efficiency of combined cycle operation can be enhanced in a number of different ways and it is important for two main reasons: (a) for economic power generation and (b) to reduce the environmental impact of power generation. Flexibility in the use of fuels is also important for a number of reasons. Planning for reliable power generation during extreme weather is one reason for a diverse fuel portfolio. Furthermore, the fuel cost is the primary driver to produce the lowest cost power. While fuel costs vary with time and location, relative costs for coal, natural gas and diesel fuel for million Btu's have been approximately in the ratio $2.5/3/17.0. While diesel power plant is extremely efficient (up to 48%) it utilizes the most expensive fuel. For cost-effective electricity production, it takes both a reasonably cheap and abundant fuel. Fuel cost comprises approximately 75%–80% of the production cost of electricity for coal plants. The cost of fuel in a similar gas plant is roughly 90% of the production cost of electricity. Thus, both high efficiency and fuel costs are equally important to produce electricity at a reasonable cost.

It makes sense to have electric utility plant capable of operating with a diverse fuel supply. As indicated in Duke energy report [10], 49,500 MW utility plant of Duke energy has achieved this objective well (source: https://sustainabilityreport.

duke-energy.com/downloads/2017-DukeSR.pdf). Diversity in fuel supply is very important for long-term sustainability of power plants [10]. While coal can be extremely important for Base Load and during extreme weather, in recent years, due to environmental concerns, coal has been substituted by gas around the world. Numerous large-scale and very efficient gas-operated combined cycle plants are operated around the world. Five biggest gas-based combined cycle power plants [11] are in eastern Europe and Asia. The world's most efficient combined cycle plant is EDF Bouchain [12]. The 605-MW unit has a net efficiency of 62.22%. While the transition from coal to natural gas has been carried out smoothly in many parts of the world, this transition has not met many requirements including carbon emission in many parts of the world. The power industry has, therefore, sought other fuel options to make transition from coal more suitable for economic and environment needs. In recent years, a mixture of fuels and pure hydrogen has found more interest in the power industry in order to make use of the infrastructure of existing coal power plants and further reduce $CO_2$ emission from these power plants.

### 2.3.1   COAL-NATURAL GAS COFIRING

An option attracting the interest of some power utilities is that of cofiring natural gas in coal-fired boilers [13]. This technique can be instrumental in improving operational flexibility and reducing emissions. It avoids the intermittency issues of renewables such as solar power, but clearly retains some of the disadvantages associated with fossil fuels in general. Both coal- and gas-based generation are vital in powering many of the world's developed and emerging economies. Both are used to provide a secure and uninterrupted supply of electricity, needed to ensure that economies and societies can develop and prosper. In some countries, coal provides much of the power, whereas in others gas dominates. However, there are many instances where the national energy mix includes combinations of the two. Each brings its own well-documented advantages and disadvantages, but recent years have seen a growing interest in means by which these two fuels might be combined in an environmentally-friendly and cost-effective way.

Although coal-fired power plants in many countries already use natural gas for start-up and warming operations, the amounts used are often quite limited. Changing market conditions are increasingly forcing many plants to find new means of operation so as to keep plants clean, efficient, and economically viable. Replacing part of the coal feed with gas and burning the two together in the plant's steam generator can help enhance both fuel and operational flexibility, which is vital in today's often challenging electricity markets. Cofiring with gas seems a promising option for at least some existing coal-fired plants, and a number of utilities are considering (or are even in the process of) converting their coal plants for cofiring, enabling them to continue operating using a mix of coal and natural gas. Importantly, some conversions allow the ratio of coal-to-gas to be varied, providing a useful degree of flexibility in terms of fuel supply and plant operation.

Potentially, gas can be added to an existing system in a number of ways. Some replace a portion of the main coal feed, whereas others use gas as a means for minimizing emissions of species such as $NO_x$. The amount of gas used for some types of application will normally be less than if the plant is converted to actually cofire gas in the true sense. This allows much greater volumes to be fed directly into a boiler and burned simultaneously with the coal feed. Depending on the individual plant and operational requirements, the existing unit can be configured in a number of ways so that cofiring becomes a practical proposition. It may be a case of replacing oil-fired igniters or warm-up guns with gas-fired equivalents, one of the simplest options. But should a plant operator wish to consistently put even more gas through his plant, it is likely that gas firing will need to be incorporated into the main burner system; there are dual or multi-fuel burners suitable for this, available commercially from a number of suppliers [13].

Cofiring offers many advantages. It allows possible adaptation/reuse of existing infrastructure and control systems. It enhances operational and fuel flexibility and cost savings can be achieved by switching to the cheapest fuel at the given time. Some emission control upgrades may be reduced in scope, delayed, or avoided. Cofiring can provide a significant reduction in the minimum unit load achievable and less coal throughput reduces wear and tear on pulverizers and coal handling systems, as well as associated operation and maintenance (O&M) costs. Reduced coal throughput generates lesser solid and liquid plant wastes and emissions to air.

As with any technology, there will be drawbacks. An obvious requirement is that the coal-fired plant has an adequate source of natural gas available at an acceptable price. If the plant already uses gas in some way, the existing infrastructure may be adequate. If not, additional supply and control equipment may be required. Depending on the overall length and any local constraints, costs for a new gas pipeline can be considerable. A major attraction often cited for cofiring is the low price of natural gas. Although this is currently the case in the USA, gas is much more expensive and less readily available in some other economies. Even in the USA, there are concerns that prices could increase significantly in the future as political and environmental pressures on hydraulic fracturing and investments in gas export facilities could drive up the price of gas, closer to those seen in Europe. Higher gas prices could cancel out any advantages and cost savings provided by cofiring.

In the USA, the Environmental Protection Agency (EPA) has suggested that, under some circumstances, cofiring could be an alternative to applying partial carbon capture and storage (CCS) to coal-fired power plants. The EPA has advised that new emission standards could be met by cofiring ~40% natural gas in highly efficient supercritical pulverized coal power plants. However, some industry observers think that more than this level would be needed and that, for this to be achievable, the boiler would need to be specifically designed to operate in this manner. Various technical issues will need to be considered when a switch to cofiring is contemplated. For example, there can be significant impacts on heat transfer within the boiler. The heat transfer characteristics of natural gas and coal flames are markedly different and, in some cases, original heat transfer surfaces may be inadequate for increased natural gas firing. If appropriate measures are not taken, problems with metallurgy can arise and major plant components run the risk of becoming unreliable. Gas supply may also be an issue. In some locations, supply restrictions may limit the maximum amount available. Furthermore, seasonal restrictions may apply, giving priority to other applications such as home heating.

The study of Haugen [13] examined a Hybrid Gas/Coal Concept (HGCC), which focuses on achieving power generation with high-efficiency and load cycling capability combined with carbon dioxide ($CO_2$) capture. The HGCC concept consists of combining an 88 MWe gas combustion turbine and a 263 MWe ultra-supercritical (USC) coal boiler with 51 MW of energy storage capacity as batteries. The HGCC concept is unique and presents a strong business case because it is flexible, efficient, innovative, resilient, demand responsive and has potential for brownfield retrofits. Other benefits include (a) the COE (carbon monoxide equivalent) from the HGCC is competitive with both USCPC and IGCC even though those plants are larger and have a natural economy of scale advantage; (b) HGCC offers significant improvement in the areas of ramp rate, turndown, and start-up flexibility (cold and warm) compared to USCPC and IGCC; (c) HGCC components are commercially available today and (d) redesigned coal boiler firing allows improved ramp rates and turndown when compared to the USCPC.

As more, non-dispatchable renewables are added to the power generation portfolio, utilities respond by adjusting the commitments to combustion-type generating resources. This has required coal units to transition from base load operation to frequent cycling at certain times of the year. The Hybrid Gas/Coal Concept (HGCC) utilizes three distinct and unique approaches to maximizing cycling flexibility (turndown and ramp rate). In order of decreasing flexibility, the concept incorporates (a) energy Storage System (ESS) (batteries)—51 MW gross, (b) combustion Turbine (GE 6F.03)—88 MW gross or (c) USC Boiler/Steam Turbine Cycle—263 MW gross. The HGCC concept offers new firing system for the combustion system. This new firing system allows the boiler minimum load to be reduced by 20%. When the plant is called upon to begin operation from a cold start, the start-up order would be (a) ESS: immediate, (b) Combustion turbine: 30 minutes to full load and (c) USC Boiler Steam Cycle: 6–9 hours to full load from cold start, approximately 3 hours and 40 minutes from warm start. The overall plant turndown when ESS is considered is approximately 7.6–1.

Even slight changes in fuel price can result in significant swings in production costs, and this can create market opportunities for utilities that have both gas- and coal-fired assets. Cofiring can be a possible option, allowing pricing and market conditions to drive the fuel choice and mix. Cofiring can offer increased fuel flexibility and potentially this can provide significant fuel and operational savings. Substituting some coal input with gas is considered to be a low-risk option, allowing utilities to better meet changing market requirements. A number of utilities have already adopted gas cofiring and others are considering converting some of their coal plants such that they can operate on a mix of the two. Increasingly, many of these plants are now expected to operate in non-baseload modes, for which they were not designed. They need to adapt and evolve to survive in the longer term, and the ability to operate on fuels that they were not originally designed for will be an important factor. Cofiring can help reduce emissions, improve operational flexibility and allow faster start-ups, bringing plants on line more quickly and cleanly.

Apart from the USA, there are a limited number of plants in countries such as Indonesia and Malaysia that cofire coal and gas and also new projects in development. However, a major factor is the availability of a reliable supply of gas at a suitable price. In some locations, although affordable for limited application, gas is too expensive for bulk use. Coal is often cheaper and more easily available. For any new potential project, as with coal-solar hybrids, various economic, operational and environmental factors need to be considered on a case-by-case basis. But, compared to solar, opportunities for cofiring are less restricted as there are many more locations where both coal and natural gas are readily available.

## 2.3.2 BIOMASS-GAS COFIRING

Co-firing renewable and fossil fuels is a step taken to move away from using fossil fuels as it decreases the rate of fossils used. For power plant owners co-firing allows to continue using the existing infrastructure retrofitting the boilers with fraction of investments compared to purchasing of new boiler [14]. Biomass and coal co-firing already is widely used in industry. Co-firing biomass and natural gas is a relatively new concept as it involves burning solid fuel together with gaseous fuel. Co-firing pulverized biomass with natural gas has been studied in laboratory experiments by Costa and Casaca [15] and Golec and Bocian [16]—both studies reported technical problems to increase the ratio of pulverized biomass above 20% on energy basis. It can be concluded that obstacles to the introduction of higher biomass ratios are technology-related—biomass and natural gas (or renewable gaseous fuel) co-firing technology and environmental impacts should be studied further. Important technical challenges associated with biomass co-firing are delivery, preparation and handling of fuel; ash deposition; fuel conversion; pollutant formation; increased corrosion rates in components facing high temperatures; utilization of fly ash; impacts on SCR systems and formation of striated flows.

Firing pulverized solid fuel is the most efficient way to fire solid fuel—the overall surface of fuel for oxidant to react with increases. Due to the viscoelastic nature of wood preparation of fine particles requires increasing amounts of energy. Rittinger model states that the energy required for size reduction is proportional to the new surface area generated [17]. Priyanto et al. researched that typical mechanical energy consumption for milling wood chips to fine 1 mm particles lies between 30% and 76% of the HHV of common wood chips (19 MJ/kg) [18], Karinkanta et al. analyzed fine grinding of wood and found dry pulverization of 4–6 mm pine sawdust to the size of 500 μm in hammer roller mill consumed approximately 0.5 kW/kg [19].

GHG emissions produced by combustion of natural gas in district heating plants, electricity plants, co-generation plants or for domestic cooking do not represent GHG emissions from the whole life cycle of natural gas. The life cycle of natural gas includes pre-production, extraction, gathering, processing, transmission, storage, distribution—these steps produce significant $CO_2$ and $CH_4$ emissions. Co-firing biomass and natural gas is a step to decrease the rate of fossil fuel use but biomass size reduction for co-firing is an energy-intensive process. Using data from published research on energy consumption for woody biomass size reduction possible GHG emissions in the production of 0.5 and 0.1 mm woody biomass particles were calculated by Kazulis et al. [14] to investigate if fossil fuel use in the process of woody biomass fuel preparation for co-firing with natural gas results in overall GHG emission savings. Latvia's grid electricity fixed $CO_2$ emission factor was used in the calculations determining GHG emissions from electricity use. Results show that GHG emissions from woody biomass preparation in both chosen sizes are smaller than GHG emissions from the amount of natural gas that woody biomass particles would substitute however knowing the actual $CO_2$ emissions from electricity would give more accurate results. Results also show that in accordance with Rittinger's law achieving smaller sizes requires increasing energy.

Preparation of pulverized biomass is energy-intensive process. In accordance with Rittinger's law achieving smaller sizes require increasing energy. The results of this study show that if fossil fuels are used to power the machines used for biomass preparation then (taking into account Latvia's fixed $CO_2$ factor for grid electricity) GHG emissions from woody biomass particles preparation in 0.5 mm sizes will be 16.89 kg $CO_2$/MWh and in 0.1 mm sizes will be 93.43 kg $CO_2$/MWh. It was found in the literature that emissions from natural gas are ranging from 266.90 to 586.16 kg $CO_2$/MWh. GHG emissions from biomass preparation are smaller in both cases. The accuracy of data is crucial, fixed $CO_2$ emission factor for the use of grid electricity will not yield accurate real-time results. The final steps of pulverization are the ones consuming increasing energy. Finding optimal pulverization size is crucial for energy saving and GHG emission reduction. If biomass preparation is done during the abundance periods of solar, wind or hydro energy it could be done with zero GHG emissions and stored for later utilization.

### 2.3.3 COAL BIOMASS COFIRING

Coal and biomass hybrid power generation has three methods, direct co-firing, biomass gasification and co-firing, and parallel power generation of coal boiler and biomass boiler. Usually, cofiring uses pulverized coal-fired boiler or circulating fluidized bed boiler. The cost varies with different methods. Direct co-firing can use existing coal power units, so it is more economical than specialized biomass power generator. Biomass gasification is to turn biomass materials into combustible gas through chemical reaction at high temperature. Then the gas is sent to the boiler and co-fire with coals. Gasification can improve power generation efficiency and reduce pollution emission. Coal biomass cofiring power system can make full use of the equipment and system of existing fuel-power plant, such as boiler, steam turbine and auxiliary system. It only needs to be upgraded and equipped with a biomass fuel processing system, so the initial investment is low. Coal biomass cofiring power generation project doesn't require additional areas outside the power plant, while specialized *biomass power generation* project requires new floor space. Biomass cofiring power generation project can make full use of the existing electricity and heating supply ***market*** of the coal-fired power plant.

The continuous operation of a specialized biomass power plant depends entirely on the supply of *biomass fuel*, while coal biomass cofiring power plants have two fuel sources, and reduce the risk of fuel shortage. Therefore, power plants adopting coal and biomass co-firing have stronger bargaining power. Compared with specialized biomass power generation, biomass and coal co-firing have lower investment and operation cost. Applying a coal biomass hybrid power system in large and high-efficiency coal power plant is an economical way to reduce $CO_2$ emission. Coal biomass cofiring power generation can achieve high efficiency by making full use of the large capacity and high steam volume of coal power plant. The power generation efficiency can achieve the highest level of coal-fired power plant.

Biomass hybrid power generation offers several other benefits over biomass direct combustion. It uses the existing large-capacity and large-efficiency generator set of the coal-fired power plant and achieves high power generation efficiency of 40%–46%. For example, the coal biomass cofiring power plant based on cogeneration units can achieve power generation efficiency as high as 70%. Coal biomass cofiring power system can use the existing desulfurization and denitrification facilities of the coal-fired plant, so as to lower the pollution emission like $SO_2$, $NO_x$ and smoke. As long as coal and biomass are used in reasonable proportions, coal biomass cofiring power generation has mature technology and wide application. Coal biomass cofiring power system adopts two kinds of fuel. It reduces the power plants' reliance on biomass fuel and increases their bargaining power. It can guarantee the purchase of biomass fuel and reduce cost. Finally, the coal biomass power system is constructed by transformation of the coal-fired power plant. It can make full use of the communal facilities of the power plant and save investment.

Coal-biomass cofiring power generation is a smart way to deal with the agriculture wastes in rural areas. Besides, it can reduce the consumption of coals, and reduce the power generation cost. By these means, we can gradually relive our reliance on fossil fuel and enlarge the proportion of renewable energy. The resource base of co-fired power (combustion of coal and biomass in the same boiler) depends on the availability of coal and the various biomass feedstocks, as well as the proximity of biomass sources to coal-fired power plants. The U.S. has 261 billion tons of recoverable coal reserves; based on an average annual coal production rate of 1.5 billion tons, the estimated recoverable reserves alone will provide the U.S. with coal for the next 150+ years. The three primary types of biomass that can be used for co-fired power are agricultural residues, forest residues and thinnings, and herbaceous and woody energy crops. A comparison of the geographies of biomass resources and existing coal-fired power plants indicates high potential for co-firing agricultural residues with coal plants in Iowa and Illinois. There is also potential for co-firing in the southeast U.S., which has large forest resources and a potential for productive herbaceous or woody energy crops. Conversely, several mid-Atlantic states with large coal resources and installed capacity of coal-fired facilities such as Ohio, Pennsylvania, West Virginia, and New York lack adequate agricultural and forest residues to support significant co-firing capacity. If the incentives for co-firing were large enough, however, co-firing could grow in these states because co-fired facilities would be able to compete with the forest products industry for standing timber.

Coal and biomass co-firing have been tested at a number of plants in the U.S. As of 2011, direct co-firing has been tested at a minimum of 38 plants. Many of these tests occurred in the late 1990s and early 2000s; very few remain in operation today. Dedicated biomass-to-energy power plants that are not tied to coal-fired power production are currently operational in California, Oregon, and other regions with sufficient feedstock availability. Many of these plants were installed in the 1980s and have been operational since. Despite the low capacity of existing co-fired power plants, the Energy Information Administration's (EIA) Annual Energy Outlook (AEO) for 2011 shows significant growth in biomass co-firing for electricity generation through 2035. Based on the AEO projection, co-fired electricity generation peaks in 2024 at just over 30 billion kWh, nearly a factor of 20 greater than levels in 2011 [20]. The EIA projected increase in biomass co-firing for electricity generation is driven by a combination of state renewable portfolio standard (RPS) requirements and the projected low cost of feedstocks.

The environmental profile of this analysis is based on a life cycle analysis (LCA) that accounts for a full list of metrics, including air emissions and resource consumption. It is based on a 550 MW, coal-fired power plant with a pulverized coal (PC) boiler that is retrofitted to co-fire biomass at a 10% feedstock share by energy. The coal feedstock is Illinois No. 6 coal, and two types of biomass feedstocks, hybrid poplar (HP) and forest residue are considered. HP is a dedicated energy crop that requires land preparation, cultivation, and harvesting, while forest residue is merely a byproduct of the forest product industry. The co-firing of Illinois No. 6 coal and HP reduces life cycle (LC) greenhouse gas (GHG) emissions by only 1.0% (from 1,118 kg $CO_2$e to 1,107 kg $CO_2$e/MWh). The $CO_2$ emissions from the combustion of biomass are carbon neutral, but the HP supply chain also includes land transformation, fertilizer production and use, and other ancillary processes that produce significant GHG emissions. The co-firing of Illinois No. 6 coal and forest residue reduces the LC GHG emissions by 6.6% (from 1,119 to 1,044 kg $CO_2$e/MWh). In contrast to HP, the acquisition of forest residue does not produce GHG emissions from land use or cultivation.

Co-firing increases the LC lead (Pb), volatile organic compounds (VOC), and particulate matter (PM) emissions. Co-firing reduces LC $SO_2$ emissions if forest residue is used, but not if HP is used. Co-firing leads to reductions in LC emissions of carbon monoxide (CO), nitrogen oxides ($NO_x$), and mercury (Hg) [21]. The costs of co-firing were evaluated using an LC costing approach. The retrofit of an existing PC plant to co-fire HP at a 10% share of feedstock energy increases the cost of electricity (COE) from $30.9 to $40.4/MWh (a 31% increase). If forest residue is co-fired instead of HP, the increase in COE is only 14%. The capital costs of the co-fired systems account for a small share (approximately 8%) of the COE because this analysis assigns capital costs only to new equipment, not existing equipment. The key drivers of cost uncertainty are the feedstock prices for coal and biomass. The technical barriers to the implementation of co-fired systems include biomass supply variability as well as higher-than-expected decreases in boiler efficiencies, equipment fouling, and co-product degradation. Regulatory uncertainty is a key non-technical risk associated with coal-biomass co-fired systems. More details on coal-biomass cofiring are given in my previous books [22,23], particularly the book on "Energy and Fuel systems integration" [23].

### 2.3.4 Power Generation with NG-Hydrogen Mixture and Hydrogen

Experts note hydrogen—the most abundant and lightest of elements—is odorless and nontoxic, and it has the highest energy content of common fuels by weight, which means it can be used as an energy carrier in a full range of applications, from power generation to transportation and industry. Though it is not found freely in nature and must be extracted (produced, or "reformed") via a separate energy source (such as power, heat, or light), the hydrogen industry is today well-established in sectors that use it as a feedstock. Increasingly, however, hydrogen is being considered the missing link in the energy transition as key technologies to produce it using renewable electricity, such as proton exchange membrane electrolyzers and fuel cells, reach technical maturity and economies of scale.

### 2.3.4.1 Characteristics of Cofiring with Hydrogen

In comparison with other alternative fuels, hydrogen is the most effective to reduce or eliminate harmful vehicle emissions and their environmental impact. Hydrogen is abundant in the atmosphere, water and solids (fossil fuels and biomass). Unfortunately, most hydrogen is present in the form of compounds and needs to be separated or recovered. Currently, the most used technique for hydrogen generation is steam–methane reforming. Electrolysis of water using excess power from renewable and nuclear sources is an upcoming technology. In the study by Shadidi et al. [24], the effect of adding hydrogen as a fuel on the performance and exhaust emissions of spark ignition (SI) and compression ignition (CI) engines was investigated. The study and the associated literature resulted in the following assertions.

When hydrogen is used as an additive in spark ignition (SI) and compression ignition (CI) engines, it will reduce the volumetric efficiency of the engine because the lower heat value (LHV) of hydrogen (120 MJ/kg) is higher than those of diesel (43.6 MJ/kg) and gasoline (43.4 MJ/kg). This reduction in volumetric efficiency will reduce engine power and torque. In SI and CI engines, hydrogen as a fuel decreases brake thermal efficiency (BTE) because hydrogen has the high molecular thermal capacity and due to the fact that with the addition of hydrogen in ICE, the progress of the combustion phase changes and the combustion efficiency is reduced. Hydrogen burns quickly and has a nine-fold faster flame speed in a diesel dual-fuel engine. The heat release rate rises as the load rises and hydrogen substitution rises. For this reason, adding hydrogen to diesel fuel in CI engines increases brake-specific fuel consumption (BSFC). By injecting hydrogen at varied ratios, the equivalent brake-specific fuel consumption (BSFC) increases compared to what observed when using gasoline due to the reduction in engine power (when hydrogen injection methods that lower engine power is utilized).

Adding hydrogen to internal combustion engines reduces the amount of carbon monoxide emissions from CI and SI engines [24]. Due to lack of carbon, combustion of increasing mass content of hydrogen in the fuel will result in lower rate of hydrocarbon synthesis. Furthermore, a high hydrogen flame raises the cylinder's pressure and improves combustion efficiency. Because of its high diffusion coefficient, pre-combustion hydrogen produces a more homogenous flammable mixture and improves oxygen availability. As a result of these factors, the amounts of carbon monoxide (CO) and unburned hydrocarbon (UHC) produced by internal combustion engines are lowered. When hydrogen is used as a fuel in internal combustion engines, the H/C rate rises, resulting in a decrease in combustion time and an improvement in combustion efficiency. Furthermore, the presence of hydrogen in combustion fuel also reduces $CO_2$ emissions. In fact, the combustion of pure hydrogen results in no $CO_2$ emission.

Shadidi et al. [24] pointed out that Hydrogen is regarded as a suitable fuel for combustion because of its fast flame speed, low ignition energy required, and high adiabatic temperature. These features contribute to a rise in the temperature of the working fluid in the cylinder as well as to an increase in NOx. The high coefficient of hydrogen emission and the greater access of fuel to oxygen will increase the homogeneity of the flammable mixture and the amount of H/C in the total fuel, which will reduce the soot in diesel engines. With the usage of hydrogen in the majority of internal combustion engines, hazardous exhaust pollutants are reduced, and engines' overall performance improves. When considering its environmental and economic benefits, hydrogen is a clean and sustainable energy source.

### 2.3.4.2   Large Scale NG-Hydrogen Combustion

In recent years, major power equipment manufacturers are developing gas turbines that can operate on a high-hydrogen-volume fuel in order to reduce carbon emission by power industry. Significant efforts are pushed by companies like Mitsubishi Hitachi Power Systems (MHPS), GE Power, Siemens Energy, and Ansaldo Energia to develop 100% hydrogen-fueled gas turbines. MHPS, a joint venture between Japanese giants Mitsubishi Heavy Industries and Hitachi, wants to make hydrogen-fired gas turbines a key facet of a global $CO_2$-free hydrogen society using renewable energy by 2050. While natural gas will continue to be a transition fuel and provide a significant role to address variability from renewables, the next phase of development will involve the storage of electricity using hydrogen. Hydrogen's production from renewables through electrolysis—which uses excess renewable power to split a water molecule—allows for the renewable hydrogen to be stored and used later in a combined cycle gas turbine (CCGT). Since 1970, MHPS has fired 29 gas turbine units with hydrogen content ranging between 30% and 90%, tests that have spanned over 3.5 million operating hours [24]. A major issue with pure hydrogen combustion is the high $NO_x$ emission. This needs to be reduced without sacrificing efficiency. Furthermore, hydrogen has a higher flame speed compared to natural gas. MHPS is also trying to reduce the risk of combustion oscillation and "flashback" (backfire) in the fuels containing higher hydrogen concentration.

MHPS solution to the above issues was to develop a "diffusion combustor" based on the company's dry low-$NO_x$ (DLN) technology [25,26] that injects fuel to air. The combustor reduces $NO_x$ using steam or water injection, but it retains a relatively wide range of stable combustion, even if fuel properties fluctuate up to 90% hydrogen. MHPS showed [24] that at 30% hydrogen, the technology can handle an output equivalent of 700 MW (in combined cycle mode with a turbine inlet temperature of 1,600°C) as well as reduce carbon emissions by about 10% compared to a conventional CCGT. MHPS is currently piloting a project to *convert one of three units at Vattenfall's 1.3-GW Magnum combined cycle plant in the Netherlands* to renewable hydrogen by 2023. The project in Groningen, which entails modifying a 440-MW M701F gas turbine, will refine the combustion technology to stay within the same $NO_x$ envelope as a natural gas power plant but do it burning 100% hydrogen, without steam or water injection [27]. The project is key to MHPS's vision to provide customers with gas turbines that could be upgraded to 100% hydrogen capability. Major issue with hydrogen combustion is the supply of hydrogen. This will need on-site electrolysis plant operated by low cost excess power generated from renewables.

Siemens is also committed to developing hydrogen combustion technology. It estimates to unveil a 25–50-MW hydrogen-burning gas technology within a few years. Siemen thinks that due to battery degradation owing to cycling, hydrogen combustion has significant future in the power industry. They, however, recognize that generating power from hydrogen has several major drawbacks. At least 60% of hydrogen gas turbines under development by an assortment of manufacturers use DLN combustor technology, which is a challenge because of the way gas turbines have evolved over the years. Power conversion also needs additional equipment and water. Running electrolysis to produce 50 MW for 1 hour at a CCGT running at 50% efficiency could require 175 MW of renewable power and 3,400 kg (more than 14,000 gallons) of hydrogen. This raises the question of affordability of hydrogen power. Due to these drawbacks, the best thinking of industry is that hydrogen power could prove more economical as short-term (3 or 4 hours a day) renewable support. Shadidi et al. [24] pointed out that Siemens has addressed technical challenges concerning $NO_x$ emissions and flashback risk control and appears to have made some gains in determining the components and materials needed for high-temperature hydrogen combustion and has incorporated additive manufacturing for burners. Siemens' test families of hydrogen turbines from 4 to 560 MW don't have very good dry, low-emissions technology capabilities yet, but the conventional capabilities are quite high. Initial focus on commercialization will likely be on units smaller than 70 MW. Siemen hopes to develop gas turbines that can handle 20% hydrogen by 2020, and 100% hydrogen by 2030.

Italian engineering firm Ansaldo Energia also appears to be making gains on its hydrogen gas technology. The company is able to burn hydrogen alone or in combination with natural gases and to do it safely and efficiently. Shadidi et al. [24] pointed out that the company already offers fuel-flexible advanced gas turbine combustion systems. Their latest GT26 F-Class and GT36 H-Class gas turbine equipment leverages the Sequential Environmental (SEV) combustion system platform and has been designed with an unrivaled ability to burn the largest range of [natural gas and hydrogen] blended fuel mixture for new power plants. It also offers a hydrogen fuel flexibility retrofit solution for the currently installed base of F-class gas turbines.

General Electric (GE), is focused on combustion systems for both aero-derivative and heavy-duty gas turbines that are capable of operating with increased levels of hydrogen [24]. Aero-derivative gas turbines can be configured with a single annular combustor (SAC), which can operate on a variety of fuels, including process fuels and fuel blends with hydrogen, and GE says there are more than 2,500 gas turbines configured with this combustion system. GE has also developed two combustor configurations for heavy-duty turbines for higher hydrogen content: the single-nozzle, which is available on B- and E-class turbines, and the multi-nozzle quiet combustor for E- and F-class turbines. These have been installed on 1,700 turbines.GE also boasts several projects that use high hydrogen content. One is at the Daesan refinery in South Korea, which has operated on a 70% hydrogen fuel since 2002 to the present time using 6B.03 gas turbine. ENEL's 2010-inaugurated Fusina plant in Italy used an 11.4-MW GE-10 gas turbine to operate on fuel that was more than 97.5% hydrogen by volume [28–31].

### 2.3.4.3 Hydrogen in Combined Cycles

Gas turbines with a combustion system for hydrogen operation [24,28–40] offer a low carbon solution to support the stability of the energy grid. As mentioned earlier, there are zero $CO_2$ emissions produced in pure hydrogen combustion. Since hydrogen combustion occurs at higher temperature resulting in high $NO_x$ emission, most efforts are directed toward solving this issue. For example, micro-mix combustion is used to implement miniaturized diffusive combustion to combust hydrogen with low emissions. With miniaturized diffusive combustion, local flame hot spots, which are caused by arising stoichiometric conditions of hydrogen, are reduced substantially with an increase in the local mixing intensity. This higher quality mixing reduces emissions of $NO_x$ with a more balanced flame profile. Micro-mix combustion was also studied to develop an adaptive combustor which uses different mixtures of fuels including hydrogen, kerosene and methane [25].

The major use of hydrogen in power industry is oriented toward the use of hydrogen in fuel cells and combustion with gas turbines. In recent years, hydrogen combined cycles have garnered significant attention. The WE-Net Program of Japan predicts the implementation of the Hydrogen-Fueled Combustion Turbine Cycle (HFCTC) as a new energy source for the power sector. The program conducted a configuration and performance study of the HFCTC. Thermodynamic analysis of the efficiency of power units which use hydrogen as a fuel showed that for power units of more than 10 MW, steam-turbine hydrogen units were the preferred method to fuel cell power units. Experimental $H_2/O_2$-steam generators, with a capacity of up to 25 MW(t), were created and the peculiarities of their use in steam-turbine hydrogen power units were examined [24,28–40]. In order to avoid $CO_2$ emissions, a novel integrated gasification combined cycle (IGCC) system was proposed with a steam-injected $H_2/O_2$ cycle and $CO_2$ recovery. The results showed that the new system has a lower energy penalty for separating and recovering $CO_2$, at an efficiency decrease of less than 1% point. Two options for $H_2$ production through fossil fuels were presented and their performances were evaluated when integrated with $H_2/O_2$ cycles. This investigation was also performed for advanced technology (TIT = 1,350°C) and of futuristic technology (TIT = 1,700°C). The analysis indicated that due to high hydrogen production cost, thermal efficiency of 60% is required [36]. The present system fell short by 10% even with futuristic technology. This means that the efficiency should be increased by about 10% points, which is equivalent to an efficiency about 20% higher than the most efficient contemporary power plant units. Meeting this requirement poses a very serious technological challenge. Moreover, when increasing the working medium temperature at the turbine inlet to 1,700°C, it is essential to implement a new, non-traditional approach to both the conceptual design of the system (configuration and working parameters) and detailed construction solutions.

Numerous models for hydrogen in combined cycles have been developed. The Graz cycle—proposed by prof. H. Jericha from Technical University of Graz—is an original combination of the Joule and Rankine cycles. In the high parameters area, the Joule cycle is utilized in a semi-closed configuration, coupled with the Rankine cycle, which operates in the low parameters area [34,35]. The Rankine cycle plays here simultaneously the role of heat sink for the Joule cycle. The hydrogen combustion chamber is the high-temperature source of heat. The original idea which distinguishes the Graz cycle from other cycles is applying the extraction of the partially cooled working medium from the Joule cycle and using it as a working fluid in the Rankine cycle. An increase in efficiency is obtained here due to a significant decrease in the compression work of the working medium in the Joule cycle (steam compressor). The Toshiba cycle is one of a group of steam cycles with direct combustion—direct hydrogen-fired Rankine steam cycles. The Toshiba cycle is also called the MORITS cycle—Modified Rankine Cycle Integrated Turbine System. It consists of four turbine parts, where two of them—the first and the last (HHP and LPT)—do not have combustion chambers in front. The regenerator (heat recovery boiler) is located before the last turbine part, where superheated steam is produced. The Westinghouse cycle is a variant of the steam cycle with direct combustion of hydrogen with oxygen. Compared to the Toshiba cycle, it does not possess the turbine part used in the highest pressure region. It can be classified as a kind of new Rankine cycle with single reheat. In contrast to the other cycles, in modified new Rankine Cycle (MNRC cycle) the heat recovery steam generator (HRSG) is placed after the last turbine stage group in the low-pressure zone. The layout of the cycle allows another reheat stage to be added before the low-pressure turbine stage group. As a result, a very high thermal efficiency of the cycle is achieved. A common feature of the similar cycles is that only one working medium is used—steam. This is made possible by replacing the external firing (as in the Rankine steam cycle) with direct firing (similar to gas turbines or piston engines). The main assumption made here is the stoichiometric combustion of a hydrogen and oxygen mixture. The combustion takes place inside a stream of cooling steam, which reduces the combustion temperature to 1,700°C. It is also assumed that hydrogen and oxygen at the ambient temperature are available at a pressure level that allows them to be supplied to the combustor. This means that hydrogen would be provided as a cryogenic liquid and that cryogenic energy could be utilized for pure oxygen production in an air-separator unit.

While the above described theoretical combined hydrogen cycles have significant merit for zero emission [32,33,37–39], evidence also exists for real life zero emission hydrogen-fueled power station. Hadidi et al. [24] pointed out that Fusina Hydrogen Power Station is a hydrogen-fueled power station located in Fusina, near Venice in the Veneto region of Italy. It is the first commercial-scale power station in the world fueled by pure hydrogen. The power station is operated by Enel. The Fusina project was launched in 2004. Construction of the power station started in April 2008, and it became operational in August 2009. It was inaugurated on 12 July 2010 [28–31]. The plant is located adjacent to the Andrea Palladio Power Station. Fusina Hydrogen Power Station has an installed capacity of 12 MW. An additional 4 MW could be generated in the Andrea Palladio Power Station through the reuse of steam produced by the hydrogen-fueled turbine. The power station is equipped with a General Electric combined-cycle gas engine. The hydrogen is provided from Versalis cracker and the adjacent petrochemical facility of Porto Marghera [28–31].

While large-scale power plants with pre-combustion $CO_2$ capture, e.g., integrated gasification combined cycles and integrated reforming combined cycles, convert the primary fuel into hydrogen-rich syngas, which can be burned in a combined cycle to achieve the maximum conversion efficiency into electricity, the use of hydrogen-rich syngas in gas turbines poses a number of issues. The most important issue to be tackled concerns the mitigation of $NO_x$ emissions, which becomes critical due to the very high hydrogen flame temperature. This issue is relevant not only for adapting machines originally designed for natural gas but also for developing new machines for hydrogen-rich fuels. In diffusive flame combustors, the flame tends to be close to the stoichiometric conditions, and hence, its temperature must be mitigated by diluting the fuel with inert species, such as water (steam) or nitrogen. This dilution causes a significant decrease in the plant's efficiency. On the other hand, in lean premixed combustors, the flame temperature is directly limited by the large excess of air and no dilution is required. However, realizing a stable premixed hydrogen flame is not straightforward because of its high flame speed demanding high air velocities to obtain short mixing times and high turbulence rates. As a drawback, premixed combustors may suffer from high-pressure drops. For this reason, gas turbine manufacturers are currently investigating different combustor geometries in order to obtain the same $NO_x$ emissions and combustor pressure drops achieved in natural gas–fueled lean premixed combustors [28–31,34–40].

## 2.4 ADVANCED SUPERCRITICAL COAL POWER PLANTS WITH CCS TECHNOLOGY

Nowadays, fossil fuels are still the backbone of power generation. In 2019, fossil fuels generated 62.7% of the total global electricity [41]. By energy source, coal contributes 36.4% of the total electricity generation [41]. Due to pressing environmental and climate concerns, decarbonization of the electric power industry has become imperative. The IEA scenarios in the 2020 World Energy Outlook predicts the reduction of $CO_2$ emissions from the power industry by as much as 60% between 2020 and 2030 [42]. Three-fourths of this reduction will be achieved from the rapid decline of conventional coal-fired generation. However, the share of advanced supercritical coal power plants is expected to rise in order to meet the ongoing growth in global electricity demand [43]. Supercritical steam cycles paired with carbon capture and storage (CCS) technologies are going to be integrated into future clean coal power plants [44]. By 2030, most of the advanced coal power plants will be upgraded with carbon capture, utilization, and storage (CCUS) technologies.

Subcritical power plants (SUBC) achieve thermal efficiency in the range between 34% and 40% (based on coal LHV) with the global average efficiency around 36%. The most efficient old-fashioned coal plant in recent history has been the Longview Plant in West Virginia [45,46]. This plant recently received top honors in being rated the best Heat-Rate plant of 2016. The 2.1 Billion dollars, 700-MW coal-fired station located in Maidsville, West Virginia had a number of problems following its construction and commissioning in December 2011. Persistence and diligence efforts for good engineering and hard work allowed the Longview plant to reach the pinnacle of top efficiencies in the U.S., achieving a Heat-Rate of 8,999 Btu's/kWh. This is an overall thermal efficiency of 38%. This includes low loads, high loads and plant startups for actual operation. Actual operating efficiencies are always slightly lower than design efficiencies because a power plant rarely operates at the most efficient load point.

In 1960 Exelon's Eddystone Station Unit #1 was the world's most efficient Ultra-Supercritical Power Plant [47]. It was coal-fueled and designed for a thermal efficiency of 41%. This plant was a pace-setting facility. Low-cost power production made possible by its high efficiency and low fuel cost. Now decommissioned, the plant remained in operation for over 50 years. Given its importance to the industry, the Eddystone facility is now an American Society of Mechanical Engineers (ASME) historical landmark [47]. The newest coal-fueled Ultra-Supercritical Power Plant is the John Turk Plant owned by American Electric Power near Fulton, Arkansas. This plant was commissioned in 2012 and was featured in POWER Magazine's, August 2013 issue. The Turk Plant uses coal fuel and is designed for an operating efficiency of 40%. Another relatively new coal plant is the KCP&L Iatan #2 facility. Also successful and operating at an overall efficiency of about 38%. Longview, Turk, Cliffside and Iatan are amongst the cleanest coal plants in the world with Flue Gas Desulfurization, minimal NOx and particulate emissions.

Modern supercritical power plants (SC) reach efficiencies between 42% and 45%. Ultra-supercritical power plants (USC) employ advanced metal alloys to withstand extreme steam conditions and achieve even higher efficiencies. A record-high net efficiency of 47.5% was achieved by the RDK Block 8 unit in Germany [48,49]. This is due to the elevated steam conditions: Superheat and reheat steam temperatures of 600°C/620°C and steam pressures of up to 275 bar. Advanced ultra-supercritical power plants (A-USC) are expected to enter operation in the next decade and will approach 50% net electricity generation efficiency [50] with the use of advanced metal alloys capable of withstanding steam temperatures and pressures over 700°C and 350 bar [51,52]. These advanced alloys (superalloys) are being developed by adding chromium (Cr), nickel (Ni), cobalt (Co), vanadium (V), wolfram (W), and molybdenum (Mo) to ferritic steels to obtain higher temperature- and corrosion-resistance [53,51].

Metallurgical advances have made USC plants a practical commercial reality. The addition of >10% chromium (Cr) to ferritic steels and >19% Cr to *austenitic steels* dramatically reduced steam-side oxidation problems [51,52,54,55]. At the same time, *feedwater* chemistry is managed by oxygenated water treatment and full condensate polishing is required. Furthermore, furnace enclosure steels of T-12 and T-23 are used rather than the T-2 steels of *supercritical* boilers [51,52,54,55]. Metallurgical advances are being pursued not only for the >1,112°F (600°C) projects of today but also for the advanced ultrasupercritical (A-USB) projects of tomorrow when steam temperatures of 1,400°F (760°C) are targeted [52]. Today the limit appears to be 1,150°F (620°C) using 9Cr–12Cr for creep strength enhancement of ferritic steels [8]. *Chromium alloys* have been essential in USB systems. Advanced austenitic alloys with significant nickel content are required to achieve and manage steam temperatures of 1,300°F (~700°C) and above [52,54]. These materials issues are explored more completely by Zhang [53] and Retzlaff and Ruegger [54]. Retzlaff and Ruegger, in particular, discuss the use of 12CrMoVCbN steel for *turbine* rotors and other applications, along with 10CrMoVCbN steel and 10CrMoVCb [54]. The availability of exotic stainless steels permits the use of higher steam temperatures in power generation.

Supercritical plants provide 10% efficiency gain and a $CO_2$ emission reduction of more than 20% compare to subcritical plants. For example, a conventional coal-fired power plant generates electricity at 36% thermal efficiency while having specific emissions of around 1,000 kg $CO_2$/MWh$_{el}$. An USC unit with 46% thermal efficiency generates 28% (0.46/0.36 = 1.28) more electricity per unit of fuel heat input than the subcritical unit, whereas the emissions are 781 kg $CO_2$/MWh$_{el}$ (1,000/1.28 = 781), a 21.9% reduction. A-USC coal power plants could achieve even lower $CO_2$ emissions, around 700 kg/MWh [56]. Further $CO_2$ emission reductions, down to 100 kg/MWh, would be possible only with the implementation of post-combustion carbon capture and storage (CCS) for the treatment of flue gases released during fossil fuel combustion.

### 2.4.1 ASSESSMENT OF CCS TECHNOLOGY

The Global Status of CCS 2020 [57] reports 26 commercial CCS facilities currently in operation with a total capture and storage capacity of 40.7 million tons of $CO_2$ per year (Mtpa). CCS units are being used in the following industries: Natural gas processing (30.5 Mtpa), power generation (2.4 Mtpa), hydrogen production (2.2 Mtpa), fertilizer production (1.8 Mtpa), methanol and ethanol production (1.6 Mtpa), oil refining (1.4 Mtpa), iron and steel production (0.8 Mtpa). At present, two capture demonstration projects are up and running in the power generation sector: the Boundary Dam power plant and the Petra Nova power plant [58]. Both are using amine-based post-combustion capture, applied on one coal-fired unit each, and the captured carbon is transported via pipelines to enhanced oil recovery fields. Enhanced oil recovery (30.7 Mtpa) and storage in dedicated geological formations (10 Mtpa) are the two types of carbon storages used in these industries. By 2050, the CCS sector is envisioned to grow to a total global installed capacity of 5,600 Mtpa of $CO_2$ [10,57].

The estimated costs of CCS projects for fossil fuel power plants span over wide ranges of values, as reported in the relevant literature [59,60]. The costs of CCS depend on the fuel type, the costs of labor, materials, operation and maintenance, the carbon capture technology, the costs of transport and storage, and the type of project (greenfield or retrofit). The levelized cost of electricity (LCOE) is between 61 and 87 US$/MWh in power plants without CCS and between 94 and 163 US$/MWh for power plants with CCS [59,60]. The LCOE could be reduced to between 61 and 139 US$/MWh when the captured $CO_2$ is sold to enhanced oil recovery projects instead of simply storing it in geological formations. The cost of captured $CO_2$ is between 33 and 58 US$/t$CO_2$, while the cost of avoided $CO_2$ (including compression, transport, and storage) is between 44 and 86 US$/t$CO_2$ [11,14]. Post-combustion CCS for combined cycle natural gas turbines (CCGT) or SCPC power plants with oxy-fuel combustion are predicted to operate with similar costs. Slightly higher costs were estimated for coal-based integrated gasification combined cycles (IGCC) with pre-combustion CCS [60]. The specific costs of $CO_2$ transport are between 2 and 15 US$/t$CO_2$, depending on the pipeline capacity, type (onshore or offshore), and length [60,61]. The specific costs of $CO_2$ storage are estimated between 1 and 18 US$/t$CO_2$, depending on the storage type (depleted oil/gas field, geological formation, or ocean storage) and the potential of using EOR credits. The worldwide $CO_2$ storage capacity is 400 billion tons in discovered capacities in oil and gas fields (depleted or for EOR projects) and 12,000 billion tons in potential (estimated) storage capacities in geological (saline) formations [62]. The $CO_2$ storage capacities are such that exceeds the global net-zero emission scenario. It is estimated that the CCS industry would be cost-effective with carbon prices between 40 and 80 US$/$tCO_2$ [57]. In future cost of CCS can be further reduced if $CO_2$ is utilized like in EOR instead of just being sequestered.

CCS technology offers vast potentials for $CO_2$ reduction in the power industry, in the cement, iron, and steel production industries, and in the oil, natural gas, and chemical processing industries [63]. Different countries are developing their own legal and technical frameworks for future large-scale CCS implementation. Lee et al. [64,65] stressed the importance of developing reliable methodologies for quantifying $CO_2$ emission reduction through CCS at national basis. Nasirov et al. [66] analyzed decarbonization possibilities in developing countries and concluded that wind and solar energy are future of electricity generation, but still 15% of the electricity will be from coal by 2050. Kumar Shukla et al. [67] reviewed the clean coal potentials for the power industry in India. They concluded that post-combustion CCS is the solution for achieving a 30% $CO_2$ emission reduction by 2030 in India. Yun et al. [68] analyzed four scenarios for the power industry in South Korea. They concluded that the CCS coal scenario offers a good perspective in terms of greenhouse gases emissions control and electricity prices. Markewitz et al. [69] studied the potential of CCS technology for the cement industry, which contributes 5% of the global $CO_2$ emissions. They concluded that CCS can remove 70%–90% of the $CO_2$ emissions from the cement industry in Germany at avoidance costs between 77 and 115 €/$tCO_2$. Toktarova et al. [70] analyzed different pathways for the decarbonization of the steel industry in Sweden. They showed that top gas recycling blast furnaces and electric arc furnaces fitted with CCS technology could reduce $CO_2$ emissions by 83% in 2045. Adu et al. [71] studied the $CO_2$ avoidance costs for post-combustion CCS technology integrated in coal-fired and natural gas combined cycle power plants. They concluded that, at 90% $CO_2$ capture efficiency, the $CO_2$ avoidance costs are $72/$tCO_2$ for the coal-fired plant and $94/$tCO_2$ for the natural gas combined cycle.

CCS is an energy-intensive technology affecting substantially the plant performance. In coal-fired power plants, previous studies have reported net efficiency losses in the range between 7% and 11%-pts [72–74] and electricity output penalties between 300 and 400 kWh/$tCO_2$ [75]. Vu et al. [72] compared the techno-economical aspects of carbon capture on USC steam cycles with air-combustion and oxy-fuel combustion. They concluded that oxy-fuel CCS power plants offer an advantage over air-combustion CCS power plants. The net efficiency loss and levelized cost of electricity was 7%-pts and $59/MWh in the first case, whereas 10%-pts and $64/MWh in the second case. Liebenthal et al. [73] estimated the net efficiency loss at 10.94%-pts for a CCS retrofit project on a supercritical power plant achieving 45.5% net efficiency at the design point. Li and Liang [76] assessed a CCS retrofit project for an existing USC power plant with 1,000 MW capacity and estimated the efficiency loss at 8.6%-pts for 90% capture rate and 6%-pts for a 50% capture rate. Xu et al. [77] estimated that modified boiler structures, waste heat recovery, and steam bleed turbines could be used to reduce the efficiency loss from 12.65% down to 8.79%-pts. Jackson and Brodal [78] found that the efficiency loss can be reduced through the

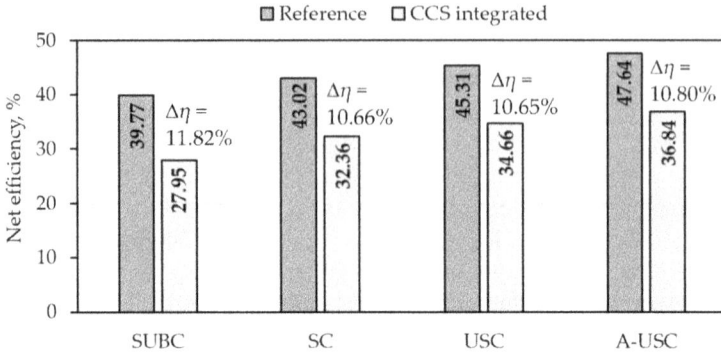

**FIGURE 2.1**   Reference and CCS integrated power plants net efficiencies [48].

optimization of the CCS compression process. Compressor designs with multiple impellers per stage and variable pressure ratios are proposed. A typical comparison of efficiency of subcritical and supercritical coal power plants with and without CCS is illustrated in Figure 2.1.

### 2.4.2   FUTURE OF SUPERCRITICAL COAL POWER PLANTS WITH CCS TECHNOLOGY

The study by Tramosljika et al. [48] makes a step forward by assessing the impact of CCS technology on four different generations of coal-fired power plants including the subcritical (SUBC), the supercritical (SC), the ultra-supercritical (USC), and the advanced ultra-supercritical (A-USC) steam cycle using performance simulation analysis. The simulation approach consists of using appropriate mass and energy balance equations to the power plant components. The obtained results from the simulation code are validated against results found in the literature. Particular attention is given to the power plant-CCS interface quantities and their effects on the efficiency loss and electricity penalty. The performance of the A-USC power plant is assessed through the combined effects of reboiler heating duty and temperature, the compression and intercooling strategy, and the advanced heat integration. The analysis is concluded with the comparison between the performance of present-state and future upgraded CCS technology.

The advanced ultra-supercritical coal-fired power plants are expected to enter operation in the next decade. These steam cycles will use advanced steel alloys and highly efficient steam generators and turbines to generate electricity from steam conditions in excess of 700°C and 350 bar. The estimated thermal efficiency of future planned A-USC units is 50% gross and 47.6% net. Relatively to subcritical units, the net efficiency of A-USC is 8%-pts higher while $CO_2$ emissions are 16.5% lower. However, $CO_2$ emissions per unit of generated electricity would still be around 700 $kgCO_2/MWh_{el}$. Post-combustion $CO_2$ capture and storage can be successfully integrated into A- USC power plants to reduce $CO_2$ emissions by 90% or more, down to 70 $kgCO_2/MWh_{el}$. However, the power plant performance and electricity output are largely affected by the energy requirements in CCS units. The study by Tramosljika et al. [48] analyzed the interface factors between the CCS unit and the power plant affecting the A-USC steam cycle efficiency. Calculations showed that $CO_2$ absorption by 30 wt% MEA solution result with net efficiency losses in the range between 9.7% and 13.4%-pts depending on the reboiler heating duty and temperature. MEA-based blends, piperazine-activated absorbents, chilled ammonia, and mixed salts could be used instead to reduce the efficiency losses by 7.8%–11.7%-pts.

The influence of the number of intercooling stages and $CO_2$ inlet pressures have also been analyzed. For an eight-stage compression process, seven intercooling stages reduce the net efficiency loss by 0.14%–0.18%-pts relatively to four stages. The $CO_2$ pressure at the compression inlet has an even stronger influence. Between inlet pressures of 1 and 2 bar, there is a net efficiency gain of 0.5%-pts while between inlet pressures of 1 and 5 bar the net efficiency gain is 1.2%-pts. Advanced heat integration by recycling of low-temperature waste heat from the CCS unit is found to bring an additional efficiency gain of 0.4%–0.6% pts to the steam cycle. Present-day CCS technology causes a net efficiency loss of 10.8%-pts and an electricity output penalty of 362.3 $kWh_{el}/tCO_2$ in the A-USC baseline scenario. Future CCS technology is expected to develop in the direction of reduced energy demand and improved $CO_2$ absorption rates. Enhanced $CO_2$ absorbents, improved compression strategies, and advanced heat integration could reduce the net efficiency loss and the electricity output penalty down to 7.2%-pts and 241.7 $kWh_{el}/tCO_2$, a 30% improvement over existing CCS technology. From a technical perspective, post-combustion CCS is ready for integration into the fossil fuel electric power industry. However, the electricity penalties and the financial downsides caused by CCS are the principal reasons dissuading electric utilities from implementing CCS at large scale. Legal and financial aspects concerning CCS and the carbon emission trading systems need further upgrades to pave the way towards the power industry decarbonization, for which CCS emerges as the solution. As mentioned before there is also significant momentum towards converting CCS technology to CCUS technology with both potential direct and indirect utilization of $CO_2$.

The share of advanced supercritical coal power plants is expected to rise in order to meet the ongoing growth in global electricity demand [48,57]. Supercritical steam cycles paired to carbon capture and storage (CCS) technologies are going to be integrated into future clean coal power plants [48,57]. By 2030, most of the advanced coal power plants will be upgraded with carbon capture, utilization, and storage (CCUS) technologies [43,44,79]. Advanced ultrasupercritical (AUSC) plants are the only solution to generate electricity in thermal power plants in the most efficient way with minimum pollution to the environment. AUSC plants, however, need to use suitable materials to match the high-temperature and high-pressure steam and *flue* gas conditions. An advanced firing system using pure oxygen instead of air (Oxyfuel combustion) eliminates thermal Nox production while the heat loss toward air heating results in less fuel consumption, higher flame temperatures, and easy $CO_2$ sequestration. The literature studies also focus on improving the integrated gasification fuel cell (IGFC) concepts, the chemical looping concepts, the indirect *coal combustion GT cycles*, the *supercritical* $CO_2$ Brayton cycle, and the bottoming/topping cycles as a means to extract additional energy from the process. Other analyses seek to replace the working fluid for reducing parasitic losses that are intrinsic to using water. Supercritical/ultrasupercritical plants with OT boilers may go for a sliding pressure operation with the possibility of saving some energy [43,44,79].

## 2.5  ADVANCED INDUSTRIAL GAS TURBINES FOR COMBINED CYCLE POWER GENERATION

The new trends for power generation around the world are the *combined cycle power plants* based on the new advanced technology gas *turbines*. The new advanced gas turbines are pushing the limits of technology in the areas of material science due to the very high firing temperatures, and of aerodynamics due to the very *high-pressure ratios* developed in the compressors. The dry low $NO_x$ *combustors* are also pushing the technology in the areas of combustion and flame stability. The concept that combined cycle power plants must be totally *base-load* operated is now a myth, as plant loads must vary during the day. This type of operation requires online total condition monitoring to ensure that the entire plant is operated at its optimum efficiency through the large operating range.

The last 20 years have seen extensive growth in *gas turbine* technology. The growth is spearheaded by the growth of materials technology, new coatings, new cooling schemes and the growth of combined cycle power plants. This, in conjunction with the increase in *compressor pressure ratio* from 7:1 to as high as 45:1 and firing temperatures from 1,400°F to 2,700°F (760°C–1,482°C), has increased the simple cycle gas turbine thermal efficiency from about 15% to over 45%. The growth of the pressure ratio and firing temperature have occurred in parallel, as both are necessary to achieve the optimum thermal efficiency. The increase in pressure ratio increases the gas *turbine* thermal efficiency when accompanied by an increase in turbine firing temperature. The increase in the pressure ratio increases the overall efficiency at a given temperature; however, increasing the pressure ratio beyond a certain value at any given firing temperature can actually result in lowering the overall cycle efficiency.

In recent years, the efficiencies of gas turbine are in the 45%–50% range, which translates to a heat rate of 7,582–6,824 Btu/kW-h. The limiting factor for most *gas turbines* has been the turbine *inlet temperature*. With new cooling schemes using steam or conditioned air and breakthroughs in blade *metallurgy*, higher turbine temperatures have been achieved. The new gas turbines have fired *inlet temperatures* as high as 2,600°F (1,427°C), and pressure ratios of 40:1 with efficiencies of 45% and above. The new advanced gas *turbines* are pushing the technology envelope in pressure (up to 588 psia, 40 Bar), temperature (2,700°F, 1,482°C), low $NO_x$ *combustion systems* (less than 9 ppm), and innovative material technology (single crystal blades). While the advanced gas turbines produce more power, use less fuel, provide higher combined cycle efficiencies, and reduce emission levels significantly, their advantages have been eclipsed by their lower availability (up to 10% lower), lower life of nozzles and blades (averaging 15,000 hours), higher degradation rate (5%–7% in the first 10,000 hours of operation) and instability of low $NO_x$ *combustors*. New advanced turbines are run at higher firing temperatures, are physically larger in size, have larger throughput (airflows and fuel flows), and have higher loadings (pressure and expansion ratios, fewer *airfoils*, larger diameters) than previous *gas turbine* designs. The large size of these gas turbines is one inherent cause of a lower availability and reliability as it takes much longer to do the various inspections and overhauls. Since new advanced turbines are run at higher pressure ratios (as high as 30:1), they create a very narrow operating margin (surge-choke). Thus any deposits on the blade could lead to degraded performance and surge in the compressors. The close tolerances between the casing and the *compressor blades* lead sometimes to excessive rubs. New advanced gas turbines are pushing the temperature envelope. The technologies (design, materials, and coatings) required to achieve the benefits are more complex to concurrently meet gas *turbine performance, emissions, and* life requirements [80,81,37]. The cost of hardware and subsequent cost of ownership have increased due to the complex designs, increased size, and higher throughput in the advanced machines. *Gas turbine* operation has become more complex and computer-driven, thus requiring new and different skill sets for staffing in plants.

The gas turbine consists of three major components; gas turbine compressor, combustor and gas expander. The advanced gas *turbines* are all digitally controlled and incorporate online condition monitoring. The addition of new online monitoring requires new and smart instrumentation. The use of pyrometers to sense blade metal's temperatures is being introduced. The blade metal temperatures are the real concern, not the exit gas temperature. The use of *dynamic pressure transducers* for the detection of surge and other flow instabilities in the compressor and also in the combustion process especially in the new low $NO_x$ combustors is being introduced. Accelerometers are being introduced to detect high-frequency excitation of the blades; this prevents major failures in the new highly loaded *gas turbines*. The use of pyrometers in control of the advanced gas turbines is being investigated. Presently, all turbines are controlled based on *gasifier turbine* exit temperatures or power turbine exit temperatures. By using the blade metal temperatures of the first section of the turbine, the *gas turbine* is being controlled at its most important parameter, the temperature of the first-stage nozzles, and blades. In this manner, the turbine is being operated at its real maximum capability.

The use of dynamic pressure transducers gives early warning of problems in the compressor. The very high pressure in most of the advanced gas turbines causes these compressors to have a very narrow operating range between surge and *choke*. Thus, these units are very susceptible to dirt and blade vane angles. The early warning provided by the use of dynamic pressure measurement at the compressor exit can save major problems encountered due to tip stall and surge phenomenon. The use of a dynamic pressure transducer in the *combustor* section, especially in the low $NO_x$ combustors, ensures that each combustor can be burning evenly. This is achieved by controlling the flow in each combustor can until the spectrums obtained from each combustor can match. This technique has been used and found to be very effective and ensures the smooth operation of the turbine [80,82,37].

Performance monitoring not only plays a major role in extending life, diagnosing problems, and increasing the time between overhauls; but it can also provide major savings on fuel consumption by ensuring that the turbine is being operated at its most efficient point. Performance monitoring requires an in-depth understanding of the equipment being tested. The development of algorithms for a complex train needs careful planning, understanding of the machinery, and process characteristics. In most cases, help from the manufacturer of the machinery would be a great asset. For new equipment, this requirement can and should be part of the bid requirements. For plants with already installed equipment, a plant audit to determine the plant machinery status is the first step. Over the life cycle of gas turbine plant, the initial cost runs about 8% of the total *life cycle cost*, the operational and maintenance cost is about 17%, and the fuel cost is about 75%.

Advanced GT often operates in DLN (dry low NO$_x$) regime. In this regime, the combustor operates in a diffusion mode at low loads (<50% load) and in a premixed mode at higher loads. Advanced GTs with DLN combustion technology may require the ability to both heat and cool fuel gas during startup. As GTs proceed to higher outputs and efficiencies, the fuel gas supply pressure requirements typically exceed 30 bar and can sometimes exceed 35 bar for can-type *combustion systems*. These supply pressures often require supplemental fuel gas compression. During startup at low fuel flow rates, operation of supplemental compressors can result in significant increases in fuel gas temperature due to compression heat. However, DLN *combustors* typically require "cold" fuel for initial startup. As a result, a startup or pilot cooler may be required downstream of the supplemental fuel *gas compressor*. Later on, as the GT ramps up in load, heated fuel would be required for full pre-mix operation and the cooler would need to be isolated or bypassed. On projects with widely varying fuel gas supply pressures, supplemental compressors can often be bypassed, even during startup. Therefore, both a dewpoint heater and a startup/pilot cooler would be required for the fuel gas supply system to meet startup fuel gas requirements under all scenarios. The advanced *gas turbine* requires heat resistant materials as well as advanced cooling technologies. For this purpose, some advanced materials for *turbine blades* and vanes, and thermal barrier coatings are developed. For the rotating blade material, MGA1400 alloy was developed. MGA1400 is a nickel-base super alloy and can be used for DS casting as well as for conventional casting. For the stationary vane material, MGA2400 alloy was developed. MGA2400 is also a nickel-base super alloy that has excellent resistance against *thermal fatigue*, oxidation, and *hot corrosion* as well as high creep strength. It also has good weldability for the settlement of accessory parts and repair [80,81,37].

Typically, advanced GTs use natural gas as the primary fuel. Duel-fuel capability, using distillate oil, is attractive to many customers with interruptible gas supply contracts. However, dual-fuel capability adds complexity to already very complicated combustion systems and controls. A difficult challenge is to achieve a switchover from one fuel to another at a reasonably high GT power level [82]. One of the most critical challenges to the erection of advanced GTs has been the implementation of field modifications. On many occasions, the OEMs used unscheduled outages, not only to correct a problem but also to implement a number of design changes based on lessons learned on other sites. This process created a "ripple effect" requiring additional changes in the plant control software and start sequences. To meet strict *emissions requirements*, all advanced GT combustion systems operate with DLN combustion systems. The combustion process between 70% and 100% load takes place in premix mode with a lean equivalence ratio, which creates a lower localized flame temperature and therefore lower NO$_x$. The premix combustion operation, in which fuel and air are mixed in advance of the combustion process, is less stable than *diffusion flame*. It is common practice to use a pilot operating in diffusion mode to provide stability and inhibit the excessive combustion-related pressure fluctuations inherent in the lean premix operation.

The *combustion system* operation from diffusion mode at low loads to full premix mode at *base load* takes place in several complicated steps and stages, requiring very close control of fuel flow and exhaust temperature. The process is sensitive to ambient conditions, combustion-associated instabilities, and even physical *combustor* dimensions, due to manufacturing or assembly tolerances. Currently, each GT is individually adjusted to meet the performance guarantees and emissions requirements without combustion oscillations. This practice has become a standard feature of GT commissioning. However, this activity has an adverse impact on the EPC contractor. The execution schedule is extended to perform a water wash of the compressor prior to the tuning process and to install and remove temporary instrumentation for full-blown GT performance testing. Because emissions limits must be met at all ambient conditions, adjustments made in the field might modify the performance correction curves for ambient temperature.

The latest advanced F-, G-, H-, and J-Class combustion turbines (CT) require new approaches for optimized combined-cycle (CC) designs than did previous generations of F-Class machines [83]. These machines, along with various market demands, resulting in challenges in overall facility design. Power block and balance-of-plant designs today must incorporate features to accommodate current market conditions and numerous additional demands on the designs, such as fast-start, base load, and cycling operation; high ramp rates; high efficiency; high reliability; lower emissions; and lower life-cycle costs. Other impacts that these larger CTs with increased capacity have on CC designs include: integrating the latest in CT steam cooling or turbine cooling air (TCA); sizing heat recovery steam generators (HRSGs) and steam turbines (STs); material selection for higher steam temperatures; as well as addressing inherent complexities of larger equipment. Today's CC plants are also trending larger in efforts to obtain economies of scale with the installed cost.

Opportunity exists with advanced CTs to increase CC efficiency by integrating heating and cooling systems between both thermodynamic cycles, including fuel gas performance heating, steam cooling of combustor transitions; and energy recovery with the TCA system. Advanced CTs place even greater design demands on the systems providing (a) fuel gas performance heating system, (b) steam cooling of combustor transitions and (c) energy recovery with the TCA system. The HRSG and ST designs must also be considered for the higher CT exhaust gas flow volumes and temperatures, resulting in greater steam-generating capabilities at higher temperatures and pressures. This presents design challenges in specifying equipment for CC configurations. Many factors influence the design of today's HRSGs, including increasing CT exhaust gas flows and temperatures. As such, design demands on HRSGs are increasing. Among the more significant modifications are (a) transition inlet duct design, (b) thermal design for fast start, (c) purge credit and stack dampers for fast start and (d) for matching HRSG steam production and the ST startup [84].

## 2.6 ADVANCED POWER GENERATION BY SUPERCRITICAL $CO_2$ THERMODYNAMIC CYCLES

Much of the recent efforts is focused on improving the overall efficiency and economics of electric power generation. To that end, according to department of energy there are three primary areas of focus for R&D to improve electric power generation efficiency: (a) increasing the fraction of the energy in the heat source that can be harvested for use in the thermal power cycle; (b) increasing the intrinsic efficiency of the thermal power cycle; and (c) decreasing the parasitic power requirement for the balance of plant (BOP). The first two focus areas cannot be pursued in isolation as they are often antagonistic. For example, recuperative heat exchange within the thermal power cycle can often lead to a higher cycle efficiency but this may be at the expense of decreasing the amount of heat that can be transferred into the cycle and lowering the overall process efficiency [85].

Most of the thermal power cycles in commercial operation are either air-breathing direct-fired open Brayton cycles (i.e., gas turbines) or indirect-fired closed Rankine cycles which use water as a working fluid (typical in pulverized coal and nuclear power plants). Within each group are a myriad of potential configurations that vary in size and complexity [86]. For any application, the best thermal power cycle will depend on the specific nature of the application and heat source. In addition to these conventional thermal power cycles, cycles based on other working fluids can be considered. In particular, the Brayton cycle based on supercritical carbon dioxide ($sCO_2$) as the working fluid is an innovative concept for converting thermal energy to electrical energy.

Numerous studies have shown that these $sCO_2$ power cycles have the potential to attain significantly higher cycle efficiencies than either a conventional steam Rankine cycle or even the state-of-the-art ultra-supercritical (USC) steam Rankine cycle [87–89]. Higher cycle efficiency will automatically lead to lower fuel cost, lower water usage, and in the case of fossil fuel heat sources, lower greenhouse gas (GHG) emissions. Further, the $sCO_2$ cycles operate at high pressures throughout the cycle, resulting in a working fluid with a high density which may lead to smaller equipment sizes, smaller plant footprint, and therefore lower capital cost. Achieving the full benefits of the $sCO_2$ cycle will depend on overcoming a number of engineering and materials science challenges that impact both the technical feasibility of the cycle as well as its economic viability.

According to the department of energy, the main R&D challenges arise from the very factors that lead to higher cycle efficiency. These include the use of: (a) elevated pressures throughout the cycle; (b) large duty heat exchangers to minimize the energy lost in cooling the working fluid; and (c) $CO_2$ as the working fluid. R&D will be needed to develop high-efficiency $CO_2$ expansion turbines. These turbines offer the promise of relatively small size because of the low turbine pressure ratio and the high density of the working fluid, but this will be partially offset by the much higher mass flow rates required and the corrosion properties of high-pressure $CO_2$. R&D will be required on seals, bearings, and materials, particularly in applications having elevated turbine inlet temperatures. R&D will also be needed to develop low-cost heat exchangers that are able to attain large heat transfer duties with small temperature differences between the hot and cold sides of the exchanger and with a small pressure drop. This will require R&D into compact heat exchanger designs, assessment of materials for suitability given the temperatures and pressures required, and advances in manufacturing techniques [85].

Power cycles using $sCO_2$ as the working fluid take on two primary configurations relevant to power generation: (a) an indirect-fired closed Brayton cycle that is applicable to advanced fossil fuel combustion, nuclear, and solar applications; and (b) a semi-closed, direct-fired, oxy-fuel Brayton cycle well-suited to fossil fuel oxy-combustion applications with $CO_2$ capture. In the supercritical state, carbon dioxide can be used efficiently throughout the entire Brayton cycle. Closed Brayton cycles for power generation have a commercial history dating to the late 1930s when the first of a series of power generation plants using the closed Brayton cycle made use of air as a working fluid. These closed-air Brayton units operated from the late 1930s through the 1970s using a variety of energy sources including coal, gas, and waste heat. These recuperated Brayton cycle installations achieved good reliability and relatively high efficiencies as compared to other power generation technologies of the era.

The range of potential applications for the indirect $sCO_2$ Brayton cycle is broad since it can be used in essentially any application that currently uses a Rankine cycle. Generally, the operating conditions where the recompression $sCO_2$ Brayton cycle attains its highest efficiency require a large degree of thermal recuperation. This reduces the heat loss in the $CO_2$ cooler and allows the heat source to heat the maximum amount of working fluid and hence, generate the maximum amount of power output. A potential disadvantage of this high degree of recuperation is that the temperature increase of the $CO_2$ in the heat source is relatively low. If the hot source operates across a wide temperature range it will create challenges in maintaining high cycle efficiency without discarding a significant portion of the available hot source energy. Many of the promising applications for indirect $sCO_2$ Brayton cycles have heat sources that have a narrow temperature range. Examples include applications with nuclear, solar, and geothermal heat sources. In each of these cases, the $sCO_2$ Brayton cycle operating state can be configured to utilize the maximum amount of energy available from the hot source. When the hot source temperature range is large, more complex modifications to the cycle are generally required. This may entail a higher degree of process-level heat integration, or reduction in cycle recuperation to increase the amount of hot source energy that can be utilized in the cycle, employing a more complex cascade cycle configuration, or possibly using a combined cycle process in which the $sCO_2$ Brayton cycle serves as the topping cycle and a Rankine cycle is used as a bottoming cycle. Conceptual designs have been proposed for each of these alternatives [90–92].

s-$CO_2$ cycle is versatile and can be applied to numerous types of heat sources. According to the department of energy, Table 2.1 provides a listing of the major categories of applications for the $sCO_2$ Brayton cycle [86], the expected cycle configuration, the peak temperature for the working fluid, and the major benefits the $sCO_2$ Brayton cycle may potentially demonstrate in each application. The principal benefit of the $sCO_2$ Brayton cycle is its potential for an increase in both cycle and process efficiency compared to processes that employ Rankine cycles. There are several secondary benefits to increase process efficiency. These include a reduction in the thermal input needed to generate a fixed amount of power which lowers the size and capital cost; and for some applications, it also lowers the fuel usage and operating costs. Increasing process efficiency also diminishes the environmental footprint of the process by reducing water usage and in the case of fossil fuel applications, reducing greenhouse gas emissions. As shown in Table 2.1, Closed Brayton cycles using supercritical $CO_2$ are actively being pursued as a high-efficiency cycle for the next generation of power blocks by a diverse set of technology areas including nuclear

**TABLE 2.1**

**Potential Applications for sCO$_2$ for Power Conversion [85]**

| Application | Cycle Type | Motivation | Size [MWe] | Temperature (°C) | Pressure [MPa] |
|---|---|---|---|---|---|
| Nuclear | Indirect sCO$_2$ | Efficiency, Size, Water Reduction | 10 – 300 | 350 – 700 | 20 – 35 |
| Fossil fuel (PC, CFB, …) | Indirect sCO$_2$ | Efficiency, Water Reduction | 300 – 600 | 550 – 900 | 15 – 35 |
| Concentrating solar power | Indirect sCO$_2$ | Efficiency, Size, Water Reduction | 10 – 100 | 500 – 1,000 | 35 |
| Shipboard propulsion | Indirect sCO$_2$ | Efficiency, Size | <10 – 10 | 200 – 300 | 15 – 25 |
| Shipboard house power | Indirect sCO$_2$ | Efficiency, Size | <1 – 10 | 230 – 650 | 15 – 35 |
| Waste heat recovery | Indirect sCO$_2$ | Efficiency, Size, Simple Cycles | 1 – 10 | <230 – 650 | 15 – 35 |
| Geothermal | Indirect sCO$_2$ | Efficiency | 1 – 50 | 100 – 300 | 15 |
| Fossil fuel (Syngas, nat gas) | Direct sCO$_2$ | Efficiency, Water Reduction, CO$_2$ Capture | 300 – 600 | 1,100 – 1,500 | 35 |

power generation, fossil fuel power generation, concentrated solar power, shipboard power, waste heat recovery, and geothermal power generation. Some of these applications are described in more detail in relevant chapters on nuclear heat, geothermal heat, solar heat and waste heat. This chapter focused on the use of s-CO$_2$ cycle for fossil fuel applications. The use of s-CO$_2$ cycle as a power block within the marine industry can be carried out either as part I ship propulsion system or for on-board power. Mendez and Rochau [93], noted that the US Navy is interested in gas turbine-generator sets with a power rating between 20 and 30 MWe, which could be a promising application for sCO$_2$ technology. The US Navy has previously explored the use of Echogen's sCO$_2$ system within marine applications, with the results suggesting fuel consumption could be reduced by 20% [94].

A direct comparison of the conventional Rankine cycle with the RCBC (Recuperated closed Brayton cycle) is difficult because the Rankine cycle is an established and mature technology and has undergone a century of development and refinement. The state-of-the-art in Rankine cycles today is the ultra-supercritical (USC) cycle having the main steam pressure of 250–290 bar and temperature of 600°C with a reheat temperature of 620°C. Since there are no commercial-scale power plants based on the RCBC, any comparison must be based on assumptions about the operating point. Although the nature of these two cycles is different, they both exhibit an increase in efficiency as the turbine inlet temperature increases. However, the magnitude of that increase will be different for the two cycles and hence each cycle will have a range of turbine inlet temperatures over which its efficiency is higher than the other cycle. There have been some limited comparisons of the performance of these two power cycles in the literature [85,95] and they consistently show that the RCBC has a higher cycle efficiency at moderate to high values of

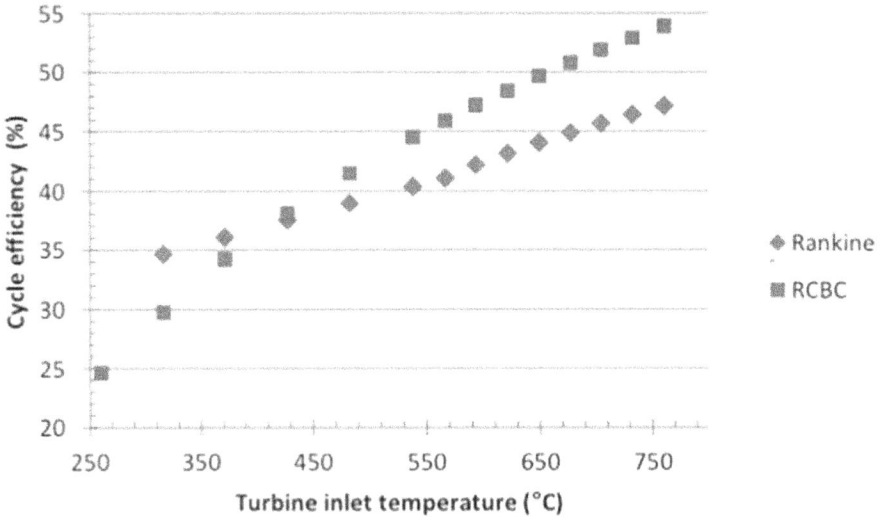

**FIGURE 2.2** Cycle efficiency of Rankine versus RCBC [85].

the turbine inlet temperature. The exact value of the turbine inlet temperature where the RCBC attains a higher efficiency will vary depending on the selected cycle configurations and assumptions used for the operating state for the RCBC. Figure 2.2 shows the results of a systems analysis performed at NETL comparing the RCBC with a Rankine cycle having a single reheat. In this analysis, the turbomachinery efficiencies for the two cycles were made equal. The results show the same trend as in prior studies and show that the RCBC has a higher efficiency than the Rankine cycle when the turbine inlet temperature exceeds approximately 425°C.

Just like the potential for higher process efficiency, other potential benefits of the s-CO$_2$ Brayton cycle remain to be demonstrated. For example, high density of working fluid should allow size reduction of some of the unit operations and thereby reduction in \$/kWe cost. Unfortunately, not all of the properties of the sCO$_2$ Brayton cycle lend themselves to size and cost reductions. For example, the sCO$_2$ Brayton cycle is more complex than the Rankine cycle, requires compressors instead of feedwater pumps, and requires recuperators having larger heat duties than the heat source. Another potential benefit is that the sCO$_2$ Brayton cycle may prove to be more practical than the Rankine cycle for air cooling in locations where water cooling is not available [96]. Finally, the cycle facilitates CO$_2$ capture from the combustion process and it can be applied to direct-heating applications, with potential for high efficiency and capture of process water.

A broad potential for power generation and propulsion of supercritical CO$_2$ technology is outlined in an excellent recent review by White et al. [97]. A market analysis conducted by Sandia National Laboratories projects an LCOE for an sCO$_2$ system between 44.8 and 56.1 \$/MWh for power ratings between 100 and 300 MWe [98], but these values are not linked to a specific application. Supercritical CO$_2$ power systems were originally conceived to overcome the limitations of steam power plants.

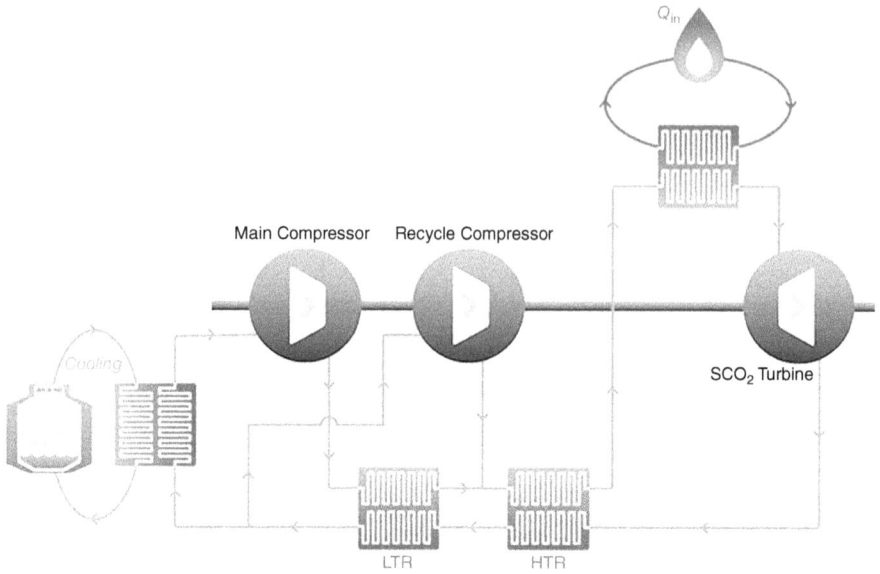

**FIGURE 2.3**   Indirect-fired supercritical $CO_2$ recompression Brayton cycle [85].

While $sCO_2$ power concepts regained popularity in the early 2000s for nuclear applications, the need for integration of carbon capture system with fossil fuel-driven power plants, required operational flexibility to accommodate the increasing penetration of renewable energy sources and its low footprint facilitated the use of $sCO_2$ cycles for base-load power generation applications. As mentioned earlier, fossil-fueled $sCO_2$ power plants are commonly classified based on the heat addition method [99]: in indirect heating (see Figure 2.3), the $sCO_2$ power block is a closed loop system where heat addition and rejection take place through heat exchangers; in direct heating (see Figure 2.4), the $CO_2$ loop is an open system in which the heat input is from combustion processes involving fuels and oxidants circulating together with $CO_2$. Hence, the $sCO_2$ unit acts as a typical bottoming heat to power system, such as steam or organic Rankine cycles. The bottoming cycle approach is also applicable for waste heat recovery.

## 2.6.1   INDIRECT $sCO_2$ HEATING—HEAT TO POWER

Heat to power generation through bottoming $sCO_2$ cycles aims at the recovery of thermal energy from the topping process to maximize the power output. This goal differs from maximizing cycle efficiency, which is typical of applications in which heat is generated at a cost, i.e. nuclear or fuel combustion. With the exception of coal-fired power plants, in indirect heating applications, the heat stream comes as a by-product of the main process. For this reason, highly recuperative cycles are not preferred for waste heat recovery. Instead, split cycles, that divert part of the $CO_2$ flow downstream of the compressor directly to the heater rather than to the recuperators, are more suited [100–102]. Bottoming heat to power cycles based on

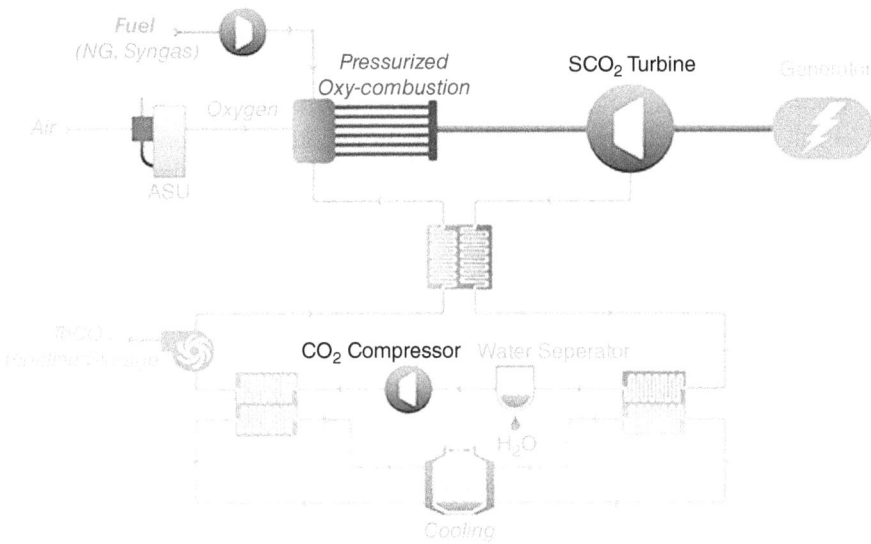

**FIGURE 2.4**  Oxygen-fueled directly-fired supercritical $CO_2$ cycle [85].

steam or organic working fluids are mature technology. A number of providers offer commercial solutions even in the kilowatt power range. Based on literature and business cases developed for industry, $sCO_2$ technology becomes particularly competitive for high-grade heat sources (>350°C) and large-scale applications (>0.5 MWe). $sCO_2$ cycle can tackle high-temperature sources, even beyond the operating ranges of ultra-supercritical (USC) steam power cycles (600°C–620°C), with favorable economies of scale and lower footprint. Potential applications for $sCO_2$ power units are gas turbines, reciprocating internal combustion engines and waste heat from energy-intensive industries.

The global gas turbine market was valued at more than $6 billion in 2019. With the additional installed capacity of 43.7 GW worldwide during 2020–2026 [103]. The 250–500 MW segment is considered the backbone of the gas turbine power generation industry [104] with China and India are the hot spots for demand [105]. In a gas turbine, waste heat is totally concentrated in the turbine exhaust in the temperature range of 350°C–600°C [106] and Supercritical $CO_2$ cycles could be employed as an alternative to current bottoming steam sections in greenfield or retrofitted combined cycle gas turbine (CCGT) power plants.

Genset power units are composed of an electrical generator driven by a reciprocating internal combustion engine which can be fueled with natural gas, diesel, gasoline, biofuels etc. The segment above 750 kVA is considered the backbone of the genset industry. The main market driver for gensets is the need for continuous and uninterrupted power supply, especially for data centers and those industries impacted by the digital revolution [107]. About one-third of fuel energy input to genset is wasted through the engine exhaust at temperatures ranging from 350°C up to 670°C [108]. While supercritical $CO_2$ power cycles can be employed as an alternative to the current bottoming ORC sections, considering the size of the reciprocating internal combustion engines, the power output is expected to be below 1 MWe. This aspect, together with a possible low capacity factor, makes the economic feasibility of the $sCO_2$ retrofit problematic.

Waste heat potential of industry at temperature above 300°C is 11.4% at the global level and 8.7% at the European level [109,110]. Iron and steel, non-ferrous metals (i.e. aluminum) and non-metallic minerals (i.e. glass and cement) have the greatest potential for this level of waste heat [111–114]. Common elements of these industrial processes are the large size of the application, the coexistence of convective and radiative mechanisms in the heat source (mostly due to furnaces), the uninterrupted nature of the process and the need to comply with emission trading systems regulations. These aspects are in favor of a waste-heat recovery retrofit with $sCO_2$ systems, which unlike steam or organic Rankine cycles, would be more compact and able to fully tackle the high-temperature recovery opportunity of these energy-intensive industries.

Indirect $sCO_2$ heating is also being considered using coal as source. Coal power still contributes 37% of today's world electricity supply and is mostly used for base-load power generation [115]. Decarbonization and flexibility are the two drivers for retrofitting or upgrading existing coal-fired power stations with $sCO_2$ technology. The majority of the published research has focused on cycle analysis and thermo-economic assessments, including carbon capture systems [116–118]. There is strong agreement in the literature that the $sCO_2$ 'boiler' is the major technological bottleneck. Compared to a steam boiler at the same duty, a coal-fired $sCO_2$ system has larger heat transfer surfaces due to the higher mass flow rates, lower heat transfer coefficients (3–5 kW/m²K [119]) and lower pressure drop requirements [120]. Furthermore, Carbon dioxide enters the heater near supercritical conditions and exits the heater at temperatures higher than the 620°C of USC boilers [121]. This poses challenges in the selection of material due to high-temperature corrosion issues both inside (due to $sCO_2$) and outside (due to flue gas) of the boiler. Non-stationary load profiles, requirement of different control strategies [122] and the cost limitations are also problematic.

## 2.6.2  DIRECT sCO₂ HEATING

The $sCO_2$ Brayton cycle can also be configured for direct heating which increases its range of potential applications. The most promising application areas for direct cycles are with fossil fuel sources. Generally, process efficiency improvement dictates the nature of specific application. This is most often connected to the improvement in cycle efficiency. Direct cycles also provide an intrinsic method to capture the water generated during combustion as liquid water which will partially offset the water withdrawal in a water-cooled application. Oxy-fired direct cycles for fossil fuel applications have the additional benefit of facilitating $CO_2$ capture which is important due to legal limitation of $CO_2$ emissions from new coal-fired power plants to 1,400 lb $CO_2$/MWh-gross [123]. Direct heating $sCO_2$ power cycles (see Figure 2.4) are open-loop internal combustion engines underpinned by variants of the Joule–Brayton cycle and in which the heat input typically results from an oxy-fuel combustion using either natural gas or syngas from a coal gasification process.

Hence, this concept is applicable and is being developed both for gaseous and solid fossil fuels. The $sCO_2$ heater found within indirect cycles is replaced by a combustor in which the heat of the oxy-combustion is diluted with $CO_2$ entering the combustor after a regenerative heating at temperatures around 750°C. The direct heating not only allows higher cycle temperatures than indirect $sCO_2$ cycles but can also deliver optimal performance at higher pressure ratios. As a result, at the turbine inlet (200+ bar and 1,100+°C) the working fluid has a higher power density compared to indirect heating cycles. Besides the theoretical efficiency advantages resulting from high turbine inlet temperature and cycle pressure ratio, the most interesting aspect of these advanced cycle architectures is the capability to perform a carbon sequestration together with power conversion (without additional equipment). This can be achieved because products of oxy-combustion are primarily $CO_2$ and water; water can be separated downstream of the cooler; the maximum cycle pressure exceeds conventional $CO_2$ pipelines (110–150 bar [124]). Hence, the $CO_2$ can be separated downstream of the compressor such that the $CO_2$ flow at the inlet of the high-pressure side of the recuperator is constant, even if the $CO_2$ loop is open.

A breakthrough in the field of direct heating $sCO_2$ fossil power generation was the invention of the Allam–Fetvedt cycle commonly referred to as the Allam cycle. The concept was patented in 2011 [137] and is being up-scaled to full-scale demonstration through initiatives led, partly or fully, by eight Rivers Capital both for natural gas (50 MWth NET Power's plant in La Porte, Texas (US) [138]) and coal-fired (300 MWe 'Allam Cycle Zero Emission Coal Power' project co-funded by US DOE [125]) power generation. The Allam cycle is a simple regenerative, semi-closed $sCO_2$ cycle in which the imbalance between residual enthalpy at the low-pressure turbine exhaust and the heat required to raise the temperature of the high-pressure $CO_2$ flow prior to combustion is compensated through an external heat addition at the recuperator. This external regenerative heat is provided either by the air separation unit (ASU) in the gas-fired version of the cycle or by the coal gasifier in the coal-fired Allam cycle [126,127]. Since the invention of the Allam cycle, alongside thermodynamic and techno-economic studies [128–130], a large body or research is currently aiming to tackle a number of issues related to direct heating $sCO_2$ cycle. These include $CO_2$ impurities due to fuel, nitrogen, water; corrosion aspects in coal gasification; $CO_2$ corrosion at high and low temperatures; combustion dynamics in high-density flow [131–133]; turbine blade film cooling and erosion due to impurities [134]; recuperator erosion due to impurities and stress magnitude due to high cycle pressure ratios; alternative control strategies to turbine throttling [135]; materials [136]; levelized costs of electricity in comparison with other oxy-fuel power generation concepts [124] and others.

For fossil-fueled applications, using the $CO_2$ stream as a heat carrier of an oxy-fuel combustion not only allows high cycle efficiency (up to 55.1% based on lower heating value of natural gas) but inherently provides the capability to further sequestrate the $CO_2$ generated from combustion. These features have triggered a strong industrial interest that is leading to the first full-scale demonstrations of $sCO_2$ Allam–Fetvedt power cycles [137,138]. However, the high cycle pressure ratios and turbine inlet temperatures of direct-fired $sCO_2$ introduce challenges related to combustion in high-density flows, turbine film cooling, material erosion and corrosion.

### 2.6.3 FUTURE TRENDS

Power cycles operating with supercritical carbon dioxide (sCO$_2$) have advantages of high thermal efficiencies using heat-source temperatures ranging between approximately 350°C–800°C, a simple and compact physical footprint and good operational flexibility [139–141]. These advantages make them promising candidates for future energy applications where their adoption could lower levelized costs of electricity compared to existing technologies. While significant research has addressed a number of issues on technology, significant hurdles including the successful demonstration of the technology at an appropriate industrial scale remain.

Advanced turbine-based cycles like supercritical CO$_2$-based (sCO$_2$) power cycles have shown the potential for increased heat-to-electricity conversion efficiencies, high power density, and simplicity of operation compared to existing steam-based power cycles. The sCO$_2$ power cycle utilizes small turbomachinery, is fuel- and/or heat-source neutral, and efficient. An advanced turbine system for supercritical CO$_2$ cycle is illustrated in Figure 2.5. The U.S. DOE continues its efforts to push the limits of turbine performance in response to the nation's increasing power supply challenges by focusing on the underlying factors affecting combustion, aerodynamics/heat transfer, and materials for advanced turbines and turbine-based power cycles. Temperature continues to be the barrier to increasing turbine efficiency. Research being pursued to allow turbines to operate in excess of 3,100°F, with low NO$_x$ emissions, increased power output, and efficiencies over 65%. Some of the technologies that will enable this transformational jump in capabilities to include ceramic matrix composites (CMCs) for airfoils and combustion components, advanced low-NO$_x$ micro-mixer combustion system that can efficiently fire multiple fuels at different loads while keeping emissions low, and pressure gain combustion. Pressure gain combustion is an alternate form of combustion that increases pressure through the combustor compared to standard combustion techniques that result in a pressure loss. Integrating this technology into a combustion turbine could provide further performance increases.

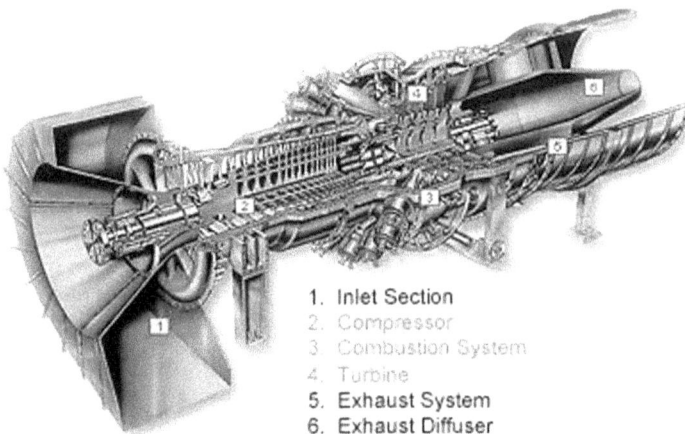

1. Inlet Section
2. Compressor
3. Combustion System
4. Turbine
5. Exhaust System
6. Exhaust Diffuser

**FIGURE 2.5** Advanced turbine system for supercritical CO$_2$ cycle [98]. (Courtesy of Siemens Westinghouse.)

Supercritical $CO_2$ power cycles using advanced turbomachinery could offer efficiency and performance improvements for some fossil energy cycles. The turbines for these cycles are unique in that they will have high power density, lower peripheral speeds, high blade loading, and high shaft speeds, all of which will factor into the final turbine designs. The high pressure, relatively high temperature, uncertainty of the $CO_2$ state near the critical point, and high power density create design challenges for the supercritical $CO_2$ turbomachinery. Advanced Ultrasupercritical (AUSC) power cycles offer efficiency and performance improvements for steam turbines operating above 700°C and 220 bar main steam temperature and pressure. Steam turbine designs for these cycles and higher will have turbomachinery challenges related to materials of construction, shaft end seals, turbine blade seals, and turbine control/ bypass valves.

NETL supports [142] the development of $sCO_2$ technology, including the Supercritical Transformational Electric Power (STEP) program, which includes the design, construction, and operation of a 10 MWe pilot scale $sCO_2$ facility. DOE is working with partners such as the Gas Technology Institute, Echogen Power Systems, Southwest Research Institute, and General Electric to develop and mature the technology at the pilot scale. The program goals are to facilitate the commercialization of $sCO_2$ Brayton Cycles and spur the development of necessary designs, materials, components, operation and control systems, sensors, understanding, and characterization needed for larger scale $sCO_2$ power conversion systems. The program goal is also to develop highly efficient and lower cost indirectly-heated power cycles that surpass the performance of advanced ultrasupercritical steam and provide the technology base for directly-heated power cycles using more advanced fossil energy conversion systems based on $sCO_2$ cycles. The direct-fired cycle can facilitate carbon capture by producing a high purity stream of carbon dioxide that is ready for use/reuse or storage. The $sCO_2$ Technology supports research and development (R&D) in five key technology areas: Turbomachinery, Recuperators, Materials, Advanced Concepts for Direct-Fired Cycles, and Systems Integration & Optimization.

Turbomachinery needs to be demonstrated at an industrial scale (i.e., ≥10 MWe) as well as for small-scale applications (i.e., <1 MWe). More R& D is needed to operate turbomachinery beyond the existing design boundaries. Further investigation to characterize compressor operation near the critical point and introduction of suitable design solutions are needed. Finally, experimental tests are necessary to provide suitable validation or introduce modifications to existing tools for turbomachinery. Achieving recuperator cost and performance targets is vital to enabling commercial implementation of the $sCO_2$ power cycle. This requires optimization of heat transfer at both local and system levels. Further work is needed to develop designs and manufacturing processes that lead to relatively low-cost, high-performance heaters that can withstand thermal cycling and fatigue and can satisfy environmental constraints in different applications. Work is also needed to further withstand high-temperature pressure differentials and flow passage design that maximizes heat transfer performance and reduces pressure drop both for heat exchangers and air and water-coupled coolers. More work is needed to improve strength, durability and environment compatibility of materials under different operating conditions. Manufacturing and fabrication issues related to high-temperature materials and coatings also need to be further evaluated. Some of the advanced concepts of s-$CO_2$ cycle need to be further evaluated. These include R&D to develop high-pressure and high-temperature oxy-fuel combustors to be used with $CO_2$ as the diluent, as well as the integration of the combustors with the turbomachinery for use in directly fired s$CO_2$ power cycles. The challenges associated with this R&D include understanding combustion kinetics and dynamics under these conditions, combustion stability, flow path design, thermal management, pressure containment, and definition of turbine inlet conditions. Finally, more investigation and development are needed for overarching control architectures based on multi-variable control approaches that integrate the power block with the heat source and heat sink. This can provide opportunities for better integration and synergistic operation of the overall energy system [139–141].

## 2.7 CHEMICAL LOOPING COMBINED CYCLE POWER PLANTS

While significant efforts are being made to reduce carbon emission in the atmosphere during combustion based power generation using carbon capture strategy, most of them are accompanied by energy penalty which results in enhanced cost. Recent studies indicate that the traditional $CO_2$ capture technologies such as amine-based absorption integrated with natural gas (NG)-based power plants result in energy penalties of 7.6%–8.4%-points [143–145]. It increases fuel cost as well as capital cost due to increase in the plant size required to provide a certain power output. In addition, the costs associated with transport and storage increase the overall CCS costs significantly [146]. The higher fuel use for this purpose counteracts some of the environmental benefits of $CO_2$ capture. Therefore, it is necessary to develop energy systems integrated with carbon capture facilities, which are highly efficient and low in energy penalty.

Chemical looping combustion (CLC) is an innovative process with inherent $CO_2$ capture at minimal energy penalty [147]. *In this process*, the energy penalty is only for compression to storage conditions due to segregated handling of the fuel and the oxidizer. CLC process involves two interconnected reactors: a fuel reactor (FR), where the OC (oxygen carrier, generally a transition metal oxide) reduction by the fuel takes place producing $CO_2$ and steam ($H_2O$), and an air reactor (AR), where the OC is oxidized by the incoming air. The steam is condensed to obtain a pure stream of $CO_2$ ready for storage. The heat released in the AR maintains the thermal balance of the system and produces a high-temperature outlet stream that is used to drive a power cycle. The generalized reactions in the two reactors are shown below.

$$\text{Reduction} \quad 4MeO + CH_4 \rightarrow 4Me + 2H_2O + CO_2 \tag{2.1}$$

$$\text{Oxidation} \quad 4Me + 2O_2 \rightarrow 4MeO \tag{2.2}$$

A number of references have given comprehensive details of the CLC process [148,149]. The power generation systems based on CLC are attractive due to the absence of the $CO_2$ separation step. However, the overall net electrical efficiency of CLC systems is limited by relatively low operating temperatures. The maximum operating temperature of the OC used in the CLC reactors is well below the inlet temperature that can be achieved by modern gas turbines (GTs). Lower turbine inlet temperature (TIT) restricts the net electrical efficiency similar to that of a natural gas combined cycle (NGCC) plant integrated with $CO_2$ capture facility. Significant efforts in the literature have been made to improve efficiency of CLC-based power generation processes. Ishida and coworkers [147,150,151] in a series of articles reported the efficiency of CLC power generation system with a GT cycle at 50.2%, similar system with air saturation with the efficiency of 55.2% and finally the system with $CO_2$ generation of 0.33 kg-$CO_2$/kWh having efficiency of 60%. The efficiencies reported in these studies are promising; however, the studies were carried out considering simple power cycles with different strategies such as air saturation, which is generally not the case. These systems require a large amount of water, which is generally irrecoverable and will result in corrosion when the flue gases are cooled below dew temperature. Furthermore, the energy penalty for $CO_2$ compression to high pressures (HPs) was also excluded. Naqvi et al. [152] introduced a steam cycle (SC) to recover the heat from the GT exhaust gases. A two-pressure heat recovery steam generator (HRSG) was used to produce the steam for power generation in steam turbines (STs). In addition, a three-stage-intercooled compression system for compressing $CO_2$ to pressures up to 200 bar was also included in the plant. They reported the net plant efficiency at full load to be 52.2%.

Naqvi and Bolland [153] proposed a multi-pressure CLC plant with single- and double-reheat systems of the AR exhaust at the same oxidation temperature. They reported that the net electrical efficiency for a single-reheat system reaches above 53% at an oxidation temperature of 1,200°C. The double-reheat system resulted in a slight efficiency improvement over the single-reheat system. The efficiency of about 51%–52% was also reported by Hassan et al. [154], Ekström et al. [155] and Porrazzo et al. [156]. Zerobin and Pröll [157] developed a process model of a pressurized CLC system and compared the performance with a simple gas turbine combined cycle (GTCC) plant consisting of a single-pressure HRSG. The study showed that the efficiency of the CLC system can be significantly improved by additional firing after CLC reactors. Petriz-Prieto et al. [158] investigated 15 different configurations consisting of three CLC systems, three power generation systems, and two OCs. The power generation systems include an SC, a steam-injected gas turbine (STIG), and a humid air turbine (HAT) cycle. The OCs considered in this study were nickel-based and iron-based. The average efficiencies reported for Ni-based OC plants: CLC with SC, CLC with STIG, and CLC with HAT were 45.92%, 47.4%, and 53.21%, respectively. Farooqui et al. [159] investigated the syngas production by $H_2O/CO_2$ splitting in a chemical looping unit and burning with oxygen from an air separation unit. The flue gas is expanded in a GT followed by heat recovery in an SC. The efficiency reported for the plant integration was 50.7%.

The studies reported above used dual circulating fluidized bed reactors in their CLC systems. While fluidized bed reactors under high pressure present some challenges, several studies have successfully used them in a variety of CLC systems [160–163]. While these studies suggest that dual fluidized bed reactors in CLC system are technically feasible, more demonstration under high pressure is needed. In addition, there are several alternative chemical looping configurations under development to simplify pressurized operations such as packed beds [164], gas switching reactors [165], rotating reactors [166], and internally circulating reactors [167]. These developments should accelerate the development of pressurized chemical looping technology. In recent years large-scale CLC process with moving bed reactors is also tested [168].

While turbine inlet temperature, the pressure drop in the reactor, gas leakage between the two reactors, etc. all have significant influences on the net electrical efficiency [157], the TIT is known to have the largest influence on net electrical efficiency [169]. As mentioned earlier, in typical CLC systems, the TIT is limited by the reactor temperature, which is maintained between 800°C and 1,200°C [170]. This reactor temperature corresponds to the combustor outlet temperature (COT) in GTs. The TIT is defined at the first GT rotor and will be lower than the COT due to blade cooling. Increasing the COT beyond the aforementioned range is beneficial for the overall net electrical efficiency of a CLC plant. However, higher temperatures result in attrition and agglomeration of the OC material. Furthermore, thermal sintering of the OC material occurs at about 70% of the melting point [171]. This means that materials with a high melting temperature and high attrition and agglomeration resistance need to be developed [172]. One way to match COT with that of modern GT is to introduce a combustion chamber downstream of the AR. The fuel is burnt in the high-temperature oxygen-depleted air that raises the COT to the desired level.

Khan et al. [173] added a combustor (COMB) to the CLC combined cycle plant upstream of the GT to raise the COT beyond the achievable CLC temperature and examine its effect on the plant's overall electrical and carbon capture efficiencies. NG or hydrogen were used as fuel in the added COMB. The study also considered the effects of steam injection and $O_2$-depleted air circulation to mitigate $NO_x$ formation due to an increase in the COT. The flow sheet of NGCC plant with COMB is shown in Figure 2.6. In this process, air is compressed and introduced into the COMB, whereas the fuel is preheated before injecting into the COMB. The high-temperature flue gases are expanded in a GT to near atmospheric pressure. The heat contained by the flue gas at the turbine outlet is further recovered in an SC. For simplicity, a single GT with a single HRSG and ST system is considered. The plant specifications and the main assumptions used are outlined by Khan et al. [173]. The effect of varying COT on the plant net electrical efficiency and carbon capture efficiency is shown in Figure 2.7. The variation in COT is achieved by varying the amount of NG burnt in the COMB. The results show that as the COT is increased, the net electrical efficiency increases, whereas the carbon capture efficiency decreases significantly. The improvement in net electrical efficiency is due to the reduction in power consumption of the AC and increased power generation by the STs with respect to the thermal input. The corresponding carbon capture efficiency is only 59.25%, which is low compared to the base case CLC combined cycle power plant (100% capture).

Khan et al. [173] also examined the effect of varying the AR outlet temperature on the net electrical and carbon capture efficiencies. The temperature was varied from 1,000°C to 1,300°C by controlling the air flow rate in the AR (1,111–721 kg/s). Moreover, the COT was kept constant at 1,416 °C by controlling the NG flow rate into the COMB (12.39–2.13 kg/s). The results showed that with an increase in the exhaust temperature the net electrical efficiency fell from 56.1% to 53.9%, whereas the carbon capture efficiency increased from 54.38% to 86.94%. The fall in the net

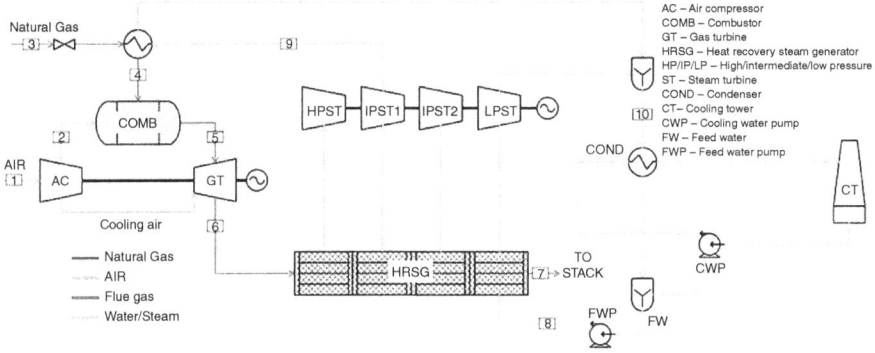

**FIGURE 2.6** Process flow sheet of a NGCC plant with COMB [173].

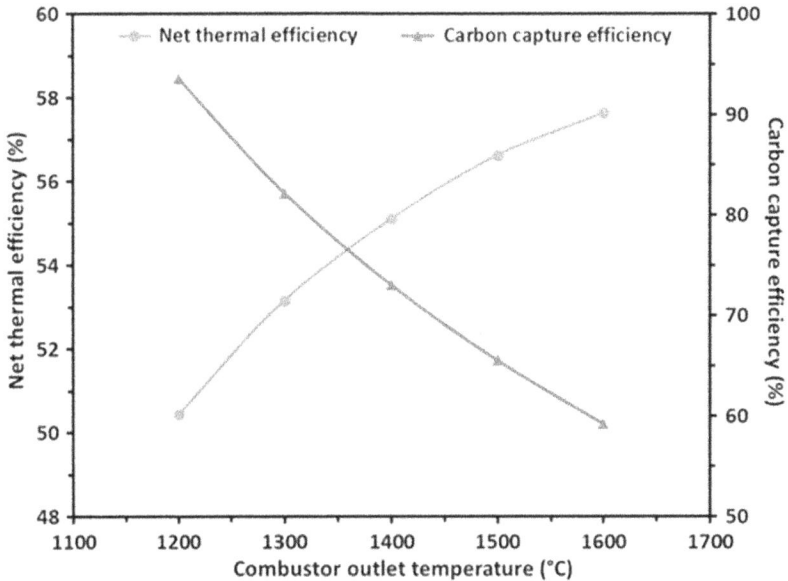

**FIGURE 2.7** Effect of COT [173].

electrical efficiency is due to the reduction in the mass flow rate of the AR exhaust. The two main reasons as to why fuel combustion in CLC is less efficient than fuel combustion in the added COMB are that (a) energy recovery from the FR outlet is less efficient due to the lower temperature of that stream and (b) more $CO_2$ compression is required. The rise in the carbon capture efficiency is due to the reduction in the NG flow rate into the COMB.

Khan et al. [173] also considered the process flow diagram of CLC integrated with a combined cycle plant and the CLC combined cycle plant with an additional COMB. Figure 2.8 presents the CLC integrated with a combined cycle plant with an additional COMB. The fuel supply, steam injection, and exhaust gas recirculation

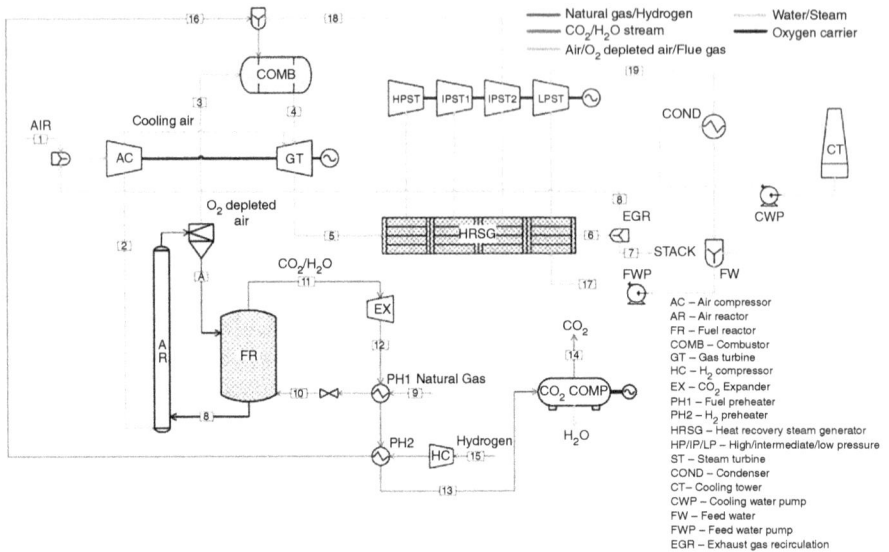

**FIGURE 2.8** Process flow sheet of a CLC plant with an additional COMB [173].

arrangements are also shown. The main objective of adding a COMB is to increase the COT beyond the achievable CLC temperature. Different fuel flowrates (either NG or $H_2$) are fed to the added COMB to vary the COT and evaluate its effect on the plant performance. Generally, a higher COT requires the greater use of EX blade cooling to mitigate the thermal stresses. The effect of changes in the amount of blade cooling on turbine efficiency is greatly dependent on the compressor pressure ratio and COT [174]. The film cooling method is extensively used for blade cooling [175]. In this method, compressor air is extracted and supplied into the blades.

The maximum blade material temperature allowed is usually between 800°C and 900°C [175]. However, due to the application of thermal barrier coatings on turbine blades, the temperature could go higher. Using the blade cooling correlation, a single-stage blade cooling is required for the COT up to 1,200°C [176]. For the COT up to 1,400°C and 1,600°C, blade cooling in two and three stages is required, respectively. The available GT module is used for both the NG and $H_2$ fuel as it is robust to changes in fuel type and composition [177]. The NG is split from the same supply line as the CLC unit and preheated using the FR exhaust before injecting into the COMB. The AR exhaust is used as the oxidizer in the COMB. Khan et al. [173] examined the effects of AR outlet temperature, plant performance with an additional COMB, plant performance with NG-fired COMB, and plant performance with $H_2$-fired COMB. In each case, the effects of several system variables on net thermal efficiency and carbon capture efficiency were examined. The study concludes: The energy penalty for carbon capture in an NGCC power plant with post-combustion capture ranges from 7.6% to 8.4%-points. This includes the energy required for the separation of $CO_2$ from the flue gas and the compression to a supercritical state. In CLC systems, the only direct energy penalty aside from a small pressure drop is for $CO_2$ compression. Despite this, due to the limitation in the maximum CLC operating temperature, the CLC plant is about 8.78% less efficient than the NGCC plant without carbon capture. When this limitation is overcome using an additional COMB after the CLC unit, the energy penalty can be as low as 4.53%-points when the fuel source is $H_2$ from an advanced $H_2$ production process. If NG is used in the added COMB, the energy penalty reduces to 2.9%-points at the expense of a lower carbon capture efficiency (72%).

When it is assumed that the CLC outlet temperature is limited to 1,150°C, the efficiency of the resulting plant is only 49.4% compared with 58.2% for the NGCC benchmark. Additional NG firing after the CLC unit to increase the COT to the same level as the benchmark plant (1,416°C) reduced the energy penalty to only 2.9%-points while achieving 72% $CO_2$ capture. The addition of COMB unit in the CLC plant can be operated with both NG and $H_2$. An efficient $H_2$ production unit is very desirable for the overall performance of the CLC system with COMB. For a highly efficient $H_2$ production process with integrated $CO_2$ capture with an efficiency of 90%, the overall plant energy penalty reduces to only 4.5%-points with 100% $CO_2$ capture. The CLC plant with an additional COMB can also benefit from continued improvements in GT technology to allow for a higher COT. In the case with a 90% $H_2$ production efficiency, a COT of 1,600°C produced a further efficiency gain of 2.98%-points. NO$x$ can be controlled by steam dilution and exhaust gas recirculation without serious effects on plant performance. Overall, the results show that adding a COMB and increasing the TIT significantly increase the net electrical efficiency and enable future gains from advanced GTs allowing for very high COT. Additional firing with NG or $H_2$ can be adapted according to $CO_2$ pricing trends in the future. Thus, a CLC combined cycle plant with an additional COMB has the potential to become a commercial technology and merits further research.

### 2.7.1 $H_2$-Powered Chemical Looping Power Generation System

In the study by Ajiwibowo et al. [178], a power-to-gas system coupled with a chemical looping combustion combined-cycle system is proposed to provide base and intermediate load power from the unused electricity from the grid. Enhanced process integration was employed to achieve optimal heat and exergy recovery. The study focuses on the design of a system consisting of a power-to-gas conversion method (for storage) and an $H_2$-powered chemical looping combustion power generation system.

Numerous efforts have been made to integrate fuel cells, especially SOEC with other energy conversion technologies. Cinti et al. proposed and investigated an integrated SOEC and Fischer-Tropsch system to produce methane from surplus renewable energy [179]. Energy and exergy evaluation of a SOEC-methanation system is also evaluated by Luo et al. [180]. The study by Ajiwibowo et al. [178] tries to focus on the effort to propose an efficient energy system which comprises of SOEC and chemical looping combustion (CLC) combined cycle power generation to produce baseload power from hydrogen as a sustainable zero-emission energy source. The study focuses on the effort to propose an efficient energy system that comprises SOEC and chemical looping combustion (CLC) combined cycle power generation to produce base and intermediate loads power from both natural gas and renewable energy-based hydrogen. The produced $H_2$ from the renewable energy source is fed to

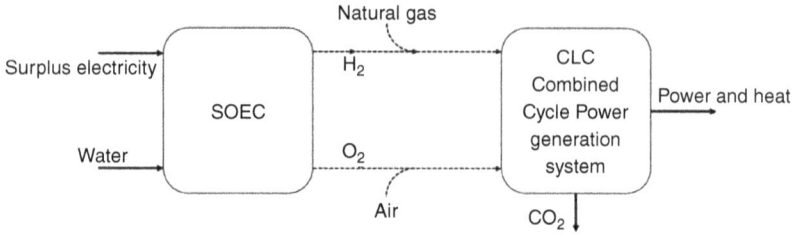

**FIGURE 2.9**  General overview of the integrated system [178].

the CLC power generation system where it is mixed with natural gas. In this way, the environmental burden caused by GHG can be decreased as the process is producing less $CO_2$ compared to traditional natural gas-fired power generation system. In addition, the inherent fluctuation characteristic of renewable energy can be stabilized, producing a stable and reliable power generation to the grid.

Figure 2.9 depicts a simplified process flow diagram for the proposed system. The system itself consists of an SOEC as the $H_2$ producer module and a chemical looping combustion combined cycle (CLCCC) as an electric power generation module. Theoretically, the CLCCC could act as the main electric power generator providing a stable electrical power output that is suitable for base and intermediate loads. The $CO_2$ emitted from the combustion process is basically separated. This leads to low $CO_2$ emission and potentially zero $CO_2$ emission if $H_2$ is used. For this system, a dual fuel system is considered where natural gas is considered, while the $H_2$ produced from the electricity from renewable energy sources or surplus electricity from the grid becomes additional fuel. Theoretically, the produced $H_2$ is consumed without being stored; therefore, the flow of natural gas to CLCCC decreases accordingly. Although a storage infrastructure could also be considered for $H_2$ in such a system, it was not considered in the study by Ajiwibowo et al. [178]. The generated electric power from the system is assumed stable, without being influenced by the fluctuation that comes with renewable energy sources and grid surplus electricity.

Overall, surplus electricity from the grid is used to convert $H_2O$ into $H_2$ and $O_2$ via the SOEC process. Afterwards, the $H_2$ is directed and fed directly to the CLCCC module. On the other hand, additional $O_2$ is also being fed into the CLCCC power system along with air, which potentially leads to higher combustion temperature. Table 2.2 describes the main assumptions and parameters used for the developed CLCCC system.

In order to achieve the highest energy efficiency for the system, an enhanced process integration methodology is utilized. This approach primarily focuses on heat and exergy recovery in the system via heat exchanger integration and compression [181,182]. Waste heat from hot downstream processes is utilized to support the heat requirements of upstream processes recuperatively. Exergy is also elevated in cold streams via compression. It is a proven methodology that has been demonstrated by various works for producing electricity or hydrogen from various sustainable sources, especially biomass [183–185].

Basically, the CLCCC is somewhat similar to a traditional combined cycle with a gas turbine, but the combustor in such system is replaced by two chemical looping

**TABLE 2.2**

**Details on the Parameters and Assumptions Used in the CLCCC [178]**

| Component/System | Value |
|---|---|
| Solid composition | 70% metal oxide |
| | 15% SiC |
| | 15% $Al_2O_3$ |
| Generator efficiency | 98% |
| Compressor isentropic efficiency | 90% |
| Turbine isentropic efficiency | 90% |
| Fuel flow rate | 6 kg/s |
| OT inlet temperature | 1,400°C |
| RT inlet temperature | 800°C – 900°C |
| ST inlet temperature | 700°C |
| Operating pressure | 15–35 bar |
| ST inlet pressure | 250 bar |

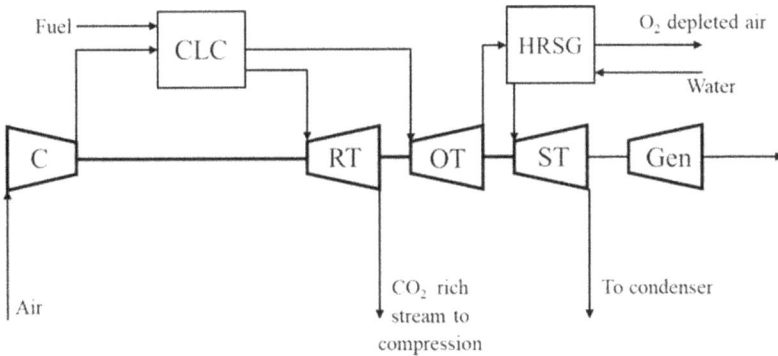

**FIGURE 2.10**   CLCCC power generation system [178] ST - Steam Turbine; C - Combustor.

combustion reactors that act as the heat source for the downstream turbines. As opposed to the traditional method, the produced fuel gas that is rich in $CO_2$ could be directly separated as it does not produce other by-products except $CO_2$ and $H_2O$. Thus, the separation of $CO_2$ is significantly less energy intensive than the traditional separation method. The schematic diagram of CLCCC is shown in Figure 2.10. In a dual fuel scenario, $H_2$ and $CH_4$ (natural gas) are considered to be fuels for this system, with the key assumption of a reactor design that would support the use of these two fuels. Two fuel gas streams coming out of the CLC process, namely, the reducer gas and the oxidizer gas, are expanded via the reducer turbine (RT) and the oxidizer turbine (OT). Afterwards, the $CO_2$-rich stream leaving the RT is directly separated by condensation and then compressed and stored. On the other hand, the high-temperature gas leaving the OT is used to generate steam for generating more power via a steam cycle.

## 2.7.2 Chemical Looping to Treat $CO_2$

Chemical looping can also be used to treat byproducts of combustion. Loutzenhiser and coworkers at Georgia Tech. are leveraging solar energy to reverse the combustion process and produce synthesis gas (mixtures of hydrogen, carbon monoxide, and small amounts of carbon dioxide), which can be converted into fuels such as kerosene and gasoline. The researchers have demonstrated that the technology works with zinc oxide, but they are searching for materials that can speed up the reactions and reduce the temperature of the first step in chemical looping. One wants something that can reduce at the lowest possible temperature in the high-temp stage and is capable of taking the oxygen from the carbon dioxide or the water vapor in the second step. Recently, the group achieved promising results with mixed ionic electronic conducting materials. Now they are trying to tune these materials to break apart either the $CO_2$ molecules or the water vapor molecules at lower temperatures. If commercialized, one can pull carbon dioxide from the air and use the sun to convert it with water into a long-term storage medium that could be shipped and used around the world without changes to transportation infrastructure.

## 2.7.3 Hydrogen and Electricity Production by Biomass Calcium Looping Gasification

Shaikh et al. [186] analyzed combined cycle biomass calcium looping gasification (CLGCC-H system) to generate hydrogen and electricity. The process simulation Aspen Plus was used to conduct a techno-economic analysis of the CLGCC-H system. The appropriate detailed models were set up for the proposed system. Furthermore, a dual fluidized bed was optimized for hydrogen production at 700°C and 12 bar. For comparison, calcium looping gasification with the combined cycle for electricity (CLGCC) was selected with the same parameters. The system exergy and energy efficiency of CLGCC-H reached as high as 60.79% and 64.75%, while the CLGCC system had 51.22% and 54.19%. The IRR and payback period of the CLGCC-H system, based on economic data, were calculated as 17.43% and 7.35 years, respectively. However, the CLGCC system has an IRR of 11.45% and a payback period of 9.99 years, respectively. The results show that the calcium looping gasification-based hydrogen and electricity coproduction system has a promising market prospect in the near future.

The calcium looping gasification with the combined cycle for electricity generation (CLGCC) and calcium looping gasification with the combined cycle for hydrogen and electricity generation (CLGCC-H) schemes mainly comprise flowing prime blocks in sequence, i.e., dual fluidized bed (gasifier and combustor), ASU, waste heat recovery (WHR), heat recovery steam generator (HRSG), gas cleaning (scrubber), acid gas removal unit (Selexol), elemental sulfur recovery unit (Claus), gas turbine (GT), and steam turbine (ST). The block diagrams of CLGCC and CLGCC–H are shown in Figures 2.11 and 2.12. The basic concept of calcium looping gasification with CaO bed material is illustrated in Figure 2.13. Pine dust was used as biomass. Other relevant parameters and operating details are described by Shaikh et al. [186]. The study once again shows the usefulness of chemical looping technique for electricity production.

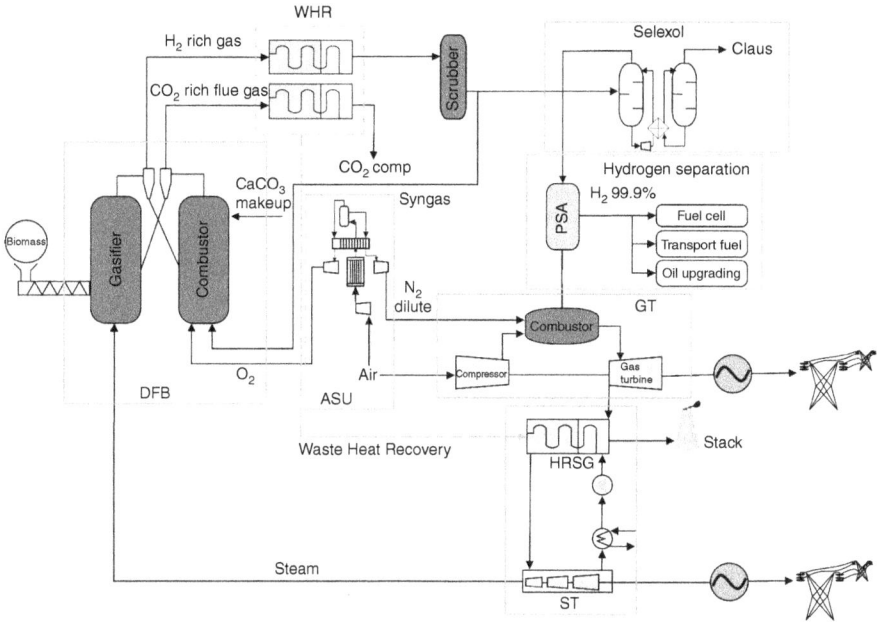

**FIGURE 2.11**  The calcium looping gasification with the combined cycle for hydrogen and electricity generation (CLGCC-H) [186].

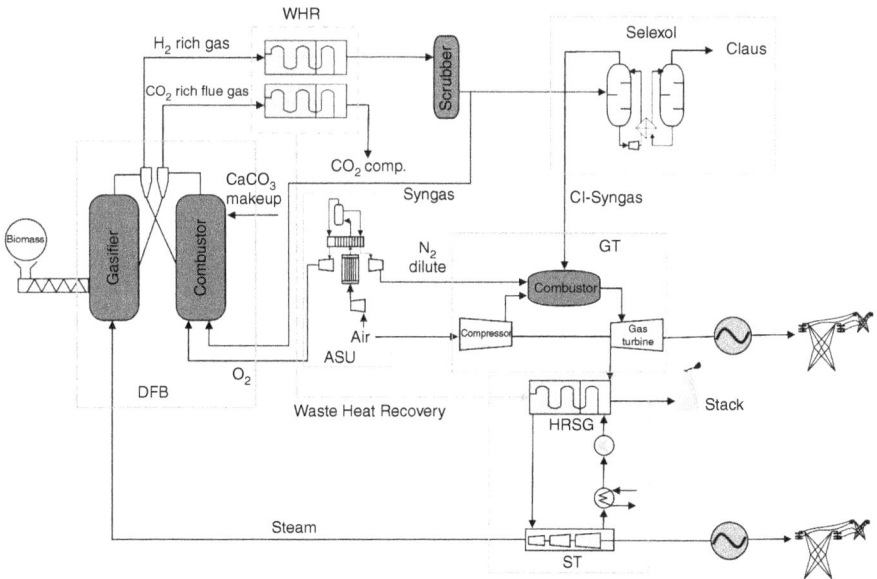

**FIGURE 2.12**  The calcium looping gasification with the combined cycle for electricity generation (CLGCC) [186].

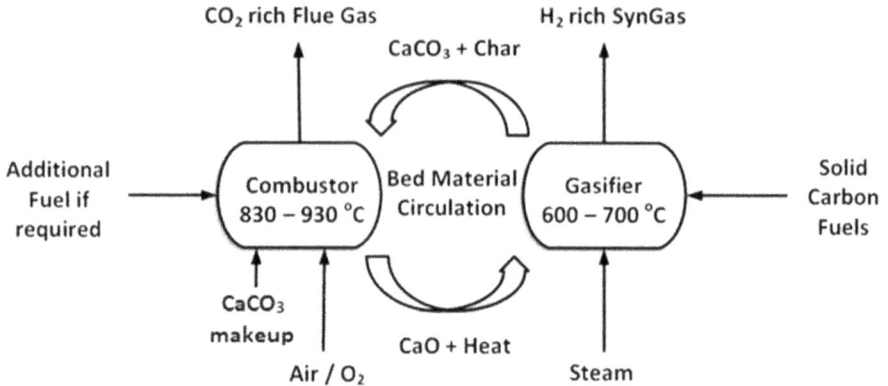

**FIGURE 2.13** Basic concept of calcium looping gasification with CaO as bed material [186].

## REFERENCES

1. Leach F., Kalghatgi G., Stone R., and Miles P. The scope for improving the efficiency and environmental impact of internal combustion engines. *Transport Eng* 2020;1:100005, www.elsevier.com/locate/treng.
2. Kalghatgi G.T. Is it really the end of internal combustion engines and petroleum in transport? *Appl Energ* 2018;225:965–974. Doi: 10.1016/j.apenergy.2018.05.076.
3. Transport and Environment. E-fuels too inefficient and expensive for cars and trucks but may be a part of aviation's climate solution study, 2017 Available from: https://www.transportenvironment.org/press/e-fuels-too-inefficient-and- expensive-cars-and-trucks-may-be-part-aviations-climate-solution-%E2%80%93 (accessed: 24 April 2019).
4. Akihama K., Takatori Y., Inagaki K., Sasaki S., and Dean A. M. *Mechanism of the Smokeless Rich Diesel Combustion by Reducing Temperature*, Detroit, MI. SAE Technical Paper 2001-01-0655; 2001, Doi: 10.4271/2001-01-0655, ISSN: 0148-7191 e-ISSN: 2688-3627.
5. Manente V., Johansson B., and Cannella W. Gasoline partially premixed combustion, the future of internal combustion engines? *Int J Eng Res* Doi: 10.1177/1468087411402441, Published Online: 6 June 2011. Published Print: June 2011.
6. Stanton D.W. *Light Duty Efficient, Clean Combustion*, Final Report by Cummins, Inc., to the U.S. Department of Energy, Report No. DE-FC26-07NT43279; 3 June 2011. http://www.osti.gov/scitech/servlets/purl/1038535/.
7. Upatnieks A., Mueller C.J., and Martin G.C. *The Influence of Charge-Gas Dilution and Temperature on DI Diesel Combustion Processes Using a Short- Ignition-Delay, Oxygenated Fuel.* SAE Paper 2005-01-2088, SAE Transactions, 114, No. 4; 2005.
8. Eismark E. *Method for Controlling a Combustion Process in a Combustion Engine.* European Patent Application EP 1 674 690 A1; 2006.
9. EPA. *Clean Diesel Combustion.* EPA 420-F-04-023. Washington, DC: Environmental Protection Agency; 2004, http://www.epa.gov/otaq/technology/420f04023.pdf.
10. A report by Duke Energy. Source: https://sustainabilityreport.duke-energy.com/downloads/ 2017-DukeSR.pdf.
11. Gas-fired-the five biggest natural gas power plants in the world. A website report in Power Technology, 14 April 2014. https://www.power-technology.com/analysis/featuregas-fired-the-five-biggest-natural-gas-power-plants-in-the-world-4214992/

12. Larson A. Worlds most-efficient combined cycle plant-EDF Bouchain. A website report by power magazine, 1 September 2017. https://www.powermag.com/worlds-most-efficient-combined-cycle-plant-edf-bouchain/

13. Haugen C. Coal-based power plants of the future—hybrid coal/gas combustion boiler concept with post combustion carbon capture (HGCC). A report by Barr Engineering, Contract Number: 89243319CFFE000017, Minneapolis, MN.

14. Kazulis V., Vigants H., Veidenbergs I., and Blumberga D. Biomass and natural gas co-firing – evaluation of GHG emissions, change. *Energ Procedia* 2018;147:558–565.

15. Costa M., and Casaca C. Co-combustion of biomass in a natural gas fired furnace. *Combust Sci Technol* 2003;175(11):1953–1977.

16. Golec T., and Bocian P. *Low Emission Pulverized Biomass Fuel Combustion Systems.* Austria, Graz: ERA-NET Bioenergy – Project FutureBioTec Technologies for clean biomass combustion; 20 September 2012.

17. Niedzwiecki L. *Energy Requirements for Comminution of Fibrous Materials – Qualitative Chipping Model.* Smaland: Linnaeus University School of Engineering; 2011.

18. Priyanto D.E., Ueno S., Hashida K., and Kasai H. Energy-efficient milling method for woody biomass. *Adv Powder Technol* 2017;28(7):1660–1667.

19. Karinkanta P., Ammala A., Illikainen M., and Niinimaki J. Fine grinding of wood – Overview from wood breakage to applications. *Biomass Bioenerg* 2018;113:31–44.

20. Annual Energy Review *U.S. Energy Information Administration Office of Energy Statistics U.S. Department of Energy.* Washington, DC 20585, DOE/EIA-0384(2011); 2011 September.

21. Yuan Y., He Y., Tan J., Wang Y., Kumar S., and Wang Z. Co-combustion characteristics of typical biomass and coal blends by thermogravimetric analysis. *Front Energ Res* 2021;9:753622. Doi: 10.3389/fenrg.2021.753622.

22. Shah Y.T. *Hybrid Power.* New York: CRC publishing co., Taylor and Francis; 2021.

23. Shah Y.T. *Energy and Fuel System Integration.* New York: CRC publishing co., Taylor and Francis; 2014.

24. Shadidi B., Najafi G., and Yusaf T. A review of hydrogen as a fuel in internal combustion engines. *Energies* 2021;14:6209. Doi: 10.3390/en14196209.

25. Karakurt A., Khandelwal B., Sethi V., and Singh R. Study of Novel Micromix Combustors to be used in Gas Turbines; using Hydrogen, hydrogen Methane, Methane and Kerosene as a fuel. *48th AIAA/ASME/SAE/ASEE Joint Propulsion Conference & Exhibit*, Atlanta, Georgia; 30 July–01, August 2012.

26. Gazzani M., Chiesa P., Martelli E., Sigali S., and Brunetti I. Using hydrogen as gas turbine fuel: Premixed versus diffusive flame combustors. *J Eng Gas Turbines Power* May 2014;136:051504-1.

27. MHPS successfully tests large-scale high-efficiency gas turbine fueled by 30% hydrogen mix – will contribute to reducing $CO_2$ emissions during power generation. A website report by Mitsubishi power, 19 January 2018, p. 116. https://power.mhi.com/news/20180119.html

28. First hydrogen power plant in Italy. *Alter Energ* 21 August 2009. Retrieved 16 October 2011.

29. Fusina: achieving low NOx from hydrogen combined-cycle power. A website report by Power Engineering International. PennWell Corporation; 1 October 2010. Retrieved 16 October 2011. https://www.powerengineeringint.com/world-regions/europe/fusina-achieving-low-nox-from-hydrogen-combined-cycle-power/

30. ENEL: First hydrogen-fuelled power now on line in Vanice, Media relation website report by Enel, Power, Tradefair group Rome, Italy; 1 October 2010. Retrieved 16 October 2011. https://www.enel.com/media/explore/search-press-releases/press/2009/08/enel-first-hydrogen-fuelled-power-now-on-line-in-venice-

31. Svetlana K., and Nigel T. Enel to start major plant conversion to coal 2011. A website report by Reuters; 12 July 2010. Retrieved 16 October 2011. https://www.reuters.com/article/enel-idAFLDE66B1CQ20100712

32. Rao A.D., editor. *Combined Cycle Systems for Near-Zero Emission Power Generation.* Salt Lake City, Utah: Woodhead Publishing Limited; 2012, 338 p.

33. http://www.energynomics.ro/wp-content/uploads/2019/06/03.-Cristian-Athanasovici-Kawasaki.pdf.

34. Milewski J. Hydrogen utilization by steam turbine cycles. *J Power Technol* 2015;95(4):258–264. Journalhomepage:papers.itc.pw.edu.pl.

35. Sanz W., Braun M., Jericha H., and Platzer M.F. Adapting the zero-emission Graz cycle for hydrogen combustion and investigation of its part load behaviour. *Proceedings of ASME Turbo Expo 2016: Turbomachinery Technical Conference and exposition GT2016*, Seoul, South Korea; 13–17 June, 2016.

36. Mitsugi C., Harumi A., and Kenzo F. We-net: Japanese hydrogen program. *Int J Hydrogen Energ* 1998;23(3):159–165.

37. Proctor D. Efficiency improvements mark advances in gas turbines. A website report in POWER magazine, 3 January 2018. https://www.powermag.com/efficiency-improvements-mark-advances-in-gas-turbines/

38. Rao, A., editor. *Combined Cycle Systems for Near-Zero Emission Power Generation.* Netherland: Elsevier; 2012. eBook ISBN: 9780857096180.

39. Faizal M., Chuah L.S., Lee C., Hameed A., Lee J., and Shankar M. Review of hydrogen fuel for internal combustion engines. *J Mech Eng Res Develop (JMERD)* 2019;42(3):35–46. Doi: 10.26480/jmerd.03.2019.35.46.

40. Adams M. Hydrogen in combined cycles. A website report by SoftInWay Inc., Burlington, MA; 13 April 2021. https://blog.softinway.com/hydrogen-in-combined-cycles/

41. BP Statistical Review of World Energy 2020, 69th ed., 2020. Available from: https://www.bp.com/content/dam/bp/business-sites/en/global/corporate/pdfs/energy-economics/statistical-review/bp-stats-review-2020-full-report.pdf (accessed: 1 November 2020).

42. IEA World Energy Outlook. 2020. Available from: https://www.iea.org/reports/world-energy-outlook-2020/achieving-net-zero-emissions-by-2050#abstract (accessed: 1 November 2020).

43. Jayarama Reddy P. *Clean Coal Technologies for Power Generation.* Boca Raton, FL: CRC Press; 2014.

44. Nomoto H. Advanced ultra-supercritical pressure steam turbines and their combination with carbon capture and storage systems (CCS). In: Tanuma T., editor. *Advances in Steam Turbines for Modern Power Plants.* Cambridge: Woodhead Publishing; 2017, pp. 501–519.

45. Larson A. Longview power plant rehabilitation results in most efficient U.S. coal plant. A website report, Power magazine, 2016. https://www.powermag.com/longview-power-plant-rehabilitation-results-efficient-u-s-coal-plant/

46. Young C. The most efficient old fashioned coal plant in recent history has been the Longview Plant in West Virginia. A website report by the State Journal, 10 July, 2017. https://ieefa.org/resources/ieefa-us-longview-coal-plant-one-americas-best-and-newest-has-just-gone-bankrupt

47. Silvestri G.J. Jr., Eddtstone station, 325 MW generating unit 1. A brief history, a website report by ASME, 2003. https://www.asme.org/wwwasmeorg/media/resourcefiles/aboutasme/who%20we%20are/engineering%20history/landmarks/226-eddystone-station-unit.pdf

48. Tramošljika B., Blecich P., Bonefačić I., and Glažar V. Advanced ultra-supercritical coal-fired power plant with post-combustion carbon capture: Analysis of electricity penalty and $CO_2$ emission reduction. *Sustainability* 2021;13:801. Doi: 10.3390/su13020801, https://www.mdpi.com/journal/sustainability.

49. Rheinhafen-Dampfkraftwerk block 8 achieved a 47.5% net thermal efficiency to world-class level. Available from: https://www.world-energy.org/article/1198.html (accessed: 20 November 2020).
50. Tumanovskii A.G., Shvarts A.L., Somova E.V., Verbovetskii E.K., Avrutskii G.D., Ermakova S.V., Kalugin R.N., and Lazarev M.V. Review of the coal-fired, over-supercritical and ultra-supercritical steam power plants. *Therm Eng* 2017;64:83–96.
51. Di Gianfrancesco A. *Materials for Ultra-Supercritical and Advanced Ulttra-Supercritical Power Plants*. Duxford: Woodhead Publishing; 2017.
52. Viswanathan R., Coleman K., and Rao U. Materials for ultra-supercritical coal-fired power plant boilers. *Int J Press Vessel Pip* 2006;83:778–783.
53. Zhang D. *Ultra-Supercritical Coal Power Plants: Materials, Technologies and Optimisation*. Cambridge: Woodhead Publishing; 2013.
54. Retzlaff K.M., and Ruegger W.A. *Steam Turbines for Ultrasupercritical Power Plants*. GER-3945. Schenectady, NY: GE Power Systems General Electric Company; 1996.
55. Viswanathan R., Purgert R., and Rao U. Materials technology for advanced coal power plants. A report by DOE grant DE-FG26-01NT-41175 and Ohio Coal Research Lab; 1999. https://www.phase-trans.msm.cam.ac.uk.
56. Rasheed R., Javed H., Rizwan A., Sharif F., Yasar A., Tabinda A.B., Ahmad S.R., Wang Y., and Su Y. Life cycle assessment of a cleaner supercritical coal-fired power plant. *J Clean Prod* 2021;279:123869.
57. Global Status of CCS. Melbourne: Global CCS Institute; November 2020. Available from: https://www. globalccsinstitute.com/resources/global-status-report/ (accessed: 31 December 2020).
58. Mantripragada H.C., Zhai H., and Rubin E.S. Boundary Dam or Petra Nova—Which is a better model for CCS energy supply? *Int J Greenh Gas Control* 2019;82:59–68.
59. Rubin E.S., Davison J.E., and Herzog H.J. The cost of $CO_2$ capture and storage. *Int J Greenh Gas Control* 2015;40:378–400.
60. IPCC special report on carbon dioxide capture and storage. In: Metz B., Davidson O., de Coninck H.C., Loos M., and Meyer L.A., editors. *Intergovernmental Panel on Climate Change*. Cambridge: Cambridge University Press; 2005. Available from: https://archive. ipcc.ch/report/srccs/ (accessed: 31 December 2020).
61. *ZEP: The Costs of $CO_2$ Capture, Transport and Storage, Post-demonstration CCS in the EU, European Technology Platform for Zero Emission Fossil Fuel Power Plants.* Brussels; 2009. Available from: https://zeroemissionsplatform.eu/wp-content/uploads/Overall-CO2-Costs-Report.pdf (accessed: 31December 2020).
62. Pale Blue Dot Energy. Global storage resource classification assessment, 2020. Available from: https://www.globalccsinstitute.com/resources/publications-reports-research/global-storage-resource-assessment-2019-update/ (accessed: 31 December 2020).
63. Leung D.Y.C., Caramanna G., and Maroto-Valer M.M. An overview of current status of carbon dioxide capture and storage technologies. *Renew Sustain Energy Rev* 2014;39:426–443.
64. Lee B.J., Lee J.I., Yun S.Y., Hwang B.G., Lim C.-S., and Park Y.-K. Methodology to calculate the $CO_2$ emission reduction at the coal-fired power plant: $CO_2$ capture and utilization applying technology of mineral carbonation. *Sustainability* 2020;12:7402.
65. Lee B.J., Lee J.I., Yun S.Y., Lim C.-S., and Park Y.-K. Economic evaluation of carbon capture and utilization applying the technology of mineral carbonation at coal-fired power plant. *Sustainability* 2020;12:6175.
66. Nasirov S., O'Ryan R., and Osorio H. Decarbonization tradeoffs: A dynamic general equilibrium modeling analysis for the Chilean power sector. *Sustainability* 2020;12:8248.

67. Kumar Shukla A., Ahmad Z., Sharma M., Dwivedi G., Nath Verma T., Jain S., Verma P., and Zare A. Advances of carbon capture and storage in coal-based power generating units in an Indian context. *Energies* 2020;13:4124.
68. Yun T., Kim Y., and Kim J.-Y. Feasibility study of the post-2020 commitment to the power generation sector in South Korea. *Sustainability* 2017;9:307.
69. Markewitz P., Zhao L., Ryssel M., Moumin G., Wang Y., Sattler C., Robinius M., and Stolten D. Carbon capture for $CO_2$ emission reduction in the cement industry in Germany. *Energies* 2019;12:2432.
70. Toktarova A., Karlsson I., Rootzén J., Göransson L., Odenberger M., and Johnsson F. Pathways for low-carbon transition of the steel industry—a Swedish case study. *Energies* 2020;13:3840.
71. Adu E., Zhang Y., Liu D., and Tontiwachwuthikul P. Parametric process design and economic analysis of post-combustion $CO_2$ capture and compression for coal- and natural gas-fired power plants. *Energies* 2020;13:2519.
72. Vu T.T., Lim Y.I., Song D., Mun T.-Y., Moon J.-H., Sun D., Hwang Y.-T., Lee J.-G., and Park Y.C. Techno-economic analysis of ultra-supercritical power plants using air- and oxy-combustion circulating fluidized bed with and without $CO_2$ capture. *Energy* 2020;194:116855.
73. Liebenthal U., Linnenberg S., Oexmann J., and Kather A. Derivation of correlations to evaluate the impact of retrofitted post- combustion $CO_2$ capture processes on steam power plant performance. *Int J Greenh Gas Control* 2011;5:1232–1239.
74. Stepczynska-Drygas K., Łukowicz H., and Dykas S. Calculation of an advanced ultra-supercritical power unit with $CO_2$ capture installation. *Energy Convers Manage* 2013;74:201–208.
75. Lucquiaud M., and Gibbins J. On the integration of $CO_2$ capture with coal-fired power plants: A methodology to assess and optimize solvent-based post-combustion capture systems. *Chem Eng Res Des* 2011;89:1553–1571.
76. Li J., and Liang X. $CO_2$ capture modelling for pulverized coal-fired power plants: A case study of an existing 1 GW ultra-supercritical power plant in Shandong, China. *Sep Purif Technol* 2012;94:138–145.
77. Xu C., Li X., Liu X., and Li J. An integrated de-carbonization supercritical coal-fired power plant incorporating a supplementary steam turbine, process heat recovery and a modified boiler structure. *Appl Therm Eng* 2020;178:115532.
78. Jackson S., and Brodal E. Optimization of the energy consumption of a carbon capture and sequestration related carbon dioxide compression processes. *Energies* 2019;12:1603.
79. Storm R. The most efficient thermal power generation plants in America. A website report, Williamson College of the Trades, 10 May 2018. https://www.williamson.edu/2018/05/the-most-efficient-thermal-power-generation-plants-in-america/
80. Boyce M.P. Advanced industrial gas turbines for power generation. In: *Combined Cycle Systems for Near-Zero Emission Power Generation*. Washington, DC: O'Reilly, pp. 44–102; 2012, Doi: 10.1533/9780857096180.44, ISBN 9780857090133.
81. Boyce M.P. An overview of gas turbines. In: *Gas Turbine Engineering Handbook*, 4th ed. The Netherland: Elsvier; 2011, November eBook ISBN 9780123838438.
82. Soares, C. Chapter 3: Gas turbine configurations and heat cycles. In: Soares C., editor. *Gas Turbines: A Handbook of Air, Land and Sea Applications*. Amsterdam: Elsevier; 2008.
83. Zachary J., Kallianpur V., and So B. Integration of modern F, G, H and J class in combined cycle applications: An EPC contractor perspective. *Proceedings of the ASME Turbo Expo 2016: Turbomachinery Technical Conference and Exposition. Volume 3: Coal, Biomass and Alternative Fuels; Cycle Innovations; Electric Power; Industrial and Cogeneration; Organic Rankine Cycle Power Systems*, Seoul, South Korea; 13–17 June, 2016. V003T08A006. ASME. Doi: 10.1115/GT2016-57022.

84. Giermak E.A., Gaikwad R., Warren S., and Sargent & Lundy. Advanced technology combustion turbines in combined-cycle applications. A website report in POWER Engineering, report by Clarion Energy Content Directors, September 2017. https://sargentlundy.com/whitepaper/advanced-technology-combustion-turbines-combined-cycle-applications/

85. Quadrennial Technology Review 2015. *Supercritical Carbon Dioxide Brayton Cycle Chapter 4: Technology Assessments*. DOE Report. Washington, DC: Department of Energy; 2015.

86. White C. Analysis of Brayton cycles utilizing supercritical carbon dioxide - revision 1. DOE/NETL-4001/070114, In Preparation. Available from: https://www.netl.doe.gov/energy-analyses/temp/AnalysisofBraytonCyclesUtilizingSupercriticalCarbonDioxide_070114.pdf.

87. Subbaraman G., Mays J.A., Jazayeri B., Sprouse K.M., Eastland A.H., Ravishankar S., and Sonwane C.G. Energy systems, Pratt and Whitney Rocketdyne, ZEPS plant model: A high efficiency power cycle with pressurized fluidized bed combustion process. *2nd Oxyfuel Combustion Conference*, Queensland, Australia; September 2011, http://www.ieaghg.org/docs/General_Docs/OCC2/Abstracts/Abstract/ occ2Final00143.pdf.

88. Kacludis A., Lyons S., Nadav D., and Zdankiewicz E. Waste heat to power (WH2P) applications using a supercritical $CO_2$-based power cycle. *Presented at Power-Gen International 2012*, Orlando, FL; December 2012.

89. Shelton, W.W., Weiland, N., White, C., Plunkett, J., and Gray, D. Oxy-coal-fired circulating fluid bed combustion with a commercial utility-size supercritical $CO_2$ power cycle. *5th International Symposium - Supercritical $CO_2$ Power Cycles*, San Antonio, TX; 29–31 March, 2016. http://www.swri.org/4org/d18/sco2/papers2015/104.pdf.

90. Kimzey G. *Development of a Brayton Bottoming Cycle using Supercritical Carbon Dioxide as the Working Fluid*. EPRI; 2012, http://www. swri.org/utsr/presentations/kimzey-report.

91. Ahn Y., Baea S.J., Kima M., Choa S.K., Baika S., Lee J.I., and Cha J.E. Cycle layout studies of S-$CO_2$ cycle for the next generation nuclear system application. *Transactions of the Korean Nuclear Society Autumn Meeting*, Pyeongchang, Korea; 30–31 October, 2014.

92. Bae S.J., Lee J., Ahn Y., and Lee J.I. Preliminary studies of compact Brayton cycle performance for small modular high temperature gas- cooled reactor system. *Ann Nucl Energ* 2015;75. http://www.sciencedirect.com/science/article/pii/S0306454914003727.

93. Mendez C., Rochau G. *sCO Brayton Cycle: Roadmap to sCO Power Cycles NE Commercial Applications*. Tech. Rep. SAND-2018-6187. Albuquerque, NM: Sandia National Laboratories; 2018. Doi: 10.2172/1452896.

94. Persichilli M., Kacludis A., Zdankiewicz E., and Held T. *Supercritical CO Power Cycle Developments and Commercialization: Why sCO Can Displace Steam Power-Gen India & Central Asia*. A paper Presented at Power-Gen India & Central Asia 2012, Pragati Maidan, New Delhi, Indi; 19–21 April, 2012.

95. Fleming D., Conboy T., Pasch J., Rochau G., Fuller R., Holschuh T., and Wright S. *Scaling Considerations for a Multi-Megawatt Class Supercritical $CO_2$ Brayton Cycle and Path Forward for Commercialization*. SAND2013-9106; November 2013, http://prod.sandia.gov/techlib/ access-control.cgi/2013/139106.pdf.

96. Conboy T.M., Carlson M.D., and Rochau G.E. Dry-cooled supercritical $CO_2$ power for advanced nuclear reactors. *J Eng Gas Turb Power* August 2014;137:012901. https://www.researchgate.net/publication/270772560_Dry-Cooled_Supercritical_CO_2_Power_for_Advanced_Nuclear_Reactors.

97. White M.T., Bianchi G., Chai L., Tassou S.A., and Sayma A.I. Review of supercritical $CO_2$ technologies and systems for power generation. *Appl Thermal Eng* 2021;185:116447. ISSN 1359-4311, Doi: 10.1016/j.applthermaleng.2020.116447, https://www.sciencedirect.com/science/article/pii/S1359431120339235.

98. Drennen T.E. *sCO Brayton System Market Analysis: Tech. Rep.* Albuquerque, NM: Sandia National Laboratories; 2020, pp. SAND2020-3248.

99. Poerner M., and Rimpel A. Waste heat recovery In: Brun K., Friedman P., and Dennis R., editors. *Fundamentals and Applications of Supercritical Carbon Dioxide (SCO) Based Power Cycles.* Woodhead Publishing; 2017, pp. 255–267. Doi: 10.1016/B978-0-08-100804-1.00010-4.

100. Crespi F., Gavagnin G., Sánchez D., and Martínez G.S. Supercritical carbon dioxide cycles for power generation: A review. *Appl Energ* 2017;195:152–183. Do10.1016/j.apenergy.2017.02.048.

101. Marchionni M., Bianchi G., and Tassou S.A. Techno-economic assessment of Joule-Brayton cycle architectures for heat to power conversion from high-grade heat sources using CO in the supercritical state *Energy* 2018;148:1140–1152. Doi: 10.1016/j.energy.2018.02.005.

102. Wright S.A., Davidson C.S., and Scammell W.O. Thermo-economic analysis of four sCO waste heat recovery power systems. *The 5th International Supercritical CO Power Cycles Symposium*, San Antonio, TX; 28–31 March, 2016.

103. Global Market InsightsGas turbine market. 2020. www.gminsights.com/industry-analysis/gas-turbine-market (accessed: 10 June 2020).

104. Slade S., and Palmer C. Worldwide gas turbine forecasts. 2020. www.mtt-eu.com/wp-content/uploads/WORLDWIDE-GAS-TURBINE-FORECAST-Turbomachinery-Magazine.pdf (accessed: 10 June 2020).

105. Mordor intelligence Gas turbine market. 2020. www.mordorintelligence.com/industry-reports/gas-turbine-market (accessed: 10 June 2020).

106. Brooks F.J. GE gas turbine performance characteristics. 2020. http://ncad.net/Advo/CinerNo/ge6581b.pdf (accessed: 10 June 2020).

107. Global Market Insights Global gas generator sets market. 2020. www.gminsights.com/industry-analysis/gas-generator-sets-market?utm_source=prnewswire.com&utm_medium=referral&utm_campaign=Paid_prnewswire (accessed: 10 June 2020).

108. Shi L., Shu G., Tian H., and Deng S. A review of modified organic Rankine cycles (ORCs) for internal combustion engine waste heat recovery (ICE-whr). *Renew Sustain Energ Rev* 2018;92:95–110. Doi: 10.1016/j.rser.2018.04.023.

109. Forman C., Muritala I.K., Pardemann R., and Meyer B. Estimating the global waste heat potential. *Renew Sustain Energ Rev* 2016;57:1568–1579. Doi: 10.1016/j.rser.2015.12.192.

110. Bianchi G., Panayiotou G.P., Aresti L., Kalogirou S.A., Florides G.A., Tsamos K., Tassou S.A., and Christodoulides P. Estimating the waste heat recovery in the European Union industry *Energy Ecol Environ* 2019;4(5):211–221. Doi: 10.1007/s40974-019-00132-7.

111. Brough D., and Jouhara H. The aluminium industry: A review on state-of-the-art technologies, environmental impacts and possibilities for waste heat recovery *Int J Thermofluids* 2020;1–2. Article 100007. Doi: 10.1016/j.ijft.2019.100007.

112. Karellas S., Leontaritis A.-D., Panousis G., Bellos E., and Kakaras E. Energetic and exergetic analysis of waste heat recovery systems in the cement industry *Energy* 2013;58:147–156. Doi: 10.1016/j.energy.2013.03.097.

113. Zhang H., Wang H., Zhu X., Qiu Y.-J., Li K., Chen R., and Liao Q. A review of waste heat recovery technologies towards molten slag in steel industry *Appl Energ* 2013;112:956–966. Doi: 10.1016/j.apenergy.2013.02.019.

114. Jouhara H., Khordehgah N., Almahmoud S., Delpech B., Chauhan A., and Tassou S.A. Waste heat recovery technologies and applications *Therm Sci Eng Prog* 2018;6:268–289. Doi: 10.1016/j.tsep.2018.04.017.

115. IEA Electricity generation by fuel and scenario, 2018–2040; 2020. https://www.iea.org/data-and-statistics/charts/electricity-generation-by-fuel-and-scenario-2018-2040 (accessed: 10 June 2020).

116. Le Moullec Y. Conceptual study of a high efficiency coal-fired power plant with CO capture using a supercritical CO Brayton cycle *Energy* 2013;49:32–46. Doi: 10.1016/j.energy.2012.10.022.

117. Park S., Kim J., Yoon M., Rhim D., and Yeom C. Thermodynamic and economic investigation of coal-fired power plant combined with various supercritical CO Brayton power cycle. *Appl Therm Eng* 2018;130:611–623. Doi: 10.1016/j.applthermaleng.2017.10.145.

118. Li Z., Liu X., Shao Y., and Zhong W. Research and development of supercritical carbon dioxide coal-fired power systems. *J Therm Stresses* 2020:1–30. Doi: 10.1007/s11630-020-1282-6.

119. Xu J., Liu C., Sun E., Xie J., Li M., Yang Y., and Liu J. Perspective of s-CO power cycles. *Energy* 2019;186, Article 115831. Doi: 10.1016/j.energy.2019.07.161.

120. Thimsen D., and Weitzel P., Challenges in designing fuel-fired sCO heaters for closed sCOBrayton cycle power plants. *The 5th International Supercritical CO Power Cycles Symposium*, San Antonio, TX; 28–31 March, 2016.

121. Mecheri M., and Moullec Y.L. Supercritical CO Brayton cycles for coal-fired power plants *Energy* 2016;103:758–771. Doi: 10.1016/j.energy.2016.02.111.

122. Cagnac A., Mecheri M., and Bedogni S. Configuration of a flexible and efficient sCO cycle for fossil power plant. *3rd European Supercritical CO Conference*, Paris, France; 19–20 September, 2019. Doi: 10.17185/duepublico/48907.

123. EPA web site. Carbon pollution standards for new, modified and reconstructed power plants; August 2015. https://www.epa.gov/cleanpowerplan/carbon-pollution-standards-new-modified-and-reconstructed-power-plants#rule summary.

124. Ferrari N., Mancuso L., Davison J., Chiesa P., Martelli E., and Romano M.C. Oxy-turbine for power plant with CO capture, *Energ Procedia* 2017;114:471–480. Doi: 10.1016/j.egypro.2017.03.1189.

125. *Rivers Capital Allam Cycle Zero Emission Coal Power Plant*. Tech. Rep. DOE Grant Proposal 89243319CFE000015; 2019. https://netl.doe.gov/coal/tpg/coalfirst/DirectSupercriticalCo2.

126. Allam R., Martin S., Forrest B., Fetvedt J., Lu X., Freed D., Brown G.W., Sasaki T., Itoh M., and Manning J. Demonstration of the Allam cycle: An update on the development status of a high efficiency supercritical carbon dioxide power process employing full carbon capture *Energy Procedia* 2017;114:5948–5966. Doi: 10.1016/j.egypro.2017.03.1731.

127. Allam R.J., Fetvedt J.E., Forrest B.A., and Freed D.A. The oxy-fuel, supercritical CO Allam cycle: New cycle developments to produce even lower-cost electricity from fossil fuels without atmospheric emissions. *Proceedings of ASME Turbo Expo 2014: Turbomachinery Technical Conference and Exposition*, Düsseldorf, Germany; 16–20 June, 2014, Vol. V03BT36A016, pp. GT2014-26952. Doi: 10.1115/GT2014-26952.

128. Weiland N.T., and White C.W. Techno-economic analysis of an integrated gasification direct-fired supercritical CO power cycle. *Fuel* 2018;212:613–625. Doi: 10.1016/j.fuel.2017.10.022.

129. Luo J., Emelogu O., Morosuk T., and Tsatsaronis G. Exergy-based investigation of a coal-fired Allam cycle. *Energy* 2019;137:01018. Doi: 10.1051/e3sconf/201913701018.

130. Penkuhn M., and Tsatsaronis G. Exergy analysis of the Allam cycle. *The 5th International Supercritical CO Power Cycles Symposium*, San Antonio, TX; 28–31 March, 2016.

131. Banuti D.T., Shunn L., Bose S., and Kim D. Large eddy simulations of oxy-fuel combustors for direct-fired supercritical CO power cycles. *The 6th International Symposium on Supercritical CO Power Cycles*, Pittsburgh, PA; 27–29 March, 2018.

132. Abdul-Sater H., Lenertz J., Bonilha C., Lu X., and Fetvedt J. A CFD simulation of coal syngas oxy-combustion in a high-pressure supercritical CO environment. *Proceedings of ASME Turbo Expo 2017: Turbomachinery Technical Conference and Exposition*, Charlotte, NC; Vol. V04AT04A051, 26–30 June, 2017. Doi: 10.1115/GT2017-63821.

133. Barak S., Pryor O., Ninnemann E., Neupane S., Vasu S., Lu X., and Forrest B. Ignition delay times of oxy-syngas and oxy-methane in supercritical CO mixtures for direct-fired cycles *J Eng Gas Turbines Power* 2020;142(2). Doi: 10.1115/1.4045743.

134. Rogalev A., Grigoriev E., Osipov S., and Rogalev N. The design approach for supercritical CO gas turbine *AIP Conf Proc* 2019;2189(1), Article 020018. Doi: 10.1063/1.5138630.
135. Weiland N., Dennis R., Ames R., Lawson S., and Strakey P. Fossil energy. In: Brun K., Friedman P., and Dennis R., editors. *Fundamentals and Applications of Supercritical Carbon Dioxide (SCO) Based Power Cycles.* Woodhead Publishing; 2017, pp. 293–338, Doi: 10.1016/B978-0-08-100804-1.00012-8 Ch. 12.
136. Strakey P., Dogan O., Holcomb G., and Richards G. Technology needs for fossil fuel supercritical CO power systems. *The 4th International Symposium on Supercritical CO Power Cycles*, Pittsburgh, PA; 9–10 September, 2014.
137. Allam R.J., Palmer M., Brown G.W., Fetvedt J.E., and Forrest B.A. *System and Method for High Efficiency Power Generation Using a Carbon Dioxide Circulating Working Fluid.* Patent US20110179799A1. Washington, DC: Rivers Capital LLC, Palmer Labs LLC; 2011.
138. Freed D., Forrest B., Rafati N., Fetvedt J., McGroddy M., *and* Allam R. Progress update on the Allam cycle: Commercialization of Net Power and the Net Power demonstration facility. *14th Greenhouse Gas Control Technologies Conference Melbourne*; 2018, pp. 21–26. https://papers.ssrn.com/sol3/papers.cfm?abstract_id=3366370#references-widget.
139. Ahn Y., Bae S.J., Kim M., KukCho S., Baik S., IkLee J., and EunCha J. Review of supercritical $CO_2$ power cycle technology and current status of research and development. *Nucl Eng Technol* October 2015;47(6):647–661.
140. Zhu Q. Innovative power generation systems using supercritical $CO_2$ cycles. *Clean Energ* December 2017;1(1):68–79. Doi: 10.1093/ce/zkx003.
141. Geng C., Shao Y., Zhong W., and Liu X. Thermodynamic analysis of supercritical $CO_2$ power cycle with fluidized bed coal combustion. *J Combust Coal Biomass Combust* 2018; Hindawi, Research Article, Open Access Volume, Article ID 6963292. Doi: 10.1155/2018/6963292.
142. Supercritical $CO_2$ power cycles for FE, 2017. A website NETL report, Pittsburgh, PA https://netl.doe.gov.
143. Diego M.E., Bellas J.-M., and Pourkashanian M. Gas turbines for carbon capture. *Appl Energ* 2018;215:778.
144. Sanchez Fernandez E., Goetheer E.L.V., Manzolini G., Macchi E., Rezvani S., and Vlugt T.J.H. Thermodynamic assessment of amine based $CO_2$ capture technologies in power plants based on European benchmarking task force methodology. *Fuel* 2014;129:318.
145. Ystad P.A.M., Bolland O., and Hillestad M. Simulation and performance comparison for $CO_2$ capture by Aqueous solvents of N-(2-Hydroxyethyl) piperazine and another five single amines. *Energy Proc.* 2012;23:33.
146. Rubin E.S., Davison J.E., and Herzog H.J. The cost of $CO_2$ capture and storage. *Int J Greenh Gas Control* 2015;40:378.
147. Ishida M., Zheng D., and Akehata T. Evaluation of a chemical-looping-combustion power-generation system by graphic exergy analysis. *Energy* 1987;12:147.
148. Hossain M.M., and de Lasa H.I. Chemical-looping combustion (CLC) for inherent $CO_2$ separations—A review. *Chem Eng Sci* 2008;63:4433.
149. Li J., Zhang H., Gao Z., Fu J., Ao W., and Dai J. $CO_2$ capture with chemical looping combustion of gaseous fuels: An overview. Energ Fuels 2017;31:3475.
150. Ishida M., and Jin H. A new advanced power-generation system using chemical-looping combustion. *Energy* 1994;19:415.
151. Ishida M., and Jin H. $CO_2$ recovery in a power plant with chemical looping combustion. *Energ Convers Manage* 1997;38:S187.
152. Naqvi R., Wolf J., and Bolland O. Part-load analysis of a chemical-looping combustion (CLC) combined cycle with $CO_2$ capture *Energy* 2007;32:360.
153. Naqvi R., and Bolland O. Multi-stage chemical looping combustion (CLC) for combined cycles with $CO_2$ capture. *Int. J. Greenh. Gas Control* 2007;1:19.

154. Hassan B., Ogidiama O.V., Khan M. N., and Shamim T.J. Energy and exergy analyses of a power plant with carbon dioxide capture using multi- stage chemical looping combustion. *Energy Resour. Technol.* 2016;139:032002.

155. Ekström C., Schwendig F., Biede O., Franco F., Haupt G., de Koeijer G., Papapavlou C., and Røkke P.E. Techno-economic evaluations and benchmarking of pre-combustion $CO_2$ capture and oxy-fuel processes developed in the European ENCAP project. *Energ Proc* 2009;1:4233.

156. Porrazzo R., White G., and Ocone, R. Techno-economic investigation of a chemical looping combustion based power plant. *Faraday Discus*, 2016:192:437.

157. Zerobin F., and Pröll T. Potential and limitations of power generation via chemical looping combustion of gaseous fuels. *Int J Greenh Gas Control* 2017;64:174.

158. Petriz-Prieto M.A., Rico-Ramirez V., Gonzalez-Alatorre G., Gómez-Castro F.I., and Diwekar U.M. A comparative simulation study of power generation plants involving chemical looping combustion system. *Comput Chem Eng* 2016;84:434.

159. Farooqui A., Bose A., Ferrero D., Llorca J., and Santarelli M. Farooqui A., Bose A., Ferrero D., Llorca J., and Santarelli M. Techno-economic and exergetic assessment of an oxy-fuel power plant fueled by syngas produced by chemical looping $CO_2$ and $H_2O$ dissociation. *J CO$_2$ Util* 2018;27:500.

160. Wang S., Wang G., Jiang F., Luo M., and Li H. Chemical looping combustion of coke oven gas by using $Fe_2O_3$/CuO with $MgAl_2O_4$ as oxygen carrier. *Energ Environ Sci* 2010;3:1353.

161. Xiao R., Chen L., Saha C., Zhang S., and Bhattacharya S. Pressurized chemical looping combustion of coal using an iron ore as oxygen carrier in a pilot scale unit. *Int J Greenh Gas Control* 2012;10:363.

162. Breault R.W. *Handbook of Chemical Looping Technology*. Weinheim: John Wiley & Sons; 2018.

163. Ryu H.-J., Lee D., Jo S.-H., Lee S.-Y., and Baek J.-I. Efficiency improvement of chemical looping combustion combined cycle power plants. *Conf. of 14th Greenhouse Gas Control Technologies*, Melbourne; 2018, p. 8.

164. Noorman S., van Sint Annaland M., and Kuipers H. Packed bed reactor technology for chemical looping combustion. *Ind Eng Chem Res* 2007;46:4212.

165. Zaabout A., Cloete S., Johansen S.T., van Sint Annaland M., Gallucci F., and Amini S. Ind Experimental demonstration of a novel gas switching combustion reactor for power production with integrated $CO_2$ capture. *Eng Chem Res* 2013;52:14241.

166. Håkonsen S.F, and Blom R. Chemical looping combustion in a rotating bed reactor- finding optimum process conditionsnfor prototype reactor. *Environ Sci Technol* 2011;45:9619.

167. Osman M., Zaabout A., Cloete S., and Amini S. Internal circulating fluidized bed reactor for syngas production using chemical looping reforming. *Chem Eng J* 2018. Doi: 10.1016/J.CEJ.2018.10.013.

168. Fan L.S. *Chemical Looping Partial Oxidation-Gasification, Reforming and Chemical Syntheses*. Cambridge: Cambridge University Press; 2017.

169. Consonni S., Lozza G., Pelliccia G., Rossini S., and Saviano F. Chemical looping combustion for combined cycles with $CO_2$ capture. *J Eng Gas Turbines Power* 2006;128:525.

170. Adanez J., Abad A., Garcia-Labiano F., Gayan P., and De Diego L.L.F. Progress in chemical looping combustion and reforming technologies. *Prog Energ Combust Sci* 2012;38:215.

171. Khan M.N., and Shamim T. Thermodynamic screening of suitable oxygen carriers for a three reactor chemical looping reforming system. *Int J Hydrogen Energ* 2017;42:15745.

172. Baek J.-I., Ryu J., Lee J.B., Eom T.-H., Kim K.-S., Yang S.-R., and Ryu C.K. Highly attrition resistant oxygen carrier for chemical looping combustion. *Energ Proc* 2011;4:349.

173. Khan M.N., Cloete S., and Amini S. Efficiency improvement of chemical looping combustion combined cycle power plants. *Energ Technol* 27 August 2019;7(11). Doi: 10.1002/ente.201900567.

174. Besharati-Givi M., and Li X. *ASME 2015 Power Conference*, ASME, New York, 2015, p. V001T09A009.

175. Moon S.W., Kwon H.M., Kim T.S., Kang D.W., and Sohn J.L. A novel coolant method for enhancing the performance of the gas turbine combined cycle. *Energy* 2018;160:625.

176. Kim T.S., and Ro S.T. The effect of gas turbine coolant modulation on the part load performance on the combined cycle plant. *Proc Inst Mech Eng A J Power Energ* 1997;211:443.

177. Nord L.O., Anantharaman R., and Bolland O. Design and off-design analyses of pre-combustion $CO_2$ capture process in a natural gas combined cycle plant. *Int J Greenh Gas Control* 2009;3:385.

178. Ajiwibowo M.W., Darmawan A., and Aziz M. Chemical looping combustion power generation system for a power-to-gas scheme. InTech open access paper; 2019. Doi: 10.5772/intechopen.85584.

179. Cinti G., Baldinelli A., Di Michele A., and Desideri U. Integration of solid oxide electrolyzer and Fischer-Tropsch: A sustainable pathway for synthetic fuel. *Appl Energ* 2016;162:308–320.

180. Luo Y., Wu X., Shi Y., Ghoniem A.F., and Cai N. Exergy analysis of an integrated solid oxide electrolysis cell-methanation reactor for renewable energy storage. *Appl Energ* 2018;215:371–383.

181. Aziz M. Power generation from algae employing enhanced process integration technology. *Chem Eng Res Desig* 2016;109:297–306.

182. Aziz M., Oda T., and Kashiwagi T. Integration of energy-efficient drying in microalgae utilization based on enhanced process integration. *Energy.* 2014;70:307–316.

183. Zaini I.N., Nurdiawati A., and Aziz M. Cogeneration of power and $H_2$ by steam gasification and syngas chemical looping of macroalgae. *Appl Energ* 2017;207:134–145.

184. Aziz M. Combined supercritical water gasification of algae and hydrogenation for hydrogen production and storage. *Energ Procedia* 2017;119:530–535.

185. Darmawan A., Hardi F., Yoshikawa K., Aziz M., and Tokimatsu K. Enhanced process integration of black liquor evaporation, gasification, and combined cycle. *Appl Energ* 2017;204:1035–1042.

186. Shaikh A.R., Wang Q., Han L., Feng Y., Sharif Z., Li Z., Cen J., and Kumar S. Techno-economic analysis of hydrogen and electricity production by biomass calcium looping gasification. *Sustainability* 2022;14:2189. Doi: 10.3390/su14042189, https://www.mdpi.com/journal/sustainability.

# 3 Advanced Nuclear Power

## 3.1 INTRODUCTION

Nuclear power was grown quickly in the 1970s and 1980s reaching a global installed capacity of 400 GWe at the present time. In an increasingly carbon-constrained future, nuclear power is becoming recognized as an integral part of the world's low-carbon energy solution. New construction in nuclear power has been most successful with State support as seen in China and South Korea, with China on track to double its nuclear power capacity from 27 to 54 GWe by 2020 [1–4]. Today, NPPs constructions have shifted from the West to the East with 4 out of 5 reactors under construction based in Asia and Eastern Europe. Between 2011 and 2018, 48 new reactors were grid-connected of which 29 units assumed operations in China [1–4].

Currently, there are 54 reactors under construction led by China with 11 reactors, India 7, Russia 6, South Korea 5 and the UAE with 4 units. Countries with two reactors under construction include Bangladesh, Belarus, Japan, Pakistan, Slovakia, Ukraine, and the USA. Argentina, Brazil, Finland, France and Turkey all have one reactor under construction. While today's global nuclear-installed capacity stands at 400, 169 GWe of that nuclear capacity will need replacing by 2030 to maintain the status quo in global nuclear capacity [1]. This could be sustained by the continued build-out of NPPs in Asia, Eastern Europe and the Middle East but recent delays in nuclear builds in Olkiluoto (Finland), Flamanville (France) and Georgia (USA) make large NPP build-outs in the West less likely. Instead, the West could diversify NPP builds with small modular reactor (SMR) and non-LWR reactor (i.e. High-Temperature Gas Reactors - HTGRs, Molten Salt Reactors - MSRs etc.) deployments [2–4].

Some studies project Chinese nuclear power capacity to expand to as much as 500 GWe should China aim to arrest global temperature rise to 1.5°C. Canada is also reinvesting in nuclear power and has announced a $12 Billion program to refurbish four CANDU reactors as well as announcing a $1.2 B program to retrofit Canadian Nuclear Laboratory at Chalk River in readiness for siting the first non-PWR based SMR. In the near term, NuScale is on track to build a FOAK (first of a kind) PWR-based SMR in Idaho National Labs by 2026 in an effort to reduce build times, enhance safety and reduce build costs. Despite the resurgence in interest in NPPs, it is tempered by the fact that 169 GWe of nuclear capacity is scheduled to come offline between 2020 and 2030 should no new reactor life extension be granted. Whether State-backed 1 GWe nuclear reactors and commercially financed SMRs could cover the retirement of older NPPs remains to be seen [1–4].

While nuclear energy still faces many obstacles to its social acceptance in many countries, the technology for nuclear power continues to make progress. Nuclear power production does not significantly emit greenhouse gases. However, large-scale nuclear power production is expensive and inefficient. Safety remains a major concern. In recent years, the advances in nuclear power are strictly focused on third and

DOI: 10.1201/9781003328087-3

fourth generation nuclear reactors. This chapter briefly examines these advances. Unlike large-scale and centralized nuclear power plants of the past, in recent years, the focus has been on small modular and distributed nuclear power plants. Besides its cost, SMR offers numerous advantages. This chapter briefly examines recent advances made globally in the design and construction of SMR. As shown, the use of SMR for cogeneration is also gaining public acceptance.

Significant advances are also made in the applications of nanotechnology for the improvement of the safety and efficiency of nuclear power production. These advances are briefly evaluated in this chapter. Nano materials have shown to be very useful in many facets of nuclear power production including their use in the treatment of nuclear waste. The direct conversion of radio isotope heat source to electricity by thermoelectricity or TPV technology has gained some momentum. This is briefly discussed in this chapter. More details on TPV technology and thermoelectricity are given in Chapters 7 and 8 respectively. Finally, the advances in the use of the supercritical $CO_2$ cycle for nuclear power production are briefly outlined.

## 3.2   BRIEF REVIEW OF GLOBAL PROGRESS ON GENERATION III ADVANCED NUCLEAR POWER REACTORS

Third-generation reactors have (a) a more standardized design for each type to expedite licensing, reduce capital cost and reduce construction time, (b) a simpler and more rugged design, making them easier to operate and less vulnerable to operational upsets, (c) higher availability and longer operating life—typically 60 years, (d) further reduced possibility of core melt accidents, (e) plant requires no active intervention for (typically) 72 hours following the shutdown, (f) stronger reinforcement against aircraft impact than earlier designs, to resist radiological release, (g) higher burn-up to use fuel more fully and efficiently, and reduce the amount of waste and (h) greater use of burnable absorbers ('poisons') to extend fuel life. Here we briefly review some of the new designs that have been evolved in various countries [5–7].

The greatest departure from most designs now in operation is that many incorporate passive or inherent safety features which require no active controls or operational intervention to avoid accidents in the event of a malfunction and may rely on gravity, natural convection or resistance to high temperatures. Another departure is that most are designed for load-following and for modular construction. This means that many small components are assembled in a factory environment (offsite or onsite) into structural modules weighing up to 1,000 tons, and these can be hoisted into place. This method speeds up construction. For example, the AP1000 footprint is very much smaller; the concrete and steel requirements are lower and it has a modular construction. At Sanmen and Haiyang in China, where the first AP1000 units were grid connected in August 2018, the first module lifted into place weighed 840 tons.

The Westinghouse AP1000 is a two-loop PWR which has evolved from the smaller AP600. SNPTC and SNERDI in China have jointly developed a passively safe 1,500 MWe (4,040 MWt) two-loop design from the AP1000, the CAP1400, or Guohe One, with 193 fuel assemblies and improved steam generators, operating at 323°C outlet temperature. The CAP1400 project may extend to a larger, three-loop

CAP1700 or CAP2100 design if the passive cooling system can be scaled to that level [5–7]. Areva NP developed a large (4,590 MWt, typically 1,750 MWe gross and 1,630 MWe net) European pressurized water reactor (EPR). In the US, the first unit (with 80% US content) was expected to be grid connected by 2020. It is now known as the Evolutionary PWR (EPR). In this reactor output electricity is at 60 Hz instead of the original design's 50 Hz. Areva NP with EdF developed a 'new model' EPR, the EPR NM or EPR2, "offering the same characteristics" as the EPR but with simplified construction and significant cost reduction—about 30%. The **Atmea1** has been developed by the Atmea joint venture established in 2007 by Areva NP and Mitsubishi Heavy Industries to produce an evolutionary 1,100–1,150 MWe net (3,150 MWt) three-loop PWR using the same steam generators as EPR. The first units are likely to be built at Sinop in Turkey. Together with German utilities and safety authorities, Areva NP has also developed another evolutionary design, the Kerena, a 1,290 MWe gross, 1,250 MWe net (3,370 MWt) BWR with 60-year design life formerly known as SWR 1000.

The advanced boiling water reactor (ABWR) is derived from a General Electric design in collaboration with Toshiba. Two examples built by Hitachi and two by Toshiba have been in commercial operation in Japan (1,315 MWe net), with another two under construction there and two in Taiwan. GE Hitachi Nuclear Energy's ESBWR is an improved design "evolved from the ABWR" but that utilizes passive safety features including natural circulation principles. The ESBWR is more innovative, with lower building costs due to modular construction, lower operating costs, 24-month refueling cycle and a 60-year operating lifetime. Mitsubishi's large APWR—advanced PWR of 1,538 MWe gross (4,451 or 4,466 MWt)—was developed in collaboration with four utilities. South Korea's APR1400 advanced PWR design has evolved from the US System 80+ with enhanced safety and seismic robustness and was earlier known as the Korean Next Generation Reactor. It was chosen for the United Arab Emirates (UAE) nuclear program on the basis of cost and reliable building schedule, and four units are under construction there, with the first expected online in 2020 [5–7].

In Canada, CANDU-9 (925–1,300 MWe) was developed from the CANDU-6 also as a single-unit plant. It had flexible fuel requirements. Some of the innovation of the CANDU-9, along with experience in building recent Korean and Chinese units, was then put back into the Enhanced CANDU-6 (EC6). EC6 is presented as a third-generation design based on Qinshan Phase III in China and is under consideration for new build in Ontario and overseas. Versatility of fuel is a claimed feature of the EC6 and its derivatives. India is developing the Advanced Heavy Water Reactor (AHWR) as the third stage in its plan to utilize thorium to fuel its overall nuclear power program. The AHWR is a 300 MWe gross (284 MWe net, 920 MWt) reactor moderated by heavy water at low pressure. The AEC says that "the reactor is manageable with modest industrial infrastructure within the reach of developing countries."

Gidropress late-model VVER-1000/V-392 units with enhanced safety (AES-92 & -91 power plants) have been built in India and China. Two more (V-466B variant) were planned for Belene in Bulgaria. The third-generation AES-2006 plant with VVER-1200 (V-392M or V-491) reactors of 3,212 MWt is an evolutionary development of the AES-92 and AES-91 plants with the VVER-1000, with longer operating lifetime (60 years for non-replaceable equipment), greater power, and greater efficiency (34.8% net instead of 31.6%) and 60 GWd/t burn-up. In Europe, the V-491 technology is being called the Europe-tailored reactor design, MIR-1200 (Modernized International Reactor) or AES-2006E, with some Czech involvement. The basic Gidropress reactor is V-510. It has upgraded pressure vessel, increased power to 3,312 MWt and 1,255 MWe gross (nominally 1,300, hence VVER-1300), improved core design still with 163 fuel assemblies to increase cooling reliability, larger steam generators, further development of passive safety with 72-hour grace period requiring no operator intervention after shutdown, lower construction and operating costs, and 40-month construction time. Gidropress has also developed the VVER-600/V-498 for sites such as Kola, where larger units are not required. It will have 60-year life and is capable of load-following. The Hualong One has 177 fuel assemblies 3.66 m long, 18–24 month refueling interval. It has three coolant loops delivering 3,050 MWt, 1,170 MWe gross, 1,090 MWe net (CNNC version). It is also being built in Pakistan. OKBM's VBER-300 PWR is a 295–325 MWe unit (917 MWt) developed from naval power plants and was originally envisaged in pairs as a floating nuclear power plant. It is designed for 60-year life and a 90% capacity factor. It is now planned to develop it as a land-based unit with Kazatomprom, with a view to exports, and the first unit will be built in Kazakhstan [5–7].

High-temperature gas-cooled reactors (Graphite-moderated) [8] use helium as a coolant at up to 950°C, which either makes steam conventionally (Rankine cycle) or directly drives a gas turbine for electricity and a compressor to return the gas to the reactor core (Brayton cycle). Fuel is in the form of TRISO particles less than a millimeter in diameter. China's HTR-PM is being built at Shidaowan in Shandong province. Fast neutron reactors (Not moderated, cooled by liquid metal) [9] are configured with a conversion or breeding ratio of more than 1 (*i.e.* more fissile nuclei are produced than are fissioned). India's 500 MWe prototype fast breeder reactor at Kalpakkam is fueled with uranium-plutonium oxide (the reactor-grade Pu being from its existing PHWRs) and with a thorium blanket to breed fissile U-233. The first (and probably only Russian) BN-800, a new more powerful (789 MWe, 880 MWe gross, 2,100 MWt) fast neutron reactor from OKBM was grid-connected at Beloyarsk in December 2015. The BN-1200 is being designed by OKBM for operation with MOX fuel initially and dense nitride U-Pu fuel subsequently, in closed fuel cycle. Today's PRISM is a GE Hitachi design for compact modular pool-type reactors with passive cooling for decay heat removal. PRISM is suited to operation with dry cooling towers due to its high thermal efficiency and small size. Westinghouse is developing a lead-cooled fast reactor (LFR) design with flexible output to complement intermittent renewable feed to the grid. Its high-temperature capabilities will allow industrial heat applications.

## 3.3   PROPOSED ADVANCED GENERATION IV NUCLEAR REACTORS

There were originally six technologies chosen for generation IV reactors, but the development of one has gone in two directions. There is also a novel design by Terra Power Company. Here we examine these in some details.

### 3.3.1 Gas-Cooled Fast Reactor (GFR)

Like other helium-cooled reactors which have operated or are under development, GFRs will be high-temperature units—typically 800°C–850°C. They employ similar reactor technology to the VHTR, suitable for power generation, thermochemical hydrogen production or other process heat applications. The reference GFR unit is 2,400 MWt/1,200 MWe with a core outlet temperature of 850°C, large enough for breakeven breeding, with thick steel reactor pressure vessel and three 800 MWt loops. The high core outlet temperature places significant demands on the fuel to operate continuously with the high power density required for good neutron economy in the core—this is recognized as the greatest challenge in the development of the GFR system. For electricity, an indirect cycle with helium will be on the primary circuit, in the secondary circuit the helium gas will directly drive a gas turbine (Brayton cycle), and a steam cycle will comprise the tertiary circuit. It would have a self-generating (breeding) core with a fast neutron spectrum and no fertile blanket. Robust nitride or carbide fuels would include depleted uranium and any other fissile or fertile materials such as ceramic pins or plates, with plutonium content of 15%–20%. As with the SFR, used fuel would be reprocessed onsite and all the actinides recycled repeatedly to minimize the production of long-lived radioactive waste [10–13]. An alternative GFR design has lower temperature (600°C–650°C) helium cooling in a primary circuit and supercritical $CO_2$ at 550°C and 20 MPa in a secondary system for power generation. This reduces the metallurgical and fuel challenges associated with very high temperatures.

The GFR system is a high-temperature helium-cooled fast-spectrum reactor with a closed fuel cycle. It combines the advantages of fast-spectrum systems for long-term sustainability of uranium resources and waste minimization (through fuel multiple reprocessing and fission of long-lived actinides), with those of high-temperature systems (high thermal cycle efficiency and industrial use of the generated heat, for hydrogen production for example). The GFR (see Figure 3.1) uses the same fuel recycling processes as the SFR and the same reactor technology as the VHTR. Therefore, its development approach is to rely, in so far as feasible, on technologies developed for the VHTR for structures, materials, components and power conversion systems. Nevertheless, it calls for specific R&D beyond the current and foreseen work on the VHTR system, mainly on the core design and safety approach. The reference design for GFR is based around a 2,400 MWth reactor core contained within a steel pressure vessel. Figure 3.2 shows the reactor core located within its fabricated steel pressure vessel surrounded by main heat exchangers and decay heat removal loops. The whole of the primary circuit is contained within a secondary pressure boundary, the guard containment. The main research needs, according to the China Academy of Sciences, are fuels, materials and thermal-hydraulics.

### 3.3.2 Lead-Cooled Fast Reactor (LFR)

The LFR is a flexible fast neutron reactor which can use depleted uranium or thorium fuel matrices and burn actinides from LWR fuel. Liquid metal (Pb or Pb-Bi eutectic) cooling is at atmospheric pressure by natural convection (at least for decay

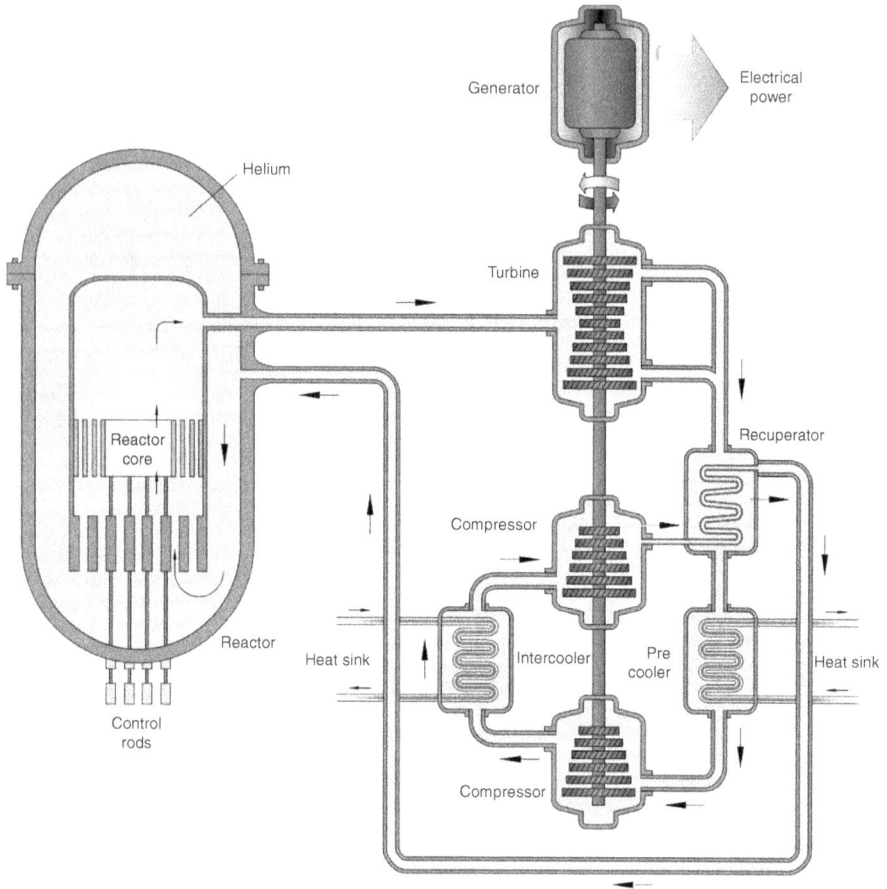

**FIGURE 3.1**   Gas-cooled fast reactor [10].

heat removal). Fuel is metal or nitride, with full actinide recycle from regional or central reprocessing plants. A wide range of unit sizes is envisaged, from factory-built "battery" with 15–20 year of life for small grids or developing countries, to modular 300–400 MWe units and large single plants of 1,400 MWe. Operating temperature of 550°C is readily achievable but 800°C is envisaged with advanced materials to provide lead corrosion resistance at high temperatures which would enable thermochemical hydrogen production. A two-stage development program leading to industrial deployment is envisaged: by 2025 for reactors operating with relatively low temperature and power density, and by 2040 for more advanced higher-temperature designs [10–13]. The lead-cooled fast reactor (see Figure 3.3) is a nuclear reactor design that features a fast neutron spectrum and molten lead or lead-bismuth eutectic coolant. Molten lead or lead-bismuth eutectic can be used as the primary coolant because lead and bismuth have low neutron absorption and relatively low melting points. Neutrons are slowed less by interaction with these heavy nuclei (thus not being neutron moderators) and therefore, help make this type of reactor a fast-neutron reactor. The coolant does, however, serve as a neutron reflector, returning some escaping

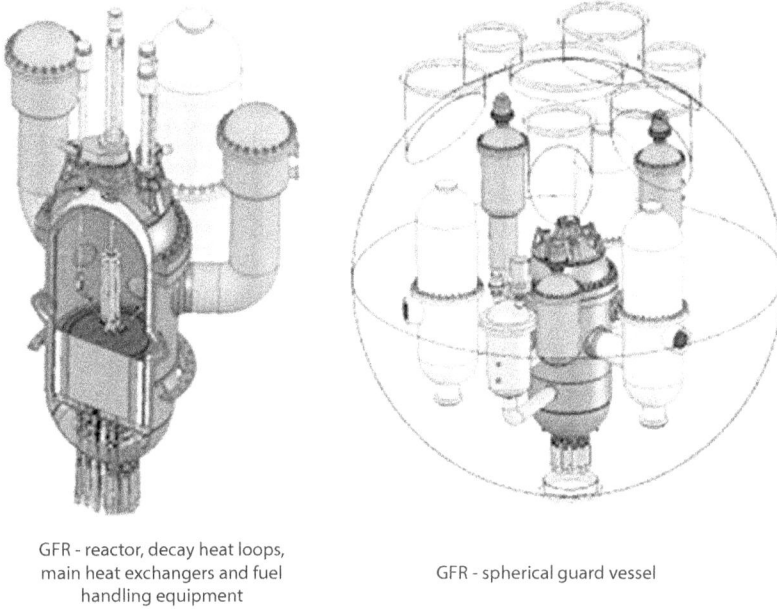

GFR - reactor, decay heat loops,
main heat exchangers and fuel
handling equipment

GFR - spherical guard vessel

**FIGURE 3.2** Details of gas-cooled reactor [10,12].

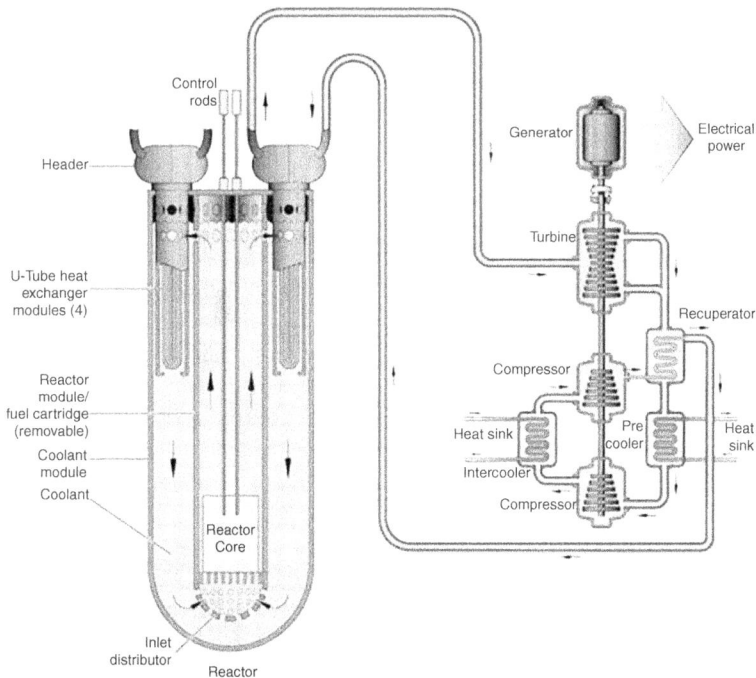

**FIGURE 3.3** Lead cooled fast reactor scheme [10].

neutrons to the core. Fuel designs being explored for this reactor scheme include fertile uranium as a metal, metal oxide or metal nitride.

The concept is generally very similar to a sodium-cooled fast reactor, and most liquid-metal reactors have used sodium instead of lead. Few lead-cooled reactors have been constructed, except for some Soviet nuclear submarine reactors in the 1970s, but a number of proposed new nuclear reactor designs are lead-cooled. The lead-cooled reactor design has been proposed as a generation IV reactor. The main research needs, according to the China Academy of Sciences, are fuels and materials.

### 3.3.3 MOLTEN SALT REACTOR (MSR)

MSR now has two variants: one a fast reactor with fissile material dissolved in the circulation fuel salt; the other with solid particle fuel in graphite and the salt functioning only as coolant. In what is considered to be a normal MSR, the uranium fuel is dissolved in the fluoride salt coolant which circulates through graphite core channels to achieve some moderation and an epithermal neutron spectrum. The reference plant is up to 1,000 MWe. Fission products are removed continuously and the actinides are fully recycled, while plutonium and other actinides can be added along with U-238, without the need for fuel fabrication. Coolant temperature is 700°C at very low pressure, with 800°C envisaged. A secondary coolant system is used for electricity generation, and thermochemical hydrogen production is also feasible [10–13]. Compared with solid-fueled reactors, MSR systems have lower fissile inventories, no radiation damage constraint on fuel burn-up, no requirement to fabricate and handle solid fuel or solid used fuel, and a homogeneous isotopic composition of fuel in the reactor. These and other characteristics may enable MSRs to have unique capabilities and competitive economics for actinide burning and extending fuel resources.

Molten salt reactors (MSR) (see Figure 3.4) use molten fluoride or chloride salts as a coolant. The coolant can flow over solid fuel like other reactors or fissile materials can be dissolved directly into the primary coolant so that the fission directly heats the salt. MSRs are designed to use less fuel and produce shorter-lived radioactive waste than other reactor types. They have the potential to significantly change the safety posture and economics of nuclear energy production by processing fuel online, removing waste products and adding fresh fuel without lengthy refueling outages. Their operation can be tailored for the efficient burn up of plutonium and minor actinides, which could allow MSRs to consume waste from other reactors. The system can also be used for electricity or hydrogen production. Terrapower and Elysium Industries in the US are researching a molten chloride-salt fast reactor to burn the US spent-fuel inventory accumulated from LWR operation. Many other reactor-vendor startups are also working on MSR designs with much pre-licensing activity occurring in Canada.

### 3.3.4 SODIUM-COOLED FAST REACTOR (SFR)

The sodium-cooled fast reactor (SFR) uses liquid metal (sodium) as a coolant instead of water that is typically used in U.S. commercial power plants. This allows for the coolant to operate at higher temperatures and lower pressures than current reactors— improving the efficiency and safety of the system. The SFR also uses a fast neutron

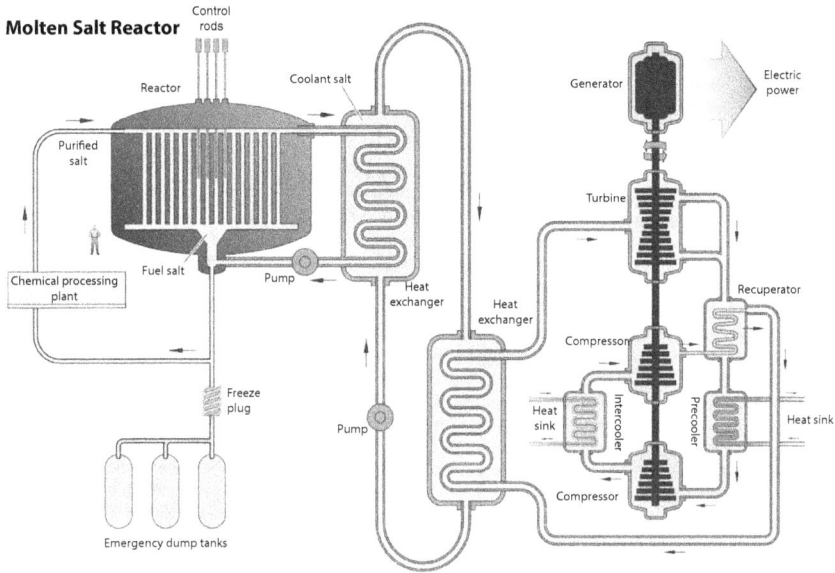

**FIGURE 3.4** MSRs have a closed fuel cycle that can be tailored for the efficient burn-up of plutonium and minor actinides [10].

spectrum, meaning that neutrons can cause fission without having to be slowed down first as they are in current reactors. This could allow SFRs to use both fissile material and spent fuel from current reactors to produce electricity considerably more efficiently than thermal spectrum reactors with once-through fuel cycles [10–13].

The sodium-cooled fast reactor (SFR) (see Figure 3.5) system features a fast-spectrum, sodium-cooled reactor and a closed fuel cycle for efficient management of actinides and conversion of fertile uranium. The SFR is designed for the management of high-level wastes and, in particular, management of plutonium and other actinides. Important safety features of the system include a long thermal response time, a large margin to coolant boiling, a primary system that operates near atmospheric pressure, and an intermediate sodium system between the radioactive sodium in the primary system and the water and steam in the power plant. With innovations to reduce capital cost, the SFR can serve markets for electricity. The SFR utilizes depleted uranium as the fuel matrix and has a coolant temperature of 500°C–550°C enabling electricity generation via a secondary sodium circuit, the primary one being at near atmospheric pressure. The main research needs, according to the China Academy of Sciences, are fuels and advanced recycle options.

### 3.3.5 Supercritical Water-Cooled Reactor (SCWR)

Supercritical fluids are those above the thermodynamic critical point, defined as the highest temperature and pressure at which gas and liquid phases can co-exist in equilibrium. They have properties between those of gas and liquid. For water the critical

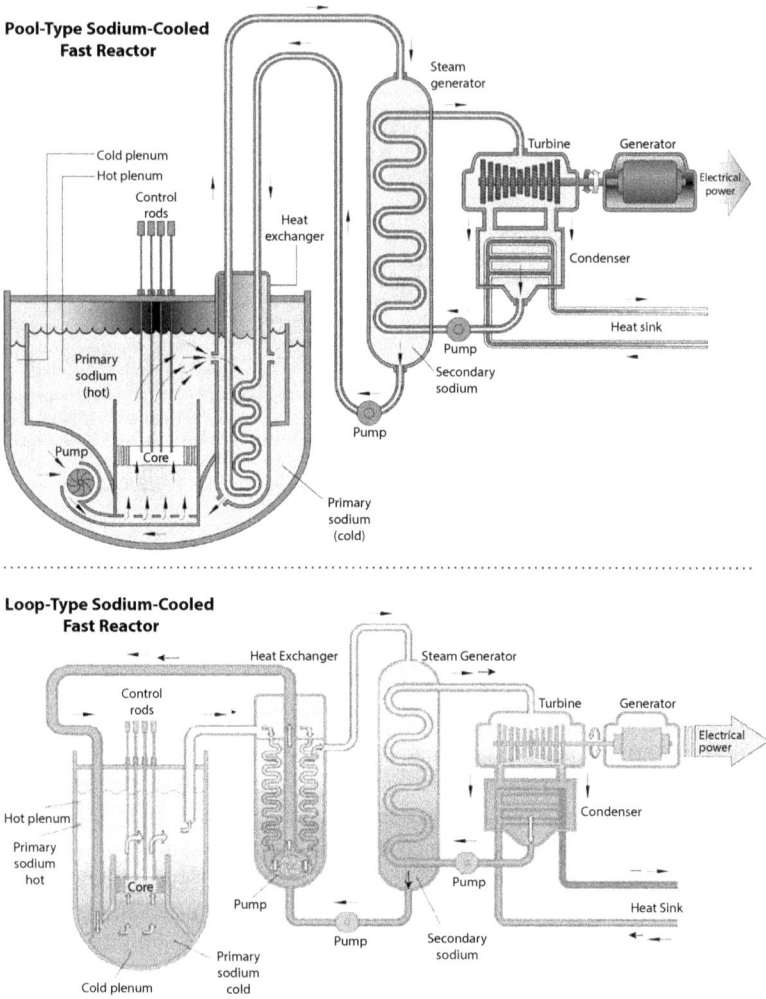

**FIGURE 3.5**   Pool-type and loop-type sodium-cooled fast reactors [10].

point is at 374°C and 22 MPa, giving it a "steam" density one-third that of the liquid so that it can drive a turbine in a similar way to normal steam. This is a very high-pressure water-cooled reactor which operates above the thermodynamic critical point of water (374°C, 22 MPa) to give a thermal efficiency about one-third higher than today's light water reactors from which the design evolves. The supercritical water (25 MPa and 510°C–550°C) directly drives the turbine, without any secondary steam system, simplifying the plant. Today's supercritical coal-fired plants use supercritical water around 25 MPa which have "steam" temperatures of 500°C–600°C and can give 45% thermal efficiency. At ultra-supercritical levels (30+ MPa), 50% thermal efficiency may be attained. Over 400 such plants are operating worldwide. China's Waigaoqiao 3 (2 × 1,000 MWe) near Shanghai operates at 600°C and 27.6 MPa with 44.5% thermal efficiency [10–13].

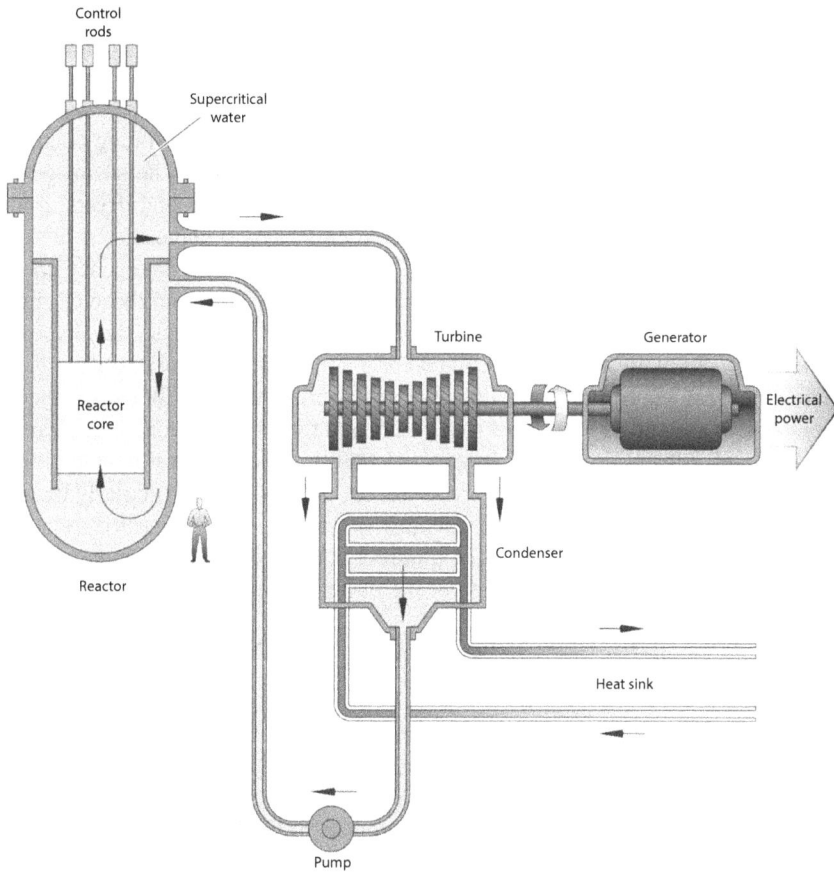

**FIGURE 3.6** SCWRs are high-temperature, high-pressure, light-water-cooled reactors that operate above the thermodynamic critical point of water (374°C, 22.1 MPa) [10].

The reactor (see Figure 3.6) core may have a thermal or a fast-neutron spectrum, depending on the core design. The concept may be based on current pressure vessel or on pressure tube reactors, and thus use light water or heavy water as moderator. Unlike current water-cooled reactors, the coolant will experience a significantly higher enthalpy rise in the core, which reduces the core mass flow for a given thermal power and increases the core outlet enthalpy to superheated conditions. For both pressure vessel and pressure-tube designs, a once-through steam cycle has been envisaged, omitting any coolant recirculation inside the reactor. As in a boiling water reactor, the superheated steam will be supplied directly to the high-pressure steam turbine and the feed water from the steam cycle will be supplied back to the core. Thus, the SCWR concepts combine the design and operation experiences gained from hundreds of water-cooled reactors with those experiences from hundreds of fossil-fired power plants operated with supercritical water (SCW). In contrast to some of the other Generation IV nuclear systems, the SCWR can be developed incrementally step-by-step from current water-cooled reactors.

SCWR designs have unique features that offer many advantages compared to state-of-the-art water-cooled reactors. SCWRs offer increases in thermal efficiency relative to current-generation water-cooled reactors. The efficiency of an SCWR can approach 44% or more, compared to 34%–36% for current reactors. Reactor coolant pumps are not required. The only pumps driving the coolant under normal operating conditions are the feed water pumps and the condensate extraction pumps. The steam generators used in pressurized water reactors and the steam separators and dryers used in boiling water reactors can be omitted since the coolant is superheated in the core. Containment, designed with pressure suppression pools and with emergency cooling and residual heat removal systems, can be significantly smaller than those of current water-cooled reactors. The higher steam enthalpy allows to decrease the size of the turbine system and thus to lower the capital costs of the conventional island.

These general features offer the potential of lower capital costs for a given electric power of the plant and of better fuel utilization, and thus a clear economic advantage compared with current light water reactors. However, there are several technological challenges associated with the development of the SCWR, and particularly the need to validate transient heat transfer models (for describing the depressurization from supercritical to sub-critical conditions), qualification of materials (namely advanced steels for cladding), and demonstration of the passive safety systems.

### 3.3.6 Very High-Temperature Gas Reactor (VHTR)

High-Temperature Gas Reactors (HTGRs) have been an understudy for over 50 years. The first helium-cooled, graphite-moderated HTGR operated at Peach Bottom, USA from 1966 to 1974. The helium coolant is inert, single phase and has no reactivity feedback; while the large graphite core gives the reactor superior thermal inertia which absorbs the decay heat after shutdown, making HTGRs passively safe. HTGRs, like MSRs, have very high-temperature outputs from 750°C to 1,000°C allowing the supply of industrial heat without $CO_2$ emissions. Later this year, China is set to commission an experimental air-cooled high-temperature gas reactor (HTR-PM) which uses two reactor modules (500 MWth) to supply a common turbine outputting 200 MWe. Successful demonstration of the HTR-PM will lead to the development of HTR-PM600 outputting 600 MWe of power [10–13]. HTGRs are suited for inland deployment where water may be scarce for LWR cooling.

The Very-High-Temperature Reactor (VHTR) is a graphite-moderated, helium-cooled reactor with a once-through uranium fuel cycle. The VHTR system is designed to be a high-efficiency system that can supply process heat to a broad spectrum of high-temperature and energy-intensive, nonelectric processes. The system may incorporate electricity-generating equipment to meet cogeneration needs. The system also has the flexibility to adopt uranium/ plutonium fuel cycles and offer enhanced waste minimization. Thus, the VHTR offers a broad range of process heat applications and an option for high-efficiency electricity production, while retaining the desirable safety characteristics offered by modular high-temperature gas-cooled reactors. The reference reactor is a 600 MWth core connected to an intermediate heat exchanger for the delivery of process heat. The reactor core can be a prismatic block core such as the operating Japanese HTTR, or a pebble-bed core such as the operating Chinese HTR-10. For hydrogen production, the system supplies heat that could be used efficiently by the thermochemical sulfur-iodine process.

**Very High Temperature Reactor Prismatic Core**

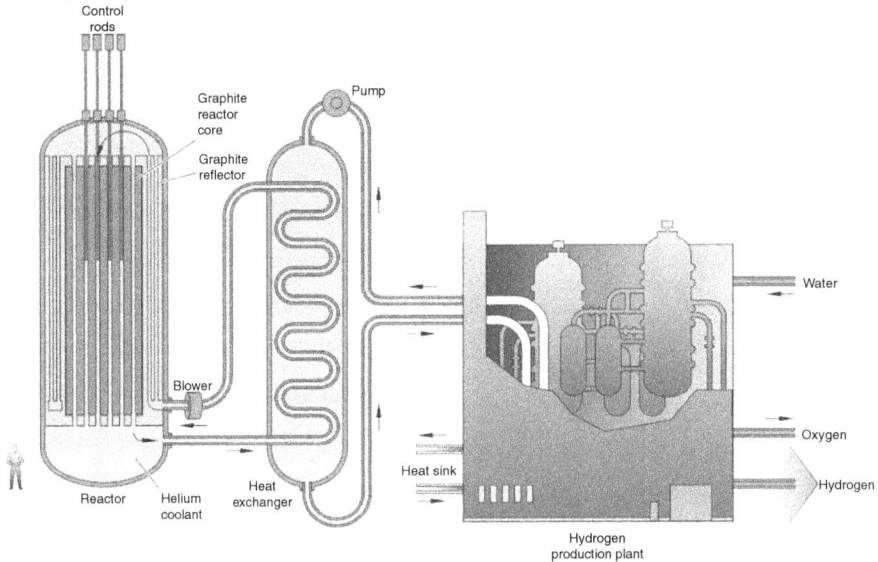

FIGURE 3.7 VHTRs offer a broad range of process heat applications and an option for high-efficiency electricity production [10].

The very high-temperature reactor (see Figure 3.7) is cooled by flowing gas and is designed to operate at high temperatures that can produce electricity extremely efficiently. The high-temperature gas could also be used in energy-intensive processes that currently rely on fossil fuels, such as hydrogen production, desalination, district heating, petroleum refining, and ammonia production. Very high-temperature reactors offer impressive safety features and can be easy to construct and affordable to maintain. The main research needs, according to the China Academy of Sciences, are fuels, materials and hydrogen production. A summary of generation IV reactors under development is given in Tables 3.1 and 3.2.

### 3.3.7 NOVEL GEN IV NUCLEAR REACTOR BY TERRA POWER

TerraPower, based in Bellevue, Wash. is poised to break ground on a test reactor *in cooperation* with the China National Nuclear Corp [11–20]. TerraPower nuclear reactor is safer and more fuel efficient. The reactor's fuel can't easily be used for weapons, and the company claims that its reactor will generate very little waste. If the reactor was left unattended, it wouldn't suffer a calamitous mishap. The TerraPower reactor is a new variation on a design that was conceived some 60 years ago by a Russian physicist, Savely Feinberg. Feinberg designed a breed-and-burn reactor which featured a slowly advancing wave of nuclear fission through a fuel source, like a cigar that takes decades to burn, creating and consuming its fuel as the reaction travels through the core [19,20].

TWR will be able to use depleted uranium, which has far less U-235 and cannot reach criticality unassisted. TerraPower's solution is to arrange 169 solid uranium

**TABLE 3.1**
**A Generation IV Reactor Designs [13]**

Summary of Designs for Generation IV Reactors [21]

| System | Neutron Spectrum | Coolant | Temperature (°C) | Fuel Cycle | Size (MW) | Example Developers |
|---|---|---|---|---|---|---|
| VHTR | Thermal | Helium | 900–1,000 | Open | 250–300 | JAEA (HTTR), Tsinghua University (HTR-10), Tsinghua University & China Nuclear Engineering Corporation (HTR-PM), X-energy |
| SFR | Fast | Sodium | 550 | Closed | 30–150, 300–1,500, 1,000–2,000 | TerraPower (TWR), Toshiba (4S), GE Hitachi Nuclear Energy (PRISM), OKBM Afrikantov (BN-1200), China National Nuclear Corporation (CNNC) (CFR-600), Indira Gandhi Centre for Atomic Research(Prototype Fast Breeder Reactor) |
| SCWR | Thermal or fast | Water | 510–625 | Open or closed | 300–700, 1,000–1,500 | |
| GFR | Fast | Helium | 850 | Closed | 1,200 | Energy Multiplier Module |
| LFR | Fast | Lead | 480–800 | Closed | 20–180, 300–1,200, 600–1,000 | Rosatom (BREST-OD-300) |
| MSR | Fast or thermal | Fluoride or chloride salts | 700–800 | Closed | 250, 1,000 | Seaborg Technologies, TerraPower, Elysium Industries, Moltex Energy, Flibe Energy (LFTR), Transatomic Power, Thorium Tech Solution (FUJI MSR), Terrestrial Energy (IMSR), Southern Company |
| DFR | Fast | Lead | 1000 | Closed | 500–1,500 | Institute for Solid-State Nuclear Physics |

**TABLE 3.2**
**Generation IV Reactor Designs under Development by GIF [11]**

| | Neutron Spectrum (Fast/Thermal) | Coolant | Temperature (°C) | Pressure[a] | Fuel | Fuel Cycle | Size (MWe) | Use |
|---|---|---|---|---|---|---|---|---|
| Gas-cooled fast reactors | Fast | Helium | 850 | High | U-238[b] | Closed, on site | 1,200 | Electricity & hydrogen |
| Lead-cooled fast reactors | Fast | Lead or Pb-Bi | 480–570 | Low | U-238[b] | Closed, regional | 20–180[c] 300–1,200 600–1,000 | Electricity & hydrogen |
| Molten salt fast reactors | Fast | Fluoride salts | 700–800 | Low | UF in salt | Closed | 1,000 | Electricity & hydrogen |
| Molten salt reactor - advanced high-temperature reactors | Thermal | Fluoride salts | 750–1,000 | Low | UO₂ particles in prism | Open | 1,000–1,500 | Hydrogen |
| Sodium-cooled fast reactors | Fast | Sodium | 500–550 | Low | U-238 & MOX | Closed | 50–150 600–1,500 | Electricity |
| Supercritical water-cooled reactors | Thermal or fast | Water | 510–625 | Very high | UO₂ | Open (thermal) closed (fast) | 300–700 1,000–1,500 | Electricity |
| Very high temperature gas reactors | Thermal | Helium | 900–1,000 | High | UO₂ prism or pebbles | Open | 250–300 | Hydrogen & electricity |

*Idaho National Laboratory.*
[a]*high = 7–15MPa.*
[b]= *with some U-235 or Pu-239.*
[c]*'battery' model with long cassette core life (15–20years) or replaceable reactor module.*

fuel pins into a hexagon. When the reaction begins, the U-238 atoms absorb spare neutrons to become U-239, which decays in a matter of minutes to neptunium-239 and then decays again to plutonium-239. When struck by a neutron, Pu-239 releases two or three more neutrons, enough to sustain a chain reaction. It also releases plenty of energy; after all, Pu-239 is the primary isotope used in modern nuclear weapons. Pu-239 won't accumulate in the TWR; instead, stray neutrons will split the Pu-239 into a cascade of fission products almost immediately. This allows the benefits of a closed fuel cycle without the expense and *proliferation* risk of enrichment and reprocessing plants typically required to get them. Enough fuel for between 40 and 60 years of operation could be in the reactor from the beginning. The reactor could be buried below ground, where it could run for an estimated 100 years. TerraPower described the concept of its main reactor design as a "Generation IV, liquid sodium-cooled fast reactor." The major benefit of such reactors is high fuel utilization in a manner that does not require *nuclear reprocessing* and could eventually eliminate the need to enrich uranium. In other words, the reactor breeds the highly fissile plutonium fuel it needs right before it burns it. Yet the "traveling wave" label refers to something slightly different from the slowly burning, cigar-style reactor (see Figure 3.8). In the TWR, an overhead crane system will maintain a reaction within a ringed portion of the core by moving pins into and out of that zone from elsewhere in the core, like a very large, precise arcade claw machine [11–20]. Molten sodium can move more heat out of the core than water, and it's actually less corrosive to metal pipes than hot water is. But it's a highly toxic metal, and it's violently flammable when it encounters oxygen [17,18].

The TWR's design features some of the same safety systems standard as nuclear reactors. In the case of an accident in any reactor, control rods crafted from neutron-absorbing materials like cadmium plummet into the core and halt a runaway chain reaction that could otherwise lead to a core meltdown. Such a shutdown is called a scram. Scramming a reactor cuts its fission rate to almost zero in a very

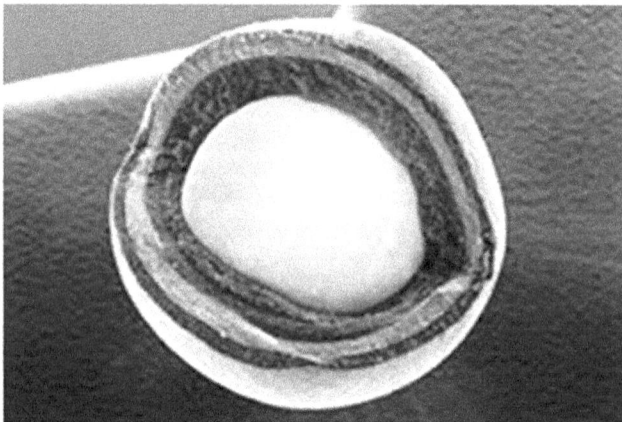

FIGURE 3.8  TRISO coated particle fuel.

short time, though residual heat can still cause a disaster. Because the TWR burns its fuel more efficiently, the TerraPower also claims it will produce less waste. Even if TerraPower's reactor succeeds wildly, it will take 20 years or more for the company to deploy large numbers of TWRs [17,18].

### 3.3.7.1 Sodium Fast Reactor (Natrium)

Natrium is a modification of Terra Power reactor. Natrium combines a molten sodium reactor with a 1 GWh molten salt energy storage system. Sodium offers a 785-K temperature range between its solid and gaseous states, nearly 8× that of water's 100-K range [14–16]. Without requiring pressurization, which increases risk and cost, sodium can absorb large amounts of heat. As an element rather than a compound, it is not at risk of disassociating at high temperature as water does. Sodium is also non-corrosive. Natrium is fueled by High-Assay, Low Enriched Uranium (HALEU) as its fuel. HALEU is enriched to contain between 5% and 20% of uranium, which can be produced from spent fuel from other nuclear power plants. Plant sites are expected to be smaller and 4× more efficient than conventional plants. Natrium *control rods* descend using only gravity in case of plant failure. Power output is a constant 345 MWe as heat. The plant is designed to run at 100% output, 24/7.

Liquid sodium used in the reactor has a higher boiling point and can absorb a lot more heat than water, which means high pressure does not build up inside the reactor. TerraPower's cooling system does not rely on any outside energy source to operate in the event of an emergency shutdown of a reactor. Instead, its system works via the hot air rising from natural circulation within the system, called a reactor vessel air cooling system, or RVACS. Such "passive" cooling systems, which rely on things like natural circulation or gravity, are at the heart of advanced reactor designs and reduce the risk and severity of accidents. This type of passive safety system has been demonstrated on a large scale. Natrium technology has the ability to store heat in tanks of molten salt for future use, "much" like a battery. The technology is similar to that used in solar plants. TerraPower also plans to build smaller than conventional ones, which is "important" for lowering the capital cost. The size also makes TerraPower plants attractive to utility companies looking to invest in a nuclear power plant to plug into their existing power grid, TerraPower's plants [17,18] have the capacity to generate enough power to operate a small city, which is the "sweet spot" for utility companies. Natrium technology, which utilizes radioactive uranium fuel (which powers nuclear reactors) much more efficiently and more completely than conventional plants.

## 3.4  SUPERCRITICAL $CO_2$ CYCLE

As mentioned earlier, the goal of Generation IV reactors [22] is to achieve higher core outlet temperature (500°C–900°C) compared to water cooled reactor (app. 300°C) [23]. This increases the thermal efficiency and reduces the cost of the power block. The gas-cooled fast reactor, and very high-temperature reactor, provide core outlet temperatures in excess of 850°C and employ a closed-loop helium Brayton cycle for the power generation cycle. However, the high operating temperatures present challenges in material selection. In general, the heat generated in the core is removed directly (by working fluid) or indirectly (where working fluid goes through another heat exchange process) which is eventually converted to electricity by a power cycle. The work of Dostal [24] was conducted with the focus of identifying power cycles that could provide comparable efficiencies at lower temperatures. The conclusion was that $sCO_2$ cycles operating with a maximum temperature of 550°C could achieve comparable efficiencies to a helium Brayton cycle operating at a temperature of 850°C, making $sCO_2$ a promising candidate for any reactor with core outlet temperatures in excess of 500°C. The study also found that the sodium-cooled fast reactor (SFR) and lead-cooled fast reactor (LFR) are the most promising applications for $sCO_2$ cycles. Sienicki and Moisseytsev [25] noted that both reactors provide core outlet temperatures in the region of 500°C–550°C, and that the temperature match between the temperature rise of the heat-transfer fluid in the primary heater and across the turbine are well matched. This enables $sCO_2$ cycles to achieve higher efficiencies than an equivalent steam cycle. A smaller footprint of s-$CO_2$ cycle also reduces capital cost and LCOE. whilst having a smaller footprint, leading to a potential reduction in capital cost and LCOE. Li et al. [26] verified this conclusion and recommended that lead cooled fast reactor (LFR) to be the new frontier of research. More evaluations of historical development and thermodynamic and techno-economic modeling are given by Li et al. [26], Sienicki and Moisseytsev [25] and more recently by Wu et al. [27].

Moisseytsev and Sienicki [28] reported the design of an LFR, named the secure transportable autonomous reactor with liquid metal coolant (STAR-LM), coupled to a $sCO_2$ recompression cycle. The study showed the turbine inlet temperature to be 540°C, with a net power output of 179 MWe by power cycle and estimated cycle thermal efficiency of 45%. Later, Sienicki et al. [29] reported on the design of a smaller reactor, referred to as SSTAR, with a net power output of 20 MWe. The system employed a direct heat exchange between the lead coolant and the $CO_2$, and the study indicated that at a turbine inlet of 550°C a thermal cycle efficiency of 44% can be obtained. One major purpose of this study was to demonstrate the viability of SMR for developing nations in a sustainable manner [29], a sentiment supported by the IEA's 2015 nuclear technology map [30]. The study also showed that the small modular reactor market could be a good proving ground for $sCO_2$ technology. Chang et al. [31] reported the pre-conceptual design of an SFR coupled to a recompression $sCO_2$ cycle with a net power of 95 MWe. Simulations also estimated that for a turbine inlet temperature of 471.5°C a cycle thermal efficiency of 39.1% could be obtained. Moisseytsev and Sienicki [32] found that while more complex power cycle layouts did not improve this efficiency, reducing the minimum temperature to 20°C, and operating a condensation cycle, could improve the efficiency by up to 4%; although this would require a heat sink. Sienicki and Moisseytsev [25] reported the detailed design of a 100 MWe small modular SFR reactor with a turbine inlet temperature of 517°C with an estimated cycle thermal efficiency of 42.3%.

The researchers at Korea Advanced Institute of Science and Technology have investigated the use of $sCO_2$ power cycles with SFRs [33], in addition to water-cooled reactors [34] and high-temperature gas-cooled reactors [35]. Yu et al. [36] reported the design of a micro modular reactor, named KAIST MMR, in which the core is directly cooled by the supercritical $CO_2$ and the intermediate heat exchanger is removed. The

KAIST MMR is designed to have a thermal power of 36.2 MWth and a 20-year lifetime without refueling. Tokyo Institute of Technology pursued $sCO_2$ cycles for application within gas-cooled reactors, by replacing helium with $sCO_2$ [37]. This reduced temperature from 850°C to 650°C while retaining a similar thermal efficiency. The study also supported the findings of Dostal [24]. Muto and Kato [38] investigated a dual expansion cycle for fast reactors and high-temperature gas-cooled reactors.

Li et al. [39] proposed a conceptual design of 10 MWe LFR, integrated with a recompression $sCO_2$ cycle. With help of thermo-economic modeling, the study predicted a thermal efficiency in the range of 36.7% and 44.5% and a cost of electricity between 50 and 55 $/MWh for a turbine inlet temperature of 550°C. In France, $sCO_2$ was considered as a power cycle for the advanced sodium technological reactor for industrial demonstration (ASTRID) project [40,41]. For the plant, which has a total thermal power of 1,500 MWth, simulations suggested that a net plant efficiency of 42.2% could be obtained for turbine inlet conditions of 25 MPa and 515°C [41]. However, as of August 2019, the ASTRID program has been canceled [42]. In the Czech Republic, the SUSEN test loop was developed with a view toward testing $sCO_2$ components with applications focused on gas-cooled fast reactors [43]. This test loop was also utilized within the $sCO_2$-HeRo project, which focused on the proof of concept for a small-scale heat removal safety backup system for a light-water reactor [44]. Linares et al. [45] explored the suitability of $sCO_2$ powercycles for the DEMO demonstration power plant, which investigated fusion applications within Europe [45]. While it appears that steam is still the power cycle of choice for that plant [46], recent simulations suggest that $sCO_2$ cycles may outperform steam providing that the heat-source temperature is above 460°C [21].

Wu et al. [27] summarized specific challenges relating to nuclear applications of $sCO_2$ cycle which include (a) understanding the interaction between the $sCO_2$ system and the reactor coolant system; (b) understanding the effect of the dynamic behavior of the reactor core on the $sCO_2$ cycle; and (c) conducting a comprehensive safety analysis. While the interest in using the $s\text{-}CO_2$ cycle for Generation IV reactors has increased, due to cost and safety considerations related to nuclear power, $sCO_2$ technology will first need to be successfully demonstrated in other applications.

## 3.5  SMALL NUCLEAR POWER REACTORS

As nuclear power generation has become established since the 1950s, the size of reactor units has grown from 60 MWe to more than 1,600 MWe, with corresponding economies of scale in operation. At the same time, there have been many hundreds of smaller power reactors built for naval use (up to 190 MW thermal) and as neutron sources, yielding enormous expertise in the engineering of small power units. The International Atomic Energy Agency (IAEA) defines 'small' as under 300 MWe, and up to about 700 MWe as 'medium'—including many operational units from the 20th century. Together they have been referred to by the IAEA as small and medium reactors (SMRs). However, 'SMR' is used more commonly as an acronym for "small modular reactor," designed for serial construction and collectively to comprise a large nuclear power plant. A subcategory of very small reactors—vSMRs—is proposed for units under about 15 MWe, especially for remote communities [47–53].

Small modular reactors (SMRs) are defined as nuclear reactors generally 300 MWe equivalent or less, designed with modular technology using module factory fabrication, pursuing economies of series production and short construction times. Small modular reactors can encompass both LWR and non-LWR designs, although for the next 10 years, nearly all deployable SMRs will be of the LWR variety which have accumulated 50+ years of operational experience and thus are easier to license compared to non-LWR designs [47–53]. SMR may be built independently or as modules in a larger complex, with capacity added incrementally as required. Economies of scale are envisaged due to the numbers produced. There are also moves to develop independent small units for remote sites. Small units are seen as a much more manageable investment than big ones whose cost often rivals the capitalization of the utilities concerned. An additional reason for interest in SMRs is that they can more readily slot into brownfield sites in place of decommissioned coal-fired plants, the units of which are seldom very large—more than 90% are under 500 MWe, and some are under 50 MWe. In the USA coal-fired units retired over 2010–2012 averaged 97 MWe, and those expected to retire over 2015–2025 average 145 MWe.

Generally, modern small reactors for power generation, and especially SMRs, are expected to have greater simplicity of design, economy of series production largely in factories, short construction times, and reduced siting costs. Most are also designed for a high level of passive or inherent safety in the event of malfunction. Also many are designed to be emplaced below ground level, giving a high resistance to terrorist threats. A 2010 report by a special committee convened by the American Nuclear Society showed that many safety provisions necessary, or at least prudent, in large reactors are not necessary in the small designs forthcoming. This is largely due to their higher surface area to volume (and core heat) ratio compared with large units. It means that a lot of the engineering for safety including heat removal in large reactors is not needed in small reactors. Since small reactors are envisaged as replacing fossil fuel plants in many situations, the emergency planning zone required is designed to be no more than about 300 m radius.

The World Nuclear Association indicates several attractive features of SMR [47–53]. It has small power and compact architecture which enables modularity of fabrication (in-factory) and facilitates implementation of higher quality standards. Lower power also leads to a reduction of the source term as well as smaller radioactive inventory in a reactor. The reactor has the potential for underground or underwater location which provides more protection from natural (*e.g.* seismic or tsunami according to the location) or man-made (*e.g.* aircraft impact) hazards. The modular design and small size lend themselves to having multiple units on the same site. The reactor has a lower requirement for access to cooling water—therefore suitable for remote regions and for specific applications such as mining or desalination. The reactor has less reliance on active safety systems and additional pumps, as well as AC power for accident mitigation. Finally, it is easy to remove or decommission smaller reactor at the end of the lifetime.

The IRIS program, led by Westinghouse, sought to fit the separate parts of reactor core, heat exchangers, control rods and pressurizer into a single, compact module. NuScale is on track to build a 12-unit-plant supplying 720 MWe at Idaho National Laboratory (INL). Other SMR projects within an 8-years deployment time frame include the Argentine CAREM (32 MWe); South Korean SMART (100 MWe + desalination as bonus), the Chinese ACP-100 (100 MWe) from CNNC and the Russian KLT-40S with two barge-mounted 35 MWe units, one of which was loaded with fuel in November 2018. SMR applications include providing high-reliability power for mission-critical infrastructure, black-start capability for micro-grids and steam generation for district heating and desalination [50]. Factory fabrication of modular components aims to reduce build times from 8 to 3 years and the smaller size would mean a lower initial capital outlay, making SMRs accessible to more countries.

SMRs are ideally suited for the fourth-generation reactors outlined earlier. It has the lowest technological risk, and FNR can be smaller, simpler and with a longer operation before refueling. Russia already has two SFRs in operations which can produce more fuel than it burns or destroys long-lived radioactive waste. This versatility stems from sodium having an extremely low neutron capture cross section resulting in a surplus of neutrons in the core that can be used to either burn actinide waste or bred more fissile fuel. Apart from the SFR, Russia is developing a lead-cooled fast reactor (BREST-300) [51,52]. Using lead as a coolant is safer than using sodium (which is more chemically reactive) even though lead is not as neutron-efficient. To compensate for the higher neutron absorption of lead, uranium-nitride fuel with a higher uranium loading per unit volume is planned for use in LFRs. Molten Salt Reactors can fulfil a variety of missions including fuel breeding and high-temperature operations for supplying industrial heat. The Chinese Government is investing US$3.3 billion to build the first liquid-fuel MSR prototype. Successful demonstration of the LF-1 demonstrator could pave the way for an MSR Breeder using thorium-232 as the fertile material for breeding uranium-233 fuel [53].

There are a number of small modular reactors coming forward requiring fuel enriched at the top end of what is defined as low-enriched uranium (LEU)—20% U-235. The US Nuclear Infrastructure Council (NIC) has called for some of the down blending of military HEU to be only to about 19.75% U-235, so as to provide a small stockpile of fuel which would otherwise be very difficult to obtain (since civil enrichment plants normally cannot go above 5%). A reserve of 20 tons of high-assay low-enriched uranium (HALEU) has been suggested. The NIC said that the only supply of fuel for many advanced reactors under development would otherwise be foreign-enriched uranium. "Without a readily available domestic supply of higher enriched LEU in the USA, it will be extremely difficult to conduct research on advanced reactors, potentially driving American innovators overseas." In 2019 the DOE contracted with Centrus Energy to deploy a cascade of large centrifuges to produce HALEU fuel for advanced reactors. Urenco USA has announced its readiness to supply HALEU from a dedicated production line at its New Mexico plant. DOE is supporting Kairos Power for the Hermes Reduced-Scale Test Reactor, Westinghouse for the eVinci microreactor; BWXT Advanced Technologies for the BWXT Advanced Nuclear Reactor (BANR); Holtec for its SMR-160; and Southern Company for its Molten Chloride Reactor Experiment, a 300 kWt reactor project to provide data to inform the design of a demonstration molten chloride fast reactor (MCFR) using TerraPower's technology. NuScale has said that it aims to deploy its SMR technology in the UK with UK partners so that the first of its units could be in operation by the mid-2020s. In September 2017 the company released its five-point UK SMR action plan. Rolls-Royce submitted a detailed design to the government for a 220 MWe SMR unit.

Ontario Ministry of Energy is focused on nine designs under 25 MWe for off-grid remote sites. All had a medium level of technology readiness and were expected to be competitive against diesel. Ontario distinguishes 'grid scale' SMRs above 25 MWe from these (very) small-scale reactors. In November 2019 CNL announced that Kairos Power, Moltex Canada, Terrestrial Energy and Ultra Safe Nuclear Corporation (USNC) had been selected as the first recipients of support under its Canadian Nuclear Research Initiative (CNRI). This is designed to accelerate SMR deployment by enabling research and development on particular projects and connecting global vendors of SMR technology with the facilities and expertise within Canada's national nuclear laboratories. In October 2020 Ontario Power Generation (OPG) announced that it would take forward engineering and design work with three developers of grid-scale SMRs—GE Hitachi (GEH), Terrestrial Energy and X-energy—to support remote area energy needs. The focus is on GEH's 300 MWe BWRX-300, Terrestrial's 192 MWe Integral Molten Salt Reactor, and X-energy's 80 MWe Xe-100 high-temperature SMRs.

The most advanced small modular reactor project is in China, where Chinergy is starting to build the 210 MWe HTR-PM, which consists of twin 250 MWt high-temperature gas-cooled reactors (HTRs). China is also developing small district heating reactors of 100–200 MWt capacity which may have a strong potential evaluated at around 400 units. Urenco has called for European development of very small—4 MWe—"plug and play" inherently-safe reactors based on graphite-moderated HTR concepts. Already operating in a remote corner of Siberia are four small units at the Bilibino co-generation plant. These four 62 MWt (thermal) units are an unusual graphite-moderated boiling water design with water/steam channels through the moderator [47–53]. Also in the small reactor category are the Indian 220 MWe pressurized heavy water reactors (PHWRs) based on Canadian technology, and the Chinese 300–325 MWe PWR such as those built at Qinshan Phase I and at Chashma in Pakistan, and now called CNP-300. The Nuclear Power Corporation of India (NPCIL) is now focusing on 540 and 700 MWe versions of its PHWR and is offering both 220 and 540 MWe versions internationally. Another significant line of development is in very small fast reactors of under 50 MWe. Some are conceived for areas away from transmission grids and with small loads; others are designed to operate in clusters in competition with large units. In December 2019, CEZ in the Czech Republic said it was focusing on 11 SMR designs including these seven: Rosatom's RITM-200, GE Hitachi Nuclear Energy's BWRX-300, NuScale Power's SMR, China National Nuclear Corporation's ACP100, Argentina's CAREM, the South Korean SMART, and Holtec International's SMR-160. The SMR reactors being operated, under construction or in advanced stage of development are illustrated in Table 3.3 [47–49].

## 3.6   ROLE OF NANOTECHNOLOGY IN NUCLEAR POWER

Nanotechnology has played a significant role in improving efficiency, cost, and safety of nuclear power in a variety of ways. Here we briefly examine the contribution of nano technology in various aspects of nuclear power production and related topics.

**TABLE 3.3**
**Global assessment of SMR [47]**

| Name | Capacity | Type | Developer |
|---|---|---|---|
| **Small Reactors Operating** | | | |
| CNP-300 | 300 MWe | PWR | SNERDI/CNNC, Pakistan & China |
| PHWR-220 | 220 MWe | PHWR | NPCIL, India |
| EGP-6 | 11 MWe | LWGR | at Bilibino, Siberia (cogen, soon to retire) |
| KLT-40S | 35 MWe | PWR | OKBM, Russia |
| RITM-200 | 50 MWe | Integral PWR, civil marine | OKBM, Russia |
| **Small Reactor Designs under Construction** | | | |
| CAREM25 | 27 MWe | Integral PWR | CNEA & INVAP, Argentina |
| HTR-PM | 210 MWe | Twin HTR | INET, CNEC & Huaneng, China |
| ACP100/Linglong One | 125 MWe | Integral PWR | CNNC, China |
| BREST | 300 MWe | Lead FNR | RDIPE, Russia |
| **Small reactors for near-term deployment – development well advanced** | | | |
| VBER-300 | 300 MWe | PWR | OKBM, Russia |
| NuScale | 77 MWe | Integral PWR | NuScale Power + Fluor, USA |
| SMR-160 | 160 MWe | PWR | Holtec, USA + SNC-Lavalin, Canada |
| SMART | 100 MWe | Integral PWR | KAERI, South Korea |
| BWRX-300 | 300 MWe | BWR | GE Hitachi, USA |
| PRISM | 311 MWe | Sodium FNR | GE Hitachi, USA |
| Natrium | 345 MWe | Sodium FNR | TerraPower + GE Hitachi, USA |
| ARC-100 | 100 MWe | Sodium FNR | ARC with GE Hitachi, USA |
| Integral MSR | 192 MWe | MSR | Terrestrial Energy, Canada |
| Seaborg CMSR | 100 MWe | MSR | Seaborg, Denmark |
| Hermes prototype | <50 MWt | MSR-Triso | Kairos, USA |
| RITM-200M | 50 MWe | Integral PWR | OKBM, Russia |
| RITM-200N | 55 MWe | Integral PWR | OKBM, Russia |
| BANDI-60S | 60 MWe | PWR | Kepco, South Korea |
| Xe-100 | 80 MWe | HTR | X-energy, USA |
| ACPR50S | 60 MWe | PWR | CGN, China |
| Moltex SSR-W | 300 MWe | MSR | Moltex, UK |

### 3.6.1 USE OF NANOTECHNOLOGY FOR WATER CONSUMPTION REDUCTION AND REACTOR EFFICIENCY ENHANCEMENT

Nuclear power plants are water-intense operations and rely on conductive heat transfer to convert nuclear energy to grid-ready electricity. The most common Western reactors are pressurized water reactors (PWRs) in which water is heated by pumping it through the reactor core and then pumping the hot water to a steam generator. This water flows through piping called the primary system and is kept in the liquid state by applying very high pressure. The heat from pressurized hot water in the primary system is transferred to cooled water in the secondary system to generate steam with no direct contact between the two water streams. The steam is then directed via piping to drive a turbine, which turns an electric generator, thus completing the cycle of converting nuclear energy to readily usable electricity for the grid. After passing through the turbines, the steam is captured and condensed for recycling. This reclaimed water can then be sent back through the steam generator. However, a significant amount of the energy of this steam is lost to the atmosphere via a third system of cooling water that is used to condense the steam. The overall process is thus accompanied by significant amount of both water and energy loss.

The water loss can be reduced using nanomaterials called core-shell phase change nanoparticles. In these particles, the center is made out of one material, and the outer skin is made out of another material. The phase change component of the name refers to the fact that the particle center changes from a liquid to a solid when it goes through a hot to cold environment. These particles are mixed into the water used for transporting the thermal energy generated within the reactor. Once mixed into the reactor water, the particle cores melt as the water picks up thermal energy from the reactor. During this time, the shell remains solid at the reactor temperature. Thus, as the water leaves the reactor it carries with it tiny particles containing bundles of liquid thermal energy wrapped in a solid core. The notion is that as these particles travel to the cooling tower, they solidify and dissipate their heat into the surrounding water, thus decreasing the amount of water needed to convert the thermal energy created by the reactor to steam for turning turbines. Since particles do not vaporize, they are much more easily retained for recycling. The Electric Power Research Institute (EPRI) is currently working with scientists at Argonne National Laboratory to commercialize these particles and has suggested that this technology could decrease power plant water requirements by as much as 20% [54].

Cooling down fuel rods is a critical technical challenge in nuclear reactors. Currently improving the heat transfer using nanofluids (liquid plus nano particles) is investigated. Solid metals have larger thermal conductivity when it is nano in size. Nanoparticles are usually made of oxides, metals, carbon nanotubes or carbides. The heat transfer in nanofluid has been extensively studied in numerous applications. Nanofluids have unique features or properties which are totally different from conventional solid-liquid mixtures making them potentially useful for heat transfer enhancement in nuclear reactors. They have a high specific surface area and therefore more heat transfer surface between particles and fluids. They have high dispersion stability with the predominant Brownian motion of particles. They require reduced pumping power as compared to pure liquid to achieve equivalent heat transfer intensification and allow system miniaturization because of the reduced clogging. Their adjustable thermal conductivity, surface wettability and other properties allow varying particle concentrations to suit different applications [10].

A lab at the Massachusetts Institute of Technology (MIT) is evaluating another novel approach to improve the thermal efficiency of pressurized water reactors. About two-thirds of the roughly 100 nuclear power plants in the United States are pressurized water reactors (PWRs). Pressurized water reactors place the water in direct contact with the fuel rods of the nuclear reactor. However, bubbles that form on the surfaces of the fuel rods reduce heat transfer efficiency by insulating the rods from the water. As the temperature of the metal surface rises, at a certain point the coolant's ability to remove heat drops dramatically. Exceed that heat-removal limit and the metal surface can overheat and even fail. Buongiorno and Hu at MIT found a way to raise that limit so that more heat can be extracted more quickly-a change which would increase the output of all PWRs [55]. Their approach calls for replacing the pure water coolant with a nanofluid in which nano-scale particles, each with a diameter of 1–100 nm (a few billionths of a meter) are added to the water.

Buongiorno and Hu at MIT used very fine, hot stainless steel wires to replicate the hot metal surface of the fuel rods. They submerged the wires in two tanks, one containing water and the other a nanofluid, and then steadily increased the temperature of the wires while taking photographs and measurements. They found that using the nanofluid rather than pure water raised the heat-removal limit by as much as 70%. Calculations based on that finding suggest that replacing the water coolant with the nanofluid in a 1,000 MWe nuclear plant could push the plant's output up to 1,200 MWe. More importantly, the necessary concentration of particles is low-just 0.1% by volume or less. Photographs provided some insights into why the nanofluid works better than the water does. In both tests, as boiling begins, bubbles formed along the hot wire. At relatively low temperatures, the bubbles were spaced out, and the sections of wire between them were in direct contact with the fluid and kept reasonably cool. But as temperatures rose, things go bad for the wire in the water. Bubbles begin to crowd together until the wire becomes covered with a continuous layer of vapor which did not conduct heat very well [56]. The wire became glowing hot and eventually broke. Images taken with a scanning electron microscope showed why the wire submerged in the nanofluid fared better. After boiling occurred, the wire in the water was still smooth, but the wire in the nanofluid had become coated with nanoparticles. That rough, porous coating encouraged the formation of bubbles, and when the nanoparticles were made of "water-friendly" materials such as alumina-the coating actually pushed newly formed bubbles away, thus prohibiting the formation of layer of vapor on the wire. The researchers have now performed preliminary experiments using their nanofluid coolants at MIT's Nuclear Research Reactor. According to Hu, initial results have been very promising. If all goes well, they hope that within a few years the use of nanofluid coolants will make today's PWRs significantly more productive. And if nanofluids were used in place of water in emergency cooling systems, they could cool down overheating surfaces more quickly, improving plant safety in all types of reactors [55].

There are also potential safety applications of having nanofluids capable of quickly transporting large quantities of thermal energy. One proposal calls for the use of nanofluids in standby coolant stored in Emergency Core Cooling Systems (ECCS). The ECCS are independent, standby systems designed to safely shut down a reactor in the case of an accident or malfunction. One ECCS component is a set of pumps and backup coolant to be sprayed directly onto reactor rods. Such systems are critical in preventing a loss of coolant accident (LOCA) from spiraling out of control. Because ECCS have backup reservoirs of coolant, technologies that make this backup coolant more effective at removing heat from the reactor could improve the safety of reactors. Because nanofluids can increase the heat transfer efficacy of water by 50% or more, some researchers have suggested that they may also be useful in emergency scenarios [57]. Generally speaking, nuclear power plants consume about 400 gallons of water/megawatt-hour (MWh). Their coal and natural gas counterparts consume approximately 300 and 100 gallons/MWh, respectively [58,59]. Thus, nuclear power plants stand to gain considerably by becoming more water-efficient. A highly conducting and stable nanofluid with exciting newer applications can be used for enhancing the efficiency of safety systems in future Nuclear Power Plants [60].

There are some challenges in the use of nano particles for efficient heat removal, particularly at the commercial level. Scaling up particle production to the large volumes of particles necessary for implementation in a power plant is expensive and labor-intensive. New synthesis infrastructures may be necessary for the large-scale production of these tiny particles. More cost savings for the production of particles are needed. A rough cost estimate can be made using commercially available alumina nanoparticles, as these particles have been tested extensively in the heat transfer literature. The typical water usage for the nuclear power plant may range from 10 to 17 million gallons per day. Assuming the scaled-up discounted price of $100/kg, and that particles could be easily recovered and recycled, loading a nuclear power plant with a 0.1% volume fraction of alumina nanoparticles would cost about $14.7–$25 million per power plant. This is a substantial initial investment. Naturally, if nanoparticles were to cost $10/kg, then particle outfitting costs of $1.5–$2.5 million per nuclear plant could be achieved which can be recovered over time by the expected 2%–4% increase in plant efficiency [58,59]. Besides cost, testing must be done to ensure that the long-term application of these particles does not threaten the operational safety of the plant. Potential pitfalls include increased corrosion, system clogging, and nanoparticle leakage into wastewater. Testing for these issues will require personnel with different sets of expertise. Eventually, EPA will have to validate the results. None of these potential roadblocks is trivial. However, while the challenges seem large, it is encouraging to see potential applications of nanotechnology in nuclear power plants.

### 3.6.2 NANOTECHNOLOGY TO PREVENT MATERIAL CRACKING

High levels of pressure, temperature and radiation can create an environment where materials used in a nuclear power plant can easily crack. In order to avoid cracks, it is important to use materials which can withstand these extreme conditions. To accomplish this goal, designing a material from its atomic level can be a possibility. Nanostructured metals and composites suggest a way to accomplish this goal because they comprise interfaces that attract, absorb and annihilate the line and defects [61]. Controlling radiation-induced defects via interfaces plays a crucial role in the removal of damages and also transmits the stability in nano-particles under certain circumstances where bulk materials demonstrate void swelling and/or embrittlement [61,62]. Steel is the widely used material in nuclear power plants (for control rods). The utilization of nano-particles in the steel will help in enhancing the mechanical, thermal, and physical properties of the steel. The cracks initiation and structural failure of the steel are due to the cyclic loading. This delivers a considerable impact on the nuclear power plant's life cycle. The addition of copper nano-particles in steel minimizes the surface unevenness which results in the confinement of the amount of stress risers and thereby reduction in fatigue cracking [62,63].

### 3.6.3   URANIUM-OXIDE NANOCRYSTALS FOR NANO-FUEL

As it is well known, in its enriched form, uranium dioxide is the major component of nuclear fuel. When depleting its progenies, it is also widely used in many fields, including radiation shielding and catalysis, *etc.* Nanostructured nuclear fuel is expected to possess superior properties, therefore fabrication of nano-fuel may become a hot topic in nanoscience [64]. Recently, Wu et al. [65] synthesized high-quality, colloidal uranium oxide nanocrystals by thermal decomposition of uranyl acetylacetonate in a mixture solution of oleic acid (OA), oleylamine (OAm), and octadecene. This work is of high significance because this $UO_2$ nanocrystal could be developed as a candidate for potential nano-fuels. Furthermore, this microstructure can hopefully enhance the thermal and radiation stability of the fuel and consequently improve its burn-up [66]. Wang et al. [67] synthesized sphere-shaped $UO_2$ nanoparticles (average diameter 100 nm) consisting of 15 nm nano- crystals and single crystalline $U_3O_8$ nanorods (diameter 80–100 nm and length 500–1,500 nm) under hydrothermal conditions. The $UO_2$ nanoparticles could be converted to porous $U_3O_8$ aggregates through thermal treatment in air. Another merit of porous uranium oxides might lie in their applications in the fabrication of transmutation fuel. More research is still much needed to test the application potentials of nanostructured $UO_2$ and $U_3O_8$ [68].

### 3.6.4   NANOPOROUS MATERIALS FOR THE SEPARATION OF HIGH-LEVEL LIQUID WASTE (HLLW)

To minimize the long-term radiological risk and facilitate the management of high-level liquid waste (HLLW), a partitioning of the long-lived minor actinides (MA Am, Cm) and some specific fission products (FPs) such as Cs, Sr, Tc and the platinum group elements are much more desirable. Compared to uranium and plutonium, the minor actinides are significantly less abundant in the spent fuel, so the scale of the separation process for minor actinides from HLLW should be considerably smaller than that of the main separation process such as PUREX. Nanoporous materials based solid-phase extraction still has their advantages such as being acid and alkali resistant, high temperature enduring and radiation resistant in processing HLLW [69–74].

In order to develop a potential direct separation process, a novel macroporous silica-based co-polymer resin was prepared [69,70]. Subsequently, the copolymer was immobilized in porous silica particles with pore size from 50 to 600 nm. Based on the so-called functionalized macroporous silica, Wei and co-workers developed the MAREC (Minor Actinides Recovery from HLLW by Extraction Chromatography) process [70–72]. Zhang et al. synthesized several other macroporous silica-based supramolecular recognition polymeric composites [73]. As the processing of HLLW normally represents harsh chemical conditions such as high nitric acid concentration and extremely high level of radiation, the chemical and radiation stability of the impregnated nanoporous silica is particularly important. According to the stability test by Wei and his co-workers [74], the silica-based support was significantly stable against $\gamma$-radiation and nitric acid.

### 3.6.5   Nanomaterials for Nuclear Waste Disposal
###            and Environment Remediation

CNTs exhibit many noteworthy properties such as strong tensile strength, large elastic modulus, high heat conductivity and electrical conductibility and large surface area [75]. These advantages make CNTs an ideal supporting material for solid- phase extraction of radionuclides. Wang et al., for example, firstly reported the sorption of lanthanides and actinides from $NaClO_4$ aqueous solution by using multiwall carbon nanotubes (MWCNTs). It was found that MWCNTs are an effective sorbent for Eu(III) [76], Am(III) [77] and Th(IV) [78,79], and the sorbent after nuclides sorption is very stable due to the strong complexation of sorbates on the MWCNTsboxymethyl cellulose grafted MWCNT (MWCNT-g-CMC) by using plasma induced grafting method.

Besides CNTs, mesoporous materials, porous molecular sieves with pore diameter between 2 and 50 nm, such as SBA-15 (Santa Barbara Amorphous-15), are also important nanomaterials that have the advantages of large surface area, well-defined pore size, excellent mechanical resistance, non-swelling, excellent chemical stability and radiation tolerance, as well as extraordinarily wide possibilities of functionalization [80,81]. These advantages also make the ordered mesoporous carbon and silica compounds attractive for nuclear waste disposal. Many other compounds are also found to be useful for nuclear waste disposal [82–88]. In addition, magnetite nanoparticles (MNPs) have received much attention recently to remove radioactive metal ions due to the easy separation of sorbent from solution by adding a magnetic field.

### 3.6.6   Nanomaterial-Based Sensors

Uranium is a toxic element which might be released into the environment by increasing the development of nuclear energy. Severe damaging effects of uranium on human health have been reported [89,90] and the maximum contamination level in drinking water is defined as 150 nM by US EPA [91].

Recently, Lee et al. [92] synthesized uranyl-specific colorimetric sensors using the uranyl-specific DNAymes and gold nanoparticles (AuNPs) by labeled or label-free methods. In the labeled method, the presence of $UO_2^{2+}$ dissociated very stable DNAymes- functionalized AuNP aggregates, whose color was purple, releasing red individual AuNPs. The detection limit of this method was 50 nM after 30 minutes contact period. The label-free method made use of different adsorption properties of single-stranded and double-stranded DNA on AuNP. The presence of $UO_2^{2+}$ induced cleavage of substrate by DNAyme and the formed single-stranded DNAs can be adsorbed on the surface of AuNPs, which prevents AuNP aggregation in a NaCl solution. Detection limits were as low as 1 nM after 6 minutes of reaction time. Both methods exhibit much lower detection limits than the maximum contamination level in drinking water, have excellent selectivity over other metal ions, and operate at room temperature.

### 3.6.7   Use of Nano Powders for Radiation Resistance

The structure and properties peculiarities of the nanocrystalline powders give the opportunity to design new nuclear energy industry materials. It was shown experimentally, that the addition of 5%–10% uranium dioxide nanocrystalline powder to traditional coarse powder allows to decrease the sintering temperature or to increase the fuel tablets size of grain. Similarly, neutron-absorbing tablets of control-rod can be made from nanopowder of dysprosium hafnate instead of presently used boron carbide. The powders in a nanocrystalline state allow to sinter and to generate compact tablet with 8.2–8.4 $g/cm^2$ density for the automatic defense system of a nuclear reactor. The dysprosium hafnate ceramics can last up to 18–20 years instead 4–5 years for boron carbide. In order to increase the radiation-damage stability of the fuel element jacket, heat-resistant ferrite-martensite steel strengthened by $Y_2O_3$ nanocrystalline powder addition has been suggested. Nanopowder with the size of particles 560 nm and crystallite size of 9 nm can be prepared by chemical coprecipitation method. In order to make the container for transport and provisional disposal of exposed fuel from nuclear reactor lighter, a new boron-aluminum alloy called as boral has been developed. This composite armed with nanopowders of boron-containing materials and heavy metals oxides can replace burnt-up corrosion-resistant steels [93–99].

Petrunin [96] pointed out that the small number of atoms in nanopowders allows a significant increase in the surface energy and resistance to extreme synthesis conditions. Many differences between the atomic arrangement of nanoparticles and the conventional coarse-grained materials have been revealed by Petrunin [97]. He showed that details of the atomic structure of nanopowders influences the peculiarity of the mechanical, electrical, magnetic, optical and chemical properties. These specific properties of the nanopowders provide the possibility of new materials development for nuclear energy. Ceramic oxides are the most suitable material for the nuclear energy industry. Nanocrystalline $UO_x^{2+}$ powders were prepared from uranyl nitrate by Petrunin et al. [99]. In order to satisfy a high-level radiation and long time usage, the grain size of fuel pellet $UO_2$ material must be increased. This was done using nanocrystalline powder of $UO_2$. The study showed that Lanthanides and their oxides are promising as effective materials for nuclear control rods. Radiological studies showed that the mixed oxides of the lanthanide $Ln_2O_3$- $MeO_2$ (where Ln – Dy, Gd; Me - Ti, Zr, Hf) with the fluorite structure have high radiation resistance.

The features of the structure and properties of nanopowders provide the possibility of developing new materials for nuclear power industry and improving materials already used in this industry. It has been experimentally shown that the use of nanopowders is promising for the modernization of the technology of a uranium-containing fuel for nuclear reactors and for the improvement of its parameters. Similar prospects have been demonstrated for the modernization of neutron absorbing pellets of controlling rods by changing of boron carbide to dysprosium hafnate. These powders in the nanocrystalline state can be sintered and compact pellets can be obtained. To increase the radiation resistance of construction materials for nuclear power plants, it is proposed to increase the strength of ferritic-martensitic steels by additions of finely dispersed (nanocrystalline) powders [93–99].

## 3.7   DIRECT CONVERSION OF HEAT FROM RADIO ISOTOPE TO ELECTRICITY BY THERMOELECTRICITY AND TPV

A radioisotope thermoelectric generator (RTG) is a nuclear electric generator which includes the natural decay of a radioactive atom, usually plutonium 238 in the form of plutonium dioxide 238$PuO_2$. As they disintegrate, radioactive atoms release heat,

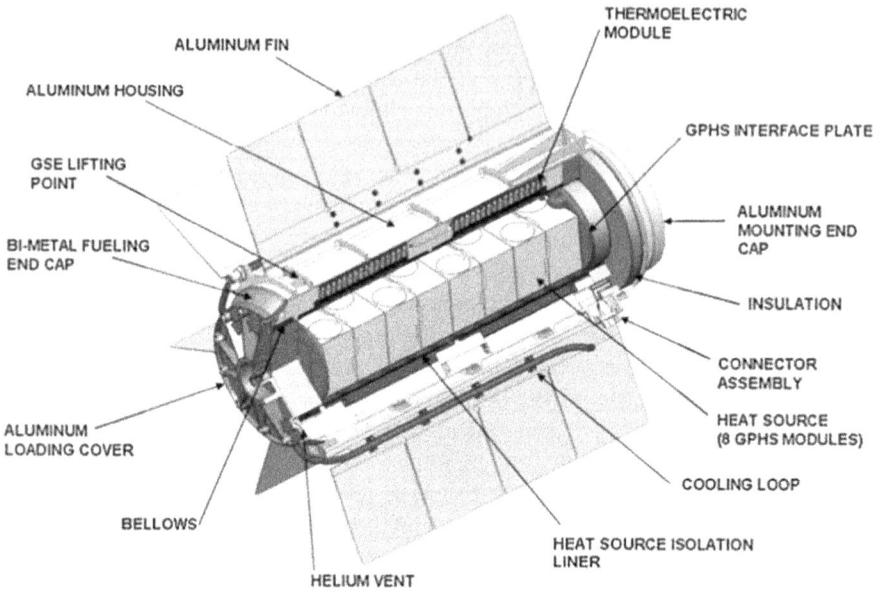

**FIGURE 3.9** Cutting view of a Multi-Mission Radioisotope Thermoelectric Generator MMRTG [105].

some of which is directly converted into electricity [100]. The first RTG was developed by Mound Laboratories in 1954 [101]. It is now well used for space applications. The first RTG launched into space by the USA was the SNAP (Systems for nuclear auxiliary power units) 3B in 1961 [102]. In 2010, the USA launched 41 RTGs on 26 space systems [103]. The RTGs used in the US space program initially included SiGe TE materials installed in the General Purpose Heat Source–Radioisotope Thermoelectric Generator (GPHS-RTG) GPHS-RTG, later succeeded by the lead telluride alloys, or TAGS, used in multi-mission RTGs (MMRTGs), illustrated in Figure 3.9. This MMRTG was developed with the use of new skutterudite thermoelectric materials in order to achieve higher efficiency and lower degradation rates, which are important for long-term missions to the outer planets [104].

RTG can also be used for remote areas. One of the first terrestrial uses of RTGs was in 1966 by the U.S. Navy for powering environmental instrumentation at Fairway Rock, a small uninhabited island in Alaska. These systems were developed for the supply of power to equipment requiring a stable and reliable power source, over several years and without maintenance. Examples of these would be power supplies for systems located in isolated or inaccessible environments, like lighthouses, navigation beacons and weather stations. Similarly, between 1960 and 1980, the Soviet Union built many unmanned lighthouses and navigation beacons, equipped with about 1,000 RTGs [106].

RTG can also be used in the medical domain. In 1966, as shown in Figure 3.10, small plutonium cells (very small RTGs fed with Pu238) were used in implanted pacemakers to ensure a very long battery life [107]. Many companies like ARCO,

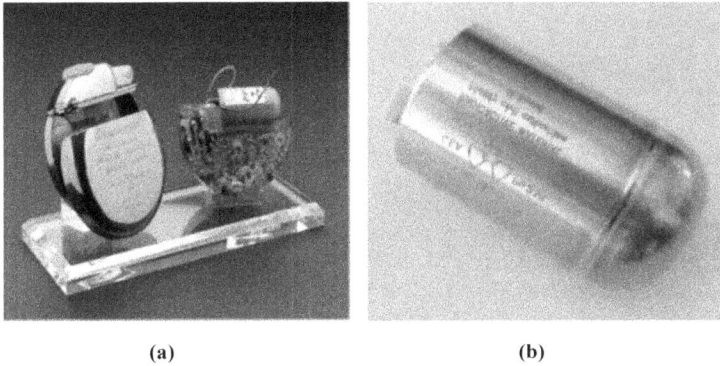

(a)                                        (b)

**FIGURE 3.10**   (a) pacemaker and (b) RTG battery [108].

Medtronic, Gulf General Atomic among others have manufactured nuclear-powered pacemakers [108]. Recent developments of lithium batteries, however, have taken away the market of nuclear batteries [109]. The use of RTG in remote locations and medical domain is not well flourished due to the risks involved in using radioisotopes.

RTPVs can also be seen as an alternative technology for other energy-generating technologies in deep space missions, like radioisotope thermoelectric generators (RTGs), or Stirling radioisotope generator (SRG), where classical PV technologies cannot be used because of a lack of usable sunlight [110,111]. While RTG shows excellent reliability, the technology is quite heavy and shows a low efficiency which leads to an increased cost. SRG, on the other hand, has typically a higher efficiency; however, a redundant system is required because of the lower reliability which results in low mass-specific power. The use of TPV for energy production in near-deep space missions could be a good alternative because it is a static system, so the influence of vibrations is small. The unproven longevity is a serious challenge for RTPV for the use in near-deep space missions. An important concern is the possible contamination of the TPV *optical surfaces* by material evaporating from the hot surfaces; also the operation at temperatures of more than 50°C for an expected mission life of 14 years will have an impact on the longevity.

The heat source in RTPV is a so-called *general purpose heat source* (GPHS) which is a stackable, compact unit designed to deliver over 600°C. Convertible photons are transmitted through a tandem filter and converted into electricity by low band gap (0.6 eV) InGaAs MIMs. For this application, only TPV cells based on III–V semiconductors can be used because of the necessity for radiation hardness. Experimental results demonstrated a ~18% *convertor* efficiency including housing thermal losses. By means of neutron radiation tests, the degradation of the MIM, the filter, and other components have been tested. The results show 20% degradation in the MIM efficiency after a 14-year mission which is comparable to the 14% degradation of an RTG. An improvement of the filters could reduce the degradation in efficiency. More details on TPV and TEG technologies are given in Chapters 7 and 8 respectively [110,111].

## REFERENCES

1. Nuclear power in the world today. 2022, March. A website report. https://world-nuclear. org›current-and-future-generation.
2. World Energy Outlook. Paris: OECD International Energy Agency; 2021.
3. OECD International Energy Agency Statistics. Paris: IEA; 2021.
4. World nuclear performance report. London: World Nuclear Association; 2021.
5. Advanced nuclear power reactors. 2021, April. A website report by https://world-nuclear.org/information-library/nuclear-fuel-cycle/nuclear-power-reactors/advanced-nuclear-power-reactors.aspx
6. Generation III reactor. Wikipedia, the free encyclopedia, last edited 29 August, 2022. https://en.wikipedia.org/wiki/Generation_III_reactor
7. Nian V. Global developments in advanced reactor technologies and international development. *World Engineers Summit – Applied Energy Symposium & Forum: Low Carbon Cities & Urban Energy Joint Conference*, WES-CUE 2017, Energy Procedia, Singapore; 19–21 July 2017, Vol. 143, pp. 605–610.
8. High temperature gas reactor. Wikipedia, the free encyclopedia, last edited 10 August, 2022. https://en.wikipedia.org/wiki/High-temperature_gas_reactor
9. Fast neutron reactors. 2021, March. A website report by World Nuclear Association, London, England. https://world-nuclear.org/information-library/current-and-future-generation/nuclear-power-in-the-world-today.aspx
10. Advanced nuclear reactors: Technology overview and current issues. 2019, April 18. A report by congressional research service, prepared for members and committee of congress. Congressional Research Service https://crsreports.congress.gov R45706.
11. Generation IV nuclear reactors. 2022. A website report by https://world-nuclear.org/information-library/nuclear-fuel-cycle/nuclear-power-reactors/generation-iv-nuclear-reactors.aspx
12. PDF. Overview of Gen-IV developments and generation IV... – SNETP. 2021. https://snetp.eu›Presentation_Kamil-Tuček-1.
13. Generation IV reactor. Wikipedia, the free encyclopedia, last edited 1 October, 2022. https://en.wikipedia.org/wiki/Generation_IV_reactor
14. Natrium TM reactor and integrated energy storage. 2021. A report by Terra Power, a nuclear innovation company, Bellevue, Washington, DC.
15. The Natrium™ Program – TerraPower. 2021, May 18. A website report by https://www.terrapower.com/natrium-program-summary/
16. Natrium: Providing flexible clean energy at a competitive cost. 2021. A website report https://natriumpower.com.
17. Terra Power. Wikipedia, the free encyclopedia, last edited 26 September, 2022. https://en.wikipedia.org/wiki/TerraPower
18. Clifford C. How Bill Gates' company terra power is building next-generation nuclear power. A website report Terra Power, Bellevue, Washington, DC. Published, updated 8 April 2021. https://www.cnbc.com/2021/04/08/bill-gates-terrapower-is-building-next-generation-nuclear-power.html
19. Patel S. The allure of TRISO nuclear fuel explained. 2021, May 1. A website report in power https://www.powermag.com/the-allure-of-triso-nuclear-fuel-explained/
20. Feltus M. *TRISO Fuels, a Presentation at Generation IV International Forum.* Washington, DC: Department of Energy; 2019, December 18.
21. Stepanek J., Entler S., Syblik J., Vesely L., Dostal V., and Zacha P. Parametric study of S-$CO_2$ cycles for the DEMO fusion reactor. *Fusion Eng Des* 2020;160(September):111992. Doi: 10.1016/j.fusengdes.2020.111992.
22. Generation IV International Forum (GIF). *Technology Roadmap Update for Generation IV Nuclear Energy Systems*, Tech. Rep. Paris: Nuclear Energy Agency; 2014, pp. 1–66.

23. Ahn Y., Bae S.J., Kim M., Cho S.K., Baik S., Lee J.I., and Cha J.E. Review of supercritical $CO_2$ power cycle technology and current status of research and development. *Nucl Eng Technol* 2015;47(6):647–661. Doi: 10.1016/j.net.2015.06.009.

24. Dostal V. *A Supercritical Carbon Dioxide Cycle for Next Generation Nuclear Reactors (Phd)*. Cambridge, MA: Massachusetts Institute of Technology; 2004, p. 317.

25. Sienicki J.J., and Moisseytsev A. Nuclear power. In: Brun K., Friedman P., and Dennis R., editors. *Fundamentals and Applications of Supercritical Carbon Dioxide ($SCO_2$) Based Power Cycles*. Woodhead Publishing; 2017, pp. 339–391. Doi: 10.1016/B978-0-08-100804-1.00013-X.

26. Li M.J., Zhu H.H., Guo J.Q., Wang K., and Tao W.Q. The development technology and applications of supercritical $CO_2$ power cycle in nuclear energy, solar energy and other energy industries. *Appl Therm Eng* 2017;126:255–275. Doi: 10.1016/j.applthermaleng.2017.07.173.

27. Wu P., Ma Y., Gao C., Liu W., Shan J., Huang Y., Wang J., Zhang D., and Ran X. A review of research and development of supercritical carbon dioxide Brayton cycle technology in nuclear engineering applications. *Nucl Eng Des* 2020, March;368:110767. Doi: 10.1016/j.nucengdes.2020.110767.

28. Moisseytsev A., and Sienicki J. *Development of a Plant Dynamics Computer Code for Analysis of a Supercritical Carbon Dioxide Brayton Cycle Energy Converter Coupled to a Natural Circulation Lead-Cooled Fast Reactor*, Tech. Rep. Argonne, IL: Argonne National Laboratory; 2006, p. ANL-06/27.

29. Sienicki J., Moisseytsev A., Yang W., Wade D., Nikiforova A., Hanania P., Ruy H., Kulesaz K., and Kim S. *Status Report on the Small Secure Transportable Autonomous Reactor (Sstar)/lead-Cooled Fast Reactor (LFT) and Supporting Research and Development*, Tech. Rep. Argonne, IL: Argonne National Laboratory; 2008, p. ANL-GenIV-089.

30. International Energy Agency. *Technology Roadmap: Nuclear Energy*, Tech. Rep. International Energy Agency; 2015, pp. 1–57., Paris, France

31. Chang Y., Finck P., and Grandy C. *Advanced Burner Test Reactor Preconceptual Design Report*, Tech. Rep. Argonne, IL: Argonne National Laboratory; 2006, p. ANL-ABR-1.

32. Moisseytsev A., and Sienicki J.J. Investigation of alternative layouts for the supercritical carbon dioxide Brayton cycle for a sodium-cooled fast reactor. *Nucl Eng Des* 2009;239(7):1362–1371. Doi: 10.1016/j.nucengdes.2009.03.017.

33. Ahn Y., and Lee J.I. Study of various Brayton cycle designs for small modular sodium-cooled fast reactor. *Nucl Eng Des* 2014;276(9):128–141. Doi: 10.1016/j.nucengdes.2014.05.032.

34. Yoon H.J., Ahn Y., Lee J.I., and Addad Y. Potential advantages of coupling supercritical $CO_2$ Brayton cycle to water cooled small and medium size reactor. *Nucl Eng Des* 2012;245:223–232. Doi: 10.1016/j.nucengdes.2012.01.014.

35. Bae S.J., Lee J., Ahn Y., and Lee J.I. Preliminary studies of compact Brayton cycle performance for small modular high temperature gas-cooled reactor system. *Ann Nucl Energ* 2014;75:11–19. Doi: 10.1016/j.anucene.2014.07.041.

36. Yu H. Hartanto D., Moon J., and Kim Y. A conceptual study of a supercritical $CO_2$-cooled micro modular reactor. *Energies* 2015;8(12):13938–13952. Doi: 10.3390/en81212405.

37. Kato Y., Nitawaki T., and Muto Y. Medium temperature carbon dioxide gas turbine reactor. *Nucl Eng Des* 2004;230:195–207. Doi: 10.1016/j.nucengdes.2003.12.002.

38. Muto Y., Kato Y. Optimal cycle scheme of direct cycle supercritical $CO_2$ gas turbine for nuclear power generation systems. *Challenges on Power Engineering and Environment - Proceedings of the International Conference on Power Engineering 2007*, Chytey, China, ICOPE 2007; 2007, pp. 86–92.

39. Li M.J., Xu J.L., Cao F., Guo J.Q., Tong Z.X., and Zhu H.H. The investigation of thermo-economic performance and conceptual design for the miniaturized lead-cooled fast reactor composing supercritical $CO_2$ power cycle. *Energy* 2019;173:174–195. Doi: 10.1016/j.energy.2019.01.135.

40. Alpy N., Cachon L., Haubensack D., Floyd J., Simon N., Gicquel L., Rodriguez G., and Avakian G. Gas cycle testing opportunity with ASTRID, the French SFR prototype. *The 3rd International Symposium on Supercritical $CO_2$ Power Cycles*, Boulder, CO; 24–25 May, 2011.

41. Floyd J., Alpy N., Moisseytsev A., Haubensack D., Rodriguez G., Sienicki J., and Avakian G. A numerical investigation of the s$CO_2$ recompression cycle offdesign behaviour, coupled to a sodium cooled fast reactor, for seasonal variation in the heat sink temperature. *Nucl Eng Des* 2013;260(7):78–92. Doi: 10.1016/j.nucengdes.2013.03.024.

42. Reuters. France drops plans to build sodium-cooled nuclear reactor. 2020. https://www.reuters.com/article/us-france-nuclearpower-astrid/france-dropsplans-to-build-sodium-cooled-nuclear-reactor-idUSKCN1VK0MC (accessed: 16 June 2020).

43. Hajek P., and Frybort O. Experimental loop s-$CO_2$ SUSEN. *The 4th International Symposium on Supercritical $CO_2$ Power Cycles*, Pittsburgh, PA; 9–10 September 2014. Doi: 10.1017/CBO9781107415324.004, arXiv:arXiv:1011.1669v3.

44. Vojacek A., Hacks A., Melichar T., Frybort O., and Hájek P. Challenges in supercritical $CO_2$ power cycle technology and first operational experience at CVR. *2nd European Supercritical $CO_2$ Conference*, Essen; 30–31 August, 2018, p. 100. Doi: 10.17185/duepublico/46075.

45. Linares J.I., Cantizano A., Moratilla B.Y., Martín-Palacios V., and Batet L. Supercritical $CO_2$ Brayton power cycles for DEMO (demonstration power plant) fusion reactor based on dual coolant lithium lead blanket. *Energy* 2016;98:271–283. Doi: 10.1016/j.energy.2016.01.020.

46. Rovira A., Sánchez C., Montes M.J., and Muñoz M. Proposal of optimized power cycles for the DEMO power plant (EUROfusion). *Fusion Eng Des* 2019;148(May):111290. Doi: 10.1016/j.fusengdes.2019.111290.

47. Small nuclear power reactors. 2021, December. A website report by https://world-nuclear.org/information-library/nuclear-fuel-cycle/nuclear-power-reactors.aspx

48. Small modular reactor. Wikipedia, the free encyclopedia, last edited 10 October, 2022. https://en.wikipedia.org/wiki/Small_modular_reactor

49. Small nuclear power reactors. A website report by IAEA, 4 November 2021 https://www.world-nuclear.org/information-library/nuclear-fuel-cycle/nuclear-power-reactors/small-nuclear-power-reactors.aspx

50. Ingersoll D.T., Houghton Z.J., Bromm R., and Desportes C. Nuscale small modular reactor for co-generation of electricity and water. *Desalination* 2014;340:84–93.

51. Shadrin A.Y., Dvoeglazov K.N., Kascheyev V.A., Vidanov V.L., Volk V.I., Veselov S.N., Zilberman B.Y., and Ryabkov D.V. Hydrometallurgical reprocessing of brest-Od-300 mixed uranium-plutonium nuclear fuel. *Procedia Chem* 2016;21:148–155.

52. Russia's Brest reactor now scheduled for 2026. 2018, November 1. Nuclear Engineering International [Internet]. Available from: https://www.neimagazine.com/news/newsrussias-brest-reactor-now-scheduled-for-2026-6803677.

53. Ho M., Obbard E., Burr P.A., and Yeoh G. A review on the development of nuclear power reactors. *2nd International Conference on Energy and Power*, ICEP2018, Sydney Australia, also Energy Procedia; 13–15 December 2018, 2019, Vol. 160, pp. 459–466.

54. Multifunctional Nanoparticles for reducing cooling tower water consumption. Palo Alto, CA: Electric Power Research Institute, 2012.

55. Buongiorno J., Hu L.W., Kim S.J., Hannink R., Truong B., and Forrest E. Nanofluids for enhanced economics and safety of nuclear reactors: An evaluation of the potential features, issues and research gaps. *Nucl Technol* 2008;162:80–91.

56. Kim S.J., Bang I.C., Buongiorno J., and Hu L.W. Surface wettability change during pool boiling of nanofluids and its effect on critical heat flux. *Int J Heat Mass Transf* 2007, September;50(19–20):4105–4116.

57. Taylor R., Coulomb S., Otanicar T., Phelan P., Gunawan A., Lv W., Rosengarten G., Prasher R., and Tyagi H. Small particles, big impacts: A review of the diverse applications of nanofluids. *J Appl Phys* 2013;113:011301. Doi: 10.1063/1.4754271.

58. Green K. Nuclear power and its water consumption secrets, a report by Monarch Partnership January 30, 2019, Wallington, UK, also see reports on water use and nuclear power plants, 2013 by Nuclear energy Institute, Washington, DC.

59. Styles, III, J.H. *Nuclear Power Plant Water Usage and Consumption*. Submitted as coursework for PH241, Palo Alto, CA: Stanford University; 2017, March 19.

60. Mair L.O. Nuclear power and nanomaterials: Big potential for small particles by federation of American scientists; 2015, February 19. www.cnn.com.

61. Beyerlein I.J. Radiation damage tolerant nanomaterials, Elsevier. *Mater Today* 2013, November;16(11):443–449.

62. Bandhu D., and Kumar R. A review on usage of nano-particles in nuclear power plants. *Int J Res-Granthaalayah* 2017, February;5(2). ISSN-2350-0530(O), ISSN-2394-3629(P) ICV (Index Copernicus Value) 2015: 71.21 IF: 4.321 (CosmosImpactFactor), 2.532 (I2OR) InfoBase Index IBI Factor 3.86.

63. Patel Abhiyan S. An overview on application of nanotechnology in construction industry. *Int J Innov Res Sci Eng Technol* 2013, November;2(11):6094–6098. ISSN: 2319-8753.

64. Shi W.-Q., Yuan L.-Y., Li Z.-J., Lan J.-H., Zhao Y.-L., and Chai Z.-F. Nanomaterials and nanotechnologies in nuclear energy chemistry. *Radiochim Acta* 2012;100:727–736. Doi: 10.1524/ract.2012.1961.

65. Wu H., Yang Y., and Cao Y.C. Synthesis of colloidal uranium- dioxide nanocrystals. *J Am Chem Soc* 2006;128:16522.

66. Mantoura S. Uranium dioxide: Nano goes nuclear. *Nat Nano- tech.* 2006. Doi 10.1038/nnano.2006.202.

67. Wang Q., Li G.D., Xu S., Li J.X., and Chen J.S. Synthesis of uranium oxide nanoparticles and their catalytic performance for benzyl alcohol conversion to benzaldehyde. *J Mater Chem* 2008;18:1146.

68. Akiyama K., Zhao Y.L., Sueki K., Tsukada K., Haba H., Nagame Y., Kodama T., Suzuki S., Ohtsuki T., Sakaguchi M., Kikuchi K., Katada M., and Nakahara H. Isolation and characterization of light actinide metallofullerenes. *J Am Chem Soc* 2001;123:181.

69. Hoshi H., Wei Y.Z., Kumagai M., Asakura T., and Morita Y. Separation of trivalent actinides from lanthanides by using R-BTP resins and stability of R-BTP resin. *J Alloys Compd* 2006;408:1274.

70. Wei Y.Z., Zhang A.Y., Kumagai M., Watanabe M., and Hayashi N. Development of the MAREC process for HLLW partitioning using a novel silica-based CMPO extraction resin. *J Nucl Sci Technol* 2004;41:315.

71. Wei Y.Z., Kumagai M., Takashima Y., Modolo G., and Odoj R. Studies on the separation of minor actinides from high-level wastes by extraction chromatography using novel silica-based ex- traction resins. *Nucl Technol* 2000;132:413.

72. Usuda S., Liu R.Q., Wei Y.Z., Xu Y.L., Yamazaki H., and Wakui Y. Evaluation study on properties of a novel R-BTP extraction resin – from a viewpoint of simple separation of minor actinides. *J Ion Exch* 2010;21:35.

73. Zhang A.Y., Hu Q.H., and Chai Z.F. Synthesis of a novel macro- porous silica-calix[4] arene- crown polymeric composite and its adsorption for alkali metals and alkaline-earth metals. *Ind Eng Chem Res* 2010;49:2047.

74. Hoshi H., Wei Y.Z., Kumagai M., Asakura T., and Morita Y. Separation of trivalent actinides from lanthanides by using R-BTP resins and stability of R-BTP resin. *J Alloys Compd* 2006;408:1274.

75. Belloni F., Kuetahyali C., Rondinella V.V., Carbol P., Wiss T., and Mangione A. Can carbon nanotubes play a role in the field of nuclear waste management? *Environ Sci Technol* 2009;43:1250.

76. Tan X.L., Xu D., Chen C.L., Wang X.K., and Hu W.P. Adsorption and kinetic desorption study of 152+154 Eu(III) on multiwall carbon nanotubes from aqueous solution by using chelating resin and XPS methods. *Radiochim Acta* 2008;96:23.

77. Wang X.K., Chen C.L., Hu W.P., Ding A.P., Xu D., and Zhou X. Sorption of 243 Am(III) to multiwall carbon nanotubes. *Environ Sci Technol* 2005;39:2856.

78. Chen C.L., Li X.L., Zhao D.L., Tan X.L., and Wang X.K. Adsorption kinetic, thermodynamic and desorption studies of Th(IV) on oxidized multi-wall carbon nanotubes. *Colloids Surf A* 2007;302:449.

79. Chen C.L., Li X.L., and Wang X.K. Application of oxidized multi- wall carbon nanotubes for Th(IV) adsorption. *Radiochim Acta* 2007;95:261.

80. Darmstadt H., Roy C., Kaliaguine S., Kim T.W., and Ryoo R. Surface and pore structures of CMK-5 ordered mesoporous carbons by adsorption and surface spectroscopy. *Chem Mater* 2003;15:3300.

81. Lee J.F., Thirumavalavan M., Wang Y.T., and Lin L.C. Monitoring of the structure of mesoporous silica materials tailored using different organic templates and their effect on the adsorption of heavy metal ions. *J Phys Chem C* 2011;115:8165.

82. Vidya K., Dapurkar S.E., Selvam P., Badamali S.K., and Gupta N.M. The entrapment of $UO_2^{2+}$ in mesoporous MCM-41 and MCM-48 molecular sieves. *Micropor Mesopor Mater* 2001;50:173.

83. Vidya K., Gupta N.M., and Selvam P. Influence of pH on the sorption behaviour of uranyl ions in mesoporous MCM-41 and MCM-48 molecular sieves. *Mater Res Bull* 2004;39:2035.

84. Li S.J., Tian G., Geng J.X., Jin Y.D., Wang C.L., Li S.Q., Chen Z., Wang H., and Zhao Y.S. Sorption of uranium(VI) using oxime-grafted ordered mesoporous carbon CMK-5. *J Hazard Mater* 2011;190:442.

85. Fryxell G.E., Lin Y.H., Fiskum S., Birnbaum J.C., Wu H., Kemner K., and Kelly S. Actinide sequestration using self-assembled monolayers on mesoporous supports. *Environ Sci Technol* 2005;39:1324.

86. Lin Y.H., Fiskum S.K., Yantasee W., Wu H., Mattigod S.V., Vorpagel E., Fryxell G.E., Raymond K.N., and Xu J.D. Incorporation of hydroxypyridinone ligands into self-assembled monolayers on mesoporous supports for selective actinide sequestration. *Environ Sci Technol* 2005;39:1332.

87. Yousefi S.R., Ahmadi S.J., Shemirani F., Jamali M.R., and Salavati-Niasari M. Simultaneous extraction and preconcentration of uranium and thorium in aqueous samples by new modified meso-porous silica prior to inductively coupled plasma optical emission spectrometry determination. *Talanta* 2009;80:212.

88. Yuan L.Y., Liu Y.L., Shi W.Q., Liu Y.L., Lan J.H., Zhao Y.L., and Chai Z.F. High performance of phosphonate-functionalized mesoporous silica for U(VI) sorption from aqueous solution. *Dalton Trans* 2011;40:7446.

89. Craft E., Abu-Qare A., Flaherty M., Garofolo M., Rincavage H., and Abou-Donia M.J. Depleted and natural uranium: Chemistry and toxicological effects. *Toxicol Environ Health B Crit Rev* 2004;7:297.

90. Zhou P., and Gu B. Extraction of oxidized and reduced forms of uranium from contaminated soils: Effects of carbonate concentration and pH *Environ. Sci. Technol.* 2005;39:4435.

91. Wei H., Li B., Li J., Dong S., and Wang E. DNAzyme-based colorimetric sensing of lead ($Pb^{2+}$) using unmodified gold nanoparticle probes. *Nanotechnology* 2008;19:095501.

92. Lee J.H., Wang Z., Liu J., and Lu Y. Highly sensitive and selective colorimetric sensors for uranyl ($UO_2^{2+}$): Development and comparison of labeled and label-free DNAzyme-gold nanoparticle systems. *J Am Chem Soc* 2008;130:14217.

93. Petrunin V.F. Development of nanomaterials for nuclear energetics. *Conference of Physics of Non-equilibrium Atomic Systems and Composites*, PNASC 2015, 18–20 February, 2015 and *the Conference of Heterostructures for Microwave, Power and Optoelectronics: Physics, Technology and Devices (Heterostructures)*, Physics Procedia; 19 February, 2015, Vol. 72, pp. 536–539.

94. Bhushan B., editor. *Springer Handbook of Nanotechnology*, 3rd ed. Berlin, Germany: Springer, pp. 1–1961; 2010.

95. Morokhov I.D., Petinov V.I., Trusov L.I., and Petrunin V.F. Structure and properties of fine metallic particles. *Soviet Phys Uspehy* 1981;24(4):295–316.

96. Petrunin V.F. Development of nanomaterials and nanotechnology in atomic industry (in Russian). *Nuclear Phys Eng* 2011;2(3):196–204.

97. Petrunin V.F. Structural characterization of ultra dispersed (nano-) materials as intermediate between amorphous and crystalline state. *Nanostruct Mater* 1999;12:1153–1156.

98. Popov V.V., Petrunin V.F., and Korovin S.A. Regularites of formation of nanocrystalline particles in titanium subgroup dioxides. *Russian J Inorganic Chem* 2010;55(10):1515–1520. Also in Strikhanov M.N., editor. *Nuclear Power Engineering. Problems. Solutions.* Berlin, German: Social Forecasts and Marketing Center, Springer; 2011.

99. Petrunin V.F., Popov V.V., Grechishnikov S.I., and Korovin S.A. Possibilities of the applications of nano powders in nuclear power industry (in Russian). *Nucl Phys Eng* 2013;4(6):555–563.

100. Yang J., and Caillat T. Thermoelectric materials for space and automotive power generation. *MRS Bull* 2006;31:224–229.

101. Jordan K., and Birden J. *Thermal Batteries Using Polonium-210. (Information Report).* Miamisburg, OH: Mound Lab.; 1954.

102. Cataldo R. Spacecraft power system considerations for the far reaches of the solar system. In: *Outer Solar System*. Cham: Springer Science and Business Media LLC; 2018, pp. 767–790.

103. Cataldo R.L.B. U.S. space radioisotope power systems and applications: Past, present and future. 2011. Available from: https://ntrs.nasa.gov/search.jsp?R=20120000731 (accessed: 18 June 2018).

104. Holgate T., Bennett R., Hammel T., Caillat T., Keyser S., and Sievers B. Increasing the Efficiency of the multi-mission radioisotope thermoelectric generator. *J Electron Mater* 2014;44:1814–1821.

105. Jaziri N., Boughamoura A., Müller J., Mezghani B., Tounsi F., and Ismail M. A comprehensive review of thermoelectric generators: Technologies and common applications. *Energ Rep* 2019;6 (7):264–287.

106. Ruo J. What future for radioisotope thermoelectric generators (RTG)? *Physics* a website report 2017. http://large.stanford.edu/courses/2017/ph241/ruffio1/

107. Prutchi D. Nuclear Pacemaker-s. 2005. Available from: http://www.prutchi.com/pdf/implantable/nuclear_pacemakers.pdf (accessed: 9 June 2019).

108. Facts about pacemakers. Available from: http://osrp.lanl.gov/pacemakers.shtml (accessed: 14 June 2018).

109. Prelas M., Boraas M., Aguilar F.D.L.T., Seelig J.-D., Tchouaso M.T., and Wisniewski D. *Potential Applications Potential Applications for Nuclear Batteries*. Cham: Springer Science and Business Media LLC; 2016, Vol. 56, pp. 285–305.

110. Stelmakh V., Fisher P.H., Wang X., Chan W., and Walker R. Radioisotope thermophotovoltaic generator design and performance estimates for terrestrial applications. *Proceedings of the 2017 25th International Conference on Nuclear Engineering ICONE25*, Shanghai, China; 2–6 July 2017. https://hdl.handle.net/1721.1/124353.

111. Wang X., Chan W., Stelmakh V., and Fisher P. Radioisotope thermophotovoltaic generator design and performance estimates for terrestrial applications. *2017 25th International Conference on Nuclear Engineering*, Shanghai, China; 2–6 July 2017, 2 July 2017, Vol. 3, Nuclear Fuel and Material, Reactor Physics and Transport Theory; Innovative Nuclear Power Plant Design and New Technology Application. Doi: 10.1115/icone25-66607. Version: Final published version, ISBN 978-0-7918-5781-6.

# 4 Advanced Power Generation from Geothermal Heat

## 4.1 INTRODUCTION

The molten core of the Earth, about 4,000 miles down, is roughly as hot as the surface of the sun, over 6,000°C, or 10,800°F. This heat is continuously replenished by the decay of naturally occurring radioactive elements, at a flow rate of roughly 30 TW, almost double all human energy consumption. That process is expected to continue for billions of years. The ARPA-E project AltaRock Energy estimates that just 0.1% of the heat content of Earth could supply humanity's total energy needs for 2 million years. There's enough energy in the Earth's crust, just a few miles down, to power all of human civilization for generations to come. The easiest way to tap this energy is to make direct use of the heat where it breaks the surface, in hot springs, geysers, and fumaroles (steam vents near volcanic activity). The warm water can be used for bathing or washing, and the heat for cooking. Using geothermal energy this way has been around since the earliest humans, going back at least to the Middle Paleolithic.

Slightly more sophisticated is tapping into naturally occurring reservoirs of geothermal heat close to the surface to heat buildings. In the 1890s, the city of Boise, Idaho, tapped one to create the US's first district heating system, whereby one central source of heat fed into multiple commercial and residential buildings. After that came digging deeper and using the heat to generate electricity. The first commercial geothermal power plant in the US was opened in 1960 in Geysers, California; there are more than 60 operating in the US today. Figure 4.1 illustrates the evolution of geothermal energy and its use as outlined by DOE GeoVision report [1]. In order to evaluate the potential for geothermal energy to contribute to America's energy future, the U.S. Department of Energy's Geothermal Technologies Office initiated the GeoVision analysis which evaluated opportunities for successful geothermal deployment based on three key objectives [1,2] (a) increased access to geothermal resources, (b) reduced costs and improved economics for geothermal projects, and (c) improved education and outreach about geothermal energy.

The *GeoVision* analysis used rigorous quantitative models to assess geothermal deployment potential under scenarios that considered a range of technologies, market conditions, and barriers. The analysis determined that achieving all three key objectives can reduce risk and costs for geothermal developers, increase growth potential for geothermal energy, and provide the United States with secure, flexible energy that offers economic benefits to the geothermal industry and environmental benefits nationwide. The analysis projected that, through technology improvements, geothermal

DOI: 10.1201/9781003328087-4

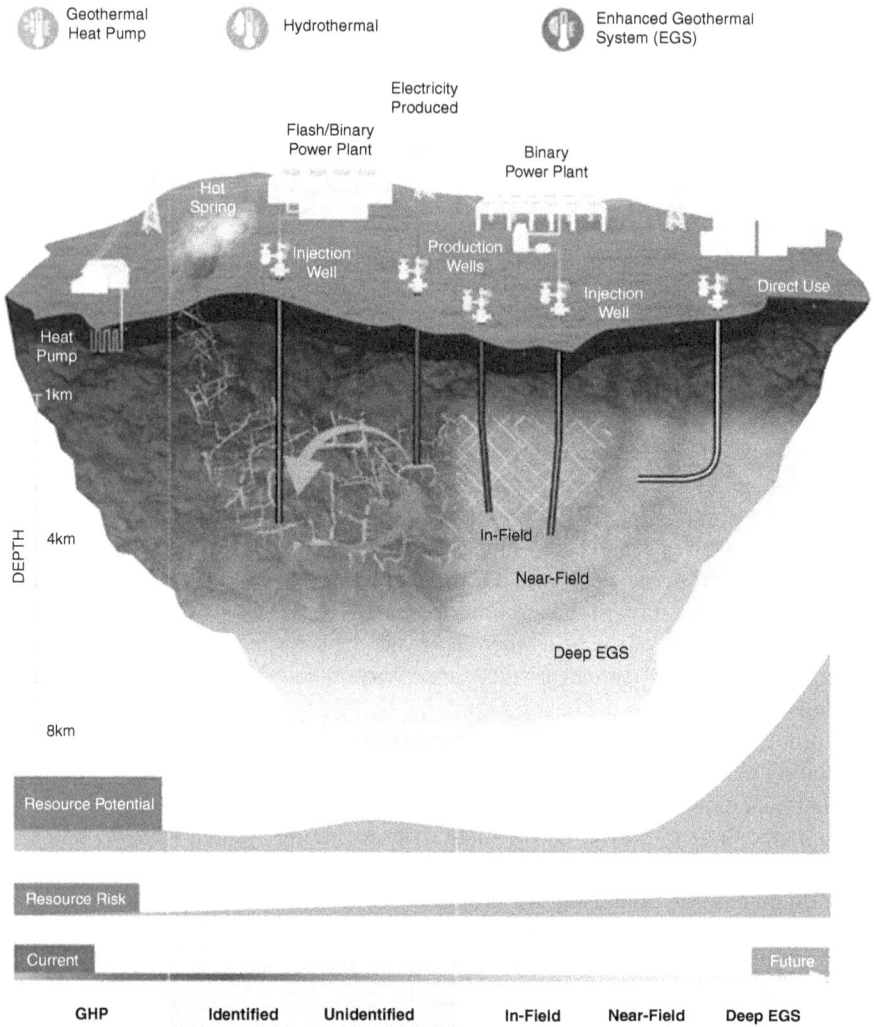

**FIGURE 4.1** Evolution of various forms of geothermal energy (DOE) [1].

electricity generation capacity has the potential to increase to more than 60 GW by 2050—providing 8.5% of all U.S. electricity generation. If power from geothermal energy can be generated economically and reliably from hotter, drier and deeper rock, it can be a perfect complement to wind and solar. It is renewable, dispatchable, and inexhaustible. It can run as baseload power around the clock, including at night, or "load follow" to complement renewables' fluctuations. It is available almost everywhere in the world, a reliable source of domestic energy and jobs that, because it is largely underground, is resilient to most weather (and human) disasters. Geothermal energy does not emit pollutants and GHG, and it can be simultaneously used for electricity and other heating needs such as district heating, industrial process heating, etc.

In order to realize geothermal energy's full potential, stakeholders must reduce risk and costs by overcoming significant technical and non-technical barriers. The *GeoVision* analysis calculated the opportunities for increasing geothermal deployment by reducing these barriers. Such increased deployment can leverage the capabilities and unique features of geothermal energy, including [2]: (a) secure, "always-on" renewable electricity generation with flexible and load-following capabilities that provide essential services to support the grid of the future, (b) nationwide, affordable solutions for electricity generation and for heating and cooling at residential, commercial, and district levels, (c) existing commercial technologies that are already proven in the market, augmented by innovative technologies with vast potential to increase electricity generation and heating and cooling solutions, (d) economic benefits to the geothermal industry and environmental benefits for the nation, and (e) revenue potential for federal, state, and local stakeholders, as well as royalty potential for leaseholders.

Geothermal energy is a sustainable renewable source of energy [3–5], which can lower fossil-fuel dependence and their environmental impact [6–8]. Electricity generation from geothermal heat can be carried out at large, medium, and small scales. The installed capacity of geothermal power has been increasing rapidly in the last few years [9] (from 6.8 GW in 1995 to an estimate of 21.4 GW in 2020 [10]). The Geo Vision analysis confirmed that improving the tools, technologies, and methodologies used to explore, discover, access, and manage geothermal resources would reduce costs and risks associated with geothermal developments and facilitate access to previously untapped sources of geothermal energy. In addition, optimizing permitting timelines alone could double geothermal capacity by 2050.

A worldwide review of published data shows that the average electric conversion efficiency of geothermal systems is about 12%, while the highest values reach was 21% [11]. For this purpose, it is worth noting that several factors influence the system efficiency, including the conversion technology (i.e., dry steam, single, double and triple flash, binary and hybrid systems), the desired output (e.g., electricity production, combined heat and power generation, and polygeneration), the size (large, medium, small, micro-scale), the ambient conditions, the temperature, the flow rate and the physic–chemical properties of the geothermal source [11]. Technologies that can be adopted for geothermal power extraction depend on the thermodynamic state of the resource.

At high temperatures, dry steam systems are usually adopted [10]. They represent a well-established technology and account for 23% of the global geothermal capacity. Furthermore, single and double flash steam power systems are the most diffused technology (42% for single flash and 19% for double flash) [10]. Finally, binary cycles represent the most recent solution that accounts for the remaining 16% of the total installed capacity in operation [12]. In this context, binary organic Ranking cycles (ORCs) play a crucial role [11,13–15]. ORCs represent an attractive and flexible technology for small applications owing to their capability to convert low enthalpy heat sources (i.e., waste heat [16–20] and renewable sources [21–23]) to electricity. They provide lower maintenance costs, better partial load performance, faster start-up and stop procedures, and higher safety and lifetime (up to 30 years) [6,24].

Geothermal ORC does not need an intermediate thermal oil circuit, as is usually present in other ORC applications [25,26]. This gives geothermal ORC a significant advantage in terms of performance, cost and architecture complexity [27].

The electrical efficiency of ORC is affected by the thermal level of the external source, the working fluid, the type of expander and operating conditions. While the system architecture can vary significantly with the specific application, its electrical efficiency can reach as high as 20% [28]. Generally, for high-temperature geothermal systems, electric power can be higher than 50 $MW_{el}$; for low-temperature ORC systems power production is lower and costs are higher. More research is being performed to improve the performance of ORC; some of which are described here. Ahmadi et al. recently provided a comprehensive review on this topic [6].

The ORC market has experienced significant growth since the early 2000s, with an average yearly capacity between 75 and 200 MWe. In 2017, the ORC market was estimated to be between USD 359 million and 402 million per year, including the sales of equipment and direct engineering services [29]. In the early 2000s, small ORC units presented a much higher cost per kWe, but units of less than 500 kWe did not represent more than 2% of the total installed capacity [29]. Recently, several low-cost and low-temperature ORC engines have entered the market to serve the needs of residential and small industries. French company Enogia provides units of 10 and 20 kWe, suitable for heat source temperatures between 80°C and 120°C using R245fa and a cycle efficiency ranging between 5% and 8%, depending on the hot source and cold sink conditions [30]. Zuccato company from Italy provides a unit of 30 kWe, with a cycle efficiency of 8.5% [31]. Infinity Turbine company developed small ORC engines starting from 5 kWe [32], and Orcan Energy company uses ORC in waste heat recovery with cycle power between 20 and 100 kWe [33].

The entrance of multiple competitive companies for small-scale ORC has allowed the reduction in relevant costs. In 2018, the global ORC market size was valued at USD 498.7 million, which is expected to grow at a compound annual growth rate of 9.7% over a forecast period up to 2025 [29]. The geothermal application segment has dominated the ORC market due to the large capacity of geothermal projects compared to other application segments such as biomass and waste heat recovery. There are very few reports concerning the cost of a geothermal ORC application, with a rough estimation being USD ~2,000–3,000/kWe installed [34]. ORC systems are also attractive for application in microgrids because of their easy-to-install configuration, equipment longevity, the potential use for co-generation applications, environmental cost benefits, and autonomous operation [35].

Once geothermal heat reaches the surface, it can be used for a wide variety of purposes. Figure 4.2 illustrates various uses of geothermal energy as a function of its temperature [1]. Depending on how hot the resource is, it can be exploited by numerous industries. Virtually any level of heat can be used directly, to run fisheries or greenhouses, to dry cement, or (the really hot stuff) to make hydrogen. To make electricity, higher minimum heat is required. The older generation of geothermal power plants used steam directly from the ground, or "flashed" fluids from the ground into steam, to run a turbine. Flash plants require heat of at least 200°C. The newer, "binary" plants run fluids from the ground past a heat exchanger and then use the heat to flash steam (meaning the underground water isn't boiled directly and there's no air or water pollution). Binary plants can generate electricity from around 100°C up. Getting the heat to the surface is the trick. For that purpose, it's useful to think of geothermal energy technology as falling into five broad categories: (a) conventional

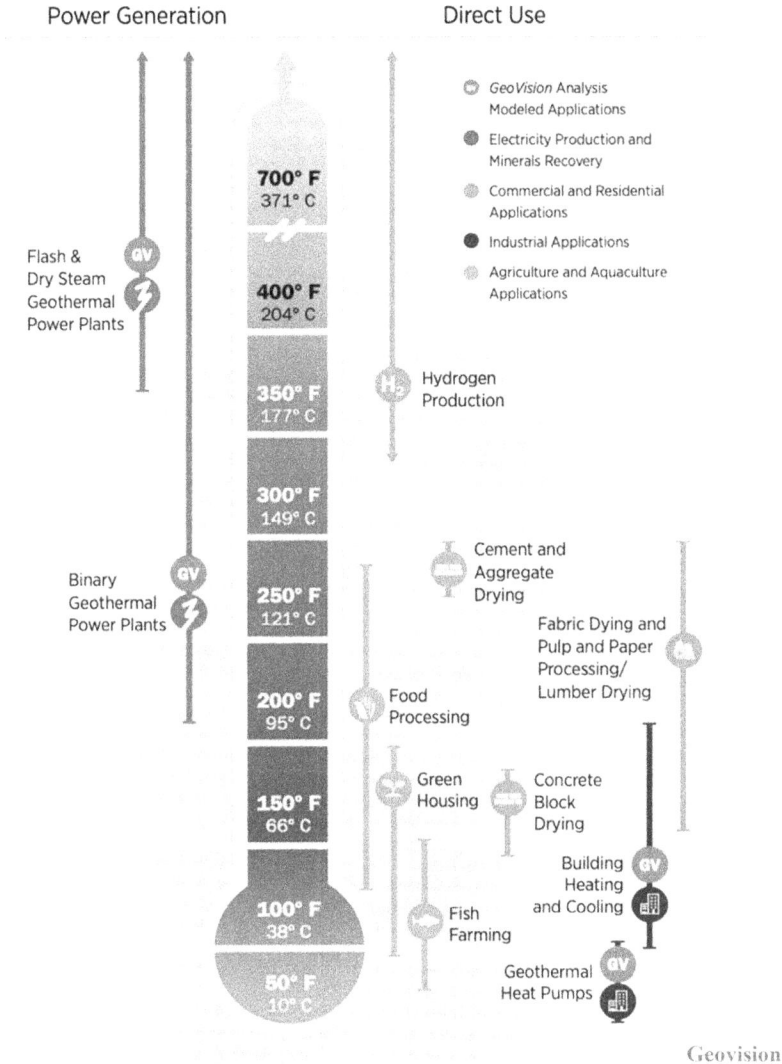

Power Generation

Direct Use

GeoVision Analysis
Modeled Applications

Electricity Production and
Minerals Recovery

Commercial and Residential
Applications

Industrial Applications

Agriculture and Aquaculture
Applications

700° F
371° C

400° F
204° C

Flash &
Dry Steam
Geothermal
Power Plants

350° F
177° C

Hydrogen
Production

300° F
149° C

Binary
Geothermal
Power Plants

250° F
121° C

Cement and
Aggregate
Drying

Fabric Dying and
Pulp and Paper
Processing/
Lumber Drying

200° F
95° C

Food
Processing

150° F
66° C

Green
Housing

Concrete
Block
Drying

100° F
38° C

Fish
Farming

Building
Heating
and Cooling

50° F
10° C

Geothermal
Heat Pumps

Geovision

**FIGURE 4.2** Use of geothermal heat as function of its temperature [1].

hydrothermal resources, (b) enhanced geothermal systems, (c) super hot rock geothermal, (d) advanced geothermal systems, and (e) supercritical high enthalpy geothermal power. Here we examine each category and its use in power generation.

## 4.2 CONVENTIONAL HYDROTHERMAL RESOURCES

In a few select areas (parts of Iceland, or California), water or steam heated by Earth's core rises through relatively permeable rock, full of fissures and fractures, only to become trapped under an impermeable caprock. These giant reservoirs of

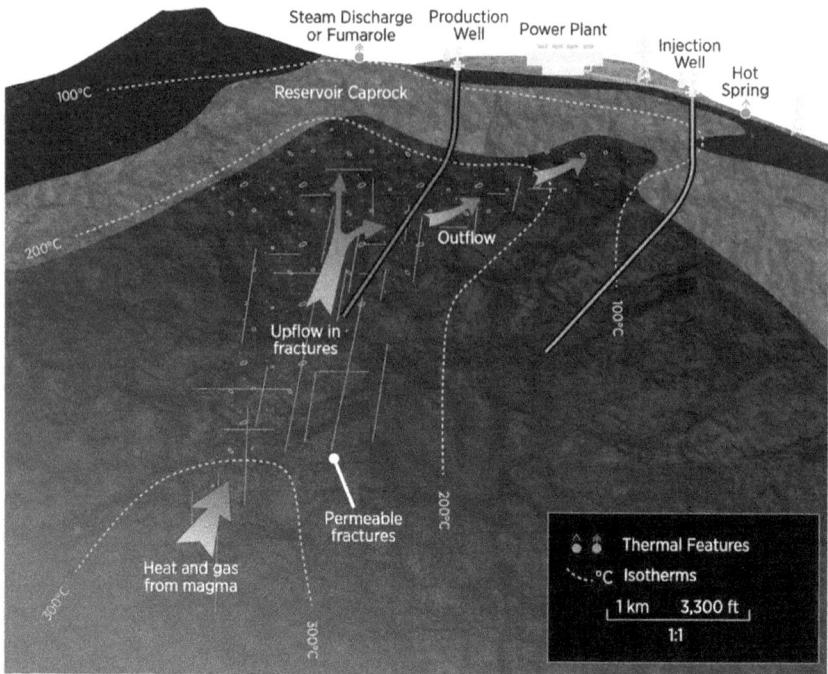

**FIGURE 4.3**    Mechanisms of geothermal heat extractions [1].

pressurized hot water often reveal themselves on the surface through fumaroles or hot springs. Figure 4.3 illustrates the mechanisms of geothermal heat extraction [1]. Once a reservoir is located, exploratory wells are drilled until a suitable location can be located for production well. The hot water that rises through that well can range from just over ambient temperature to 370°C, depending on the field. Once the heat is extracted from them, the fluids are cooled and returned to the field via an injection well, to maintain pressure. Almost all conventional geothermal projects make use of high-quality hydrothermal resources.

It is difficult to identify hydrothermal reservoirs; since hot springs and fumaroles remain the only reliable way to identify them. Exploration and characterization of new fields are expensive and uncertain. Another problem is that they are extremely geographically concentrated. In the US, as shown in Figure 4.4, geothermal electricity from hydrothermal resources is mostly located in California, Nevada, Hawaii, and Alaska, where tectonic plates are grinding beneath the surface [1]. The global geothermal electricity fleet has an average capacity factor—time spent running relative to maximum capacity—of 74.5%, and newer plants often exceed 90%. Geothermal energy is the only renewable energy source that provides continuous baseload power. As of the end of 2019, global installed geothermal electric capacity, dispersed across 29 countries, **reached 15.4 GW**, with the US in the lead. Figure 4.5 illustrates the historical distribution of geothermal power generation in the top five countries. The largest availability of geothermal reservoirs is between 100°C and 220°C, known as

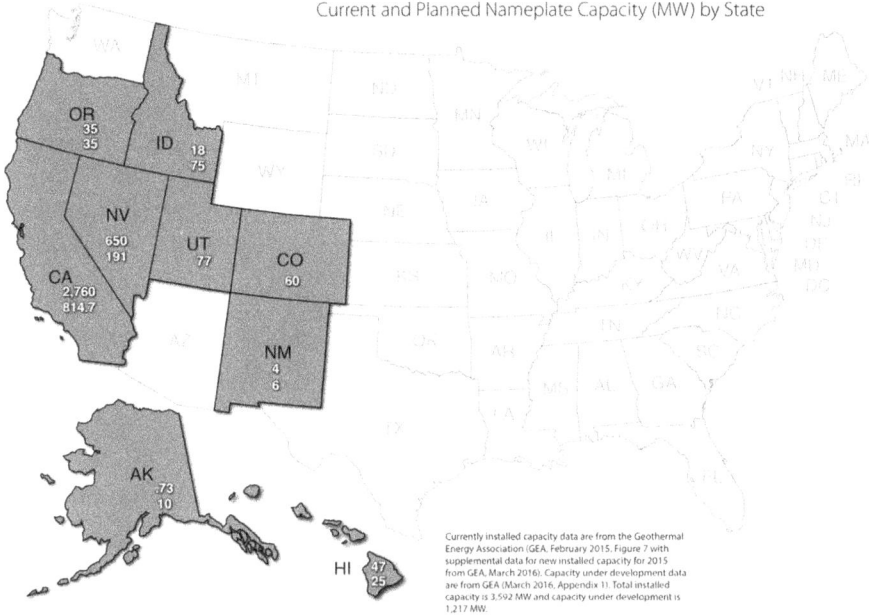

**FIGURE 4.4** Locations of geothermal power in the U.S [225]. (Source: NREL report.)

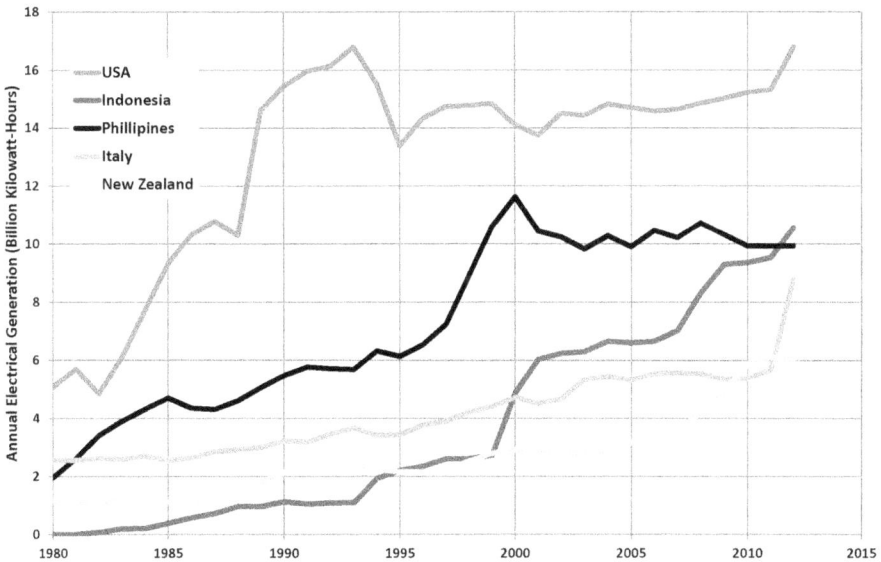

**FIGURE 4.5** Trends in the top five geothermal electricity-generating countries, 1980–2012 (US EIA) [37].

medium-temperature geothermal resources [36]. Unfortunately, most of the big, well-explored, well-characterized fields have been tapped out, at least with conventional technology. Therefore, geothermal energy that relies on high-quality hydrothermal resources remains a niche solution which is difficult to standardize and scale.

Low-temperature geothermal is still a new technology, compared to its sister resource High-T geothermal. Most power plants are moderate to high in the temperature category [24]. Many companies have proposed development plans for Low-T areas, though few have actually begun drilling. It seems that most of the countries currently harnessing high-temperature geothermal (U.S., Iceland, and the Philippines) are currently exploring Low-T resources as well. United Technologies Corp. in 2006, replaced the diesel generation system (producing power at 30 cents/kWh [27]) with a low-temperature power plant at Chena Hot Springs in Alaska to generate electricity at temperatures as low as 75°C. While this power plant is a role model for low-temperature power plant, it is less efficient and puts out electricity at 6 cents/kWh, as opposed to high-temperature power plant which can sell electricity at 3–3.5 cents/kWh. Alaska is the home to the world's lowest temperature geothermal plant to date.

Low-T sources are economical when they are "piggy-back" on the oil and gas industry. In the U.S, nearly 3,000 L of hot water are produced from oil and gas wells which requires expensive disposal. "Co-produced fluid" technologies eliminate waste while providing electricity to be used on-site, or sold to the grid. The use of ORC for waste fluids can result in energy savings and diminished greenhouse gas emissions. Geothermal plants require high upfront costs. The use of the existing infrastructure of oil and gas fields eliminates the need for *hydraulic fracturing* and drilling. This concept is even more powerful when infrastructure from oil and gas production is no longer in use [38]. The first successful generation of electricity from co-production (funded by the U.S. DOE) occurred in 2008 at the Rocky Mountain Oilfield Testing Center in the U.S. The binary geothermal power generation system used low temperatures to generate 150–250 gross kW of power [39]. Low-temperature and high-temperature geothermal power plants swap fuel for either thermal oils or water-based steam, thus cutting back emission levels to only about 1% of the $CO_2$ released by comparable fossil fuel plants. The plant also generated and released only 1%–3% of the sulfur compounds produced and emitted by coal and oil-fired power plants [29]. This approach also provides some secondary benefits. $CO_2$ emission level caused by the fuel transportation to the plant is also avoided. While a geothermal plant requires a considerable area under ground for its infrastructure, its imprint on the surface area is minimal. Farming, wildlife grazing, and recreational activities have not been shown to suffer any negative effects from geothermal power plant operations [30].

### 4.2.1 Low-Temperature Organic Rankine Cycle

Low-temperature geothermal sources are generally near the surface because the geothermal gradient caused by primordial heat and radioactive decay increases the underground temperature by about 2.5°C–3°C per 100m. As mentioned earlier, power generation from geothermal resources is usually accomplished through either the use of a single flash, dual flash or binary flash power plant [31,40]. Conventional steam turbines require fluids in excess of 150°C. Because water drawn from a Low-T geothermal resource [41] is not at or near the boiling point of water, it is not possible to 'flash' these fluids to produce steam. Instead, the heat energy held within the water is transferred to a secondary fluid (thermal oil or silicone-based oil), which has a much lower boiling point, and it is then this fluid which is 'flashed' to produce steam to drive a turbine. Binary plants such as organic Rankine cycle (ORC) and Kalina cycle are currently considered as appropriate solutions to the geothermal energy utilization within low-temperature range [10,42].

Unlike high-temperature power generation, low-T generates less efficient and low voltage power generation which results in higher equipment cost and smaller profit [43]. For ORC, significant literature on working fluid selection, component design, cycle optimization, and economic evaluation is available. Zhai et al. [44] studied the influence of working fluid properties on the performance of geothermal ORC systems. The study indicated that the optimized evaporation temperatures were almost the same for the selected working fluids and those with double bonds or cyclic structure delivered higher thermal efficiencies. By using zeotropic working fluids in geothermal ORC systems, Heberle et al. [45] and others [46] found that the use of zeotropic fluids increased cost and thermal efficiency by up to 15% compared to pure fluids. Imran et al. [47] conducted multi-objective optimization of the evaporator in a geothermal ORC system and the Pareto front solutions were obtained to indicate the trade-off between pressure drop and cost. Sauret and Rowlands [48] presented the rationale for using radial-inflow turbines in geothermal ORC systems. Walraven et al. [49] compared the performances of ORC systems with different cooling technologies and found that the mechanical-draft wet cooling towers showed better economics.

ORC system which generally includes a pump, an evaporator, an expander and a condenser can be operated in a number of different configurations which include superheated and recuperated systems. Superheater is included in ORC systems to maintain the working fluid in the superheated region after heat absorption and to prevent liquid droplet formation in the expander. Roy et al. [50] presented an analysis of non-recuperative ORC with a superheating process using R12, R123, R134a, and R717 as working fluids, and the results revealed that the system with R123 was the best choice for converting low-grade heat to power. Algieri and Morrone [51] observed twofold improvement in performance by superheating. Recuperator is often used to preheat the working fluid at the pump outlet by recovering part of the heat released during desuperheating at the expander outlet. This can reduce the amount of heat required in the evaporator and increase thermal efficiency. Ventura and Rowlands [52] investigated the performance of recuperated ORC systems with various working fluids for different heat sources. While the inclusion of a recuperator could generally enhance the performance by increasing the specific power production, there was a threshold pressure, above which the recuperator had no positive influence on the ORC system. Saturated and superheated ORC systems were compared to evaluate the influence of superheater on the system performance. The study revealed that superheating can improve the system performance with R1234yf and isobutane, whose critical temperature is relatively low; the maximum decrease in payback period reached 73% and 60%. However, a deterioration in thermo-economic performance was observed for working fluids with higher critical temperatures, e.g., R245fa, R1233zd, and isopentane, with the maximum increment on the payback period of 32%, 33%, and 60%, respectively. While recuperation presents advantages such as thermal-efficiency enhancements and cooling water savings, these did not compensate for the costs associated with the installation of the additional heat exchanger, which consequently makes recuperation less favorable in most cases. The comparison results indicated that an increment up to 8% on the payback period was observed when the recuperator was included in the system. A significant decrease in the payback period, up to a maximum of 90%, was achieved by switching from subcritical to transcritical ORC systems. This indicates that transcritical ORC systems offer a promising option for power generation in geothermal applications.

Under a fixed payback period, a higher net power output indicates the ORC geothermal plant could achieve a higher profit in the long-term operation. After selecting a system payback period of 10 years as an example, it was found that R245fa outperforms the working fluids in saturated (subcritical) and transcritical ORC systems, while isobutane was the most suitable fluid for superheated (subcritical) systems. The inclusion of recuperator resulted in reductions in the maximum power output by 0.6%, 2.5%, and 0.7%, in saturated, superheated and transcritical cases, respectively, relative to non-recuperated cycle systems. Recuperator was favored in superheated and transcritical configurations, in which a decrease of 4% and 6% of SIC were observed. For the saturated-cycle system, however, the SIC increased by 14%. The study showed that transcritical cycles provide a power output that is nearly 20% higher than equivalent subcritical systems; in addition, lower SICs (by 9% and 7%, respectively) can be achieved by recuperated cycles relative to saturated and superheated systems. For a geothermal heat source ranging from 140°C to 200°C, the study indicated that non-recuperated transcritical-cycle systems with a working fluid whose critical temperature is close to the heat-source temperature typically deliver the maximum power output under a fixed payback period. This suggests that these configurations have the potential to offer more profit in long-term operation of the ORC geothermal plant beyond the project payback period.

Astolfi et al. [53,54] carried out both thermodynamic and economic assessment of binary ORC power plants for medium- and low-temperature geothermal sources. The studies indicated that working fluids with critical temperature close to the heat source temperature were optimal for subcritical configurations; superheating was not profitable and supercritical configurations performed better from the economic point of view. Transcritical operation can improve systems' thermodynamic performance as it can achieve a better thermal match between the working fluid and the heat source, hence reducing the exergy losses in the heat exchange processes. Transcritical ORC system, however, gave poor economics [55]. Lecompte et al. [56], showed that while transcritical ORC system improved net power output by 31.5%, its specific investment cost increased by 72.8%. Several studies examined the effects of ORC configuration on its thermodynamic and economic performances. Zare [57] showed that the system with internal heat exchanger (IHE) was superior from the thermodynamic point of view while simple system showed the best economic performance. Walraven et al. [49] showed that the transcritical and multi-pressure subcritical cycles were the best options with exergy efficiency of up to 50%. Zhang et al. [58] showed that subcritical ORC with R123 yielded the highest exergy efficiency of 54% and transcritical ORC with R125 provided the lowest levelized electricity

### TABLE 4.1
### Working-Fluid Candidates with Key Properties [43]

| Working Fluid | Critical Temperature (°C) | Critical Pressure (kPa) | Normal Boiling Point (°C) | Molecular Weight (kg/kmol) |
|---|---|---|---|---|
| Isopentane | 187.2 | 3,378 | 27.8 | 72.1 |
| R1233zd | 165.6 | 3,571 | 18.3 | 130.5 |
| R245fa | 154.0 | 3,651 | 15.1 | 134.1 |
| Isobutane | 134.7 | 3,629 | −11.7 | 58.1 |
| R134a | 101.1 | 4,059 | −26.1 | 102.0 |
| R1234yf | 94.7 | 3,382 | −29.5 | 114.0 |

cost (LEC) of 0.056 $/(kWh). Vetter et al. [59] demonstrated that transcritical ORC with propane achieved a thermal efficiency of 10%, which was higher than that of subcritical cycles.

Song et al. [60] compared different ORC configurations (subcritical and transcritical, saturated and superheated, simple and regenerative cycles) with different organic fluids (isobutane, isopentane, R134a, R245fa, R1233zd, and R1234yf) for the exploitation of a 40 kg/s geothermal source at 180°C. The relevant properties of these fluids are described in Table 4.1. A bi-variable optimization was adopted based on the exergy efficiency maximization and payback period minimization. The comparison between optimized configurations demonstrated that transcritical systems presented a lower payback period (up to 90%) compared to subcritical arrangements for all the investigated working fluids. Furthermore, the authors demonstrated that when the payback period is fixed, the transcritical ORCs provide higher net power, larger long-term economic profit, and lower specific investment cost compared to subcritical systems. The advantages of transcritical configurations were also confirmed for geothermal source temperatures between 140°C and 200°C.

Preißinger et al. [61] observed similar results of a significant increase in the gross power and a noticeable decrease in the payback period for transcritical configuration compare to subcritical configuration for geothermal source temperatures between 100°C and 190°C and different working fluids. Transcritical organic Rankine cycles (ORCs) were selected for the investigation by Morrone and Algieri [62], owing to their high performance, low payback period and efficient heat exchange with low and medium-temperature heat sources [58,60,63–65]. Since, organic fluids present usually low critical pressures, transcritical configurations are very attractive for these systems due to the possibility to guarantee proper performance without operating in extreme and dangerous conditions [13].

Astolfi et al. [53] highlighted that transcritical systems guarantee lower specific costs compare to subcritical ORCs when the ratio between the critical temperature of the organic fluid and the geothermal source thermal level ranges between 0.8 and 0.9. Higher energy production and efficiency in a transcritical system overcomes its larger investment costs. The study also demonstrated that supercritical ORCs offer better energy and economic performance for high geothermal sources [53].

The proposed unit was able to work at partial loads to fulfil the energy demand of the final users, to increase the system flexibility and to respond efficiently to load variations [27]. The internal regeneration was considered to maximize the electric power and achieve higher efficiency with respect to the mean value of typical geothermal plants (close to 12%) and isobutane was selected as working fluid owing to its good performance and low specific costs [66–69]. Isobutane is considered one of the organic fluids with zero ODP (ozone depletion potential), suitable for supercritical ORC applications [39]. Even though the relatively high pressures lead to higher operating costs, the integrated system reaches higher global efficiency for high thermal level geothermal sources.

Maloney et al. [70] found that high pressure values are more appropriate for high expander inlet temperatures to maximize the system efficiency. The study showed that in the range of temperatures 180°C–240°C isobutane revealed good energy performance in transcritical ORCs. The study also showed that the best isobutane cycles are supercritical for the highest values of temperature of the geofluid investigated (180°C and higher). This was supported by Toffolo et al. [71] and others [72,73] who showed that from an energetic and exergetic point of view [74,75] isobutane represents one of the most cost-effective fluids for the exploitation of high-temperature geothermal sources.

In the studies carried out by Askari et al. [76] and others [77,78], a lumped thermodynamic model was developed to predict the performance of the geothermal-driven integrated system at full and part loads and to select the suitable ORC system and the corresponding operating conditions, and to calculate the energy production on an hourly basis over the whole year. The proper system configuration was defined adopting a multi-variable optimization and different operating strategies (i.e., thermal-driven, electric-driven and mixed-mode) were compared [76–78]. The payback period and net present value were compared for the different operating strategies. The influence of the maximum temperature on the ORC behavior was evaluated through a parametric investigation for both simple and regenerative ORC arrangements and the proper system configuration was defined by adopting a multi-variable optimization using "minimum distance" method. The purpose was to find the most suitable apparatus with electric power and efficiency. The optimized system obtained in this manner consisted of a regenerative 101.4 $kW_{el}$ ORC unit with 17.4% electric efficiency and a geothermal heat exchanger that guarantees 249.5 kWt.

## 4.2.2 ORC WITH COGENERATION

Combined heat and power (CHP) generation presents several advantages compared to conventional separate electric and thermal production [79–80]. CHP provides lower costs, emissions, fuel consumption, and higher decentralized generation. The flexibility of geothermal-driven CHP systems based on ORCs has been receiving great attention from both the scientific community and industry [32–35,81]. In particular, a great amount of energy and techno-economic analysis and optimization approaches are usually proposed for medium and large-scale geothermal power plants (electric power higher than 200 $kW_{el}$) and for different system arrangements [32,82,83]. The objective functions combine energy parameters (i.e., first and second law analysis

[84–86], electric power [81], heat exchanger areas [84,87], etc.) and economic purposes (e.g., specific cost of investment and levelized cost of energy, energy production cost [88–90]).

The study by Morrone and Algieri [62] examined an innovative transcritical geothermal-driven ORC system for small-scale applications. In this study, the ORC unit was coupled with a traditional geothermal heat exchanger for the direct production of thermal energy. The sub-systems worked in parallel to satisfy the electric and thermal demands of a small commercial center. In a parallel arrangement, the heat was delivered to both the thermal user and the ORC at a high temperature and at a lower flow rate [83,91]. In a series design, the same geothermal flow passed through the ORC and subsequently through the heat exchanger [83,91]. As a consequence, the ORC unit significantly limited the thermal level of the heat generation in series arrangements [92,93]. According to the literature [83], the parallel configuration is the most suitable solution for the supply and return temperatures higher or close to 80°C and 60°C [83], respectively, typical of space heating applications [6]. On the other hand, the series configuration is recommended for applications with lower temperatures (supply and return temperature close to 50°C and 30°C [22,27,83]) suitable for soil warming, swimming pools, de-icing, etc. [6]. Furthermore, the parallel configuration guarantees better ORC electric performance, owing to the lower condensation temperatures compared to series arrangements and higher flexibility.

An innovative high-energy geothermal CHP demonstrator in parallel configuration was developed within the Fongeosec project [94,95]. In this project, the geothermal source was used to feed a 5.5 $MW_{el}$ subcritical ORC and a 15 $MW_{th}$ district heating network in Alsace (France) [96]. Parallel arrangements are also available in Bavaria (Germany) with Traunreut and Oberhaching geothermal plants. The global electric and thermal power are equal to 8.4 $MW_{el}$ and 52.0 $MW_{th}$, respectively, while the net electric process efficiency ranges between 8.9% and 13.1% [97]. A lot of research effort is currently focused on organic Rankine cycles for CHP applications operating in subcritical conditions and driven by relatively low-temperature sources.

The adoption of a transcritical ORC configuration with superheated conditions at the expander inlet has also been considered. Transcritical systems showed higher performance compared to subcritical configurations [51,39] and provided a more efficient thermal exchange between the geofluid and organic fluid and lower irreversibility and exergy destruction [98]. Another benefit was the component downsizing due to the higher density of fluid. However, transcritical configurations present higher investment costs than subcritical arrangements owing to their higher operating pressures and larger size of some components (turbines, pumps and evaporators) [61,53]. Nevertheless, transcritical ORCs appear a promising technology to also exploit geothermal sources from an economic perspective [53,61,66].

As a test case, the application of the integrated CHP system to a commercial center located in Phlegraean Fields (Southern Italy) was considered by Algieri [13] where high-temperature geothermal aquifers are present at a very shallow depth. The payback time was used to evaluate the economic viability. The system consisted of an organic Rankine cycle (ORC) for the electric production and a heat exchanger (HEX) which was able to provide the thermal energy. The two components work in parallel and simultaneously and the total geothermal mass flow rate can be arranged between the ORC and HEX units according to the user energy demand. For this purpose, the geofluid was pumped to the two components from the geothermal reservoir and afterwards was reinjected into the ground. The adopted parallel configuration guaranteed high flexibility to the CHP system, owing to the possibility to split the geothermal mass flow rate in the ORC and HEX units, depending on the hourly user energy request [92]. Furthermore, the parallel arrangement permitted one to meet the high-temperature heat demand for the space heating and also when the ORC condensation temperature was set to low values to maximize the electric power and efficiency [83,92]. The integrated system was able to exchange the electric energy with the grid and a traditional natural gas packaged firetube boiler was adopted to meet the thermal load when the geothermal production was not sufficient. Figure 4.6 illustrates a schematic of an integrated ORC-CHP system.

In order to examine economic viability, the costs of the different components were estimated through the bare module cost methodology and three operating strategies were compared: electric- and thermal-driven and mixed strategies. For each operating mode, the yearly electric and thermal energy balances were evaluated on an hourly basis. In the first two cases, the priority was given to the fulfilment of the heat and electricity request, respectively; as a consequence, the geothermal mass flow rate was divided between HEX and ORC sub-systems on the basis of the privileged load. In the mixed operation, the thermal-driven strategy was adopted when the heat demand was present in order to minimize the heat dumping. On the other hand, the geofluid was completely exploited in the ORC apparatus when the thermal request was absent to maximize the electric efficiency and production. The results highlighted that mixed-mode represents the best solution according to both energy and economic points of view. Particularly, the integrated system provides an electric and thermal self-consumption higher than 90%, while thermal and electric

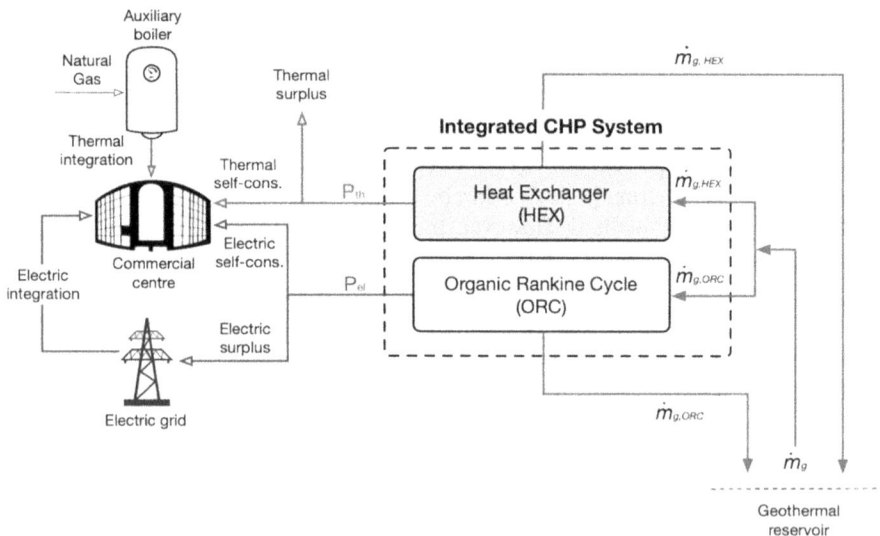

**FIGURE 4.6**   Simplified scheme of the integrated geothermal-driven CHP system [62].

strategies revealed lower performance. Furthermore, the mixed strategy assured no heat dumping and the highest equivalent hours of operation—i.e., 8,095—whereas thermal- and electric-driven strategies showed 2,977 and 6,028 hours, respectively. The payback time lower than 5 years confirmed that the proposed system represented a viable technical solution for small-scale and combined energy production. Finally, the geothermal exploitation in the suggested integrated apparatus provided a yearly $CO_2$ emission reduction of more than 250 tons and a greenhouse gas saving of more than 300 tons of $CO_2$ equivalent per year.

### 4.2.3  ORC with Autonomous Polygeneration Microgrids

The study by Kyriakarakos et al. [99] investigated techno-economical use of small-scale (<15 kWe) low-temperature (<150°C) geothermal Organic Rankine Cycle (ORC) technology for polygeneration microgrids in order to fully cover essential energy needs of small settlements. A schematic representation of autonomous poly-generation microgrid is illustrated in Figure 4.7. The investigation aimed at con-cluding whether geothermal ORC in microgrid concept integration is technically feasible, cost-effective, and attractive for future investments, thus constituting an alternative solution to the commonly applied photovoltaics (PVs), wind turbines, and hybrid configurations with battery pack energy storage. In order to analyze the ORC system for low-T geothermal sources, small-scale ORC performance under varied conditions was introduced based on extensive experimental results already published [100,101]. These studies presented a complete analysis of the performance of a two-stage, small-scale (~10 kW) ORC operating at temperatures up to 140°C that favors its application for geothermal small-scale microgrid integration. Scroll expenders were used as expansion machines [102].

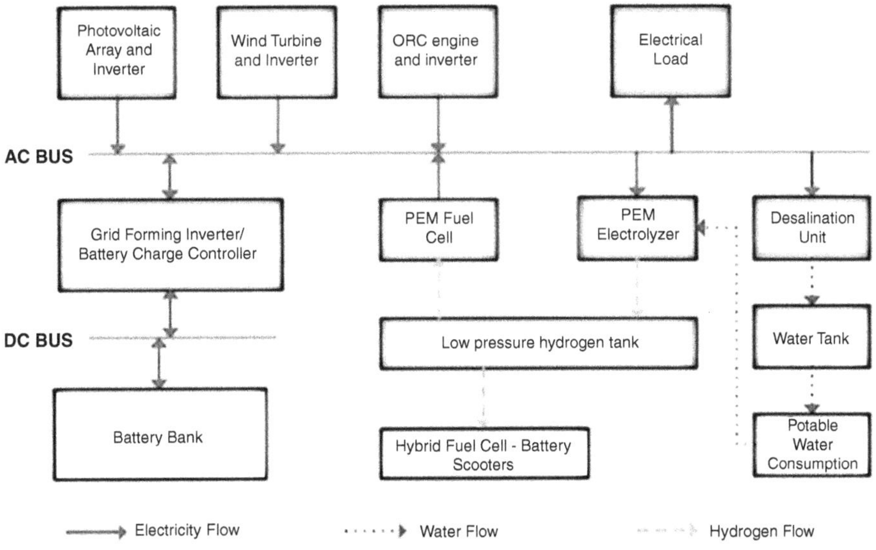

**FIGURE 4.7**   Schematic representation of autonomous polygeneration microgrid [99].

The autonomous polygeneration microgrid (APM) concept was first presented almost a decade ago [103]. It aims to holistically meet the needs of an off-grid community in terms of electrical loads, space heating and cooling, potable water production through desalination, and the use of hydrogen as fuel for transportation in the most cost-effective manner possible. At the same time, the use of renewables minimizes the carbon footprint of the community. In principle, any power source can be used in an APM. Photovoltaics (PVs) and wind turbines have been investigated extensively since PVs can be installed practically anywhere in the world and wind turbines in areas with sufficient wind potential.

In addition to the initial research, advanced energy management systems were also developed based on fuzzy logic [104], a combined approach of petri nets and fuzzy cognitive maps [105], a multi-agent system enabling demand-side management [106], and eventually an investigation of transitioning to a decentralized topology in the energy management system from the centralized one [107]. Other artificial intelligence paradigms were also investigated successfully, such as game theory [108] and fuzzy Q-learning [109]. The biggest improvement was observed when migrating from an on–off approach for the various subsystems to variable load operation. Kyriakarakos et al. [110] also investigated various components of the APM. Hybrid capacitors were investigated as an alternative to batteries, with higher utilization of the hydrogen subsystem obtained positive results. These results led to the realization of a hybrid capacitors bank [111]. Dimitriou et al. [112] developed and experimentally analyzed a reverse osmosis desalination unit incorporating energy recovery. Since commercial lithium-ion batteries present high efficiencies, high depth of discharge, extended lifetimes, reasonable cost, and warranties of 10 years [113], they are prime candidates for use in polygeneration microgrids. Battery disposal and recycling, however, faced a sustainability challenge in the developing world [114].

Chandrasekharam and Bundschuh [115] showed that microgrid integration by ORC makes a competent solution for low-enthalpy geothermal resources (<150°C) occurring at shallow depths. The main reason is that it exhibits higher conversion efficiency and maturity [116] compared to all alternative technical solutions (e.g., thermoelectric generator, Stirling engines, etc.). The thermal efficiency of the ORC is in the order of 5%–6% when heat is supplied at 100°C, which is about 30% of Carnot efficiency [117,118]. Most of the ORC systems currently installed are in the range of a few hundred kWe to a few MWe [119]. In the field of geothermal electricity generation, an investigation into relevant literature reveals few systems that include autonomous geothermal electricity generation mostly in the MWe range [120,121], with few cases of geothermal-based isolated grids starting from 50 kWe [122,123]. In Bologna, a 11 kWe capacity unit using R134a and a prototype four-cylinder piston expander was constructed and tested, exploiting a geothermal well with a temperature of 65°C with a net efficiency of 4.4% [9]. Another experimental ORC with a capacity of 1 kWe combined with a geothermal power plant of 85°C–105°C was developed in Germany, where two different working fluids, R245fa and R1233zd(E), were compared for the efficiency of the ORC operation [124]. Eyerer et al. [27] developed a regenerative ORC–CHP plant with a twin-screw expander operating at a constant heat source temperature of 135°C, which resembled a geothermal source, and was tested in full and partial load operation. The plant reached a very high operational range and high efficiency in part-load operation. Finally, three units with a capacity of a few kWe were developed, realized, and tested extensively at the Agricultural University of Athens [101,117,125].

There are some concerns about the use of geothermal energy. H2S and geothermal water release can cause contamination of air and existing surface water [126]. H2S is generally treated by caustic scrubbing followed by oxidation, adsorption, and catalytic conversion in order to produce elemental sulfur [127] which can be used as fertilizer [128]. After proper treatment based on its chemical constituents, geothermal water can be used for irrigation [129].

Distributed geothermal power generation using ORC and microgrid topology is used all over Europe and developing nations [130]. The possibility of using existing low-T geothermal wells for power and heat can find significant use for agriculture in developed nations. In remote rural areas of Sub-Saharan Africa where access to electricity is 31% [131], electricity is provided through autonomous systems mainly in the form of microgrids [132]. In the developing world, the application of small-scale ORC units in microgrids utilizing low-enthalpy geothermal energy resources can provide a low-cost alternative for providing energy access. Thus, geothermal power generation with ORC under a microgrid topology could play a key role in fulfilling full basic power demand in small communities in both the developed and developing world. The technology will, however, have to compete with other renewable energy technologies.

In the past energy in remote areas were provided by PV and wind energy sources. Since these are intermittent, large packs of batteries were needed to provide energy storage. The development of low-T geothermal sources with ORC provided another source of energy. The power produced by ORC engines depends on the temperature of the source. In general, it was found that raising temperature from 95°C to 120°C, nearly doubles power output. Thus at high temperatures, battery pack storage is not needed and the overall cost of energy supply and the operation and maintenance costs of microgrids are reduced.

In the developing world, many development-aid financing programs in the past decades focused on providing autonomous PV-battery systems but did not include any provisions for long-term maintenance of these systems. This proved to be of high importance since the lack of funds to exchange the battery banks was and still is one of the most common reasons for stranded non-operational autonomous systems in Sub-Saharan Africa [133]. The current trend is also to facilitate productive uses of energy applications while providing electrification to communities in order to increase economic activity and ultimately the income of the local population [134]. Agricultural loads such as milling for on-site production of flour, ice making, or water pumping coupled with a tank can be scheduled anytime throughout the day. This can further decrease the need for large battery banks, since optimal scheduling is feasible, increasing the applicability and cost-effectiveness of geothermal ORC engine use [135]. Public-private partnership in financing helps the system.

Finally, as far as the environmental sustainability of the system is concerned, the geothermal ORC system needs a much lower capacity battery bank. Moreover, it is not oversized in terms of installed power, which is necessary while using intermittent renewable energy sources. This has a direct positive impact in terms of its sustainability in comparison with the large intermittent and battery pack-based systems.

Low-T geothermal sources that are needed for ORC are also easy to find and have secondary positive attributes to agricultural communities as mentioned before.

### 4.2.4 KALINA BINARY CYCLES

In order to replace the previously used Rankine Cycle as a bottoming cycle for a combined-cycle energy system as well as for generating electricity using low-temperature heat resources, Alexander I. Kalina designed a new power cycle in which ammonia–water is used as a working fluid [136,137]. The first version of the Kalina cycle is characterized by a second condenser, after the separator, at one intermediate pressure, allowing an additional degree of freedom in the composition of the boiling mixture and allowing the distillation unit to operate at a pressure lower than the maximum one [136,138]. A further difference concerns the recuperative heat exchanger, which, in the Kalina scheme is placed downstream of the turbine. In these situations (medium-low temperatures heat sources application and small power conversion system) the plant layout may be simplified and the cycle has a single main condenser, at the lowest cycle temperature, and the separator is placed after the evaporator [139].

Water and ammonia have similar molecular weight but quite different thermophysical characteristics as presented in Table 4.2. When mixed, water and ammonia form a nonazeotropic mixture. In such a mixture, ammonia is the more volatile component of the two, having a lower boiling and condensing temperature than water. For the thermodynamic properties of the ammonia-water mixture used in the Kalina cycle, a basic model is assumed. In the gas phase, above water $T_{sw}$ is the saturation temperature, it is believed that the overheated mixture serves as an ideal solution of superheated ammonia and water vapor (Figure 4.8). In the gas phase, when the temperature is between the pure water saturation temperature $T_{sw}$ and the mixture dew point $T_d$, the water portion is assumed at the considered pressure to be in a meta-stable vapor state. Similarly in the liquid region, we assume a meta-stable liquid state for ammonia between the saturation temperature of pure ammonia $T_{sa}$ and the bubble point of the mixture $T_b$. We have a saturated steam mixture with ammonia mass fraction $X_g$ and a saturated liquid mixture with ammonia mass fraction $X_f$ in the wet vapor mixture area $T_d > T > T_b$. In the liquid region below the bubble point of the $T_b$ mixture, a Gibbs excess function is assumed for moving away from ideal-solution conduct [140]. The Kalina cycle, as shown in Figure 4.9,

---

**TABLE 4.2**
**Properties of Water and Ammonia [136]**

| Name | Chemical Formula | Molecular Weight (kg/kmol) | Critical Temperature (°C) | Critical Pressure (MPa) | Boiling Point at 1.013 Bar (°C) | Freezing Point at 1.013 Bar (°C) |
|------|------------------|----------------------------|---------------------------|-------------------------|----------------------------------|-----------------------------------|
| Ammonia | $NH_3$ | 17.0 | 101.1 | 4.059 | −33.3 | −77.7 |
| Water | $H_2O$ | 18.0 | 373.9 | 22.064 | 100.0 | 0.0 |

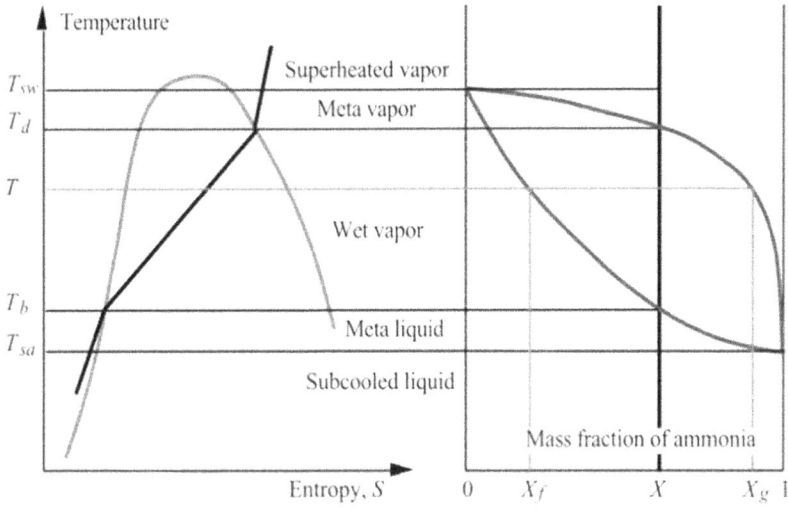

**FIGURE 4.8**   The various states on T-S and T-X diagrams [136].

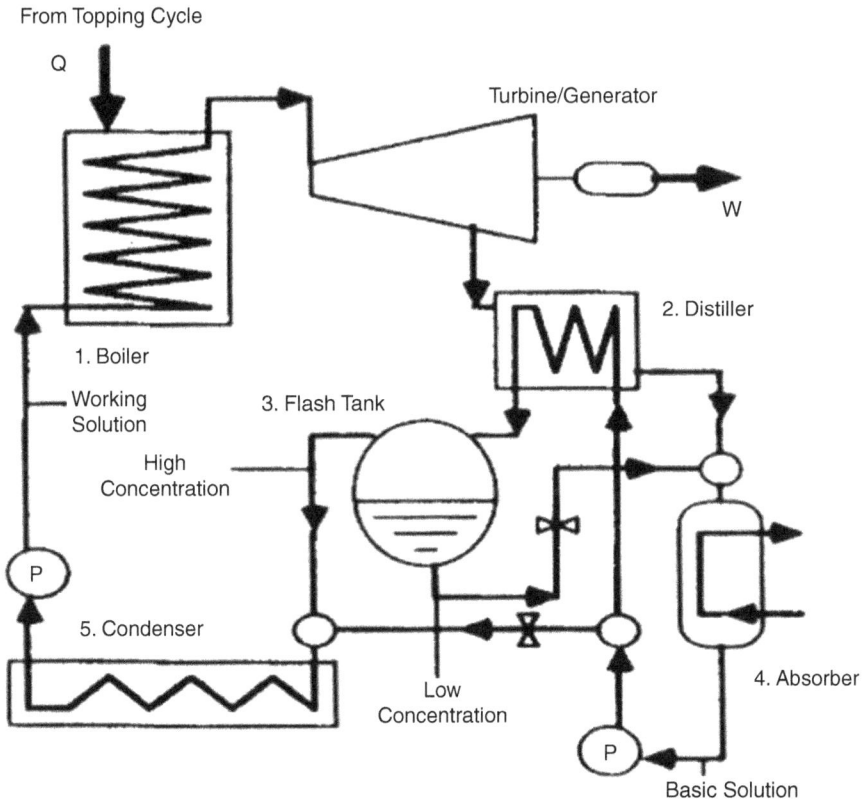

**FIGURE 4.9**   Schematic diagram of the Kalina Cycle [136,141].

consists of six main components: boiler, turbine distiller, absorber, flash tank, and condenser. Except for the absorber and flash tank, these components are the same as those of the Rankine cycle [141]. The cycle operates as follows:

The binary mixture, the ammonia/water mixture, boils at a variable temperature. More heat is extracted from the boiler and delivered to the power cycle. The superheated mixture is expanded to the turbine back pressure. The pressure drop through the turbine may be greater than that of the Rankine cycle. The waste heat from the turbine exhaust is used to distil (separate) the low temperature boiling fluid (ammonia) from the high temperature boiling fluid (water), using the significant difference in volatility between the two fluids. The high- concentration (strong solution) and low-concentration (weak solution) fluids are separated in the flash tank. The low-concentration solution is condensed. The high-concentration solution is condensed at a relatively high pressure. This pressure is higher than the turbine back pressure [141].

In theory, the Kalina cycle can help convert approximately 45% of a direct-fired system's heat input to electricity and up to 52% for a combined-cycle plant (a gas turbine produces exhaust, which enables a steam turbine to produce electricity). This compares with about 35% and 44%, respectively, for the steam cycle [142]. Moreover, the Kalina cycle cycles can give up to 32% more power in the industrial waste heat application compared to a conventional Rankine steam cycle. However, the Kalina cycle in small direct-fired biomass-fueled cogeneration plants does not show better performance than a conventional Rankine steam cycle [143]. When both cycles are used as a "bottoming" cycle with the same thermal boundary conditions, it can be found when the heat source is below 1,100 1 F (537°C), the Kalina cycle may show 10%–20% higher second law efficiencies than the simple Rankine cycle [144].

Jonsson [145] investigated the Kalina cycles as bottoming processes for natural gas-fired gas and gas–diesel engines. It was shown that the Kalina cycle has a better thermodynamic performance than the steam Rankine cycle for this application. All simulated Kalina cycle configurations generated more power than the steam cycles, except for one simple Kalina cycle configuration compared with a dual-pressure steam cycle. The best Kalina bottoming cycle could generate 40%–50% more power than a single-pressure steam cycle and 20%–24% more power than a dual-pressure steam cycle. A Kalina bottoming cycle could add 6%–8% points in efficiency to the gas engines, while a single-pressure steam bottoming cycle could add about 5% points. For the gas–diesel engines, the efficiency augmentation was 4%–7% points for the Kalina bottoming cycles, 4%–5% points for a single-pressure steam cycle and 4%–6% points for a dual-pressure steam cycle.

The adoption of the Kalina cycle to a certain heat source and a certain cooling fluid sink has one degree of freedom more than the ORC cycle, as the ammonia–water composition can be adjusted as well as the system high and low-pressure levels [146]. Nihaj and Shan [136] indicated that in a particular case in the Republic of Croatia, the geothermal source has a higher temperature (175°C), therefore, ORC in which isopentane is used as the working fluid has better both the thermal efficiency (the First Law efficiency) and the exergetic efficiency (the Second Law efficiency): 14.1% vs. 10.6% and 52% vs. 44% [147].

## 4.3 ENHANCED GEOTHERMAL SYSTEMS (EGS)

Conventional hydrothermal systems are limited to specialized areas where heat, water, and porosity come together. But those areas are limited. There's plenty of heat stored down in all that normal, solid, nonporous rock. Enhanced geothermal systems drill down into solid rock, inject water at high pressure through one well, fracture the rock to let the water pass through, and then collect the heated water through another well. Figure 4.10 illustrates the mechanism of EGS. It is important

Heated fluids are recovered at the surface for energy production

Power Plant

Production Well

Injection Well

Fluids are injected into the earth for continuous energy recovery

Heated fluid is produced back to the surface

Injection creates fractures resulting in an EGS reservoir

FIGURE 4.10  Mechanism of EGS [1].

to note that the line between a conventional hydrothermal resource and a resource that requires EGS is not sharp. There are many gradations and variations between wet/porous and dry/solid. This is tantamount to having a supply curve, where the variables are temperature, depth, well permeability, and reservoir permeability, and everything that exist between the two extremes. In simple terms, as the resource gets deeper and the rock becomes hotter and less porous, the engineering difficulty of accessing it increases [148].

The basic idea has always been that EGS would start off within existing hydrothermal reservoirs, where fields are relatively well-characterized. Then, as it learned, honed its technology, and brought down costs, it would branch out from "in field" into "near field" resources—solid rock adjacent to reservoirs, at similar depth. Eventually it would be able to venture farther out into new fields and deeper into hotter rock. In theory, EGS could eventually be located almost anywhere in the world. EGS startups like Fervo are growing quickly and bigger established companies are running profitable EGS projects today.

The engineering challenges become more daunting, as the targets get deeper and drier. There are PR challenges as well. Injecting fluids into the ground in order to fracture rock is known as "fracking" in the oil and gas business, and it has not got favorable public acceptance because of its linkage to localized earthquakes. In fact, there are numerous US states and countries where it is banned. Still, if the engineering and marketing challenges can be overcome, the prize is almost unthinkably large. Assuming an average well depth of 4.3 miles and a minimum rock temperature of 150°C, the GeoVision study estimates a total US geothermal resource of at least 5,157 GW of electric capacity—around five times the nation's current installed capacity. Alternatively, according to DOE, using EGS for direct heat could provide the US with 15 million terawatt-hours-thermal (TWhth) compared to a total US annual energy consumption of 1,754 TWhth for residential and commercial space heating. This EGS-based resource is theoretically sufficient to heat every US home and commercial building for at least 8,500 years. There is even more heat deeper down, 6 miles and further [148].

### 4.3.1 Ogachi, Japan, HDR Project for $CO_2$ Sequestration

The Ogachi HDR project was conducted during the same time period as the one at Hijiori. Its success as an HDR project was limited owing to lack of support for a multi-well/multi-reservoir system, such as the Hijiori project. In 1990 the original well OGC-1 was drilled to 1,000 m depth into a 228°C uplifted mylonitized granodiorite basement characterized by two major faults. This well was used to stimulate both fault zones. In 1992 a production well OGC-2 was drilled directionally to intercept both reservoirs; its true length was 1,100 m. A circulation test resulted in only 3% recovery of injected water in OGC-2. After further hydrofracs, first to OGC-2 and then to both wells simultaneously, the rate of recovery improved in steps to 10%–25%. In 1999 a new well OGC-3 was drilled to about 1,300 m into the new fractures. A flow test from OGC-1 to OGC-3 indicated connections at three locations at 770, 950, and 970 m depths.

While planned long-term flow test using the three Ogachi wells was never carried out, the wells were later used to study the concept of *carbon dioxide sequestration* in geothermal formations. The basic concept involved forcing down in a well a mixture of water saturated with dissolved $CO_2$ into a porous formation. Under the reservoir pressures and temperatures, the $CO_2$ exists in a *supercritical state* with enhanced ability to dissolve minerals from the formation. As the fluid migrates through the reservoir, chemical processes including degassing and precipitation occur in various parts of the formation. At Ogachi, experiments were carried out using the existing wells. The first test was run in 2006 with 6.3 kg/s of neutralized river water being injected at a pressure of 15 MPa into OGC-1 and 0.67 kg/s produced via OGC-2; the produced fluid was at 127.5°C. For these experiments, OGC-3 was an observation well. Upon reaching steady conditions, $CO_2$ was introduced at the injection well as centimeter-size cubes of dry ice. While this test did not prove fixation of carbon, the next text carried out in 2007 into OGC-2 indicated that the injected $CO_2$ precipitated as carbonates within a few days. The tests improved the understanding of $CO_2$–rock interactions in a non-aqueous, hot formation. The results also indicated that $CO_2$ may be a useful working fluid in an EGS system, instead of water. Generally, $CO_2$-EGS reservoir behave in three zones each having different modes of interaction with the formation. The inner most zone 1 is devoid of any natural water since it is assumed to be both displaced by and dissolved into the $CO_2$ stream. While $CO_2$-water *supercritical* ionic aqueous mixture would not remove minerals from the rock, it would remove loosely bound water molecules in some minerals which then result in improved permeability of rock. Hot $CO_2$ is recovered from this zone. Zone 2 surrounds zone 1 and in this zone increased permeability between $CO_2$ and water occurs. Finally, in the most outer zone 3, $CO_2$ is dissolved in water and it is considered as sealing calcite zone. This proposed mechanism needs experimental verification.

## 4.4  SUPER-HOT-ROCK GEOTHERMAL (HDR)

At the far horizon of EGS is "super hot rock" geothermal, which seeks to tap into extremely deep, extremely hot dry rock. This is the third category of geothermal resources. Superhot rock geothermal energy is a visionary technology deserving of investment, and yet almost entirely unrecognized in the decarbonization debate. It has the potential to meet long-term demands for zero-carbon, always-on power, and can generate hydrogen for transportation fuel and other applications. When water exceeds 373°C and 220 bars of pressure, it becomes "supercritical," There are two important things about supercritical water. First, its *enthalpy* is much higher than water or steam, meaning it holds anywhere from 4 to 10 times more energy per unit mass. Second, it is so hot that it almost doubles the *Carnot efficiency* of its conversion to electricity. In short, supercritical condition provides more energy and electricity. This means an individual geothermal project at 400°C would have about 50 MW capacity, compared to the roughly 5 MW capacity of an EGS project at 200°C—42% hotter, ten times the power. One can get more power out of three wells on a 400°C project than you can out of 42 EGS wells at 200°C, using less fluid and a fraction of the physical footprint [2]. Experience to date shows that the hotter geothermal gets, the more competitive its power price, to the point that super-hot EGS could

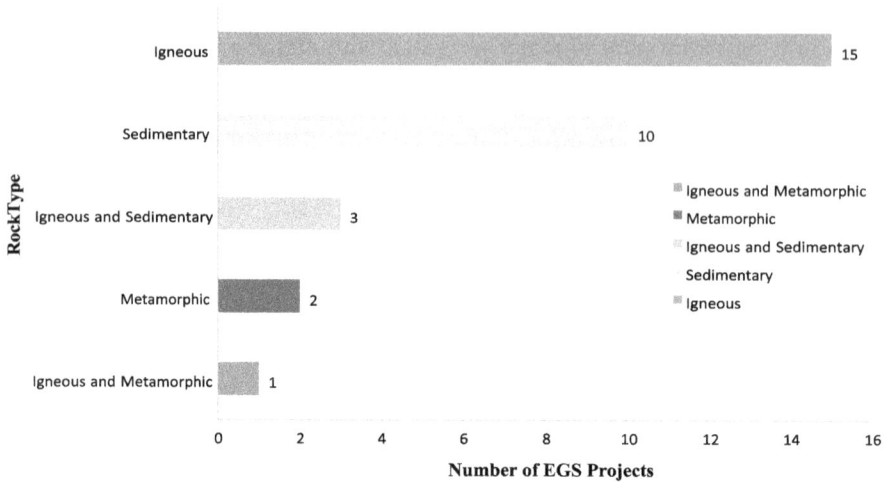

**FIGURE 4.11**   EGS projects classified on the basis of rock types [149].

be the cheapest baseload energy available. For example, levelized cost of electricity ($/MWh) for EGS project can be within 80–120 compared to 85 for solar PV, 96 for advanced nuclear, 99 for natural gas combined cycle and 56 for onshore wind. However, it could be as low as 46 for super hot EGS [1,2].

Superhot rock geothermal could have a few distinct advantages over other energy sources. As shown in Figure 4.11, EGS projects are generally classified based on rock types [149]. It is projected to be affordable, requiring little area to produce large amounts of energy (high energy density) due to the very large amount of energy that can be produced per well. Superhot rock is expected to produce five to ten times as much energy as the power produced from one of today's commercial geothermal wells. Superhot rock geothermal "mines" deep at very high temperature and heat in the Earth's crust. This contrasts with today's small (~15 GW globally) commercial geothermal industry that typically depends on upwelling of hot groundwater at locations with high near-surface heat. Superhot rock geothermal injects water into superhot dry crystalline rock by opening existing fractures at a depth where water is so hot it possesses properties of both liquids and gasses, allowing injected water to travel rapidly through existing rock fractures and gather very large volumes of heat energy. Production wells bring this steam energy to the surface to produce power in electric turbines and/or to generate hydrogen.

In order to increase the heat transfer between rock and water, water can be pumped into the rock causing it to *hydraulically fracture*, or the rock can be *control-exploded* first. The explosion should create small rocks that can heat the water into steam more effectively. Care should be given so that the explosion does not create cracks that can allow the water or steam to leave the reservoir except through the provided well. HDR system was used for the first time experimentally in Los Alamos, New Mexico, the USA in 1970. The experiment was then followed by similar projects in Australia, France, Germany, Japan, and the United Kingdom. These reservoirs are generally hot *impermeable rocks* at depths shallow enough to be accessible (<3,000 m). To extract heat from such formations, the rock must be fractured and a fluid circulation system developed. Although hot dry rock resources are virtually unlimited in magnitude around the world, only those at shallow depths are currently economical.

The engineering challenges for super hot rock geothermal are very significant. These include new casings and cements development; knowledge of water chemistry at high heat; development of materials that resist corrosion and high heat; and improved drilling techniques. There are even new, "non-contact drilling" methods being developed, including AltaRock's, *frickin' lasers* ("millimeter waves," technically) technique. While no well is currently producing electricity from supercritical water, several past wells (in Hawaii and California's Salton Sea, e.g.) have encountered supercritical water and there are exploratory projects using supercritical water in Japan, Italy, Mexico, including several other counties.

HDR is a condition where *water is not naturally present at the site*. The magma only heats dry rock on top of it. In order to tap heat from the dry rock, two wells can be drilled into the rock. One well is used to carry water from the surface down into the HDR. Once the water is heated, steam created is then channeled up through the second well into a *turbine* above the surface. The rest of the system is similar to the ones described earlier; i.e. direct dry steam, flash steam, or ORC. Hot dry rock (HDR) is an extremely abundant source of *geothermal energy* that is difficult to access. A vast store of thermal energy is contained within hot – but essentially dry and impervious crystalline *basement* rocks found almost everywhere deep beneath Earth's surface. HDR contains vast energy with low environment impact.

The feasibility of mining heat from the deep Earth was proven in two separate HDR reservoir flow demonstrations—each involving about 1 year of circulation—conducted by the Los Alamos National Laboratory between 1978 and 1995. These groundbreaking tests took place at the Laboratory's *Fenton Hill* HDR test site in the *Jemez Mountains* of north-central *New Mexico*, at depths of over 8,000 ft (2,400 m) and rock temperatures in excess of 180°C. The results of these tests demonstrated conclusively the engineering viability of the revolutionary new HDR geothermal energy concept. The two separate reservoirs created at Fenton Hill are still the only truly confined HDR geothermal energy reservoirs flow-tested anywhere in the world. Although these tests demonstrated that HDR systems could be constructed, the flow rates and energy extractions rates did not justify the cost of the wells.

This technology has been tested extensively with multiple deep wells drilled in several field areas around world including the USA, Japan, Australia, France, and the UK with investment of billions of research funds. HDR system project was tested by France and Germany in *Soultz-sous-Forêts*. While preliminary work started in 1986 through 1997, significant advances were made in 2003 when the already existing wells were deepened to 5.1 km. Stimulations were done to create another reservoir. During circulation tests in 2005–2008 water was produced at a temperature of about 160°C with low water loss. Construction of a power plant was begun. The power plant started to produce electricity in 2016, it was installed with a gross capacity of 1.7 MWe. The 1.7 MW test plant is purely a demonstration plant. In comparison, normal geothermal power plant development typically involves initial plants from 10 to 100 MW. More work is needed to make HDR technology commercially viable.

While limited success has been achieved, so far not enough rock fractures and steam generation are achieved to have successful and continuous large-scale power production. No commercial projects are ongoing or likely due to the high cost and limited capacity of the engineered reservoirs, associated wells, and pumping systems. For this technology to successfully compete with other energy sources, drilling costs would have to drop drastically or new approaches that result in much more extensive, complex, and higher rate flow paths through actual fracture networks would have to be established.

The Fenton Hill tests clearly demonstrated the advantages of a fully engineered HDR reservoir over naturally occurring hydrothermal resources, including EGS. With all the essential physical characteristics of the reservoir—including rock volume, fluid capacity, temperature, etc.—established during the engineered creation of the reservoir zone, and the entire reservoir volume enclosed by a hyper-stressed periphery of sealed rock, any variations in operating conditions are totally determined by intentional changes made at the surface. In contrast, a natural hydrothermal "reservoir"—which is essentially open and therefore unconfined (having boundaries that are highly variable)—is inherently subject to changes in natural conditions. It is, however, true that less confined, more complex, lower pressure, and more pervasively fractured natural systems support much higher well flow rates and low-cost development of energy generation. This makes them commercially more viable.

Another advantage of an HDR reservoir is that its confined nature makes it highly suitable for load-following operations, whereby the rate of energy production is varied to meet the varying demand for electric power—a process that can greatly increase the economic competitiveness of the technology. This concept was evaluated and proven near the end of Phase II of Fenton Hill test. Such load-following operations could not be implemented in a natural hydrothermal system or even in an EGS system because of the unconfined volume and boundary conditions. Load following almost never improves economics for geothermal development because the fuel cost is effectively paid up front, so delaying use just hurts the economics. Normal geothermal systems have also (by necessity) been applied to follow loads but this kind of generation increases maintenance costs and generally reduces revenue (in spite of the higher prices for some of the load). The experiments at Fenton Hill have clearly demonstrated that HDR technology is unique, not only with respect to how the pressurized reservoir is created and then circulated but also because of the management flexibility it offers.

Hot Wet Rock (HWR) hydrothermal technology makes use of hot fluids found naturally in basement rock; but such HWR conditions are rare. By far the bulk of the world's geothermal resource base (over 98%) is in the form of basement rock that is hot but dry—with no naturally available water. This means that HDR technology is applicable almost everywhere on Earth (hence the claim that HDR geothermal energy is ubiquitous). On the other hand, an uneconomic resource is actually just energy storage and is not useful. Typically, the temperature in those vast regions of the accessible crystalline basement rock increases with depth. This geothermal gradient, which is the principal HDR resource variable, ranges from less than 20°C/km to over 60°C/km, depending upon location. The concomitant HDR economic variable is the cost of drilling to depths at which rock temperatures are sufficiently high to permit the development of a suitable reservoir. The advent of new technologies for drilling hard crystalline basement rocks, such as new PDC (polycrystalline diamond compact) drill bits, drilling turbines or fluid-driven percussive technologies (such as Mudhammer) may significantly improve HDR economics in the near future.

It is important to distinguish between HDR and EGS. Some sources describe the permeability of the Earth's basement rock as a continuum ranging from totally impermeable HDR to slightly permeable HWR to highly permeable conventional hydrothermal. However, this continuum concept is not technically correct. A more appropriate view would be to consider impermeable HDR rock as a separate state from that of the continuum of permeable rock—just as one would consider a completely closed faucet as distinct from one that is open to any degree, whether the flow be a trickle or a flood. In the same way, HDR technology should be regarded as totally distinct from EGS. Unfortunately it is not easy to open the faucet to obtain significant flow.

Recently Seattle-based *AltaRock Energy* announced the results of a path-breaking comprehensive technical and economic feasibility study, completed in collaboration with Baker Hughes, an energy technology company, and the University of Oklahoma, demonstrating the superior energy density and competitive economics of an engineered geothermal system (EGS) resource in high temperature (>400°C) impermeable rock at the Newberry Volcano near Bend, Ore. This result is a major step in developing the first SuperHot Rock (SHR) geothermal resource in the United States. AltaRock defines SuperHot Rock as EGS in high-temperature rock above 400°C. SuperHot Rock development targets energy densities per well as high as 5–10 times that of both conventional EGS and hydrothermal developments in the 200°C–250°C range.

The analysis carried out by AltaRock energy and its collaborators indicated that SuperHot Rock resources could achieve a competitive Levelized Cost of Electricity (LCOE) of <$0.05/kWh. In comparison, a conventional EGS resource of 200°C–230°C would produce power at an LCOE >$0.10/kWh. The significant cost difference between the two systems results from much higher energy density – SuperHot Rock being 5–10 times higher per well than conventional ESG wells – with one-tenth the water requirements and surface area and infrastructure based on conventional EGS use cases. SuperHot Rock geothermal, will require development of engineered reservoirs in deep basements where hotter 'supercritical' temperatures can yield up to ten times more energy than a conventional geothermal well.

The extreme case of superhot substance is magma which is partially molten rock at very high temperatures (> 600°C). It is found at depths of 3–10 km and deeper. This great depth inhibits accessibility. Magma has a temperature that ranges from 700°C to 1200°C and it possesses the largest source of geothermal energy.

## 4.5 ADVANCED GEOTHERMAL SYSTEMS (AGS)

As indicated above, a typical geothermal plant works by drilling down into a naturally occurring underground reservoir of hot water, normally in volcanic regions. This hot water is pumped to the surface and turns into steam when it returns to the atmospheric pressure, driving an electricity generating turbine. The steam is then cooled and the cold water is pumped back into the underground reservoir to be reheated. A so-called low enthalpy geothermal energy (LEGE) plant has similar principles but does not utilize underground hot water reservoirs-it simply pumps water down to a depth of 1.5 km and uses hot rock to heat the water, which is then pumped back to the surface to power an ORC electricity-generating system. LEGE has never taken off because the amount of energy needed to pump the heated water to the surface and later reinject it equates to 50% or more of the electricity the plant generates.

On the other hand, a fourth category of technologies named as advanced geothermal systems has emerged recently, which holds out promise that geothermal power could someday be accessible anywhere. AGS refers to a new generation of "closed loop" systems, in which no fluids are introduced to or extracted from the Earth; there's no fracking. Instead, fluids circulate underground in sealed pipes and boreholes, picking up heat by conduction and carrying it to the surface, where it can be used for a tunable mix of heat and electricity. Closed-loop geothermal systems have been around for decades, but a few startups have recently amped them up with technologies from the oil and gas industry. Drill a deep hole anywhere on the planet and the temperature will rise about 30°C every kilometer down. In certain volcanic hotspots, temperature gradients reach 60°C or more per kilometer. So drill to a depth of 3–5 km—as the oil & gas industry sometimes does—and the temperature of the rock will potentially be hundreds of degrees Celsius.

Eavor-Loop is a closed-loop geothermal energy extraction system. It is a buried well system consisting of two vertical wells several kilometers (generally about 1.5 miles apart) deep connected by an extended system of interconnected multilaterals. Multilaterals are drilled connecting the vertical wells, parallel to each other in a kind of radiator design, to maximize surface area and soak up as much heat as possible. (Precise lateral drilling is borrowed from the shale revolution, and from the oil sands.) The multilaterals are not cased with steel, but rather a polymer in the drilling mud is used to seal the inside of the pipe from the surrounding rock. This is done to increase heat extraction efficiency in the horizontal plane by exposing the system to a larger volume of rock. A benign working fluid is circulated within the completely contained system, which harvests heat from within the earth and brings it to the surface where it can be used for commercial heating applications or electricity generation. Circulation is added by the *thermosiphon* effect, where temperature differences between the upflowing and downflowing working fluid cause the system to circulate without pumps. The key feature of Eavor's geothermal technology is the closed-loop nature of the system. This mitigates many of the risks associated with traditional geothermal technologies, which use wells to produce brine from *subsurface aquifers*. As a closed-loop system that operates in complete isolation from the surrounding environment, the system requires no *fracking*, produces no *greenhouse gas emissions*, poses no earthquake or subsidence risk, no water use, produces no brine or solids, and no *aquifer contamination*.

Adjusting the pressure inside the Eavor loop using a valve on the surface can change the speed of the cycle, and thus raise or lower the electricity output, converting a baseload system into a dispatchable one that is able to generate energy on demand, according to the power and balancing needs of the local grid. In simple terms, the output of a project can be increased by connecting multiple parallel horizontal wells—or multilaterals—to the up- and downpipes, creating a giant underground radiator, as well as increasing the size of the ORC on the surface. Power output is anywhere between 3 and 8 MW for one Eavor-Loop [with 8–12 multilaterals for a single loop], but one can actually stack up more of these Eavor-Loops in pretty tight formation... and drill a lot of them from the same drill path. We could have projects where we have the drilling rig, basically sitting in one close area for 5 years, doing nothing but drilling. With ten multilaterals that go out several kilometers and come back several kilometers, that could be 50 km of drilling right there for one Eavor-Loop. But you could have ten Eavor-Loops from the same drilling location. Drilling has improved such that it not only can control the direction and everything else, but it can do it with speed and cheapness. No one had previously connected horizontal eight-inch boreholes "toe to toe" from wells that started several kilometers apart, mainly because there was no reason to do so. The magnetic ranging technology that makes this possible was traditionally used by oil & gas companies to avoid running into other boreholes. As long as we're within 130 m of the other well, one can pick it up on this magnetic-ranging technology, and then fine-tune the direction for the last little bit prior to intersection.

An Eavor-Loop can act as baseload (always-on) power, but it can also act as flexible, dispatchable power—it can ramp up and down almost instantaneously to complement variable wind and solar energy. It does this by restricting or cutting off the flow of fluid. As the fluid remains trapped underground longer, it absorbs more and more heat. So, unlike with solar, ramping the plant down does not waste (curtail) the energy. The fluid simply charges up, like a battery, so that when it's turned back on it produces at above nameplate capacity. This allows the plant to "shape" its output to match almost any demand curve. Directional drilling in high temperatures, above 150°C or so, remains difficult, with equipment prone to melting. As rock becomes harder, equipment must also be hardened to additional vibrations. And electronics need to be better insulated. The Eavor-lite project is only mining heat of about 70°C. (It was not intended to be commercially viable.) To make geothermal work, Eavor and other companies will need to master going deeper and hotter. Temperature distribution of geothermal sources in the U.S. is illustrated in Figure 4.12.

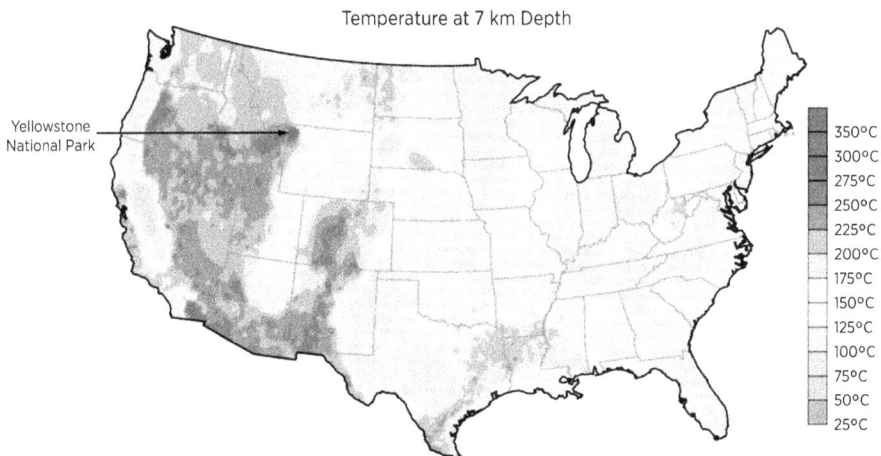

Temperature at 7 km Depth

Yellowstone National Park

350°C
300°C
275°C
250°C
225°C
200°C
175°C
150°C
125°C
100°C
75°C
50°C
25°C

FIGURE 4.12   Temperature distribution of geothermal sources in U.S. [1].

Without the parasitic load of a pump, Eavor can make profitable use of relatively low heat, around 150°C, available almost anywhere about a mile and a half down. The Eavor-Lite Demonstration Project is a full-scale prototype of the firm's technology suite located near Rocky Mountain House, Alberta. Construction began in August 2019 and was completed in February 2020. The system consists of two vertical wells connected toe to toe, with one multilateral well drilled off the horizontal section, parallel to the original well. The project was not intended to be commercially viable, but was designed to demonstrate critical elements of the technology at the lowest cost. It has shown that the lateral wells can be precisely targeted, the thermosiphon effect works, and the plant's costs and output can be reliably predicted in advance.

The firm announced the official completion and third-party validation of the project on February 5, 2020. In January 2020, Eavor Yukon announced a partnership with Carmacks Development Corporation (CDC), owned by the *Little Salmon Carmacks First Nation*, to establish Eavor-Loop technology in the Yukon. The company has three or four commercial plants in various stages of planning. In May 2020, Evor Technologies entered into a letter of intent with *Enex Power Germany GmbH* to form a geothermal project development company to construct Eavor-Loop heat and power projects within Enex's existing geothermal license area in Bavaria, Germany. In May 2020, the firm announced the launch of second-generation technology featuring a new demonstration site, Eavor-Long, in locations across Canada, France, Norway and the United States. In France and the Netherlands Eavor will provide heat and in Japan, electricity. Eavor believes it will be able to provide gigawatts of baseload and dispatchable renewable energy anywhere in the world for less than $50/MWh by the end of the decade, making its technology cost-competitive with natural gas and coal.

### 4.5.1 EAVOR-LOOP 2.0

A recent change in Eavor loop design that will reduce the physical footprint and enable even more precise drilling. Instead of the two vertical wells being located at a distance, they will be right next to each other. Lateral wells branch out from them, staying parallel until they meet at the end. Like so: With the wells so close to one another, they can use "magnetic ranging" (with a transmitter in one well and a receiver in another) to remain at a fixed distance from one another. Meeting at the end is easier than meeting in the middle. As for land use, after the initial drilling, the only part that technically needs to be aboveground is the air cooler that cools the water before it descends. Power lines, even the electric generator itself, could be underground. And if there's a water cooler rather than an air cooler, that too could be underground. Theoretically, one could have zero surface footprint. Since all Eavor needs to work is hot rock, which is pretty reliably located beneath almost any site in the world, it avoids the need for expensive exploration and modeling.

At the moment, Eavor-Loops could be built in a sufficiently deep sedimentary basin anywhere in the world, but the company is collaborating with France's Drillstar on a "percussion hammer" drilling technology that would enable kilometers-deep boreholes to be dug in harder igneous rock and at greater depths with hotter temperatures. This is the technology where the traditional oil and gas industry invested a lot of money. There were about ten of these different [hard-rock drilling technology] projects around the world. The company has been testing it up in Norway in nice hard gneiss, which is basically granite and the results for the rates of penetration are very encouraging. On top of this, the company recently came up with an Eavor Loop design that only requires one above-ground site, rather than the two envisaged in the original "daisychain" Eavor-Loop 1.0, and so avoids the need for a second site the right distance away. It's like having the daisychain [design] folded in half, so that you're drilling the up and the down well from the same drilling pad, sort of like two pitchforks doing along and then they meet at the tips. These two wells can go down close together, and we will always know where the other well is. It's not like coming at it blind toe to toe where you only 'see' the other well at the last moment. This new design is called Ever- Loop 2.0.

When one is in sedimentary basins, one generally drills the horizontal section to stay in [the sedimentary rock] formation. But once one is in the igneous basement rock and gets down to the multilateral section, one might as well go the same distance and go deeper because that's hotter and more efficient. This technique does not cause earthquakes like fracking because in fracking water is forcing down the well at such high pressure that three kilometers down it's actually breaking the rocks apart and at the same time sand is flushed through to keep the cracks open enough to allow oil or gas to flow out. This is very different to drilling eight-inch boreholes that are hermetically sealed with Eavor's patented rock-pipe technology, which permeates into the rock and solidifies in the pore spaces (without the need for traditional steel casing). It doesn't cause earthquakes, doesn't contaminate the groundwater and there are no $CO_2$ emissions.

Eavor Loop is expensive and capital intensive and requires long-term commitments from major players and politicians. There are three different types of electrical grid where Eavor's economics should be attractive, First, anywhere with a high enough feed-in tariff [such as the aforementioned project with Enex in Germany]; secondly, islands and city-states, where electricity prices are typically high, and there is not enough land to do wind and solar; and third, those with power markets that place a value on load-following capacity and ancillary services. Eavor can also produce heat for industry or district heating.

## 4.6 SUPERCRITICAL HIGH ENTHALPY GEOTHERMAL POWER

This can be considered as the fifth method of geothermal energy recovery and power production. High enthalpy geothermal systems have been harnessed for electrical power generation for over 100 years. While the classification scheme for low, medium, and high enthalpy systems varies for different authors [150–155], reservoir temperatures above 150°C–225°C are generally considered to be able to provide high enthalpy fluids. Most of the developed geothermal systems so far have a temperature range of 150°C–300°C. Generally, active volcanic centers and elevated temperature gradient undergrounds resulting from high heat flow caused by shallow intrusions of magma produces geothermal systems with temperature higher than 250°C. Most of the unexploited energy sources are igneous-related geothermal systems. Magmatic systems in the United States are believed to contain much more thermal energy than all known hydrothermal systems in the same region [156–158]. High enthalpy system in Iceland can produce a tenfold increase in energy output for a single well than a conventional hydrothermal system [159]. Most high enthalpy systems are associated with supercritical conditions which for water is temperature greater than 374°C and pressure greater than 221 bar.

Supercritical conditions are generally associated with the brittle–ductile transition zone where magmatically dominated fluids are found in the hotter plastic rock and hydrothermal fluids circulate through the overlying cooler brittle rock [160]. In this zone, any fractures generated can be sealed by the retrograde solubility of quartz [161–163]. Watanabe et al. [164] suggest that there is not a step-function decrease in permeability associated with the brittle–ductile transition, and that potentially exploitable resources may occur in nominally ductile granitic crust at temperatures of 375°C–460°C and depths of 2–6 km. Friðleifsson et al. [165] showed that much deeper and hotter wells carrying high enthalpy fluids under supercritical conditions have high energy productivity and high rates of mass transport because of their much higher ratio of buoyancy forces relative to viscous forces [166]. This results in more efficiency and more attractive economic conditions. While supercritical temperatures have been found in volcanic areas, higher temperature can also be achieved by drilling deeper at the depth of about 3.6 km [167] where boiling point-depth curve results in the critical point of water. The presence of a sealing horizon also allows the pressure to exceed hydrostatic conditions. High enthalpy improves well's productivity and sustainability.

Recently, there have been several initiatives focused on identifying the potential opportunities and challenges associated with supercritical and high enthalpy conditions. Several potential opportunities for international collaborations are identified [168]. The initiatives are both exploratory and production in nature. The main purpose has been to demonstrate viability of recovering energy from supercritical conditions. A brief summary of wells with supercritical conditions are summarized in Table 4.3. Some of these wells were "dry," indicating very low permeability. All of these wells experienced serious issues with regard to rock-physical and fluid properties, leading to challenges related to drilling, completion, and fluid handling. Previous studies, however, indicate that a number of serious issues have been encountered while trying to handle and utilize fluids from geothermal reservoirs at temperature and pressure conditions exceeding the supercritical conditions of water. Early experiments in high enthalpy geothermal fields clearly identified bottlenecks in terms of exploration, drilling, completing, and monitoring. For example, in highly fractured reservoirs, wells were drilled with total circulation loss, and sometimes a very low rate of penetration (ROP) and high bit wear. The additives to the drilling fluid made fluid sometime coagulate, blocking the drill string and eventually leading to stuck pipe. Due to the high temperatures and acidic reservoir fluids, drill string fatigue and corrosion was observed, sometimes leading to breakage [169–171]. While a number of remedies were performed including improved drilling, completion, and cementing practices, monitoring of subsurface conditions by measurements while drilling (MWD) etc., as pointed out by Reinsch et al. [172], supercritical geothermal power development requires better resource assessment and exploration methods, understanding of in-situ rock and fluid properties as well as drilling and completion technologies,

**TABLE 4.3**

**List of Wells with Reported Temperatures Above the Critical Point of Water (374.15°C) [172]**

| Country | Site (Depth) | P/T | Permeability/Fractures |
|---|---|---|---|
| Italy | Sasso-22 (4,092 m) | 380°C at 3,970 m during stop of drilling | Highly fractured down to bottom |
| | San Pompeo 2 (2,966 m) | >400°C, >240 bar | Fractured in 2,930 m, no fractures between 2,300 and 2,930 m |
| | Carboli 11 (3,455 m) | 427°C @ 3,328 m | |
| | San Vito 1 (3,045 m) | 419°C BHT (1 week after end of drilling) | Permeability available |
| Iceland | NJ-11 (2,265 m) | >380°C, possibly 220 bar | Permeability, low vertical permeability |
| | RN-17 (3,082.4 m) | 320°C–380°C BHT from extrapolated temperature logs and fluid chemistry information | Permeability about 300 m above total depth |
| | IDDP-1 (2,104 m) | Wellhead: 450°C, 140 bar, BHT: molten magma | Permeability, production 10–12 kg/s steam |
| | K-36 (2,501 m) | Possibly superheated conditions at >2 km | Permeability available |
| | K-39 (2,865 m) | Freshly quenched silicic glass, 385.6°C at 2,822 m in drill pipe shortly after end of drilling | Feed zone in 100 m distance from intrusion |
| | IDDP-2 (4,659 m) | 427°C, 340 bar, 6 days after end of circulation | Permeability indicated |
| Japan | Kakkonda, WD-1a (3,729 m) | 500°C | No permeability, little fracture density in the ductile part |
| US | Wilson No 1 (3,672 m) | 325°C unequilibrated (30 hours after end of circulation), 400°C from fluid inclusions, high-pressure fluid zone during drilling, 489 bar indicated by mud weight | Steam entry at 3,631, in sidetrack A2 (3,762 m) 3.5 kg/h steam entry at 3762, pressure declining |
| | Prati-32 (3,396 m) | 400°C @ 3,352 m | Fluid entry at 3,352 m, 10.6 kg/s steam at a normalized pressure of 6.9 bar |
| | IID-14 (2,073 m) | 390°C, 207 bar at TD | Total circulation loss below 2,033 m |
| | KS-2 (2,440 m) | 342°C and 146 bar @ 1676 m, supercritical conditions extrapolated at TD | Permeability below 1,965 m; 4 kg/s steam @ 12 bar wellhead pressure during flow testing |
| | KS-13 (2,488 m) | 1050°C | Injectivity above the intrusion |

*(Continued)*

**TABLE 4.3** *(Continued)*
**List of Wells with Reported Temperatures Above the Critical Point of Water (374.15°C) [172]**

| Country | Site (Depth) | P/T | Permeability/Fractures |
|---|---|---|---|
| | Lanipuna-1 (2,557 m) | BHT: ≥363°C 32 hours after end of circulation | Low permeability, conductive temperature gradient |
| Mexico | H-8 (2,300 m), H-11 (2,376 m), H-12 (2,984 m), H-26 (2,546 m), H-27 (2,584 m), H29 (2,186 m), H-32 (2,186 m) | >380°C estimated, young intrusions at H-12 and H-26 | Circulation loss close to bottom, overall low permeability |
| Kenya | MW-01 (2,198 m) | 391°C | Permeable zones identified at 1,200–1,400, 1,600–1,750, and 1,900–2,008 m depth |
| | MW-04 (2,118 m) | Fresh quenched glassy cuttings during drilling, 390°C during flow testing, 140 bar shut-in, 20 bar flowing conditions | Low permeability above molten magma; permeability noted between 1,700 and 1,800 m depth |
| | MW-06 (2,172 m) | Fresh quenched glassy cuttings, 325°C | Low permeability above molten magma; permeability noted between 1,802 and 2002 m depth |

*Source:* From Utilizing Supercritical geothermal systems: a review of past ventures and ongoing research activities.

logging and monitoring instruments and strategies, simulation tools and field testing to gain better knowledge of downhole conditions. Many new investigations in Japan, New Zealand, Mexico and Europe [173] are underway to further understand the drilling of a well under supercritical high enthalpy conditions. Reinsch and co workers [172–175] have identified seven such projects which are (a) Iceland Deep Drilling project; (b) Krafla Magma Testbed project; (c) Japan Beyond Brittle project; (d) DESCRAMBLE project; (e) New Zealand Hotter and Deeper project; (f) GEMex project and (g) Newberry Deep Drilling project.

The current phase of the Iceland Deep Drilling project (IDDP-2) involves deepening of the RN-15 well in the Reykjanes geothermal field from its original depth of 2,507 m to a depth of ≈5 km (Figure 4.13). The well was completed on January 25, 2017, at a depth of 4,659 m, where an unequilibrated bottom-hole temperature of 427°C was recorded together with a fluid pressure of 340 bar [176]; several permeable zones were encountered below 3,000 m. The well will be used to carry out numerous studies including petrographic analysis of retrieved core and cuttings samples to characterize the lithology and alteration of the well, running (as conditions permit) a comprehensive suite of downhole well logs, injecting cold water into the

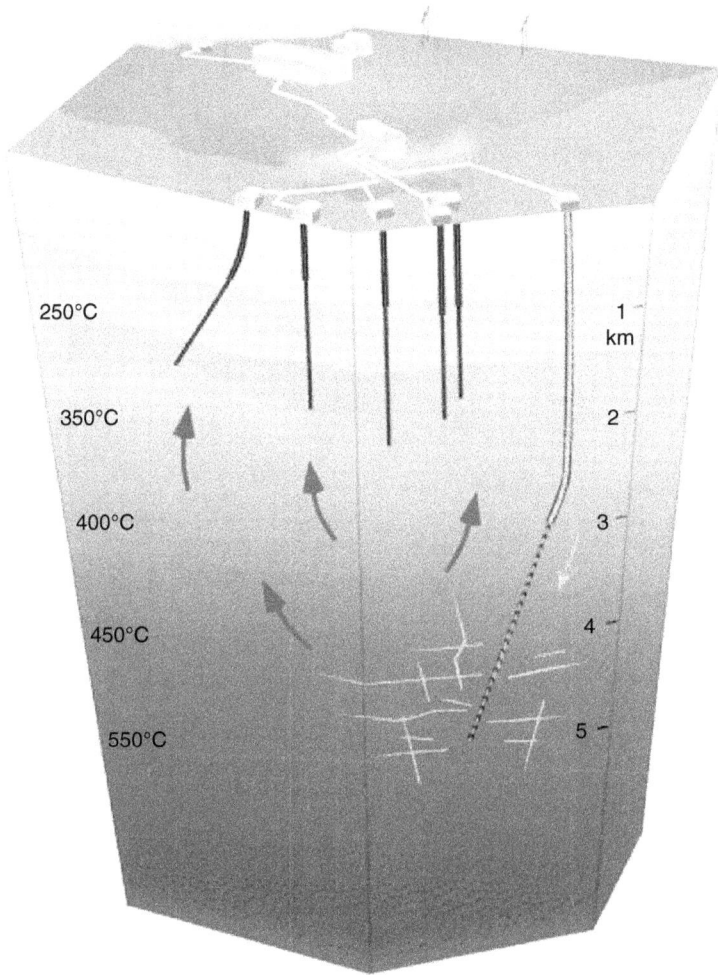

**FIGURE 4.13** Conceptual model of the Reykjanes geothermal field showing the existing conventional geothermal wells (solid lines) and the IDDP-2 well (hollow and dotted lines) [172].

completed well to stimulate fracture permeability etc. The ultimate objective of the DEEPEGS project in Iceland is to deliver steam for electrical power generation [177]. The main goal of the Krafla Magma Testbed project was to closely observe, sample, and manipulate the transition zone from host rock to magma in order to test the concept of directly harnessing magmatic systems (e.g., Chu et al. [178]). An improved understanding of the roots of geothermal systems gained from dedicated research wells will be used to explore the potential for direct energy extraction from magma. Knowledge gained from the project will also help to establish a holistic model of volcanic systems and hence allow more reliable eruption forecasts for populated regions worldwide [179]. The Japan Beyond Brittle project (JBBP) was initiated to investigate the feasibility of creating enhanced geothermal systems in the brittle–ductile

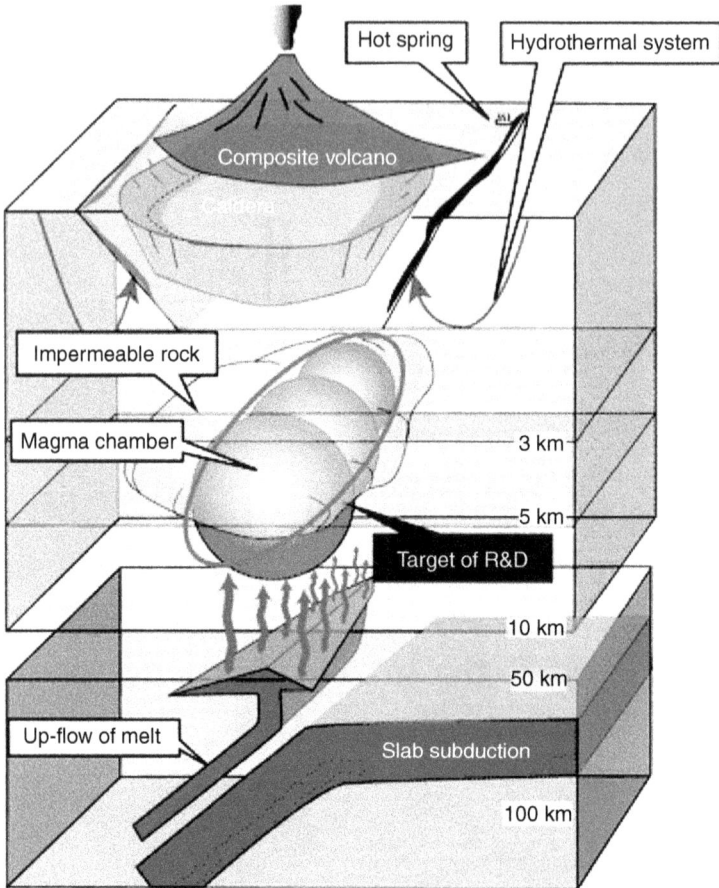

**FIGURE 4.14** Conceptual model of the JBBP project for supercritical geothermal systems in northern Honshu, Japan. These systems lie above shallow magma chambers that are associated with Miocene and younger caldera complexes in the Tohoku region [172].

transition zone [180–182]. Anticipated advantages of such a system include a potentially very large geothermal energy resource that could result in economic energy extraction, simpler reservoir design and control, reduced parasitic fluid losses, and reduced induced seismicity. Tohoku area of northern Honshu in Japan has been identified as a promising target for this effort because the data from geophysical surveys suggest the presence of shallow magma chambers that would provide a widespread source of heat (Figure 4.14) [183]. Studies of other young uplifted and exhumed plutons in Japan (e.g., Bando et al. [184]) support the idea that supercritical conditions of 400°C–500°C can be found at depths of 3–5 km in association with cooling and fractured young magmatic intrusions. Current work is focused on identifying a field site where a deep well can be drilled into such a supercritical system.

One of the objectives of the DESCRAMBLE project (Drilling in dEep, Super-CRitical AMBient of continental Europe), funded by the European Commission was to deepen the existing Venelle-2 well in the Larderello geothermal field in Italy from 2.2 km (and 350°C) to a depth of 3–3.5 km. The goal was to characterize and test the deep high-temperature resource below the currently exploited reservoir horizons (which is expected to have a temperature of ≈450°C, e.g., Büsing et al. [186]; Liotta and Ranalli [187]; Stamnes et al. [188]). The main research objectives of DESCRAMBLE were to improve drilling and monitoring methods and to develop better ways to physically and chemically characterize deep crustal fluids and rocks [177]. Research efforts in New Zealand have included a study of the deep (5–7 km) geothermal resources in the Taupo Volcanic Zone, which are estimated to have temperatures >400°C and a potential of 10 GWe [189,190]. Newman et al. [191] used full tensor 3D MT modeling to provide evidence of deep-seated electrically conductive plumes to 10 km depth. Bannister et al. [192] conducted a passive seismic broadband survey of the region to elucidate changes in crustal velocity structure between 3 and 8 km depth. One of the goals of these surveys was to develop an integrated image of the brittle–ductile transition zone and identify potential deep drilling targets. Additional work has been conducted on the geochemical links between geothermal and magmatic fluids in this region [193]. GEMex, a joint geothermal research project launched by Europe and Mexico aimed to assess two unconventional geothermal sites in Mexico: EGS development at Acoculco and a super-hot resource in Los Humeros. The project's objectives include increasing knowledge of these two particular sites as well as using advanced techniques to identify deep structures to reduce drilling risks, developing stimulation methods to increase permeability, and carefully evaluating social and environmental risks associated with the development of these types of resources. In a second stage, drilling into the explored reservoir is anticipated [194,195]. Finally, The Newberry Deep Drilling project (NDDP) aimed at drilling into the brittle–ductile transition zone at the Newberry Volcano, central Oregon state, USA. The main focus of the project was to study the heat and mass transfer in the crust in view of natural hazards and geothermal energy resource utilization. The Newberry Volcano contains one of the largest geothermal reservoirs in the western United States with a conductive thermal anomaly at ≈320°C at 3,000 m depths [196]. It is assumed that at depths <5,000 m very high temperatures can be reached [197].

There has been extensive work on conceptual models and numerical simulations of supercritical geothermal systems. Fournier [160] developed a comprehensive conceptual model of the link between magmatic and hydrothermal systems that includes sharp thermal and fluid-pressure gradients across the brittle–ductile transition near the supercritical temperature. Early work focused on developing simulators that could handle temperatures ranging from those found in hydrothermal systems up to those of magmatic systems (e.g., Hayba and Ingebritsen [198]). Ingebritsen et al. [199] provided an overview of numerical modeling of magmatic hydrothermal systems, looking at the transfer of heat and metals from magma bodies into the overlying crust through fluid–rock interaction. Yano and Ishido [200] simulated flow to a well from a supercritical reservoir and noted that complex behavior might be expected due to nonlinear changes in compressibility and fluid viscosity. Norton and Dutrow [201] suggested that magma–hydrothermal processes should be thought of as complex dynamical systems whose behavior near the supercritical region is likely chaotic. Fournier [161] suggested that zones of increased permeability might be episodic features associated with elevated strain rates. Watanabe et al. [202] noted that the high heat capacity of supercritical fluids allows for more effective heat mining from high-temperature rocks.

More recent works which include works of Croucher and O'Sullivan [203], O'Sullivan et al. [204,205], Magnusdottir and Finsterle [206] and Gunnarsson and Aradóttir [207] have focused on regions with supercritical conditions, which pose challenges for many conventional reservoir model simulators due to rapid changes in the physical properties of water. Weis and Driesner [208] noted many similarities between supercritical geothermal systems and porphyry copper systems. Driesner et al. [209] describe a modular numerical simulator platform capable of simulating fluid flow and heat transfer while also capturing geo-mechanical and geochemical processes. Scott et al. [210,211] applied this modeling approach to evaluate processes in the supercritical zone in the IDDP-1 well. Scott et al. [212] modeled the temporal evolution of high enthalpy geothermal systems associated with shallow intrusions and observed that host rock permeability and composition, intrusion depth, intrusion geometry, and strain rate all play important roles in the thermal structure. Scott et al. [213] noted that, for saline hydrothermal systems, the depth of the magmatic intrusions powering geothermal systems impacts the efficiency of heat transfer. Additional modeling work by Carcione and Poletto [214] and Carcione et al. [215,216] have focused on developing improved geophysical models to capture the changing physical properties associated with the brittle–ductile transition (BDT) and supercritical conditions. Farina et al. [217] applied this algorithm to simulate full waveforms, demonstrating that discontinuities associated with the transition to supercritical conditions and the presence of magmas can be seismically observable. These tools can be used for seismic characterization in conjunction with passive seismic, exploration seismic, and seismic-while-drilling methods [218,219].

A critical challenge confronting the commercial utilization of supercritical geothermal systems is the need for improved drilling systems, well completion methods, power plants, logging tools, and characterization methods that can withstand high temperatures and aggressive fluids. Several European Commission funded projects have addressed and are currently addressing these issues. These projects include The HiTI project [220,221], the development and testing of a high-temperature distributed temperature sensing (DTS) cable [174,175], the MultiSensor memory tool that records temperature, pressure, fluid flow and casing collar locations, high-temperature borehole televiewer and resistivity logging tools, and new Na/Li geothermometers [221,222], and high-temperature tracers [223,224]. The DESCRAMBLE project developed a slick-line temperature and pressure logging tool by SINTEF that can withstand downhole conditions of 450°C and 450 bar for up to 8 hours. The tool helped characterizing supercritical systems [225], and these methods have been employed at the IDDP sites in Iceland. One approach involves adaption the seismic-while-drilling method to geothermal systems [218,219]. In addition, new laboratory set-ups were developed to investigate rock properties at supercritical conditions (e.g., Kummerow and Raab [226,227]). The objective of the GeoWell project is to develop reliable, cost-effective, and environmentally safe well completion and monitoring technologies to accelerate the development of geothermal resources for power generation in Europe and worldwide. In order to further promote the successful development of supercritical geothermal systems in future, three main areas of future international collaboration: (a) data sharing; (b) coupled process modeling; and (c) underground field laboratories are needed.

## 4.7  SUPERCRITICAL $CO_2$ CYCLE

Supercritical $CO_2$ power cycles can be used in recovery of geothermal power. In existing geothermal plants, a brine solution is pumped deep underground where it extracts heat from the surrounding rock and is heated up to around 100°C–200°C. The hot brine is then returned to the surface where it is converted to electricity by a power block, which is typically a closed-loop organic Rankine cycle (ORC) with indirect heating between the hot brine and the organic fluid. In principle, $sCO_2$ power cycles could replace the ORC system [228,229]. Another option is the operation of a closed-loop direct cycle where the brine is replaced with $sCO_2$ and the hot, high-pressure $CO_2$ leaving the well is expanded within a turbine. This direct cycle can lead to more effective geothermal extraction because the power required to drive the cycle is reduced due to a stronger thermosiphon effect [230].

Adams et al. [231] investigated the use of $CO_2$ as the heat-carrier fluid. Subsequently, Adams et al. [229] demonstrated that direct $sCO_2$ cycles produced more power at lower reservoir depths, significantly more power at higher reservoir depths, and that for indirect systems a trans-critical $sCO_2$ power block could outperform an ORC system operating with the refrigerant R245fa. Green Fire Energy company is developing the ECO2G, which is a direct closed-loop $sCO_2$ system for geothermal applications [232]. Recently, they demonstrated the technology at the Coso Geothermal Field. In this study [233] emphasis was placed on testing the heat exchanger technology operating with $sCO_2$, and hence an expansion valve was used in place of the turbine. Based on thermo-economic study of closed-loop direct s-$CO_2$ cycle, Glos et al. [230] suggested that LCOEs of 0.20 and 0.12 $/kWh could be obtained for 52 and 157 MWe systems respectively for installation at brownfield sites where there are existing wells.

Geothermal sources have a much lower temperature than nuclear reactors and coal-fired power plants. Generally, the $tCO_2$ Rankine cycle is more suitable for low-grade energies. A relatively low condensation temperature is required to ensure the operation of a $tCO_2$ power system. However, it may be difficult to condense $CO_2$ if the ambient temperature is high. Few investigations have evaluated the feasibility of the $sCO_2$ cycle for the utilization of geothermal energy. Ruiz-Casanova et al. [234] compared four different configurations and reported that intercooling could reduce the mass flow of $CO_2$ and thus decline the compression work. Regeneration slightly increased the mass flow rate of $CO_2$, resulting in an increase in the power output. Among the four configurations, the performance of the regenerative $sCO_2$ cycle with intercooling was the best, with optimal thermal and exergy efficiencies of 11.51% and 52.49%, respectively.

For high-temperature applications, special attention should be paid to the corrosivity of $CO_2$ and impurities in the system materials. Some materials, such as chromium oxide and alumina can form a protective oxide film on the surface and have good compatibility with $CO_2$. Haynes 230 [235] or 617 alloy [236] can be used for high thermal fatigue resistance. Several MW-level $CO_2$ experimental systems, which concentrated on simple and recuperated configurations, have been built. More experimental systems need to be developed to validate the theoretical results. Furthermore, during the development of experimental system, technical issues can be discovered, and expertise can be gained aiding future engineering development.

Compared with the conventional steam Rankine cycle, the $sCO_2$ cycle has low critical pressure, high density, high heat transfer rate, high specific power, and small size [237], thereby making it suitable for various heat sources. Besides geothermal systems, s-$CO_2$ cycle can be used for numerous other applications. Table 4.4 shows the operating conditions of the $sCO_2$ cycles for these applications along with their use for geothermal energy. The table articulates various advantages and disadvantages of each application.

### 4.7.1 $CO_2$ Plume Geothermal System (CPG)

The CPG system described McDonnell et al. [185] and Randolph and Saar [239] involve the pumping of supercritical $CO_2$ into natural high porosity and permeability reservoirs. The patent, on which the concept is based, includes the technology and the use of supercritical $CO_2$ for energy generation purposes in the form of electricity or for the purpose of exchanging heat [240,241]. The injected $CO_2$ is heated by the geothermal heat flux from Earth's interior. At the production side of the system the $CO_2$ is utilized either for electricity production through a turbine connected to a generator, or in a binary cycle through a heat exchanger to provide energy for electricity generation (binary system), and/or direct use of the thermal energy in district heating systems (Figure 4.15). In either case, the heated carbon dioxide will increase its enthalpy which can be converted through a mechanical system into power or energy. In return carbon dioxide is stored and sequestered for long term in the reservoir.

The possibility of a geothermal system coupled with $CO_2$ sequestration has been investigated previously [242]. The study has determined that such a system can provide three times greater heat extraction rates compared to conventional water-based systems [239]. In a steam turbine high-pressure high temperature steam is used to power the system. The higher the pressure and temperature the more energy is stored in the steam. At a supercritical state, the fluid has both gaseous and liquid-like characteristics. In the case of water, the triggering pressure and temperature are 22.06 MPa and 373.95°C [243]. However, $CO_2$ ($P_{cr} = 7.37$ MPa, $T_{cr} = 30.97$°C) has a much lower pressure/temperature ratio. A supercritical $CO_2$ power cycle was used to extract energy from a high-pressure low enthalpy reservoir [244]. A numerical simulation was used to assess the available thermal energy. The details of the simulation technique are summarized in [185,239,245]. The model [239] was constructed as a reference for a five-spot well configuration with 1 km² map view area and 707 m between the injection and the production wells. For the model, the reservoir thickness was set at 305 m. For the modelled scenario, the heat extraction rate was calculated as 47.0 MW, averaged over a 25-year period. The amount of sequestered $CO_2$ was calculated according to the occupiable space in the reservoir. Secondly, a natural 5% fluid loss was expected using $CO_2$ as the working fluid in tight reservoirs. Randolph and Saar [239] calculated this with a 7% loss based on the assumption that fluid loss is higher in naturally permeable formations [246]. More details on model and model calculations are given by McDonnell et al. [185] and Randolph and Saar [239].

**TABLE 4.4**

**Comparison of Operating Conditions for s-CO$_2$ Cycle for Various Heat Sources [238]**

| Application | Operating Conditions Temperature | Pressure (MPa) | Cycle Configuration | Advantages | Challenges |
|---|---|---|---|---|---|
| Nuclear reactor | 300°C–600°C | 12–25 | Recompression cycle, double recompression cycle | High thermal efficiency, compact size, simple system layout, intercooling and reheating can improve performance | High operating pressure, system safety, material corrosion |
| Coal-fired power plant | 600°C–700°C | 20–35 | Recompression cycle with reheating and intercooling, partial-split arrangement of the heater, cascade system consisting of top and bottom cycles | High thermal efficiency, compact size, air cooling is possible, integration with carbon capture system (CCS) | Complicated layout, high operating temperature, turbine manufacturing |
| Waste heat recovery | 320°C–570°C | 15–25 | Recompression cycle, recompression cycle with split expansion, recuperated cycle, split recuperated cycle | Compact size, environmentally friendly, synergic energy utilization with multi-heat sources | High cost, small-scale turbine design and manufacturing |
| Solar power system | 450°C–900°C | 25–30 | Recompression cycle, recuperated cycle, recompression with reheating and intercooling | Environmentally friendly, integration with thermal storage, air cooling is possible | Robustness under dynamic working conditions, material corrosion under high-temperature conditions |
| geothermal power system | 60°C–145°C | 12–16 | Recuperated cycle, recuperated cycle with intercooling | Environmentally friendly, suitable for low-temperature heat source | Sensitive to ambient temperature, state at the turbine inlet close to the critical point |

expansion device
(turbine)

generator

to district space/water
heating (optional)

cold cooling cold
air device air
(optional)

compressor

$CO_2$
from
emitter

cold
$CO_2$

$CO_2$
from
emitter

heat exchanger
(optional)

> 1 km

caprock / trap (low permeability)

geothermal reservoir
(high porosity and permeability)

$CO_2$ plume

salty $H_2O$,
hydrocarbons,
or other

permanent
$CO_2$ storage

$CO_2$

**geothermal heat flow**

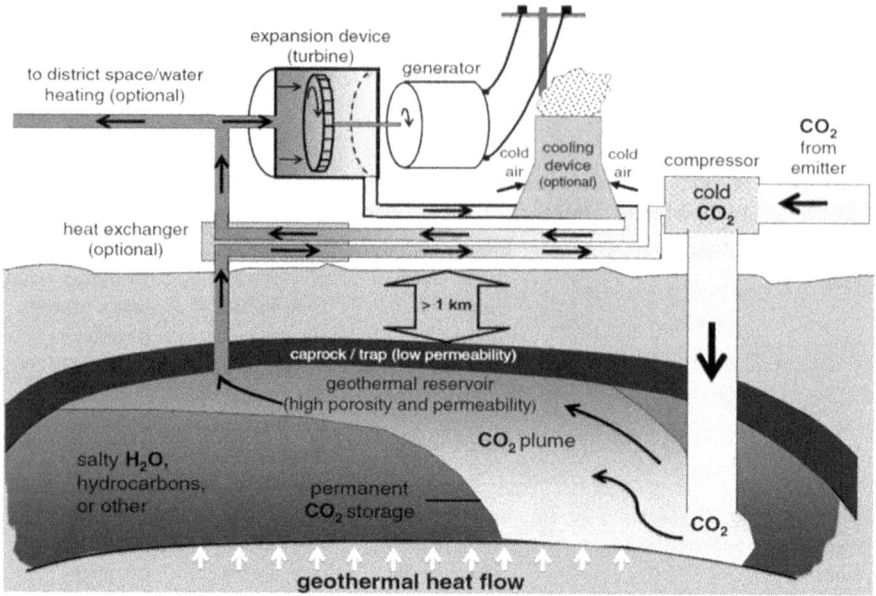

**FIGURE 4.15**  Schematic representation of the $CO_2$ plume geothermal (CPG) system show-ing the possibility of $CO_2$ sequestration and scenarios with optional elements for electricity production, or district heating systems. The electricity generator is driven by a turbine main-tained by the flow of high pressure vaporized $CO_2$ gas. The waste heat can be further utilized through a heat exchanger for district heating or cooling purposes [185].

McDonnell et al. [185] used Germany as a case study for this concept. Germany is the most energy-intensive country, and it also has an untapped potential for geother-mal energy in the northern as well as the western regions. The $CO_2$ plume geother-mal system using supercritical carbon dioxide as the working fluid can be utilized in natural high porosity (10%–20%) and permeability ($2.5 \times 10^{-14}$ to $8.4 \times 10^{-16} m^2$) reservoirs with temperatures as low as 65.8°C. The feasibility of the project was assessed based on market conditions and policy support in Germany as well as the geologic background of sandstone reservoirs near industrialized areas (Dortmund, Frankfurt) and the possibility of carbon capture integration and $CO_2$ injection. The levelized cost of electricity for a base case resulted in € 0.060/kWh. Optimal system type was assessed in a system optimization model. The project had a potential to sup-ply 6,600/12,000 households with clean energy (electricity/heat) and sequester car-bon dioxide at the same time. The concept thus offers some commercial possibilities.

## 4.8  ROLE OF NANO TECHNOLOGY ON GEOTHERMAL POWER GENERATION

Geothermal energy generation is also enhanced by nanotechnology. In conventional geothermal energy production, cold fluids are injected into naturally heated hot rocks usually found over 1,500 m below the earth surface. The heated fluid is then extracted and used to generate electricity. Nanotechnology is now helping to make geothermal energy more practical by allowing efficient energy production closer to the surface and at lower temperatures. The heat-retaining properties of the fluid are also being enhanced with nanoparticles. Still on energy generation, it has been demonstrated that sunlight, concentrated on nanoparticles, can produce steam with high energy efficiency. The stem can be used to run power plants in developing countries. Sheets of nanotubes have also been used to build "thermocells" that generate electricity when the sides of the cell are at different temperatures. By wrapping these nanotube sheets around hot pipes such as the exhaust pipe of cars, electricity can be generated from heat that is usually wasted.

Geothermal power is renewable and produces almost no pollution. Unlike solar and wind power, it provides steady base-load power. But in conventional geothermal power production, hot rock needs to heat water to 300°F [149°C] or hotter. That hot rock might be found 5,000 feet beneath the ground in a few places, but typically it's much deeper. Nanomaterials may help make geothermal more practical by allowing efficient energy production closer to the surface at lower temperatures. Some metal-organic nano materials can absorb 30% of their weight in organic compounds. This could help drive turbines with organic compounds at lower temperatures. The nano-material coating can prevent the corrosive materials from corrosion. They decrease the frictional, tear and wear losses thus they increase the efficiency of the system. They are extremely light weighted but are harder than materials like steel, iron etc. Plenty of raw materials are available for manufacture and they are durable for a number of years. Nanotechnology is, however, very expensive and its production cost is high and the impact of nanotechnology on the environment is still unknown.

Heat transfer processes play a key role in the efficiency of geothermal energy systems [247]. Enhancement in thermal performance of thermal devices applied in these systems leads to improved efficiency of the overall system. Applying nanofluids as heat transfer fluids is an attractive idea to achieve higher efficiencies [248]. In addition to improvement in heat transfer, using nanofluids can reduce the size of systems. Diglio et al. [249] investigated the effect of using nanofluid as a heat carrier instead of conventional fluid, mixture of water and ethylene glycol, on a borehole heat exchanger. In this study, various nano particles such as graphite, alumina, aluminum, etc were used in low concentrations. The volumetric concentrations of nanofluids were between 0.1% and 1%. It was observed that applying nanofluids can result in reduction in borehole thermal resistance. Since the type of nano structures influences the thermophysical specifications [250], the results were dependent on the kind of nanofluids. In addition, it was observed that using Ag nanofluids resulted in the highest convective heat transfer and pressure drop. By considering both heat transfer and pressure drop, it was concluded that using Cu-based nanofluids led to the most reduction in the length of borehole heat exchanger.

In another study carried out by Deneshipour et al. [251], two nanofluids including $Al_2O_3$/water and CuO/water were used in geothermal borehole heat exchanger as circuit fluids. The volumetric concentration varied between 0% and 6% to evaluate the impact of concentration. Results of the study revealed that by using CuO/water nanofluid, higher heat extraction was achievable compared with $Al_2O_3$/water nanofluid; however, the pressure loss was higher in the case of CuO/water nanofluid.

Convective heat transfer coefficient increased with concentration of nano particles. Jamshidi et al. [252] used $Al_2O_3$ nano particles in volumetric concentrations between 0% and 0.5%. The results showed that increase in the concentration resulted in improved heat flux. Smaller nano particles gave better heat transfer. The study also showed that nanofluid with 0.5 vol% conc. can result in approximately 18% increase in energy obtained from the earth.

The extracted heat from the earth in geothermal systems depends on numerous factors such as working fluid, its flow rate and well specification. In a study carried out by Sui et al. [253], $Al_2O_3$/water nanofluid was employed as working fluid in a geothermal system. The results showed that an increase in flow rate led to lower temperature of returning flow and an increase in extracted heat. The use of nanofluid resulted in increased temperature of returning fluid. At low mass flow rate, the effect of nanofluid on extracted heat was not significant, while the increase in mass flow rate had a significant effect.

While nanoparticles improve heat transfer, their sedimentation at the bottom of the borehole is of concern. The study of Sun et al. [254] indicates that high velocity of fluid can reduce sedimentation. Using nanofluids in heat exchangers and boreholes can reduce the size of structures and improve the efficiency of the systems. In general, Nanotechnology can improve mechanical properties and heat transfer capacity in many parts of geothermal power production [255]. More work is needed to find optimum nanofluid systems.

## REFERENCES

1. *GeoVision-Harnessing the Heat Beneath Our Feet*. DOE report, DOE/EE-1306 Washington, DC: US Department of Energy; 2019, May.
2. Roberts D. @drvolts Geothermal energy is poised for a big breakout "An engineering problem that, when solved, solves energy." 2020, October 21. https://www.vox.com›2020/10/21›renewable-energy.
3. Carlos de Oliveira Matias J., Godina R., and Pouresmaeil E. Sustainable energy systems: Optimization and efficiency. *Appl. Sci.* 2020;10:4405.
4. Amelio M., and Morrone P. *Residential Cogeneration and Trigeneration. In Current Trends and Future Developments on (Bio-) Membrane.* Amsterdam: Elsevier; 2020, pp. 141–175. ISBN 978-0-12-817807-2.
5. Algieri A., Andiloro S., Tamburino V., and Zema D.A. The potential of agricultural residues for energy production in Calabria (Southern Italy). *Renew Sustain Energ Rev* 2019;104:1–14.
6. Ahmadi A., El Haj Assad M., Jamali D.H., Kumar R., Li Z.X., Salameh T., Al-Shabi M., and Ehyaei M.A. Applications of geothermal organic Rankine cycle for electricity production. *J Clean Prod* 2020;274:122950.
7. Joci'c N., Müller J., Požar T., and Bertermann D. Renewable energy sources in a post-socialist transitional environment: The influence of social geographic factors on potential utilization of very shallow geothermal energy within heating systems in Small Serbian Town of Ub. *Appl Sci* 2020;10:2739.
8. Yu G., and Yu Z. Research on a coupled total-flow and single-flash (TF-SF) system for power and freshwater generation from geothermal source. *Appl Sci* 2020;10:2689.
9. Bianchi M., Branchini L., Pascale A.D., Melino F., Ottaviano S., Peretto A., Torricelli N., and Zampieri G. Performance and operation of micro-ORC energy system using geothermal heat source. *Energ Procedia* 2018;148:384–391.

10. Anderson A., and Rezaie B. Geothermal technology: Trends and potential role in a sustainable future. *Appl Energ* 2019;248:18–34.
11. Zarrouk S.J., and Moon H. Efficiency of geothermal power plants: A worldwide review. *Geothermics* 2014;51:142–153.
12. Bertani R. Geothermal power generation in the world 2010–2014 update report. *Geothermics* 2016;60:31–43.
13. Algieri A. Energy exploitation of high-temperature geothermal sources in volcanic areas—A possible ORC application in phlegraean fields (Southern Italy). *Energies* 2018;11:618.
14. Nami H., and Anvari-Moghaddam A. Geothermal driven micro-CCHP for domestic application—Exergy, economic and sustainability analysis. *Energy* 2020;207:118195.
15. Ehyaei M.A., Ahmadi A., El Haj Assad M., and Rosen M.A. Investigation of an integrated system combining an organic Rankine cycle and absorption chiller driven by geothermal energy: Energy, exergy, and economic analyses and optimization. *J Clean Prod* 2020;258:120780.
16. Li P., Han Z., Jia X.; Mei Z., Han X., and Wang Z. An improved analysis method for Organic Rankine Cycles based on radial-inflow turbine efficiency prediction. *Appl Sci* 2018;9:49.
17. Gutierrez J.C., Valencia Ochoa G., and Duarte-Forero J. Regenerative organic Rankine cycle as bottoming cycle of an industrial gas engine: Traditional and advanced exergetic analysis. *Appl Sci* 2020;10:4411.
18. Alshammari F., Karvountzis-Kontakiotis A., Pesiridis A., and Minton T. Radial expander design for an engine organic Rankine cycle waste heat recovery system. *Energ Procedia* 2017;129:285–292.
19. Bellos E., and Tzivanidis C. Investigation of a hybrid ORC driven by waste heat and solar energy. *Energ Convers Manag* 2018;156:427–439.
20. Peris B., Navarro-Esbrí J., Molés F., Martí J.P., and Mota-Babiloni A. Experimental characterization of an organic Rankine cycle (ORC) for micro-scale CHP applications. *Appl Therm Eng* 2015;79:1–8.
21. Flores R.A., Aviña Jiménez H.M., González E.P., and González Uribe L.A. Aerothermodynamic design of 10 kW radial inflow turbine for an organic flashing cycle using low-enthalpy resources. *J Clean Prod* 2020;251:119713.
22. Kaczmarczyk M., Tomaszewska B., and Operacz A. Sustainable utilization of low enthalpy geothermal resources to electricity generation through a cascade system. *Energies* 2020;13:2495.
23. Pili R., Eyerer S., Dawo F., Wieland C., and Spliethoff H. Development of a non-linear state estimator for advanced control of an ORC test rig for geothermal application. *Renew Energ* 2020;161:676–690.
24. Dong L., Liu H., and Riffat S. Development of small-scale and micro-scale biomass-fuelled CHP systems—A literature review. *Appl Therm Eng* 2009;29:2119–2126.
25. Morrone P., Algieri A., and Castiglione T. Hybridisation of biomass and concentrated solar power systems in transcritical organic Rankine cycles: A micro combined heat and power application. *Energ Convers Manag* 2019;180:757–768.
26. Algieri A., and Morrone P. Techno-economic analysis of biomass-fired ORC systems for single-family combined heat and power (CHP) applications. *Energ Procedia* 2014;45:1285–1294.
27. Eyerer S., Dawo F., Wieland C., and Spliethoff H. Advanced ORC architecture for geothermal combined heat and power generation. *Energy* 2020;205:117967.
28. Drescher U., and Brüggemann D. Fluid selection for the organic Rankine cycle (ORC) in biomass power and heat plants. *Appl Therm Eng.* 2007;27:223–228.
29. Macchi E., and Astolfi M. *Organic Rankine Cycle (ORC) Power Systems: Technologies and Applications.* Cambridge: WoodheadPublishing; 2016. ISBN 978-0-08-100510-1.

30. Rivera Diaz A., Kaya E., and Zarrouk S.J. Reinjection in geothermal fields—A worldwide review update. *Renew Sustain Energ Rev* 2016;53:105–162.
31. El Haj Assad M., Bani-Hani E., and Khalil M. Performance of geothermal power plants (single, dual, and binary) to compensate for LHC-CERN power consumption: Comparative study. *Geotherm Energ* 2017;5:17.
32. Lee I., Tester J.W., and You F. Systems analysis, design, and optimization of geothermal energy systems for power production and polygeneration: State-of-the-art and future challenges. *Renew Sustain Energ Rev.* 2019;109:551–577.
33. Heberle F., and Brüggemann D. Thermoeconomic analysis of hybrid power plant concepts for geothermal combined heat and power generation. *Energies* 2014;7:4482–4497.
34. Lecompte S., Huisseune H., Van den Broek M., De Schampheleire S., and De Paepe M. Part load based thermo-economic optimization of the organic Rankine cycle (ORC) applied to a combined heat and power (CHP) system. *Appl Energ* 2013;111:871–881.
35. Eller T., Heberle F., and Brüggemann D. Transient simulation of geothermal combined heat and power generation for a resilient energetic and economic evaluation. *Energies* 2019;12:894.
36. Hettiarachchi H.M., Golubovic M., Worek W.M., and Ikegami Y. Optimum design criteria for an organic Rankine cycle using low-temperature geothermal heat sources. *Energy* 2007;32:1698–1706. Doi: 10.1016/j.energy.2007.01.005.
37. Geothermal power. Wikipedia, the free encyclopedia, last edited 3 October, 2022. https://en.wikipedia.org/wiki/Geothermal_power
38. Calise F., Cappiello F.L., Dentice d'Accadia M., and Vicidomini M. Thermo-economic analysis of hybrid solar-geothermal polygeneration plants in different configurations. *Energies* 2020;13:2391.
39. Lecompte S., Ntavou E., Tchanche B., Kosmadakis G., Pillai A., Manolakos D., and De Paepe M. Review of experimental research on supercritical and transcritical thermodynamic cycles designed for heat recovery application. *Appl Sci* 2019;9:2571.
40. Valdimarsson P. Geothermal power plant cycles and main components. *2011, United Nation University, Geothermal Training Program Organized by UNP-GTP and LaGeo*, in Santa Tecla, El Salvador; 16–22 January, 2011, pp. 1–24.
41. Helston C. Low temperature geothermal power- Energy British Columbia. 2012, May. A website report http://www.energybc.ca/lowtempgeo.html.
42. Hijriawan M. Pambudi N.A., Biddinika M.K., Wijayanto D.S., Kuncoro I.W., Rudiyanto B. and Wibowo K.M. Organic Rankine cycle (ORC) in geothermal power plants. *J Phys Conf Ser* 2019;1402:044064.
43. Saadon S., and Islam S.M.S. A recent review in performance of organic Rankine cycle (ORC), *Intech Open* 2019. Doi: 10.5772/intechopen.89763.
44. Zhai H., Shi L., and An Q. Influence of working fluid properties on system performance and screen evaluation indicators for geothermal ORC (organic Rankine cycle) system. *Energy* 2014;74:2–11. Doi: 10.1016/j.energy.2013.12.030.
45. Heberle F., Preißinger M., and Brüggemann D. Zeotropic mixtures as working fluids in organic Rankine cycles for low-enthalpy geothermal resources. *Renew Energ* 2012;37:364–370. Doi: 10.1016/j.renene.2011.06.044.
46. Oyewunmi O.A., and Markides C.N. Thermo-economic and heat transfer optimization of working-fluid mixtures in a low-temperature organic Rankine cycle system. *Energies* 2016;9:448, Doi: 10.3390/en9060448.
47. Imran M., Usman M., Park B.S., Kim H.J., and Lee D.H. Multi-objective optimization of evaporator of organic Rankine cycle (ORC) for low temperature geothermal heat source. *Appl Therm Eng* 2015;80:1–9. Doi: 10.1016/j.applthermaleng.2015.01.034.
48. Sauret E., and Rowlands A.S. Candidate radial-inflow turbines and high-density working fluids for geothermal power systems. *Energy* 2011;36:4460–4467. Doi: 10.1016/j.energy.2011.03.076.

49. Walraven D., Laenen B., and D'haeseleer W. Minimizing the levelized cost of electricity production from low-temperature geothermal heat sources with ORCs: Water or air cooled? *Appl Energ* 2015;142:144–153. Doi: 10.1016/j.apenergy.2014.12.078.

50. Roy J.P., Mishra M.K., and Misra A. Performance analysis of an organic Rankine cycle with superheating under different heat source temperature conditions. *Appl Energ* 2011;88:2995–3004. Doi: 10.1016/j.apenergy.2011.02.042.

51. Algieri A., and Morrone P. Comparative energetic analysis of high-temperature subcritical and transcritical organic Rankine cycle (ORC). A biomass application in the Sibari district. *Appl Therm Eng* 2012;36:236–244. Doi: 10.1016/j.applthermaleng.2011.12.021.

52. Ventura C.A., and Rowlands A.S. Recuperated power cycle analysis model: Investigation and optimisation of low-to-moderate resource temperature organic Rankine cycles. *Energy* 2015;93:484–494. Doi: 10.1016/j.energy.2015.09.055.

53. Astolfi M., Romano M.C., Bombarda P., and Macchi E. Binary ORC (organic Rankine cycles) power plants for the exploitation of medium–low temperature geothermal sources–Part A: Thermodynamic optimization. *Energy* 2014a;66:423–434. Doi: 10.1016/j.energy.2013.11.056.

54. Astolfi M., Romano M.C., Bombarda P., and Macchi E. Binary ORC (organic Rankine cycles) power plants for the exploitation of medium–low temperature geothermal sources–Part B: Techno-economic optimization. *Energy* 2014b;66:435–446. Doi: 10.1016/j.energy.2013.11.057.

55. Oyewunmi O.A., Ferré-Serres S., Lecompte S., van den Broek M., De Paepe M., and Markides C.N. An assessment of subcritical and trans-critical organic Rankine cycles for waste-heat recovery. *Energ Procedia* 2017;105:1870–1876. Doi: 10.1016/j.egypro.2017.03.548.

56. Lecompte S., Lemmens S., Huisseune H., Van den Broek M., and De Paepe M. Multi-objective thermo-economic optimization strategy for ORCs applied to subcritical and transcritical cycles for waste heat recovery. *Energies* 2015;8:2714–2741. Doi: 10.3390/en8042714.

57. Zare V. A comparative exergoeconomic analysis of different ORC configurations for binary geothermal power plants. *Energ Convers Manage* 2015;105:127–138. Doi: 10.1016/j.enconman.2015.07.073.

58. Zhang S., Wang H., and Guo T. Performance comparison and parametric optimization of subcritical organic Rankine cycle (ORC) and transcritical power cycle system for low-temperature geothermal power generation. *Appl Energ* 2011;88:2740–2754. Doi: 10.1016/j.apenergy.2011.02.034.

59. Vetter C., Wiemer H.J., and Kuhn D. Comparison of sub-and supercritical Organic Rankine Cycles for power generation from low-temperature/low-enthalpy geothermal wells, considering specific net power output and efficiency. *Appl Therm Eng* 2013;51:871–879. Doi: 10.1016/j.applthermaleng.2012.10.042.

60. Song J., Loo P., Teo J., and Markides C.N. Thermo-economic optimization of organic Rankine cycle (ORC) systems for geothermal power generation: A comparative study of system configurations. *Front Energ Res* 2020, February 07. Doi: 10.3389/fenrg.2020.00006.

61. Preißinger M., Heberle F., and Brüggemann D. Advanced organic Rankine cycle for geothermal application. *Int J Low Carb Tech* 2013;8:i62–i68.

62. Morrone P., and Algieri A. Integrated geothermal energy systems for small-scale combined heat and power production: Energy and economic investigation. *Appl Sci* 2020;10:6639. Doi: 10.3390/app10196639 www.mdpi.com/journal/applsci.

63. Li J., Liu Q., Ge Z., Duan Y., and Yang Z. Thermodynamic performance analyses and optimization of subcritical and trans critical organic Rankine cycles using R1234ze€ for 100–200°C heat sources. *Energ Convers Manag* 2017;149:140–154.

64. Gao T., and Liu C. Off-design performances of subcritical and supercritical Organic Rankine Cycles in geothermal power systems under an optimal control strategy. *Energies* 2017;10:1185.
65. Yang M.-H., and Yeh R.-H. Economic performances optimization of the transcritical Rankine cycle systems in geothermal application. *Energ Convers Manag* 2015;95:20–31.
66. Manente G., Toffolo A., Lazzaretto A., and Paci M. An organic Rankine cycle off-design model for the search of the optimal control strategy. *Energy* 2013;58:97–106.
67. Nasruddin N., Dwi Saputra I., Mentari T., Bardow A., Marcelina O., and Berlin S. Exergy, exergoeconomic, and exergoenvironmental optimization of the geothermal binary cycle power plant at Ampallas, West Sulawesi, Indonesia. *Therm Sci Eng Prog* 2020;19:100625.
68. Dumont O., Dickes R., De Rosa M., Douglas R., and Lemort V. Technical and economic optimization of subcritical, wet expansion and transcritical organic Rankine cycle (ORC) systems coupled with a biogas power plant. *Energ Convers Manag* 2018;157:294–306.
69. Garg P., Orosz M.S., and Kumar P. Thermo-economic evaluation of ORCs for various working fluids. *Appl Therm Eng* 2016;109:841–853.
70. Moloney F., Almatrafi E., and Goswami D.Y. Working fluid parametric analysis for regenerative supercritical organic Rankine cycles for medium geothermal reservoir temperatures. *Energ Procedia* 2017;129:599–606.
71. Toffolo A., Lazzaretto A., Manente G., and Paci M. A multi-criteria approach for the optimal selection of working fluid and design parameters in organic Rankine cycle systems. *Appl Energy* 2014;121:219–232.
72. Akbari A.D., and Mahmoudi S.M.S. Thermoeconomic analysis & optimization of the combined supercritical $CO_2$ (carbon dioxide) recompression Brayton/organic Rankine cycle. *Energy* 2014;78:501–512.
73. Heberle F., and Brüggemann D. Exergy based fluid selection for a geothermal organic Rankine cycle for combined heat and power generation. *Appl Therm Eng* 2010;30:1326–1332.
74. Wang X., Levy E.K., Pan C., Romero C.E., Banerjee A., Rubio-Maya C., and Pan L. Working fluid selection for organic Rankine cycle power generation using hot produced supercritical $CO_2$ from a geothermal reservoir. *Appl Therm Eng* 2019;149:1287–1304.
75. Gawlik K., and Hassani V. Advanced binary cycles: Optimum working fluids. *Proceedings of the IECEC-97 Thirty-Second Intersociety Energy Conversion Engineering Conference* (Cat. No.97CH6203), Honolulu, HI; 27 July–1 August, 1997, Vol. 3, pp. 1809–1814.
76. Askari I., Calise F., and Vicidomini M. Design and comparative techno-economic analysis of two solar polygeneration systems applied for electricity, cooling and fresh water production. *Energies* 2019;12:4401.
77. Amber K., Day T., Ratyal N., Kiani A., and Ahmad R. Techno, economic and environmental assessment of a Combined Heat and Power (CHP) system—A case study for a university campus. *Energies* 2018;11:1133.
78. Ghaem Sigarchian S., Malmquist A., and Martin V. Design optimization of a complex polygeneration system for a hospital. *Energies* 2018;11:1071.
79. Chen X., Si Y., Liu C., Chen L., Xue X., Guo Y., and Mei S. The value and optimal sizes of energy storage units in solar-assist cogeneration energy hubs. *Appl Sci* 2020;10:4994.
80. Li D., Xu X., Yu D., Dong M., and Liu H. Rule based coordinated control of domestic combined micro-CHP and energy storage system for optimal daily cost. *Appl Sci* 2017;8:8.
81. Fiaschi D., Lifshitz A., Manfrida G., and Tempesti D. An innovative ORC power plant layout for heat and power generation from medium- to low-temperature geothermal resources. *Energ Convers Manag* 2014;88:883–893.

82. Li T., Zhu J., and Zhang W. Comparative analysis of series and parallel geothermal systems combined power, heat and oil recovery in oilfield. *Appl Therm Eng* 2013;50:1132–1141.

83. Van Erdeweghe S., Van Bael J., Laenen B., and D'haeseleer W. Comparison of series/parallel configuration for a low-T geothermal CHP plant, coupled to thermal networks. *Renew Energ* 2017;111:494–505.

84. Mohammadzadeh Bina S., Jalilinasrabady S., and Fujii H. Energy, economic and environmental (3E) aspects of internal heat exchanger for ORC geothermal power plants. *Energy* 2017;140:1096–1106.

85. Marty F., Serra S., Sochard S., and Reneaume J.M. Exergy analysis and optimization of a combined heat and power geothermal plant. *Energies, MDPI* 2019;12 (6):1175. Doi: 10.3390/en12061175.hal-02102937.

86. Pastor-Martinez E., Rubio-Maya C., Ambriz-Díaz V.M., Belman-Flores J.M., and Pacheco-Ibarra J.J. Energetic and exergetic performance comparison of different polygeneration arrangements utilizing geothermal energy in cascade. *Energ Convers Manag* 2018;168:252–269.

87. El-Emam R.S., and Dincer I. Exergy and exergoeconomic analyses and optimization of geothermal organic Rankine cycle. *Appl Therm Eng.* 2013;59:435–444.

88. Toselli D., Heberle F., and Brüggemann D. Techno-economic analysis of hybrid binary cycles with geothermal energy and biogas waste heat recovery. *Energies* 2019;12:1969.

89. Van Erdeweghe S., Van Bael J., Laenen B., and D'haeseleer W. Design and off-design optimization procedure for low-temperature geothermal organic Rankine cycles. *Appl Energ* 2019;242:716–731.

90. Astolfi M., La Diega L.N., Romano M.C., Merlo U., Filippini S., and Macchi E. Techno-economic optimization of a geothermal ORC with novel "Emeritus" heat rejection units in hot climates. *Renew Energ* 2020;147:2810–2821.

91. Van Erdeweghe S., Van Bael J., Laenen B., and D'haeseleer W. Optimal configuration for a low-temperature geothermal CHP plant based on thermoeconomic optimization. *Energy* 2019;179:323–335.

92. Wieland C., Meinel D., Eyerer S., and Spliethoff H. Innovative CHP concept for ORC and its benefit compared to conventional concepts. *Appl Energ* 2016;183:478–490.

93. Habka M., and Ajib S. Evaluation of mixtures performances in organic Rankine cycle when utilizing the geothermal water with and without cogeneration. *Appl Energ* 2015;154:567–576.

94. Marty F., Serra S., Sochard S., and Reneaume J.-M. Simultaneous optimization of the district heating network topology and the organic Rankine cycle sizing of a geothermal plant. *Energy* 2018;159:1060–1074.

95. Marty F., Serra S., Sochard S., and Reneaume J.-M. Economic optimization of a combined heat and power plant: Heat vs electricity. *Energ Procedia* 2017;116:138–151.

96. Ademe. Fongeosec—Conception d'un Démonstrateur de Centrale Géothermique Haute Enthalpie. Available from: https://www.ademe.fr/fongeosec (accessed: 3 September 2020).

97. Eyerer S., Schiffechner C., Hofbauer S., Bauer W., Wieland C., and Spliethoff H. Combined heat and power from hydrothermal geothermal resources in Germany: An assessment of the potential. *Renew Sustain Energ Rev* 2020;120:109661.

98. Cau G., and Cocco D. Comparison of medium-size concentrating solar power plants based on parabolic trough and linear Fresnel collectors. *Energ Procedia* 2014;45:101–110.

99. Kyriakarakos G., Ntavou E., and Manolakos D. Investigation of the use of low temperature geothermal organic Rankine cycle engine in an autonomous polygeneration microgrid. *Sustainability* 2020;12:475. Doi: 10.3390/su122410475 www.mdpi.com/journal/sustainability.

100. Manolakos D., Kosmadakis G., Ntavou E., and Tchanche B. Test results for characterizing two in-series scroll expanders within a low-temperature ORC unit under partial heat load. *Appl Therm Eng* 2019;163:114389.
101. Ntavou E., Kosmadakis G., Manolakos D., Papadakis G., and Papantonis D. Experimental testing of a small-scale two stage organic Rankine cycle engine operating at low temperature. *Energy* 2017;141:869–879.
102. Kosmadakis G., Mousmoulis G., Manolakos D., Anagnostopoulos I., Papadakis G., and Papantonis D. Development of open-drive scroll expander for an organic Rankine cycle (ORC) engine and first test results. *Energ Procedia* 2017;129:371–378.
103. Kyriakarakos G., Dounis A.I., Rozakis S., Arvanitis K., and Papadakis G. Polygeneration microgrids: A viable solution in remote areas for supplying power, potable water and hydrogen as transportation fuel. *Appl Energ* 2011;88:4517–4526.
104. Kyriakarakos G., Dounis A.I., Arvanitis K., and Papadakis G. A fuzzy logic energy management system for polygeneration microgrids. *Renew Energ* 2012;41:315–327.
105. Kyriakarakos G., Dounis A.I., Arvanitis K., and Papadakis G. A fuzzy cognitive maps–petri nets energy management system for autonomous polygeneration microgrids. *Appl Soft Comput.* 2012;12:3785–3797.
106. Kyriakarakos G., Piromalis D.D., Dounis A.I., Arvanitis K., and Papadakis G. Intelligent demand side energy management system for autonomous polygeneration microgrids. *Appl Energy* 2013;103:39–51.
107. Karavas C.-S., Kyriakarakos G., Arvanitis K.G., and Papadakis G. A multi-agent decentralized energy management system based on distributed intelligence for the design and control of autonomous polygeneration microgrids. *Energ Convers Manag* 2015;103:166–179.
108. Karavas C.-S., Arvanitis K., and Papadakis G. A game theory approach to multi-agent decentralized energy management of autonomous polygeneration microgrids. *Energies* 2017;10:1756.
109. Kofinas P., Dounis A., and Vouros G. Fuzzy Q-learning for multi-agent decentralized energy management in microgrids. *Appl Energ* 2018;219:53–67.
110. Kyriakarakos G., Piromalis D.D., Arvanitis K., Dounis A.I., and Papadakis G. On battery-less autonomous polygeneration microgrids: Investigation of the combined hybrid capacitors/hydrogen alternative. *Energ Convers Manag* 2015;91:405–415.
111. Karavas C.-S., Arvanitis K., Kyriakarakos G., Piromalis D.D., and Papadakis G. A novel autonomous PV powered desalination system based on a DC microgrid concept incorporating short-term energy storage. *Sol Energ.* 2018;159:947–961.
112. Dimitriou E., Mohamed E.S., Kyriakarakos G., and Papadakis G. Experimental investigation of the performance of a reverse osmosis desalination unit under full- and part-load operation. *Desalin Water Treat* 2014;53:3170–3178.
113. Beltran H., Ayuso P., and Pérez E. Lifetime expectancy of Li-ion batteries used for residential solar storage. *Energies* 2020;13:568.
114. Haefliger P., Mathieu-Nolf M., Lociciro S., Ndiaye C., Coly M., Diouf A., Faye A.L. Sow A., Tempowski J., Pronczuk J., and Junior A.P.F. Mass lead intoxication from informal used lead-acid battery recycling in Dakar, Senegal. *Environ Health Perspect* 2009;117:1535–1540.
115. Chandrasekharam D., and Bundschuh J. *Low-Enthalpy Geothermal Resources for Power Generation.* Boca Raton, FL: CRC Press; 2008.
116. Garcia S.I., Garcia R.F., Carril J.C., and Garcia D.I. A review of thermodynamic cycles used in low temperature recovery systems over the last two years. *Renew Sustain Energ Rev* 2018;81:760–767.
117. Kosmadakis G., Manolakos D., and Papadakis G. Experimental investigation of a low-temperature organic Rankine cycle (ORC) engine under variable heat input operating at both subcritical and supercritical conditions. *Appl Therm Eng* 2016;92:1–7.

118. Lu Y., Roskilly A., Yu X., Tang K., Jiang L., Smallbone A., Chen L., and Wang Y. Parametric study for small scale engine coolant and exhaust heat recovery system using different organic Rankine cycle layouts. *Appl Therm Eng* 2017;127:1252–1266.

243. Pioro I., and Mokry S. Thermophysical properties at critical and supercritical pressures. In: Belmiloudi A., editor. *Heat Transfer - Theoretical Analysis, Experimental Investigations and Industrial Systems*. London: IntechOpen Ltd.; 2011, January 28. ISBN 978-953-307-226-5.

119. Ismail B.I. Introductory chapter: Power generation using geothermal low-enthalpy resources and ORC technology. In: Ismail B., editor. *Renewable Geothermal Energy Explorations*. Rijeka, Croatia: IntechOpen; 2019. Doi: 10.5772/intechopen.84390.

120. VanderMeer J.B., and Mueller-Stoffels M. *Wind-Geothermal-Diesel Hybrid Micro-Grid Development: A Technical Assessment for Nome*, AK. Master's Thesis. Oldenburg: University of Oldenburg; 2014.

121. Kaplan U., Sfar R., and Shilon Y. Small scale geothermal power plants with less than 5.0 MW capacity. *Bull Hydrogéol* 1999;17:433–440.

122. Welch P., Boyle P., Giron M., and Sells M. Construction and startup of low temperature geothermal power plants. *Geotherm Resour Council Trans* 2011;35:1351–1356.

123. Schochet D.N. Case histories of small scale geothermal power plants. *Proceedings of the 2000 World Geothermal Congress*, Tohoku, Japan; 28 May–10 June, 2000, pp. 2201–2204.

124. Welzl M., Heberle F., and Brüggemann D. Experimental evaluation of nucleate pool boiling heat transfer correlations for R245fa and R1233zd(E) in ORC applications. *Renew Energ* 2020;147:2855–2864.

125. Manolakos D., Papadakis G., Kyritsis S., and Bouzianas K. Experimental evaluation of an autonomous low-temperature solar Rankine cycle system for reverse osmosis desalination. *Desalination* 2007;203:366–374.

126. Shortall R., Davidsdottir B., and Axelsson G. Geothermal energy for sustainable development: A review of sustainability impacts and assessment frameworks. *Renew Sustain Energ Rev* 2015;44:391–406.

127. Finster M., Clark C., Schroeder J., and Martino L. Geothermal produced fluids: Characteristics, treatment technologies, and management options. *Renew Sustain Energ Rev* 2015;50:952–966.

128. Boswell C.C., and Friesen D.K. Elemental sulfur fertilizers and their use on crops and pastures. *Nutr Cycl Agroecosyst* 1993;35:127–149.

129. Shah M., Sircar A., Varsada R., Vaishnani S., Savaliya U., Faldu M., Vaidya D., and Bhattacharya P. Assessment of geothermal water quality for industrial and irrigation purposes in the Unai geothermal field, Gujarat, India. *Groundw Sustain Dev* 2019;8:59–68.

130. Antics M., and Sanner B. Status of geothermal energy use and resources in Europe. *Proceedings of the Proceedings European Geothermal Congress*, Unterhaching; 30 May–1 June, 2007.

131. IEA. *World Energy Outlook 2017*. Paris: OECD Publishing; 2017.

132. Dagnachew A.G., Lucas P.L., Hof A.F., Gernaat D.E., De Boer H.-S., and Van Vuuren D.P. The role of decentralized systems in providing universal electricity access in Sub-Saharan Africa—A model-based approach. *Energy* 2017;139:184–195.

133. Graffy E.A. Sparking a worldwide energy revolution: Social struggles in the transition to a post-petrol world—By Kolya Abramsky. *Rev Policy Res* 2012;29:309–311.

134. Kyriakarakos G., and Papadakis G. Microgrids for productive uses of energy in the developing world and blockchain: A promising future. *Appl Sci* 2018;8:580.

135. Kyriakarakos G., Balafoutis A., and Bochtis D. Proposing a paradigm shift in rural electrification investments in sub-saharan Africa through agriculture. *Sustainability* 2020;12:3096.

136. Shan A.N.M.N.U. A review of Kalina cycle. *Int J Smart Energ Technol Environ Eng* 2020, September;1(1). http://globalpublisher.org/journals-1007/.

137. Kalina A.I. Combined-cycle system with novel bottoming cycle. *J Eng Gas Turbines Power* 1984, October;106(4):737–742.

138. Kalina A.I. Combined cycle and waste heat recovery power systems based on a novel thermodynamic energy cycle utilizing low-temperature heat for power generation. *1983 Joint Power Generation Conference: GT Papers*, Indianapolis, Indiana; 1983.

139. Bombarda P., Invernizzi C.M., and Pietra C. Heat recovery from diesel engines: A thermodynamic comparison between Kalina and ORC cycles. *Appl Therm Eng* 2010, February;30(2–3):212–219.

140. Kalina A.I., Tribus M., and E1-Sayed Y. M. A theoretical approach to the thermophysical properties of two-miscible-component mixtures for the purpose of power-cycle analysis. ASME, Paper 86-WA/HT-54, ASME Winter Annual Meeting, Anaheim, CA; 1986.

141. Park Y.M., and Sonntag R.E. A preliminary study of the Kalina power cycle in connection with a combined cycle system. *Int J Energ Res* 1990;14(2):153–162.

142. Jurgen R.K. The promise of the Kalina cycle: Using an ammonia-water mixture, the Kalina steam cycle may permit thermal-mechanical-electrical energy conversion efficiencies of 45 percent. *IEEE Spect* 1986;23(4):68–70. ieeexplore.ieee.org.

143. Thorin E. Power cycles with ammonia-water mixtures as working fluid analysis of different applications and the influence of thermophysical properties. Ph. D. thesis, Royal Institute of Technology, KTH, Stockholm, Sweden, 2000.

144. El-Sayed Y.M., and Tribus M. A theoretical comparison of the Rankine and Kalina cycles. *ASME Pub AES* 1985;1:97–102.

145. Jonsson M. *Advanced Power Cycles with Mixtures as the Working Fluid*, Ph. D. thesis. Stockholm, Sweden: Royal Institute of Technology, KTH; 2003.

146. Valdimarsson P., Ing S., and Eliasson L. Factors influencing the economics of the Kalina power cycle and situations of superior performance. *A paper presented at International Geothermal Conference*, Raykjavik, Iceland, 2003.

147. Guzovi Z., Lon Car D., and Ferdelji N. Possibilities of electricity generation in the Republic of Croatia by means of geothermal energy. *Energy* 2010;35:3429–3440.

148. DiPippo R. Chapter 22 - Enhanced geothermal systems—Projects and plants. In: DiPippo R., editor *Geothermal Power Plants*, 4th ed. Butterworth-Heinemann; 2016, pp. 609–656. ISBN 9780081008799. Doi: 10.1016/B978-0-08-100879-9.00022-7.

149. Breede K., Dzebisashvili K, Liu X., and Falcone G. A systematic review of enhanced (or engineered) geothermal systems: Past, present and future. *Geotherm Energ* 2013;1(4):1–27.

150. Axelsson G., and Gunnlaugsson E. Geothermal utilization, management and monitoring. In: *Long-Term Monitoring of High-and Low Enthalpy Fields Under Exploitation*. Morioka: World Geothermal Congress Short Courses; 2000, pp. 3–10.

151. Benderitter Y., and Cormy G. Possible approach to geothermal research and relative costs. In: Dickson M., and Fanelli M., editors. *Small Geothermal Resources: A Guide to Development and Utilization*. New York: UNITAR; 1990, pp. 59–70.

152. Hochstein M. Classification and assessment of geothermal resources. In: Dickson M., and Fanelli M., editors. *Small Geothermal Resources*. Rome: UNITAR/UNDP Centre for Small Energy Resources; 1990, pp. 31–59.

153. Kaya E., Zarrouk S.J., and O'Sullivan M.J. Reinjection in geothermal fields: A review of worldwide experience. *Renew Sustain Energ Rev* 2011;15(1):47–68. Doi: 10.1016/j.rser.2010.07.032.

154. Muffler P., and Cataldi R. Methods for regional assessment of geothermal resources. *Geothermics*. 1978;7(2–4):53–89. Doi: 10.1016/0375-6505(78)90002-0.

155. Nicholson K. *Geothermal Fluids*. Berlin: Springer; 1993. Doi: 10.1007/978-3-642-77844-5.

156. Smith R., and Shaw H. Igneous-related geothermal systems. In: White D., and Williams D., editors. *Assessment of Geothermal Resources of the United States—1975, U.S. Geological Survey Circular*. Vol. 726; 1975, pp. 58–83. Doi: 10.2172/860709.

157. Smith R., and Shaw H. Igneous-related geothermal systems. In: Muffler L., editors. *Assessment of Geothermal Resources of the United States, 1978, U.S. Geological Survey Circular*; 1979, Vol. 790, pp .12–17. Doi: 10.2172/6870401.

158. Tester J.W., Anderson B.J., Batchelor A.S., Blackwell D.D., DiPippo R., Drake E.M., Garnish J., Livesay B., Moore M.C., Nichols K., Petty S., Toksöz M.N., and Veatch Jr R.W. *The Future of Geothermal Energy in the 21 Century Impact of Enhanced Geothermal Systems (EGS) on the United States.* Cambridge, MA: MIT Press; 2006. http://www1.eere.energy.gov/geothermal/egs_technology.html.

159. Friðleifsson G.O., and Elders W.A. The Iceland deep drilling project: A search for deep unconventional geothermal resources. *Geothermics* 2005;34:269–285. Doi: 10.1016/j.geothermics.2004.11.004.

160. Fournier R.O. Hydrothermal processes related to movement of fluid from plastic into brittle rock in the magmatic-epithermal environment. *Econ Geol* 1999;94(8):1193–1211. Doi: 10.2113/gsecongeo.94.8.1193.

161. Fournier R.O. The transition from hydrostatic to greater than hydrostatic fluid pressure in presently active continental hydrothermal systems in crystalline rock. *Geophys Res Lett* 1991;18(5):955–958. Doi: 10.1029/91GL00966.

162. Saishu H., Okamoto A., and Tsuchiya N. The significance of silica precipitation on the formation of the permeable-impermeable boundary within earth's crust. *Terra Nova* 2014;26(4):253–259. Doi: 10.1111/ter.12093.

163. Tsuchiya N., and Hirano N. Chemical reaction diversity of geofluids revealed by hydrothermal experiments under sub- and supercritical states. *Isl Arc* 2007;16(1):6–15. Doi: 10.1111/j.1440-1738.2007.00554.x.

164. Watanabe N., Numakura T., Sakaguchi K., Saishu H., Okamoto A., Ingebritsen S.E. and Tsuchiya N. Potentially exploitable supercritical geothermal resources in the ductile crust. *Nat Geosci* 2017;10(2):140–144. Doi: 10.1038/NGEO2879.

165. Friðleifsson G.O., Albertsson A., Stefansson B., Gunnlaugsson E., and Adalsteinsson H. Deep unconventional geothermal resources: A major opportunity to harness new sources of sustainable energy. *Proceedings, 20th World Energy Conference*, World Energy Council, Rome; 2007, p. 21.

166. Elders W.A., Friðleifsson G., and Albertsson A. Drilling into magma and the implications of the Iceland Deep Drilling project (IDDP) for high-temperature geothermal systems worldwide. *Geothermics* 2014a;49:111–118. Doi: 10.1016/j.geothermics.2013.05.001.

167. White D.E. *Hydrology, Activity, and Heat Flow of the Steamboat Springs Thermal System.* Washington, DC: USGS, Washoe County, Nevada. U.S. Geological Survey Professional Paper 458-C; 1968, p. 109.

168. Dobson P., Asanuma H., Huenges E., Poletto F., Reinsch T., and Sanjuan B. Supercritical geothermal systems—review of past studies and ongoing research activities. *Proceedings, 42nd Workshop on Geothermal Reservoir Engineering*, Stanford University, Stanford; 2017, p. 13.

169. Gunnlaugsson E., Armannsonn H., Thorhallsson S., and Steingrimsson B. Problems in geothermal operation. Scaling and corrosion. *Short Course VI on Utilization of Low- and Medium Enthalpy Geothermal Resources and Financial Aspects of Utilization, Organized by UNU-GTP and Lageo*, in Santa Tecla, El Salvador; 2014.

170. Miller R. Chemistry and materials in geothermal systems. In: Casper L., and Pinchback T., editors. *Geothermal Scaling and Corrosion, STP717.* ASTM International; 1980, p. 7. Doi: 10.1520/STP30061S.

171. Sanada N., Kurata Y., Nanjo H., Ikeuchi J., and Kimura S. Corrosion in acidic geothermal flows with high velocity. *Proceedings 20th New Zealand Geothermal Workshop*, Auckland, New Zealand; 1998, pp. 121–126.

172. Reinsch T., Dobson P., Asanuma H., Huenges E., Poletto F., and Sanjuan B. Utilizing supercritical geothermal systems: A review of past ventures and ongoing research activities. *Geotherm Energ* 2017;5:16. Doi 10.1186/s40517-017-0075-y.

173. Reinsch T., Huenges E., Bruhn D., Thorbjörnsson I., Gavriliuc R., and van Wees J. Geothermal R&D, new projects and perspectives for basic scientific research. *Proceedings, European Geothermal Congress 2016*, Strasbourg, France; 2016, p. 4.

174. Reinsch T., and Henninges J. Temperature-dependent characterization of optical fibres for distributed temperature sensing in hot geothermal wells. *Meas Sci Technol* 2010;21(9):094022. Doi: 10.1088/0957-0233/21/9/094022.

175. Reinsch T., Henninges J., and Ásmundsson R. Thermal, mechanical and chemical influences on the performance of optical fibres for distributed temperature sensing in a hot geothermal well. *Environ Earth Sci* 2013;70(8):3465–3480. Doi: 10.1007/s12665-013-2248-8.

176. Friðleifsson G.O., and Elders W.A. The Iceland deep drilling project geothermal well at Reykjanes successfully reaches its supercritical target. *Geotherm Resour Counc Bull* 2017;46:30–33.

177. Friðleifsson G.O., Bogason S.G., Stoklosa A.W., Ingolfsson H.P., Vergnes P., Thorbjörnsson I., Peter-Borie M., Kohl T., Edelmann T., Bertani R., Sæther S., and Palsson B. Deployment of deep enhanced geothermal systems for sustainable energy business. *Proceedings, European Geothermal Congress 2016*, Strasbourg, France; 2016, p. 8.

178. Chu T., Dunn J., Finger J., Rundle J., and Westrich H. The magma energy program. *Geotherm Resour Counc Trans* 1990;14:567–577.

179. Sigmundsson F., Eichelberger J., Papale P., Ludden J., Dingwell D., Mandeville C., Pye S., Markússon S., Árnason K., and Ingólfsson H. Krafla magma testbed. *GEORG Geothermal Workshop*, Reykjavik, Iceland; 2016.

180. Asanuma H., Muraoka H., Tsuchiya N., and Ito H. The concept of the Japan Beyond-Brittle project (JBBP) to develop EGS reservoirs in ductile zones. *Geotherm Resour Counc Trans* 2012;36:359–364.

181. Asanuma H., Soma N., Tsuchiya N., Kajiwara T., and Yamada S. Concept of development of supercritical geothermal resources in Japan. *Proceedings, 2015 International Conference on Geothermal Energy in Taiwan*, Taipei, Taiwan; 2015, pp. 66–68.

182. Muraoka H., Asanuma H., Tsuchiya N., Ito T., Mogi T., and Ito H. The Japan beyond-brittle project. *Sci Drill.* 2014;17:51–59. Doi: 10.5194/sd-17-51-2014.

183. Tsuchiya N., Yamada R., Uno M. Supercritical geothermal reservoir revealed by a granite–porphyry system. *Geothermics* 2016;63:182–194. Doi: 10.1016/j.geothermics.2015.12.011.

184. Bando M., Bignall G., Sekine K., and Tsuchiya N. Petrography and uplift history of the Quaternary Takidani Granodiorite: Could it have hosted a supercritical (HDR) geothermal reservoir? *J Volcanol Geotherm Res* 2003;120(3–4):215–34. Doi:10.1016/S0377-0273(02)00399-2.

185. McDonnell K. Molnár L., Harty M., and Murphy F. Feasibility study of carbon dioxide plume geothermal systems in Germany–Utilising carbon dioxide for energy. *Energies* 2020;13:2416. Doi: 10.3390/en13102416 www.mdpi.com/journal/energies.

186. Büsing H., Niederau J., and Clauser C. Pressure-enthalpy formulation for numerical simulations of supercritical water/steam systems applied to a reservoir in Tuscany, Italy. *Proceedings, European Geothermal Congress 2016*, Strasbourg, France; 2016, p. 8.

187. Liotta D., and Ranalli G. Correlation between seismic reflectivity and rheology in extended lithosphere: Southern Tuscany, inner northern Apennines, Italy. *Tectonophysics.* 1999;315(1–4):109–122, Doi: 10.1016/S0040-1951(99)00292-9.

188. Stamnes Ø.N., Røed M.H., Hjelstuen M., Kolberg S., Knudsen S., Vedum J., and Halladay N. Development of a novel logging tool for 450°C geothermal wells. *Proceedings, European Geothermal Congress 2016*, Strasbourg, France; 2016, p. 6.

189. Bignall G. Hotter and deeper: New Zealand's research programme to harness its deep geothermal resources. *Proceedings, World Geothermal Congress 2010*, Bali, Indonesia; 2010, p. 3.

190. Bignall G., and Carey B. A deep (5 km?) geothermal science and drilling project for the Taupo Volcanic Zone—who wants in? *Proceedings New Zealand Geothermal Workshop 2011*, Auckland, New Zealand; 2011, p. 5.

191. Newman G., Lindsey N., Gasperikova E., Bertrand E., and Caldwell T. The importance of full impedance tensor analysis for 3D magnetotelluric imaging the roots of high temperature geothermal systems: Application to the Taupo Volcanic Zone, New Zealand. *Proceedings, World Geothermal Congress 2015*, Melbourne, Australia; 2015, p. 5.

192. Bannister S., Bourguignon S., Sherburn S., and Bertrand T. 3-D seismic velocity and attenuation in the Central Taupo Volcanic Zone, New Zealand: Imaging the roots of geothermal systems. *Proceedings, World Geothermal Congress 2015*, Melbourne, Australia; 2015, p. 17.

193. Bégué F., Deering C.D., Gravley D.M., Chambefort I., and Kennedy B.M. From source to surface: Tracking magmatic boron and chlorine input into the geothermal systems of the Taupo Volcanic Zone, New Zealand. *J. Volcanol Geotherm Res* 2017. Doi: 10.1016/j.jvolgeores.2017.03.008.

194. Bruhn D. GEMex: Cooperation Europe–Mexico for the development of unconventional geothermal systems. In: *Tagungsband, Der Geothermiekongress*; 2017, p. 1.

195. López Hernandez A., Garduno-Monrroy V., Vargas-Medina J., Romo-Jones J., Prol-Ledesma R., Bruhn D., and Flores-Armenta M. GEMex-cooperación Mxico-Europa para la investigación de sistemas geotérmicos mejorados y sistemas geotérmicos super-calientes. *GEOS*. 2016.

196. Cladouhos T.T., Petty S., Swyer M.W., Uddenberg M.E., Grasso K., and Nordin Y. Results from Newberry Volcano EGS demonstration, 2010–2014. *Geothermics* 2016;63:44–61. Doi: 10.1016/j.geothermics.2015.08.009.

197. ICDP. Newberry deep drilling project (NDDP). 2017. http://www.icdp-online.org/fileadmin/icdp/projects/doc/nddp/Newberry_ICDP_Workshop_Call_final-1.pdf.

198. Hayba D.O., and Ingebritsen S.E. *The Computer Model Hydrotherm, a Three-Dimensional Finite-Difference Model to Simulate Ground-Water Flow and Heat Transport in the Temperature Range of 0 to 1,200°C*. US Geological Survey Water-Resources Investigations Report, 94-4045; 1994, p. 85. http://pubs.er.usgs.gov/publication/wri944045.

199. Ingebritsen S.E., Geiger S., Hurwitz S., and Driesner T. Numerical simulation of magmatic hydrothermal systems. *Rev Geophys* 2010;48:RG1002. Doi: 10.1029/2009RG000287.

200. Yano Y., and Ishido T. Numerical investigation of production behavior of deep geothermal reservoirs at super-critical conditions. *Geothermics* 1998;27(5–6):705–721. Doi: 10.1016/S0375-6505(98)00041-8.

201. Norton D.L., and Dutrow B.L. Complex behavior of magma-hydrothermal processes: Role of supercritical fluid. *Geochimica et Cosmochimica Acta*. 2001;65(21):4009–4017. Doi: 10.1016/S0016-7037(01)00728-1.

202. Watanabe K., Niibori Y., and Hashida T. Numerical study of heat extraction from super-critical geothermal reservoir. *Proceedings, World Geothermal Congress 2000*, Kyushu, Tohoku, Japan; 2000, p. 5.

203. Croucher A.E., and O'Sullivan M.J. Application of the computer code TOUGH2 to the simulation of supercritical conditions in geothermal systems. *Geothermics*. 2008;37(6):622–634. Doi: 10.1016/j.geothermics.2008.03.005.

204. O'Sullivan J., Kipyego E., Croucher A., Ofwona C., and O'Sullivan M. A supercritical model of the Menengai geothermal system. *Proceedings, World Geothermal Congress 2015*, Melbourne, Australia; 2015, p. 9.

205. O'Sullivan J., O'Sullivan M., and Croucher A. Improvements to the AUTOUGH2 supercritical simulator with extension to the air-water equation-of-state. *Geotherm Resour Counc Trans*. 2016;40:921–929.

206. Magnusdottir L., and Finsterle S. An iTOUGH2 equation-of-state module for modeling supercritical conditions in geothermal reservoirs. *Geothermics*. 2015;57:8–17. Doi: 10.1016/j.geothermics.2015.05.003.

207. Gunnarsson G., and Aradóttir E.S.P. The deep roots of geothermal systems in volcanic areas: Boundary conditions and heat sources in reservoir modeling. *Transp Porous Media* 2014;108(1):43–59. Doi: 10.1007/s11242-014-0328-1.

208. Weis P., and Driesner T. The interplay of non-static permeability and fluid flow as a possible pre-requisite for supercritical geothermal resources. *Energ Procedia* 2013;40:102–106. Doi: 10.1016/j.egypro.2013.08.013.

209. Driesner T., Weis P., and Scott S. A new generation of numerical simulation tools for studying the hydrology of geothermal systems to "supercritical" and magmatic conditions. *Proceedings, World Geothermal Congress 2015*, Melbourne, Australia; 2015, p. 4.

210. Scott S., Driesner T., and Weis P. Geologic controls on supercritical geothermal resources above magmatic intrusions. *Nat Commun* 2015a;6. Doi: 10.1038/ncomms8837.

211. Scott S., Driesner T., and Weis P. Hydrology of a supercritical flow zone near a magmatic intrusion in the IDDP-1 well—insights from numerical modeling. *Proceedings, World Geothermal Congress 2015*, Melbourne, Australia; 2015b, p. 5.

212. Scott S., Driesner T., and Weis P. The thermal structure and temporal evolution of high-enthalpy geothermal systems. *Geothermics* 2016;62:33–47. Doi: 10.1016/j.geothermics.2016.02.004.

213. Scott S., Driesner T., and Weis P. Boiling and condensation of saline geothermal fluids above magmatic intrusions. *Geophys Res Lett* 2017;44(4):1696–16705. Doi: 10.1002/2016GL071891.

214. Carcione J.M., and Poletto F. Seismic rheological model and reflection coefficients of the brittle–ductile transition. *Pure Appl Geophys* 2013;170(12):2021–2035. Doi: 10.1007/s00024-013-0643-4.

215. Carcione J.M., Poletto F., Farina B., and Craglietto A. Simulation of seismic waves at the earth's crust (brittle–ductile transition) based on the Burgers model. *Solid Earth* 2014;5(2):1001–1010. Doi: 10.5194/se-5-1001-2014.

216. Carcione J.M., Poletto F., Farina B., and Craglietto A. The Gassmann–Burgers model to simulate seismic waves at the earth crust and mantle. *Pure Appl Geophys* 2016;174(3):849–863. Doi: 10.1007/s00024-016-1437-2.

217. Farina B., Poletto F., and Carcione J.M. Seismic wave propagation in poro-viscoelastic hot rocks. *Proceedings, European Geothermal Congress 2016*, Strasbourg, France; 2016, p. 7.

218. Poletto F., Corubolo P., Schleifer A., Farina B., Pollard J., and Grozdanich B. Seismic while drilling for geophysical exploration in a geothermal well. *Geotherm Resour Counc Trans*. 2011a;35:1737–1741.

219. Poletto F., Corubolo P., Farina B., Schleifer A., Pollard J., Peronio M., and Bohm G. Drill-bit SWD and seismic interferometry for imaging around geothermal wells. *SEG Technical Program Expanded Abstracts*, San Antonio 2011 Annual Meeting, San Antonio, TX; 2011b, pp. 4319–4324.

220. Ásmundsson R., Pezard P., Sanjuan B., Henninges J., Deltombe J.L., Halladay N., Lebert F., Gadalia A., Millot R., Gibert B., Violay M., Reinsch T., Naisse J.M., Massiot C., Azaïs P., Mainprice D., Karytsas C., and Johnston C. High temperature instruments and methods developed for supercritical geothermal reservoir characterisation and exploitation—the HiTI project. *Geothermics* 2014;49(0):90–98. Doi: 10.1016/j.geothermics.2013.07.008.

221. Sanjuan B., Millot R., Brach M., Asmundsson R., and Giroud N. Use of a new sodium/lithium (Na/Li) geothermometric relationship for high-temperature dilute geothermal fluids from Iceland. *Proceedings, World Geothermal Congress 2010*, Bali, Indonesia; 2010, p. 12.

222. Sanjuan B., Millot R., Ásmundsson R., Brach M., and Giroud N. Use of two new Na/Li geothermometric relationships for geothermal fluids in volcanic environments. *Chem Geol* 2014;389:60–81. Doi: 10.1016/j.chemgeo.2014.09.011.

223. Gadalia A., Braibant G., Touzelet S., and Sanjuan B. Tracing tests using organic compounds in a very high temperature geothermal field, Krafla (Iceland). Final Report, No. RP-57661-FR, BRGM. Mainland France: BRGM; 2010.

224. Juliusson E., Markusson S., and Sigurdardottir A. Phase-specific and phase-partitioning tracer experiment in the Krafla reservoir, Iceland. *Proceedings, World Geothermal Congress 2015*, Melbourne, Australia; 2015, p. 12.

225. van Wees J.D., Hopman J., Dezayes C., Vernier R., Manzella A., Bruhn D., Scheck-Wenderoth M., Flovenz O., Hersir G.P., Halldosdottir S., and Liotta D. IMAGE: The EU funded research project integrated methods for advanced geothermal exploration. *Proceedings World Geothermal Congress 2015*, Melbourne, Australia; 2015, p. 11.

226. Kummerow J., and Raab S. Temperature dependence of electrical resistivity-Part I: Experimental investigations of hydrothermal fluids. *Energ Procedia*. 2015a;76:240–246. Doi: 10.1016/j.egypro.2015.07.854.

227. Kummerow J., and Raab S. Temperature dependence of electrical resistivity—Part II: A new experimental set-up to study fluid-saturated rocks. *Energ Procedia*. 2015b;76:247–255. Doi: 10.1016/j.egypro.2015.07.855.

228. Conboy T.M., Wright S.A., Ames D.E., and Lewis T.G. *$CO_2$-Based Mixtures as Working Fluids for Geothermal Turbines*, Tech. Rep. Albuquerque, New Mexico: Sandia National Laboratories; 2012, p. SAND2012– SAND4905.

229. Adams B.M., Kuehn T.H., Bielicki J.M., Randolph J.B., and Saar M.O. A comparison of electric power output of $CO_2$ plume geothermal (CPG) and brine geothermal systems for varying reservoir conditions. *Appl Energ* 2015;140:365–377. Doi: 10.1016/j. apenergy.2014.11.043.

230. Glos S., Grotkamp S., and Wechsung M. Assessment of performance and costs of $CO_2$ based next level geothermal power (NLGP) systems. *3rd European Supercritical $CO_2$ Conference*, Paris, France; 19–20 September, 2019, p. 109. Doi: 10.17185/ duepublico/48876.

231. Adams B.M., Kuehn T.H., Bielicki J.M., Randolph J.B., and Saar M.O. On the importance of the thermosiphon effect in CPG ($CO_2$ plume geothermal) power systems. *Energy* 2014;69:409–418. Doi: 10.1016/j.energy.2014. 03.032.

232. Higgins B.S., Muir M.P., Eastman A.D., Oldenburg C.M., and Pan L. Process modeling of a closed-loop $SCO_2$ geothermal power cycle. *The 5th International Supercritical $CO_2$ Power Cycles Symposium*, San Antonio, TX; 29–31 March, 2016. Doi: 10.1007/ s13398-014-0173-7.2, arXiv:arXiv:1011. 1669v3.

233. Amaya A., Scherer J., Muir J., Patel M., and Higgins B. Green Fire energy closed loop geothermal demonstration using supercritical carbon dioxide as working fluid. *45th Workshop on Geothermal Reservoir Engineering*, Stanford, CA; 10–12 February, 2020, p. SGP-TR-21.

234. Ruiz-Casanova E., Rubio-Maya C., Pacheco-Ibarra J. J., Ambriz-Díaz V.M., Romero C. E., and Wang X. Thermodynamic analysis and optimization of supercritical carbon dioxide brayton cycles for use with low-grade geothermal heat sources. *Energ Convers Manag* 2020;216:112978. Doi: 10.1016/j.enconman.2020.112978.

235. Haynes International. *Haynes 230 Alloy*. Kokomo, Indiana; 2007.

236. Li X., Kininmont D., Le Pierres R., and Dewson S.J. Alloy 617 for the high temperature diffusion-bonded compact heat exchangers. *Proceedings of ICAPP*, Anaheim, CA, ICAPP; 2008.

237. Feng Y., and Wang J. Review of supercritical carbon dioxide brayton cycle research. *Energ Conserv* 2019;(2):91–100. Doi: 10.16643/j.cnki.14-1360/td.2019.02.044.

238. Wang E., Peng N., Zhang M. System design and application of supercritical and transcritical $CO_2$ power cycles: A review. *Front Energ Res* 2021, November 10. Doi: 10.3389/fenrg.2021.723875.

239. Randolph J.B., and Saar M.O. Coupling carbon dioxide sequestration with geothermal energy capture in naturally permeable, porous geologic formations: Implications for $CO_2$ sequestration. *Energ Procedia* 2011;4:2206–2213.

240. Saar M.O., Randolph J.B., and Kuehn T.H. *Carbon Dioxide-Based Geothermal Energy Generation Systems and Methods Related Thereto.* U.S. Patent US8,316,955 B2. Minneapolis: University of Minnesota; 2012, November 27.

241. Saar M.O., Randolph J.B., and Kuehn T.H. *Carbon Dioxide-Based Geothermal Energy Generation Systems and Methods Related Thereto.* Europe Patent 2406562. Minnesota: University of Minnesota; 2014, December 17.

242. Garapati N., Randolph J.B., Valencia J.L., and Saar M.O. $CO_2$-Plume Geothermal (CPG) heat extraction in multi-layered geologic reservoirs. *Energ Procedia* 2014;63:7631–7643.

243. Pioro I., and Mokry S. Thermophysical properties at critical and supercritical pressures. In: Belmiloudi A., editor. *Heat Transfer - Theoretical Analysis, Experimental Investigations and Industrial Systems.* London: IntechOpen Ltd.; 2011, January 28. ISBN 978-953-307-226-5.

244. Jiang P.X., Zhang F.Z., and Xu R.N. Thermodynamic analysis of a solar–enhanced geothermal hybrid power plant using $CO_2$ as working fluid. *Appl Ther Eng* 2017;116:463–472.

245. Pruess K. The Tough codes-a family of simulation tools for multiphase flow and transport processes in permeable media. *Vadose Zone J.* 2004;3:738–746.

246. Pruess K. Enhanced geothermal systems (EGS) using $CO_2$ as working fluid-a novel approach for generating renewable energy with simultaneous sequestration of carbon. *Geothermics* 2008;35:351–367.

247. Wang K., Yuan B., Ji G., and Wu X. A comprehensive review of geothermal energy extraction and utilization in oilfields. *J Pet Sci Eng* 2018;168:465–477. Doi: 10.1016/J. PETROL.2018.05.012.

248. Bobbo S., Colla L., Barizza A., Rossi S., Fedele L., and Nazionale C. Characterization of nanofluids formed by fumed $Al_2O_3$ in water for geothermal applications. *Int Compress Eng Refrig Air Cond High Perform Build Conf* 2016;2284:1–9.

249. Diglio G., Roselli C., Sasso M., and Jawali Channabasappa U. Borehole heat exchanger with nanofluids as heat carrier. *Geothermics* 2018;72:112–123. Doi: 10.1016/J. GEOTHERMICS.2017.11.005.

250. Ahmadi M.H., Hajizadeh F., Rahimzadeh M., Shafii M.B., and Chamkha A.J. Application GMDH artificial neural network for modeling of $Al_2O_3$/water and $Al_2O_3$/ Ethylene glycol thermal conductivity 2018;36:773–782.

251. Daneshipour M., and Rafee R. Nanofluids as the circuit fluids of the geothermal borehole heat exchangers. *Int Commun Heat Mass Transf* 2017;81:34–41. Doi: 10.1016/J. ICHEATMASSTRANSFER.2016.12.002.

252. Jamshidi N., and Mosaffa A. Investigating the effects of geometric parameters on finned conical helical geothermal heat exchanger and its energy extraction capability. *Geothermics* 2018;76:177–189. Doi: 10.1016/J.GEOTHERMICS.2018.07.007.

253. Sui D., Langåker V.H., and Yu Z. Investigation of thermophysical properties of nanofluids for application in geothermal energy. *Energ Procedia* 2017;105:5055–5060. Doi: 10.1016/J.EGYPRO.2017.03.1021.

254. Sun X.H., Yan H., Massoudi M., Chen Z.H., Wu W.T., and Sun X.H. Numerical simulation of nanofluid suspensions in a geothermal heat exchanger. *Energies* 2018;11:919. Doi: 10.3390/en11040919.

255. Ahmadi M.H., Ramezanizadeh M., Nazari M.A., Lorenzini G., Kumar R., and Jilte R. Applications of nanofluids in geothermal: A review mathematical modelling of engineering problems. *Int Informat Eng Technol Associat* 2018, December;5(4):281–285.

# 5 Advanced Solar Thermal Power Systems

## 5.1 INTRODUCTION

The sun is the most important energy source available to us. Outside the Earth's atmosphere, the average power of the solar radiation perpendicular to the main direction of the sun rays is of the order of 1.36 kW/m$^2$. This quantity, which is traditionally called the solar constant, is not a constant and varies inversely proportional to the square of the distance from Earth to the sun. It fluctuates about 6.9% during the year. Incoming radiation from the sun passes through the Earth's atmosphere, whereby it is partially absorbed or scattered by components of the atmosphere, such as aerosols, gases, and particles. About 30% of solar radiation is reflected without being absorbed by the atmospheric components or surface of Earth, while the amount of scattering depends on the atmospheric conditions. In quantitative terms, the amount of solar energy reaching the Earth's surface, "ground-level," is about 885 million terawatt hours (TWh, or 3.06 10$^{24}$ joules) per year [1–4]. In 2013, the annual anthropogenic energy consumption was 13,541 Mtoe (million tons of oil equivalent) [2], (0.157 TWh or 5.67 10$^{20}$ J) per year, just a fraction of a percentage point of the solar energy reaching the Earth's surface.

The physical impact of the sun and solar radiation on Earth affects, among other things, the currents and tides, the weather, and the biosphere. Thus, the sun is not only a direct renewable and practically inexhaustible source of energy but also the ultimate origin of most other renewable energy resources; wind energy, ocean energy, as well as biomass. It is worth noting that biomass was the original energy material source for fossil fuels, coal, shale, oil, and gas, albeit a million years in the making. This leaves geothermal and nuclear energy as the only sources of energy available on Earth that are independent of the sun. The intensity of the direct solar radiation reaching the surface of the Earth is not geographically uniform. An approximate upper limit is 1 kW/m$^2$, although in many instances and regions of the world it is much less than this. The regions of the world that receive the most amount of direct solar radiation are typically located within the latitudes of ±40°, in what is sometimes referred to as the "sunbelt." In these regions, the annual energy per unit received by a surface that is kept at all instances perpendicular to the main direction of the sun rays is usually equal to or larger than 1,800 kWh/m$^2$/year. The main advantages of solar energy are that it is abundant, clean and renewable; it is geographically distributed and consequently free of geopolitical tensions and it has high energy content. The disadvantages are that it has relatively low surface density, it is intermittent and it is variable [1–4].

The spectral distribution of the direct solar radiation just outside the Earth atmosphere is equivalent to that of a blackbody at 5,777 K, ranging from near-ultraviolet through to near-infrared, with most photons and energy in the visible spectrum.

DOI: 10.1201/9781003328087-5

The maximum percentage of heat that can be transformed into work for a given ambient temperature is known as the exergy of the heat. According to the second law of thermodynamics, this exergy increases with the temperature at which the heat is delivered and decreases with increasing ambient temperatures. The fact that solar radiation outside the Earth's atmosphere can be considered as heat delivered from a thermal reservoir at 5,777 K implies that more than 94% of this heat can be transformed into work. At ground level, because of the interactions with the atmosphere, the spectral distribution of the direct solar radiation can no longer be considered equivalent to that of a blackbody. However, its exergy content is still very high and between 84% and 96% according to most authors [1–4]. In addition, of being intermittent due to the day and night cycles, the solar radiation on Earth, at ground level, is also variable. The amount of direct solar radiation reaching the ground depends on the atmospheric conditions in general and on cloud cover in particular. This variability, however, is significantly lower than the variability of other renewable energy sources, such as wind, and decreases with increased time intervals. All solar energy systems are designed to maximize the advantages provided by the sun as an energy source and to minimize the disadvantages.

As shown in this and subsequent two other chapters solar energy can be converted to power in a number of different ways. In this chapter, we specifically focus on the conversion of solar thermal energy to power by thermodynamic cycles which are normally designated as CSP (or CST) or concentrated solar power. While hydropower has risen from 715 GW in 2004 to 1,114 GW in 2017, solar PV has risen from 2.6 GW in 2004 to 402 GW in 2017 and wind power has risen from 48 GW in 2004 to 539 GW in 2017, the growth in CSP has been more modest (0.4 GW in 2004 to 4.9 GW in 2017). This growth is still more than that for geothermal power and biopower. Furthermore, CSP appears to have significant future potential, particularly in hybrid form. The United States and Spain are the major producers of solar thermal power. Figures 5.1 and 5.2 illustrate the growth of CSP in these and other countries.

Concentrating solar thermal (CST) technologies collect and concentrate radiation from the sun to transform it into high-temperature thermal energy. This thermal energy can later be used for a plethora of high-temperature thermal applications, such as heating and cooling, process heat, material processing, electricity production, or chemical processes. In most CST (CSP) technologies, once the solar radiation is transformed into thermal energy, instead of using it when the transformation takes place, the thermal energy can be stored as such and used when it makes the most sense to do it, such as to maximize economic returns. The high-temperature thermal energy derived from the solar radiation collected and concentrated in a CST system can also be hybridized; that is, mixed with thermal energies derived from other heat sources such as biomass or fossil fuels. The hybridization with biomass is particularly interesting because it will produce a renewable energy system capable of delivering thermal energy to run an industrial process or produce electricity 24 hours a day, all days of the year. Thus, the concentration of solar radiation, its intermediate transformation into high-temperature thermal energy, the possibility of storing this thermal energy to dispatch it when

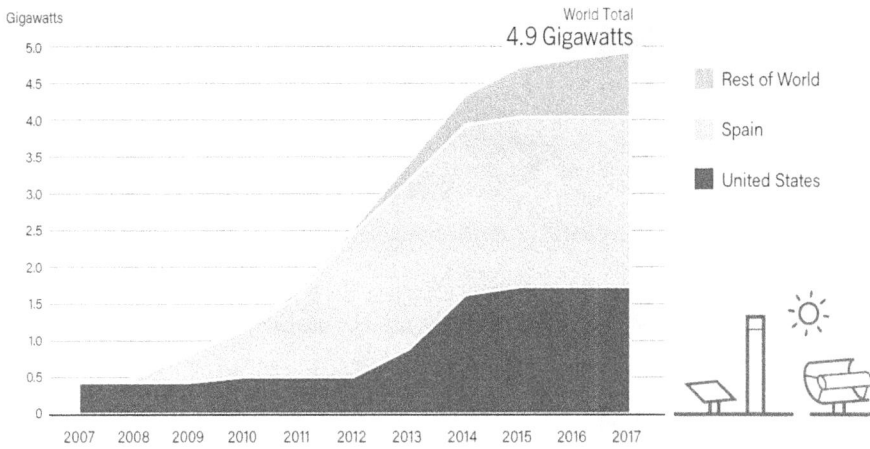

**FIGURE 5.1** Concentrating solar thermal power global capacity by country and region over the period 2007–2017 [5,6].

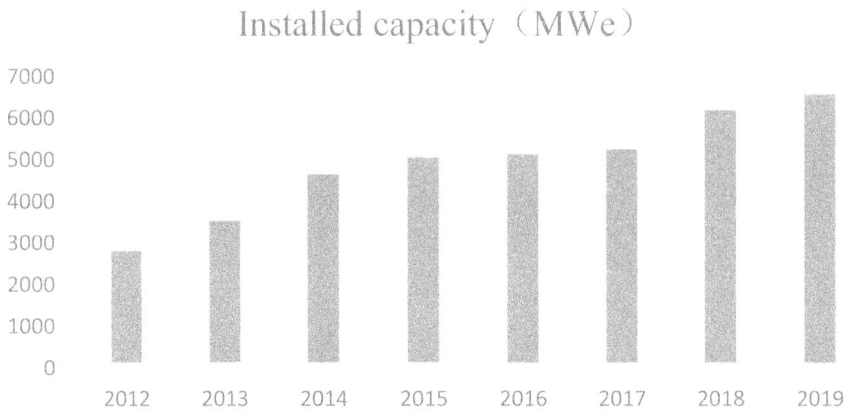

**FIGURE 5.2** Global capacity of CSP [7]. From 2012 to 2019, the global cumulative installed capacity of CSP

needed, and the possibility of hybridizing it with thermal energy from other heat sources are the defining characteristics of CST technologies. Those that set them apart from the rest of renewable energy technologies. More often high temperature concentrated solar energy is converted to solar power as illustrated in Figure 5.3. In recent years significant efforts are made to improve the efficiency of solar thermal power. LCOE of CSP (CST) power compared to other forms of power generation is illustrated in Table 5.1.

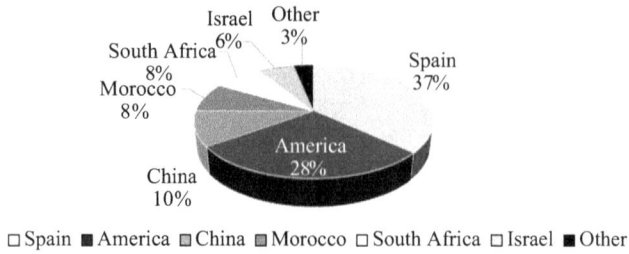

**FIGURE 5.3**  Schematic diagram of CSP (CST) generation technology system [7]. The percentage contributions of the total installed CSP capacity of the world by different countries at the end of 2019.

**TABLE 5.1**
**LCOE Estimates (c/kWh) [5,6,8,9–13]**

| Technology | Europe | USA | China |
|------------|--------|-----|-------|
| CSP (CST) | 17.6–43.10 | 17.6–43.10 | 16.70–40.50 |
| PV | 8.80–22.00 | 9.70–20.25 | 6.95–13.15 |
| Wind | 6.25–10.30 | 5.40–11.95 | 4.30–8.20 |
| Hydro | 8.80 | 7.95 | 2.65 |
| Coal | 10.55–14.95 | 6.20 | 3.10–3.45 |

## 5.2  THERMAL EFFICIENCY AND THE NEED FOR CONCENTRATION

Due to the relatively low surface density of direct solar radiation, to deliver a large amount of power a CST system needs a large area of solar radiation collecting surfaces. The efficient transformation of solar radiation into thermal energy, however, is best done, in receivers with a small surface area, since thermal losses are proportional to the area. Because of this, all CST systems collect and concentrate solar radiation using surfaces that are quite different from the receiver surfaces where the transformation of concentrated solar radiation into thermal energy takes place and is much less expensive per unit area. Traditionally, in a CST system, reflective surfaces are used to collect and concentrate the direct solar radiation upon the receiver. However, refractive surfaces can also be used.

Typically, the receiver of a CST system consists of an absorber material, or absorber coated material, upon which the concentrated solar radiation impinges and is transformed into thermal energy. A large fraction of this thermal energy is transformed into useful energy, for example, in the form of the increase in enthalpy of a working fluid. The rest is used to heat up the absorber material to its operating temperature and after that dissipated as thermal losses. The thermal losses are proportional to the receiver area and the temperature difference between the receiver area and the ambient. Conductive and convective losses are linearly proportional to this temperature difference, while radiative losses are proportional to the difference of the fourth power of these temperatures. Because of this, as the operating temperature increases, radiative losses become more and more dominant. This is the reason why, for low- to mid-temperature CST systems, the system designers make a significant effort to keep convection and conduction losses at bay, using instant vacuum technologies to reduce these losses to a minimum, while for very high-temperature CST systems where the dominant losses are radiative these techniques are seldom used since they are ineffective in reducing radiative losses.

In order to increase the thermal efficiency, that is, the ratio of the useful energy delivered by the receiver to the concentrated solar radiation impinging upon it, the system designer can increase the concentration, that is, the ratio between the total area of the solar collecting surfaces and the total area of the surface of the receiver, decrease the operating temperature of the receiver, thereby, reducing the temperature difference with the ambient or use specially engineered materials and techniques to increase the absorption of the solar radiation in the receiver and to minimize its thermal losses, such as spectrally selective absorber coatings, evacuated tubes, and materials and meta-materials with the appropriate thermal properties. To achieve high thermal efficiencies while operating at a high temperature almost inevitably requires achieving high concentration ratios. How much direct solar radiation can be concentrated is limited by the second law of thermodynamics. The limit is related to the fact that as seen from the surface of Earth the sun subtends a finite solid angle. This implies, that direct solar radiation is not completely parallel, but distributed over a cone of directions around the direction from the observer to the center of the solar disk defined by a half-angle of 4.58 mrad. Traditional image-forming concentrators, such as parabolic troughs or parabolic dishes fall short of the concentration limits for linear and point-focusing concentrators, respectively. The concentration of the best possible parabolic dish concentrator is of the order of 25% of the maximum 3D concentration limit. To approach the limits of concentration with actual physical devices, one has to use new types of concentrators that do not form images and are adequately known as non-imaging concentrators.

The second law of thermodynamics limits the conversion of thermal energy into work. According to this law, the efficiency of the conversion of thermal energy into work increases with the temperature at which the thermal energy is delivered to the heat engine. In a CST system designed to produce work, the overall light-to-work efficiency is the product of two efficiencies: (a) the thermal efficiency with which the direct solar radiation from the sun is transformed into useful thermal energy; and (b) the efficiency with which the useful thermal energy is subsequently transformed into work by a heat engine. While the efficiency at which the direct solar radiation is transformed into useful thermal energy decreases with the increase in operating temperature, the efficiency at which the useful thermal energy is transformed into work increases with it. The overall light-to-work efficiency is, therefore, the product of two functions with opposite tendencies with regard to the variation of the operating temperature of the receiver and as such it must have an optimum; that is, a value of the operating temperature of the receiver for which the light-to-work efficiency is a maximum. Since higher the concentration, higher the temperature at which the receiver can operate with a given amount of thermal losses, the optimum operating temperature increases with concentration.

## 5.3 MAIN COMMERCIALLY AVAILABLE SOLAR CONCENTRATING TECHNOLOGIES

Currently, there are four main commercially available technologies to concentrate the direct solar radiation into a receiver, where this radiation will be transformed into thermal energy. These are graphically illustrated in Figure 5.4. Two of these solar concentrating technologies are linear-focusing technologies and the other two are point-focusing technologies. All of them are imaging technologies but can be combined with non-imaging secondary reflectors to further increase the concentration they can achieve. A comparison of these technologies is illustrated in Table 5.2 [14–22]. Figure 5.5 illustrates the distribution of the use of these four technologies to produce solar thermal energy worldwide under different sets of conditions.

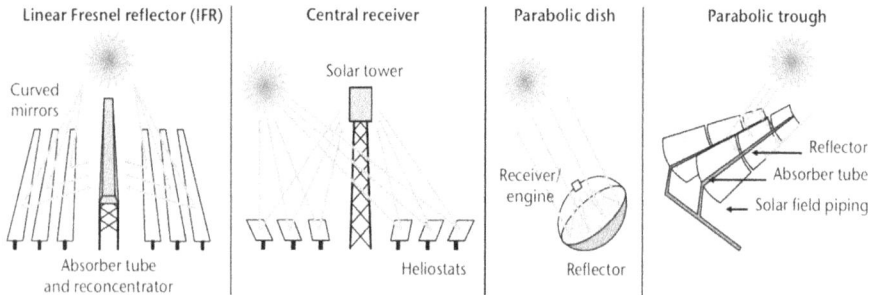

**FIGURE 5.4**  Main CSP technologies [5,57].

**TABLE 5.2**
**Comparison of CSP Technologies [5,8,25–32]**

| CSP Type | Operating Temperature (C) | Ratio of Solar Concentration | Thermal Storage Suitability | Average Annual Efficiency (%) | Land Use Efficiency (Total Area/Power) |
|---|---|---|---|---|---|
| Parabolic Trough | 20–400 | 15–45 | Suitable | 15 | 3.9 |
| Linear Fresnel Reflector | 50–300 | 10–40 | Suitable | 8–11 | 0.8–1 |
| Solar Trough | 300–1,000 | 150–1,500 | Highly suitable | 17–35 | 5.4 |
| Parabolic Dish | 120–1,500 | 100–1,000 | Difficult | 25–30 | 1.2–1.6 |

## Percentage of worldwide STE plants by technology

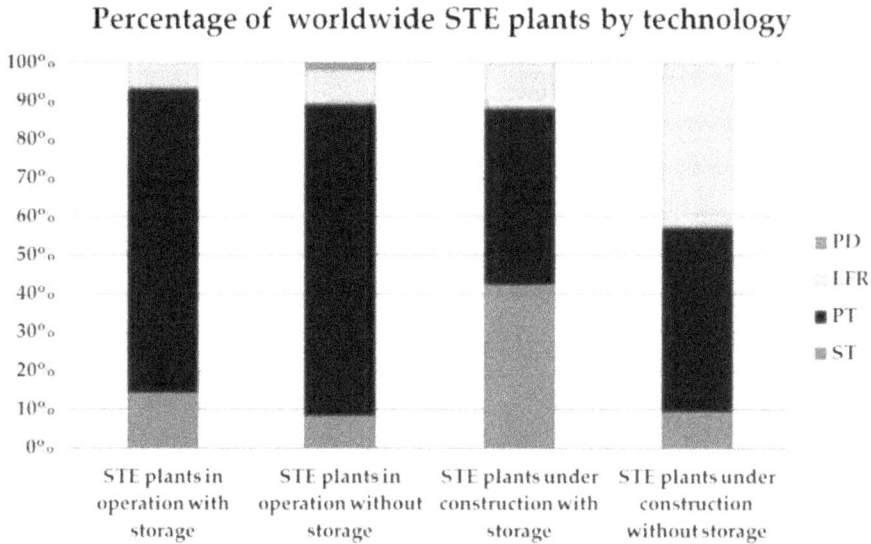

FIGURE 5.5   STE worldwide capacity categorized by technology and with/without storage [5,23,24].

## 5.4   LINE FOCUS SOLAR CONCENTRATORS

Line focus solar concentrators have reflectors that concentrate solar radiation onto a linear receiver. The two dominant concentrator technologies are parabolic troughs and linear Fresnels.

### 5.4.1   PARABOLIC TROUGH

A conventional parabolic trough solar concentrator comprises a parabolic reflector, with reflector support structure, pylons with joint attachments to allow single-axis solar tracking, pier foundations, a receiver fixed to the reflector support structure and pipework to convey the heat transfer fluid to the receiver and to its storage or utilization point. The troughs are laid out in parallel rows as a solar field with spacing between the rows to minimize shading of the reflectors while allowing sufficient access for maintenance and minimizing the pipework and parasitic pumping energy for the heat transfer fluid. The heat transfer fluid normally enters at one end of the trough and leaves at the other, although it might also have a tube-in-tube arrangement to allow entry and exit at one end (Figure 5.6). The troughs are normally installed to maximize the annual energy output, with the collector aligned from solar north to south, allowing tracking from east to west. The troughs may also be installed to maximize the energy output at solar noon in winter, with the collector aligned from east to west. Commercially available trough solar concentrators achieve concentration ratios within the 50–80 range.

Parabolic trough technology needs to increase performance to achieve a solar field system with improved efficiency and lower operation and maintenance (O&M) costs [34].

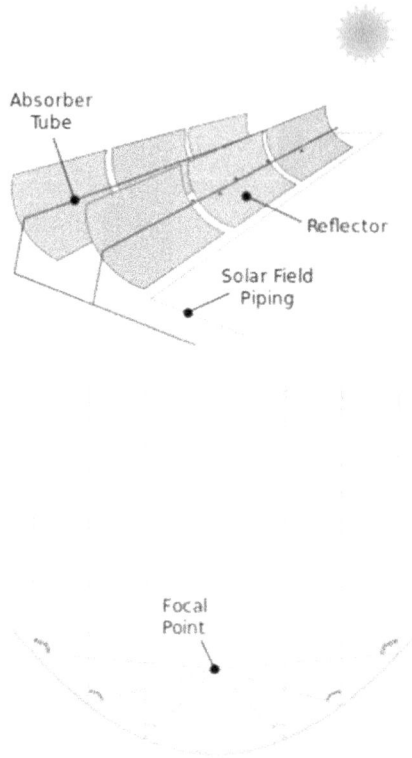

**FIGURE 5.6** Details of Parabolic trough concentrator [33].

The performance of the STE plant is mainly related to the optical and thermal properties of the collector system and to the maximum operation temperature. The commonly used oil-based HTFs, however, are restricted to operation temperatures of 400°C, and therefore limit the overall plant efficiency. The key innovation to overcome this limitation is the use of high-temperature HTF, such as molten salts or direct steam generation (DSG), which enables operation temperatures up to 550°C. On the other hand, project execution and solar field construction costs can be significantly reduced, with a potential reduction of levelized cost of electricity (LCOE), at higher operational temperatures in a molten salt power plant with thermal storage unit, of above 20% compared with a non-storage standard HTF plant and 15% to an oil-based HTF plant with thermal storage. Parabolic systems sometimes use thermocline tanks to address some of the energy storage issues. Cylindrical mirrors focus the sun's rays on an absorbent tube. These mirrors then collect sunlight along one axis, while the receivers that contain the heat transfer fluid are hit with focused sunlight. As a result, this technology enables the use of natural gas as a secondary source of fuel. French company ALTO Solution designs and develops a parabolic trough to concentrate solar power either for the production of process heat or for producing electricity. The startup's *Sol.CT* technology also includes an ultra-high performance concrete structure that reduces optical errors. The approximation of a near-perfect parabola makes such mirrors cheaper compared to thermally curved mirrors. In addition, parabolic troughs offer higher resistance to wind.

The main two challenges in parabolic trough receivers are cost-reducing measures (including manufacturing, installation, and maintenance) and durability. Cost reduction can be afforded in several ways: using cheaper materials without affecting performance and durability; increasing tube length from actual 4–6 m or even longer sizes and with new simpler and innovative designs such as re-evacuable tubes or dynamic vacuum systems. Longer pipes present a higher useful surface, less costly welding between pipes, larger collectors that require fewer servomotors and controls, and so on. The main degradation problem observed in receiver tubes, operating in commercial solar power plants, is vacuum loss in the inner annulus produced by hydrogen diffusion from oil-based HTF through the metal pipe wall. Getters have been extensively used to try to reduce hydrogen partial pressure in vacuum annulus, but the 25 years expected durability is not achieved. Moving from oil-based HTF systems to molten salts or DSG can reduce vacuum degradation problems and can help to reduce receiver tube cost [34].

The cost of Parabolic trough can also be optimized by using selective coating on the reflector. Double selective coated receivers have shown significant economic performance improvement over traditional receivers in parabolic trough CSPs, particularly in high temperatures and low irradiation conditions. The LCOE results of the power plants utilizing synthetic oil and MS are similar to the AEP results. The literature indicates that the LCOE of MS CSP plants with double-selective-coated receivers in Phoenix, Sevilla, and Tuotuohe can be decreased by 6.9%, 8.5%, and 11.6%, respectively. The main components of a heat collecting element or solar receiver are metallic pipe; glass pipe; selective absorber; antireflective coating (ARC); getters; glass metal welding; and bellows. The glass cover is connected to the steel pipe using metallic expansion bellows to compensate for different linear thermal expansion of glass and steel when solar receiver is working at nominal temperature. The glass metal welding is used to connect the glass cover and the flexible bellows, and it is shielded from concentrated solar radiation to avoid the thermal and mechanical stress that could affect welding performance and durability. The heat thermal fluid (HTF) is pumped into the metallic pipe to collect thermal energy absorbed by the selective absorber and vacuum is established in the annulus to reduce thermal losses.

Glass cover is used in heat collector elements (HCEs) to reduce convective thermal losses and to preserve selective absorber from the outdoor exposition. The glass pipe, concentric to metallic absorber pipe, needs to have high solar transmittance and a low thermal expansion coefficient to provide good thermal shock resistance, close to the thermal expansion coefficient of metallic absorber pipe at operating temperature. This reduces the mismatch between linear expansion coefficients of glass and metal pipes. Vacuum in annulus between glass and metallic pipes reduces thermal losses by conduction-convection and avoids thermal oxidation of the selective absorber deposited on the metallic pipe. This annulus is sealed with a glass-metal welding, and a metallic bellow is used to compensate for the mismatch expansion coefficient between glass and metal pipes. The glass material used as a glass cover in HCE has to be borosilicate glass. This glass is characterized by the presence of substantial amounts of silica ($SiO_2$) and boric oxide ($B_2O_3 > 8\%$) as glass network formers. Borosilicate glasses have a solar transmittance from 0.91 to 0.93, good thermal shock resistance, and good outdoor durability.

The glass jackets of receiver tubes are coated with a film on both sides (inner and outer) to reduce the reflection losses in the glass, thereby increasing the optical efficiency of the receiver tube. This film, known as ARC (antireflective coating) has to satisfy two requirements (related to coating thickness, and refractive index value) to obtain the destructive interference of light which is reflected at the glass coating interface and at the coating air interface. The material most widely used as ARC on glass is silicon dioxide ($SiO_2$). Its porous nature, necessary to increase the glass transmittance, is the weak spot of this material in terms of durability. The pores easily absorb water and other volatile organic components, increasing the refractive index of the coating and lowering the transmittance. Another consequence of the porous structure is the weak mechanical performance coming from the weak binding force between the silica particles and substrate, as well as between the particles [35]. The technology most widely used for preparing the ARC on glass receiver tubes is the sol gel dip-coating method. Efforts are also made to develop multifunctional coatings that work not only as ARC but also as easy-to-clean or self-cleaning coating.

HTF is pumped into the receiver metallic pipes along collector lines to be heated by solar concentrated radiation. Hence, the metallic pipe needs to have good thermal conductivity, good mechanical property and good corrosion resistance regarding HTF employed in the system. Steel is commonly used for this purpose. Carbon and low-alloyed stainless steels have better thermal conductivity than high-alloyed austenitic stainless steel, and they are cheaper but these steels have lower corrosion resistance than austenitic stainless steel. Receiver thickness increases with the maximum operative pressure (100 bar), and this leads to a thicker tube (>4.5 mm) which increases the cost of the technology [36]. When synthetic oil and organics such as biphenyl/diphenyl oxide systems are used as HTF, austenitic AISI 321L stainless steel is preferred compared to other austenitic stainless steel due to its lower hydrogen permeation. Hydrogen diffusion needs to be prevented because it causes HTF thermal degradation and it accumulates in vacuum annulus [37]. Typical wall thickness employed is 2 mm.

Selective absorber is mainly responsible for optical and thermal efficiency of a receiver tube. It absorbs concentrated solar radiation that reaches the metallic pipe, and the collected energy is transferred to the HTF. Receiver tube thermal losses by conduction and convection are reduced with vacuum environment within metal and glass pipes and radiative losses are minimized by the low thermal emittance of the selective absorber. Absorber collects solar radiation and has to transfer heat to the HTF, hence increasing wall thickness and decreasing steel heat transfer coefficient leads to a higher absorber temperature [38]. Selective absorbers are degraded due to thermal stress, oxidation and diffusion. Absorber degradation impacts strongly solar field performance. Air leakage at the glass to metal seal appears to be the primary cause of failure, resulting in a loss of vacuum and oxygen exposure of the cermet coating. Archimede Solar Energy sputtered cermet shows a relatively good stability in the air at high temperature up to 450°C and good optical properties, [39]. Huiyin Group, in China, has developed a sputtered selective absorber cermet with good optical performance, and it has been successfully used in Fresnel systems without thermal degradation. A quite promising absorber has been used for flat collectors and medium temperature applications produced by sol-gel deposition technology that uses a $CuMnO_4$ spinel mixed oxide [40]. This absorber is stable in air at temperatures higher than 1,000°C, and it is almost transparent to IR radiation. Combining this spinel absorber with an IR reflector stable at high temperature, such as platinum or b-tungsten, it is possible to prepare a selective absorber which is stable in air at temperatures of 550°C.

Receiver tubes have vacuum in the annulus between metallic and glass pipes, with pressure usually lower than $10^3$ mbar, to reduce conductive convective thermal losses and to preserve selective absorber from air oxidation. Annulus space is sealed using a glass metal welding that attaches borosilicate glass cover with a metallic ring that is finally welded to a bellow installed to compensate dilatation mismatch between glass and steel pipes. Vacuum maintenance is one of the most critical issues in solar receivers' design, construction, and maintenance cost. Vacuum degradation leads to a dramatic increase in thermal losses due to conduction/convection; it can produce selective absorber degradation resulting in the need for expensive tube replacement. Most usual problems in vacuum maintenance, observed in plant operation, have been glass metal welding failure and hydrogen diffusion from HTF to vacuum annulus. Hydrogen diffusion problem can be solved by using molten salts or DSG instead of oil-based systems, using hydrogen diffusion barriers on steel pipe or removing or reducing hydrogen content from HTF using filters with any material that adsorbs hydrogen or venting the fluid system on a regular basis so that the gas does not migrate through the steel receiver wall [37]. The problem of mechanical stresses in the seal can be resolved using Kovar or cheaper metal alloys such as SS AISI 430 [41,42].

In recent years, a common practice among manufacturers of receiver tubes is to incorporate so-called "hydrogen getters" in- side the vacuum space of each HCE. These materials absorb any traces of hydrogen gas that might be formed over time as a result of the thermal degradation of the HTF. The getter consists of a material that adsorbs hydrogen to form a hydride and thereby, removes hydrogen from the annular volume. The limitation of this method is the finite capacity of the getter for hydrogen. Once the getter saturates, it cannot adsorb additional hydrogen allowing the concentration of hydrogen in the annulus to increase. Getters are also expensive. Another method to remove hydrogen from the annulus is to locate hydrogen-permeable membrane as a barrier between the annular volume and ambient air. The membrane is most commonly a thin layer of palladium that is selectively permeable to hydrogen. At elevated temperatures, hydrogen permeates through the membrane from the annulus to ambient air where it reacts with oxygen to form water. This method works in principle, but practical implementation results in failure of the palladium membrane due to corrosion of the membrane or hydrogen embrittlement when operating at the design temperature, so it is not used in new generation receiver tubes. Probably, the most obvious solution to solve the vacuum maintenance problem is to design solar receivers that can be evacuated during operation on demand. These receivers can be continuously evacuated in plant operation in a dynamic system or re-evacuated when hydrogen partial pressure is high enough.

A new absorber tube has been developed by ARIES INGENIERIA company, which presents a continuous open chamber in which the dynamic vacuum is held and the internal steel tube and external glass tube are totally independent [43]. Vacuum level and gas composition in- side the chamber can be maintained continuously with minimal thermal losses. ARIES claims that vacuum does not have to be continuously created; on the contrary, the operation would have a very large periodicity with little energy consumption. Furthermore, minimal maintenance is achieved by the use of conventional pumping systems [44]. Huiyin Group has developed a serviceable HCE [45] that provides the capability to evacuate vacuum tubes during operation in a solar plant and regenerate getters located in a "getter box." It presents a reusable nozzle to connect a vacuum pump to the vacuum tube under operation in a power plant when hydrogen pressure increases.

The bellow is the airtight connection between the absorber and the glass envelope. Depending on the specific receiver design, the bellow might also have to support the weight of the glass envelope. The main degradation problems observed in metallic bellows are due to the process known as intergranular stress corrosion cracking (ISSC). This corrosion process is produced by a susceptible material in an aggressive medium, particularly during the welding process. Bellows are not in contact with HTF and one bellow part is inside the vacuum chamber, so it is protected against any aggressive medium. The other bellow surface is exposed to air and it is necessary to avoid the formation of an aggressive environment. Both solar receiver tube ends are usually covered with an insulating material to reduce thermal losses and it is critical to select a suitable insulating material without chlorides traces and very low humidity content to preserve the receiver from aggressive mediums that could lead to ISSC. Residual stresses can be introduced by cold deformation and forming, welding, heat treatment, machining, and grinding. Recommended materials for bellows are stainless steels stabilized with titanium (Grade 321) or niobium (Grade 347) that are less susceptible to suffer ISSC. Bellow design is generally optimized to obtain maximum active length at operating temperature. Currently, HCE producers publicize values close to 97% at 400°C.

## 5.4.2 LINEAR FRESNEL

A conventional linear Fresnel solar concentrator comprises reflectors that may be flat or slightly parabolic, with a reflector support structure, pier foundation, frame with joint attachments to allow single-axis solar tracking and a receiver fixed above, but independent of the reflector support structure by having independent reflectors rather than a more continuous reflector surface. While there may be some loss of optical efficiency from having independent reflectors, there may be other advantages that contribute to the overall collector efficiency or cost-effectiveness such as pipework to convey the heat transfer fluid to the receiver and to its storage or utilization point. The individual reflectors are laid out parallel to the ground such that each reflector has a different focal length to its receiver. The spacing of the reflectors is close to minimize the discontinuity in the reflective area or aperture. The width of the reflectors is optimized to allow access to the reflectors for maintenance, while not being too large to complicate the support structure or tracking (Figure 5.4). While a reflector may be placed so that it can be associated with more than one receiver, access to the receiver for maintenance also needs to be considered. The alignment of linear Fresnel systems is the same as for parabolic troughs. Commercially available linear Fresnel solar concentrators achieve concentration ratios within the 30–70 range. Linear Fresnel reflectors use long, thin segments of mirrors to focus sunlight onto a fixed absorber located at a common focal point of the reflectors. This concentrated energy is transferred through the absorber into some thermal fluid (this is typically oil capable of maintaining liquid state at very high temperatures). The fluid then goes through a *heat exchanger* to power a *steam generator.* As opposed to traditional LFRs, the CLFR utilizes multiple absorbers within the vicinity of the mirrors. A compact linear Fresnel reflector (CLFR) —also referred to as a concentrating linear Fresnel reflector—is a specific type of linear Fresnel reflector (LFR) technology. They are named for their similarity to a *Fresnel lens*, in which many small, thin lens fragments are combined to simulate a much thicker simple lens.

The reflectors are located at the base of the system and converge the sun's rays into the absorber. A key component that makes all LFRs more advantageous than traditional parabolic trough mirror systems is the use of "Fresnel reflectors." These reflectors make use of the Fresnel lens effect, which allows for a concentrating mirror with a large aperture and short focal length while simultaneously reducing the volume of material required for the reflector. This greatly reduces the system's cost since sagged-glass parabolic reflectors are typically very expensive. However, in recent years thin-film nanotechnology has significantly reduced the cost of parabolic mirrors. Fresnel reflectors help lower energy costs as they incorporate a narrow design, allowing for less windage compared to other types of solar concentrators. Moreover, fresnel reflectors provide higher peak power and provide a more uniform load distribution. This design additionally eliminates the need for massive concrete foundations and ensures low-resource utilization. ALSOLEN, a Moroccan startup works on solar thermal power plants using fresnel reflectors. Apart from integrating power and thermal storage, the power plants designed by this company also provide a boost to energy efficiency, for example, by using absorption chiller machines. The company also builds mobile factories that support the manufacturing processes of the solar field.

A major challenge that must be addressed in any solar concentrating technology is the changing angle of the incident rays (the rays of sunlight striking the mirrors) as the sun progresses throughout the day. In order to extend the efficiency of Fresnel collector, the first try goes toward taking proper care of the incoming "etendue." The first suggestion is to use multiple receivers instead of one. This concept was baptized as compact LFR, or CLFR for short. Compact signifies that less spacing can now be left between mirrors for ground loss reduction in larger primaries. This concept adds a degree of freedom. That of sending the reflected radiation to one or the other receiver, thereby controlling blocking. The second concept comes from the idea of placing the mirrors on an "etendue" conserving curve. The "etendue"-conserving curve uses an extra degree of freedom, that of raising/ lowering the height of each mirror with respect to its neighbors and thereby further reducing (even eliminating) shading and blocking.

The reflectors of a CLFR are typically aligned in a north-south orientation and turn about a single axis using a computer-controlled solar tracker system. This allows the system to maintain the proper angle of incidence between the sun's rays and the mirrors, thereby optimizing energy transfer.

The absorber is located at the focal line of the mirrors. It runs parallel to and above the reflector segments to transport radiation into some working thermal fluid. The basic design of the absorber for the CLFR system is an inverted air cavity with a glass cover enclosing insulated steam tubes, as shown in Figure 5.7a. This design has been demonstrated to be simple and cost effective with good optical and thermal performance. For optimum performance of the CLFR, several design factors of the absorber must be optimized. First, heat transfer between the absorber and the thermal fluid must be maximized. This relies on the surface of the steam tubes being selective. A selective surface optimizes the ratio of energy absorbed to energy emitted. Acceptable surfaces generally absorb 96% of incident radiation while emitting only 7% through infra-red radiation. Electro-chemically deposited black chrome is generally used for its ample performance and ability to withstand high temperatures. Second, the absorber must be designed so that the temperature distribution across the selective surface is uniform. Non-uniform temperature distribution leads

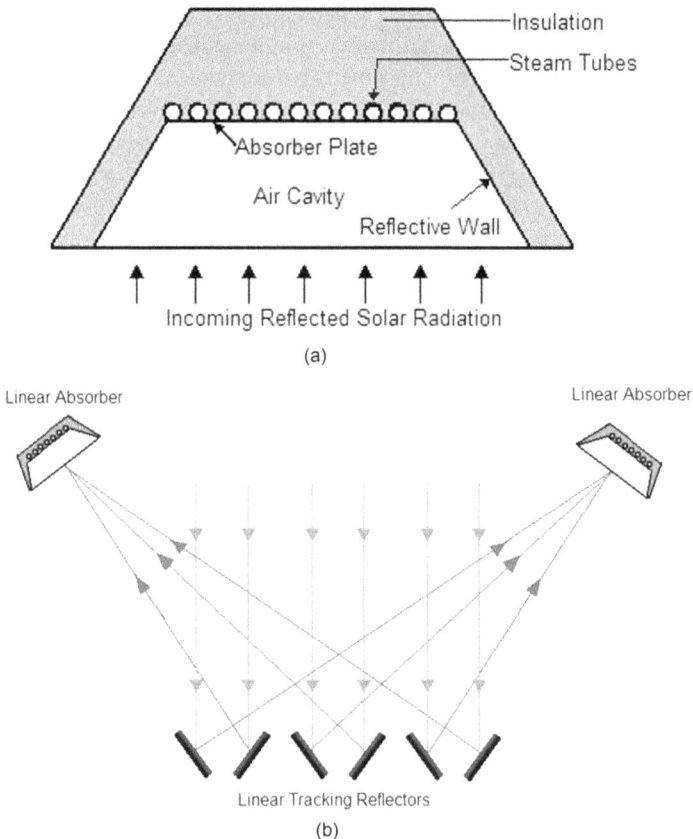

FIGURE 5.7    (a) Compact linear Fresnel reflector absorber; (b) CLFR alternating inclination [46].

to accelerated degradation of the surface. Typically, a uniform temperature of 300°C (573 K; 572°F) is desired. Uniform distributions are obtained by changing absorber parameters such as the thickness of insulation above the plate, the size of the aperture of the absorber and the shape and depth of the air cavity.

As opposed to the traditional LFR, the CLFR makes use of multiple absorbers within the vicinity of its mirrors. These additional absorbers allow the mirrors to alternate their inclination, as illustrated in Figure 5.7b. This arrangement is advantageous for several reasons. First, alternating inclinations minimize the effect of reflectors blocking adjacent reflectors' access to sunlight, thereby improving the system's efficiency. Second, multiple absorbers minimize the amount of ground space required for installation. This in turn reduces the cost to procure and prepare the land. Finally, having the panels in close proximity reduces the length of absorber lines, which reduces both thermal losses through the absorber lines and the overall cost of the system. The receiver may be atmospheric or in an evacuated environment. For very high temperatures (above 400°C up to 600°C), selective coatings for atmospheric receivers (tubular or not) with guaranteed long-term stability/durability are still problematic. For the evacuated tubular receiver, a sort of standard in the market today, the 70-mm receiver [47,48] is often used. In order to reduce the number of rays that get lost without reaching the receiver, a number of options can be adopted. Beside the CPC-type concentrator [49,50] there are several other second stage-type concentrators that can be combined with a primary concentration stage. Examples are compound elliptical concentrator (CEC) [49,50,51], tailored-edge ray concentrator (TERC) [51,52], simultaneous multiple surface (SMS) [51,53], and aplanatic optics [51,54]. In theory, it can be shown that TERCs provide the highest possible concentration. However, they must be severely truncated since by definition they extend from the secondary to the primary, and the resulting full shadow would make them useless. Another possible solution is an SMS single receiver solution with a very large primary (20 m) designed for the 70 mm evacuated tube and to solve the gap problem (losses in gap) at the same time reducing the losses through the fact in the other solutions some rays still need to traverse several times the outer glass envelope. This SMS solution strives for having the largest number of rays going only once through the glass envelope. The "etendue" is conserved through the SMS solution development condition, of simultaneous primary and secondary optimization.

However, it should be noted that the SMS solution has a second-stage concentrator with a size that is typically larger than other types, as, for instance CECs [55]. A larger second stage means larger shadow over the primary for off-axis incidence angles. The last configuration [56] proposes a second stage of the CEC type and some of the other features of the preceding solution. It is a multiple [2] receiver solution, but the two receivers are placed on a single tower. The idea is to join two asymmetric solutions. The second stage is composed of two macro focal CECs [57], asymmetric, and containing a V-groove section that is designed to eliminate the possibility of rays escaping through the gap between the receiver and outer glass tube. The total primary width is now 26 m. These advanced concepts accomplish to provide much higher concentration values (more than the double of what parabolic troughs provide). This means less thermal losses at higher temperatures, themselves associated with larger efficiency conversion from heat to electricity and a number of other advantages like substantially larger apertures, thus a substantial reduction of rows in a collector field, less HTF, less components, less parasitic losses and thermal system losses, and less O&M. The advanced LFR solutions shown to correspond to very large primaries, larger than 20 m. One with 26 m, implies, when compared with 6 m aperture PTs in collector field, to a potential reduction of the number or rows by a factor larger than 4. This is a substantial reduction of the number of pipes, valves, bents, HTF fluid volume, thermal and parasitic pumping power losses, and so on, also with benefits on O&M tasks and costs. There are also inherent advantages of LFR systems, more compact (tighter ground cover) and with fixed receivers, a substantial simplification, since it eliminates the need for flexible hosings or swivel-type joints, present in all PT fields. Other ones are the lowest cost associated with flat mirrors, simple tracking mechanisms, better response to high winds. Compare to parabolic troughs, LFR has an extra penalty from effects arising from the cosine of incidence angle. However, the advantages of the advanced LFR concepts compensate for this. Thus, LFRs have a great potential for cost reduction but some inherent problems to solve as well. These problems are mainly related, as mentioned earlier, the optical losses due to the cosine effect. In this sense, future developments will be strongly connected with more advanced designs and configurations, seeking maximum concentration, and maximum efficiency in order to compensate for the aforementioned losses.

As regard to the practical applications of the LFR concept, *Areva Solar* (Ausra) built a linear Fresnel reflector plant in New South Wales, Australia. Initially a 1 MW test in 2005 was expanded to 5 MW in 2006. AREVA Solar also built the 5 MW *Kimberlina Solar Thermal Energy Plant* in Bakersfield, California in 2009. This was the first commercial linear Fresnel reflector plant in the United States. The factory was planned to be capable of producing enough solar collectors to provide 200 MW of power per month. In March 2009, the German company *Novatec* Biosol constructed a Fresnel solar power plant known as PE 1. The solar thermal power plant used a standard linear Fresnel optical design (not CLFR) and had an electrical capacity of 1.4 MW. The commercial success of the PE 1 led *Novatec Solar* to design a 30 MW solar power plant known as PE 2. PE 2 has been in commercial operation since 2012. From 2013 on *Novatec Solar* developed a molten salt system in cooperation with *BASF*. The largest CSP system using Compact linear Fresnel reflector technology is the 125 MW Reliance Areva CSP plant in India. In China, a 50 MW commercial-scale Fresnel project using molten salt as its heat transfer medium has been under construction since 2016. After the grid connection in 2019, it now seems to operate successfully until present time.

### 5.4.3 Options for Heat Transfer Fluids for Line Focus Concentrators

The global efficiency of an STE plant mainly depends on the efficiency of both the PCS (power conditioning system) and the solar field. Working fluids affect plant efficiency by the pressure drop in the solar field circuit and the heat transfer coefficient in the receiver tubes. The pressure drop has a direct impact on the plant's electricity consumption because the pumping power demanded by the solar field is the main fraction of the overall electricity consumption of the plant. The high value of the heat transfer coefficient is essential for the integrity of the receiver tubes. High-temperature difference between the irradiated and the nonirradiated parts of the tubes created by low heat transfer coefficient can contribute to the bending and even breakage of the tubes [58]. Sixty-three solar thermal electric (STE) plants with parabolic trough collectors (PTCs) were fully operational during mid-2015. Sixty-one of these plants were using thermal oil as a working fluid in the solar field. The use of thermal oil in the solar field is usually known as "heat transfer fluid (HTF) technology" because the thermal oil transfers the thermal energy delivered by the solar field to the thermal energy storage (TES) and to the steam generating system producing the steam for the plant power conversion system (PCS).

Thermal oils are preferred because of their affordable price, low vapor pressure, good thermal stability, and long lifetime if the working conditions recommended by the manufacturers are fulfilled. However, they are not perfect because they can cause environmental contamination in case of leaks, create fire hazards when there is a leak and have limited working temperature which can limit efficiency. In the case of electricity production using LFR, the most common HTFs are also thermal oils, for energy delivered up to 390°C, providing a much higher solar-to-electricity conversion efficiency [59]. In southern Iberian Peninsula locations, these systems can have a yearly conversion efficiency of about 15%. Thermal oil as HTF usually requires a heat exchanger, as in solar-thermal electricity (STE) applications to produce steam for the Rankine cycle (steam turbine). In many of the present solar power plants for STE: the thermal energy storage is connected to the thermal oil loop by means of a heat exchanger. Energy storage is a highly desirable feature and sets STE completely apart from PV electricity production. Many hours of storage can be provided in this fashion, allowing for nominal power production for many hours on demand, even with no sunshine: thus STE is said to be truly dispatchable. Three other innovative working fluids for both PT and LFR investigated so far are liquid-water/steam (the so-called direct steam generation, DSG, technology); molten salts, composed of ternary or binary salt mixtures (potassium and sodium nitrates mainly) and pressurized gases, like $CO_2$, $N_2$ or air. These three alternative working fluids have advantages and disadvantages when they are compared to thermal oils traditionally used in PTC. For example, while DSG has a simpler plant configuration, higher steam temperature, no fire hazard and no pollutant, it lacks a cost-effective thermal storage system, presents the complexity of solar field control and higher pressure in solar field piping. While molten salts result in cheaper thermal storage, no fire hazard and no pollutant, it produces freezing hazard, complex solar field design and higher electricity self-consumption. Finally, while pressurized gas produces higher steam temperature, cheaper thermal storage and no fire hazard and no pollutant, it results in lower heat transfer coefficient, presents the complexity of solar field control and produces higher pressure in solar field piping and higher pumping power.

### 5.4.3.1 Direct Steam Generation (DSG Technology)

In the DSG process liquid water is pumped to the solar field inlet to be preheated, evaporated, and converted into saturated or superheated steam (depending on the solar field design) as it circulates through the receiver tubes. The test facility implemented at the Plataforma Solar de Almería (PSA) within the DISS-phase I project

[14] was the first life-size DSG test facility available in the world to study the DSG process under real solar conditions. STE plant with PTC and thermal oil as working fluid is composed of four main subsystems: solar field, TES, steam generator, and PCS. When DSG is used in the solar field, the steam generator is no longer needed because the steam for the PCS is directly provided by the solar field. The removal of the steam generator has twofold benefits: the simplification of the plant configuration and a lower investment cost because the steam generating system is quite expensive. Another important benefit of DSG is the possibility to increase the temperature of the steam provided to the PCS, thus increasing the overall plant efficiency because of better thermodynamic parameters of the Rankine cycle. DSG has no thermal limitation and steam at temperatures even higher than 500°C can be produced in the solar field. The non-pollutant nature of water is another benefit of DSG, because water leaks have neither fire nor contamination hazards, while oil leaks are very dangerous for people and for the environment.

The lack of a suitable and cost-effective TES is a significant barrier to the commercial deployment of DSG STE plants. Dispatchability is the main benefit of STE plant when they are compared with wind farms or photovoltaic (PV) plants. Since dispatchability is provided by the TES systems, the lack of a competitive TES for DSG is a significant constraint of this alternative working fluid. Solar plants using thermal oil can be provided with TES systems based on molten salts so that the thermal energy is stored as sensible heat (i.e., temperature increase) of a binary mixture of molten salts, the so-called solar salt (i.e., 40% of potassium nitrate and 60% of sodium nitrate). This type of TES system had been implemented in 21 STE plants with PTC at the end of 2015. TES systems using molten salts are cost-effective and they provide the STE plants with an excellent degree of dispatchability. However, this type of TES system is not suitable for solar plants with DSG because the steam provided by the solar field will condense as it trespasses its thermal energy to a storage media. Another constraint of DSG is the complexity of the solar field control because uneven two-phase flow (i.e., liquid water and steam) conditions in parallel rows of PTC are likely to happen during solar radiation transients [20,21]. Therefore, control devices must be installed at each row to avoid flow instabilities. This problem does not happen in solar fields using thermal oil. Although DSG imposes higher pressures in the solar field circuit this can be overcome by the use of thicker steel pipes and mechanically more resistant fittings. The working pressures above 100 bar are not recommended for large DSG solar fields because water leaks at flanges and nonwelded connections will often be likely to occur due to the daily thermal cycles at the solar field.

It has been experimentally proven that the best configuration for collector rows with DSG is one with the end of the evaporating section separated from the beginning of the superheating section by a liquid-water/steam separator. With this configuration, the feed water flow at the inlet is higher than the steam flow required at the outlet, thus assuring an excess of liquid water at the end of the evaporating section. This excess of liquid water is recirculated to the inlet of the collector row and mixed with the feed water coming from the solar field feed pump. The first commercial DSG plant was installed in Thailand and put into operation in 2010 [15]. It is the plant Thai Solar One (TSE-1), with a nominal power of 5 MWe. The DSG technology, including TES with PCM, is already available for small IHP applications demanding saturated steam. The use of DSG requires additional research on control systems and operation strategies for superheated steam production without water/steam separators inserted in the solar field, optimization of solar field control and development of TES systems with PCM suitable for large storage capacities (about 1 GWh). The most common heat transfer fluids for LFR are water (under pressure), steam (saturated and superheated), and thermal oils. Conventional LFRs have been offered to produce water or saturated steam for operating temperatures up to 270°C. They are typically associated with a Rankine cycle for electricity production or with a direct thermal application such process heat, air conditioning/refrigeration, and desalination. PTs have been used for all these applications as well.

### 5.4.3.2 Molten Salts

The replacement of thermal oil by molten salts has several benefits. The first one is the possibility to work at higher temperature in the solar field (550°C–560°C) thus overcoming the thermal limit imposed by current thermal oils. This higher temperature at the solar field not only increases the efficiency of the PCS thermodynamic cycle, but it also enhances the integration of the TES into the plant because the heat exchangers (HXs) acting as interface between the TES and the solar field are no longer needed. Additionally, the investment cost of the TES is significantly cut down because the inventory of storage media is reduced by 64% due to the higher temperature difference between the cold and hot tanks (from 290°C to 550°C) instead of 90°C (from 290°C to 380°C) with thermal oil. Molten salts have high density, high heat capacity, high thermal stability, and very low vapor pressure even at elevated temperatures, which are important features for a heat transfer fluid. However, due to corrosion issues with carbon steel at temperatures higher than 400°C, molten salts (sodium and potassium nitrates) require the use of stainless steel in piping, vessels, and elements [60]. Another benefit of molten salts is the avoidance of the fire and contamination hazards that can occur by leakage of thermal oil in piping and vessels. Parasitic pumping losses are lower with molten salt because the volumetric flow and the pressure drop are lower than with thermal oil.

The most outstanding problem with molten salts, however, is the significant freezing hazard due to the high melting point of the current salts. For the salt used in commercial STE plants the freezing and melting processes take place in the temperature range from 220°C (melting) to 240°C (freezing). This high freezing temperature introduces a significant hazard of salt plug formation in the pipes when the plant is not operating and the ambient temperature is low. Once a salt plug is formed, the first problem is the difficulty to find its location. This search for the location of the plug can take a lot of time, and it could even be an impossible task when the circuit layout is very complex (e.g., a high number of valves, pipe fittings, and accessories). The design and implementation of an efficient heating system is the only way to prevent salt from freezing and to avoid these problems. In other options, EHT or impedance systems are the most developed. The need for an anti-freezing system in all the solar field piping, instruments, and receiver tubes makes the solar field design more complex and expensive. Three main salts commercially used for solar applications are Hitec XL (Sodium nitrate Potassium nitrate Calcium nitrate) with a melting point of 140°C and maximum working temperature of 500°C, Hitec (Sodium nitrate Potassium nitrate Sodium nitrite) with a melting point of 142°C and maximum working temperature of 538°C and Solar salt (Sodium nitrate Potassium nitrate) with melting point of 240°C and maximum working temperature of 593°C [18].

Two different electricity-based concepts are used in solar plants for heating pipelines and components in salts systems: impedance heating and mineral insulated heat tracing. Impedance heating directly heats the pipeline by flowing electrical current through the pipeline wall (Joule effect) by direct connection to a low voltage, high current source from a dual-winding power transformer. This concept heats uniformly the pipe and it is best suited for long straight runs of pipe without components, and it is critical that piping and instruments installed in an impedance system are not grounded. Even though high currents are flowing through the pipes, there is no safety hazard for personnel because the voltage is low. Impedance heating is the option used nowadays for the receiver pipes of PTCs using molten salt as working fluid because the glass cover surrounding the inner steel pipe makes the use of heat tracing unfeasible. Mineral-insulated electrical heat tracing (EHT) is nowadays the most common option in molten salt applications to maintain the temperature of pipelines (not the receiver tubes) and components above the salt freezing point, typically at 290°C. This option represents no risk for the instrumentation or personnel but requires a more careful design and installation than an impedance system. The electrical consumption of impedance systems in the steady state at 290°C is higher than in heat tracing systems, mainly due to higher electrical losses in the feed cables and power transformer [17].

At the end of 2015, there were 22 commercial STE plants using molten salt in the thermal storage system, and only one plant was using molten salt as working fluid in the solar field. This plant is the 5 MWe Archimede solar plant, owned by ENEL and built at Priolo Gargallo near Syracuse in Sicily, Italy. The solar field of this plant has inlet- outlet temperatures of 290°C/550°C, and it is composed of nine parallel loops with six 100-m and 590 m² PTCs in each loop. Since the high melting point is the main disadvantage associated with molten salts when used in solar fields with PTCs, new systems which include binary mixture of potassium and sodium nitrates and ternary mixtures of potassium, lithium and sodium nitrates have been reported. It has been proven at laboratory scale that the heat capacity and thermal conductivity can be significantly improved by adding a small percentage (1.0 wt.%) of silica or alumina nanoparticles [61]. Overall economic benefits of these different mixtures are not as yet clear. Often the molten salts are directly used as HTF and as means to store energy, as it is done at the Central Receiver plant of Torressol/Gema- solar [62]. Molten salts are used in this fashion, using the advanced LFR technology.

### 5.4.3.3 Compressed Gases

The use of pressurized gases is another option to replace thermal oil in PTCs. They are safe and clean fluids from an environmental viewpoint because they do not pose the fire hazards and environmental constraints associated with thermal oils, and they are able to work at higher temperatures in the solar field without thermal stability problems. Additionally higher temperature would increase power conversion efficiency and higher temperature differences between solar field inlet and outlet would enhance and make more efficient the integration of TES systems based on latent heat (e.g., two-tank molten salt systems). This in turn will also increase the storage capacity per volume and thus reduce the amount of storage medium. However, pressurized gasses also have some disadvantages when compared to thermal oils. The lower density of compressed gases reduces the heat transfer coefficients in the receiver tubes and increases the required pumping power. The only way to reduce the negative impact of a lower fluid density is to increase working pressure in the piping and associated components (i.e., HXs and vessels) of solar system. This would, however, affect the circuit pressure drop [16]. The evaluation of the technical and economic feasibility of pressurized gases as HTF was experimentally addressed by CIEMAT in the gas-cooled solar collectors project [19]. This project included the analysis of several working gases (helium, $CO_2$, $N_2$, and air) and the design, construction, and testing at PSA of the first-of-its-kind experimental solar facility composed of two 50-m parabolic-trough collectors connected to a complete hydraulic circuit using pressurized gases. The two solar collectors can be connected in series or in parallel, depending on the intended outlet temperature (in series for 525°C, or in parallel for 400°C), while the nominal inlet temperature was 225°C in both cases. The solar field was connected to a two-tank molten salt TES in order to evaluate the performance of not only the solar collectors with pressurized gas but also the overall performance of the system composed of the solar field and the TES.

The results of these [16,19,22] and other studies indicate that $CO_2$ is probably the best candidate for using pressurized gas as heat transfer fluid, especially when working at supercritical conditions ($T > 31$°C and $P > 7.4$ MPa). However, $CO_2$ also has some problems. In presence of water, $CO_2$ could react to form carbonic acid ($H_2CO_3$), which is corrosive to carbon steel which is widely used in solar plants. Also, supercritical $CO_2$ at temperatures higher than 400°C is incompatible with the graphite sealing used in the ball joints installed at the end of the receiver tubes of PTCs to allow their thermal expansion and rotation [19]. $N_2$ presents fewer corrosion issues than either $CO_2$ or air and can be obtained on site in a simple way from compressed air by means of a nitrogen generator. Therefore, nitrogen could be the most feasible HTF among the proposed gases for PTCs. The overall costs of a large solar field with nitrogen as HTF will be higher than that of a synthetic-oil field due to the increasing cost of distributed blowers and HXs. On the other hand, higher temperature differences in TES imply a lower molten salts inventory and therefore the cost of storage will be lower for the $N_2$ plant. Moreover, the cost of synthetic oil and the corresponding equipment (pumps, expansion vessels, conditioning system, and so on) will be replaced by a lower cost equipment for nitrogen. With the optimized solar field design composed of many modules connected in parallel to a molten-salt circuit, the overall plant efficiency for nitrogen is similar to the efficiency of a traditional STE plant with thermal oil. The study by Biencinto et al. [59] proved that similar annual performances can be attained for parabolic plants using synthetic oil and compressed $N_2$. However, there are significant uncertainties concerning the investment cost for the $N_2$ plant because of the high number of blowers (177) and HXs (531) required in the solar field. In order to improve this technology, future works should investigate new gases as heat transfer fluids with better thermo-hydraulic properties than $N_2$. Also, development of commercial PTCs with larger aperture areas and bigger receiver tubes could help to reduce the number of HXs per collector loop. While the use of gases as HTF is potentially very interesting, it brings an added problem, that of the substantially smaller heat transfer coefficient within the receiver, requiring higher pressures and higher temperatures. The usefulness of the concept is more apparent for much higher temperatures (i.e., from 600°C above). This means that they are better suited to be combined with central receiver (three-dimensional, 3D) type optics due to their capacity to deliver substantially higher concentration values. In the final analysis, none of the three options, water and steam, molten salts, and pressurized gases is the perfect in short term as a replacement for synthetic oils in PTCs in commercial plants. All of them have some issues and challenges such as: (a) DSG needs latent-heat thermal storage systems with phase-change materials and mass-flow control devices cheaper than conventional control valves (b) molten salts needs new salt mixtures with lower melting point and good thermal conductivity and nonrotating devices to absorb thermal expansion of receiver tubes and (c) compressed gas needs new gases mixtures (i.e., water and gas) with good thermos-hydraulic behavior and needs to address incompatibility between $CO_2$ and graphite.

### 5.4.3.4  Comparison of Molten Salt Heating Fluids for PT and LFR

As mentioned above, a concern with salts is the freezing/fusion point. Significant more demonstration with molten salt is needed before full-scale commercialization. In Europe demonstration plants such as that of PCS at Casaccia, Italy or the EMSP in Evora, Portugal, and the solar direct molten salts loop next to an existing power plant at Priolo Gargalo, Sicily, Italy are a part of that effort [63–65]. A combination of molten salts and LFR technology simplifies some of the unsolved issues, because of its much simpler possible mechanical configurations and arrangements compared with parabolic trough technology. LFR technology operates with fixed receivers in contrast with those in standard PT collectors, where potential problems arising from movable joints or flexible hoses must be addressed. LFR technology places the evacuated receivers high above the ground (usually 8 m or more) and that makes gravity-assisted salts drainage a simpler matter. LFR technology reduces considerably the length of the piping loop needed, especially not just because of high concentration, as explained for advanced LFR technology, but also because of solutions like the one described in the following, where a double receiver is placed on a single tower. Finally, LFR technology can be implemented easily with a slight pending gradient, to facilitate draining operations. For all the mentioned reasons advanced LFR technology has been gathering renewed interest since it can supply both electricity (using molten salts as HTF) or process heat (using water/steam as HTF) at a lower cost than today's STE standard. In terms of energy storage concepts, there will also be an opportunity for evolution; for instance, change from two-tank configurations to one-tank only and hybrid solutions with salts and different solid materials as storage media with extensive ongoing research to validate new materials.

High concentration is a very important feature for potential energy delivery cost reduction. Advanced LFR concentrators are uniquely placed to deliver high concentration and thus stand as a very interesting way to produce low-cost electricity or simple low-cost heat for direct use. LFR technology has a much smaller presence in the commercial plant market than its rival, linear PT technology. One of the reasons, is that while LFR has been perceived as potentially cheap, but associated with low-efficiency conversion from solar to electricity. These perceptions have relegated LFR to applications like industrial process heat or other niche markets. The new ideas around advanced LFR technologies are a breakthrough and stand to completely change that perception, showing the true potential of LFR with advanced optics to operate at much higher temperatures as well and in conditions that are much easier to install, operate, and control. A renewed interest in the technology will thus have a serious impact on the CSP market. Advanced LFR technologies with molten salts will become a contender for large-scale competitive electricity production.

## 5.5 POINT-FOCUS SOLAR CONCENTRATORS

Point-focus systems have reflectors that concentrate solar radiation onto a central receiver that is effectively a point compared to the reflector. The two dominant technologies are parabolic dishes (Figure 5.4) and central receiver systems, known as solar towers. The solar tower (Figure 5.4) has independent reflector facets, known as heliostats, rather than a more continuous reflector surface. While there may be some loss of optical efficiency from having heliostats, there may be other advantages that contribute to the overall collector efficiency or cost-effectiveness.

### 5.5.1 PARABOLIC DISH

A conventional parabolic dish collector (Figure 5.8) comprises a dish parabolic reflector, with reflector support structure, pier foundation, a receiver fixed to the reflector support structure, pylons with joint attachments to allow double-axis solar tracking and pipework to convey the heat transfer fluid to the receiver and to its storage or utilization point. The parabolic reflectors may be continuous or comprise discrete elements conforming to a parabolic shape. The receiver is fixed to the reflector support structure so that both the dish and receiver track the sun. Its shape is part of a *circular paraboloid*, that is, the surface generated by a *parabola* revolving around its axis. The parabolic reflector transforms an incoming *plane wave* travelling along the axis into a *spherical wave* converging toward the focus. Conversely, a spherical wave generated by a *point source* placed in the *focus* is reflected into a plane wave propagating as a *collimated beam* along the axis. The size of the receiver needs to be optimized to minimize the shadow it, and its support structure might create on the reflector. The mass of the receiver needs to be optimized to minimize the mass that needs to track the sun. The most common consideration is for a parabolic dish to have a Stirling engine placed at the receiver. Alternatively, the receiver might have a heat transfer fluid to drive an independent process or heat engine. One of the world's largest dish collectors is illustrated in Figure 5.8c. The dishes are laid out as a solar field with spacing to minimize collisions and shading of the collectors while allowing sufficient access for maintenance and minimizing the pipework and parasitic pumping energy for the heat transfer fluid. Commercially available dish concentrators achieve concentration ratios of more than 2,000 [67].

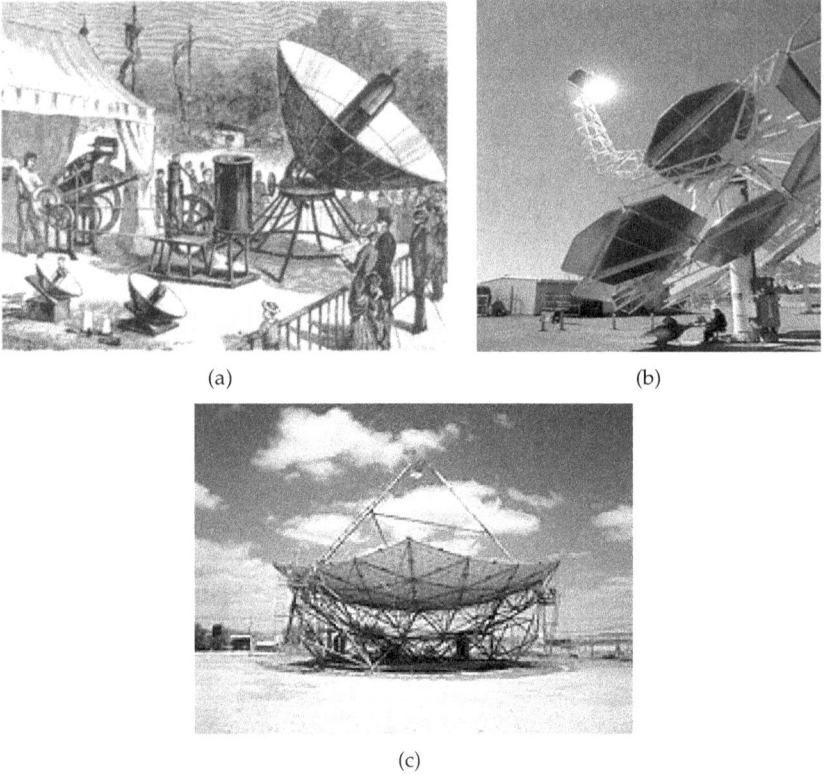

(a)                                                          (b)

(c)

**FIGURE 5.8**  Solar parabolic collector. (a) First concept; (b) Modern concept; (c) largest design [5,66].

Dish Stirling technology intends to produce low-cost energy among the various renewable energy technologies, especially in large-scale production and hot zones. The Stirling system uses a large reflective parabolic dish and focuses the sunlight on the dish, where the receiver captures the heat and turns it into usable thermal energy. Usually, the dish is combined with the Stirling engine in the Dish-Stirling system [68,69]. This also creates rotational kinetic energy that converts into electricity using an electric generator. Swedish company Azelio's technology eliminates using water for cooling or steam production, as the Stirling engine converts the heat into a mechanical movement to drive the generators directly. This results in higher solar to electricity conversion rate. Azelio also uses multiple suppliers of solar concentrators globally and integrates its own software to control them. There are a few variations of dish technology as described below.

### 5.5.1.1  Focus-Balanced Reflector

It is sometimes useful if the *center of mass* of a reflector dish coincides with its *focus* (see Figure 5.9). This allows it to be easily turned so it can be aimed at a moving source of light, such as the Sun in the sky, while its focus, where the target is

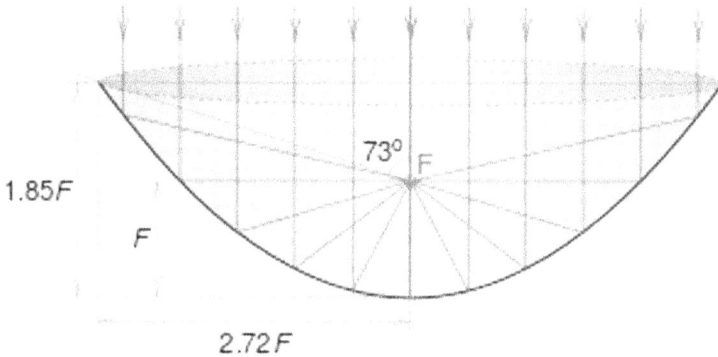

**FIGURE 5.9**   An oblique projection of a focus-balanced parabolic reflector [66].

located, is stationary. The dish is rotated around *axes* that pass through the focus and around which it is balanced. If the dish is *symmetrical* and made of uniform material of constant thickness, and if $F$ represents the focal length of the paraboloid, this "focus-balanced" condition occurs if the depth of the dish, measured along the axis of the paraboloid from the vertex to the plane of the rim of the dish, is 1.8478 times $F$. The radius of the rim is 2.7187 $F$. The angular radius of the rim as seen from the focal point is 72.68°.

### 5.5.1.2   Scheffler Reflector

The focus-balanced configuration (see above) requires the depth of the reflector dish to be greater than its focal length, so the focus is within the dish. This can lead to the focus being difficult to access. An alternative approach is exemplified by the Scheffler Reflector. This is a paraboloidal mirror which is rotated about axes that pass through its center of mass, but this does not coincide with the focus, which is outside the dish. If the reflector were a rigid paraboloid, the focus would move as the dish turns. To avoid this, the reflector is flexible, and is bent as it rotates so as to keep the focus stationary. Ideally, the reflector would be exactly paraboloidal at all times. In practice, this cannot be achieved exactly, so the Scheffler reflector is not suitable for purposes that require high accuracy. It is used in applications such as *solar cooking*, where sunlight has to be focused well enough to strike a cooking pot, but not to an exact point.

### 5.5.1.3   Off-Axis Reflectors

A circular paraboloid is theoretically unlimited in size. Any practical reflector uses just a segment of it. Often, the segment includes the *vertex* of the paraboloid, where its *curvature* is greatest, and where the *axis of symmetry* intersects the paraboloid. However, if the reflector is used to focus incoming energy onto a receiver, the shadow of the receiver falls onto the vertex of the paraboloid, which is part of the reflector, so part of the reflector is wasted. This can be avoided by making the reflector from a segment of the paraboloid which is offset from the vertex and the axis of symmetry.

## 5.5.2 Heliostat Field-Central Receiver

A conventional heliostat field-central receiver (see Figure 5.10) solar concentrator, comprises heliostat reflectors, that may be flat or slightly parabolic, with reflector support structure, pier foundation, a receiver mounted on a tower, pedestals with joint attachments to allow double-axis solar tracking and pipework to convey the heat transfer fluid to the receiver and to its storage or utilization point. The heliostat reflectors are placed in a solar field surrounding the tower. The receiver and tower need to be optimized to minimize the shadow they might create on the solar field. The solar field needs to be optimized in terms of the heliostat size, closeness to minimize the discontinuity in the reflective area, or aperture, spacing to minimize collisions, while also allowing sufficient access for maintenance. The system layout needs to be optimized to minimize the pipework and parasitic pumping energy for the heat transfer fluid. Commercially available heliostat field, central receiver solar concentrators, achieve concentration ratios within the 500–800 range. The systems using molten salts (40% *potassium nitrate*, 60% *sodium nitrate*) as the *working fluids* are now in operation. These working fluids have high *heat capacity*, which can be used to store the energy before using it to boil water to drive turbines. These designs also allow power to be generated when the sun is not shining.

As discussed by Ho and Iverson [71,72], unique challenges associated with high-temperature receivers include the development and use of geometric designs (e.g., dimensions and configurations), materials, heat-transfer fluids, and processes that maximize solar irradiance and absorptance, and minimize heat loss, and have high reliability at high temperatures over thousands of thermal cycles. Advantages of direct heating of the working fluid include reduced exergetic losses

**FIGURE 5.10**   PS10 solar power tower [70].

through intermediate heat exchange, while advantages of indirect heating include the ability to store the heat transfer media (e.g., molten salt and solid particles) for energy production during nonsolar hours. Ho and Iverson [71,72] show that a high concentration ratio on the receiver and reduced radiation losses are critical to maintain high thermal efficiencies at temperatures above 650°C. Reducing the convective heat loss is less significant, although it can yield a several percentage point increase in thermal efficiency at high temperatures (note that the convective heat loss in cavity receivers can be a factor of two or more greater than that in external receivers because of the larger absorber area [73]). Increasing the solar absorptance, and/or decreasing the thermal emittance, can also increase the thermal efficiency. The advantage of a tower-type solar station is that sunlight on the central receiver tower focuses on a smaller area, removing the need for pumping coolant through the pipeline around a large field of heliostats. This results in higher temperatures of the working fluid in the receiver and also improves steam performance. The relatively low cost makes combinations of solar panels and gas turbine power plants more profitable for isolated places. A list of solar power towers in the world is illustrated in Table 5.3.

## 5.6   SOLAR POWER TOWERS USING ADVANCED CYCLES

In recent years several advances are made to improve the performance of solar tower thermal power generation using advanced cycles [74,75]. Higher efficiency power cycles are being pursued to reduce the levelized cost of energy from concentrating solar power-tower technologies [76]. These cycles, which include combined air-Brayton, supercritical-$CO_2$ ($sCO_2$) Brayton, and ultra-supercritical steam cycles, require higher temperatures than those previously achieved using central receivers. Current central receiver technologies employ either water/steam or molten nitrate salt as the heat transfer and/or working fluid in subcritical Rankine power cycles. The gross thermal-to-electric efficiency of these cycles in currently operating power-tower plants is typically between 30% and 40% at turbine inlet temperatures <600°C. At higher input temperatures, the thermal-to-electric efficiency of the power cycles increases following Carnot's theorem. However, at temperatures greater than 600°C, molten nitrate salt becomes chemically unstable, producing oxide ions that are highly corrosive [77], which results in significant mass loss [78]. Here we examine the implementation issues of a few of these advanced cycles.

### 5.6.1   SUPERCRITICAL STEAM RANKINE CYCLES

SSRCs operate at pressures greater than 22.1 MPa in a range of temperatures only moderately higher (600°C–720°C, up to 760°C in the case of advanced ultra-supercritical cycles) than those of the state-of-the-art solar power towers and have been identified as a promising option to increase the efficiency and reduce the cost of the electricity generated by CSP plants in the near or medium term. The Sunshot Initiative of the US Department of Energy [79] identified the possibility to adapt current molten salt and direct steam generation (DSG) solar power towers to supercritical steam cycles operating in the range of 600°C–700°C. The long-term scenario for

**TABLE 5.3**
**List of Solar Power Towers [70]**

| Name | Developer/Owner | Completed | Country | Town | Height (m) | Height (ft) | Collectors | Installed maximum Capacity (MW) | Yearly Total ENERGY production (GWh) |
|---|---|---|---|---|---|---|---|---|---|
| ACME Solar Tower | ACME Group | 2011 | India | Bikaner | 46 | 150 | 14,280 | 2.5 | |
| Ashalim Power Station | Megalim Solar Power | 2018 | Israel | Negev Desert | 260 | 853 | 50,600 | 121 | 320 |
| Cerro Dominador Solar Thermal Plant | Acciona (51%) and Abengoa (49%) | 2021 | Chile | Calama | 243 | 820 | 10,600 | 110 | |
| Crescent Dunes Solar Energy Project | Solar Reserve | 2016 | United States | Tonopah | 200 | 656 | 10,347 | 110 | 500 |
| Dahan Power Plant | Institute of Electrical Engineering of Chinese Academy of Sciences | 2012 | China | Dahan | 118 | 387 | 100 | 1 | |
| Gemasolar Thermosolar Plant | Torresol Energy | 2011 | Spain | Sevilla | 140 | 460 | 2,650 | 19.9 | 80 |
| Greenway CSP Mersin Solar Tower | Greenway CSP | 2013 | Turkey | Mersin | 60 | 200 | 510 | 1 (5 MWt) | |
| Haixi 50 MW CSP Project | Luneng Qinghai Guangheng New Energy | 2019 | China | Haixi Zhou | 188 | 617 | 4,400 | 50 | |
| Hami 50 MW CSP Project | Supcon Solar | 2019 | China | Hami | 180 | 590 | | 50 | |
| Ivanpah Solar Power Facility (3 towers) | BrightSource Energy | 2014 | United States | Mojave Desert | 139.9 | 459 | 173,500 | 392 | 650 |
| Jemalong CSP Pilot Plant | | 2017 | Australia | Jemalong | 5×27 | 5×89 | 3,500 | 1.1 (6 MWt) | |
| Jülich Solar Tower | German Aerospace Center | 2008 | Germany | Jülich | 60 | 200 | 2,000 | 1.5 | na, demonstrator |
| Khi Solar One | Abengoa | 2016 | South Africa | Upington | 205 | 673 | 4,120 | 50 | 180 |

**TABLE 5.3 (Continued)**
**List of Solar Power Towers [70]**

| Name | Developer/Owner | Completed | Country | Town | Height (m) | Height (ft) | Collectors | Installed maximum Capacity (MW) | Yearly Total ENERGY production (GWh) |
|---|---|---|---|---|---|---|---|---|---|
| *Mohamed bin Rashid Al Maktoum Solar Park* | *ACWA Power* | 2020 | United Arab Emirates | *Saih Al-Dahal*, Dubai | 262.44 | 861 | | 1 (5–6 MWt) | na, demonstrator |
| *National Solar Thermal Test Facility* | *U.S. Department of Energy* | 1978 | United States | Mojave Desert | 60 | 200 | | | |
| *Ouarzazate Solar Power Station* | *Moroccan Agency for Sustainable Energy* | 2009 | Morocco | Ouarzazate | 250 | 820 | 7,400 | 150 | 500 |
| *PS10 solar power plant* | *Abengoa Solar* | 2007 | Spain | Sanlúcar la Mayor | 115 | 377 | 624 | 11 | 23.4 |
| *PS20 solar power plant* | *Abengoa Solar* | 2009 | Spain | Sanlúcar la Mayor | 165 | 541 | 1,255 | 20 | 48 |
| *Qinghai Gonghe CSP* | | 2019 | China | Gonghe | 210 | 689 | | 50 | 156.9 |
| *Shouhang Dunhuang 10 MW Phase I* | | 2018 | China | Dunhuang | 138 | 453 | 1,525 | 10 | |
| *Shouhang Dunhuang 100 MW Phase II* | *Beijing Shouhang IHW* | 2018 | China | Dunhuang | 220 | 722 | 12,000 | 100 | 390 |
| *Sierra SunTower (2 towers)* | | 2010 | United States | Mojave Desert | 46 | 150 | 24,000 | 5 | na, demolished |
| *Sundrop Farms* | *Aalborg CSP* | 2016 | Australia | Port Augusta | 127 | 417 | 23,712 | 1.5 | |
| *Supcon Solar Delingha* | *Supcon Solar* | 2016 | China | Delingha | 200 | 656 | | 50 | 146 |
| *Supcon Solar Delingha 10MW (2 towers)* | *Supcon Solar* | 2013 | China | Delingha | 100 | 328 | | 10 | |
| *The Solar Project* | *U.S. Department of Energy* | 1981 | United States | Mojave Desert | 100 | 328 | 1,818 later 1,926 | 7, later 10 | na, demolished |

2020 analyzed by NREL [80] has already considered a solar power tower plant using an SSRC. Kolb [81] explored the benefits of a new generation of molten salt CSP plants with SSRCs operating in the range 600°C–650°C, finding potential LCOE reductions of up to 8% compared to current sub-critical molten salt plants. Singer et al. [82] studied the cost reduction potential of ultra- supercritical steam cycles at 350 bar and up to 720°C with a thermal efficiency of 55% coupled with a solar tower using tubular receivers with different liquid metals and salt mixtures as HTMs. According to their estimates, the LCOE could be reduced by up to 15% compared to the current molten salt CSP plants. Singer et al. [83,84] also compared the potential of two innovative receiver concepts namely internal direct absorption and beam down with tubular receivers using different HTMs and ultra-supercritical steam cycles, estimating potential LCOE reductions from 7.2% (direct absorption receiver with chloride salt) to 0.5% (beam down with molten nitrate salts) in comparison to molten salt power towers of 2016. Peterseim and Veeraragavan [85] compared three solar power towers using an advanced steam cycle with sub-critical parameters and two supercritical solar power towers: the first, a hybrid solar-natural gas configuration with state-of-the-art molten salts and steam parameters of 280 bar and 620°C and the second, a solar-only plant with a precommercial, advanced molten salt and steam parameters of 620°C at 280 bar. They found that the LCOE could be reduced by about 4.3% with the third option, which requires the development of a commercial molten salt mixture, stable at about 700°C. The authors in [86] analyzed the requirements, in terms of materials technology, for the use of advanced ultra-supercritical steam cycles operating at steam conditions up to 760°C with up to 35% improved efficiency compared to superheated steam cycles of 2016.

The development of SSRC solar power towers poses several challenges. First, working at high temperature and pressure increases the costs of materials and equipment. The use of SSRC also requires upscaling of current power tower plants. The largest solar power tower built today is Crescent Dunes (Nevada, USA) with a rated power of 110 MW and 10 full-hour molten salt thermal energy storage (TES); on the other hand, the smallest commercially available supercritical steam turbines are in the range of the 250 MW. Hybrid solar-fossil fuel options may provide a path to overcome the current gap [85]. Increasing the temperature of receivers higher than current 560°C will require the use of other HTMs like solid particles, liquid metals, air, or supercritical steam. Also, the materials that can withstand the demanding conditions of 600°C–700°C during the lifetime of the plant will be required for the receiver, heat exchangers [87], and other equipment. SSRC can be configured with either direct or indirect supercritical steam generation. In the DSG configuration, water enters the solar receiver at low temperature and super-critical pressure and is heated to temperatures above 550°C. No additional heat exchanger or steam generator is required unless the plant includes an indirect-type TES. The main concern for the development of direct supercritical steam receivers is the demanding requirements in terms of materials and design to withstand the high-pressure and high-temperature conditions of the steam [71]. An early 2010s project in CSIRO [78,88] demonstrated the feasibility of such receivers, with a pilot project (309 kW thermal power) generating supercritical steam at pressures from 22.5 to 23.5 MPa and temperatures of 570°C. Another option to integrate solar power towers with supercritical steam is to use a high-temperature HTM. In this case, the concentrated solar flux is used to increase the temperature of the HTM and a high-pressure HTM steam generator is used to generate the supercritical steam feeding the turbine. The overall configuration of the plant can be very similar to current molten salt plants. Potential HTMs are molten salt mixtures, liquid metals, solid particles, and so on.

## 5.6.2 Supercritical $CO_2$ Brayton Cycle

The key components of a CSP plant are the solar collector, the solar receiver and the power block, although thermal-energy storage is also a key component to decouple the availability of the sun and the demand for power [89]. As shown earlier, cumulatively, the global installed capacity of CSP has grown five-fold since 2010, and in that time has seen a drop in LCOE from 0.346 to 0.182 \$/kWh [90]. However, compared to other technologies CSP is still in its infancy, and is associated with higher LCOE values compared to other renewable technologies, such as solar PV(0.068 \$/kWh) [90]. Much of the drop in LCOE for CSP systems can be attributed to reductions in the costs of solar collector fields, which represent around 40% of the total capital cost [91]. In contrast, there have not been significant developments in the power block technology, which remains based on the steam Rankine cycle. The deployment of a $sCO_2$ power block, in place of steam, could offer a number of benefits. These are well analyzed by Silva-Perez [74]. Firstly, $sCO_2$ power cycles are capable of achieving higher thermal efficiencies at temperatures relevant to CSP applications, which enables $sCO_2$ to produce the same amount of energy using a smaller solar field. The possibility for more compact turbomachinery, a simpler cycle layout, and a smaller physical footprint could also reduce both the thermal mass and complexity of the power block. This can improve the cycle's response time during intermittent operation. The system should also allow a more compact thermal energy storage system because the lack of phase change within the primary heat-addition process reduces the pinch point within the primary heater [91,96].

Supercritical $CO_2$ closed loop Brayton cycles have the potential to provide higher efficiency than the sub-critical steam cycles used in CST (CSP) power plants and equivalent or higher than supercritical steam cycles operating in the same temperature range. The system also has the advantage of lower operating pressure and greater compactness. In addition, the heat rejection temperature ranges of $sCO_2$ cycles make them appropriate for the use of dry-cooling. Sensible heat TES can be easily integrated because the $CO_2$ presents a single phase in all the cycle processes. These characteristics reveal a potential to develop highly efficient and compact CSP plants with the potential to reduce capital, operational and maintenance costs of the power block, enabling a significant reduction in the LCOE of CSP plants. This motivation is supported by the Gen3 CSP roadmap [92], developed within the US SunShot program, which maps out the pathway toward the next generation of CSP systems.

While $sCO_2$-CSP systems have been reviewed by Yin et al. [93], the identification of optimal cycle layouts and operating conditions for CSP applications by thermodynamic modelling and optimization remains a major research topic. Although different $sCO_2$ cycle configurations exist, the selection of the most appropriate one for integration in CSP towers depends on the cycle efficiency, the temperature difference, and the complexity which is directly associated to its cost. Chacartegui et al. [94] compared two configurations of $sCO_2$ Brayton cycles simple and recompression and an $sCO_2$-ORC combined cycle, integrated with central receiver systems. Their analysis, which used a simple solar tower model, concluded that the recompression cycle provides the best performance. Other analyses using an effectiveness model for the recuperator [95] suggested that the performance of the partial cooling and recompression cycles were similar. A later study using a conductance model for the recuperators [96] showed that the partial cooling cycle provides higher efficiency than the recompression cycle up to high values of the recuperator conductance. Assuming that the cost of the equipment increases with the recuperator conductance, the partial cooling configuration is better for low- and medium-conductance values. In addition, they also found advantages for the partial cooling cycle regarding the design and operation of the solar receiver. The simple $sCO_2$ Brayton cycle has a lower efficiency than the partial cooling and the recompression cycle, but has the advantage of its simplicity for near-term implementation. The study of Iverson et al. [97] includes the transient part-load response of the cycle and the identification of necessary research for the successful implementation of $sCO_2$ Brayton cycles in CSP power plants, including the development of turbines, bearings, seals, heat exchanger designs, and materials. Padilla et al. [98,99] compared the thermal and exergetic performance of four cycle configurations: simple, recompression, partial cooling, and recompression with main compression intercooling configurations. The results indicated that the latter has the best performance with a thermal efficiency of about 47% at temperature greater than 700°C.

Neises and Turchi [96] compared simple recuperated, re-compression and partially cooled cycles for CSP applications in terms of thermodynamic performance. Subsequently, the authors considered the economic performance and concluded that the partially-cooled cycle had a LCOE that was 6.2% lower than either the simple or recompression cycle [100]. Binotti et al. [101] compared cycles for a CSP application with temperatures up to 800°C and found that the compression cycle with main compression inter-cooling outperformed either the recompression cycle or partially-cooled cycles. Finally, Crespi et al. [49] assessed the overnight capital costs of $sCO_2$ for CSP applications and concluded that the partially-cooled cycle is a promising candidate when both thermodynamic performance and capital costs are considered. Their results also suggest that very complex cycles may be unsuitable, even though they have high thermal efficiencies.

Ultimately, all these studies point towards the recompression and partially-cooled cycles as most promising candidates $sCO_2$-CSP systems. Alongside thermodynamic modelling, there is a need to demonstrate the technology at a commercial scale. Of this, the developments under the SunShot program are the most notable, which target an LCOE of 0.06 $/kWh for CSP systems, with a thermal efficiency of 50%, and under which a 10 MWe $sCO_2$ turbine is being tested up to 750°C and 250 bar under reduced flow conditions [102,103]. Another notable development is the Shouhang-EDF demonstration plant in China, which involves retrofitting a 10 MWe steam power plant with a $sCO_2$ powerblock by the end of 2020. The plant employs a recompression cycle with intercooling and preheating with an estimated net efficiency of 35.6%, and while the current maximum turbine inlet temperature is limited to 468°C, the project eventually aims to achieve higher temperatures [104]. Finally, it is worth noting that optimal locations for CSP plants are typically in hot and arid regions where there may be limited availability of water. As such, it is necessary to rely on dry cooling for the heat-rejection process, which leads to compressor inlet temperatures in the region of 50°C once the approach temperature in the cooler is considered. Whilst $sCO_2$ cycles are suitable for dry cooling [91], this increase in compressor temperature moves the compression process away from the critical point, somewhat negating the low compressor work promised by operating with $sCO_2$, and may also require increased compressor inlet pressures to maximize efficiency [105]. To this end, $CO_2$-blends have been proposed to increase the critical point of the working fluid, not only moving the compression process closer to the critical point but also facilitating the use of a transcritical cycle. Manzolini et al. [106] studied CO2-blends for CSP applications, and suggest that for turbine inlet temperatures of 550°C and 700°C $CO_2$-blends could increase thermal efficiency by 2% and reduce LCOE by10% compared to a conventional steam cycle. Ranjan and coworkers at Georgia Tech's and Purdue University are developing a new breed of heat exchanger that can withstand extremely high temperatures and pressures. The group has developed a process for inexpensively fabricating a high-temperature composite material into complicated 3-D shapes. In addition to making solar power more competitive, the heat exchangers could also be used with $sCO_2$ to boost efficiency in fossil fuel power plants.

### 5.6.2.1  Integration of Solar Power Towers and Supercritical $CO_2$ Cycles

Solar power towers and $sCO_2$ cycle can be integrated in a number of different ways. These are well analyzed by Silva-Perez [74]. The main advantage of direct receivers, where the working fluid is the same at the solar receiver and power cycle, is that it eliminates the need for intermediate heat exchangers, thus avoiding the thermal and exergy losses and the cost associated to this equipment. There are three main options for a direct $sCO_2$ receiver: tubular, pressurized volumetric, and fluidized bed, small particle-gas receiver. Of these three options, only the first seems to be sufficiently mature for its commercial deployment in the near to medium term. Tubular receivers would operate at pressures of up to 30 MPa. Ortega and Christian [107] established the design requirements for tubular $CO_2$ receivers operating at supercritical conditions. They have also developed a coupled optical-thermal-fluid model [108] and performed a structural and creep-fatigue evaluation of such receiver [109]. Their results show that thermal efficiencies close to 85% can be achieved at the receiver when using appropriate aiming-point strategies to obtain adequate flux pro- files on the receiver surface and flow patterns with high recirculation of the working fluid. Besarati et al. [110] proposed a direct $CO_2$ solar receiver based on compact heat exchanger (CHE) technology.

The second option is the indirect receivers for the $sCO_2$ cycle, which allows an integration of $sCO_2$ receivers with storage. Since thermal [111] storage of supercritical fluids is not viable [112], the use of different HTMs as working fluid appears as the best option for the integration of solar power towers with storage and $sCO_2$ cycles. The concentrated solar flux is used to increase the temperature of the HTM and a high-pressure HTM-$CO_2$ heat exchanger is used to generate the high-temperature supercritical $CO_2$ feeding the $sCO_2$ turbine. The overall configuration of the plant can be very similar to molten salt plants. Potential HTMs are, again, molten salt mixtures, liquid metals, and solid particles in different receiver configurations. Ho et al. [113] reviewed several high-temperature designs for $sCO_2$ Brayton cycles and concluded that the most viable option today for indirect $CO_2$ heating and TES is the use of falling particle receivers.

The third option involves thermochemical cycles for TES coupled to $sCO_2$ cycles. Buckingham et al. [114] propose redox transitions in metal oxides and sulfur-based cycles, where energy is stored inexpensively in the form of elemental sulfur). These reactions that occur at temperatures between 500°C and 1,000°C take place at the solar receiver (moving-bed reactor). While this option has a potential to increase efficiency, the cost is uncertain. Calcium looping with the $sCO_2$ cycle is a promising option too, to be integrated with solar power towers [115]. The calcium looping has a high volumetric energy density and can produce temperatures close to 900°C, and it is well coupled to $sCO_2$ cycles and solar power towers.

### 5.6.3 COMPARISON OF SUPERCRITICAL STEAM RANKINE AND CARBON DIOXIDE BRAYTON CYCLES

A review of high-efficiency thermodynamic cycles and their applicability to CSP systems performed by Dunham and Iverson [116] concluded that steam Rankine systems may offer higher thermal efficiencies up to temperatures of about 600°C, while an $sCO_2$ recompression Brayton cycle may be the best candidate for higher temperature, with potential efficiency of about 60% at 30 MPa and above 1,000°C and wet-cooling. While the study in [117] concluded that a superheated steam Rankine cycle is both more efficient and more cost-effective than the three $sCO_2$ power cycle concepts and the SSRC considered in their analysis, some of the assumptions made in the analysis may be questionable. Furthermore, final conclusions may change with significant advances made in the field of high-temperature materials for advanced power generation and other related fields.

### 5.6.4 DECOUPLED SOLAR COMBINED CYCLES

DSCCs have been identified as a promising option for LCOE reduction that takes advantage of the high temperature achievable by means of CSP systems and the use of TES. A DSCC is the combination of a high-temperature cycle where the heat is provided by a CSP system and a lower temperature, bottoming cycle. The heat rejected during the operation of the high-temperature cycle is used to charge the TES. The energy stored in the TES can be asynchronously used to feed the bottoming cycle, thus decoupling the operation of both cycles and allowing for great operation flexibility. The DSCC concept seems "naturally" linked to the solar tower technology because of its capacity to efficiently achieve high concentration ratios and high temperature, thus taking advantage of the high exergy of the solar radiation. The DSCC concept provides great flexibility in the design of the plant. Despite its relatively recent development, the first reference to DSCCs dates from 2012 [118] and several different configurations can be found in the literature. Researchers of CENER (National Renewable Energy Center, Spain) identified and analyzed two configurations [118], both of them based on multitower solar fields, each tower having its own gas turbine (Brayton cycle). Both the multitower solar energy collection and the Brayton cycle have the potential to achieve high efficiency. In the first configuration (concept A) the heat rejected from every gas turbine is used to charge a common, single medium-temperature thermal storage system that provides thermal energy to the bottoming cycle, a single low- temperature organic Rankine cycle (ORC). The second configuration (concept B) combines the multitower system with a single high-temperature molten salts TES and a superheated Rankine cycle. In both cases, the authors consider a solar energy collection system with a heliostat field, a beam-down reflector, and a secondary concentrator coupled to a ground-based high-temperature air receiver. The authors analyze two cases ("conservative" and "optimistic") for each configuration.

Agalit et al. [76] propose a decoupled configuration with a Brayton cycle and a superheated-steam Rankine cycle and two TES systems. The solar field with a high-temperature air receiver (800°C–1,200°C) is connected to the gas turbine and a high-pressure regenerator/packed-bed storage system using natural (quartzite) rocks. The system includes a fuel combustor which can be used to complement or replace the solar energy input. This configuration is a modification of the SUNSPOT concept [119] which includes only the low-pressure TES and is conceived for partially decoupled operation (the Rankine cycle would operate during nighttime). Crespo [120] proposes a DSCC hybrid configuration, with a large fuel-driven Brayton cycle and a molten salt heat exchanger to recover the exhaust thermal energy of the air turbine, combined with a molten salt CSP plant operating at about 560°C. The storage system can be charged either from the solar receiver or from the gas turbine exhausts under the same conditions. The generation of electricity from the steam cycle is completely decoupled either from the gas turbine or from the solar part.

These studies indicate that the combination of a high-temperature power cycle with a bottoming cycle using a TES system in place of the conventional heat recovery steam generator not only allows to a very large extent the decoupling of the operation of the bottoming cycle from that of the high-temperature cycle, but it also introduces additional degrees of flexibility in the design and operation of solar tower systems that can be used to create more flexible, reliable, and cost-effective systems. Both power cycles operate in temperature ranges only moderately superior to molten salt plants, being thus well suited to the temperature levels achievable with central receiver systems. The integration of these power systems poses, however, significant challenges both on the power cycle and on the solar energy collection system, especially in terms of the identification or development of new materials that can withstand the demanding operating conditions of these systems without increasing the costs so much that the potential efficiency increase is negatively counterweighed. According to the literature, SRRCs can be a good option for temperatures up to 600°C–650°C, while $sCO_2$ cycles would be more competitive for higher temperatures. Thermochemical storage based on ammonia dissociation, metal oxides or sulfur redox reactions, or calcium looping coupled to supercritical plants also seem to be promising long-term options. That alternative can also include DSCCs, where the heat rejected by a high-temperature cycle integrated with a solar power tower is used to charge a TES system which, in turn, feeds a bottoming cycle. For concentrated solar power (CSP) applications, the goal is to develop systems that can achieve levelized costs of electricity that are competitive with solar photovoltaic (PV). Demonstration plants are under construction as part of the SunShot and STEP programs in the US and the Shouhang-EDF plant in China.

### 5.6.5 INNOVATIVE POWER CONVERSION CYCLES WITH LIQUID

Due to the safety issues with the liquid metal storage system during the operation of IEA-SSPS and Jemalong projects noted above, researchers realized that the use of a sodium/salt binary scheme is needed [121,122]. It seems necessary to minimize the volume of sodium by restricting its use to the solar receiver loop and then use an intermediate loop with a different heat transfer fluid (e.g., solar salt) for the storage and heat transfer to the steam generator. With this strategy, the system will take advantage of the main benefits of sodium and other liquid metals regarding efficiency, size, and aperture area of the solar receiver, where the heat transfer rates are important, and will avoid the low heat capacity and operational risks associated with storage and exchanger involving steam [123]. However, the binary system has an additional level of complexity resulting from the extra heat transfer loop and introduces a new source of risk in case of a leakage in the sodium/salt heat exchanger because the reaction would be strongly exothermic. Another critical issue is that the potential improvement of efficiency in the receiver compared to an all-salt plant would be slightly more than 1%. Because of that, there is little chance that liquid metals replace the well-understood solar salt central receiver systems for typical Rankine subcritical cycles.

As an alternative, future technologies may address receivers based on heat pipe concepts such as the ones implemented in dish-Stirling systems for distributed generation [124]. Typically, in dishes with solarized Stirling engines, the receiver absorbs the light and transfers the energy as heat to the working gas, usually helium or hydrogen. Thermal fluid working temperatures are between 650°C and 750°C. This temperature strongly influences the efficiency of the engine. Because of the high operating temperatures, radiation losses strongly penalize the efficiency of the receiver; therefore, a cavity design is an optimum solution for this kind of system.

Two different heat transfer methods are commonly used in parabolic dish receivers to be used with Stirling engines or at solarized reactors for thermochemical applications [125]. In directly illuminated receivers, the same gas fluid used inside the engine is externally heated in the receiver through a pipe bundle. Although this is the most conventional method, a good high-pressure, high-velocity, heat-transfer gas such as helium or hydrogen must be used. In indirect receivers, one method is heat pipes, which employ a metal capillary wick impregnated with a liquid metal heated up through the receiver plate and vaporized. The wick structure distributes sodium across a solar-heated dome, and thermal energy is removed as sodium evaporates typically at a temperature range between 700°C and 850°C. The vapor then moves across the receiver and condenses in a cooler section, transferring the heat to the engine. Evaporation/condensation processes guarantee good temperature control, providing uniform heating of the Stirling engine [126,127]. Coventry et al. [240] provided insight into different conceptual level design alternatives that result from adapting solar dish technology to central receivers in solar towers with liquid-vapor phase change sodium. More experimental testing and research are necessary to assess the controllability of temperature gradients on material surface and potential failures prior to scaling up the technology. If the technical challenges regarding materials and heat transfer in liquid to vapor sodium phase change receivers are eventually solved, the technology may be used up to 850°C to power high-temperature thermo-dynamic cycles such as supercritical steam, air Brayton, and combined cycles [128]. Higher temperatures, beyond 1,000°C, might be achievable with other liquid metals such as LBE in future developments. However, it should be noted that the impact on plant efficiency is only incremental, given the convolution of efficiencies in the different subsystems, such as heliostat field, receiver, heat transfer fluid loop, heat exchangers, storage, and thermodynamic cycles.

There are various options available for heat transfer fluids providing higher tem-perature. Some examples are air streams highly charged in particles [221] and new for-mulations of carbonates and chloride-based salts. While molten salts are considered as the commercial mature option of CRS (central receiver system) plants, their relatively high corrosive nature to metal alloys remains as an issue. The corrosion characteris-tics of many other molten salts are not available in the literature [129]. Sodium and other liquid metals can be the working fluid itself in the cycle, and therefore provide disruptive schemes with high efficiency and the removal of the turbomachinery and the entire power block. Romero and Gonzalez-Aguilar [121] pointed out that there are two technologies that are retained for CSP integration studies at the system level, such as AMTEC (Alkali Metal Thermal to Electric Converter) cells and LMMHD (Liquid Metal Magneto-Hydro-Dynamic) power conversion systems. Fritsch et al. [130] have provided conceptual insight and thermodynamic analysis of these systems.

Romero and Gonzalez-Aguilar [121] have described both AMTEC and LMMHD in detail. AMTEC is an electrochemical device that uses a recirculating alkali metal (sodium or potassium) working fluid passing through a solid electrolyte in a closed circuit to produce an electron flow in an external load. Typical solids used are ceramic electrolytes such as $b^{00}$ or $P^{00}$ alumina, which are very good conductors of ions but poor conductors of electrons, due to their crystal structure. In addition, they with-stand high-temperature differences between anode and cathode surfaces. The liquid metal is driven around a closed thermodynamic cycle between a heat source and a heat sink held at different temperatures and, during the vapor phase of the cycle, the available work from the isothermal expansion of the working fluid as it passes through the electrolyte is converted directly into electric power. The AMTEC device is characterized by high potential efficiencies and no moving parts except the liquid metal itself. It accepts a heat input in a range from about 600°C to 1,000°C and pro-duces a direct current with predicted device efficiencies of 10%–30%. It can be used as a topping cycle with a bottoming Rankine cycle [130]. Loop-type heat pipe system used by AMTEC has the potential to enhance thermal transport capabilities by sepa-rating the liquid and vapor lines and thus reducing the fluid dynamic resistance at the liquid-vapor interface that results from liquid and vapor flowing in the opposite direction to one another. Since, a solar receiver is usually tilted, a thermosiphon heat pipe can be utilized, as long as the condenser section of the heat pipe is located in a higher position in the gravity field [131].

Romero and Gonzalez-Aguilar [121] point out that in an LMMHD generator, a highly electro-conductive two-phase mixture composed of a liquid metal and a gas (or vapor) moves across a magnetic field and thus generates electrical power. The two-phase flow is propelled by the expanding gas bubbles and the gas goes through the thermodynamic cycle. For LMMHD, some solar-assisted designs have been proposed since the 1980s [132]. In many cases, solar designs have been conceived for low temperatures from 80°C to 300°C [133]. At low-temperature operation of the solar assisted-LMMHD power generator, the MHD duct becomes free from many problems such as electrode and duct life, erosion-corrosion, preheating, cooling, and so on, and the overall system is free from the emission of harmful chemical species [121].

The liquid metal flow can be accelerated in a number of different ways. In a one-component flow, the liquid metal becomes partially gaseous in the receiver and then the changes in density accelerate the flow. In a two-component operation, possible combinations are alkali metals with helium or argon or lead or lead alloys with water. This mixer is similar to a direct contact heat exchanger and the theoretical efficiencies of the device can exceed 60% [134]. Liquid-vapor phase change sodium integrated receivers may reach temperatures above 800°C providing CRS plants access to the use of Brayton cycles or advanced direct conversion systems.

SHGT (solar hybrid gas turbine) systems offer a reduced cost of solar electricity, full dispatchability, simplicity of operation, and reduced water consumption. This system was analyzed both theoretically and experimentally. However, as of now, such systems are not mature. Further developments to improve efficiency and modifications for solar applications are required. While this is feasible, significantly more work is needed to provide technically proven units. The important issues are mainly associated with interfacing to external air heating, modification of the combustor system, component cooling, and system control. SHGT systems also require high-temperature receivers with outlet temperatures in the range of 1,000°C. The corresponding receiver technology is not mature yet. More work is also needed for technical readiness, upscaling and long-term operation of various concepts. Finally, a modular system design approach needs to be adopted to limit the development cost on the gas turbine and receiver side and distribute these costs over a larger number of identical units. The modular configuration needs to be carefully selected and developed. Since the molten salt mixture is most suitable for SSRC solar power towers, Peterseim and Veeraragavan [85] analyzed a hybrid configuration based on the current molten salt plants producing supercritical steam at 280 bar and 545°C with additional natural gas superheating to reach a steam temperature of 620°C. This design would not require any major technological development since only the molten salt and steam generator would need to be modified to operate at supercritical pressure.

## 5.7 ADVANCED MIRROR CONCEPTS FOR CONCENTRATING SOLAR THERMAL SYSTEMS

In CST (CSP) the use of proper mirrors is very important because they involve a significant cost, both for the initial investment and during the operation and maintenance (O&M) of the plant [135]. Solar beams are first impacted by the mirror (also known as reflector), hence its effective and efficient use is very important for the overall energy conversion process. The evolution and advances made to the reflectors are important to all the stakeholders. According to the materials roadmap published by the European Commission, the target performance in terms of reflectance for silvered reflectors in the following decades (2020–2030) is set at 95%–96%, considering a reference starting point of 94% in 2010. Mirrors are also expected to cost 25% less compared to 2010 values [136]. They are expected to last for 10–30 years under severe outdoor environments [137] or even more than 30 years [138]. The 2018 roadmap by IEA also calls for the development of lightweight, low-cost reflector optics [139].

The improvement in reflector efficiency reduces the solar field size and retains the electricity output and the investment cost for collectors or heliostats. Even small improvements of 0.5 ppt in reflectance will have a high impact on the annual revenues of the power plant. High reflectance can be achieved with silver as the reflecting surface and thin transparent front coats. Polymer films have been tested for reflectors for concentrating solar applications [140,141] because silvered-polymer films are flexible and allow any kind of collector geometry. However, their solar-weighted hemispherical reflectance values of around 94%, are still below the state-of-the-art 4 mm silvered-glass mirror reflectance value of 94.7%. In terms of specularity and durability, glass-based mirrors are superior and will be preferred in the future The reduction of the glass thickness from 4 to 1 mm boosts the reflectance by around 1 ppt. Commercially available thin-glass mirrors (around 1 mm glass thickness) achieve around 95.7% solar-weighted hemispherical reflectance. The durability of this type of material has already been widely checked [142,143]. The reflectance can be further improved and weight can be reduced by the use of ultra-thin reflectors. Prototype mirrors of more than 97% solar-weighted hemispherical reflectance have been produced with 100 mm ultra-thin glass. The lower weight allows the reduction of collector cost due to the requirement for less support material, simpler foundations, cheaper motors, and reduced transport costs. Thin-glass mirrors, however, need to be supported by a backing structure to provide stiffness against wind loads. Several substrate materials such as concrete, composite materials, aluminum, steel, and glass are being investigated for this purpose. Thin-glass mirrors <1 mm are flat and flexible, and the substrate material needs to provide the shape accurately. The bonding between the mirror and backing material (e.g., using adhesives) needs to be durable and must not introduce waviness or additional shape errors. Therefore, special care must be taken with the gluing process.

Stainless steel is the preferred material as a reflective metal because of its low cost and easy availability and for processes where energy demand is low; however, its efficiency is low. The solar-weighted hemispherical reflectance attained by stainless steel was reported to be 0.572, with a maximum hemispherical spectral value of 0.680 at 1,100 nm [144]. Different reflectance spectra were reported as a function of the polishing treatment. In particular, the reflectance of mechanically polished stainless steel containing 18% Cr and 10% Ni was found to be 8%–10% lower than with the same material polished electrochemically at 80°C in a solution of 60% $H_3PO_4$, 20% $H_2SO_4$, and 20% $H_2O$ [145]. A low-cost parabolic-trough collector incorporating a polished stainless steel reflector, named MEXSOL, was developed to be coupled to a dried hybrid energy system [146]. This work reported the average hemispherical reflectance value of 0.70 [146]. While stainless steel provides flexibility in the collector design, cost and market favorability, its challenges are high reflectance, suitable durability and low cost. Future research needs include the study of the efficiency of different approaches to increase reflectance and durability.

One of the major problems with CST plants is soiling accumulation on the reflector surface requiring expensive cleaning activities and water consumption. Anti-soiling coatings are currently receiving major coating that prevents the accumulation of dust particles on the mirror surface which may have detrimental effects on the optical properties of the reflector. The coating and reflector combination must be durable and reliable while experiencing high temperatures, sandstorms, and higher than normal exposure to both ultraviolet (UV), visible, and infrared radiation, and must prove their ability to withstand any cleaning procedures [147]. Cuddihy [148] indicated that a low soiling surface requires hardness, smoothness, hydrophobic characteristics, low surface energy, non-stickiness and easy-to-clean characteristics, which can be achieved by some appropriate anti-soiling coatings. Repelling charged dust particles by coating is also desirable. Hydrophobic and hydrophilic properties are important for the anti-soling effect. Materials used for hydrophobic coatings are mainly fluoropolymers [149]. On the other hand, water-attracting hydrophilic coatings have high surface energies [149].

For CST plants, three types of coatings; an antireflection (AR) coating, a SC coating, and a multilayer (ML) coating consisting of both AR and SC have been examined. The only solar mirror with anti-soiling coating currently marketed is dura GLARE, by the company Flabeg [150]. These coatings repel the dust particles blown on the mirrors and to reduce costs for cleaning and maintenance activities and thereby improving the performance of the solar field. Future research needs assurance of the coating efficiency over the expected lifetime of solar cells under real operating conditions.

Often the performance of the primary concentrator is improved by using secondary concentrators which redirect solar radiation reflected by the primary concentrators to the focal point or line. Their role is to increase the concentrated solar flux density and hence decrease thermal radiation losses by reducing the receiver size. Also, these components are used to increase the optical efficiency of the system by improving the flux distribution homogeneity or reducing the amount of solar flux missing the receiver due to scattering (spillage). Compound parabolic concentrators (CPCs) are typically used as secondary concentrators, either for concentrating solar radiation on line-focusing systems (named 2D-CPC) or point-focusing systems (named 3D-CPC) [151]. Normally, 2D-CPC are used in linear Fresnel collectors, and 3D-CPC is employed in STs or "beam-down" tower systems, which allow the placing of solar receiver/reactors on the ground rather than on the tower, with the concentrated irradiation entering from the top [152,153]. The secondary concentrators of the REFOS [154,155] and SOLGATE research projects [156] were installed in the Central Electrosolar de Almería 1 (CESA-1) central tower at the PSA at about 60 m height. For high power levels and high temperatures, the cooling of the front edges of the aluminum plates, in this case, was insufficient. Also, the elevated temperatures caused some glass mirrors to crack [154].

SFERA-I European project examined the durability of solar reflectors for secondary concentrators by simulating adverse environmental conditions created naturally or by human-made. The study indicated that aluminum reflectors and thin silvered-glass reflectors glued to an aluminum structure showed minimum reflectance losses and structural degradation under the operation conditions of cooled 3D secondary concentrators. The research center Fraunhofer ISE (Institute for Solar Energy Systems), from Germany, investigated several approaches to avoid silver corrosion in high-temperature reflectors for secondary concentrators of linear Fresnel collectors, where temperatures of up to 300°C can be reached [157]. This study showed that low coefficients of expansion were obtained for the dielectric barrier layers but silver causes high thermal stress on the glass due to its large coefficient of expansion. European project RAISELIFE [158] is developing a secondary mirror for CST which maintains its optical and mechanical properties for operating temperatures of up to 350°C. Acceptable stability and reduced thermal stress under high radiation flux and temperatures, and minimum interference with the adhesive material are the main challenges.

## 5.8   NEXT-GENERATION RECEIVERS

Receivers are very important for the overall efficiency of CST systems. The novel design of receivers also includes their novel applications. In this section, we briefly review advances made in the development of next generation of receivers. Ho [72] has presented an excellent review of operating principles, modeling and testing activities, thermal efficiency, advantages and other challenges related to next-generation receivers. The present summary follows this review.

### 5.8.1   PARTICLE RECEIVERS

In order to achieve higher operating temperatures (>700°C), inexpensive direct storage, and higher receiver efficiencies for concentrating solar power technologies, thermochemical reactions, and process heat falling particle receivers are extensively examined in the literature [159–176]. Falling particle receivers use solid particles that are heated either directly or indirectly as they fall through a beam of concentrated direct solar radiation. Once heated, the particles may be stored in an insulated tank and used to heat a secondary working fluid (e.g., steam, $CO_2$, or air) for the power cycle. Particle receivers have the potential to increase the maximum temperature of the heat-transfer media to over 1,000°C. Thermal energy storage costs can be significantly reduced by directly storing heat at higher temperatures in a relatively inexpensive medium (i.e., sand-like particles). Because the solar energy is directly absorbed in the particles, the flux limitations associated with tubular central receivers are significantly relaxed. The falling particle receiver appears well suited for scalability ranging from 10 to 100 MWe power-tower systems. According to Ho [72], previous studies have considered alternative particle receiver designs including free-falling [171], centrifugal [173,174,177,178], flow in tubes with or without fluidization [168,175,176,179–182], multi-pass recirculation [162,170] north- or south-facing [159,164], and face-down configurations [170]. In general, these particle receivers can be categorized as either direct or indirect particle heating receivers. Direct particle heating receivers irradiate the particles directly as they fall through a receiver, while indirect particle heating receivers utilize tubes or other enclosures to convey and heat the particles.

The most basic form of a direct particle heating receiver consists of particles falling through a cavity receiver, where the particles are irradiated directly by concentrated sunlight. The particles are released through a slot at the base of a hopper above the receiver, producing a thin sheet (or curtain) of particles falling through the receiver. A number of assessments and studies have been performed on direct free-falling particle receivers since its inception in the 1980s [159–163,165–167,170–172,183–197]. The majority of those studies focused on modeling the particle hydraulics and radiant heat transfer to falling particles. Various geometries and configurations of falling particle receivers have been considered, including north/south-facing cavity receivers as well as face-down cavity receivers with a surrounding heliostat field [161,170,190]. While earlier studies [171,198] gave thermal efficiency of 50% with a maximum temperature of 250°C, subsequent studies [199,200] with recirculating particle receiver gave 700°C temp. with thermal efficiency ranging from 50% to 80%. Results showed that the particle temperature rise and thermal efficiency were dependent on particle mass flow rate and irradiance. Higher particle mass flow rates yielded greater thermal efficiencies but lower particle temperature rise. One way to increase temperature is to recirculate particles multiple times, but this increases cost and complexity of the process. The study identified major issues with the system as: nonuniform irradiance distributions on the particle curtain, variable mass flow rates, wind impacts, particle loss through the aperture, particle elevator reliability, and wear on the receiver walls from direct flux and high temperatures (>1,000°C).

Kim et al. [166] performed tests of particles free-falling along a 3 m drop length to evaluate the influence of wind direction (induced by fans). They found that the least amount of particle loss occurred when the wind was oriented directly toward (normal to) the aperture. Air recirculation and air curtains have been proposed as a means to mitigate the impacts of wind on particle flow and to reduce convective losses [172,191,193,201–204]. Tan et al. [172,202–204] simulated the use of an aero window to mitigate heat loss and wind impacts in falling particle receivers. Ho et al. [191,193] performed experimental and numerical studies that evaluated the impact of an air curtain on the performance of a falling particle receiver. Results showed that the air curtain only reduced particle loss when particles were released near the aperture in the presence of external wind. Larger particles and mass flow rates were also shown to reduce particle loss through the aperture. Numerical results showed that the presence of an air curtain could reduce the convective heat losses, but only at higher temperatures (>600°C).

One of the issues is to achieve high temperature by increasing the residence time of particles within the concentrated sunlight. This can be achieved by inserting a porous structure or an array of obstacles to slow the downward velocity while still allowing direct absorption of concentrated solar energy [184,205]. Ho et al. performed on-sun tests of a particle receiver consisting of a staggered array of stainless steel chevron-shaped mesh structures [199] and found that while peak particle temperatures reached over 700°C near the center of the receiver, the particle temperature increase near the sides was lower due to a nonuniform irradiance distribution. While this method improved heating, reduced impacts of winds and particle loss through the aperture, they caused overheating, oxidizing and deteriorating of the stainless steel 316 mesh materials. Another obstructed flow design employed a spiral ramp along which particles flow under the influence of gravity and mechanically induced vibration [206]. This method required beam-down optics, caused restricted amount of particle flow but provided 60% thermal efficiency. Another study [207] using obstructed flow and beam-down optics investigated the system where particles are lifted upwards with a screw elevator toward an aperture. The particles are irradiated by concentrated sunlight before spilling into the hollow screw for subsequent heat exchange and reaction.

Rotating kilns were used in solar particle heating applications [175] since 1980. The general principle is to feed particles into a rotating kiln/receiver with an aperture at one end of the receiver to allow incoming concentrated sunlight. The centrifugal force of the rotating receiver causes the particles to move along the walls of the receiver while they are irradiated by the concentrated sunlight. Flamant et al. showed that these systems have a very high absorption factor (0.9–1), but low thermal efficiency (10%–30%). Wu et al. [173,174,177,178] developed a centrifugal particle receiver design and prototype that employs a similar concept. Wu et al. reported a particle outlet temperature of 900°C and a receiver efficiency of about 75% [173]. Challenges include maintaining a constant and sufficient mass flow rate of particles at larger scales, parasitic energy requirements, and system reliability.

Fluidization of solid particles in a solar receiver have been proposed for several decades. Flamant et al. [175,176] tested a fluidized-bed receiver that consisted of a vertical transparent silica tube (15 cm long 6.5 cm diameter) that was fluidized with compressed air from the bottom and irradiated at the top. The study showed that ability to convey the particles and achieve adequate mass flow rates may pose a challenge. Chinese Academy of Sciences [208–210] performed numerical and experimental studies on the thermal performance of an air receiver with silicon carbide particles in transparent quartz tubes. Results of those tests showed that the heated air reached over 600°C with minimum temperature differences between the particles and the air below 10°C. Steinfeld et al. [182] designed and tested a fluidized-bed receiver reactor that employed a vortical flow of air in a conical-shaped receiver. The particle/gas stream was introduced near the aperture, where concentrated sunlight entered the receiver and heated the swirling particles before the particles exited the receiver. A final type of fluidized particle receiver involves the use of very small carbon particles dispersed in air that flows through the receiver. Concentrated sunlight irradiates and oxidizes the carbon particles, which volumetrically heats pressurized air passing through the receiver for high-temperature Brayton cycles [211–219]. Potential advantages include the following: solar radiation is absorbed throughout the gas volume due to the large cumulative surface area of the particles; higher incident fluxes with no solid absorber that can be damaged; particles are oxidized leaving a particle-free outlet stream [214]. Challenges include the development of a suitable window for the pressurized receiver and the development of a solid-gas suspension system that maintains a uniform particle concentration and temperature within the receiver.

Besides direct heating particle receivers mentioned above, numerous efforts are made to devise indirect particle heating receivers. Ma et al. [168,181,220] proposed an indirectly heated particle receiver with particles flowing downward under the force of gravity around a staggered array of tubes within an enclosure. The tubes were irradiated by concentrated sunlight on the interior surfaces while transferring heat to the particles flowing around the exterior side of the tubes inside of an enclosure). The study showed that the heat transfer to the particles was limited in locations around the tubular structures where the particles lost contact with the heated wall surfaces. Other limitations included maintaining a sufficient mass flow and obtaining a significant penetration and uniform flux of concentrated sunlight within the tubular cavities. Advantages to this design include no loss of particles through an open aperture and reduced heat losses relative to an open cavity receiver. Flamant et al. [179,180,221] proposed and demonstrated an indirect particle receiver in which the particles are forced upward through irradiated tubes by airflow, which fluidizes the particles and increases heat transfer from the tube walls to the flowing particles. Challenges in this system include parasitic energy requirements to fluidize the particles through the receiver tubes with the sufficient mass flow to meet desired power requirements. The potential for hot spots and significant tube surface temperatures that radiate energy to the environment also exists.

The literature studies indicate that each of the particle receiver designs has promising advantages, along with challenges that need to be addressed. Directly heated particle receivers have a significant advantage over direct particle heating, but particle loss may be a problem in open cavities with significant wind effects. Indirect particle receivers have the advantage of particle containment and no particle losses, but additional heat transfer resistance between the irradiated surface and the particles is a challenge. Fluidizing the particles within tubes has been shown to enhance heat transfer. For large-scale electricity production, which will require significant particle mass flow rates, gravity-driven flow (free-falling or with obstructions) appears to be the most promising. Another promising trend is the adoption of solid particle receivers, which can operate at a temperature higher than 1,000°C, leading to higher efficiency and significant cost reduction. This new concept offers several advantages compared to conventional HTFs. First, they can replace the thermal fluids and storage medium at lower cost. They can also achieve high chemical stability under high temperatures. Moreover, with the very high temperature that can be obtained by using solid particle receivers, the opportunity of integrating CSP plants with highly efficient thermodynamic cycles is increased. This technology is still in the experimental verification stage and more research on the practical application and compatibility of the thermo-physical properties of such particles to be used as HTF and storage medium are still needed.

### 5.8.2 NOVEL HIGH-PERFORMANCE RECEIVER DESIGNS BASED ON APPLICATIONS

Additional novel receiver designs have been proposed to reduce heat losses and achieve higher efficiencies. These include designs to increase light trapping and solar absorption, and air curtains to reduce convective heat losses. Garbrecht et al. [222] proposed an external receiver with an array of numerous pyramidal structures on the exterior of the receiver that intercepted incident radiation and could create a radiation trap. Molten salt flowed from the interior of each pyramid toward the tip, and then along the sides of the pyramid. Simulations showed that for an irradiance of 1 MW/

m², this receiver design could achieve a thermal efficiency above 90%. The reflective radiative losses could be reduced to about 1%, and thermal emittance was about 3% of incident radiation. Challenges with this proposed design include achieving sufficient heat transfer at the tip of the pyramid, where the greatest flux occurs with the potential for stagnant internal flow. In addition, costs associated with the complexity of the numerous flow-through pyramidal structures may be high.

Lubkoll et al. [223] describe a spiky central receiver air preheater (SCRAP) design that employs a large number of spikes or tubes that are irradiated from concentrated sunlight. Each spike consists of an inner and outer tube through which air flows. Similar to the pyramidal designs, cold air flows through the inner tube toward the end and then flow back along the outer tube. The temperature of the outer surface of each tube is anticipated to rise from the spike tip to the interior root of each spike, where radiation is minimized. The highest cooling effect from the air occurs at each spike tip, where the irradiance is greatest. CFD simulations showed that radiative losses were only few percent of the total incident power, but convective losses were significantly higher (>16%) due to the large surface area. The thermal efficiency with an air outlet temperature of about 800°C was calculated at about 80%. Challenges include reducing the pressure drop within each spike and reducing convective losses.

Another approach to increase light trapping involves the use of alternative configurations of the tube panels for external receivers. Conventionally, the tube panels are arranged in either cylindrical or cubical fashion. Any incident radiation that is reflected is lost to the environment. By arranging the panels in a bladed configuration, reflected light can be intercepted by the surrounding panels. In addition, by introducing the cold heat-transfer fluid to the outer regions of the panels first, the hottest portions of the panels will be near the interior where radiative heat losses are reduced [224]. CFD simulations showed that the thermal efficiency was about 95% with radiative losses of less than 4% and convective losses of less than 1%. Wind effects were not considered. Challenges included structural considerations to prevent dynamic loading and fatigue from wind effects and the proper accommodation of headers and static loads in these novel bladed receiver configurations.

The concept of introducing light-trapping features at multiple length scales (fractal-like geometry) was introduced by Ho et al. [225,226]. In this concept, light-trapping features and processes can occur at the macro scale, meso scale, and micro-scale. At the macro scale (meters to tens of meters), bladed or spiky receiver geometries can be employed as described earlier. At the meso scale (millimeters to tens of centimeters), alternative shapes and arrangements for the tubes carrying the heat-transfer fluid can be designed and have been shown to increase the effective solar absorptance due to additional light trapping [225,227,228]. At the micro-scale (tenths of a millimeter or less), surface features and texturing can be used to increase light trapping and reduce thermal emittance. Combined, these features and geometries and multiple length scales may increase the thermal efficiency significantly. Challenges include reliability and costs associated with novel features, especially at the smaller scales.

Air curtains have been proposed as a way to mitigate convective heat losses from cavity receivers. Air curtains were proposed for cavity-based solid particle receivers as a means to reduce both convective heat losses and particle losses through the aperture [191,193,201–204]. Simulation results showed that the use of an air curtain had the potential to reduce convective losses by several percentage points, but testing showed that the airflow could cause particle instability [193]. The use of air curtains in dish-based cavity receivers was also investigated by Zhang et al. [229]. Different orientations of the airflow across the aperture of the cavity were investigated numerically, and optimal configurations were identified. While air curtains may reduce convective heat losses, a complete demonstration at high temperatures with economic analysis has not been performed.

### 5.8.3 Next Generation of Liquid Metal and Other High-Performance Receiver Designs for Central Tower Systems

In a solar power tower plant, the receiver is the heat exchanger where the concentrated sunlight is intercepted and transformed into thermal energy useful in thermodynamic cycles [230]. Radiant flux and temperature are substantially higher than in parabolic troughs, and this requires high-performance materials and system design with high technology. In order to maximize radiation absorption and minimize losses, the solar receiver should mimic a black body, and cavities, black-painted tube panels, or porous absorbers are used to trap incident photons. In most designs, a solar receiver is a single unit that centralizes all the energy collected by the large mirror field, and therefore high availability, thermal efficiency and durability are necessary requirements. Typical receiver absorber operating temperatures are between 500°C and 1,200°C and incident flux covers a wide range of flux density, from 300 to over 1,000 kW/m$^2$.

The solar receiver must reduce thermal and optical losses to increase its efficiency. Solar receiver is classified based on the use of intermediate absorber materials, the kind of thermal fluid used, or heat transfer mechanisms. According to the geometrical configuration, there are basically two design options, external and cavity-type receivers. In a cavity receiver, the radiation reflected from the heliostats passes through an aperture into a box-like structure before impinging on the heat transfer surface. Cavities are constrained angularly and subsequently used in polar field (north or south) layouts. External receivers can be designed with a flat plate tubular panel or a cylindrically shaped unit. Cylindrical external receivers are the typical solution adopted for surrounding heliostat fields. Receivers can be directly or indirectly irradiated depending on the absorber materials used to transfer the energy to the working fluid [231]. Directly irradiated receivers make use of fluids or particle streams that are able to efficiently absorb the concentrated flux. Particle receiver designs make use of falling curtains or fluidized beds.

The key design element in indirectly heated receivers is the radiative/convective heat exchange mechanism. There are two heat transfer options; tubular panels and volumetric surfaces. In tubular panels, the cooling thermal fluid flows inside the tube and removes the heat collected by the external black panel surface by convection. It therefore operates as a recuperative heat exchanger. Depending on the heat transfer fluid properties and incident solar flux, the tube might undergo thermo- mechanical stresses. When water is used as heat transfer fluid, since heat transfer is through the tube surface, when water is used as heat transfer fluid it is difficult to operate at an incident peak flux above 600 kW/m$^2$. However, with high thermal conductivity liquids such as sodium it is possible to reach in tubular panels operating fluxes above 1 MW/m$^2$. The contact surface can be improved by using wires, foams, or appropriately shaped materials within a volume. In volumetric receivers, highly porous structures operating as convective heat exchangers absorb the concentrated solar radiation [232]. The solar radiation is not absorbed on an outer surface, but inside the structure "volume." The heat transfer medium (mostly air) is forced through the porous structure and is heated by convective heat transfer. The maximum temperature is achieved inside the absorber. Under specific operating conditions, volumetric absorbers tend to have an unstable mass-flow distribution which is handled by an absorber material and appropriate selection of the operating conditions.

Selection of a particular receiver technology is a complex task, since operating temperature, heat storage system, and thermodynamic cycle influence the design. In general, tubular technologies allow either high temperatures (up to 1,000°C) or high pressures (up to 120 bar), but not both [233]. Directly irradiated particle receivers or volumetric absorbers allow even higher temperatures but limited pressures. Central receiver systems have a consolidated scheme when subcritical Rankine cycle is used and storage determines the dispatching economic feasibility of the project. In this case, molten nitrate salts are becoming the reference design material since nitrates are cheap and provide high storage capacity. As of 2016, solar receivers cooled with nitrate molten salts have thermal efficiencies of 88%. However, such a system has an upper temperature limit of 580°C and high melting point of nitrate molten salt requires trace heating which complicates operation and maintenance. These factors suggest the need for an alternate liquid metal medium.

Liquid sodium is extensively used in nuclear industry for liquid metal fast breeder reactors and carries extensive practical know-how in sensitive aspects such as safety and control, and hardware validation of key components such as pumps, valves, lines, and steam generators. Its boiling temperature (890°C) is substantially higher than solar salts and presents outstanding thermal conductivity. The vapor pressure at 595°C is only slightly above atmospheric pressure. Its main shortcoming is that it reacts with water and air, resulting in high maintenance costs. Melting point is relatively high (98°C), because of which heat tracing is required in the heat transfer fluid loop like in the case of molten salts.

Molten salt is cheaper than sodium by a factor of two and has a three-to-one advantage in its volumetric heat capacity, factors which are particularly important in the thermal storage subsystem. On the other hand, sodium has a five times higher heat transfer rate, which means that sodium receivers (like water/steam receivers) can be single pass. The thermal conductivity of liquid metals leads to high heat transfer coefficients. For receiving the same level of solar radiation and for the same flow velocity and tubular dimensions, molten sodium can require 57% less absorber area compare to molten salt.

LBE is another good alternative. Even though its thermal conductivity is significantly lower than sodium, it provides a wide range of temperatures for operation. However, its high density diminishes the heat capacity in storage systems and it produces large corrosion rates [234]. Furthermore, since LBE presents a high solubility limit for both nickel and copper (up to a few weight percent) at 600°C [235], inhibitors or protective layers are required to use nickel-based alloys and steels with high nickel content. A similar problem is found with molten tin above 600°C. Suitable options appear to be graphite, molybdenum, tungsten, or rhenium [236] or perhaps the use of some ceramics and refractory materials [234]. Although sodium has much better heat transfer characteristics than LBE, the stringent safety measures requirements have motivated the use of LBE temporarily in the German construction described in [237]. New approaches are, however, needed.

Apart from temperature range, liquid metals also reduce the size of heat exchangers and solar receivers due to their high thermal conductivity. Boerema et al. [238] developed a simple receiver model to determine the influences of the fluids' characteristics on receiver design and efficiency. They found that Hitec has a high Pr number, depending on temperature, while liquid sodium has a very low Pr number, with only a relatively small variation over the temperature range. The study also found for liquid sodium has a high heat transfer coefficient (an order of magnitude greater than Hitec) and a low heat capacity. High thermal conductivity of liquid metals leads to high heat transfer coefficients. It should be noticed that for the case of sodium the heat transfer coefficient decreases when the temperature increases because thermal conductivity decreases with temperature.

The efficiency and the pipe surface temperature at the receiver exit using liquid sodium are quite stable for increasing pipe diameters, whereas for Hitec both magnitudes are highly dependent on the pipe diameter. This is important as the use of larger diameter pipes reduces the number of pipes needed for the receiver and thus the manufacturing costs [238]. Compare to molten salts, at the same flux density, sodium decreases the thermal gradient of convection on the inner tube wall surface. This also reduces tube wall temperature and the risk of temperature hot spots and thus pipe stresses as well. Higher flux densities lead to smaller receivers for the same power output. The reduction of absorber surface decreases material and manufacturing costs. According to preliminary receiver performance studies, compared to solar salt, liquid sodium leads to an absolute efficiency increase of 1.1% by utilizing higher concentration ratios [238]. In addition, due to the absorber area reduction as mentioned earlier, radiation and convection losses might also decrease. Both of these factors imply higher receiver efficiencies and performance.

Early experimental solar tower facilities started operation in the 1980s with small demonstration systems between 0.5 and 10 MW [73,239,230]. The thermal fluids used in the receiver were at that time liquid sodium, saturated or superheated steam, and nitrate-based molten salts. The earlier studies also indicated that solar receivers with proper designs could work at relatively high efficiencies (above 80%) for temperatures high enough in the absorber material to heat up thermal fluids able to produce superheated steam above 500°C. A comparison of solar receivers tested at different power loads in early experimental facilities indicated that for nominal power load, the liquid sodium receiver achieves thermal efficiencies near 90. Alternative approaches are taking place to move forward with the use of liquid metals efficiently and safely. Other liquid metals like Na-K mixtures or LBE are being explored at the laboratory scale to assess performance and understand operational issues. The use of liquid metals such as sodium, Na-K, or LBE mixtures in solar receivers may lead to very compact designs of the aperture since the high conductivity improves the heat transfer through the absorber wall [240]. The absorber can also work at high fluxes, above 2 MW/m², since thermal stress is reduced and temperature gradient between the tube wall and the bulk of the fluid is minimized. Sodium is largely used for the receiver where the heat transfer properties of liquid metals are advantageous.

According to Romero, and Gonzalez-Aguilar [121], the most extensive operational experience of liquid metal solar receivers took place in Almería, Spain, within the framework of the International Energy Agency (IEA) SSPS project. The project was developed during 1982–1986 and two receivers were analyzed. The first receiver was a north-facing cavity type, having a vertical octagonal-shaped aperture. The absorber panel was a cylinder segment of 120° with 4.5 m diameter. Sodium was flowing through six horizontal parallel tubes (35 mm ID) which was winding in a serpentine from the inlet header on the bottom of the cavity to the outlet header at the top. Sodium entered the inlet header at 270°C and exited through the outlet header at 530°C. [241]. The second receiver, so-called advanced sodium receiver or ASR, was external and formed by five panels arranged to form a rectangular absorber. Each panel consisted of a tube bundle with 39 tubes (14 mm OD and 12 mm ID). Both receivers were able to produce 2.4–2.5 MW thermal power [242]. Unfortunately, a sodium fire occurred at the IEA/SSPS central receiver plant in 1986. It resulted in its shut down. Since then much development has been done to model the behavior of sodium combustion and fire extinguishing [243]. In addition, the cost of the fluid remains as another important factor against liquid metals since they are relatively expensive [234]. The cost of sodium (US $2/kg) is four times more expensive than solar salt typically used in current solar towers (60–40%wt NaNO₃- KNO₃) and LBE is about 26 times more expensive (US $13/kg).

CRS-SSPS has been the only project offering experimental data after an extended test campaign. Few other initiatives can be acknowledged from a literature survey. A pioneering project took place during the early 1980s by Rockwell International and the US Department of Energy whose results were instrumental for the design of the SSPS receivers. The development involved the construction and testing of a 3.6 m² sodium-cooled test receiver for evaluation at the Central Receiver Test Facility (CRTF) in Albuquerque, New Mexico [244]. Thermal efficiency reported was higher than 90%, with some uncertainty in flow measurement. Another solar thermal power plant based upon the use of liquid sodium as heat transfer fluid is Jemalong Solar Thermal Station developed in Forbes, New South Wales, Australia by the company Vast Solar. While in 2015 fire was handled from a tank leaking sodium in this plant, subsequently pilot plant was started in 2016. Romero and Gonzalez-Aguilar [121] have presented an excellent review for the use of next generation of liquid metal and other high-performance receiver designs for central tower systems.

## 5.9   STATUS OF COMMERCIAL CSP PLANTS

The current state-of-the-art CSP technologies for electricity production largely uses parabolic trough (PT), linear Fresnel (LF) and central receiver tower (CRT) systems. PT uses synthetic oil as heat transfer fluid and two-tank molten salt storage system; LFR uses water as the heat transfer fluid for direct steam generation without storage and CRT uses molten salt as heat transfer fluid and a two-tank molten salt storage system. These three CSP commercial technologies for electricity production typically use a Rankine cycle to transform the useful thermal energy generated in the receiver (s) into electricity. For parabolic dish systems, the most demonstrated concept uses a Stirling engine, although these systems are not fully commercially available. Parabolic troughs track the sun in one direction normally east to west and give a concentration ratio of 50–80. The receiver usually consists of an absorber metal pipe insulated inside an evacuated glass tube. The heat transfer fluid, synthetic oil, is generally heated to temperature around 400°C. The superheated steam coming out of heat exchanger provides subcritical Rankine cycle efficiency of about 34%. The thermal efficiency of Rankine cycle depends on the temperature and pressure conditions of the steam. For example, in subcritical operation with steam temperature 565/565 and pressure of 16.5 MPa, efficiency range of 34–38 can be obtained. In supercritical steam conditions with temperature of 565/585 and pressure >24.8 MPa, thermal efficiency in the range of 38–41 can be obtained. In ultra-supercritical steam conditions with temperature of 593/621 and above and pressure greater than 24.8 MPa, thermal efficiency in the range of 41–42 can be obtained. Finally, in advanced ultra-supercritical conditions of the steam with temperature of 677 and above and pressure greater than 34.5 MPa the efficiency can be greater than 42%. The Rankine cycle involves water being heated to create pressurized steam that is fed into a turbine. Expansion of the steam causes the turbine to rotate, converting thermal energy into mechanical energy that can drive an alternator or generator, to create electricity.

More than 90% of current commercial CSP power plants are based on parabolic trough technology, less than 5% based on linear Fresnel, and about 5% based on central receiver tower. In recent years, however, central receiver tower system is gaining more popularity due to its ability to get higher temperature and higher Rankine cycle efficiency. Linear Fresnel systems provide a concentration ratio of 30–70 with subcritical Rankine cycle efficiency of about 36%. The reflectors in this system also focus the sun's rays onto a fixed linear receiver by tracking the sun in one direction normally from east to west. The receiver usually consists of absorber metal pipes in an insulated cavity with a glass window. The heat transfer fluid is normally water allowing direct steam generation at higher temperatures (about 450°C) than is possible with synthetic oil in parabolic troughs. Central receiver towers are point-focusing technologies with a concentration ratio of about 500–800. A field of many mirrors, called heliostats, reflects the sun's rays onto a central receiver located on a tower. Each heliostat has two-axis tracking to direct the sun's rays toward the central receiver, which is normally an array of absorber tubes, referred to as an external receiver. The heat transfer fluid in the absorber tubes could be molten salt, which can be heated to temperatures of around 560°C, or steam. The use of cavity receivers partially insulates the absorber tubes to reduce heat losses. Due to higher superheated steam temperature, its subcritical Rankine cycle efficiency is about 38%.

The overall conversion efficiency of solar energy into electricity is dependent upon each step in the conversion process, including the thermal efficiency of the power block. The thermal efficiency of a Rankine cycle depends upon the temperature and pressure of the steam. To increase the temperature of a CSP system, the concentration ratio needs to be increased, and it follows that point-focus systems achieve this more readily than line-focus systems. Alternative power blocks to the Rankine cycle include the Brayton and Stirling cycles. The Brayton cycle uses a gas as the working fluid, which needs to be compressed rather than pumped through the cycle to the turbine. If the cycle is open, air can be used as the working fluid, with gases such as helium, argon, and supercritical carbon dioxide being preferred for a closed cycle. The Stirling cycle also uses a gas as the working fluid, but involving a reciprocating engine to compress and convert the expanding gas into mechanical energy.

The IEA 2014 roadmap reported 4 GW of installed CSP power plants worldwide with an anticipated deployment of 11 GW of CSP plants by 2020. The IEA also proposed that "achieving this roadmap's vision of 1,000 GW of installed CSP capacity by 2050 would avoid the emissions of up to 2.1 gigatons (Gt) of carbon dioxide ($CO_2$) annually." The present CSP industry exists due to the establishment of the initial trough plants at Kramer Junction in California USA, in the 1980s and deployment of main trough but also Fresnel and solar tower plants in Spain from 2005 to 2012. The 1980s trough plants saw the construction on nine solar energy generating systems (SEGS) totaling 354 MW. These plants ranged in net output from 13.8 to 80 MW and used PT as collector and thermal oil as the heat transfer fluid and possibly as fluid for thermal storage. These plants were at three different locations (Daggett, Kramer Junction, and Harper Lake) and in their more than 30 years of operation, the Kramer Junction SEGS plants in particular have demonstrated that CSP plants are very reliable, they operate beyond their originally estimated operational life and they improve their performance and reduce their operational cost with time. New plants developed in Spain since 2006, were trough technology, although some used linear Fresnel and tower technology; incorporated thermal energy storage, based on nitrate molten salts, of 4–15 hours at nominal plant capacity and had a high proportion (60%) of local content, that is, goods and services from local companies.

Troughs were the preferred technology due to the proven experience of the SEGS plants in California providing a low project risk for the finance sector, although CSP power plants using linear Fresnel and solar tower technologies were also constructed. There was widespread integration of thermal energy storage, based on the use of hot and cold tanks of molten nitrate salts. The 50 MW Andasol I plant in Granada is typical of the trough plants. The 20 MW Gemasolar Thermasol plant in Andalucía is representative of the tower plants installed in Spain. This plant operated continuously for 30 days. The best market potential for CSP power plants is in regions with direct solar insolation greater than 1,800 kWh/m²/year which occur mainly within latitudes of 40, which includes the Americas, Australia, China, India, Middle East, North Africa, and South Africa. In 2015, the emerging markets were predominantly in Morocco, South Africa, China, and Chile where the insolation coexisted with increasing demand and policy drivers for renewable energy and/or stable grid supply. While the trend to improve the performance of CSP power plants is to increase the operating temperature and thereby the concentration ratio, the opportunities for industrial process heat tend to be at lower temperatures.

## 5.10   HYBRID CSP POWER PLANTS

CSP has been hybridized with other sources for power production. Here we briefly examine a few such hybrid power sources. The details on hybrid power generation using CSP are given in an excellent review by Yousef et al. [245].

## 5.10.1 Fossil Fuels-Solar Hybrids for Power

Power generation by coal and solar energy can be integrated into a number of different ways. For example, solar cells can be added to a combine (CGT) plant. Clearly, these solar assets generate electricity, but this is fed into the grid independently of the coal or gas-fired plant. India plans to install a significant amount of solar CSP generating capacity, with some new facilities being located at existing coal-fired power plants [246,247]. A number of projects combining CSP and coal have been investigated worldwide, in the USA, Europe, Australia, South Africa, China and the Middle East. Colorado Integrated Solar Project (CISP) was the first integrated CSP-coal project [248]. In this project, PT was integrated into 49 MW coal-fired power plant, to increase the plant efficiency and reduce coal consumption, which led to a reduction in conventional plant emissions and $CO_2$ [248]. The Electric Power Research Institute also hosted a demonstration project at a Pulverized Coal Plant, to integrate a 36 MW solar concentrator field into 245 MW Escalante Generating Station in New Mexico [249].

CSP-coal integration has been also suggested as part of the development of solar thermal power generation strategy in China [250]. One of the studies evaluated the effect of using CSP to generate steam as boiler feed-water to 330 MW hybrid power plant in Changji City [251]. The study showed that, besides the reduction in gas emission, the electricity generated by integrated CSP coal could be 20%–30% cheaper than a standalone CSP power plant. In addition to the reduction in coal combustion and $CO_2$ emissions, the integration of CSP with coal reduces the total investment cost because of the sharing of main components. In a numerical study performed regarding the integration of a fossil fuel plant with a CSP plant, it was estimated that the integrated power plant can cost 72% of the overall CSP cost [249]. Moreover, in the study, it was obtained that electricity power generation can be increased up to 25% [252]. Zhang et al. [253] investigated two suggested arrangements to integrate solar tower with coal power plant. In the first scheme the solar tower was used in parallel with a coal-fired boiler to heat superheated steam, whereas in the second scheme the solar tower was used to heat sub-cooled feed water using a molten salt heat exchanger with a flue gas bypass in the boilers. Hong et al. [254] derived theoretical calculations of solar power efficiency, including exergy destruction of solar integration. They applied the derived equations to an existing integrated 330 MW coal-fired power plant to evaluate the thermal performance of converting solar energy to power. The study showed that the amount of output work for integrated power plant was increased compared with the coal-fired power system. Further studies with different configurations to integrate different CSP technologies with coal power plant are described by Yousef et al. [245].

A limited number of coal-solar projects are true hybrids. These operate under an entirely cooperative arrangement where the two sources of energy are harnessed to create separate but parallel steam paths. These paths later converge to feed a shared steam-driven turbine and generate electricity as a combined force. This form of hybrid technology integrates these two disparate forms of power so that they combine the individual benefits of each. This approach can replace a portion of coal demand by substituting its energy contribution via input from a solar field.

During daylight operation, solar energy can be used to reduce coal consumption (coal-reducing mode). As solar radiation decreases during the latter part of the day, the coal contribution can be increased, allowing the plant's boiler to always operate at full load. When solar radiation increases again, the process is reversed, with solar input gradually reducing that of coal. Alternatively, input from the solar field can be used to produce additional steam that can be fed through the turbine, increasing electricity output (solar boost). Whichever mode is adopted, the design and integration of the solar field into the conventional system are critical for the proper functioning of a hybrid plant. In principle, this form of hybrid technology can be applied to any form of conventional thermal (coal-, gas-, oil-, or biomass-fired) power plant, either existing or new build [246,247].

Xcel Energy and Abengoa Solar partnered a demonstration that used sun-tracking parabolic trough technology to supplement the use of coal. The main aim was to demonstrate the potential for integrating solar power into large-scale coal-fired power plants to increase plant efficiency, reduce the amount of coal required, and hence reduce conventional plant emissions and $CO_2$. It was also to test the commercial viability of combining the two technologies. Coal-solar hybrid technology has also been investigated in a number of other countries that maintain major coal-fired power sectors, and significant work has been undertaken in, for example, South Africa, China, and several European countries. In the Atacama desert region of Chile, as part of increasing electricity supply in the region, a 5-MW coal-solar hybrid project is being developed by Engine and Solar Power at the existing 320-MW Mejillones coal-fired power plant. In a recent development, NTPC (formerly the n Limited) announced the start of a coal-solar hybrid project (the Integrated Solar Thermal Hybrid Plant) to be developed at its Dadri power plant in India. This will be the first Indian project to use solar energy to heat boiler feedwater with the aim of increasing plant efficiency and reducing coal demand. To save costs and manpower, the project will feature the robotic dry cleaning of the solar panels [246,247].

An advantage of adding a solar thermal module to an existing coal-fired power plant is that much of the necessary infrastructure (steam cycle, etc.) and plant requirements already exist. This can make the economics more attractive than those of a stand-alone solar thermal generating unit. Solar energy input can be harnessed by parabolic troughs, compact linear Fresnel reflectors (CLFRs), or power towers. As part of a hybrid system, these raise steam that is fed into a power plant, reducing the amount of coal (or gas) required. Such thermal hybrid projects may be the most cost-effective option for large-scale use of solar energy. According to US studies carried out by EPRI, potentially, a solar trough system could provide 20% of the energy required for a steam cycle. The IEA GHG R&D Program has recently examined the integration of solar energy technologies with CCS-equipped plants [255,247].

Globally, there are around twenty hybrid solar thermal plants being developed, some based on gas and some on coal. In the USA, a major demonstration project (using parabolic trough solar collectors) was undertaken at Unit 2 of Xcel Energy's coal-fired Cameo Generating Station. The main aims were to decrease coal use, increase plant efficiency, lower $CO_2$ emissions, and to test the commercial viability of combining the two technologies. The solar component was deemed to have operated satisfactorily—coal use was lower, and overall $SO_2$, NO and $CO_2$ emissions were reduced. During operation, the system produced the equivalent of 1 MW (of the plant's 49 MW) from solar power. In Australia, the utility CS Energy is building a 44 MW solar thermal add-on to its 750 MW coal-fired supercritical Kogan Creek plant in Queensland. This is the largest solar project in the Southern Hemisphere and the world's largest linear Fresnel reflector solar CSP installation. The add-on allowed increased electricity production and avoided an estimated 35,600 $tCO_2$. A second Australian CSP project (using a Linear Fresnel reflector system with a total mirror surface of 18,500 $m^2$) is at Macquarie Generation's 2 GW coal-fired Liddell Power Station in New South Wales. This incorporates a 9.3 MW capacity Novatec solar boiler to generate steam that is fed into the existing plant, helping reduce coal requirements and plant emissions. The reduction in coal used cuts $CO_2$ emissions by 5,000 $tCO_2$ [246,247].

Coal-solar hybrid can also be applied to brown coal gasification, power production and direct conversion of coal to liquids in a number of different ways [246,254]. A major challenge is to increase the solar share of coal-solar hybrids. To date, input from the solar component has tended to be limited, often as a consequence of their application to retrofit applications on ageing plants as opposed to new build. There is some consensus that the sector needs bigger hybrid projects based on highly efficient, newly built coal-fired plants; this would provide more scope for improved efficiency and better economics. The projects developed in Chile and India will provide useful data and operational experience, hopefully encouraging further uptake of the technology. Technology development continues, aimed at reducing system costs, increasing the efficiency of solar-to-electricity conversion and minimizing the environmental footprint. Some argues that a much greater solar share (possibly up to 30%–40%) could be realistic [246,247,255].

In the case of hybrid CSP-natural gas power plants, integration can be directed to the steam cycle (SC) or gas turbine (GT). Therefore, solar energy through CSP is used to either boil water in the heat recovery steam-generator (HRSG) and then inject it into the high-pressure drum/steam turbine or heat the air leaving the compressor before combustion. Most of the existing hybrid CSP worldwide in USA, Mexico, Italy, Morocco, Algeria, Iran and Egypt are integrated with CCGT for supplying additional saturated steam to HRSG through a high-pressure drum [256]. A typical natural gas-solar hybrid system is illustrated in Figure 5.11. Examples of worldwide-integrated CSP-natural gas power plant are given by Yousef et al. [245].

Integration of CSP with natural gas can be achieved using various configurations and CSP technologies at different collection temperatures. Supplying heat to gas turbine through CSP is a technical challenge, since gas turbines operate at temperatures higher than that needed by steam turbines. On other hand, the flexible operation of open cycle gas turbine (Brayton cycle), where fuel and air can be adjusted, makes their coupling with CSP more valuable [257]. Therefore, CSP heat can be added to the cycle after compression stage and before combustion, where the adjusted fuel rate assures combustion can reach the required operating temperature. To add air steam heated in CSP directly to a Brayton cycle, pressurized receiver is the best choice to be used [258,259]. Another configuration to integrate CSP with a gas turbine is to inject steam into the combustor, by replacing or supplementing the steam generators. This

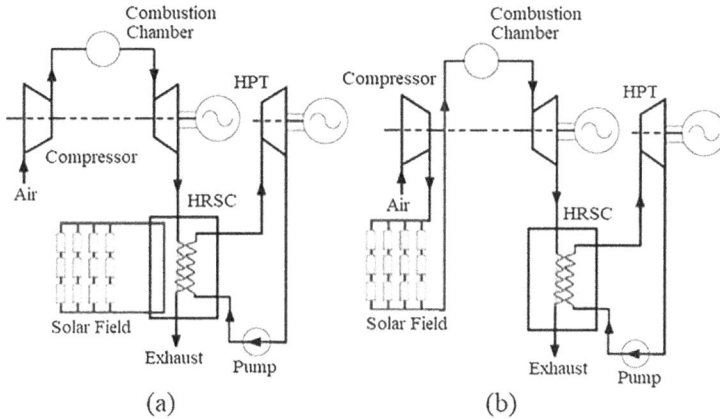

**FIGURE 5.11** Integrated CSP-natural gas configurations: (a) solar energy used to superheat high-pressure water; (b) solar energy used to preheat the air leaving from the compressor before entering the combustion chamber [245].

configuration is known as steam injection gas turbine (STIG), and it can increase the power output from the gas turbine, as well as increase solar share. This configuration does not require a high temperature from CSP, since the steam is injected to HRSG not to the gas combustion. Moreover, depending on the compression ratio of the gas turbine, STIG can use simple and less expensive CSP technology such as PT or linear Fresnel, with saturation steam temperature in a range of 200°C–300°C [260,261]. A summary of various possible configurations of integrated CSP-natural gas systems is given by Yousef et al. [92,245].

### 5.10.2 CSP-GEOTHERMAL POWER PLANT

While geothermal power plant (GPP) has a shorter construction time and lower investment cost, due to the low enthalpy of geothermal resources, its overall efficiency is low (~10%) [262]. Thus, in order to increase the overall efficiency of GPP, the integration with other power plants should be considered. There are three types of GPP, as illustrated in Figure 5.12: (a) Dry steam GPP: where the dry steam drawn from underground wells is directed to the turbine to generate electricity, (b) Flash steam GPP: where water at ~180°C temperature is drawn from the well, and converted to steam due to pressure reduction. The converted steam is separated and used for power generation and (c) Binary cycle GPP: where water in the range of 107°C–182°C is drawn from the well to boil a working fluid in an organic Rankine cycle (ORC). In order to integrate CSP with GPP, the solar energy can be directed to preheat the brine or to superheat the working fluid in the geothermal cycle, while the geothermal energy is used to preheat the feed-water in the Rankine cycle of the solar thermal power plant [260–266].

As pointed out by Yousef et al. [245], the first CSP-GPP power plant (Stillwater), located in Nevada, USA, integrated a 2 MW PT technology with a 33 MW geothermal

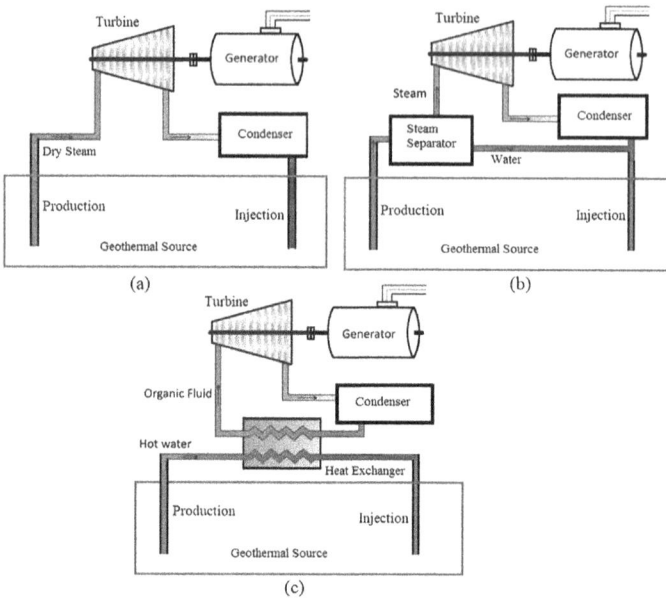

**FIGURE 5.12** Types of geothermal power plants: (a) dry steam geothermal plant; (b) flash steam geothermal plant; (c) binary cycle [245].

station and a 26 MW solar PV unit [267]. The integrated CSP-GPP combined the continuous generating capacity of binary-cycle and medium-enthalpy geothermal power with solar thermodynamic technology. The heat collected by CSP was added to the geothermal brine before it entered the ORC. This increased the temperature of the incoming geothermal fluid to the plant, as well as the plant power output. Moreover, due to the CSP-GPP integration, the capacity factor of the plant was enhanced, without increasing the nominal power [92,268].

Ciani et al. [269] modelled an integrated CSP-GPP with and without TES and compared the monthly electrical production of both solutions. The highest electrical production was achieved in the integrated CSP-GPP with TES, where an increase of 19% in annual energy production was achieved due to the improved utilization of the extra thermal power produced by CSP and stored in TES. Zhou et al. [270] used Aspen HYSYS to simulate an integrated CSP-GPP and compared its performance with stand-alone solar and GPPs in the Australian context. The simulation results showed that the integrated CPS-GPP outperformed the individual solar power plant and geothermal plant. The study also showed that at a reservoir temperature of 150°C, the cost of the electricity production could be reduced by 20% in integrated CSP-GPP, compared to the individual enhanced geothermal system.

Craig et al. [271] used IPSEpro simulation software to simulate an integrated CSP-GPP. The configuration used in the simulation increased the output power for integrated CSP-GPP by 8%, compared to the individual CSP power plant. They also found that the electrical efficiency conversion in the integrated CSP-GPP increased significantly from 1.7 to 2.5, and as a consequence resulted in a more efficient use of geothermal energy. Moreover, the overall efficiency of the integrated CSP-GPP was 3.6 times higher than the efficiency of the individual GPP. Technical and economic analyses of integrated CSP-GPP using low-temperature ORC were conducted by Ayub et al. [272]. The study showed improved LCOE by several percentages by integration. Astolfi et al. [273] evaluated the potential of CSP-GPP based on ORC. The CSP was used to provide an additional heat source for the cycle as well as increase power production. The study showed that besides improvement in LCOE, this kind of integration could increase the overall capacity factor and guarantee safe operation for the turbine in an ORC-based plant with a reduced electricity cost.

One of the efficient but challenging techniques for integrating low grade heat CSP-GPP is the multi-generation system. CSP-GPP integration for multi-generation (heat and electricity) was examined by Al-Ali and Dincer [274] and Ezzat and Dincer [275]. In the latter study, TES was also used. Both studies indicated improved performance by integration. Calise et al. [276] conducted a study using TRNSYS to simulate a highly integrated CSP-GPP system. The integrated system included ORC-PT-geothermal wells absorption, chillers-multi-effect distillation and a thermal recovery subsystem. The power generated by solar and geothermal sources was used for electricity, cooling, water heating and desalination for small communities. These and other study [277] also showed that incorporating TES into the system can maintain the grid stability, but with additional expenses in the capital expenditures. Alibaba et al. [278] investigated the conventional and advanced, exergo-environmental and exergo-economic analysis of a geothermal–solar hybrid power plant (SGHPP) based on an organic Rankin cycle (ORC) cycle. In practice, only a few CSP-GPP hybrid plants are implemented for power production. This can be explained by the high initial, operation and maintenance costs as well as the complexity of building the entire hybrid power plant. More details on the global assessment of CSP-GPP integration are discussed by Yousef et al. [245].

### 5.10.3 CSP AND BIOMASS

The integration of solar and biomass energy to generate power is recommended in the areas where these two resources are abundant such as Bangladesh, India, Greece, Southern Turkey etc. Both CSP and biomass technologies convert thermal energy to electrical power and that makes them compatible for integration; different components can be shared during this integration, lowering the investment cost [279].

Borges thermo-solar plant, located in Spain, was the first commercial integrated CSP-BIM plant in the world and has operated since December 2012. The electricity production is 98,000 MWh/year generated by PT during the day and by biomass energy at night, with a 22.5 MW net power capacity, avoiding 24,500 tons emissions [280,281]. Servet et al. [282] evaluated the integrated CSP-BIM power plant compared to the stand-alone CSP and biomass power plant. The study showed that, due to the possibility of using shared equipment, the investment cost was ~24% lower than that related to the two technologies, resulting in lower LCOE for the integrated system. Moreover, energy generation was 2.77 times higher than the individual CSP plant, and the use of biomass technology eliminated the need for TES. Kaushika et al. [283] described the integration of PT and a biomass plant using the biogas produced from the waste of the distillery plant. Two studies, [284,285], investigated the integration of PT and biomass energy in order to (a) supply heat to ORC through a biomass furnace and PT, (b) to superheat steam generated by PT and (c) to use steam from PT for biomass gasification. The conversion efficiency from solar to electrical recorded an increase of 10.5%, while the investment cost decreased by ~23.5% due to the integration [286,287]. Multi-generation systems to produce electricity, chilled and hot water by integrated PT and biomass have also been investigated [288–290]. The integration of LFR with biomass energy has also been investigated; this integration allows to increase the steam temperature up to ~500°C, leading to higher conversion efficiency [291–293].

Nixon et al. [294] evaluated the technical and economic feasibility of integrating LFR with a biomass power plant for multi-generation based on five case studies, with thermal capacities ranging from 2.0 to 10 MW. The results showed that the integrated CSP-BIM power plant can be a viable option for multi-generation (electrical power, cooling and heating) for small and medium applications. Peterseim et al. [295] investigated a number of integrated CSP-BIM systems to determine the best technical and economic combination. The integrated CSP-BIM delivered ~69% lower annual electricity generation than individual CSP systems [295]. Peterseim et al. [296] investigated the integration of a solar tower with biomass in Griffith, New South Wales. The results showed that the integrated system provided lower electricity cost at AU$155/MWh, as well as reduced the investment cost by 43% compared to stand-alone CSP systems. Alfonso-solar et al. [297] proposed a hybrid system, combining biomass and photovoltaics, to supply electricity to educational buildings.

Ravaghi-Ardebili et al. [298] modeled a biomass gasification unit using low temperature of 410°C collected by PT. In order to achieve efficiency equivalent to the efficiency of high-temperature biomass gasification, different operating conditions and parameters, along with the alternative design option in the reactor configuration were considered. Khalid et al. [299] evaluated the energy and exergy efficiencies for a hybrid solar tower and biomass integrated system to generate multi-energy for commodities heating, cooling and electrical power. Integration of solar energy and biomass showed good synergy; CSP and BIM both generated electricity using thermal energy, with a potential to deliver 100% dispatchable power supply. The integration of the CSP-biomass can improve the flexibility and competitiveness of the power plant, increasing the capacity factor of the plant, as well as decreasing the size of the solar field of the equivalent stand-alone CSP plant. Moreover, other elements, such as the turbine generator, compressed air, condenser, cooling tower, deaerator pipes, feed water, valves, control devices and many other components, can be shared, which results in lower investment cost. In addition, the biomass consumption in the hybrid power plant is reduced compared to the stand-alone biomass plant and thus, the risk associated with the biomass supply is decreased [300]. Solar thermal energy can be injected at different points and temperatures to obtain optimal steam parameters for the host plant in terms of temperature and pressure. Furthermore, CSP-biomass hybridization offers the possibility to operate on new thermodynamic cycles other than the classical solar ORC such as Brayton cycle, supercritical $CO_2$ cycle and combined ORC-externally fired gas turbine cycle for both power generation and cogeneration applications [301,302]. CSP can also be used as a clean source of heat, to produce biofuels via gasification.

## 5.10.4  CSP-Photovoltaic

PV-CSP hybrid systems can be of two ways: (a) decoupled PV-CSP in which both PV and CSP subsystems can be planned independently and integrated by the electric power system [303] and (b) coupled systems such as PV-topping technology or the spectral beam splitting (SBS) technology. In decoupled PV-CSP systems, PV panels may contribute to power production during the day as PV has lower LCOE, while the CSP plant with TES can provide the necessary power during low irradiance periods supplementing PV or operating during the night [304]. The idea behind the PV-topping system is to recover the heat dissipated from solar cells; here cells are operated at a high temperature and the dissipated heat is used to generate solar power through the CSP system [305]. While decoupled systems have begun their commercialization stage, coupled ones are still under research and face several technological challenges.

Although both PV and CSP use the same energy source, the inclusion of a storage option in the CSP allows for the continuous operation of the power plant, regardless of the solar energy availability. This integration might lead to a lower-cost solution, compared to CSP alone, especially in terms of baseload output requirements. A hybrid CSP-PV plant consists of a field of PV modules together with a CSP field, coupled with a thermodynamic cycle, usually the ORC. The CSP part of the hybrid plant can be coupled with a thermal storage facility, to improve the dispatchability and renewable penetration in the grid. Chow et al. [306] pointed out that the market of solar thermal and photovoltaic electricity generation is growing rapidly. The final success of the integrative technologies relies on the coexistence of robust product design/construction and reliable system operation/maintenance in the long run to satisfy the user needs.

Recent studies have proven that hybridization of CSP-PV can provide dispatchable and stable power at a lower production cost. Selected recent studies on CSP-PV integration are summarized by Yousef et al. [245]. According to the literature review, the hybrid CSP-PV option has significant potential for coupling the technical and economic advantages of both technologies. In fact, such an option may contribute to improve the power quality, grid stability and renewable penetration in the grid compared to PV plant alone. Moreover, both PV and CSP technologies have reached their maturity, which makes their integration more attractive and cost-effective. This hybridization is suitable for large-scale solar power plants but can also be implemented in micro-scale solar power plants in remote areas. With the low cost of PV technology, the CSP-PV hybridization can replace the complex and costly electro-chemical battery with a lower cost and well-established thermal storage system. Such configuration could achieve a minimum of LCOE of 0.12 USD/kWh.

Cocco et al. [307] aimed to demonstrate the improvement of power dispatchability that can be achieved with a suitable integration of Concentrating Solar Power (CSP) and Concentrating Photovoltaic (CPV) plants to mitigate the effects of the variability and intermittency of solar energy. Orosz et al. [308] showed that Sunlight to electricity efficiencies of Parabolic Trough Collector (PTC) plants are typically on the order of 15%, while commercial solar Photovoltaic (PV) technologies routinely achieve efficiencies of greater than 20%, albeit with much higher conversion efficiencies of photons at the band gap. Hybridizing concentrating solar power and photovoltaic technologies can lead to higher aggregate efficiencies due to the matching of photons to the appropriate converter based on wavelength. Zhai et al. [309] illustrated an optimum design method for the CSP-PV plant based on a genetic algorithm considering the operation strategy. The study showed the operation strategy of the CSP-PV system for parabolic trough CSP system and PV system which are now commercially operated. Bennett et al. [310] examined a hybrid CSP/CPV spectrum splitting solar collector for power generation. Cygan et al. [311] examined the workings of a full spectrum solar system using a hybrid CPV/CSP system to deliver variable electricity and on-demand heat. Spectral beam splitting (SBS) films are crucial for the development of hybrid systems based on photovoltaic (PV) and concentrating solar thermal (CST) technologies. In the study by Zhang et al. [312], a novel double-layer $SiN_x$/Cu SBS film was prepared via magnetron sputtering. This film was developed based on the linear Fresnel solar thermal technology used in PV/CST hybrid systems. More details on this hybrid system are also given by Yousef et al. [245].

## REFERENCES

1. Blanco M.J., and Miller S. Chapter 1 - Introduction to concentrating, solar thermal (CST) technologies. In: Blanco M., and Santigosa L., editors. *Advances in Concentrating Solar Thermal Research and Technology*. Amsterdam: Woodhead Publishing series in energy; 2017, pp. 1–25. Doi: 10.1016/B978-0-08-100516-3.00001-0.
2. Technology roadmap. Solar thermal electricity; by Paris, France: International Energy Agency IEA; 2014.
3. Key world energy statistics; by Paris, France: International Energy Agency IEA; 2015.
4. Zamfirescu C., and Dincer I. How much exergy one can obtain from incident solar radiation? *J Appl Phys* 2009;105:044911.
5. R'boac M.S., Badea G., Enache A., Filote C., Rata G.R.M., Lavric A., and Felseghi R-A. Concentrating solar power technologies. *Energies* 2019;19:1048. Doi:10.3390/en19061048 www.mdpi.com/journal/energies.
6. Toro C., Rocco M.V., and Colombo E. Exergy and thermoeconomic analyses of central receiver concentrated solar plants using air as heat transfer fluid. *Energies* 2016;9:885.
7. Wang R., and Jun M.A. Status and future development prospects of CSP. *IOP Conference Series: Earth and Environment Science* 687 021088, IOP publishing, EPPCT 2021 - an open access paper; 2021. Doi: 10.1088/1755-1315/687/1/012088.
8. Dowling A.W., Zheng T., and Zavala V.M. Economic assessment of concentrated solar power technologies: A review. *Renew Sustain Energy Rev* 2017;72:1019–1032.
9. Salvatore J. *World Energy Perspective—Cost of Energy Technologies*. London: Bloomberg New Energy Finance; 2013. Available from: https://www.worldenergy.org/wp-content/uploads/2013/09/WEC_ J1143_CostofTECHNOLOGIES_021013_WEB_Final.pdf (accessed: 17 October 2018).
10. Ehtiwesh I.A.S., Coelho M.C., and Sousa A.C.M. Exergetic and environmental life cycle assessment analysis of concentrated solar power plants. *Renew Sustain Energ Rev* 2016;56:145–155.
11. Shravanth Vasisht M., Srinivasan J., and Ramasesha S.K. Performance of solar photovoltaic installations: Effect of seasonal variations. *Sol Energ* 2016;131:39–46.
12. Zsiborács H., Hegedüsné Baranyai N., Vincze A., Háber I., and Pintér G. Economic and technical aspects of flexible storage photovoltaic systems in Europe. *Energies* 2018;11:1445.

13. International Renewable Energy Agency (IRENA). Renewable power generation costs. Available from: https://www.irena.org/-/media/Files/IRENA/Agency/Publication/2018/Jan/IRENA_2017_Power_Costs_2018.pdf (accessed: 20 June 2018).

14. Hennecke K., and Zarza E. Direct solar steam generation in parabolic troughs (DISS). Update on project status and future planning. *J. de Physique* 1999;9(3):Pr3–469.

15. Khenisi A., Krüger D., Hirsch T., and Hennecke K. Return of experience on transient behavior at the DSG solar thermal power plant in Kanchanaburi, Thailand. *Energ. Procedia* 2015, May;69:1603–1612.

16. Muñoz-Anton J., Biencinto M., Zarza E., and Díez L.E. Theoretical basis and experimental facility for parabolic trough collectors at high temperature using gas as heat transfer fluid. *Appl Energ* 2014;135(C):373–381.

17. Pacheco J.E., and Kolb W.J. Comparison of an impedance heating system to mineral insulated heat trace for power tower applications. *A Paper Presented at 1997 ASME International Solar Energy Conference*, Washington, DC; 1997.

18. Raade J., and Padowitz D. Development of Molten salt heat transfer fluid with low melting point and high thermal stability. *J Sol Energ Eng* 2011, July;133(3):031013 (6 pages). Doi: 10.1115/1.4004243.

19. Rodríguez M.M., Marquez J.M., Biencinto M., Adler J.P., and Díez L.E. First experimental results of a solar PTC facility using gas as the heat transfer fluid. *A Paper Presented at SolarPACES Conference*, Berlin, Germany; 2009.

20. Valenzuela L., Zarza E., Berenguel M., and Camacho E. Control concepts for direct steam generation process in parabolic troughs. *Sol Energ* 2005;78(2):301–311. Doi: 10.1016/j.solener.2004.05.008.

21. Valenzuela L., Zarza E., Berenguel M., and Camacho E. Direct steam generation in solar boilers. *IEEE Cont Syst Magaz* 2004;24(2):15–29.

22. Muñoz J., Zarza E., Díez L.E., Martínez-Val J.M., Lopez C., and Gavela R. The new technology of gas cooled trough collectors. *A Paper Presented at SolarPACES Conference*, Granada, Spain; 2011.

23. National Renewable Energy Laboratory (NREL) Project Listing. Available from: http://www.nrel.gov/csp/solarpaces/ (accessed: 30 October 2018).

24. CSP World Website. Available from: http://www.cspworld.org/cspworldmap (accessed: 5 January 2018).

25. Pelay U., Luo L., Fan Y., Stitou D., and Rood M. Thermal energy storage systems for concentrated solar power plants. *Renew Sustain Energy Rev.* 2017;79:82–100.

26. Giurca I., Aşchilean I., Naghiu G.S., and Badea G. Selecting the technical solutions for thermal and energy rehabilitation and modernization of buildings. *Procedia Technol* 2016;22:789–796.

27. Gaglia A.G., Lykoudis S., Argiriou A.A., Balaras C.A., and Dialynas E. Energy efficiency of PV panels under real outdoor conditionse. An experimental assessment in Athens, Greece. *Renew Energy* 2017;101:236–243.

28. Pentiuc R.D., Vlad V., and Lucache D.D. Street lighting power quality. *A Paper Presented at the 8th International Conference and Exposition on Electrical and Power Engineering (EPE)*, Iasi, Romania; 16–18 October 2014.

29. Gábor Pintér G., Hegedüsné Baranyai N., Wiliams A., and Zsiborács H. Study of photovoltaics and LED energy efficiency: Case study in Hungary. *Energies* 2018;11:790.

30. Naghiu G.S., Giurca I., Aşchilean I., and Badea G. Multicriterial analysis on selecting solar radiation concentration ration for photovoltaic panels using Electre-Boldur method. *Procedia Technol* 2016;22:773–780.

31. Corona B., Ruiz D., and San Miguel G. Life cycle assessment of a HYSOL concentrated solar power plant: Analyzing the effect of geographic location. *Energies* 2016;9:413.

32. Kumara A., Sahb B., Singh A.R., Deng Y., He X., Kumar P., and Bansal R.C. A review of multi criteria decision making (MCDM) towards sustainable renewable energy development. *Renew Sustain Energy Rev* 2017;69:596–609.
33. Parabolic trough. Wikipedia, the free encyclopedia, last edited 27 July, 2022. https://en.wikipedia.org/wiki/Parabolic_trough
34. Morales A., and San Vicente G. Chapter - A new generation of absorber tubes for concentrating solar thermal (CST) systems. In: Blanco M., and Santigosa L.R., editors. *Advances in Concentrating Solar Thermal Research and Technology.* Amsterdam: Woodhead publishing series in energy; 2017, pp. 59–74.
35. Mehner A., Dong J., Prenzel T., and Datchary W. Mechanical and chemical properties of thick hybrid solegel silica coatings from acid and base catalyzed sols. *J. Sol-Gel Sci Technol.* 2010;54:355–362.
36. Chiarappa T. Performance of direct steam generator solar receiver: Laboratory vs real plant. *Energ Procedia* 2015, May;69:328–339.
37. Moens L., and Blake D.M. Mechanism of hydrogen formation in solar parabolic trough receivers. *A Paper Presented at 14th Biennial CSP SolarPACES Symposium*, Las Vegas, NV; 4–7 March 2008.
38. Roldan M.I., Valenzuela L., and Zarza E. Thermal analysis of solar receiver pipes with super- heated steam. *Appl Energ* 2013, March;103:73–84.
39. Raccurt O., Disdier A., Bourdon D., Donnola S., Stollo A., and Gioconia A. Study of the stability of a selective solar absorber coating under air and high temperature conditions. *Energ Procedia* 2015;69:1551–1557.
40. Bayon R., San Vicente G., Maffiotte C., and Morales A. Preparation of selective absorbers based on CuMn spinels by dip-coating method. *Renew Energ* 2008;33(2008):348–353.
41. Angelantoni G. Archimede's social presentation. 2008. Available from: http://www.ice.gov.it/sedi/umbria/energia/angelantoni/Archimede%20Company%20Profile.pdf.
42. Raggi C., and Chiarappa T. *A Glass-to-Metal Joint for a Solar Receiver.* EP 2626336 A1 European Patent. Syracuse, Italy: Archemedes Solar Energy; 2013..
43. Cachafeiro H. New concepts advance solar thermoelectric energy. *Euro Photonics.* 2012 Available from: http://www.photonics.com/Article.aspx?AID1/452546.
44. HITECO. HITECO (new solar collector concept for high temperature operation in CSP applications). Brussels, Belgium, 2013. Final Report Summary EU-FP7-ENERGY reference: 256830.
45. Ven K. Receivers for parabolic troughs. *CSP Today* 2014. Available from: http://social.csptoday.com/sites/default/files/5aHUIYIN.pdf.
46. Compact linear Fresnel reflector. Wikipedia, the free encyclopedia, last edited 25 July, 2022. https://en.wikipedia.org/wiki/Compact_linear_Fresnel_reflector.
47. Schott Solar. http://www.schott.com/csp/english/schott-solar-ptr-70-receivers.html.
48. Flabegd Ultimate Trough. http://www.flabeg-fe.com/en/engineering/ultimate-trough.html.
49. Winston R., Minano J.C., Benítez P., Shatz N., and Bortz J.C. *Non-imaging Optics.* Amsterdam: Elsevier Academic Press; 2005.
50. Rabl A. *Active Solar Collectors and Their Applications.* Oxford: Oxford University; 1985.
51. Chaves J. *Introduction to Non-imaging Optics*, 2nd ed. New York: CRC Press, Taylor and Francis Group; 2016.
52. Gordon J.M., and Ries H. Tailored edge-ray concentrators as ideal second stages for Fresnel reflectors. *Appl Opt* 1993;32:2243.
53. Minano J.C. *High Efficiency Non-Imaging Optics.* United States Patent 6.6639.733B2. Madrid Spain: Light Prescriptions Innovators; 2003.
54. Canavarro D., Chaves J., and Collares-Pereira M. Infinitesimal etendue and simultaneous multiple surface (SMS) concentrators for fixed receiver troughs. *Sol. Energ.* 2013, November;97:493–504.

55. Canavarro D., Chaves J., and Collares-Pereira M. A novel compound elliptical-type concentrator for parabolic primaries with tubular receiver. *Sol. Energ.* 2016;134:383–391. Doi: 10.1016/j.solener.2016.05.027.

56. Chaves J., Canavarro D., and Collares-Pereira M. Dual asymmetric macrofocal CEC LFR solar concentrator. U.S. Patent No., WO2017131544A1, Universidade De Evora, 2017.

57. Available from: https://www.iea.org/newsroom/news/2017/january/making-freshwater-from-the-sun.html (accessed: 25 October 2018).

58. Moya E. Chapter 5 - Innovative working fluids for parabolic trough collectors. In: Blanco M., and Santigosa L.R., editors. *Advances in Concentrating Solar Thermal Research and Technology*. Amsterdam: Woodhead Publishing series in energy; 2017, pp. 75–106.

59. Biencinto M., González L., Zarza E., Díez L.E., and Muñoz-Antón J. Performance model and annual yield comparison of parabolic-trough solar thermal power plants with either nitrogen or synthetic oil as heat transfer fluid. *Energ. Conver. Manag.* 2014;87:238–249.

60. Goods S., Bradshaw R., and Chavez J. Corrosion of stainless and carbon steels in molten mixtures of industrial nitrates. Technical report SAND 94-8211. Albuquerque, NM: Sandia Lab, DOE; 1994.

61. Chieruzzi M., Cerritelli G.F., Miliozzi A., and Kenny J.M. Effect of nanoparticles on heat capacity of nanofluids based on molten salts as PCM for thermal energy storage. *Nanoscale Res. Lett.* 2013;8:448.

62. Eck M., Bahl C., Bartling K.H., Biezma A., Eickhoff M., Ezquirro E., Fontela P., Hennecke K., Laing D., Möllenhoff M. and Nölke M. Direct steam generation in parabolic troughs at 500°C- a German-Spanish project targeted on component development and system design. *A Paper Presented at the 14th International Symposium on Concentrated Solar Power and Chemical Energy Technologies*, Las Vegas, NV; 7 March 2008.

63. EMSP Facility, Portugal. http://www.en.catedraer.uevora.pt/.

64. ENEA PCS Facility, Italy. www.enea.it/en/video/the-pcs-plant.

65. Priolo Gargalo Facility, Italy. Wikipedia, Last edited 12 August, 2021. https://en.wikipedia.org/wiki/Archimede_solar_power_plant.

66. Parabolic reflector. Wikipedia, the free encyclopedia, last edited 25 March, 2022. https://en.wikipedia.org/wiki/Parabolic_reflector

67. The Big dish (solar thermal). Wikipedia, the free encyclopedia, last edited 29 January, 2021. https://en.wikipedia.org/wiki/The_Big_Dish_(solar_thermal)

68. Bloess A., Schill W.P., and Zerrahn A. Power-to-heat for renewable energy integration: A review of technologies, modeling approaches, and flexibility potentials. *Appl Energ* 2018;212:1611–1626.

69. Available from: http://analysis.newenergyupdate.com/csp-today/technology/trouble-dish-stirling-csp (accessed: 12 June 2018).

70. Solar power tower. Wikipedia, the free encyclopedia, last edited 13 July, 2022. https://en.wikipedia.org/wiki/Solar_power_tower

71. Ho C.K., and Iverson B.D. Review of high-temperature central receiver designs for concentrating solar power. *Renew Sustain Energ Rev.* 2014;29:835–846.

72. Ho C. Chapter 6 - A new generation of solid particle and other high-performance receiver designs for concentrating solar thermal (CST) central tower systems. In: Blanco M., and Santigosa L.R., editors. *Advances in Concentrating Solar Thermal Research and Technology*. Amsterdam: Woodhead publishing series in energy; 2017, pp. 107–125.

73. Falcone P.K. *A Handbook for Solar Central Receiver Design*. SAND86-8009. Livermore, CA: Sandia National Laboratories; 1986.

74. Silva-Pérez M.A. Chapter 17 - Solar power towers using supercritical $CO_2$ and supercritical steam cycles, and decoupled combined cycles. In: Blanco M., and Santigosa L.R., editors. *Advances in Concentrating Solar Thermal Research and Technology*. Amsterdam: Woodhead publishing series in energy; 2017, pp. 383–402.

75. Mehos M., Turchi C., Vidal J., Wagner M., Ma Z., Ho C., Kolb W., Andraka C., Ma Z., and Kruizenga A. *Concentrating Solar Power Gen3 Demonstration Roadmap*, Tech. Rep., NREL/TP-5500-67464. Golden, CO: National Renewable Energy Laboratory; 2017, p. 127.

76. Agalit H., Zari N., Maalmi M., and Maaroufi M. Numerical investigations of high temperature packed bed TES systems used in hybrid solar tower power plants. *Sol Energ* 2015;122:603–616. Doi: 10.1016/j.solener.2015.09.032.

77. Al-Sulaiman F.A., and Atif M. Performance comparison of different supercritical carbon dioxide Brayton cycles integrated with a solar power tower. *Energy* 2015;82:61–71. Doi: 10.1016/j.energy.2014.12.070.

78. McNaughton R. Advanced steam generating receivers for high-concentration solar collectors final report: Project Results. CSIRO, ARENA; June 2014. http://www.csiro.au/Outcomes/Energy/Renewables-and-Smart-Systems.aspx.

79. Gary J., Turchi C., and Siegel N. CSP and the DOE sunshot initiative. *System* 2011;1:8.

80. Assessment of Parabolic trough and power tower solar technology cost and performance forecast. An NREL report number NREL/SR-550-34440 prepared by Sargent and Lundy LLC consulting group , Chicago, IL, October, 2003, contract no., DE-AC36-99-GO10337.

81. Kolb G.J. *An Evaluation of Possible Next Generation High-Temperature Molten-Salt Power Towers*. Albuquerque, NM: Sandia National Laboratories; 2011; SAND2011-9320, TRN: US201205%%95, OSTI identifier 1035342.

82. Singer C., Buck R., Pitz-Paal R., and Müller-Steinhagen H. Assessment of Solar power tower driven ultrasupercritical steam cycles applying tubular central receivers with varied heat transfer media. *J Sol Energ Eng* 2010;132:041010. Doi: 10.1115/1.4002137.

83. Singer C., Buck R., Pitz-Paal R., and Müller-Steinhagen H. Economic potential of innovative receiver concepts with different solar field configurations for supercritical steam cycles. *J Sol Energ Eng* 2013;136:21009. Doi: 10.1115/1.4024740.

84. Singer C., Giuliano S., and Buck R. Assessment of improved molten salt solar tower plants. *Energ Procedia* Elsevier B.V, 2013;1553–1562. Doi: 10.1016/j.egypro.2014.03.164.

85. Peterseim J.H., and Veeraragavan A. Solar towers with supercritical steam parametersdis the efficiency gain worth the effort? *Energ Procedia* 2015;69:1123–1132. Doi: 10.1016/j.egypro.2015.03.181.

86. Siefert J.A., Libby C., and Shingledecker J. Concentrating solar power (CSP) power cycle improvements through application of advanced materials. 2016;070030. Doi: 10.1063/1.4949177.

87. Sabharwall P., Clark D., Glazoff M., Zheng G., Sridharan K., and Anderson M. Advanced heat exchanger development for molten salts. *Nuclear Eng Desig* 2014;280:42–56. Doi: 10.1016/j.nucengdes.2014.09.026.

88. Gardner W., Mcnaughton R., Kim J., and Barrett S. Development of a solar thermal supercritical steam generator. *A Paper Presented at the 50th Annual Conference, Australian Solar Energy Society,* Melbourne, Australia; 2012.

89. White M.T., Bianchi G., Chai L., Tassou S.A., and Sayma A.I. Review of supercritical $CO_2$ technologies and systems for power generation. *Appl Thermal Eng* 2021;185:116447.

90. IRENA. *Renewable Power Generation Costs in 2019*. Tech. Rep., Abu Dhabi: International Renewable Energy Agency; 2020, p. 144.

91. Turchi C.S., Stekli J., and Bueno P.C. Concentrating solar power. In: Brun K., Friedman P., and Dennis R., editors. *Fundamentals and Applications of Supercritical Carbon Dioxide (SCO₂) Based Power Cycles*. Woodhead Publishing; 2017, pp. 269–292. Doi: 10.1016/B978-0-08-100804-1.00011-6.

92. NREL. *Concentrating Solar Power Projects*. NREL Transforming Energy; 2019, Golden, Colorado.

93. Yin J.M., Zheng Q.Y., Peng Z.R., and Zhang X.R. Review of supercritical $CO_2$ power cycles integrated with CSP. *Int J Energy Res* 2020;44:1337–1369. Doi: 10.1002/er.4909.

94. Chacartegui R., De Escalona J.M.M., Sanchez D., Monje B., and Sanchez T. Alternative cycles based on carbon dioxide for central receiver solar power plants. *Appl Thermal Eng* 2011;31:872–879. Doi: 10.1016/j.applthermaleng.2010.11.008.
95. Turchi C.S., Ma Z., Neises T.W., and Wagner M.J. Thermodynamic study of advanced supercritical carbon dioxide power cycles for concentrating solar power systems. *J Sol Energ Eng* 2013;135:041007. Doi: 10.1115/1.4024030.
96. Neises T., and Turchi C. A comparison of supercritical carbon dioxide power cycle configurations with an emphasis on CSP applications. *Energ. Procedia* 2013;49:1187–1196. Doi: 10.1016/j.egypro.2014.03.128.
97. Iverson B.D., Conboy T.M., Pasch J.J., and Kruizenga A.M. Supercritical $CO_2$ Brayton cycles for solar-thermal energy. *Appl Energ* 2013;111:957–970. Doi: 10.1016/j.apenergy.2013.06.020.
98. Padilla R.V., Soo Too Y.C., Benito R., and Stein W. Exergetic analysis of supercritical $CO_2$ Brayton cycles integrated with solar central receivers. *Appl Energ* 2015;148:348–365. Doi: 10.1016/j.apenergy.2015.03.090.
99. Padilla R.V., Too Y.C.S., Beath A., McNaughton R., and Stein W. Effect of pressure drop and reheating on thermal and exergetic performance of supercritical carbon dioxide Brayton cycles integrated with a solar central receiver. *J Sol Energ Eng.* 2015;137:051012. Doi: 10.1115/1.4031215.
100. Neises T., and Turchi C. Supercritical carbon dioxide power cycle design and configuration optimization to minimize levelized cost of energy of molten salt power towers operating at 650°C. *Sol Energ.* 2019, January;181:27–36. Doi: 10.1016/j.solener.2019.01.078.
101. Binotti M., Astolfi M., Campanari S., Manzolini G., and Silva P. Preliminary assessment of $sCO_2$ cycles for power generation in CSP solar tower plants. *Appl Energ* 2017;204:1007–1017. Doi: 10.1016/j.apenergy.2017.05.121.
102. Allison T.C., Jeffrey Moore J., Hofer D., Towler M.D., and Thorp J. Planning for successful transients and trips in a 1 MWe-scale high-temperature $sCO_2$ test loop. *J Eng Gas Turbines Power* 2019;141(6):061014. Doi: 10.1115/1.4041921.
103. Moore J.J., Cich S., Towler M., Allison T., Wade J., and Hofer D. Commissioning of a 1 MWe supercritical $CO_2$ test loop. *A Paper Presented at The 6th International Supercritical $CO_2$ Power Cycles Symposium*, Pittsburgh PA, 27–29 March; 2018.
104. Moullec Y.L., Qi Z., Zhang J., Zhou P., Yang Z., Chen W., Wang X., and Wang S. Shouhang-EDF 10 MWe supercritical $CO_2$ cycle + CSP demonstration project. *A Paper Presented at 3rd European Supercritical $CO_2$ Conference*, Paris, France; 19–20 September, 2019, p. 120. Doi: 10.17185/duepublico/48884.
105. Dyreby J., Klein S., Nellis G., and Reindl D. Design considerations for supercritical carbon dioxide brayton cycles with recompression. *J Eng Gas Turbines Power* 2014;136(10):101701. Doi: 10.1115/1.4027936.
106. Manzolini G., Binotti M., Bonalumi D., Invernizzi C., and Iora P. $CO_2$ mixtures as innovative working fluid in power cycles applied to solar plants. Technoeconomic assessment. *Sol Energ.* 2019;181(3):530–544, Doi: 10.1016/j.solener.2019.01.015.
107. Ortega J.D., and Christian J.M. Design requirements for direct supercritical carbon dioxide. *Proceedings of the 9th International Conference on Energy Sustainability*, San Diego, CA; 2015, pp. 1–6.
108. Ortega J., Khivsara S., Christian J., Ho C., Yellowhair J., and Dutta P. Coupled modeling of a directly heated tubular solar receiver for supercritical carbon dioxide Brayton cycle: Optical and thermal-fluid evaluation. *Appl Thermal Eng* 2016. Doi: 10.1016/j.applthermaleng.2016.05.178.
109. Ortega J., Khivsara S., Christian J., Ho C., and Dutta P. Coupled modeling of a directly heated tubular solar receiver for supercritical carbon dioxide Brayton Cycle: Structural and creep fatigue evaluation. *Appl Thermal Eng* 2016. Doi: 10.1016/j.applthermaleng.2016.06.031.

110. Besarati S.M., Yogi Goswami D., and Stefanakos E.K. Development of a solar receiver based on compact heat exchanger technology for supercritical carbon dioxide power cycles. *J Sol Energ Eng* 2015;137:031018. Doi: 10.1115/1.4029861.
111. Ho C.K., Carlson M., Garg P., and Kumar P. Cost and performance tradeoffs of alternative solardriven S-CO2 Brayton cycle configurations. *Proceedings of the ASME 2015 Power and Energy Conversion Conference*, San Diego, CA; 2015, pp. 1–10.
112. Kelly B.D., Izygon M., and Vant-Hull L. Advanced thermal energy storage for central receivers with supercritical coolants. DE-FG36-08GO18149. 2010. Doi: 10.2172/981926.
113. Ho C.K., Conboy T., Ortega J., Afrin S., Gray A., Christian J.M., Bandyopadyay S., Kedare S.B., Singh S., and Wani P. High-temperature receiver designs for supercritical $CO_2$ closed-loop Brayton cycles. *ASME 2014 8th International Conference on Energy Sustainability*, Boston, MA: ASME; 2014. Doi: 10.1115/ES2014-6328.
114. Buckingham R., Wong B., and Brown L. Thermochemical energy storage for concentrated solar power - coupling to a high efficiency supercritical $CO_2$ power cycle. *Proceedings of the SolarPaces Conference,* Granada, Spain; 2011, pp. 0–5.
115. Chacartegui R., Alovisio A., Ortiz C., Valverde J.M., Verda V., and Becerra J.A. Thermochemical energy storage of concentrated solar power by integration of the calcium looping process and a $CO_2$ power cycle. *Appl Energ* 2016;173:589–605. Doi: 10.1016/j.apenergy.2016.04.053.
116. Dunham M.T., and Iverson B.D. High-efficiency thermodynamic power cycles for concentrated solar power systems. *Renew Sustain Energ Rev.* 2014;30:758–770. Doi: 10.1016/j.rser.2013.11.010.
117. Cheang V.T., Hedderwick R.A., and McGregor C. Benchmarking supercritical carbon dioxide cycles against steam Rankine cycles for concentrated solar power. *Sol Energ* 2015;113:199–211, Doi: 10.1016/j.solener.2014.12.016.
118. Sanchez M., Blanco M.J., García-Barberena J., and Monreal A. The potential for cost reduction ofsolar towers with decoupled combined cycles. *A Paper Presented at SolarPaces Conference,* Marrakech, Morocco; 2012.
119. Heller L., and Gauché P. Modeling of the rock bed thermal energy storage system of a combined cycle solar thermal power plant in South Africa. *Sol Energ* 2013;93:345–356. Doi: 10.1016/j.solener.2013.04.018.
120. Crespo L. STE plants: Beyond dispatchability firmness of supply and integration with VRE. *Energ Procedia* 2015;69:1241–1248. Doi: 10.1016/j.egypro.2015.03.161.
121. Romero M.J. Chapter 7 - Gonzalez-Aguilar Next generation of liquid metal and other high-performance receiver designs for concentrating solar thermal (CST) central tower systems. In: Blanco M., and Santigosa L.R., editors. *Advances in Concentrating Solar Thermal Research and Technology.* Amsterdam: Woodhead publishing series in energy; 2017, pp. 129–153.
122. Pacio J., Singer C., Wetzel T., and Uhlig R. Thermodynamic evaluation of liquid metals as heat transfer fluids in concentrated solar power plants. *Appl Thermal Eng* 2013;60:295–302.
123. Falcone P.K. *A Handbook for Solar Central Receiver Design.* SAND86-8009. Livermore, CA: Sandia National Laboratories; 1986.
124. Mancini T., Heller P., Butler B., Osborn B., Schiel W., Goldberg V., Buck R., Diver R., Andraka C., and Moreno J. Dish-stirling systems: An overview of development and status. *J Sol Energ Eng* 2003;125:135–151.
125. Diver R.B. Receiver/reactor concepts for thermochemical transport of solar energy. *J Sol Energ Eng* 1987;109(3):199–204.
126. Laing D., and Palsson M. Hybrid dish/stirling systems: Combustor and heat pipe receiver development. *J Sol Energ Eng* 2002:124.

127. Moreno J.B., Modesto-Beato M., Rawlinson K.S., Andraka C.E., Showalter S.K., Moss T.A., Mehos M., and Baturkin V. Recent progress in heat-pipe solar receivers. *SAND2001-1079. 36th Intersociety Energy Conversion Engineering Conference*, Savannah, GA; 2001, pp. 565–572.

128. Pacioa J., Fritsch A., Singer C., and Uhlig R. Liquid metals as efficient coolants for high-intensity point-focus receivers: Implications to the design and performance of next generation CSP systems. *Energ Procedia* 2014;49:647–655.

129. Vignarooban K., Xinhai X., Arvay A., Hsu K., and Kannan A.M. Heat transfer fluids for concentrating solar power systems e a review. *Appl Energ* 2015;146:383–396.

130. Fritsch A., Flesch J., Geza V., Singer C., Uhlig R., and Hoffschmidt B. Conceptual study of central receiver systems with liquid metals as efficient heat transfer fluids. *Energ Procedia* 2015;69:644–653.

131. Boo J.H., Kim S.M., and Kang Y.H. An experimental study on a sodium loop-type heat pipe for thermal transport from a high-temperature solar receiver. *Energ Procedia* 2015;69:608–617.

132. Kaushik S.C., Verma S.S., and Chandra A. Solar-assisted liquid metal MHD power generation: A state of the art study. *Heat Recov Syst CHP* 1995;15(7):675–689.

133. Branover H., El-Boher A., and Yakhot A. Testing of a low-temperature liquid-metal MHD power system. *Energ Convers Manag* 1982;22:163–169.

134. Baker R., and Tessier M. *Handbook of Electromagnetic Pump Technology*. New York: Elsevier Science Publishing Co. Inc; 1987.

135. Fernandez-García A., Sutter F., Martínez-Arcos L., Valenzuela L., and Sansom C. Chapter 2 - Advanced mirror concepts for concentrating solar thermal systems. In: Blanco M., and Santigosa L.R., editors. *Advances in Concentrating Solar Thermal Research and Technology*. Amsterdam: Woodhead publishing series in energy; 2017, pp. 29–43.

136. Commission staff working paper SEC(2011) 1609 final. *Materials Roadmap Enabling Low-Carbon Energy Technologies*. Brussels, Belgium: European Commission; 2011.

137. Kennedy C.E., Terwilliger K., and Milbourne M. *Development and Testing of Solar Reflectors*. Tech. Rep. No. NREL/CP-520-36582. Golden, CO: NREL; 2004.

138. Pitchumani R. *Concentrating Solar Power Program. SunShot Grand Challenge and Peer Review 2014*. Anaheim, CA: US Department of Energy (DOE).

139. *Technology Roadmap: Solar Thermal Electricity*. Paris, France: International Energy Agency (IEA; 2014.

140. DiGrazia M.J., Gee R., Jorgensen G.J., and Bingham C. Service life prediction for ReflecTech® mirror film. *A Paper Presented at WREF 2012, World Renewable Energy Forum*, Denver CO; 2012.

141. Almanza R., Hernández P., Martínez I., and Mazari M. Development and mean life of aluminium first-surface mirrors for solar energy applications. *Sol Energ Mater Sol Cells* 2009;93:1647–1651.

142. Kennedy C.E., and Terwilliger K. Optical durability of candidate solar reflectors. *J Sol Energ Eng Trans ASME* 2005;127(2):262–269.

143. Sutter F., Fernandez-García A., Wette J., and Heller P. Comparison and evaluation of accelerated aging tests for reflectors. *Energ Procedia* 2014;49:1718–1727.

144. Echazú R., Cadena C., and Saravia L. Estudio de materiales reflectivos para concentradores solares. *AVERMA* 2000;4:08–11.

145. Okićc M., Roušar I., Táborský Z., and Roháček K. Reflectivity and surface composition of electrochemically and mechanically polished stainless steel. *Mat Chem Phys* 1987;17(3):301–309.

146. García-Ortiz Y., Yanez-Mendiola J., and Valenzuela L. Cylindrical parabolic collectors material from low cost (stainless steel) applied to dried hybrid system. *Dyna* 2016;91(1):1–7.

147. Sarver T., Al-Qaraghuli A., and Kazmerski L.L. A comprehensive review of the impact of dust on the use of solar energy: History, investigations, results, literature and mitigation approaches. *Renew Sustain Energ Rev* 2013;22:698–733.

148. Cuddihy E.F. Theoretical considerations of soil retention. *Sol Energ Mater.* 1980;3:21–33.

149. Lorenz T., Klimm E., and Weiss K.A. Soiling and anti-soiling coatings on surfaces of solar thermal systems featuring an economic feasibility analysis. *Energ Procedia* 2014;48:749–756.

150. Schwarberg F., and Schiller M. Enhanced solar mirrors with anti-soiling coating. SolarPACES 2012. *A Paper Presented at 18th International Conference on Solar Power and Chemical Energy Systems*, Marrakech, Morocco; 11–14 September, 2012.

151. Winston R., Minano J.C., and Benitez P. *Non-imaging Optics.* Amsterdam: Elsevier Academic Press; 2005.

152. Yogev A., Kribus A., Epstein M., and Kogan A. Solar tower reflector systems: A new approach for high temperature solar plants. *Int J Hydrogen Energ* 1998;23:239–245.

153. Segal A., and Epstein M. The optics of the solar tower reflector. *Sol Energ* 2000;69:229–241.

154. Buck R., Bra¨uning T., Denk T., Pfa¨nder M., Schwarzb¨zl P., and Tellez F. Solar-hybrid gas turbine-based power tower systems (REFOS). *J Sol Energ Eng* 2002;124(1):2–9.

155. Schmitz M., Schwarzbözl P., Buck R., and Pitz-Paal R. Assessment of the potential improvement due to multiple apertures in central receiver systems with secondary concentrators. *Sol Energ* 2006;80:111–120.

156. SOLGATE. *Solar Hybrid Gas Turbine Electric Power System.* Final Publishable Report. Contract ENK5-CT-2000-00333. Brussels, Belgium: European Commission; 2006.

157. Dallmer-Zerbe K., Georg A., Graf W., Klimm E., Kühne M, and Platzer W. *High-Temperature Corrosion of Coatings for Secondary Reflectors.* Fraunhofer ISE Annual Report. Freiburg, Germany:; 2011.

158. http://cordis.europa.eu/project/rcn/200815_en.html.

159. Christian J.M., and Ho C.K. Alternative designs of a high efficiency, north-facing, solid particle receiver. *SolarPACES 2013*, Las Vegas, NV; 17–20 September 2013.

160. Falcone P.K., Noring J.E., and Hruby J.M. *Assessment of a Solid Particle Receiver for a High Temperature Solar Central Receiver System.* SAND85-8208. Livermore, CA: Sandia National Laboratories; 1985.

161. Gobereit B., Amsbeck L., Buck R., Pitz-Paal R., and Müller-Steinhagen H. Assessment of a falling solid particle receiver with numerical simulation. *SolarPACES 2012*, Marrakech, Morocco; 11–14 September 2012.

162. Ho C., Christian J., Gill D., Moya A., Jeter S., Abdel-Khalik S., Sadowski D., Siegel N., Al-Ansary H., Amsbeck L., Gobereit B., and Buck R. Technology advancements for next generation falling particle receivers. *Proceedings of the SolarPACES 2013 International Conference*, Energy Procedia; 2014, Vol. 49, pp. 398–407.

163. Hruby J.M., and Steele B.R. A solid particle central receiver for solar-energy. *Chem Eng Progres* 1986;82(2):44–47.

164. Khalsa S.S.S., Christian J.M., Kolb G.J., Röger M., Amsbeck L., Ho C.K., Siegel N.P., and Moya A.C. CFD simulation and performance analysis of alternative designs for high-temperature solid particle receivers. *A Paper Presented at ASME International Conference on Energy Sustainability*, Washington, DC; 2011.

165. Khalsa S.S.S., and Ho C.K. Radiation boundary conditions for computational fluid dynamics models of high-temperature cavity receivers. *J Sol Energ Eng-Trans ASME* 2011;133(3).

166. Kim K., Moujaes S.F., and Kolb G.J. Experimental and simulation study on wind affecting particle flow in a solar receiver. *Sol Energ* 2010;84(2):263–270.

167. Kolb G.J., Diver R.B., and Siegel N. Central-station solar hydrogen power plant. *J Sol Energ Eng-Trans ASME* 2007;129(2):179–183.

168. Ma Z.W., Glatzmaier G., and Mehos M. Fluidized bed technology for concentrating solar power with thermal energy storage. *J Sol Energ Eng-Trans ASME* 2014;136(3). 031014-1–031014-9.

169. Rightley M.J., Matthews L.K., and Mulholland G.P. Experimental characterization of the heat transfer in a free-falling-particle receiver. *Sol Energ* 1992;48(6):363–374.

170. Röger M., Amsbeck L., Gobereit B., and Buck R. Face-down solid particle receiver using recirculation. *J Sol Energ Eng* 2011;133:1–10.

171. Siegel N.P., Ho C.K., Khalsa S.S., and Kolb G.J. Development and evaluation of a prototype solid particle receiver: On-sun testing and model validation. *J Sol Energ Eng-Trans ASME* 2010;132(2).

172. Tan T.D., and Chen Y.T. Review of study on solid particle solar receivers. *Renew Sustain Energ Rev* 2010;14(1):265–276.

173. Wu W., Trebing D., Amsbeck L., Buck R., and Pitz-Paal R. Prototype testing of a centrifugal particle receiver for high-temperature concentrating solar applications. *J Sol Energ Eng Trans ASME* 2015;137(4): 041011 (7 pages), paper no: SOL-14-1127. Doi: 10.1115/1.4030657.

174. Wu W., Uhlig R., Buck R., and Pitz-Paal R. Numerical simulation of a centrifugal particle receiver for high-temperature concentrating solar applications. *Num Heat Trans Part A-Appl* 2015;68(2):133–149.

175. Flamant G. Theoretical and experimental-study of radiant-heat transfer in a solar fluidizedbed receiver. *AIChE J* 1982;28(4):529–535.

176. Flamant G., Hernandez D., Bonet C., and Traverse J.P. Experimental aspects of the thermochemical conversion of solar-energy - decarbonation of $CaCO_3$. *Sol Energ* 1980;24(4):385–395.

177. Wu W., Amsbeck L., Buck R., Uhlig R., and Ritz-Paal R. Proof of concept test of a centrifugal particle receiver. *Proceedings of the SolarPACES 2013 International Conference,* Las Vegas, NV; 2014, Vol. 49, pp. 560–568.

178. Wu W., Amsbeck L., Buck R., Waibel N., Langner P., and Pitz-Paal R. On the influence of rotation on thermal convection in a rotating cavity for solar receiver applications. *Appl Thermal Eng* 2014;70(1) 694–704.

179. Flamant G., Gauthier D., Benoit H., Sans J.L., Boissiere B., Ansart R., and Hemati M. A new heat transfer fluid for concentrating solar systems: Particle flow in tubes. *Proceedings of the SolarPACES 2013 International Conference,* Las Vegas, NV; 2014, Vol. 49, pp. 617–626.

180. Flamant G., Gauthier D., Benoit H., Sans J.L., Garcia R., Boissiere B., Ansart R., and Hemati M. Dense suspension of solid particles as a new heat transfer fluid for concentrated solar thermal plants: On-sun proof of concept. *Chem Eng Sci* 2013;102:567–576.

181. Martinek J., and Ma Z. Granular flow and heat-transfer study in a near-blackbody enclosed particle receiver. *J Sol Energ Eng* 2015;137(5):051008.

182. Steinfeld A., Imhof A., and Mischler D. Experimental investigation of an atmospheric-open cyclone solar reactor for solid-gas thermochemical reactions. *J Sol Energ Eng-Trans ASME* 1992;114(3):171–174.

183. Hruby J.M., Steele B.R., and Burolla V.P. *Solid Particle Receiver Experiments: Radiant Heat Test.* SAND84-8251. Albuquerque, NM: Sandia National Laboratories; 1984.

184. Hruby J.M. *A Technical Feasibility Study of a Solid Particle Solar Central Receiver for High Temperature Applications.* SAND86-8211. Livermore, CA: Sandia National Laboratories; 1986.

185. Evans G., Houf W., Greif R., and Crowe C. Gas-particle flow within a high-temperature solar cavity receiver including radiation heat-transfer. *J Sol Energ Eng-Trans ASME* 1987;109(2):134–142.

186. Meier A. A predictive CFD model for a falling particle receiver reactor exposed to concentrated sunlight. *Chem Eng Sci* 1999;54(13–14):2899–2905.

187. Chen H., Chen Y., Hsieh H.T., and Siegel N. CFD modeling of gas particle flow within a solid particle solar receiver. *Proceedings of the ASME International Solar Energy Conference,* Beijing, China; 2007, pp. 37–48.

188. Klein H.H., Karni J., Ben-Zvi R., and Bertocchi R. Heat transfer in a directly irradiated solar receiver/reactor for solid-gas reactions. *Sol Energ* 2007;81(10):1227–1239.

189. Martin J., and John Vitko J. *ASCUAS: A Solar Central Receiver Utilizing a Solid Thermal Carrier.* SAND82-8203. Livermore, CA: Sandia National Laboratories; 1982.

190. Gobereit B., Amsbeck L., and Buck R. Operation strategies for falling particle receivers. *A Paper Presented at ASME 2013 7th International Conference on Energy Sustainability,* ES-FuelCell2013-18354, Minneapolis, MN; 14–19 July, 2013.

191. Ho C.K., and Christian J.M. Evaluation of air recirculation for falling particle receivers. *A Paper Presented at ASME 2013 7th International Conference on Energy Sustainability,* ES-FuelCell2013-18236, Minneapolis, MN; 14–19 July, 2013.

192. Christian J.M., and Ho C.K. System design of a 1 MW north-facing, solid particle receiver. *A Paper Presented at the SolarPACES 2014 International Conference,* Beijing, China, Energy procedia; 2014.

193. Ho C.K., Christian J.M., Moya A.C., Taylor J., Ray D., and Kelton J. Experimental and numerical studies of air curtains for falling particle receivers. *A Paper Presented at ASME 2014 8th International Conference on Energy Sustainability,* ES-FuelCell2014-6632, Minneapolis, MN; June 29–July 2, 2014.

194. Knott R., Sadowski D.L., Jeter S.M., Abdel-Khalik S.I., Al-Ansary H.A., and El-Leathy A. High temperature durability of solid particles for use in particle heating concentrator solar power systems. *A Paper Presented at the ASME 2014 8th International Conference on Energy Sustainability,* ES-FuelCell2014-6586, Boston, MA; June 29–July 2, 2014.

195. Siegel N., Gross M., Ho C., Phan T., and Yuan J. Physical properties of solid particle thermal energy storage media for concentrating solar power applications. *Proceedings of the SolarPACES 2013 International Conference,* Energy procedia; 2014, Vol. 49, pp. 1015–1023.

196. Ho C.K., Christian J., Romano D., Yellowhair J., and Siegel N. Characterization of particle flow in a free-falling solar particle receiver. *A Paper Presented at the ASME 2015 Power and Energy Conversion Conference,* San Diego, CA; June 28–July 2, 2015.

197. Siegel N.P., Gross M.D., and Coury R. The development of direct absorption and storage media for falling particle solar central receivers. *ASME J Sol Energ Eng* 2015;137(4):041003–041007.

198. Ho C.K., Khalsa S.S., and Siegel N.P. Modeling on-sun tests of a prototype solid particle receiver for concentrating solar power processes and storage. *ES2009: A Paper Presented at the ASME 3rd International Conference on Energy Sustainability,* San Francisco, CA, Vol. 2; 2009.

199. Ho C.K., Christian J.M., Yellowhair J., Siegel N., Jeter S., Golob M., Abdel-Khalik S.I., Nguyen C., and Al-Ansary H. On sun testing of an advanced falling particle receiver system. *A Paper Presented at SolarPACES 2015,* Cape Town, South Africa; 13–16 October, 2015.

200. Ho C.K., Christian J.M., Yellowhair J., Armijo K., and Jeter S. Performance evaluation of a high temperature falling particle receiver. *A Paper Presented at ASME Power & Energy Conference,* Charlotte, NC; 26–30 June, 2016.

201. Kolb G.J. *Suction-Recirculation Device for Stabilizing Particle Flows within a Solar Powered Solid Particle Receiver.* United States Patent 8109265, 12/368,327. Albuquerque, NM: Sandia National Laboratories; 2012, February 7.

202. Chen Z.Q., Chen Y.T., and Tan T.D. Numerical analysis on the performance of the solid solar particle receiver with the influence of aero window. *A Paper Presented at the ASME Fluids Engineering Division Summer Conference* -2008, Pt a and B, Jacksonville, FL; 2009, Vol. 1.

203. Tan T.D., and Chen Y.T. Protection of an aero window, one scheme to enhance the cavity efficiency of a solid particle solar receiver. *HT2009: A Paper Presented at the ASME Summer Heat Transfer Conference 2009*, San Francisco, CA; 2009, Vol. 2.

204. Tan T.D., Chen Y.T., Chen Z.Q., Siegel N., and Kolb G.J. Wind effect on the performance of solid particle solar receivers with and without the protection of an aero window. *Sol Energ* 2009;83(10):1815–1827.

205. Lee T., Lim S., Shin S., Sadowski D.L., Abdel-Khalik S.I., Jeter S.M., and Al-Ansary H. Numerical simulation of particulate flow in interconnected porous media for central particle-heating receiver applications. *Sol Energ* 2015;113:14–24.

206. Xiao G., Guo K.K., Ni M.J., Luo Z.Y., and Cen K.F. Optical and thermal performance of a high temperature spiral solar particle receiver. *Sol Energ* 2014;109:200–213.

207. Ermanoski I., Siegel N.P., and Stechel E.B. A new reactor concept for efficient solar thermochemical fuel production. *J Sol Energ Eng-Trans ASME* 2013;135(3).

208. Bai F., Zhang Y., Zhang X., Wang F., Wang Y., and Wang Z. Thermal performance of a quartz tube solid particle air receiver. *Proceedings of the SolarPACES 2013 International Conference,* Las Vegas, NV; 2014 Vol. 49, pp. 284–294.

209. Wang F., Bai F., Wang Z., and Zhang X. Numerical simulation of quartz tube solid particle air receiver. *International Conference on Concentrating Solar Power and Chemical Energy Systems, SolarPACES 2014 Energy Procedia,* Beijing, China; 2015, Vol. 69, pp. 573–582.

210. Zhang Y.N., Bai F.W., Zhang X.L., Wang F.Z., and Wang Z.F. Experimental study of a single quartz tube solid particle air receiver. *International Conference on Concentrating Solar Power and Chemical Energy Systems, SolarPACES 2014 Energy Procedia,* Beijing, China; 2015, Vol. 69, pp. 600–607.

211. Abdelrahman M., Fumeaux P., and Suter P. Study of solid-gas-suspensions used for direct absorption of concentrated solar-radiation. *Sol Energ* 1979;22(1):45–48.

212. Hunt A.J. A new solar receiver utilizing a small particle heat exchanger. *Proceedings of the 14th International Society of Energy Conversion Engineering Conference*, New York, Institute of Electrical and Electronics Engineers; 1979.

213. Hunt A.J., and Brown C.T. *Solar Testing of the Small Particle Heat Exchanger (SPHER)*. Berkeley, CA: Lawrence Berkeley National Laboratory; 1982, Report no. LBL-16497.

214. Miller F.J., and Koenigsdorff R. Theoretical-analysis of a high-temperature small-particle solar receiver. *Sol Energ Mater* 1991;24(1–4):210–221.

215. Miller F.J., and Koenigsdorff R.W. Thermal modeling of a small-particle solar central receiver. *J Sol Energ Eng-Trans ASME* 2000;122(1):23–29.

216. Crocker A., and Miller F.J. Coupled fluid flow and radiation modeling of a cylindrical small particle solar receiver. *Proceedings of the ASME 6th International Conference on Energy Sustainability -2012*, Pts a and B, SanDiego, CA; 2012, pp. 405–412.

217. del Campo P.F., Miller F.J., and Crocker A. Three-dimensional fluid dynamics and radiative heat transfer modeling of a small particle solar receiver. *Proceedings of the ASME 7th International Conference on Energy Sustainability,* Minneapolis, MN; 2014, p. 2013.

218. Fernandez P., and Miller F.J. Performance analysis and preliminary design optimization of a small particle heat exchange receiver for solar tower power plants. *Sol Energ* 2015;112:458–468.

219. Kitzmiller K., and Miller F.J. Thermodynamic cycles for a small particle heat exchange receiver used in concentrating solar power plants. *J Sol Energ Eng-Trans ASME* 2011;133(3).

220. Ma Z., and Zhang R. *Solid Particle Thermal Energy Storage Design for a Fluidized-Bed Concentrating Solar Power Plant*. United States Patent 13/855092. Golden, CO: NREL; 2 April 2013.

221. Benoit H., Lopez I.P., Gauthier D., Sans J.L., and Flamant G. On-sun demonstration of a 750 degrees C heat transfer fluid for concentrating solar systems: Dense particle suspension intube. *Sol Energ* 2015;118:622–633.

222. Garbrecht O., Al-Sibai F., Kneer R., and Wieghardt K. CFD-simulation of a new receiver design for a molten salt solar power tower. *Sol Energ* 2013;90:94–106.

223. Lubkoll M., vonBackstrom T.W., Harms T.M., and Kroger D.G. Initial analysis on the novel spiky central receiver air pre-heater (SCRAP) pressurized air receiver, international conference on concentrating solar power and chemical energy systems. *Sol PACES* 2015;2014(69):461–470.

224. Ho C.K., Christian J.M., and Pye J. *Bladed Solar Thermal Receivers for Concentrating Solar Power*. United States Patent Application 14535100. Albuquerque, NM: Sandia Corporation; 6 November 2014.

225. Ho C.K., Christian J.M., Ortega J.D., Yellowhair J., Mosquera M.J., and Andraka C.E. Reduction of radiative heat losses for solar thermal receivers. *A Paper Presented at the SPIE Optics' Photonics Solar Energy Technology High and Low Concentrator Systems for Solar Energy Applications IX*, San Diego, CA; 17–21 August, 2014.

226. Ho C.K., Ortega J.D., Christian J.M., Yellowhair J.E., Ray D.A., Kelton J.W., Peacock G., Andraka C.E., and Shinde S. *Fractal Materials and Designs with Optimized Radiative Properties*. United States Patent Application 62/015052. Sandia Corporation; 2014, June 20.

227. Ortega J.D., Christian J.M., and Ho C.K. Coupled optical-thermal-fluid and structural analyses of novel light-trapping tubular panels for concentrating solar power receivers. *A Paper Presented at the SPIE Optics Photonics Solar Energy Technology High and Low Concentrator Systems for Solar Energy Applications X*, San Diego, CA; 9–13 August, 2015.

228. Yellowhair J.E., Ho C.K., Ortega J.D., Christian J.M., and Andraka C.E. Testing and optical modeling of novel concentrating solar receiver geometries to increase light trapping and effective solar absorptance. *A Paper Presented at the SPIE Optic Photonics Solar Energy Technology High and Low Concentrator Systems for Solar Energy Applications X*, San Diego, CA; 9–13 August, 2015.

229. Zhang J., Pye J.D., and Hughes G.O. Active air flow control to reduce cavity receiver heat loss. *A Paper Presented at Ninth International Conference on Energy Sustainability, ASME Power & Energy*, SanDiego, CA; June 2015.

230. Romero M., and Gonzalez-Aguilar J. Solar thermal CSP technology. *Wires Energ Environ*. 2014;2014(3):42–59.

231. Romero M., Gonzalez-Aguilar J., and Zarza E. Concentrating solar thermal power. In: Goswami Y., and Kreith F., editors. *Energy Efficiency and Renewable Energy Handbook*, 2nd ed. Boca Raton, FL: CRC Press Taylor & Francis Group; 2015.

232. Gomez-Garcia F., González-Aguilar J., Olalde G., and Romero M. Thermal and hydrodynamic behavior of ceramic volumetric absorbers for central receiver solar power plants: A review. *Renew Sustain Energ Rev* 2016;57:648–658.

233. Kribus A. Future directions in solar thermal electricity generation. *Solar Thermal Electricity Generation. Colecci on documentos CIEMAT*. Madrid, CIEMAT; 1999, 251–285, ISBN 84-7834-353-9.

234. Pacio J., and Wetzel T. Assessment of liquid metal technology status and research paths for their use as efficient heat transfer fluids in solar central receiver systems. *Sol Energ* 2013;93:11–22.

235. Heinzel A., Weisenburger A., and Müller G. Corrosion behavior of austenitic steels in liquid lead bismuth containing $10^{-6}$wt% and $10^{-8}$wt% oxygen at 400°C–500°C. *J Nucl Mater* 2014;448(1–3):163–171.

236. Weeks J. Lead, bismuth, tin and their alloys as nuclear coolants. *Nucl Eng Design* 1971;15:363–372.

237. Flesch J., Fritsch A., Cammi G., Marocco L., Fellmoser F., Pacio J., and Wetzel T. Construction of a test facility for demonstration of a liquid lead-bismuth-cooled 10kW thermal receiver in a solar furnace arrangement e SOMMER. *Energ Procedia* 2015;69:1259–1268.

238. Boerema N., Morrison G., Taylor R., and Rosengarten G. Liquid sodium versus Hitec as a heattransfer fluid in solar thermal central receiver systems. *Sol Energ* 2012;86:2293–2305.

239. Grasse W., Hertlein H.P., and Winter C.J. Thermal solar power plants experience. In: Winter C.J., Sizmann R.L., and Vant-Hull L.L, editors. *Solar Power Plants*. Berlin: Springer-Verlag; 1991, pp. 215–282. ISBN 3-540-18897-5.

240. Coventry J., Andraka C., Pye J., Blanco M., and Fisher J. A review of sodium receiver technologies for central receiver solar power plants. *Sol Energ* 2015;122:749–762.

241. The IEA/SSPS Solar Thermal Power Plants d Facts and Figures d Final Report of the International Test and Evaluation Team (ITET). Volume 1: Central Receiver System (CRS). Kesselring P., and Selvage, C.S., editors. Springer Verlag Heidelberg GmbH.

242. Becker M., and Vant-Hull L.L. Thermal receivers. In: Winter C.J., Sizmann R.L., and Vant-Hull L.L., editors. *Solar Power Plants*. Berlin: Springer-Verlag; 1991, pp. 163–197. ISBN3-540-18897-5.

243. Olivier T.J., Radel R.F., Nowlen T.K., Blanchat T.K., and Hewson J.C. *Metal Fire Implications for Advanced Reactors Part 1: Literature Review*. SAND2007-6332. Albuquerque, NM: Sandia National Laboratories; 2007.

244. Rockwell International. *Sodium Solar Receiver Experiment e Final Report*. Report number SAND82-8192. Albuquerque, NM: Sandia National Laboratories; 2013.

245. Yousef B.A., Hachicha A.A., Rodriguez I., Abdelkareem M.A., and Inyaat A. Perspective on integration of concentrated solar power plants. *Int J Low-Carbon Technol* 2021, September;16(3):1098–1125. Doi: 10.1093/ijlct/ctab034.

246. Miller K. Hybrid solar thermal integration at existing fossil generation facilities. Available from: www.eskom.co.za/AboutElectricity/Engineering/G3CSP_FossilHybridMillerFINAL.pptx Johannesburg, South Africa, Black & Veatch (South Africa); 2013, August, 31.

247. Mills S.J. *Integrating Intermittent Renewable Energy Technologies with Coal-Fired Power Plant*. CCC/189. London: IEA Clean Coal Centre; 2013 January, 2011.

248. Mills S.J. Combining solar power with coal-fired power plants, or cofiring natural gas. *Clean Energ* 2018;2:1–9.

249. Maxson A., and Libby C. *Solar Thermal Hybrid Demonstration Project at a Pulverized Coal Plant, Overview, Value, and Deliverables*. Electric Power Research Institute (EPRI); 2010, pp. 1–7., Palo Alto , CA

250. Servert J., Zhifeng W., and Martinez D.. *People's Republic of China: Concentrating Solar Thermal Power Development*. Beijing, China; January 2012. *Technical Assistance Consultant's Report*, p. 135.Project number :43356/TA 7402, CDTA.

251. Hong H., Peng S., Zhao Y., and Liu Q. Hybrid power plant in China. *Energ Procedia* 2014;49:1777–1783.

252. Pierce W., Gauché P., von Backström T., Brent A.C., and Tadros A. A comparison of solar aided power generation (SAPG) and stand-alone concentrating solar power (CSP): A South African case study. *Appl Therm Eng* 2013;61:657–662.

253. Zhang M., Du X., Pang L., Xu C., and Yang L. Performance of double source boiler with coal-fired and solar power tower heat for supercritical power generating unit. *Energy* 2016;104:64–75.

254. Hong H., Peng S., Zhang H., Sun J., and Jin H. Performance assessment of hybrid solar energy and coal-fired power plant based on feed-water preheating. *Energy* 2017;128:830–838.

255. Mills S.J. Combining solar power with co-fired power plants or cofiring natural gas. A report by IEA clean coal center, CCC/279. London; 2017, October. ISBN 978-92-9029-602-7, www.iea-coal.org.

256. Rovira A., Sánchez C., Valdés M., Abbas R., Barbero R., Montes M.J., Muñoz M., Muñoz-Antón J., Ortega G., and Varela F. Comparison of different technologies for integrated solar combined cycles: Analysis of concentrating technology and solar integration. *Energies* 2018;11:1–16.

257. Kitzmiller K., and Miller F.J. Effect of variable guide vanes and natural gas hybridization for accommodating fluctuations in solar input to a gas turbine. *J Sol Energ Eng., Transaction of ASME* 2012;134:041008-1–041008-12.

258. Quero M., Korzynietz R., Ebert M., Jiménez A.A., Del Río A., and Brioso J.A. Solugas—operation experience of the first solar hybrid gas turbine system at MW scale. *Energ Procedia* 2014;49:1820–1830.

259. Korzynietz R., Brioso J.A., del Río A., Quero M., Gallas M., Uhlig R., Ebert M., Buck R., and Teraji D. Solugas—comprehensive analysis of the solar hybrid Brayton plant. *Sol Energ* 2016;135:578–589.

260. Livshits M., and Kribus A. Solar hybrid steam injection gas turbine (STIG) cycle. *Sol Energ* 2012;86:190–199.

261. Polonsky G., and Kribus A. Performance of the solar hybrid STIG cycle with latent heat storage. *Appl Energ* 2015;155:791–803.

262. Zarrouk S.J., and Moon H. Efficiency of geothermal power plants: A worldwide review. *Geothermics* 2014;51:142–153.

263. Zhou C., Doroodchi E., and Munro I.M.B. A feasibility study on hybrid solar– geothermal power generation. *A Paper Presented at New Zealand Geothermal Workshop*, Auckland; 2011.

264. Greenhut A., Tester J., DiPippo R., Field R., Love C., Nichols K., Augustine C., Batini F., Price B., Gigliucci G., and Fastelli I. Solar–geothermal hybrid cycle analysis for low enthalpy solar and geothermal resources. *A Paper Presented at World Geothermal Congress*, Bali, Indonesia; 2010.

265. Manente G., DiPippo R., Paci M., Field R., Tester J.W., and Rossi N. Hybrid solar–geothermal power generation to increase the energy production from a binary geothermal plant. *A Paper Presented at the ASME 2011 International Mechanical Engineering Congress & Exposition*, Denver, CO; 2011.

266. Todorovic M.S., and Lic̆ina D.Z. Parametric analysis and thermodynamic limits of solar assisted geothermal co-and tri-generation systems. A report by *ASHRAE Trans* 2011;117.

267. DiMarzio G., Angelini L., Price W., Chin C., and Harris S. The still water triple hybrid power plant: Integrating geothermal, solar photovoltaic and solar thermal power generation. *Proceedings World Geothermal Congress*; 2015, pp. 1–5.

268. Hashem H. World's first commercial CSP-geothermal hybrid underway. *New Energy Update: CSP*, Reuters Events; 2014. https://analysis.newenergyupdate.com/csp-today/technology/worlds- first-commercial-csp-geothermal-hybrid-underway.

269. Ciani Bassetti M., Consoli D., Manente G., and Lazzaretto A. Design and off-design models of a hybrid geothermal-solar power plant enhanced by a thermal storage. *Renew Energ* 2018;128:460–472.

270. Zhou C., Doroodchi E., and Moghtaderi B. An in-depth assessment of hybrid solar–geothermal power generation. *Energ Convers Manag* 2013;74:88–101.

271. Turchi C., Zhu G., Wagner M., Williams T., and Wendt D. Geothermal solar hybrid designs: Use of geothermal energy for CSP feed water heating. *GRC Trans* 2014;38:817–823.

272. Ayub M., Mitsos A., and Ghasemi H. Thermo-economic analysis of a hybrid solar-binary geothermal power plant. *Energy* 2015;87:326–335.

273. Astolfi M., Xodo L., Romano M., and Macchi E. *Technical and Economical Analysis of a Solar–Geothermal Hybrid Plant Based on an Organic Rankine Cycle*. The Netherland: Elsevier; 2011.

274. Al-Ali M., and Dincer I. Energetic and exergetic studies of a multigenerational solar–geothermal system. *Appl Therm Eng* 2014;71:16–23.

275. Ezzat M.F., and Dincer I. Energy and exergy analyses of a new geothermal– solar energy based system. *Sol Energ* 2016;134:95–106.

276. Calise F., d'Accadia M.D., Macaluso A., Piacentino A., and Vanoli L. Exergetic and exergoeconomic analysis of a novel hybrid solar–geothermal polygeneration system producing energy and water. *Energ Convers Manag* 2016;115:200–220.

277. Ghasemi H., Sheu E., Tizzanini A., Paci M., and Mitsos A. Hybrid solar–geothermal power generation: Optimal retrofitting. *Appl Energ* 2014;131:158–170.
278. Alibaba M., Pourdabani R., Manesh M., Herrera-Miranda I., Gallardo-Bemal I., and Hernandez-Hernandez J. Conventional and advanced exergy-based analysis of hybrid geothermal–solar power plant based on ORC Cycle. *Appl Sci* 2020;10(15):5206, Doi: 10.3390/app10155206.
279. Chasapis D., Drosou V., Papamechael I., Aidonis A., and Blanchard R. Monitoring and operational results of a hybrid solar-biomass heating system. *Renew Energ* 2008;33:1759–1767.
280. Tang B., Xu S., Hou X., Li J., Sun L., Xu W., and Wang X. Shape evolution of silver nanoplates through heating and photo induction. *ACS Appl Mater Interf* 2013;5:646–653.
281. Cot A., Amettler A., Vall-Llovera J., Aguiló J., and Arque J.M. Termosolar Borges: A thermosolar hybrid plant with biomass. *Third International Symposium on Energy From Biomass and Waste*, Italy, Venice, Italy, CISA, Environmental Sanitary Engineering Centre; 2010, pp. 8–11.
282. Servet J., San Miguel G., and Lopez D. Hybrid solar–biomass plants for power generation; technical and economic assessment. *Global NEST J* 2011;13:266–276.
283. Kaushika N.D., Mishra A., and Chakravarty M.N. Thermal analysis of solar biomass hybrid co-generation plants. *Int J Sust Energ* 2005;24:175–186.
284. Sterrer R., Schidler S., Schwandt O., Franz P., and Hammerschmid A. Theoretical analysis of the combination of CSP with a biomass CHP-plant using ORC-technology in Central Europe. *Energ Procedia* 2014;49:1218–1227.
285. Nixon J.D., Dey P.K., and Davies P.A. Which is the best solar thermal collection technology for electricity generation in north-west India? Evaluation of options using the analytical hierarchy process. *Energy* 2010;35:5230–5240.
286. Schnatbaum L. Biomass utilization for Co firing in parabolic trough power plants. *A Paper Presented at SolarPACES Conference*, Berlin, Germany; 2009.
287. Pérez Á., and Torres N. Solar parabolic trough biomass hybrid plants: A cost- efficient concept suitable for places in low irradiation conditions. *A Paper Presented at Solar-PACES Conference*, Perpignan, France; 2010.
288. Vidal M., Martín L., and Martín M. *24th European Symposium on Computer Aided Process Engineering,* Budapest, Hungary; 2014.
289. Vidal M., and Martín M. Optimal coupling of a biomass based polygeneration system with a concentrated solar power facility for the constant production of electricity. *Comput Chem Eng* 2015;72:273-283.
290. Zhang X., Li H., Liu L., Zeng R., and Zhang G. Analysis of a feasible trigeneration system taking solar energy and biomass as co-feeds. *Energ Convers Manag* 2016;122:74–84.
291. Rojas M., Hellwig U., Peterseim J.H., and Harding J. Combining solar thermal power systems with solid fuels in new plants. *A Paper Presented at SolarPACES Conference*, Perpignan, France; 2010.
292. Peterseim J.H., Tadros A., White S., Hellwig U., and Klostermann F. Concentrated solar power/energy from waste hybrid plants—creating synergies. *A Paper Presented at SolarPACES Conference*, Marrakech, Morocco; 2012.
293. Fluri T.P., Lude S., Lam J., Morin G., Paul C., and Platzer W.J. Optimization of live steam properties for a linear Fresnel power plant. *A Paper Presented at SolarPACES Conference*, Marrakech, Morocco; 2012.
294. Nixon J.D., Dey P.K., and Davies P.A. The feasibility of hybrid solar-biomass power plants in India. *Energy* 2012;46:541–554.
295. Peterseim J.H., Hellwig U., Tadros A., and White S. Hybridisation optimization of concentrating solar thermal and biomass power generation facilities. *Sol Energ* 2014;99:203–214.
296. Peterseim J.H., Tadros A., White S., Hellwig U., Landler J., and Galang K. Solar tower-biomass hybrid plants—maximizing plant performance. *Energ Procedia* 2014;49:1197–1206.

297. Alfonso-Solar D., Vargas-Salgado C., Sánchez-Díaz C., and Hurtado-Pérez E., Small-scale hybrid photovoltaic-biomass systems feasibility analysis for higher education buildings. *Sustainability* 2020;12:9300. Doi: 10.3390/su12219300 www.mdpi.com/journal/sustainability

298. Ravaghi-Ardebili Z, Manenti F, Corbetta M., Pirola C., and Ranzi E. Biomass gasification using low-temperature solar-driven steam supply. *Renew Energ* 2015;74:671–680.

299. Khalid F., Dincer I., and Rosen M.A. Energy and exergy analyses of a solar-biomass integrated cycle for multigeneration. *Sol Energ*. 2015;112:290–299.

300. Bai Z., Liu Q., Lei J., Wang X., Sun J., and Jin H. Thermodynamic evaluation of a novel solar-biomass hybrid power generation system. *Energy Convers Manag* 2017;142:296–306.

301. Pantaleo A.M., Camporeale S.M., Miliozzi A., Russo V., Shah N., and Markides C.N Novel hybrid CSP- biomass CHP for flexible generation: Thermo-economic analysis and profitability assessment. *Appl Energ*. 2017;204:994–1006.

302. Pantaleo A.M., Camporeale S.M., Sorrentino A., Miliozzi A., Shah N., and Markides C.N. Hybrid solar-biomass combined Brayton/organic Rankine-cycle plants integrated with thermal storage: Techno-economic feasibility in selected Mediterranean areas. *Renew Energ* 2020;147:2913–2931.

303. Liu Q., Bai Z., Wang X., Lei J., and Jin H. Investigation of thermodynamic performances for two solar-biomass hybrid combined cycle power generation systems. *Energ Convers Manag* 2016;122:252–262.

304. Platzer W.J. Combined solar thermal and photovoltaic power plants–an approach to 24h solar electricity? *AIP Conference Proceedings*, AIP Publishing LLC, Penang, Malaysia; 2016, p. 070026.

305. Ju X., Xu C., Han X., Zhang H., Wei G., and Chen L. Recent advances in the PV-CSP hybrid solar power technology. *AIP Conference Proceedings*, AIP Publishing LLC, Bydgoszcz, Poland; 2017, p. 110006.

306. Chow T.T. Tiwari G.N., and Menezo C. Hybrid solar: A review on photovoltaic and thermal power integration. *Int J Photoenerg* 2012, December; 2012. Article ID 307287. Doi: 10.1155/2012/307287.

307. Cocco D., Migliari L., and Petrollese M. A hybrid CSP–CPV system for improving the dispatchability of solar power plants. *Energ Convers Manag* 2016;114:312–323, Doi: 10.1016/j.enconman.2016.02.015.

308. Orosz M., Zweibaum N., Lanc T., Ruiz M., and Morad R. (2016). Spectrum-splitting hybrid CSP-CPV solar energy system with standalone and parabolic trough plant retrofit applications. Vol. 1734, p. 070023. Doi: 10.1063/1.4949170.

309. Zhai R. Chen Y., Liu H., Wu H., and Yang Y. Optimal design method of a hybrid CSP-PV plant based on genetic algorithm considering the operation strategy. *Int J Photoenerg* 2018;2018, Open access article, by Hindawi, Article ID 8380276. Doi: 10.1155/2018/8380276.

310. Bennett K.W. *A Hybrid (CSP/CPV) Spectrum Splitting Solar Collector for POWER Generation*, Ph.D. thesis. Merced: University of California; 2018.

311. Cygan D., Abbasi H., Kozlov A., Pondo J., Winston R., Widyolar B., Jiang L., Abdelhamid M., Kirk A.P., Drees M., Miyamoto H., Elarde V., and Osowski M.L. Full spectrum solar system: Hybrid concentrated photovoltaic/concentrated solar power (CPV-CSP). *MRS Adv* 2016;1:1–6. Doi: 10.1557/adv.2016.512.

312. Zhang X., Lei D., Zhang B., Yao P., and Wang Z. SiN$_x$/Cu spectral beam splitting films for hybrid photovoltaic and concentrating solar thermal systems. *ACS Omega* 2021;6(33):21709–21718, Publication Date: August 15, 2021. Doi: 10.1021/acsomega.1c03178.

# 6 Advances in Photovoltaic Technology

## 6.1 INTRODUCTION

In solar photovoltaic (PV) technology, the power source is photovoltaic modules that convert light directly to electricity. This differs from, and should not be confused with concentrated solar power described in the previous chapter which uses heat to drive a variety of conventional generator systems. Both approaches have their own advantages and disadvantages, but to date, for a variety of reasons, photovoltaic technology has seen much wider use in the field both at distributed level (rooftop mounted) and for utility-scale power station. While the global PV installation rate is increasing at an amazing rate (see Figure 6.1), photovoltaic systems have many variations. Besides flat plat PV, the PV system can also be used in conjunction with concentrated collector as the CPV system with several options depending on the level of concentration and capturing of the selected range of solar radiation spectrum. Since solar radiation generates both electricity and heat, the resulting high temperature of the solar cell affects its efficiency. Solar cell temperature is therefore controlled by providing a coolant medium. Thus, both PV and CPV can be hybridized by having a coolant lowering the temperature of the cell. These are called PVT and CPVT technologies. A number of cooling technologies can be used for this purpose and the thermal energy collected in the coolant can also be used for a number of secondary co or multi-generation applications. This chapter covers the important aspects of all four (PV, PVT, CPV, and CPVT) variations of photovoltaic technology. Since CPVT is the most elaborate form of PV technology, it is often used as a standard for general discussion.

Solar PV is emerging as one of the most competitive renewable sources of power generation due to its rapid cost decline. A decline of 74% in total installed costs occurred between 2010 and 2018. Lower solar PV module prices and ongoing reductions in balance-of-system costs remain the main drivers of reductions in the cost of electricity from solar PV [2]. The average total installed cost of utility-scale solar PV projects has declined by between 66% and 84% in major markets during the period 2010–2018 [3]. Alongside the decrease in installed costs, the global weighted average capacity factor of utility-scale PV systems has been increasing. Between 2010 and 2018 capacity factors increased from an average of 14%–18%. There are three major drivers for these increases: (a) the trend towards greater deployment in regions with higher irradiation levels, (b) the increased use of tracking systems, and (c) improvements in the performance of systems as losses are reduced, for instance though improvements in inverter efficiency [4]. Further growth will also require new business models and technical solutions for both the supply and demand sides, such as through the use of aggregators and the deployment of new technologies, including behind-the-meter batteries and demand response [5].

DOI: 10.1201/9781003328087-6

**Global PV installation rate**

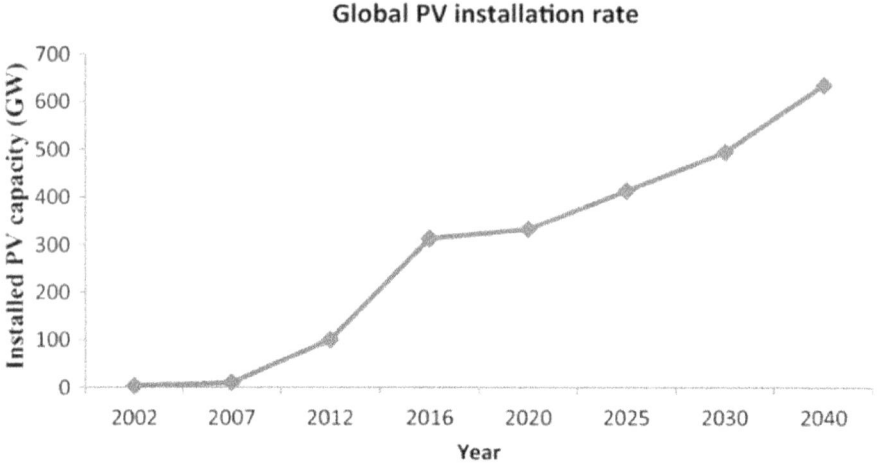

FIGURE 6.1   Global PV installation rate [11].

Solar PV power installations have been dominating the renewables industry for many years. As of the end of 2018, the global capacity of installed and grid-connected solar PV power reached 480 GW, representing a compound annual growth rate (CAGR) of nearly 43% since 2000 [6]. IRENA's REmap analysis shows that solar PV power installations could grow almost sixfold over the next 10 years, reaching a cumulative capacity of 2,840 GW globally by 2030 and rising to 8,519 GW by 2050. This implies total installed capacity in 2050 almost eighteen times higher than in 2018. The global weighted average LCOE of utility-scale PV plants is estimated to have fallen by 77% between 2010 and 2018, from around USD 0.37–USD 0.085/kWh, while auction and tender results suggest they will fall to between USD 0.08 and 0.02/kWh in 2030. Based on IRENA's REmap Case, the solar PV share of global power generation would reach 13% by 2030 and 25% by 2050. The global solar market in 2018 was dominated by Asia, accounting for over half of the world's addition of solar capacity. The European Union represented the world's second-largest solar PV market, mainly driven by Germany with 45 GW cumulative installed capacity by the end of 2018, followed by North America with 55 GW, of which the United States accounted for 90% [7]. Under the REmap scenario Asia would continue to lead global solar PV installations, with 65% of the total capacity installed by 2030.

Solar PV installations have occurred in three different ways; centralized utility grid-connected, distributed operations like rooftop installations and off-grid applications when no grid is available, or the cost of connecting to grid is too high. At a global level, around 60% of total solar PV capacity in 2050 would be utility scale, with the remaining 40% distributed and off-grid. Small and medium-sized distributed energy sources are mainly connected to the lower voltage systems near the end users. One example is rooftop PV panels. Due to net metering and fiscal incentives and falling costs, rooftop solar PV systems have significantly increased in recent years. Behind the meter storage allows consumers to store excess electricity from rooftop solar PV and consume it later when needed or sell it the grid. In 2018, distributed-scale solar PV capacity additions amounted to about 43 GV [8]. Currently, off-grid solar solutions constitute about 85% of all off-grid energy installations which include solar home systems (about 50%), solar lanterns, solar lighting systems (about 35%), rechargeable batteries (10%) and mini-grids (2%) [9]. The main cause for such growth is rapid decline in costs and increase in investments [2].

As mentioned earlier, while photovoltaic (PV) cells absorb and convert solar radiation into electricity [1–16], the main part of the collected solar radiation is transformed into heat, increasing PV-cell temperature and, therefore, reducing PV-cell efficiency. This undesirable phenomenon can be avoided by using a heat extraction device with fluid circulation. In this way, PV panels can work at a more satisfactory temperature. PV systems which combine PV modules with thermal units (involving the circulation of a fluid, e.g. air, water or others) are known as hybrid Photovoltaic/Thermal (PVT) systems and produce both electrical and thermal energy [4]. In other words, PVT installations provide higher total energy output (in comparison to PV modules which generate only electrical energy) and environmental advantages [10]. The same concept applies to CPV systems.

PVT and CPVT systems are appropriate for different kinds of applications (domestic, industrial, etc.) compare to PV and CPV systems which are mainly focused on the generation of electrical power. By placing emphasis on the working fluid, water-cooled PVT (PVT/water) and air-cooled PVT (PVT/air) systems are commonly used for water and indoor space heating, respectively [1]. The temperature of the working fluid plays a pivotal role and is associated with the type of application [1,17]. PVT and CPVT systems appropriate for different temperatures/applications have been developed: domestic hot water production, indoor space heating/cooling, desalination, pool heating, crop drying, industrial process heating and so on [1,4,10–23]. In the case of PVT applications in the building sector, it can be noted that there are two basic categories: (a) Building-Added Photovoltaic/Thermal (BA PVT) (for instance, these systems are mounted on the roof of a building), (b) Building-Integrated Photovoltaic/Thermal (BIPVT) (these systems form part of the building structure itself, e.g. façade- or roof-integrated configurations) [1,4,10–23].

As shown in Figure 6.2, PVT (and CPVT) can be characterized based on the type of collector, method of cooling, nature of coolant, etc. CPV and CPVT systems can also be subdivided into three categories based on solar distribution method and conversion sequence of energy flux as (a) the normal heat recovery CPV or CPVT, (b) the spectral beam splitting (SBS) CPV or CPVT, and (c) the energy distribution fitting (EDF) CPV or CPVT. For both PV (and PVT) and CPV (and CPVT) systems, as the system configuration is mainly affected by the CR of optical concentrators, they could be further categorized into three types: low CR system (CR < 10×) which is named as low concentration PVT system (LCPVT), medium CR system (10× < CR < 100×) which is named as medium concentration PVT system (MCPVT), and high CR system (CR > 100×) which is named as high concentration PVT system (HCPVT). Often ultra-high concentration category (UHCPVT) is also used as a possible fourth category. As shown later, these concentration differentiations result in different strategies for capturing solar energy, different cooling technologies, different coolant outgoing temperatures and different applications of thermal energy retained in the outgoing coolant.

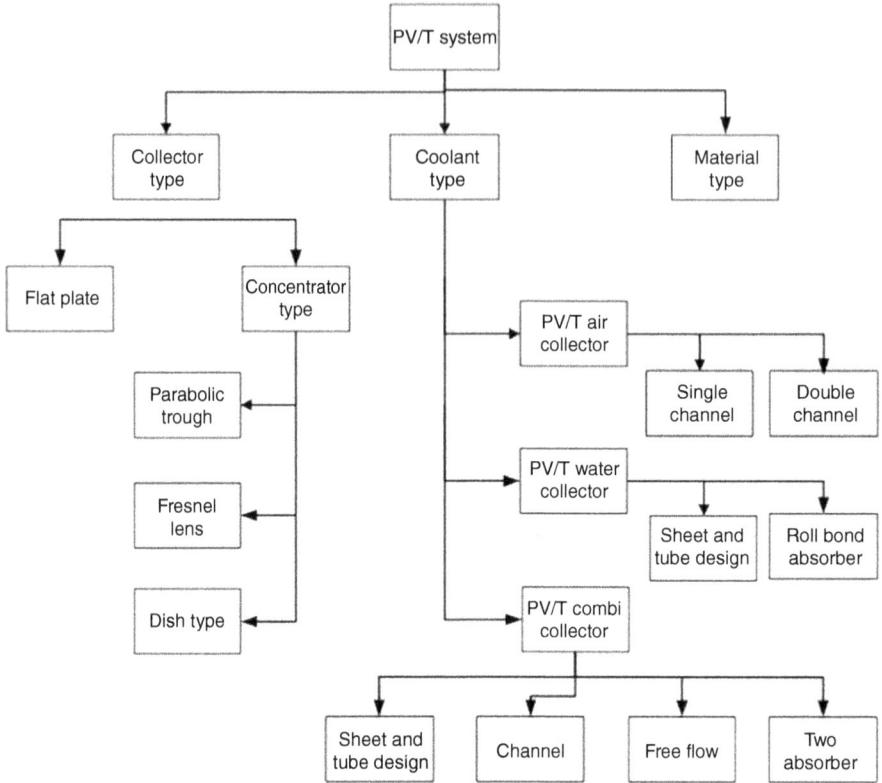

**FIGURE 6.2**  Characterization of PVT and CPVT technologies [1].

The normal CPVT system is developed based on the conventional flat-plate PVT system and separated by the efficiency of the solar concentrator. The thermal subsystem is the cooling system of PV modules and is usually active cooling on the module backside. In recent years, there are other cooling methods, such as immersion cooling, passive cooling, phase change cooling, etc., are also examined. During operation, all incident solar energy is first received by PV modules. Part of the radiation with energy near the bandgap of PV cells is converted into electricity in the PN junction, while the majority of the rest part is converted to heat through the thermalization process. After that, the cooling system absorbs the heat for further utilization such as domestic hot water or space heating. From an energy conversion point of view, this process is based on cascade utilization of solar energy.

The SBS CPVT system is based on the notion that solar energy can be utilized according to its spectral wavelength. The spectral beam of solar radiation is ranging from 200 to 4,000 nm, corresponding to photon energy 0.31–6.20 eV. As only photons with energy over the solar cell bandgap can be converted by PV cells, photons with lower energy are directed to the thermal receiver for heat generation. As bandgaps of silicon and GaAs solar cells are 1.11 and 1.43 eV, respectively, the solar radiation is usually split into ultraviolet + visible (UV + VIS) part and near-infrared (NIR) part. Thus, the SBS CPVT system can make full utilization of solar energy in a broad solar spectrum wavelength range. An additional optical element for spectral beam splitting is necessary, and the spectral beam splitter could be a dichroic filter, a liquid or solid absorptive filter, a holographic filter, a luminescent filter, etc. [24]. Compared with the normal CPVT system in which the temperature of the thermal subsystem is limited by PV cells' operating temperature, the SBS CPVT system has thermally decoupled PV and thermal subsystems, which means the thermal yield can achieve a higher temperature over the operating temperature limitation of PV cells.

The EDF CPVT system is an emerging concept. The EDF CPVT system is specially designed for nonuniform concentrated solar radiation, such as the Gaussian distribution of point-focused concentrators. This type of CPVT design aims to make full use of the spillage loss of the truncated energy profile. The edge part of concentrated illumination, which is usually discarded, is recovered in this type of CPVT. Under the nonuniform concentrated illumination, the facula area with different CRs can be adequately received by PV and thermal receivers separately. This concept is still being explored at basic research and demonstration levels.

The normal and SBS CPVT systems have been developed in the 1970s and 1980s [18], respectively, and they have been reviewed in several reports [25–38]. However, most of these papers mainly concern the examinations of flat-plate PVT systems, and discussions on CPVT systems are not quite profound and comprehensive. Only a few reviews are completely concerned with the CPVT, including the reviews of Sharaf and Orhan [26,27], Zhang et al. [28], and Ju et al. [39]. In these reviews, the scientific basics, technological fundamentals, characteristics, design considerations, advances, solar components, energy and exergy models, performance assessment methods, and application areas of CPVT systems are summarized. As an emerging concept, there is no significant research reported on EDF CPVT. Typical research was recently presented by Meng et al. [40], in which an annular PV receiver is deployed outside of a thermal receiver. Thus, the central high concentrated illumination is received by the thermal receiver, and the remaining is converted by the PV module. Reversed designs using a high concentration area for PV and a low concentration area for the thermal receiver were also proposed by Ju et al. [41] and Han et al. [42,43] for CPVT or CPV-CSP hybrid systems. Combinations of these three solar energy harness approaches may inspire novel designs and better performance, such as the combination of EDF and normal CPVT concept [41–43], and the combination of interference/absorption SBS and normal CPVT concept [44–46].

The above three types of CPVT differ greatly from each other. Most of the literature have not focused on the specific type of CPVT. Most literature is on normal CPVT where design and operation constraints are much tighter. The system designs and performance characteristics are also quite different for normal CPVT than that of SBS and EDF CPVT systems. While this chapter mainly focuses on normal CPVT systems, some analysis of SBS CPVT is presented in the following two sections.

### 6.1.1  SPECTRAL BEAM SPLITTING TECHNOLOGY

The central idea of Spectral Splitting is to irradiate the PV cells in a PVT collector only a selected range of the solar spectrum that can be converted into electricity with maximum efficiency. This is the segment of wavelengths where the spectral response SR (resp. the external quantum efficiency EQE) of a PV cell reaches its maximum, meaning that the incident photons have suitable energy in order to generate electron-hole pairs efficiently, e.g. without causing relevant heat losses within the cell due to thermalization. The remaining parts of the spectrum containing photons with less-suitable energy for electricity generation are converted into thermal energy directly. On the one hand, this is the infrared range (IR) of the spectrum, where the photon energy does not exceed the bandgap energy of the PV, while on the other hand, the photons' energy in the ultraviolet range (UV) is far beyond the bandgap energy so that it can only be converted partly into electricity. Spectral beam splitting technology thus exposes the solar cells to the selected optimum spectrum band, in which the cells express higher performance. This method rejects the light wavelength which cannot be converted to electricity before it is absorbed by the PV cell. It thus addresses the problem of battery overheating from the source and balances the uneven energy distribution on the surface of PV cells, but requires spectrum matching of the wavelength band and PV cells. Huang et al. [47] presented a comprehensive review of spectral splitting technologies used in PVT systems and outlined the challenges and opportunities for design considerations. In addition, the authors proposed a generalized technique to find an optimal nano filter for PVT and highlighted that the stability of the nano materials is still a great challenge for practical applications.

Resch and Holler [12] have reviewed design concepts for a spectral splitting CPVT receiver. Mojiri et al. [48] summarized the approaches which can reduce optical losses, by using a single device to concentrate and split the light into a single stage, as illustrated in Figure 6.3. The device can be made by a set of prisms based on a curved surface (as shown in Figure 6.3a [49]) or a combination of different dichroic concentrating mirrors (as shown in Figure 6.3b [50]. The principle is directing the different wavelength bands which are separated by the spectral beam splitter to the most efficient collector. These arrangements make the device realize spectral

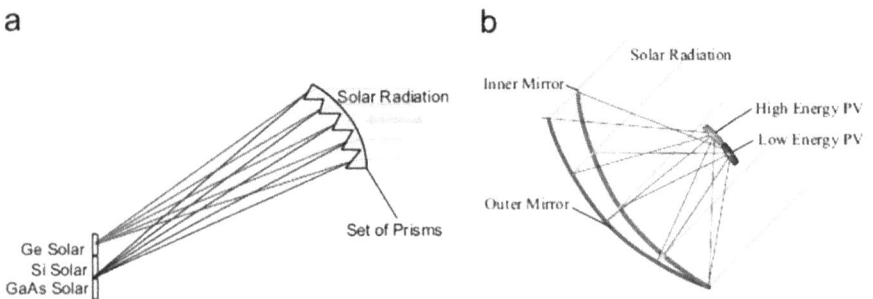

**FIGURE 6.3** Device of concentrating and splitting the spectral light into a single stage (a) A concentrating PV system with spectral separation using a set of prisms; 100 × and 17.5 × concentration levels for monochromatic and polychromatic (between 730 nm and 1,000 nm) beams respectively were achieved (b) A concentrating PV system comprising a dish-shaped 1 m² reflector made of two layers of faceted mirrors to create two focal points 20 cm off each other; Inner mirror reflects wavelengths above 650 nm and transmits below that; the outer mirror reflects all wavelengths [19].

splitting and produce concentrating effect in the meantime. The advantage is that it can reduce the number of interfaces and consequently decrease the reflection losses.

A novel spectrum splitting CPV system was invented by Taudien et al. in 2013 [51]. In this system, single and low-cost optical materials with dispersive properties, such as glass and plastic, were selected to produce the desired spectral separation. Simultaneously, reverse-ray tracing methods were utilized to optimize the shape of the solar concentrator's top and bottom interfaces, to ensure that the desired split spectrum would be concentrated in the target PV cells. This novel system had been proved to overcome the efficiency and cost limitations of current spectrum splitting CPV systems, with aggregate cell conversion which exceeded 45%. It showed that the potential of mass production simultaneously can meet the requirements for a high-efficiency low-cost CPV system.

Orosz et al. [52] retrofitted a standalone parabolic trough plant (PTC) to a spectrum-splitting hybrid Concentrator photovoltaics-concentrated PV (CSP-CPV) system. In this study, spectral filtering was adopted to transfer the sunlight which is unusable or poorly utilized for PV to a heat collection element, meanwhile, reflecting the useful sun rays to a concentrated PV receiver. According to the experimental validation and economic analysis, this hybrid CSP-PV power system expressed the capacity of higher efficiency and lower cost, with a 10% output improvement at an expected investment cost of less than $1 per additional net Watt, when compared to existing PTC plants. Meanwhile, it preserved the dispatchability of the CSP system's thermal energy storage.

### 6.1.2 CONCEPT OF MULTI-JUNCTION SOLAR CELL FOR SPECTRAL BEAM SPLITTING

As mentioned above, while the concentration of a CPV system can achieve tens to hundreds of suns, but only some specific wavelength can be utilized by traditional PV cells to convert into electricity. The spectral response is determined by the selected PV cell's material. For example, silicon solar cells absorb sunlight with a wavelength of 400–1,200 nm, whereas gallium arsenide solar cells can absorb between 400 and 900 nm [53]. The absorbable wavelength takes up only a small part of the entire sunlight spectrum, and the other sun radiation which is out of the PV cell's response wave band would turn into thermal energy that causes a continuous rise in the cell temperature and seriously affects the photoelectric conversion efficiency. Spectral separation of solar energy can address these problems, in which spectral splitting receiver and multi-junction solar cell are the specific applications. Currently, the majority of Concentrator photovoltaics systems use triple junction cells to achieve higher photoelectric conversion efficiency and lower cost. But the triple junction cell has limited efficiency, and although the spectral beam splitting system can realize high theoretical conversion efficiency, it is very costly for mass production [51].

Xiao et al. [19] pointed out that multi-junction solar cells (MJC) is a typical application of spectral beam splitting technology in the solar energy conversion field, which is caused by monolithically stacking multiple semiconductor materials with different wavelength matching [54]. Because of the characteristics of high reliability, anti-radiation performance and specific power, tandem and multi-junction solar cells are identified to have the greatest development potential in the future, and are gradually becoming the third-generation solar cells and replacing the silicon PV cells in recent years [55]. Widyolar et al. [56] simulated and optimized the hybrid solar CPV/CSP parabolic trough collector (PTC) systems (single stage and novel two-stage design) with spectral and temperature optimization, which integrated with different spectral beam splitting (SBS) approaches (ideal, interference filter, and novel integrated semi-transparent/back-reflecting solar cell filters) with different solar cells (c-Si, CdTe, GaAs, and InGaP), to maximize the solar-to-electric conversion. Yaping et al. [57] reported that out-of-band transmission in a semi-transparent triple junction cell could be 87% (ignoring grid contact shading losses). Widyolar et al. [58] reported that out-of-band reflectance of a bare GaAs cell to be 92% (closer to 83% with assumed encapsulant losses). Sun et al. [59] pointed out that typical c-Si and CdTe cells present high absorption for post-bandgap wavelengths and they cannot be used as spectrum splitting devices. Widyolar et al. [56] therefore used novel semi-transparent or back-reflecting solar cells (GaAs, InGaP) as the beam splitters for a two-stage CPV/CSP parabolic trough collector (PTC) system, as illustrated in Figure 6.4. The results of these studies showed that spectral beam splitting enhances solar-to-electric conversion by 45% over traditional PTC CSP systems. The hybrid CPV/CSP system thermally-generated electricity is at a lower cost than the current CSP by efficiently utilizing the solar spectrum. Economic analysis indicated that the installation cost of the c-Si SBS PTC system using interference filters was $2.39/W. However, the novel semi-transparent and back-reflecting solar cell beam splitters demonstrated lower conversion efficiencies than typical interference (dichroic) filter systems [56].

Xiao et al. [19] indicated that multi-junction solar cells demonstrated the highest conversion efficiency of all the photovoltaic technologies. Among the various classes of multi-junction solar cells, such as Si-base, thin film, space, high-efficiency and concentrator multi-junction solar cells, the concentrator multi-junction solar cell is the type which has been applied in CPV systems. For concentrator multi-junction solar cells, the InGaP/GaAs/Ge triple junction solar cell has been developed at the

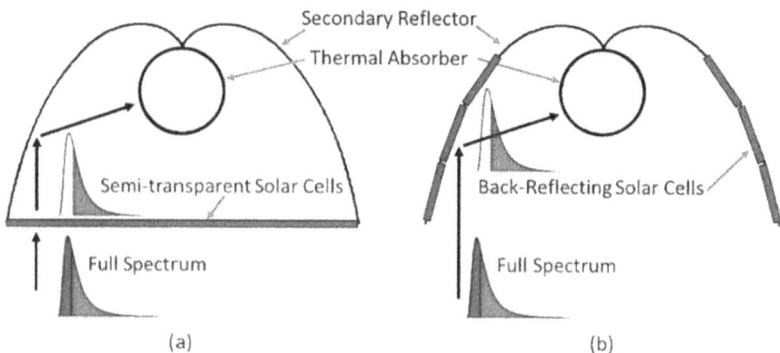

**FIGURE 6.4**  Spectral beam splitting by solar cells (purple) in semi-transparent (a) and back-reflecting (b) configurations as part of a secondary concentrator. High energy photons are absorbed and low energy photons are either (a) transmitted into the secondary concentrator or (b) reflected towards the thermal absorber [19].

beginning of 1990. The United States Spectrolab company launched the InGaP/ GaAs/Ge triple junction solar cell with the conversion efficiency in single spectrum at 26.8%, 28.3% and 29.9% respectively. The company subsequently applied the InGaP/GaAs/Ge triple junction solar cell into high concentrated photovoltaic (CPV) power system, the results show that the efficiency was up to 41.6%. Besides, GaInP/ InGaAs/InGaNAs/Ge four-junction solar cell has been produced by the utilization of epitaxial growth technology, which expresses the highest conversion efficiency in the world [60].

A novel concentrated photovoltaic-thermoelectric (CPV-TE) system using a triple-junction solar cell combined with thermoelectric cooler (TEC) and thermoelectric generator (TEG) was invented and experimented with by Teffah and Zhang [61]. The structure of the used triple junction solar cell and the band gap energy of these three $GaInP_2$/GaAs/Ge junctions is shown in Figure 6.5a. Nishioka et al. [14] examined different triple junction stack InGaP/InGaAs/Ge as shown in Figure 6.5b. In both cases, the results showed electrical power generation of the triple junction solar cell would increase by enhancing the concentration ratio, with the increase in efficiency.

Sadewasser et al. [62] conducted the heat management of $CuInSe_2$ micro-concentrator solar cells. The micro-concentrator solar cells showed a significant benefit in heat management over the conventional CPV devices. Since the devices are smaller than 200 μm, the temperature increase of micro-concentrator solar cells can be kept below 10°C above that of the regular flat panel solar cell. It was concluded that heat distribution can be improved by micro-meter size devices and thus eliminates the need for any additional cooling systems. A comprehensive review in the field of spectral beam splitting for photoelectric converting, including the application of multi-junction solar cells, has been presented by many researchers [17–20].

All these studies indicate that while utilization of spectral separation technology is cost-effective, more novel research ideas are needed to further improve CPVT cell cost, efficiency and manufacturability. There is also a lack of economical assessment of spectral separation technology. Burhan et al. [63] indicated that there is no commercial tool available that can analyze CPV performance with the utilization of multi-junction solar cells, and they proposed and techno-economically analyzed a novel standalone CPV-hydrogen system utilizing multi-junction solar cell through a micro genetic algorithm (micro-GA). More study is needed to examine the trade-off between reduced power generation cost by increasing efficiency and power needed for cooling costs.

## 6.2  MATERIALS ALTERNATIVES FOR SOLAR PV CELL

Continuous cost reduction and performance improvements through emerging and innovative technologies are still the major objectives of the solar industry. First-generation technologies, which have been evolving along the whole PV value chain, still account for the majority of global annual production [64]. Tandem and perovskite technologies also offer interesting perspectives, but in the longer term due to barriers that still need to be addressed and overcome (durability, price). Numerous types of materials and design alternatives are being investigated. These are briefly illustrated in Figure 6.6.

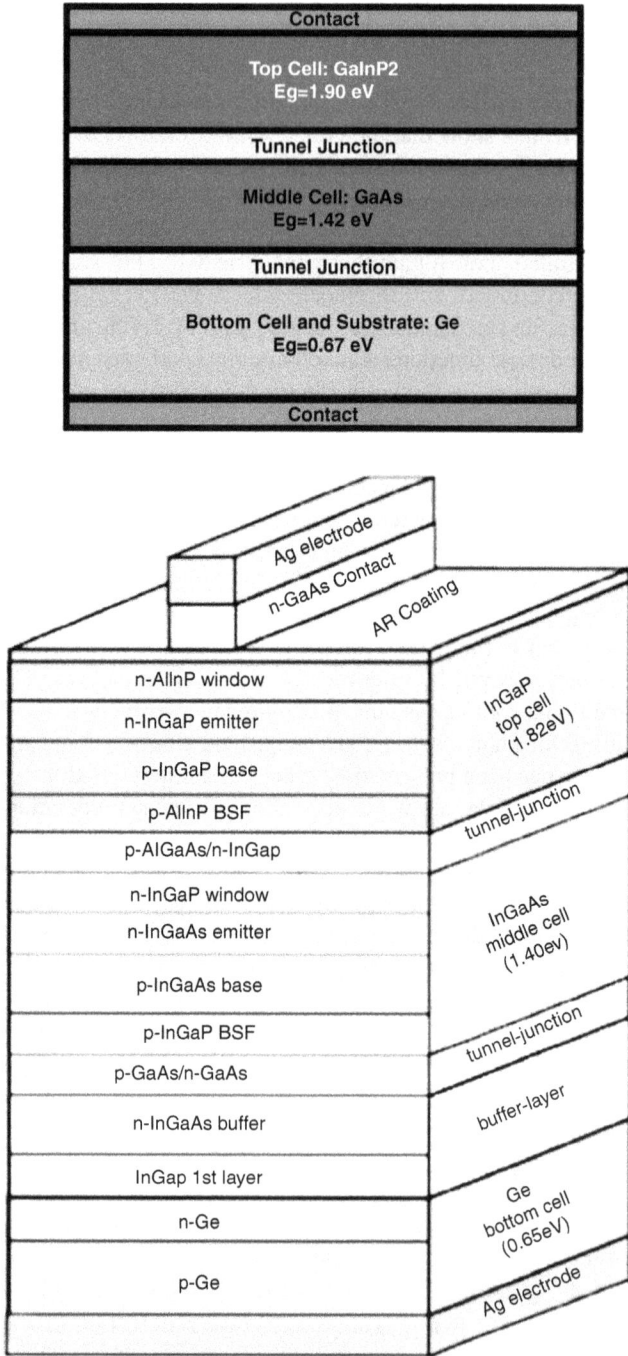

**FIGURE 6.5** (a) Structure of triple junction cell based on GaInP$_2$/GaAs/Ge [19]. (b) Structural details of a InGaP/InGaAs/Ge triple-junction PV stack [14].

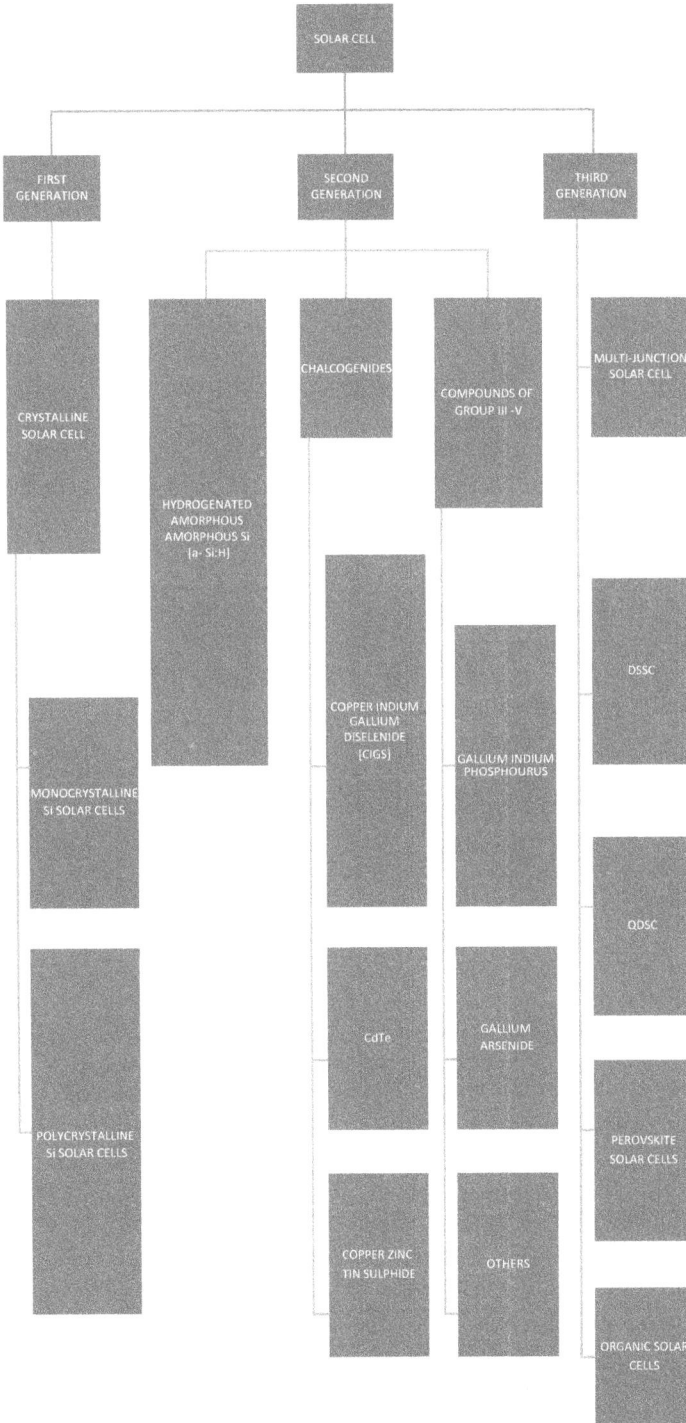

**FIGURE 6.6** Various types of solar cells and current advancements [65,66].

## 6.2.1 SILICON BASED CELLS

Silicon is one of the most dominating materials in solar PV technology. Its history and future are reviewed by Goetzberger and Hebling [67]. The first-generation PV modules were manufactured from the crystalline structure of silicon. However, this technology is constantly in R&D phase. With continuous progress, the cell efficiency during the 1980s and 2000+, increased up to 25% [68]. The University of New South Wales, also demonstrated the efficiency of 24.7% of silicon solar cells from Passivated Emitter, Rear Locally (PERL) on Flat Zone (FZ) silicon substrate and another 24.5% efficiency based on Passivated Emitter, Rear Totally (PERT) diffused silicon solar cells fabricated on Magnetically Confined Czochralski (MCZ) substrates [69]. On broad classification, monocrystalline and multi-crystalline are two basic forms of crystalline Si technology. Crystalline silicon technology has the highest commercial efficiency, and highly developed manufacturing process and maintains at least 80% of the market share of worldwide production of PV modules.

Monocrystalline material is widely used because of its high-efficiency level as compared to multi-crystalline. Zhao et al. [70] reported the honeycomb textured monocrystalline solar cells, with an efficiency of 24.4%. While a significant number of companies in China and Germany produce these modules with efficiencies in the range of 16.0%–16.9%, the SunPower Corporation [71] of the USA produces the modules with an efficiency of 20.4% (range of 20.0%–20.9% efficiency). It should be noted that generally module efficiency is somewhat smaller than cell efficiency. Zhao et al. [70] reported that multi-crystalline honeycomb textured solar cells, with an efficiency of 19.8%. Multi-crystalline solar modules are cheaper to manufacture, and they are preferred more in market. However, they are less efficient as compared to monocrystalline solar modules. The efficiency of the multi-crystalline modules produced by the companies lies in the range of 15.0%–16.9%. The highest commercial module efficiency is 16.9% from Neo Solar Power Corporation (T10) from Taiwan, whereas the maximum efficiency of monocrystalline is 20.4%. Efforts are being pursued to improve the efficiency of multi-crystalline cells [72].

Crystalline silicon (c-Si) panels belong to the first-generation solar PV panels and they hold a 95% share of worldwide PV production [64]. The economies of scale of silicon, make c-Si more affordable and highly efficient compared to other materials. The improvements in efficiency and power output of solar panels are expected to continue through 2030 [64]. While c-Si continues to hold a strong market position, there remains a lot of scope for improvement, including: (a) lowering the cost of c-Si modules for better profit margins; (b) reducing metallic impurities, grain boundaries, and dislocations; (c) mitigating environmental effects by reducing waste; and (d) yielding thinner wafers through improved material properties [73].

Amorphous silicon is the most commonly developed and a non-crystalline allotropic form of silicon. It is most popular among thin film technology but it is prone to degradation. Some of the varieties of a-Si are amorphous silicon carbide (a-SiC), amorphous silicon germanium (a-SiGe), and microcrystalline silicon (μ-Si) and amorphous silicon-nitride (a-SiN) [74]. Due to random structure, a-Si has a high band gap of 1.7 eV [75] and hence, compared to monocrystalline silicon, it has a 40 times higher rate of light absorptivity [76]. It also holds the first position in the current market amongst all thin film materials. The first amorphous thin film solar cell of ~1 μm thick was reported with an efficiency of 2.4% by Carlson and Wronski [77]. The potential of thin film solar cell is reviewed by Rech and Wagner [78] and further illustrated in the subsequent section. Stion Corporation of USA is manufacturing a-Si modules with the highest efficiency of 13.8%. China and Germany have the largest number of companies making modules using a-Si with the majority of companies producing modules in a range of 5%–9.9% efficiency. More work is needed to improve a-Si module efficiency.

## 6.2.2 NON-SI-BASED SOLAR CELLS

While currently most solar cells are made from silicon, new materials for solar cells are being developed. One of the most promising materials is perovskites, a type of mineral very good at absorbing light. The first perovskite PV devices in 2009 converted just 3.8% of the energy contained in sunlight into electricity. However, because crystals are very easy to make in the lab, their performance was quickly improved and by 2018 their efficiency had soared to 24.2%, close to silicon's lab record of 26.7% [79]. Perovskites still face some significant challenges before achieving market maturity. One of the main ones is durability. Because the crystals dissolve easily, they are not able to handle humid conditions and need to be protected by moisture through encapsulation, for instance through an aluminum oxide layer or sealed glass plates. Another challenge is to demonstrate high efficiency for larger cell areas. If these barriers can be overcome, perovskite cells have the potential to change the dynamics and economics of solar power because they are cheaper to produce than solar cells and can be produced at relatively low temperatures, unlike silicon.

Cadmium telluride solar cells are formed from cadmium and tellurium. Because of its ideal band gap of 1.45 eV, and longer stability [80], it is one of the promising materials in thin film technology. Besides some remarkable results by Compaan (81) and Schock and Pfisterer (82), efficiency of 10.6% and 11.2% was obtained on thin film 0.55 and 1 micron thick CdTe respectively by Nowshad et al. (83). In addition, CdTe on plastic foil with an efficiency of 11.4% is reported by Upadhayaya et al. [84]. In general, 15% to 16% cell efficiency has been obtained by Britt and Ferekides [85], Aramoto et al. [86] and Wu et al. [87]. In July 2011, First Solar (2011) company recorded 17.3% cell efficiency, which was confirmed by NREL. Cadmium telluride cells have achieved an efficiency of 21%, very similar to CIGS, and are characterized by good absorption and low energy losses [64]. CdTe solar cells are made through low- temperature processes, which makes their production very flexible and affordable. CdTe currently has the largest market share of all thin-film technologies [88].

Copper indium gallium selenide is an advanced researched material, in which gallium is added to copper indium (Di) selenide [89]. Until 2006, the best efficiency of CIGS cells was 20% [90] and about 13% for modules [91]. The commercial maximum efficiency of CIGS/CIS modules is 15.0%, manufactured by Mia Solé in the USA. As compared to the maximum efficiency of monocrystalline and multi-crystalline modules, CIGS/CIS modules lag by 5.4% and 1.9%. Though as compared to a-Si module, CIGS/CIS modules lead by 1.2%. CIGS cells have achieved high-efficiency levels (22.9%) comparable to commercial crystalline silicon [64]. However, manufacturing CIGS cells can be difficult due to the rarity of indium, as well as to the complex stoichiometry and multiple phases to produce them, restricting large-scale production in the near term [88].

Gallium arsenide (GaAs) is a compound semiconductor form of Gallium (Ga) and Arsenide (As). It has a similar structure like silicon cells with high efficiency and less thickness. In addition, it is lighter as compared to monocrystalline and multi-crystalline silicon cells [92]. Its energy band gap is 1.43 eV [93,94] which can be improved by alloying it with Aluminum (Al), Antimony (Sb), Lead (Pb), which in turn will form a multijunction device [95]. The Dutch Radboud University Nijmegen made a single junction GaAs cell that reached up to 28.8% efficiency [96] while Sharp [97] Company has reached up to 36.9% and Spire Corporation, has manufactured the most efficient triple-junction, GaAs cell, with an efficiency of 42.3%, which was verified by NREL [98]. There are, however, negligible commercial modules available in the market. CNT is formed by hexagonal lattice carbon [99]. A transparent conductor material made of CNT provides an excellent current. Meiller [100] believes that these cells can convert 75% of light into electricity. Shi et al. [101] reported a Titanium Dioxide ($TiO_2$)-coated CNT silicon solar cell with an efficiency of up to 15%. The record efficiency of solar cells with different materials developed over years is illustrated in Figure 6.7. A comparison of different types of solar cells is illustrated in Table 6.1 [102].

Transparent solar panels use a tin oxide coating on the inner surface of the glass panes to conduct current out of the cell. The cell contains titanium oxide that is coated with a photoelectric dye. Most conventional solar cells use visible and infrared light to generate electricity. In contrast, the innovative new solar cell also uses

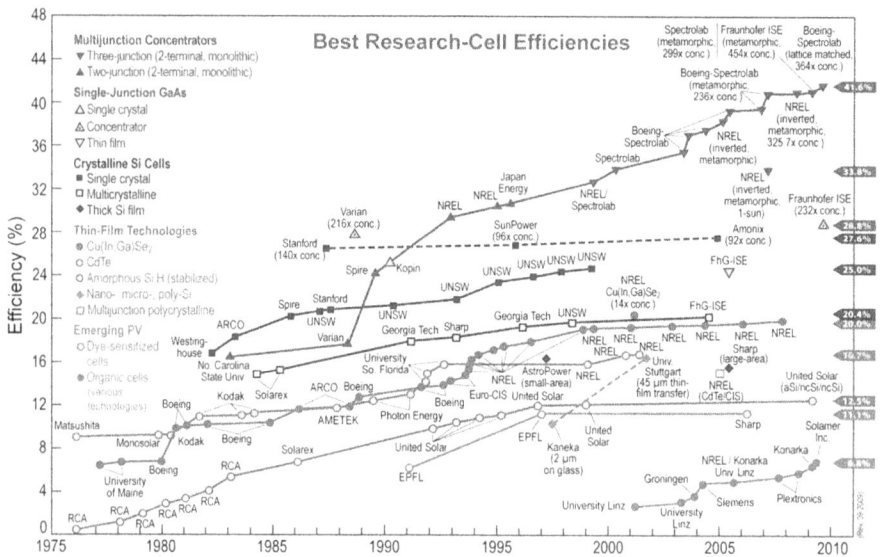

**FIGURE 6.7** Record efficiencies for different types of solar cells in the laboratory. (Source: NREL) [101,103].

**TABLE 6.1**
**Comparison of Different Types of Solar Cells [102]**

| Categories | Technology | $\eta$ (%) | $V_{OC}$ (V) | $I_{SC}$ (A) | W/m² | $t$ (μm) |
|---|---|---|---|---|---|---|
| Crystalline silicon cells | Monocrystalline | 24.7 | 0.5 | 0.8 | 63 | 100 |
| | Polysilicon | 20.3 | 0.615 | 8.35 | 211 | 200 |
| Thin film solar cells | Amorphous silicon | 11.1 | 0.63 | 0.089 | 33 | 1 |
| | CdTe | 16.5 | 0.86 | 0.029 | - | 5 |
| | CIGS | 19.5 | - | - | - | 1 |
| Multi-junction cells | MJ | 40.7 | 2.6 | 1.81 | 476 | 140 |

*Note:* Here *n*, efficiency; $V_{oc}$, open circuit voltage; $I_{sc}$, Short circuit current; *t*, film thickness.

ultraviolet radiation. They replace conventional window glass or placed over the glass. The installation surface area could be large, leading to potential uses that take advantage of the combined functions of power generation, lighting and temperature control Another name for transparent photovoltaics is "translucent photovoltaics" (they transmit half the light that falls on them). Similar to inorganic photovoltaics, organic photovoltaics are also capable of being translucent. Transparent and translucent photovoltaics comes as Non-wavelength-selective and Wavelength-selective [104]. In 2017, MIT researchers developed a process to successfully deposit transparent graphene electrodes onto organic solar cells resulting in a 61% transmission of visible light and improved efficiencies ranging from 2.8% to 4.1%. Perovskite solar cells, popular due to their promise as next-generation photovoltaics with efficiencies over 25%, have also shown promise as translucent photovoltaics. In 2015, a semi-transparent perovskite solar cell using a methylammonium lead triiodide perovskite and a silver nanowire mesh top electrode demonstrated 79% transmission at an 800 nm wavelength and efficiencies at around 12.7% [104].

## 6.3  NOVEL SOLAR CELL ARCHITECTURES

The emergence of new cell architectures has also allowed higher efficiency levels. A major driver of this shift has been the emergence of the PERC cells and their compatibility with other emerging innovations, such as half-cut cells. The most important technological shift in the market relates to bifacial cells and modules. Here we briefly examine a few novel solar cell architectures. Some of these architectures were devised to serve special applications.

### 6.3.1  PERC

A PERC cell uses advanced silicon cell architecture. Literally, it stands for *Passivated Emitter and Rear Cell*. PERL, PERT, and PERF are now usually considered as being part of the same family. PERC cells are not much different in construction from a typical monocrystalline PV cell; however, the key improvement is the integration of a back-surface passivation layer, which is a layer of material on the back of the cells that is able to improve the cell's efficiency [105]. The dielectric passivation layer contributes to the increase of efficiency by reducing electron recombination, increasing the solar cell's ability to capture light and reflecting specific wavelengths that normally generate heat out of the solar cells.

The main advantage of the PERC cell structure is that it enables manufacturers to achieve higher efficiencies than with standard solar cells which are reaching their physical limits. With the current state of the technology, it is possible to achieve up to 1% absolute gain in efficiency. The efficiency gain of implementing PERC architecture for monocrystalline cells is about 0.8%–1% absolute, while the boost for multicrystalline cells is a little lower, at 0.4%–0.8% [105]. While there are more steps in the manufacturing process, the gain in efficiency enables costs decrease, also at the system level. PERC has started only recently to enter the commercial arena but has quickly become the new industry standard for monocrystalline cells. Several factors have facilitated this remarkable progress, including the major shift of the market towards monocrystalline cells, the improvement in reliability and throughput of production tools, which has consequently improved the passivation quality of the films [105].

The structure of a PERC solar cell (see Figure 6.8) from front to rear include (a) screen-printed silver paste front contact, (b) anti-Reflective Coating (ARC), (c) silicon wafers that form the P-N junction, (d) local Aluminum Back Surface Field (Al-BSF), (e) dielectric passivation layer, (f) SiNx Capping Layer and (g) screen-printed Aluminum paste layer. The challenge behind the PERC technology is to be able to scale up the technology while controlling the process. There are two important challenges for PERC modules. The first one is related to Light Induced Degradation. LID is the effect that causes a module to lose a percentage of its power after its first exposition to light. Due to the higher doping levels commonly applied in PERC cells, the negative effect due to LID is increased with PERC technology compared to standard cells with an Al-BSF. The second challenge is around Potential Induced Degradation, particularly for polycrystalline PERC. This kind of defect can completely damage the performance of a power plant [106].

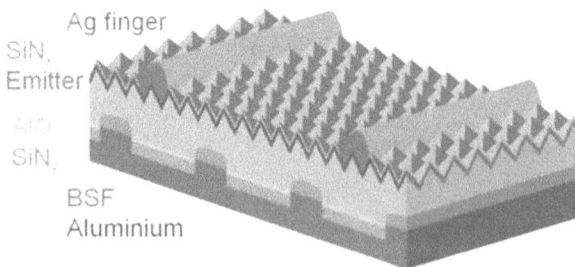

**FIGURE 6.8** Typical commercial PERC solar cell with AlO$_x$/SiN$_y$ rear passivation and local screen-printed Al rear contacts formed by laser-opening of the rear dielectric. (Reprinted from [106].)

## 6.3.2   THIN FILM SOLAR CELL

Thin film is an alternative technology, which uses less or no silicon in the manufacturing process. A more in-depth review of amorphous and crystalline thin film silicon solar cell is done by Roedern [107] and in the initial phases of the development of thin film solar cells, 10.7% efficiency was demonstrated by Yamamoto et al. [108]. A thin-film solar cell is a second-generation solar cell that is made by depositing one or more thin layers, or thin film (TF) of photovoltaic material on a substrate, such as glass, plastic or metal. These technologies generally include two main families: (a) silicon-based thin film (amorphous [a-Si] and micromorph silicon [a-Si/c-Si]; and (b) non-silicon based (perovskites, cadmium telluride [CdTe] and copper-indium-gallium- diselenide [CIGS]). These technologies can be cheaper to produce, as such they are being deployed on a commercial scale, but they have historically had lower efficiency levels. Film thickness varies from a few nanometers (nm) to tens of micrometers (μm), much thinner than thin-film's rival technology, the conventional, first-generation crystalline silicon solar cell (c-Si), that uses wafers of up to 200 μm thick. This allows thin film cells to be flexible, and lower in weight. It is used in building-integrated photovoltaics and as semi-transparent, photovoltaic glazing material that can be laminated onto windows. Other commercial applications use rigid thin film solar panels (interleaved between two panes of glass) in some of the world's largest photovoltaic power stations. Different types of thin film solar cells are illustrated in Figure 6.9.

Thin-film technology has always been cheaper but less efficient than conventional c-Si technology. However, it has significantly improved over the years. The lab cell efficiency for CdTe and CIGS is now beyond 21%, outperforming multicrystalline silicon, the dominant material currently used in most solar PV systems. Thin films degrade faster but are still expected to last 20 years. Thin film technology is often classified as emerging or third-generation photovoltaic cells and includes organic, dye-sensitized, as well as quantum dot, copper zinc tin sulfide, nanocrystal, micromorph, and perovskite solar cells. Whilst first-generation crystalline silicon-based PV has therefore consolidated its commercial status following performance improvements and cost reductions in recent years, the overall market share of thin-film technologies has been constantly decreasing since 2012. Currently, thin-film technology accounts for only 5% of the global solar PV market, while silicon-based solar modules still hold approximately 95% of the global PV module market [73].

## 6.3.3   HYBRID CELLS

Hybrid solar cells combine the advantages of both organic and inorganic semiconductors. Hybrid photovoltaics have organic materials that consist of conjugated polymers that absorb light as the donor and transport holes. Inorganic materials in hybrid cells are used as the acceptor and electron transporter in the structure. The hybrid photovoltaic devices have a potential for not only low-cost roll-to-roll processing but also for scalable solar power conversion. In hybrid solar cells, an organic material is mixed with a high electron transport material to form the photoactive layer. The two materials are assembled in a heterojunction-type photoactive layer, which can have

**FIGURE 6.9**  Different types of thin film solar cells [109].

a greater power conversion efficiency than a single material [108]. Controlling the interface of inorganic-organic hybrid solar cells can increase the efficiency of the cells. This increased efficiency can be achieved by increasing the interfacial surface area between the organic and the inorganic to facilitate charge separation and by controlling the nanoscale lengths and periodicity of each structure so that charges are allowed to separate and move toward the appropriate electrode without recombining. The three main nanoscale structures used are mesoporous inorganic films infused with electron-donating organic, alternating inorganic-organic lamellar structures, and nanowire structures.

Mesoporous films have been used for a relatively high-efficiency hybrid solar cell. The structure of mesoporous thin film solar cells usually includes a porous inorganic that is saturated with organic surfactant. The organic absorbs light and transfers electrons to the inorganic semiconductor (usually a transparent conducting oxide), which then transfers the electron to the electrode. Problems with these cells include their random ordering and the difficulty of controlling their nanoscale structure to promote charge conduction. Recently, the use of alternating layers of organic and inorganic compounds has been controlled through electrodeposition-based self-assembly. This is of particular interest because it has been shown that the lamellar structure and periodicity of the alternating organic-inorganic layers can be controlled through solution chemistry. To produce this type of cell with practical efficiencies, larger organic surfactants that absorb more of the visible spectrum must be deposited between the layers of electron-accepting inorganic.

Researchers have also been able to grow nanostructure-based solar cells that use ordered nanostructures like nanowires or nanotubes of inorganic surrounding by electron-donating organics utilizing self-organization processes. Ordered nanostructures offer the advantage of directed charge transport and controlled phase separation between donor and acceptor materials. The nanowire-based morphology offers reduced internal reflection, facile strain relaxation and increased defect tolerance. The ability to make single-crystalline nanowires on low-cost substrates such as aluminum foil and to relax strain in subsequent layers removes two more major cost hurdles associated with high-efficiency cells. There have been rapid increases in efficiencies of nanowire-based solar cells and they seem to be one of the most promising nanoscale solar hybrid technologies.

The photoactive layer can be created by mixing nanoparticles into a polymer matrix. Solar devices based on polymer-nanoparticle composites most resemble polymer solar cells. In this case, the nanoparticles take the place of the fullerene-based acceptors used in fully organic polymer solar cells. The use of nano particles over fullerenes is preferred because (a) Fullerenes production is difficult and energy-intensive while synthesis of nanoparticles is a low-temperature process, (b) limited testing of nanoparticle solar cells indicates they may be more stable than Fullerenes over time (c) nanoparticles are more absorbent than fullerenes, meaning more light can be theoretically absorbed in a thinner device, (d) nanoparticle size can affect absorption and (e) nanoparticles with size near their Bohr radius can generate two excitons when struck by a sufficiently energetic photon.

Hybrid cell efficiency must be increased to start large-scale manufacturing. Three factors affect efficiency. First, the bandgap should be reduced to absorb red photons, which contain a significant fraction of the energy in the solar spectrum. Current organic photovoltaics have shown 70% of quantum efficiency for blue photons. Second, contact resistance between each layer in the device should be minimized to offer a higher fill factor and power conversion efficiency. Third, charge-carrier mobility should be increased to allow the photovoltaics to have thicker active layers while minimizing carrier recombination and keeping the series resistance of the device low. Problems include controlling the amount of nanoparticle aggregation as the photo layer forms. The particles need to be dispersed in order to maximize interface area, but need to aggregate to form networks for electron transport. The network formation is sensitive to the fabrication conditions. Dead-end pathways can impede flow. A possible solution is implementing ordered heterojunctions, where the structure is well controlled. The nanoparticles involved are typically colloids, which are stabilized in solution by ligands. The ligands decrease device efficiency because they serve as insulators which impede interaction between the donor and nanoparticle acceptor as well as decrease electron mobility. Some, but not complete success has been had by exchanging the initial ligands for pyridine or another short chain ligand [60]. CNT may also be used as a photovoltaic device not only as an add-in material to increase carrier transport but also as the photoactive layer itself. Oxidized CNTs tend to become more metallic, and so less useful as a photovoltaic material [65].

In general, since hybrid cell technology is based on the principle of combining crystalline silicon with non-crystalline silicon [110], its manufacturing process is complex. Wu et al. [111] found a high ratio of performance to cost for hybrid cells. Based on that, Sanyo has manufactured a hybrid cell with a module efficiency of 17.8% [112]. In April 2014, Panasonic manufactured a hybrid cell that combines a thin crystalline silicon wafer coated with amorphous silicon giving better performance in low light and at high temperature with a record-breaking highest conversion efficiency of 25.6%, which was confirmed by the National Institute of Advanced Industrial Science and Technology (AIST) [113,114].

### 6.3.4 DYE-SENSITIZED SOLAR CELL

In order to cut production cost, address environmental concerns along with the improvement of efficiency, a new material technology called dye-sensitized solar cell [115] was tested. Several researchers [116,117] investigated on sensitization of wide bandgap semiconductor materials like Zinc Oxide (ZnO) by organic dyes for photo-electochemical (PEC) process. Deb et al. [118] reported the first use of $TiO_2$ in the PEC process. However, these cells are still considered as under the early stage of development. In July 2013, École Polytechnique Fédérale De Lausanne (EPFL) scientist has set a world record of 15% exceeding the power conversion efficiencies of conventional, amorphous silicon-based solar cells and they believe that this will open a new era for dye-sensitized solar cells (DSSC) [119,120] and soon it can be a good competitor to the existing materials available in the market for solar cells [121].

Dye-sensitized solar cells consist of a photo-sensitized anode, an electrolyte, and a photo-electrochemical system. Hybrid solar cells based on dye-sensitized solar cells are formed with inorganic materials ($TiO_2$) and organic materials. Hybrid solar cells based on dye-sensitized solar cells are fabricated by dye-absorbed inorganic materials and organic materials. $TiO_2$ is the preferred inorganic material since this material is easy to synthesize and acts as an n-type semiconductor due to the donor-like oxygen vacancies. However, titania only absorbs a small fraction of the UV spectrum. To enhance diffusion length (or carrier lifetime), a variety of organic materials are attached to the titania [102,108].

Mesoporous materials contain pores with diameters between 2 and 50 nm. A dye-sensitized mesoporous film of $TiO_2$ can be used for making photovoltaic cells and this solar cell is called a "solid-state dye-sensitized solar cell." The pores in mesoporous $TiO_2$ thin film are filled with a solid hole-conducting material such as p-type semiconductors or organic hole conducting material. Many organic materials have also been tested to obtain a high solar-to-energy conversion efficiency in dye synthesized solar cells based on mesoporous titania thin film. Liquid organic electrolytes contain highly corrosive iodine, leading to problems of leakage, sealing, handling, dye desorption, and maintenance. Much attention is now focused on the electrolyte to address these problems [102,108].

For solid-state dye-sensitized solar cells, the first challenge originates from disordered titania mesoporous structures. Mesoporous titania structures should be fabricated with well-ordered titania structures of uniform size (~ 10 nm). The second challenge comes from developing the solid electrolyte, which should be transparent to the visible spectrum (wide band gap) and its fabrication should be possible for depositing the solid electrolyte without degrading the dye molecule layer on titania. Finally, the LUMO of the dye molecule should be higher than the conduction band of titania. Several p-type semiconductors tend to crystallize inside the mesoporous titania films, destroying the dye molecule-titania contact. Therefore, the solid electrolyte needs to be stable during operation.

### 6.3.5 TANDEM AND MULTI-JUNCTION SOLAR CELLS

Tandem solar cells are stacks of individual cells, one on top of the other, that each selectively convert a specific band of light into electrical energy, leaving the remaining light to be absorbed and converted to electricity in the cell below. Emerging PV technologies comprise several types of tandem cells that can be grouped mainly depending on the materials used (*e.g.* organic, inorganic, hybrid) as well as on the kind of connection used [122]. The tandem cell approach has been used to fabricate the world's most efficient solar cells that can convert 46% of sunlight into electricity. Unfortunately, these devices use very expensive materials and fabrication processes, and still cannot break through the market [88]. Tandem cells, just like DSSC material technology, are also one of the new immerging ideas to increase the efficiency of solar modules. The concept is that by stacking several cells of different band gaps such that the gap energy decreases from the top and each cell converts solar spectrum at its maximum efficiency, overall it will increase the efficiency of a completely stacked cell. This tandem arrangement can be achieved with the different thin film materials, according to their band gaps [123]. Yamaguchi et al. [124,125] have reviewed the efficiency and status of multi-junction solar cells. The theoretical efficiency of a single-junction cell is around 31% [126]. The stabilized efficiency of a single junction cell is 9.3%, for double junction 12.4%, whereas 13% for triple junction was described by Guha [127]. The strategy of the double junction of Hydrogenated amorphous silicon (a-Si:H) intrinsic layers was developed by Fuji Electric & Co. (Japan) and Phototronics (Germany) (part of RWE Schott) with a stabilised lab efficiency of ~8.5% for cells and 5.5% for commercially available module efficiency [128]. Based on the concept intermediate Transparent Conducting Oxides (TCO) reflector layer for light trapping, Yamamoto et al. [129] showed the efficiency of 14.7%. Nevertheless, in June 2013, Sharp hit a world record of 44.4% triple junction solar cell, which was verified by the Fraunhofer Institute [130]. To date, their higher price and higher price-to-performance ratio have limited their use in special roles, notably in aerospace where their high power-to-weight ratio is desirable. In terrestrial applications, these solar cells are emerging in concentrator photovoltaics (CPV), with a growing number of installations around the world [102].

Tandem fabrication techniques have been used to improve the performance of existing designs. In particular, the technique can be applied to lower-cost thin-film solar cells using amorphous silicon, as opposed to conventional crystalline silicon, to produce a cell with about 10% efficiency that is lightweight and flexible. It is difficult to create "monolithically integrated" cell, where the cell consists of a number of layers that are mechanically and electrically connected. These cells are much more difficult to produce because the electrical characteristics of each layer have to be carefully matched. In particular, the photocurrent generated in each layer needs to be matched, otherwise electrons will be absorbed between layers. This limits their construction to certain materials, best met by the III–V semiconductors. The choice of materials for each sub-cell is determined by the requirements for lattice-matching, current-matching, and high-performance optoelectronic properties [102].

The majority of multi-junction cells have been produced to date using three layers. However, the triple junction cells require the use of semiconductors that can be tuned to specific frequencies, which has led to most of them being made of gallium arsenide (GaAs) compounds, often germanium for the bottom-, GaAs for the middle-, and $GaInP_2$ for the top-cell. The theoretical efficiency of MJ solar cells is 86.8% for an infinite number of pn junctions, implying that more junctions increase efficiency. Furthermore, as the cell approaches the limit of efficiency, the increased cost and complexity grow rapidly [102]. As shown in Table 6.1, MJ solar cells and other photovoltaic cells have significant differences. Physically, the main property of a MJ solar cell is having more than one pn junction in order to catch a larger photon energy spectrum while the main property of the thin film solar cell is to use thin films instead of thick layers in order to decrease the cost efficiency ratio. MJ solar panels are more expensive than others. These differences imply different applications: MJ solar cells are preferred in space and c-Si solar cells for terrestrial applications. The efficiencies of solar cells and Si solar technology are relatively stable, while the efficiency of solar modules and multi-junction technology is progressing.

### 6.3.6 BIFACIAL SOLAR CELLS

A mono facial solar cell produces photocurrent only if the face where the p-n junction has been formed is illuminated. Instead, a bifacial solar cell (BSC) is designed in such a way that the cell is active on both its faces and will produce photocurrent when either side, front or rear, is illuminated. The efficiency of bifacial solar cells, defined as the ratio of incident luminous power to generated electrical power, is measured independently for the front and rear surfaces under one or several suns (1 sun = 1,000 W/m²). The bifaciality factor (%) is defined as the ratio of rear efficiency in relation to the front efficiency subject to the same irradiance.

BSCs and modules (arrays of BSCs) were invented and first produced for space and earth applications in the late 1970s and became mainstream solar cell technology by the 2010s. It is foreseen that it will become the leading approach to photovoltaic solar cell manufacturing by 2030. Several in-depth reviews on bifacial solar

## TABLE 6.2
## Characteristics of Various Types of Bifacial Solar Cells [123]

**PERT**
- Efficiency: 19.5%–22% (front), 17%–19% (rear)
- Bifaciality: 80%–90%
- Mostly commercialized (e.g. Yingli, Trina, LG) on the n-type c-Si wafer due to longer carrier lifetime than p-type and absence of boron in the bulk material avoiding light-induced degradation (LID).

**PERL**
- Efficiency: 19.8% (front)
- Bifaciality: 80%–90%
- Mainly based on p-type c-Si wafer
- Boron is locally diffused into contact areas at the rear side

**PERC**
- Efficiency: 19.4%–21.2% (front), 16.7%–18.1% (rear)
- Bifaciality: 70%–80%
- Mostly commercialized (e.g. JA Solar, LONGi, Trina) e.g. on the p-type c-Si wafer

**IBC**
- Efficiency: 23.2%
- Bifaciality: 70%–80%
- Mainly based on n-type c-Si wafer
- No metal grid contact on the front side

**HIT**
- Efficiency: 24.7%
- Bifaciality: 95%–100%
- Mostly commercialized (e.g. Panasonic, Hanergy) on n-type c-Si waferCurrent bifacial solar cells

cells and their technology elements cover the current state-of-the-art [122,130–132]. They summarize the most common BSC designs currently being marketed and then provide a review of their technological aspects. Various bifacial PV modules with different architectures for their BSCs are currently available in the PV market. These include Passivated Emitter Rear Contact (PERC), Passivated Emitter Rear Locally-diffused (PERL), Passivated Emitter Rear Totally diffused (PERT), Heterojunction with Intrinsic Thin-layer (HIT), Interdigitated Back Contact (IBC). The characteristics of various types of bifacial solar cells are described in Table 6.2 and they are graphically illustrated in Figure 6.10.

Bifacial solar cells have been under development for decades and their manufacturing process can be considered one of the most advanced for solar modules today [133]. China retains its status as the largest manufacturer of, and end market for, bifacial modules. Worldwide demand has also increased, with countries such as the United States, Brazil and the United Kingdom increasing their use of bifacial modules for utility-scale PV plants. Based on the current market trend, bifacials are extending their geographical reach from Europe and Japan to emerging markets and across the globe [134].

One type of bifacial module is the *glass-glass module*. These are solar panels with solar cells arranged between two glass panes. They are typically applied to utility-scale systems and provide a heavy-duty solution for harsh environments (*e.g.* high temperatures, high humidity) because they are less sensitive to penetration of moisture. The technology has already been under development for decades, but its high costs and heavy weight have been a barrier to its development. In a PV module, solar cells are electrically connected to strings. This interconnection, however, can cause optical losses in the module, which affects the reliability of the product. To overcome this limitation, various industrial stringing equipment and soldering technologies are being developed, such as half-cells, solar shingles and

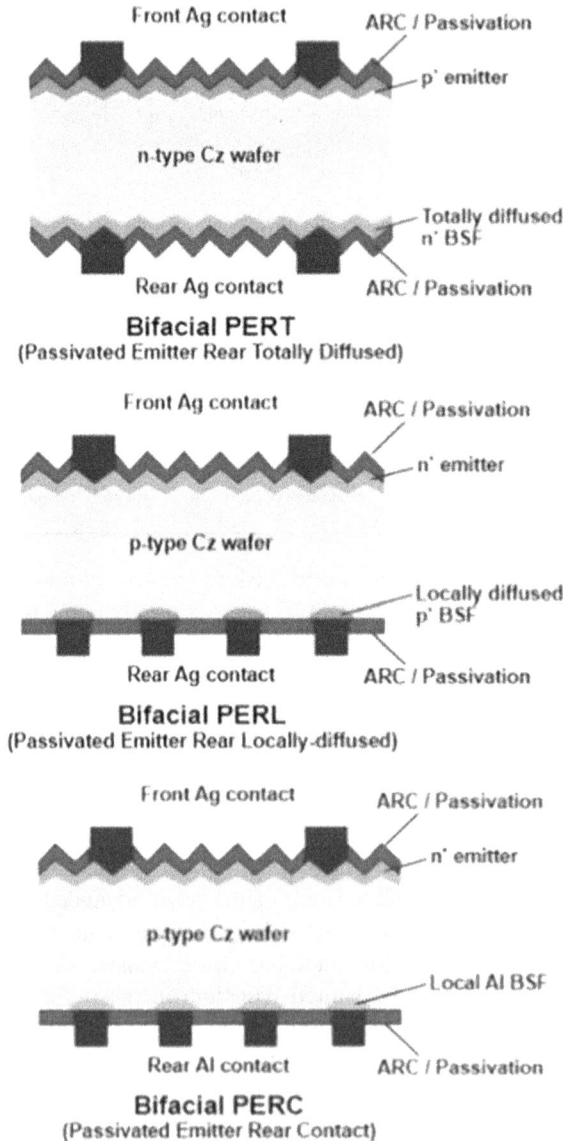

**Bifacial PERT**
(Passivated Emitter Rear Totally Diffused)

**Bifacial PERL**
(Passivated Emitter Rear Locally-diffused)

**Bifacial PERC**
(Passivated Emitter Rear Contact)

**FIGURE 6.10**   Graphical illustration of several bifacial solar cells [122].

*(Continued)*

**Bifacial IBC**
(Interdigitated Back Contact)

**Bifacial HIT**
(Heterojunction with Intrinsic Thin-layer)

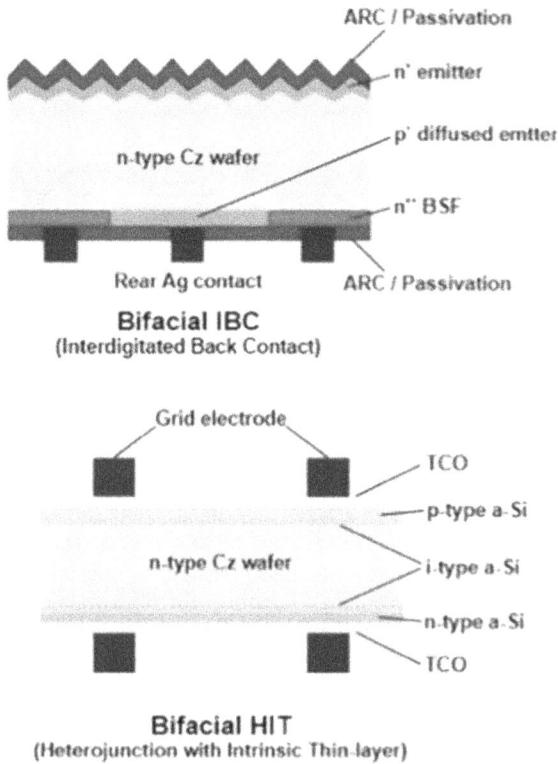

**FIGURE 6.10 (*CONTINUED*)** Graphical illustration of several bifacial solar cells [122].

multi-busbars. The growth in bifacial cells compared to mono facial solar cells is illustrated in Figure 6.11.

### 6.3.7 HALF-CELLS

In electrochemistry, a half-cell is a structure that contains a conductive electrode and a surrounding conductive electrolyte separated by a naturally occurring Helmholtz double layer. Chemical reactions within this layer momentarily pump electric charges between the electrode and the electrolyte, resulting in a potential difference between the electrode and the electrolyte. The typical anode reaction involves a metal atom in the electrode dissolved and transported as a positive ion across the double layer, causing the electrolyte to acquire a net positive charge while the electrode acquires a net negative charge [135]. The growing potential difference creates an intense electric field within the double layer, and the potential rises in value until the field halts the net charge-pumping reactions. This self-limiting action occurs almost instantly in an isolated half-cell; in applications, two dissimilar half-cells are appropriately connected to constitute a Galvanic cell. Half cells involve deliberately cutting a fully processed cell into the half with very advanced laser machines. Half-cells are being

# BIFACIAL VS. MONOFACIAL SOLAR CELLS IN THE WORLD MARKET

☒ Monofacial   ▨ Bifacial

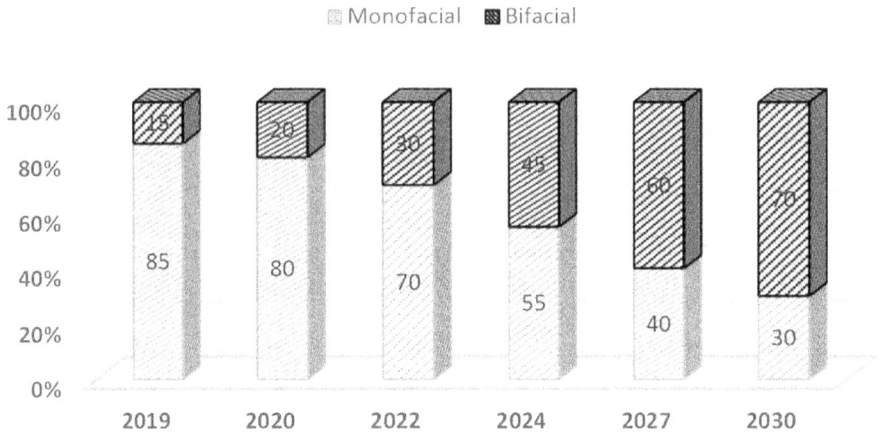

**FIGURE 6.11**  Forecast of the worldwide market shared for bifacial solar cell technology according to the International Technology Roadmap for Photovoltaic (ITRPV) -11th Ed., April 2020 [122].

adopted quickly due to their ease in manufacturing, improvement in module performance and durability, and an instant power boost of 5–6 W [133]. With the integration of PERC, half-cut technology has achieved efficiencies of up to 18% and power ratings of up to 300 W [134].

## 6.3.8   EMERGING NOVEL IDEAS ON SOLAR CELL DEVELOPMENT

Besides the ones mentioned above, there are several far-reaching emerging new ideas on the generation of electricity from solar energy that are being pursued in numerous universities and research organizations. For example, at Georgia Tech five novel ideas that are worth mentioning. These are all carried out by various professors in the engineering college at Georgia Tech university.

Yee at Georgia Tech is developing a thermo-electro-chemical converter using isothermal expansion of sodium by solar heat to directly generate electricity with no moving parts. The process is called Na-TECC, and in this process, electricity is generated from solar heat by thermally driving a sodium redox reaction on opposite sides of a solid electrolyte. The resulting positive electrical charges pass through the solid electrolyte due to an electrochemical potential produced by a pressure gradient, while the electrons travel through an external load where electric power is extracted. The new process results in improved efficiency and less heat leaking out. The goal is to reach heat-to-electricity conversion efficiency of more than 45%. The technology could be used for distributed energy applications. It can also be used with other heat sources such as natural gas, biomass, and nuclear to directly produce electricity without boiling water and spinning turbines.

Yee and researchers from Stanford University are also developing a technology called Betavoltaics that is similar to photovoltaic devices but instead of using photons from the sun, it uses high-energy electrons emitted from nuclear byproducts. This study uses strontium-90, a prevalent isotope in nuclear waste. Strontium-90 is unique because it emits two high-energy electrons during its decay process. What's more, the strontium-90s energy spectrum aligns well with the design architecture already used in crystalline silicon solar cells, so it could yield highly efficient conversion devices. In lab-scale tests with electron beam sources, the researchers have been achieving power conversion efficiencies of between 4% and 18%. With continued improvements, the betavoltaic devices could ultimately generate about 1 W of power continuously for 30 years — which would be 40,000 times more energy dense than current lithium-ion batteries. Initial applications include military equipment that requires low-power energy for long periods of time or powering devices in remote locations where changing batteries is problematic.

Tentzeris and coworkers at Georgia Tech have developed an electromagnetic energy harvester that can collect enough ambient energy from the radio frequency (RF) spectrum to operate devices for the Internet of Things (IoT), smart skin and smart city sensors, and wearable electronics. Harvesting radio waves is not brand new, but previous efforts have been limited to short-range systems located within meters of the energy source. This team is the first to demonstrate long-range energy harvesting as far as seven miles from a source. The researchers unveiled their technology in 2012, harvesting tens of microwatts from a single UHF television channel. Since then, they've dramatically increased capabilities to collect energy from multiple TV channels, Wi-Fi, cellular, and handheld electronic devices, enabling the system to harvest power in the order of milliwatts. Hallmarks of the technology include ultra-wideband antennas that can receive a variety of signals in different frequency ranges, unique charge pumps that optimize charging for arbitrary loads and ambient RF power levels and antennas and circuitry, 3-D inkjet-printed on paper, plastic, fabric, or organic materials, that are flexible enough to wrap around any surface. (The technology uses principles from origami paper folding to create "smart" shape-changing complex structures that reconfigure themselves in response to incoming electromagnetic signals.) The researchers have recently adapted the harvester to work with other energy-harvesting devices, creating an intelligent system that probes the environment and chooses the best source of ambient energy to collect. It also combines different forms of energy, such as kinetic and solar, or electromagnetic and vibration. More work is needed to scale the printing process for its commercialization.

Cola and coworkers at Georgia Tech developing the use of optical rectenna for solar cells. Rectennas, which are part antenna and part rectifier, convert electromagnetic energy into direct electrical current. Cola's team is using nanoscale fabrication techniques and different physics. Instead of converting particles of light, which is what solar cells do, this concept converts waves of light. Key to this technology are antennas small enough to match the wavelength of light (about one micron) and a super-fast diode — achieved in part by building the antenna on one of the metals in the diode. The process includes (a) carbon nanotubes are grown vertically off a substrate, (b) using atomic layer deposition, the nanotubes are coated with aluminum oxide to serve as an insulator and (c) extremely thin layers of calcium and aluminum metals are placed on top to act as an anode. As light hits the carbon nanotubes, a charge moves through the rectifier, which switches on and off to create a small direct current. The metal-insulator-metal-diode structure is fast enough to open and close at a rate of 1 quadrillion times per second. From a performance perspective, the devices currently operate just under 1% efficiency. Yet because theory matches lab experiments, the Cola group hopes to increase broad-spectrum efficiency to 40% (which compares to 20% efficiency for silicon solar cells). Other important benefits include: The optical rectenna works at high temperatures, and mass production should be inexpensive. The technology also can be tuned to different frequencies, so the rectenna can be used as a detector or in energy harvesting. The researchers are now focused on lowering contact resistance and growing the nanotubes on flexible substrates for applications that require bending.

Kippelen and coworkers at Georgia Tech are developing paper-based electronics — organic solar cells, organic light-emitting diodes (OLEDs), and organic field-effect transistors (OFETs) — fabricated on cellulose-based substrates that can be recycled easily. The use of paper for substrates has generated considerable buzz among researchers, but its high porosity and surface roughness pose challenges. Today's organic electronic components use very thin carbon-based semiconductor layers — about 1,000 times thinner than the average human hair. Because they are so thin, you need nearly atomically flat substrates where the surface is down to a nanometer. The team is using cellulose nanocrystals (CNCs), a type of wooden wunderkind material, to develop new semiconductor devices, demonstrating that CNCs are a viable alternative to traditional plastic substrates — while offering new environmental benefits. Devices made on these substrates can be easily dissolved in water, allowing semiconducting materials and metal layers to be filtered and recycled. Applications will depend on economics and performance. For CNC-based solar cells, the researchers have achieved power conversion efficiencies of 4%. Efficiencies could be increased to 10% but would require more expensive materials. So instead of paper-based solar farms becoming the norm, the group predicts low-power applications, such as computer covers and mousepads, for CNC-based solar cells. Cellulose-based OLEDs, which have performance comparable to current devices, show greater potential for market adoption. The trend in flat-panel displays is a larger size and higher resolution. Glass substrates, however, pose manufacturing and transportation problems because of their rigidity and breakability and plastic has problems at the end-of-product lifecycle. Yet with the low cost and flexibility of paper-based OLEDs, flat panel displays could be the size of a wall.

## 6.4  ADVANCES IN SOLAR CELL ARCHITECTURE

Novel solar cell architectures are also designed for specialized applications. Solar shingles, also called photovoltaic shingles, are solar panels designed to look like and function as conventional roofing materials, such as asphalt shingle or slate, while also producing electricity. There are several varieties of solar shingles, including shingle-sized solid panels that take the place of a number of conventional shingles in a strip, semi-rigid designs containing several silicon solar cells that are sized more like conventional shingles, and newer systems using various thin-film solar cell technologies that match conventional shingles both in size and flexibility. Solar shingles are manufactured by RGS Energy, CertainTeed, and SunTegra among others. A key advantage of solar shingle is that they eliminate the need for ribbon, connecting cells like roof tiles. Module aesthetics are improved, as the panels are homogeneously colored. Finally, cells for shingle modules have busbars at opposite ends and cells are sliced into several strips, which reduces the current and consequently the load on fingers. This also enables a reduction in the number of fingers as well as their thickness, which decreases shading and improves the output power of the cell [133].

Floating PV cells are attractive because there is no land occupancy, installation and decommissioning are easy with more compact space requirements and it saves water quantity and quality by restricting evaporation. The cooling of floating structure is simple and natural cooling can be increased by a water layer on the PV modules or by submerging them (the so-called SP2 (Submerged Photovoltaic Solar Panel)). In these cases, the global PV modules efficiency rises due to the absence of thermal drift, with a gain in energy harvesting up to 8%–10%. A large floating platform can be easily turned and can perform vertical tracking. Equipping a floating PV plant with a tracking system costs little extra while the energy gain can range from 15% to 25%. Floating PV also allows storage opportunities through coupling with hydroelectric basins or using compressed air energy storage. Floating solar is often installed on existing hydropower. It can control algae growth by limiting light for biological fouling. Many studies claim that solar panels over water are more efficient. The energy gain reported ranges from 5% to 15%. There are, however several challenges in installing, operating and maintaining floating PV cells. The Mooring (or anchoring) systems for floating PV need to be able to withstand wind and heavy waves, Operating on water over its entire service life, the system is required to have significantly increased corrosion resistance, particularly when installed over salt water. Operation and maintenance activities are as a general rule more difficult to perform on the water than on land [136].

Floating PV is a rapidly growing market. According to a World Bank report, as of the end of September 2018, the global cumulative installed capacity of floating PV plants was 1.1 GW [137]. Demand for floating PV is expanding, especially on islands (and other land-constrained countries), because the cost of the water surface is generally lower than the cost of land [73]. Floating solar is particularly well suited to Asia, where land is scarce but there are many hydroelectric dams with existing transmission infrastructure. Unsurprisingly, the world's top ten plants are located in Asia, namely China, Japan and the Republic of Korea. In particular, Korea has announced the completion next year of what will then become the world's largest floating solar plant, with a capacity of 102.5 MW. Other Asian countries, namely Singapore, Thailand, Viet Nam and India, are also actively pursuing floating solar project development. Europe has huge potential and demand for floating PV – especially in the Netherlands and France. Large-scale potential can also be found in opencast lignite coal pits. For instance, today the largest floating solar project is in China, with a capacity of 70 MW, and is located in a former coal-mining area of Anhui Province. Meanwhile, another plant in eastern China (in Panji District, Huainan City) has just been connected, becoming the world's largest floating solar plant, which will generate almost 78,000 (MWh) in its first year. Hydropower reservoirs and other artificial bodies of water also have enormous potential. With utility-scale power electronics and grid connection already established, coupling PV with hydropower offers significant added value [138]. For instance, the Norway-based independent power producer Statkraft announced the construction of a 2 MW floating PV plant in Albania. The company is using an innovative technology consisting of a membrane-type flotation device, 72 m in diameter, accommodating 500 kW of PV. Glass-glass modules are mounted onto special rails, in a way that the modules will be in permanent contact with a thermal membrane (designed to withstand stress and sun exposure) that sits on the water's surface. The water cools the membrane, which in turn cools the modules and enables them to produce more energy [100]. In addition, by covering the surface of the water reservoir, floating solar plants can also reduce evaporation and protect water quality from excessive algae growth [136,139]. A typical floating PV cell with a pumped hydro virtual battery concept is illustrated in Figure 6.12 [136].

Building-integrated photovoltaics (BIPV) are photovoltaic materials that are used to replace conventional building materials in parts of the building envelope such as the roof, skylights, or facades. They are increasingly being incorporated into the construction of new buildings as a principal or ancillary source of electrical power, although existing buildings may be retrofitted with similar technology. The advantage of integrated photovoltaics over more common non-integrated systems is that the initial cost can be offset by reducing the amount spent on building materials and labor that would normally be used to construct the part of the building that the BIPV modules replace. In addition, BIPV allows for more widespread solar adoption when the building's aesthetics matter and traditional rack-mounted solar panels would disrupt the intended look of the building. These advantages make BIPV one of

**FIGURE 6.12**   FPV with a pumped hydro "virtual battery" concept. (Copyright EDP S.A., photo by Pixbee) [136].

the fastest growing segments of the photovoltaic industry. The term building-applied photovoltaics (BAPV) is sometimes used to refer to photovoltaics that are a retrofit – integrated into the building after construction is complete. Most building-integrated installations are actually BAPV. Some manufacturers and builders differentiate new construction BIPV from BAPV. Typical building integrated PV panels are illustrated in Figures 6.13 and 6.14.

**FIGURE 6.13**   The CIS Tower in Manchester, England was clad in PV panels [104].

**FIGURE 6.14**   The headquarters of Apple Inc., in California. The roof is covered with solar panels [104].

There are four main types of BIPV products: (a) crystalline silicon solar panels for ground-based and rooftop power plant, (b) amorphous crystalline silicon thin film solar pv modules which could be hollow, light, red blue yellow, as glass curtain wall and transparent skylight, (c) CIGS-based (Copper Indium Gallium Selenide) thin film cells on flexible modules laminated to the building envelope element or the CIGS cells are mounted directly onto the building envelope substrate and (d) double glass solar panels with square cells inside. Building-integrated photovoltaic modules are available as flat roofs, pitched roofs, façade and photovoltaic windows (glazing) and photovoltaic stained glass [104]. The cell technologies most often used are plasmonic solar cells, perovskite solar cell technology and dye-sensitized solar cells.

BIPV solutions have several advantages. First, they are multifunctional as they can be adapted to a variety of surfaces (*e.g.* roofs, windows, walls) as an integrated solution, providing both passive and active functions. A key passive function is thermal and acoustic insulation, as with any other construction material, which is complemented by a unique active function – the PV component – which generates renewable electricity that can be directly used in the building. Other functions, also unique to BIPV systems, include the possibility of real-time thermal or lighting regulation [140]. Second, they provide a cost-efficient solution. They offer potential cost-reduction benefits related to the savings on roofing material, as well as potential efficiencies and time saving in labor/construction (i.e. with roofing and panel installation done simultaneously). Moreover, BIPV can reduce the cost of refurbishment and renovation of existing buildings and create a business case for efficient strategies. When compared to classic roofing materials, BIPV cladding is somewhat more expensive. However, when the additional revenue generated from the electricity produced is taken in to consideration, this higher cost is more than offset [140]. Finally, other advantages include versatility and design flexibility in size, shape and color. An EU-funded project, PVSITES, is currently developing a new generation of solar panels that can be part of traditional house elements like roofs, windows and glass façades.

Solar trees work very much like real ones, as they have leaf-like solar panels connected through metal branches using sunlight to make energy. Solar trees can be seen as complementary to rooftop solar systems. They are more ergonomic than solar panels, taking nearly 100 times less space to produce the same amount of electricity as a horizontal solar plant and, as such, constitute a solution for land- and space- scarce economies [67]. Solar trees are intended to bring visibility to solar technology and to enhance the landscape and architecture they complement, usually in a commercial or public context. Solar trees may build awareness and interest in solar technology and also provide shade and meeting places. There are numerous examples of solar trees some of which are graphically illustrated in Figures 6.15–6.17 [141].

Solar carports are ground-mounted solar panels that are installed so that parking lots and home driveways can be laid underneath to form a carport. They have been a very popular alternative or supplement to the classic rooftop systems, with the advantage that they can be installed entirely independently of the roof angle,

**FIGURE 6.15**   Ross Lovegrove Solar Tree on display [141].

**FIGURE 6.16** Spotlight Solar product "Lift" at net zero school in North Carolina, Sandy Grove Middle School [141].

**FIGURE 6.17** CSIR's solar power tree [141].

shape and orientation of the house. They offer a number of benefits. First, if coupled with a well-designed charging system, the electricity produced can be used for EV charging and thus reduce the costs of running the vehicle [142]. Second, they can provide energy storage enhancements by having battery storage integrated and available in the system, making the solution independent of sunshine hours. Third, unlike ground-level systems, they are easy to customize and can save space as they do not require an additional structure or land to install them on.

Solar carports are becoming more popular because they minimize energy expenses, resulting in significant savings. The revenue generated from electricity can reduce the costs of maintaining a parking lot. It also reduces the carbon footprint of parking space with simultaneous efficient use of space. The solar carport can also be adaptable and serve multiple functions such as a place for EV charging with simultaneous protection from elements. Solar carport also generates more power for the owner for a variety of other usages [142].

Agrivoltaics or agrophotovoltaics is the simultaneous use of areas of land for both solar photovoltaic power generation and agriculture. The coexistence of solar panels and crops implies a sharing of light between these two types of production, so the design of agrivoltaic facilities requires trading off such objectives as optimizing crop yield, crop quality, and energy production. Benefits include higher electricity production, higher crop yields and less water used [143]. APV is a win-win situation for both crops and solar panels. Many types of food crops, such as tomatoes, grow better in the shade of solar panels. A key advantage of solar panels is that their efficiency is increased. Cultivating crops underneath reduces the temperature of the panels, as they are cooled down by the fact that the crops below are emitting water through their natural process of transpiration [144]. In a dual-use APV technology, solar arrays are installed some five meters above the ground in order to allow the land to be accessed by farm machinery or a system where solar paneling is installed on the roofs of greenhouses. The shade produced by such a system can reduce production of some crops, but such losses may be offset by the energy produced. APV is commercialized in China and Japan. Dupraz et al. [145] calculated that the land use efficiency may increase by 60%–70% (mostly in terms of usage of solar irradiance). Photovoltaic arrays in general produce much less carbon dioxide and pollutant emissions than traditional forms of power generation for farm needs. Agrivoltaic systems have been pointed out to be good for shade-tolerant crop production. It has been postulated that agrivoltaics would be beneficial for summer crops due to the microclimate they create and the side effect of heat and water flow control. Wheat crops do not fare well in a low-light environment and are not compatible with agrivoltaics. There are three basic types of agrivoltaics that are being actively researched: solar arrays with space between for crops, stilted solar arrays above crops, and greenhouse solar arrays. The optimization requires (a) orientation of solar panels in the south for fixed or east-west panels for panels rotating on an axis, (b) spacing between solar panels for sufficient light transmission to ground crops and (c) elevation of the supporting structure of the solar panels to homogenize the amounts of radiation on the ground. Most conventional systems install fixed solar panels on agricultural greenhouses, above open fields crops or between open fields crops. It is possible to optimize the installation by modifying the density of solar panels or the inclination of the panels [146].

## 6.5  INDUSTRIAL PROGRESS ON SOLAR CELL EFFICIENCY AND POWER LEVELS

The solar PV industry is changing rapidly, with innovations occurring along the entire value chain. In recent years, a major driver for innovation has been the push for higher efficiency [147]. This is reflected by the expansion of passivated emitter and rear cell/contact (PERC) technology, which offers more efficient solar cells and as such increases the performance of solar panels. Increasing cell efficiency is key for competitive module manufacturing, as it directly decreases cell processing costs by reducing quantities required for a given output. Efficiency is also very important at the system level, with several factors explaining the push for higher- efficiency technologies. From the technical perspective, higher levels of efficiency reduce the number of modules that need to be transported to the installation site, the necessary land area and the length of wires and cables required. From a marketing perspective, companies able to offer the highest-efficiency modules are also generally perceived as having the highest level of technical expertise [147].

Solar panel **efficiency** is a measure of the amount of sunlight (irradiation) that falls on the surface of a solar panel and is converted into electricity. Due to the many advances in photovoltaic technology over recent years, the average panel conversion efficiency has increased from 15% to well over 20%. This large jump in efficiency resulted in the power rating of a standard size panel increasing from 250 W up to 370 W. Solar panel efficiency is determined by **two main factors;** the photovoltaic (PV) cell efficiency, based on the cell design and silicon type, and the total panel efficiency, based on the cell layout, configuration and panel size. **Cell efficiency** is determined by the cell structure and base silicon material used which is generally either P-type or N-type. Cell efficiency is calculated by what is known as the fill factor (FF), which is the maximum conversion efficiency of a PV cell at the optimum operating voltage and current. Svarc [148,149] puts out a yearly evaluation of industrial progress on solar panel efficiency and power. The following discussion is a summary of his recent reports. For more updates, the reader should follow his yearly evaluation of the progress made by the solar panel industry.

According to Svarc [148,149], cell design plays a significant role in panel efficiency. Key features include silicon type, multiple busbars (MBB), and passivation type (PERC). The high-cost IBC cells are currently the most efficient (20%–22%), due to the high purity N-type silicon cell base and no losses from busbar/finger shading. However, recent mono PERC cells with MBB and the latest heterojunction (HJT) cells have achieved efficiency levels well above 20%. Total Panel efficiency is measured under **standard test conditions** (STC), based on a cell temperature of 25°C, solar irradiance of 1,000 W/m$^2$ and Air Mass of 1.5. The efficiency (%) of a panel is calculated by the maximum power rating (W) at STC, divided by the total panel area in meters. Overall panel efficiency can be influenced by many factors including; temperature, irradiance level, cell type, and interconnection of the cells. Surprisingly, even the color of the protective back sheet can affect efficiency. A black back sheet might look more aesthetically pleasing, but it absorbs more heat resulting in higher cell temperature which increases resistance, this in turn slightly reduces total conversion efficiency [148,149].

**TABLE 6.3**
**Efficiency of Top Five Solar Panels [148,149]**

| Number | Make | Efficiency (%) |
|---|---|---|
| 1 | Sun power | 22.8 |
| 2. | Canadian solar | 22.5 |
| 3 | LG | 22.3 |
| 4 | Panasonic | 22.2 |
| 5 | Jinko solar | 22.2 |

Panels built using advanced 'Interdigitated back contact' or IBC cells are the most efficient, followed by heterojunction (HJT) cells, half-cut and multi-busbar monocrystalline PERC cells, shingled cells and finally 60-cell (4–5 busbar) mono cells. 60 cell poly or multi-crystalline panels are generally the least efficient and equally the lowest cost panels. In recent years, there has seen a surge in manufacturers releasing more efficient solar panels based on high purity N-type and heterojunction or HJT cells. For the first time, the efficiency of the at least top ten panels is now above 21%. The efficiency of the top five panels is illustrated in Table 6.3 [149]. As shown in this table, while Sun Power and Canadian Solar lead the pack, LG, Panasonic and Jinko are close behind. The more efficient panels using N-type cells also have a lower rate of light-induced degradation_or LID, which is as low as 0.3% power loss per year. These panels will generate close to 90% or more of the original rated capacity after 25 years, depending on the solar panel warranty details. All of these panels generate power between 355 and 410 W.

The increased efficiency has a significant commercial value since higher efficiency will pay back energy consumed and cost incurred in manufacturing solar panels in less time. Based on detailed lifecycle analysis, most silicon-based solar panels already repay the embodied energy within 2 years. However, as panel efficiency has increased beyond 20%, payback time has reduced to less than 1.5 years in many locations. Increased efficiency also means a solar system will generate more electricity over the average 20+ year life of a solar panel and repay the upfront cost sooner, meaning the return on investment (ROI) will be reduced even further [148,149].

Solar panel efficiency generally gives a good indication of performance, especially as many high-efficiency panels use higher grade N-type silicon cells with improved temperature coefficient and lower power degradation over time. Some manufacturers such as LG and SunPower even offer warranties with 90% or higher retained power output after 25 years of use. Higher efficiency panels generate more energy per square meter and thus require less overall area. This is perfect for rooftops with limited space and can also allow larger capacity systems to be fitted to any roof. In real-world use, solar panel operating efficiency is dependent on many external factors. Depending on the local environmental conditions these various factors can reduce panel efficiency and overall system performance. The main factors which affect solar panel efficiency are irradiance (W/m$^2$), shading, panel orientation, temperature, location (latitude), time of year and dust and dirt [148,149].

The factors which have the most significant impact on panel efficiency in real-world use are **irradiance, shading, orientation and temperature.** The level of

**solar irradiance**, measured in watts per square meter (W/m²), is influenced by atmospheric conditions such as clouds & smog, the latitude and time of year. Naturally, if a panel is fully shaded the power output will be very low, but partial shading can also have a big impact, not only on panel efficiency but total system efficiency. For example, slight shading over several cells on a single panel can reduce power output by 50% or more, which in turn can reduce the entire string power by a similar amount since most panels are connected in series and shading one panel affects the whole string. Therefore it is very important to try to reduce or eliminate shading if possible. Luckily there are **add-on devices** known as optimizers and micro-inverters which can reduce the negative effect of shading, especially when only a small number of panels are shaded.

The power rating of a solar panel, measured in Watts (W), is calculated under standard test conditions (STC) and measured at a cell temperature of 25°C. However, in real-world use, cell temperature generally rises well above 25°C, depending on the ambient air temperature, wind speed, time of day and amount of solar irradiance (solar energy - W/m²). During sunny weather, the internal cell temperature is often 20°C–30°C higher than the ambient air temperature, which equates to approximately 8%–12% reduction in total power output - depending on the type of solar cell and its temperature coefficient. *Conversely, extremely cold temperatures can result in an increase in power generation as the PV cell voltage increases at lower temperatures below STC. Solar panels can generate power above the nameplate power rating during very cold, sunny weather* [148,149].

According to Svarc [148,149], higher or lower cell temperature will either reduce or increase the power output by a specific amount for every degree above or below 25°C (STC). This is known as the power temperature coefficient which is measured in %/°C. Monocrystalline panels have an average temperature coefficient of −0.38%/°C, while polycrystalline panels are slightly higher at −0.40%/°C. Monocrystalline **IBC cells** have a much better (lower) temperature coefficient of around −0.30%/°C while the **best performing cells at high temperatures are HJT (heterojunction) cells** which are as low as −0.25%/°C. Power temperature coefficient is measured in %/°C - **Lower is more efficient Some of them are illustrated as [148,149]**:

1. Polycrystalline cells −0.39–0.43%/°C
2. Monocrystalline cells −0.35–0.40%/°C
3. Monocrystalline IBC cells −0.28–0.31%/°C
4. Monocrystalline HJT cells −0.25–0.27%/°C

Generally, cell temperature is 20°C–30°C higher than the ambient air temperature which equates to approximately 8%–12% reduction in power output. Note that cell temperature can rise as high as 80°C when mounted on a dark-colored rooftop during very hot 40+°C, windless days.

The most efficient solar panels on the market generally use either N-type (IBC) monocrystalline silicon cells or the another highly efficient N-type variation, heterojunction (HJT) cells. Most other manufacturers currently use the more common P-type mono-PERC cells; however, several large volume manufacturers, including JinkoSolar, Longi Solar and Trina, are now starting to shift to the more efficient N-type cells.

The efficiency of various types of PV cells are [149]:

1. Polycrystalline −15%–18%
2. Monocrystalline −16.5%–19%
3. Polycrystalline PERC −17%–19.5%
4. Monocrystalline PERC −17.5%–20%
5. Monocrystalline N-type −19%–20.5%
6. Monocrystalline N-type HJT −19%–21.7%
7. Monocrystalline N-type IBC −20%–22.6%

All manufacturers produce a range of panels with different efficiency ratings depending on the silicon type used and whether they incorporate PERC, multi busbar or other cell technologies. Very efficient panels above 21% featuring N-type cells are generally much more expensive, so if cost is a major limitation it would be better suited to locations with limited mounting space, otherwise, you can pay a premium for the same power capacity which could be achieved by using 1 or 2 additional panels. However, high-efficiency panels using N-type cells will almost always outperform and outlast panels using P-type cells due to the lower rate of light-induced degradation or LID, so the extra cost is usually worth it in the long term.

Panel efficiency is calculated by the power rating divided by the total panel area, so just having a larger size panel does not always equate to higher efficiency. However, larger panels using larger size cells increase the cell surface area which does boost overall efficiency. Most common residential panels still use the standard 6" (156 mm) square 60-cell panels while commercial systems use the larger format 72 cell panels. However, a new industry trend emerged in 2020 towards much larger panel sizes built around new larger size cells which increased panel efficiency and boosted power output up to an impressive 600 W. Common Solar panel sizes are [148,149]:

1. 60 cell panel (120 HC): Approx width 0.98 m × length 1.65 m
2. 72 cell panel (144 HC): Approx width 1.0 m × length 2.0 m
3. 96/104 cell panel: Approx width 1.05 m × length 1.60 m
4. 66 cell panel (132 HC) - Approx width 1.10 m × length 1.80 m - **New**
5. 78 cell panel (156 HC): Approx width 1.30 m × length 2.3 m - **New**

HC = Half-cut cells
A **standard size** 60-cell (1 m × 1.65 m) panel with 18%–20% efficiency typically has a power rating of 300–330 W, whereas a panel using higher efficiency cells, of the same size, can produce up to 370 W. As previously explained, the most efficient standard-size panels use high-performance N-type **IBC** or Interdigitated Back Contact cells which can achieve up to 22.6% panel efficiency and generate an impressive 380–400 W. Popular half-cut or split cell modules have double the number of cells with roughly the same panel size. A panel with 60 cells in a half-cell format is doubled to 120 cells, and 72 cells in a half-cell format have 144 cells. The half-cut cell configuration is **slightly more efficient** as the panel voltage is the same but the current is split between the two halves. Due to the lower current, half-cut panels have

lower resistive losses resulting in increased efficiency and a lower temperature co-efficient which also helps boost operating efficiency [148,149].

To decrease manufacturing costs, gain efficiency and powerful solar panels, manufacturers have started moving away from the standard 156 mm (6″) square cell wafer size in favor of larger wafer sizes. While there are a variety of various cell sizes under development with the most popular being 166, 182 and 210 mm. The larger cells combined with new larger panel formats have enabled manufacturers to develop **extremely powerful solar panels with ratings up to 670 W**. Larger cell sizes have a greater surface area and when combined with the latest cell technologies such as multi-busbar (MBB), PERC and tiling ribbon, can boost panel efficiency up to 22% [148,149].

In the solar world, panel efficiency has traditionally been the one factor most manufacturers strived to lead. In 2020 a new battle emerged to develop the world's most powerful solar panel and over the last 12 months many of the industry's biggest players announced next-generation panels with power ratings well above 500 W. However, the race for the most powerful panel really heated up last July when Trina Solar revealed a panel which will deliver an impressive 600 W. Then in August, at the SNEC PV Power Expo in China, JinkoSolar unveiled a 610 W version of their current Tiger Pro panel while Trina solar proposed a 660 W+ panel is on the horizon. Amazingly, there were close to 20 manufacturers at SNEC 2020 showcasing panels rated over 600W with the most powerful panel being the Jumbo 800 W panel from JA solar. However, this panel was incredibly large at 2.2 m high and 1.75 m wide. Panels rated from 400 to 450 W are now common which is remarkable considering only in 2019, the most powerful panes available were only just reaching 360 W [148,149].

### 6.5.1 DESIGN FOR UTILITY-SCALE SYSTEMS

The main driver for the development of larger, more powerful solar panels stems from the desire to decrease the costs of utility-scale solar farms and ultimately reduce electricity prices. Larger panels require a similar amount of connections and labor as smaller panels, thus the installation cost per kW is reduced resulting in lower overall cost or reduced LCOE. As explained below, the new high-powered panels are physically much larger in size compared to common panels found on residential rooftops. Although, those hoping to get a dozen 600 W panels on your home rooftop to get an easy 7 kW array will be a little disappointed. At this stage, most high-powered panels will only be available for commercial and utility-scale systems, plus the extra-large size is not well suited to most residential rooftops [148,149]. While the solar industry as a whole is slowly shifting to larger, higher wattage panels, the front runners in the race are Risen, Trina Solar, Astroenergy, AE Solar, and Canadian Solar (see Table 6.4).

These well-known companies all launched ultra high-power panels with ratings well above 600 W over the last year. Additionally, premium manufacturer SunPower (now Maxeon) announced 540 W panels in the next-generation 'Performance 5' series. In the past, most increases in panel power came from efficiency gains due to advances in solar cell technology. While that is partly a driver behind the massive

**TABLE 6.4**
**Most Powerful Solar Panels [148,149]**

| Make | Power (W) | Cell size (mm) | Efficiency (%) |
|------|-----------|----------------|----------------|
| Risen | 700 | 210 | 22.5 |
| Trina solar | 670 | 210 | 21.6 |
| Astroenergy | 670 | 210 | 21.6 |
| AE solar | 665 | 210 | 21.4 |
| Canadian solar | 665 | 210 | 21.4 |

**TABLE 6.5**
**New Technologies in Panels and Cells [148,149]**

1. MBB - Multi-busbars
2. PERC/PERC+ - Passivated emitter & rear cell
3. TOPCon - Tunnel-Oxide Passivating Contact
4. N-type silicon cells
5. High-density cells - Reducing inter-cell gaps

jump in panel wattage, the main factor is the new larger cell sizes being developed together with a higher number of cells per panel. These new cell formats and configurations mean the new panels are physically much larger in size. Generally, these large panels are best suited for utility-scale solar farms or large commercial installations. Traditionally, solar panels were available in two main sizes - the standard format 60 cell panels used for residential rooftops, and the larger format 72 cell commercial size panels. Besides the standard sizes, there were a few premium manufacturers such as SunPower and Panasonic producing unique 96 and 104 cell panels. The industry-standard panel size for much of the last decade was built around the 156 mm × 156 mm or 6-inch square cell format. While there are a variety of cell sizes under development, a few cell sizes have emerged as the new industry standard; these include 166, 182 and 210 mm.

Along with the different cell sizes, there is a myriad of new panel configurations built around the many cell combinations. According to Svarc [148,149] the three most popular which have emerged are 66-cell (half-cut 132), 78-cell (half-cut 156), and 84-cell (half-cut 168) panels. The extra-large 210 mm cells are also well suited to unique cell dividing formats such as 1/3 cut cells; where the square wafer is divided into three segments rather than the common half-cut or half-size cell. In order to achieve these impressive power ratings, the panels and cells have not just increased in size but cell efficiency has improved substantially using numerous new technologies (listed in Table 6.5) along with advanced rear side passivation techniques like TOPCon [148,149]. Research has found that while the use of N-type silicon is one of the simplest ways to boost efficiency but also one of the more costly methods. However, the price gap between P-type and N-type silicon is reducing as the economy of scale lower the cost of manufacturing the high-performance N-type silicon wafers.

## 6.5.2   MBB - MULTI-BUSBARS

Silicon solar cells are metallized with thin strips printed on the front and rear of a solar cell; these are called busbars and have the purpose of conducting the electric direct current (DC) power generated by the cell. Older solar cells typically had two busbars; however, the industry has moved towards higher efficiencies and busbars have increased to three (or more) in most solar cells. The increased number of busbars has several advantages: first is the high potential for cost saving due to a reduction in metal consumption for front-facing metallization [150]; second, series resistance losses are reduced by employing thin wires instead of regular ribbon [133]; and third, optimizing the width of the busbars leads to an additional rise in efficiency. A higher number of busbars leads to higher module efficiencies because of reduced internal resistance losses; this is due to the lower distance between the busbars. Finally, multi-busbar design is highly beneficial for bifacial technology, especially for improving the bifaciality for PERC cells of 90% [133].

According to Svarc [148,149], of the many cell improvements, the most common technology used to increase efficiency has been multi-busbars (MBB). Traditional ribbon busbars (5BB or 6BB) are being rapidly phased out in favor of nine or more thin wire busbars (9BB). Some manufacturers such as REC have even moved to 16 micro-wire busbars in the new Alpha panel series. Wider cells also mean more busbars can fit across the cell surface with 10 or 12 busbars cells also becoming more common. Bifacial panels featuring MBB are also growing in popularity due to the increased power output by utilizing the rear side of the panel to achieve up to 20% or more power (roughly 80 W extra). However, bifacial panels are generally only beneficial over light colored surfaces such as light sandy or rocky ground used in large MW scale solar farms located in more arid areas. To further boost panel efficiency and increase power, manufacturers such as Trina Solar have introduced techniques to eliminate the vertical inter-cell gap between cells. Removing the typical 2–3 mm vertical gaps and squeezing the cells together results in more panel surface area available to absorb sunlight and generate power. Manufacturers have developed a number of techniques to minimize or eliminate the gap with the most common being to simply reduce the cell spacing from around 2.0 to 0.5 mm. The reason for this gap was due to traditional larger ribbon busbars requiring 2.0 mm+ to bend and interconnect the front and rear of each cell. However, the transition to using much smaller wire busbars enabled the gap to be reduced significantly.

According to Svarc [148,149], LONGi Solar is another manufacturer that managed to reduce the inter-cell gap down to 0.6 mm by using what the company describes as a "smart soldering" method using integrated segmented ribbons. This new technology uses a unique triangular busbar design across the front surface of the cell, with a very thin flattened section that bends and runs behind the cell to form the interconnection. Jinko Solar, currently the world's largest panel manufacturer, developed what the company refers to as Tiling Ribbon or TR cells. Tiling Ribbon cell technology is the elimination of the inter-cell gap by slightly overlapping the cells creating more cell surface area. This in turn boosts panel efficiency and power output. The tiling ribbon technology also dramatically reduces the amount of solder required by using inter-cell compression joining methods rather than soldering. Shingled cell panels, such as those used in the Sunpower Performance series, use a similar technology where overlapping thin cell strips can be configured into larger format high-power panels.

Other manufacturers are taking a similar approach to boost efficiency by reducing the inter-cell gap as much as possible but still leaving a very small gap of around 0.5 mm or less. This effectively removes the gap without having to develop new cell interconnection techniques. Cells built on an N-type silicon base offer improved performance over the more common P-type silicon due to a greater tolerance to impurities which increases overall efficiency. In addition, N-type cells have better temperature tolerance compared to both mono and multi P-type cells. More importantly, N-type cells have a much lower rate of LID (light-induced degradation) and do not generally suffer from LeTID (light and elevated temperature induced degradation), unlike many P-type cells [148,149].

## 6.6 INNOVATIONS IN SOLAR CELL POWER MONITORING AND MAINTENANCE

An operation and management (O&M) system is a key component of a solar plant, as it ensures that the PV system will be able to maintain high levels of technical and economic performance over its lifetime [119]. In addition, the O&M phase is the longest in the lifecycle of a PV project, as it typically lasts 20–35 years. As such, ensuring the quality of O&M services is essential to mitigate potential risks. Innovations and improvements, including more data-driven solutions, are becoming increasingly important because they help O&M services to keep up with market requirements. Important trends in O&M innovation can be grouped into three main categories: (a) smart PV power plant monitoring; (b) retrofit coatings for PV modules and related maintenance measures and (c) issues related to large-scale PV plant integration to the grid.

### 6.6.1 SMART PV POWER PLANT MONITORING

Electricity generation from PV plants is limited by the variable nature of the sun's radiation. The growing penetration of PV into electricity markets creates the need for new regulations to guarantee grid stability and the correct balancing of electricity demand and supply [151]. The ability to predict PV production is therefore an essential tool to capture economies in a market with a high penetration of non-predictable energy. Current simulation models allow the prediction of energy production on an hourly basis for at least the next 48 hours. Improvements in communication procedures between devices (i.e. modules, inverters, sensors, etc.) would contribute to improving intraday forecasting, calculation of performance expectations, and exchange with the energy grid [151]. Frequent failings in communication between devices and the cloud or data center infrastructure are, however, a challenge. In order to overcome this challenge, the Internet of things (IoT) represents a valid solution for PV systems, since it provides an interoperability environment where all devices in the field are connected to each other and spontaneously show themselves as available to be connected to the system [151]. The efforts are also being made throughout the whole PV market to increase standardization of communication, which will improve the security level, options for communication, and configuration costs for solar monitoring [151].

The process of monitoring is also now replaced by intelligent systems, such as drones. Drones are becoming highly suited to the solar industry due to a wide range of surveillance and monitoring capabilities, the possibility of long-range inspection and easy control and capability to monitor large-scale solar parks in a shorter time. With the help of sensing elements, drones efficiently capture the necessary data and send them to the cloud for analysis in less time and in more accurate form [152]. The innovative monitoring systems identify the root causes of performance problems that lead to plant underperformance and unavailability. These innovations include advances in single plant and system portfolio monitoring and management, such as: (a) automated maintenance (preventive and emergency), intervention management and (re)scheduling, based on parameters such as alarms and performance data; and (b) algorithms for equipment or plant behavior and reliability predictions based on historical failure data and simulation models [153].

### 6.6.2 Maintenance Measures

One of the major maintenance issues with solar park is to keep solar PV modules cool because their performance and lifetime are reduced by the heat of the sun. In fact, the chemical reactions that cause degradation of solar modules double for every 10°C increase above ambient temperature (around 25°C) causing their lifetime, as well as their voltage output, to shrink [154]. While the use of water can be extremely effective in maintaining the equilibrium of solar panels, incorporating water-based systems into module manufacturing or installation adds cost and complexity [154]. Other approaches include applying a transparent coating of patterned silica to solar cells to capture and radiate heat from infrared rays back to the atmosphere. This was found to improve absolute cell efficiency by more than 1%. Greater improvements in efficiency can also be obtained through processes including infrared reflation and radiative transfers. Radiative transfer to the sky or to cooler areas around the panels holds the greatest promise for boosting solar module efficiency [154].

Soiling of PV lens requires regular module washing. In this context, several anti-soiling solutions are being implemented. First, robotic panel cleaning technology consists of robots moving along the array of panels; but this process is expensive. This is, however, very effective for areas with high soiling rates, insufficient water supplies and high labor costs. They are mostly deployed in the Middle East and North Africa, as they are best suited for frequent washing schedules, such as those that require washing on a weekly basis [155]. Second, sprinkler systems consist of a water filtration system and a soap dispensing system, mainly used in very dry areas to keep the panels clean with the same cleaning effect as rain. This system has two main disadvantages: (a) the large amounts of water used; and (b) the need to constantly monitor filters and soap levels, which is costly and time-consuming. Finally, anti-soiling coatings are used to treat modules so that they get dirty less quickly while making them easier to clean, and thus maintain higher performance levels for a longer [151] period of time. Currently, various anti-soiling coatings are already commercially available and have been installed, and in parallel new solutions are also being developed, mostly based on spray technologies. More discussion on anti-soiling coatings is given in Chapter 5. Ground-mounted PV plants may suffer from shading if vegetation inside and outside the plant is not properly controlled. The main innovations in this area seek to reduce the maintenance required without using pesticides or other dangerous products. These include using weed control fabrics inside the plant, under the modules and around the perimeter to help limit weed growth. These fabrics combine soil erosion control and weed control into a single product, thereby minimizing maintenance needs for green areas. Also, selecting seeds of plants with slow growth and limited height avoids the need for frequent maintenance [153].

### 6.6.3 Issues Related to Large-Scale PV Power Integration in the Grid

Large-scale PV power integration into the grid has resulted in significant growth in solar inverters demand. This growth, largely in the US, China, India, Japan, Canada and Chile will continue to grow. Germany, France and Italy were the top markets in Europe, and the Gulf Cooperation Council (GCC) countries such as Kuwait, Qatar, Saudi Arabia and the United Arab Emirates were the major markets in the Middle East [156]. China, the US and India collectively accounted for approximately 70% of global PV inverter installations in 2018. In all three countries, solar PV and inverter installations have been driven by government support policies and schemes such as net metering and feed-in-tariffs [2,3,5–7].

Solar PV systems can be categorized into three main categories according to their size: large, medium, and small. The large- or utility-scale PV power plants (LS-PVPP) are those typically generating above 100 $MW_p$, so they can be connected to high or medium voltage. This category of PV energy is typically delivered in three phases and requires some transformers parallel to the power system. The second category, which is the medium-scale PV, generates from 1 to 100 $MW_p$ and is connected to the distribution level [157]. This category can particularly be found in all sizes of buildings such as government sites, malls, and residential areas. For residential areas, small-scale PV systems are used with a generation capacity that ranges from 1 up to 1,000 $kW_p$, where it is connected to low voltage. Due to the integration of solar PV systems, voltage increase, reverse power flow, and some other issues may appear in the grid.

Some other challenges have been encountered due to the intense integration of LS-PVPPs into the power system. The variability of solar radiation is a significant issue that may lead to instability in the point-of-common-coupling (PCC), concerning the voltage and frequency responses. The control of LS-PVPP in power curtailment (PC) mode is becoming challenging due to the intermittent energy nature of the source. Reduced inertia is another challenging issue due to the absence of rotational machinery in LS-PVPPs, which is used in traditional power plants, thereby introducing instability in the grid. In order to ensure the reliability of PV system penetration in LS-PVPPs, the grid codes (GCs) must be analyzed as new GCs are needed from LS-PVPPs for better configurations as well as to improve the control stability.

Widen et al. [158], Gallo et al. [159] and Marinopoulos et al. [160] discussed some issues related to the high level of integration of PV systems into the electrical power system, such as voltage drop and network losses. From the control and implementation points of view, the LS-PVPPs were studied by Morjaria et al. [161]. Mansouri et al. [156] provided a detailed review of several other aspects of issues related to large-scale solar PV integration into the grid. These include (a) frequency stability,

(b) angle stability, (c) voltage stability, (d) active power curtailment, (e) reactive power control, (f) fault ride through, (g) maximum power point tracking technique, and (h) impacts on photovoltaic cells and panels. The review provided different strategies used by researchers to handle these issues.

## 6.7  ADVANCES IN CPV TECHNOLOGY

Unlike conventional photovoltaic systems, CPV (concentrated photovoltaics) uses lenses or curved mirrors to focus sunlight onto small, highly efficient, multi-junction (MJ) solar cells. In addition, CPV systems often use solar trackers and sometimes a cooling system to further increase their efficiency. Systems using high-concentration photovoltaics (HCPV) possess the highest efficiency of all existing PV technologies, achieving near 40% for production modules and 30% for systems. They enable a smaller photovoltaic array that has the potential to reduce land use, waste heat and material, and balance system costs. The rate of annual CPV installations peaked in 2012 and has fallen to near zero since 2018 with the faster price drop in crystalline silicon photovoltaics. In 2016, cumulative CPV installations reached 350 MW, less than 0.2% of the global installed capacity of 230,000 MW that year.

HCPV directly competes with concentrated solar power (CSP) as both technologies are suited best for areas with high direct normal irradiance, which are also known as the Sun Belt region in the United States and the Golden Banana in Southern Europe. As of 2012, CSP was more common than CPV. Figure 6.18 illustrates an example of CPV technology [162]. The key principle of CPV is the use

**FIGURE 6.18**  Concentrator photovoltaics (CPV) modules on dual axis solar trackers in Golmud, China [162].

of cost-efficient concentrating optics that dramatically reduce the cell area, allowing for the use of more expensive, high-efficiency cells and potentially a levelized cost of electricity (LCOE) competitive with Concentrated Solar Power and standard flat-plate PV technology in certain sunny areas with high Direct Normal Irradiance (DNI) [163]. While CPV technology has had difficulty penetrating the commercial market, its continued increased efficiency and resulting reduction in system cost have made it more competitive in recent years [162,164,165].

CPV is of most interest for power generation in sun-rich regions with Direct Normal Irradiance (DNI) values of more than 2,000 kWh/(m²a). CPV systems are differentiated according to the concentration factor of the technology configuration (see Table 6.6). More than 90% of the capacity publicly documented installed through the end of July 2015 was in the form of high concentration PV (HCPV) with two-axis tracking because of its economic advantage. Concentrating the sunlight by a factor of between 300× and 1,000× onto a small cell area enables the use of highly efficient but comparatively expensive multi-junction solar cells based on III–V semiconductors (e.g. triple junction solar cells made of GaInP/GaInAs/Ge). Low concentration designs – those with concentration ratios below 100× – are also being deployed. These systems primarily use crystalline silicon (c-Si) solar cells and single-axis tracking, although dual-axis tracking can also be used. While different types of solar collectors/concentrators can be used in CPV, as shown in Figure 6.19,

## TABLE 6.6
## Classification of CPV [162,164]

| Class of CPV | Typical Concentration Ratio | Tracking | Type of Converter |
|---|---|---|---|
| High Concentration PV (HCPV) | 300–1,000 | Two-axis | III-V multi-junction solar cells |
| Low Concentration PV (LCPV) | <100 | One or two-axis | c-Si or other cells |

FIGURE 6.19 Left and middle: Example of a CPV system using Fresnel lenses to concentrate the sunlight: FLATCON® concept originally developed at Fraunhofer ISE. Right: Example of a mirror-based system developed by the University of Arizona, USA [164].

Fresnel lens and mirrors as concentrating optics are very popular [164]. An anti-soiling layer consisting of a $WO_3$ photocatalyst coated on a polymethylmethacry-late (PMMA) substrate is a primary material used for the manufacturing of Fresnel lenses for CPV modules. The performance of PV modules can be increased due to the two main properties of the coating film: anti-reflectiveness and SC effect.

A key reason for the increasing number of large-scale power plants using HCPV is the significant increase in the efficiency of individual modules, which also leads to a reduction of area-related system costs. Soitec recently demonstrated a CPV module efficiency of 38.9% at Concentrator Standard Test Conditions (CSTC) [166] and efficiencies of commercially available CPV modules that exceed 30%. In addition to these higher efficiencies, tracking allows CPV systems to produce a larger amount of energy throughout the day in sunny regions, notably during the late part of the day when electricity demand peaks. At the same time and in contrast to CSP, the size of the installations can be scaled over a wide range, i.e from kW to multi MW, and in this way adapted to the local demands. Some CPV systems also disturb a smaller land area, since the trackers, with relatively narrow pedestals, are not closely packed. This makes it possible to continue to use the land for agriculture. Finally, HCPV is advantageous in hot climates in particular, since the output of the solar cells used does not decline as severely at high temperatures as that of conventional c-Si solar modules [164].

Industry reports indicate that the total capital equipment (capex) requirement, while varying by design and manufacturing process, can be lower for CPV than for traditional flat plate technologies, including c-Si. Additionally, a bottom-up analysis from NREL based on a specific HCPV system with a Fresnel lens primary optic and refractive secondary lens estimates the total capex for cells and modules in this design (assuming a vertically integrated company) to be around $0.55/$W_p$(DC), with a much lower capex for variations on the design. Most HCPV companies have their optics and cells manufactured by a third party, in which case the capital equipment requirements for the HCPV company itself can be quite low. Table 6.7 summarizes the strengths and weaknesses of CPV [162,164].

In order to produce equal or greater energy per rated watt than conventional PV systems, CPV systems must be located in areas that receive plentiful direct sunlight. This is typically specified as average DNI (Direct Normal Irradiance) greater than 5.5–6 $kWh/m^2/day$ or 2,000 $kWh/m^2/yr$. Otherwise, evaluations of annualized DNI vs. GNI/GHI (Global Normal Irradiance and Global Horizontal Irradiance) irradi-ance data have concluded that conventional PV should still perform better over time than presently available CPV technology in most regions of the world [162]. The main challenge cited by the industry is the difficulty of CPV to compete with flat-plate cSi PV modules on cost, in light of the recent drop in c-Si module prices due to the significant advantage in manufacturing scale that flat-plate currently has over CPV. CPV companies expect that this technology can compete on an LCOE basis with flat-plate PV when installed in sunny areas, but the road to scale has been difficult.

While a breadth of designs in the CPV space exist, the majority of companies are HCPV and most of those employ Fresnel primary lenses in refractive, point-focus systems. Some companies have moved towards smaller cells and higher concentra-tions in hopes of reducing costs and thermal management requirements. Almost all

## TABLE 6.7
## Strength and Weakness of CPV Technology [162,164]

| CPV Strengths | CPV Weaknesses |
| --- | --- |
| High efficiencies for direct-normal irradiance | HCPV cannot utilize diffuse radiation LCPV can only utilize a fraction of diffuse radiation |
| Low-temperature coefficients | Tracking with sufficient accuracy and reliability is required |
| No cooling water required for passively cooled systems (as is required for CSP) | May require frequent cleaning to mitigate soiling losses, depending on the site |
| Additional use of waste heat possible for systems with active cooling possible (e.g. large mirror systems) | Limited market – can only be used in regions with high DNI, cannot be easily installed on rooftops |
| Modular – kW to GW scale | Strong cost decrease of competing technologies for electricity production |
| Increased and stable energy production throughout the day due to tracking | Bankability and perception issues due to shorter track record compared to PV |
| Very low energy payback time | New generation technologies, without a history of production (thus increased risk) |
| Potential double use of land, e.g. for agriculture. Low environmental impact[1] | Additional optical losses |
| Opportunities for cost-effective local manufacturing of certain steps | Lack of technology standardization |
| Less sensitive to variations in semiconductor prices | |
| Greater potential for efficiency increase in the future compared to single-junction flat plate systems could lead to greater improvements in land area use, system, BOS and BOP costs | |

HCPV companies now operate near 500× or 1,000×. In LCPV, both the designs and concentration ratios tend to be much more varied than in HCPV, with groups even targeting building-integrated CPV (BICPV) and modules floating on the water. Trackers have also made great strides in recent years, being both more reliable and lower cost than in the past. Despite this convergence within HCPV onto similar module designs, and the recent availability of some standard components, firms continue to use their own custom components. Industry has concerns about stable growth for the future.

According to Philipps et al. [164], the literature analysis shows that CPV has the potential for reducing the LCOE, which encourages the continued development of this technology. If installations continue to grow through 2030, CPV could reach a cost ranging between €0.045 and €0.075/kWh (Figure 6.20) [164]. The system prices, including installation for CPV power plants would then be between €700 and €1,100/kWp. High efficiency is one of the key drivers to make HCPV more cost-competitive on the LCOE level. Philipps et al. [164] point out that the efficiency of III–V multi-junction solar cells is the key driver to lower the LCOE of energy produced by HCPV technology. Solar cells made by Sharp [167] and Fraunhofer

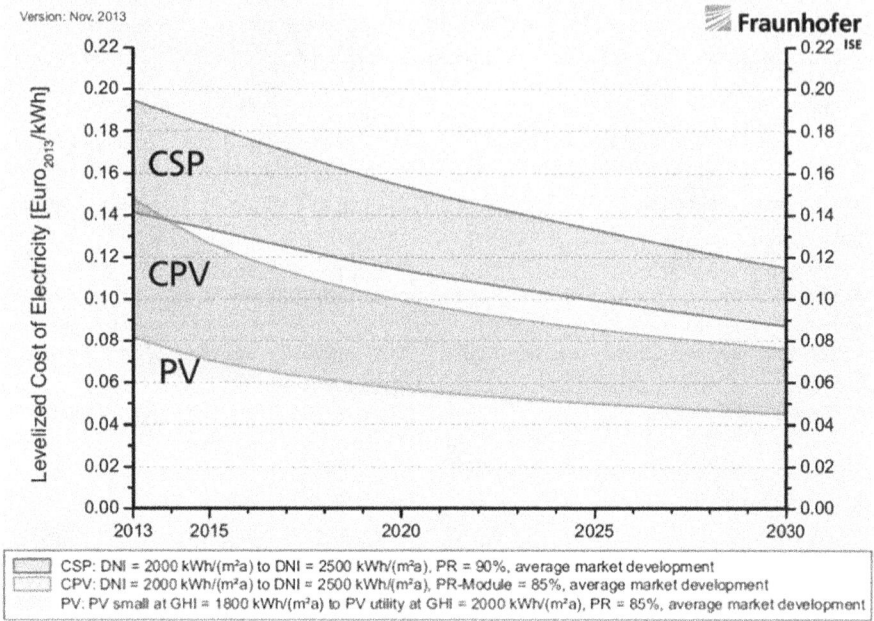

CSP: DNI = 2000 kWh/(m²a) to DNI = 2500 kWh/(m²a), PR = 90%, average market development
CPV: DNI = 2000 kWh/(m²a) to DNI = 2500 kWh/(m²a), PR-Module = 85%, average market development
PV: PV small at GHI = 1800 kWh/(m²a) to PV utility at GHI = 2000 kWh/(m²a), PR = 85%, average market development

**FIGURE 6.20** Development of the LCOE of PV, CSP and CPV plants at locations with high solar irradiation of 2,000 kWh/(m²a) -2,500 kWh/(m²a) [164].

ISE [166] achieved today's champion efficiencies of 44.4% and 46.0% for triple- and four-junction solar cells, respectively. The most common III–V multi-junction solar cell in space and terrestrial concentrator systems is a lattice-matched $Ga_{0.50}In_{0.50}$ $P/Ga_{0.99}In_{0.01}As/Ge$ triple-junction solar cell. Table 6.8 presents cell architectures that have achieved record cell efficiencies above 41%. These use different elements from the wide range of technology building blocks available for III–V multi-junction solar cells. More detailed analyses on this subject are provided by Philipps and Bett [169], Friedman et al. [170] and Luque [171]. Philipps et al. [164] also carried out the materials needed for II–V multi-junction cells. The data presented in the literature indicate that Gallium (Ga), indium (In), and germanium (Ge) are usually employed in the designs for III–V multi-junction cells employed in CPV, and these have limited global supplies. Based on the analysis the final conclusion of Philipps et al. [131] was that material availability of Ga, In, and Ge for III–V multi-junction cells would be a much more significant challenge if the cells are used for low concentration or one sun applications.

The choice of the best PV cell has a significant impact on the performance of a PVT and a CPVT system. Four generations of PV cells are being designed and developed and first two generations are commercially available. Da Silva and Fernandes [172] noted that out of monocrystalline, polycrystalline and amorphous silicon PV cells, the monocrystalline cells gave the best efficiency. Literature indicates that high-efficiency PV cells are not always the best choice [173]. Several important factors such as cell cost, heat extraction method and concentration ratio

## TABLE 6.8
## Summary of Record Concentrator Cell Efficiencies above 41% Based on III–V Multi-junction Solar Cells [162,164]

| Cell Architecture | Record Efficiency (Accredited test lab) | Institution | Comments |
|---|---|---|---|
| GaInP/GaAs//GaInAsP/ GaInAs | 46.0 @ 508 suns (AIST) | Fraunhofer ISE/Soitec/CEA | 4J, wafer bonding, lattice matched grown on GaAs and InP |
| GaInP/GaAs/GaInAs/ GaInAs | 45.7% @ 234 suns (NREL) | NREL | 4J, inverted metamorphic |
| GaInP/GaAs/GaInAs | 44.4 @ 302 suns (Fraunhofer ISE) | Sharp | 3J, inverted metamorphic |
| GaInP/GaAs/GaInNAs | 44.0% @ 942 suns (NREL) | Solar Junction | 3J, MBE, lattice matched, dilute nitrides, grown on GaAs |
| GaInP/Ga(In)As/ GaInAs | 42.6% @ 327 suns (NREL) (40.9% @ 1,093 suns) | NREL | 3J, inverted metamorphic |
| | 42.4% @ 325 Suns (NREL) (41% @ 1,000 suns) | Emcore | |
| GaInP-GaAs-wafer- GaInAs | 42.3% @ 406 suns (NREL) | Spire | 3J, epi growth lattice matched on front and inverted metamorphic on back of GaAs wafer |
| GaInP-Ga(In)As-Ge | 41.6% @ 364 suns (NREL) | Spectrolab | 3J, lattice matched, commercially available; |
| GaInP-GaInAs-Ge | 41.1% @ 454 suns (Fraunhofer ISE) | Fraunhofer ISE | 3J, upright metamorphic; commercially available from AZUR SPACE, Spectrolab |

*Note that LCPV systems mainly use c-Si solar cells.*

have significant impacts on the option of PV cells. Multijunction or non-silicon PV cells are appropriate for HCPVT systems, on the other hand, for LCPVT, common crystalline PVs are better because of their lower costs. Due to the Shockley-Queisser balance limit, the maximum efficiency of a p-n junction PV cell with 1.1 eV band gap is 33% [125] and could reach 40.7% by considering the maximum concentration ratio [174]. The material selected should have high thermal conductivity, high electrical insulation, high resistivity to thermal shocks and low cost. Rosell et al. [175] investigated the effect of adhesive on the thermal efficiency of CPVT systems and the results showed that 13.5% enhancement in the thermal efficiency of the collector is achieved by increasing 100% thermal conductivity of the adhesive. Dupeyrat et al. [176] presented a technique in which the front glazing, adhesive material, PV cells and absorber are laminated in one step and this resulted in about 600% enhancement in thermal conductivity.

### 6.7.1 SOLAR THERMAL COLLECTORS FOR PVT AND CPVT

A significant theoretical and experimental research on PV cell and flat plate PVT collector has been carried out. Based on this research, the design of two prototype and workable PVT collectors are present in Figure 6.21. For the first collector (PVT-A), the PVT absorber was manufactured using an improved technique for the lamination of single-crystalline silicon solar cells on the top of an optimized flat heat exchanger [176]. This PVT absorber was then encased in the frame of a solar collector. For the second collector (PVT-B), the PVT absorber was assembled by connecting square copper channels mechanically and thermally on the back of a conventional thin film module (CdTe). CdTe modules have a lower efficiency than sc-Si modules, but their temperature dependency is much lower as well. Flat plate collectors are also reviewed in several publications [36,37,177]. In a flat plate collector, the glass cover is used for minimizing the convection heat losses and transmittance of long wave radiations from the absorber tube. For the absorber plate, the selective coating is utilized. For cold environmental conditions, it is better to use evacuated tube collectors instead of flat plate collectors. In an evacuated tube collector, a heat pipe is surrounded by a vacuum tube. This vacuum reduces the convection and the conduction heat losses. The pipe is coupled with an absorber plate and a working fluid like methanol evaporates in the evaporator segment.

Flat plate, parabolic trough, dish and evacuated tube collectors are the most common types which are utilized in CPVT collectors. Sharan et al. [178] consider a linear CPVT unit. They examine tubular, horizontal flat plate and vertical flat plate collectors. Tubular demonstrates the best results by considering both thermal and electrical efficiencies. There are two kinds of solar thermal collectors which are used in the CPVT systems; non-concentrated (or LCPVT) solar collectors such as flat plate collectors and evacuated tube collectors and concentrated collectors (or MCPVT and HCPVT) such as dish collectors and parabolic trough collectors. In the CPVT systems, expensive PV cells are replaced with cheaper ones as mentioned before. The energy conversion efficiency increases with the increase of solar intensity. The reduction ratio for the cell number could be reached even to 1,000 and this reduction would decrease the cost per kW [179]. As mentioned earlier, CPVT collectors are classified into four levels: Low concentration ratio (1–10 suns), medium (10–80 suns),

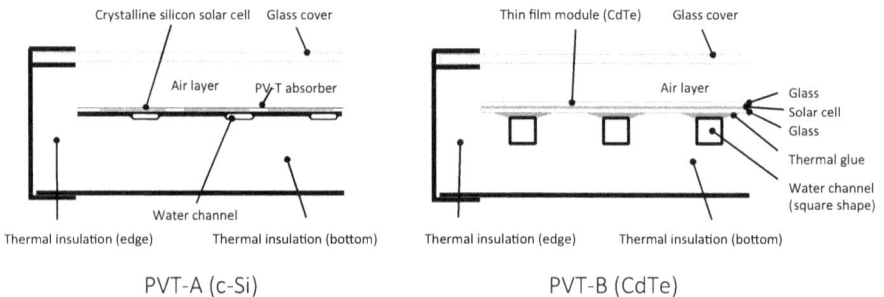

PVT-A (c-Si)  PVT-B (CdTe)

**FIGURE 6.21** Cross sections of two covered flat-plate PVT water collectors. Left: (PVT-A) with laminated c-Si solar cells. Right: (PVT-B) with CdTe thin film solar cells [11].

high (80–500 suns) and ultra-high (>500 suns). Linear Fresnel Reflectors (LFR) are the most common types utilized in CPVT systems. However, parabolic dish collector [180], central receiver systems [181] and parabolic trough collectors [182] are also used in CPVT systems. Different types of concentrating solar collectors are shown in Figure 6.22 [180–184]. Different types of solar thermal collectors with concentration ratio, operating temperature and efficiency ranges are described in Table 6.9.

Several CPVT systems have been designed and manufactured with the overall efficiency of more than 65.1% [185] and the outlet temperature of the working fluid is approximately 200°C [186]. The most important factor in the CPVT collectors is their concentration ratio. The highly concentrated CPVT systems could lead to a small cell

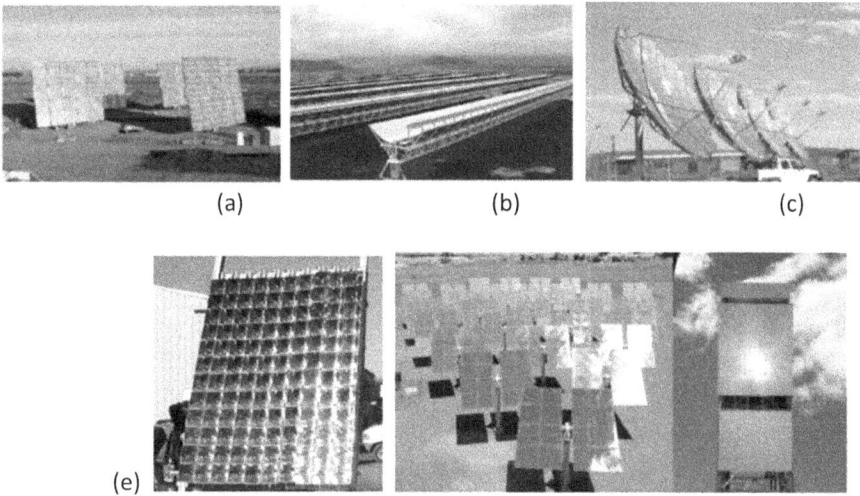

(a)            (b)            (c)

(e)

**FIGURE 6.22** (a) Fresnel lenses, (b) parabolic trough, (c) parabolic dish, (d) compound parabolic concentrators and (e) heliostat field central receiver [16].

**TABLE 6.9**
**Solar Thermal Collectors with Efficiency, Operating Temperature and Concentration Ratio Ranges [16]**

| Technology | Concentration Ratio | Operating Temp (°C) | Efficiency (%) |
|---|---|---|---|
| Flat Plate Collector (FPC) | 1 | 35–65 | 45–60 |
| Evacuated Tube Collector (ETC) | 1 | 50–100 | 30–50 |
| FPC coupled with booster reflectors | 1–3 | 40–80 | 30–50 |
| ETC coupled with booster reflectors | 1–3 | 70–120 | 40–50 |
| Compound Parabolic Concentrator (CPC) | 1–5 | 50–120 | 30–40 |
| Parabolic Trough Collector (PTC) | 40–80 | 350–450 | 14–16 |
| Linear Fresnel Reflector (LFR) | 20–50 | 200–350 | 8–10 |
| Parabolic Dish Collector (PDC) | 800–5,000 | 500–1,200 | 20–30 |
| Heliostat Field Central Receiver (HFCR) | 500–1,200 | 500–1,000 | 12–18 |

surface, so this cost-saving issue could be used for efficient multi-junction solar cells. The multi-junction solar cells have high-performance figures and efficiencies which could reach more than 40% [187–189]. For an efficient CPVT system, PV modules are expected to have efficiency above 30% [190–193]. For MCPVT and HCPVT, it is common to use dish parabolic, parabolic trough and linear Fresnel reflectors. Fresnel lens reflectors are one of the best collectors to be used in CPVT systems due to their small size, lightweight structure and low cost. Araki et al. [190,191] investigate a CPVT system using Fresnel lenses. Their collector efficiency and the nominal power generation were reported to be 25.8% and 30 kW, respectively. A review of CPVT using linear Fresnel reflectors is reported by Xie et al. [194].

CPVT system is a hybrid application of PVT and CPV collectors for achieving more performance. There are two disadvantages of PVT systems; First, generating the desired amount of electrical energy from PV cells needs high investments. Second, the thermal energy of these systems is used for only low-temperature applications. In a CPVT system, both of these demerits are covered by maintaining the PV cells at a moderated temperature and utilizing the spectrum concentration. As CPVT receivers operate under highly concentrated sunlight (>200 times), one main challenge for the design of an appropriate CPVT receiver is to remove a comparably large quantity of heat from a small area in order to keep solar cell operating temperature as low as possible, i.e. the thermal resistance ($R_{th}$) between heat transfer fluid and solar cell has to be low.

### 6.7.2   SOLAR CONCENTRATORS

CPV uses relatively cheap and suitable optical devices to concentrate the light on small and highly efficient photovoltaic solar cells [17]. Hence, the cost can be reduced by replacing the cell surface with cheaper optical devices [183] having a lower surface area. The advantages of CPV over flat plate systems include lower cost, superior efficiency, higher annual capacity factor, less materials availability issues, less toxic material use, ease of recycling, ease of rapid manufacturing capacity scale-up and high local manufacturing content [195]. These advantages are described in more detail by Swanson [195].

Solar concentrators are classified by their optical characteristics such as the concentration factor, distribution of illumination, focal shape, and optical standard. Concentration factor is also known as the number of suns, which is the ratio of the mean radiant flux density on a receiver area compared to the average normal global irradiance [196]. The classification based on the concentration factor includes [197]: (a) low concentration (LCPV): (1–40), (b) medium concentration (MCPV): (40–300), (c) high concentration (HCPV): (300–2,000). The efficiency of different PV cells is the ratio of the optimal electric power delivered by the PV cell, the area of the PV cell exposed to sunlight, and solar irradiance received by the PV [198].

LCPV is more simple compared to HCPV [199] due to its higher tracker tolerances, passive heat sinks, lower cost optics, reduced manufacturing costs, and reduced installation precision. The experimental findings by Butler et al. [200] show that LCPV has the potential to harvest more energy when using standard Si solar cells in a basic concentration configuration. However, Pérez-Higueras et al. [197]

indicated that in short term, HCPV offers a more profitable return due to its increased cell efficiency compared to flat module photovoltaic systems in the energy generation market. HCPV efficiencies obtained in laboratories have varied from approximately 37%–42% depending on materials and their structure and the number of suns used. The commercial efficiency was in general lower and ranged from 27% to 39% depending on the number of suns and multi-junction materials used in the solar cell [162]. More details are given by Pérez-Higueras et al. [197].

The solar collectors are also classified into two optical categories: (a) imaging optical concentrators, which means the image formed on the receiver by the optical concentrators [201] and (b) non-imaging optical concentrators: the receiver is not concerned with forming an image on it by optical concentrators [202].

During past decades, a lot of developments have been made in designing different models of solar concentrators. These developments are presented in an excellent review by Khamooshi et al. [17]. Here we briefly summarize existing know-how.

A Fresnel lens is a flat optical component where the bulk material is eliminated because the surface is made up of many small concentric grooves. These grooves individually act as prisms since each groove is approximated by a flat surface that reflects the curvature at that position of the conventional lens [203]. Fresnel lenses recently have been one of the best choices due to their desirable properties such as small volume, light weight, as well as mass production with low cost [194]. In early Fresnel lenses, glass was replaced by polymethylmethacrylate (PMMA), with optical characteristics almost the same as glass including good transmissivity and resistance to sunlight. It is now considered as the suitable material choice for the manufacturing of Fresnel lenses [204,205]. Figure 6.23 shows the schematic view of the conventional lens and Fresnel lens [17].

In general, the concentration of flux depends on the maximum concentration of optical flux, the real component of the refractive index, acceptance angle along the plane of the azimuth and the acceptance angle of the altitude. Fresnel lenses can be either circular or linear. Nakata et al. [206] and Harmon [207] examined circular Fresnel lens. The study by Whitfield et al. [208] discovered that the Linear Fresnel lens system has the advantage of being simple and totally enclosed yet is more costly than some of the others. The point-focus Fresnel lens has the advantage of having the potential for simple mass-produced optics but it loses efficiency at

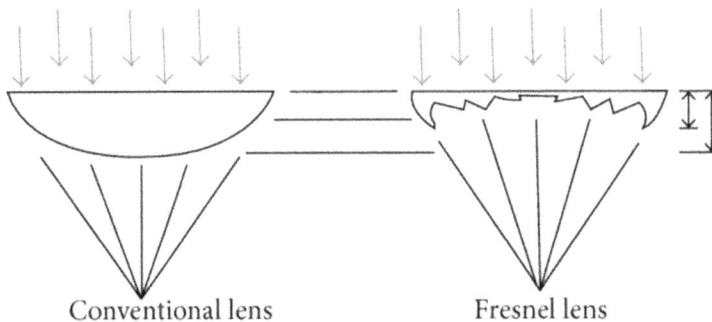

FIGURE 6.23 Conventional lens and Fresnel lens [17].

higher concentrations. The optical properties of flat linear Fresnel glass lenses are also presented by Franc et al. [209]. While Fresnel lens has small volume and light weight and can be mass produced they suffer from (a) imperfection on the edges of the facets, causing the rays to be improperly focused at the receiver; (b) possibility of lost light due to incidence on the draft facet and (c) reduction in luminance in order to minimize the upper disadvantages. As pointed out by Khamooshi et al. [17], Fresnel lens systems are used for concentrated solar energy applications for various thermal usages, thermal heating, solar cooking, photocatalytic, solar building, solar-pumped laser, lighting, and surface modification of metallic materials [203,210–212].

The solar parabolic trough collector is the most recognized technology due to its high dispatchability and low unit cost [213]. In parabolic trough concentrators, the parabolic-shaped mirror focuses sunlight on the receiver tube which is placed at the focal point of the parabola [214]. Reflectivity of the mirror, incident angle, tracking error, intercept factor, as well as absorptivity of the receiver, are the factors which can affect the performance of the parabolic trough concentrator [213]. Riffelmann et al. [215] showed that the image quality of the mirror, slope error, and collector assembly also affect the optical efficiency of a parabolic trough collector. Omer and Infield [216] indicated that the two-stage concentration of the parabolic trough collector can enhance the concentration efficiency of the parabolic trough. This design provides an efficient concentration of the incident solar radiation without any frequent tracking system. The performance of the parabolic trough collector also depends on receiver design and heat loss from the receiver [17,217–222]. The heat loss can be increased by inserting porous inserts into the inner surface of the receiver. The porous inserts increase the heat transfer rate by (a) increasing the effective fluid thermal conductivity, (b) enhancing mixing between the fluid and receiver wall, and (c) lowering thermal resistance by developing a thinner hydrodynamic boundary layer [214]. While parabolic concentrator makes efficient use of direct solar radiation, it is expensive, uses only direct radiation and has low optical and quantum efficiencies [213].

Compound parabolic concentrators (CPCs) are designed to efficiently collect and concentrate distant light sources with some acceptance angle. The geometrical concentration ratio and theoretical maximum possible concentration ratio of the CPC depend on the aperture area, receiver area, and maximum acceptance angle [223,224]. CPCs can be in both two-dimensional and three-dimensional configuration. Suzuki and Kobayashi's [225] examined the optimum acceptance angle of the 2-D CPC concentrator The results indicated that the optimum half-acceptance angle is 26° irrespective of the change in the diffuse radiation fraction.

Senthilkumar et al. [226] showed that the three-dimensional compound parabolic concentrator (3D CPC) is more efficient than the 2D CPC because of the higher concentration ratio. Yehezkel et al. [227] analyzed the losses due to reflection properties and calculated the effect of these losses on concentration ratio. Khalifa and Al-Mutawalli [228] carried out an experimental study on the effects of two-axis sun tracking on the thermal performance of CPC in two different modes; (a) a batch feeding where no flow through the collector was allowed and (b) different steady water flow rates were used. The results led us to the conclusion that the energy gain of a CPC collector can be increased by using two-axis tracking systems. Mallick et al. [229] designed a novel non-imaging asymmetric compound parabolic photovoltaic concentrator (ACPPVC) with different numbers of PV strings connected in series and demonstrated that an ACPPVC increased the maximum power point by 62% when compared to a similar non-concentrating PV panel. While for compound parabolic concentrator, most of radiation within the acceptance angle can transmit through the output aperture into receivers, it needs a good tracking system in order to get maximum efficiency [17].

Figure 6.24 illustrates two-dimensional hyperboloid concentrators. This kind of concentrator is also called the elliptical hyperboloid concentrator. The advantage of this concentrator is that it is very compact since only a truncated version of the concentrator needs to be used. Because of this factor, it is mainly used as a secondary concentrator [230]. Garcia-Botella et al. [202] found out that the one-sheet hyperbolic concentrator is an ideal 3D asymmetric concentrator as its shape does not disturb the flow lines of an elliptical disk. It also does not need a tracking system where two different acceptance angles in transversal and longitudinal directions, are needed. While the hyperboloid concentrator is very compact, it needs to introduce lens at the entrance aperture to work effectively. Ali et al. [231] carried out optical performance evaluation of a 2-D and 3-D novel hyperboloid solar concentrator.

Sellami et al. [232] designed a 3-D concentrator and coined the name Square elliptical hyperboloid (SEH). This type of concentrator can be integrated into either glazing windows or facades for photovoltaic applications. This configuration can collect both diffuse and direct beams. They also found that optical efficiency depends on the size of the SHE. The 3-D solar concentrator acquired from the hyperboloid has the ability to concentrate all the entering rays [233] such as the trumpet concentrator, which is composed of a revolution of the hyperbolic type and is considered as an ideal concentrator [234–239]. Saleh Ali et al. [236] presented a method for

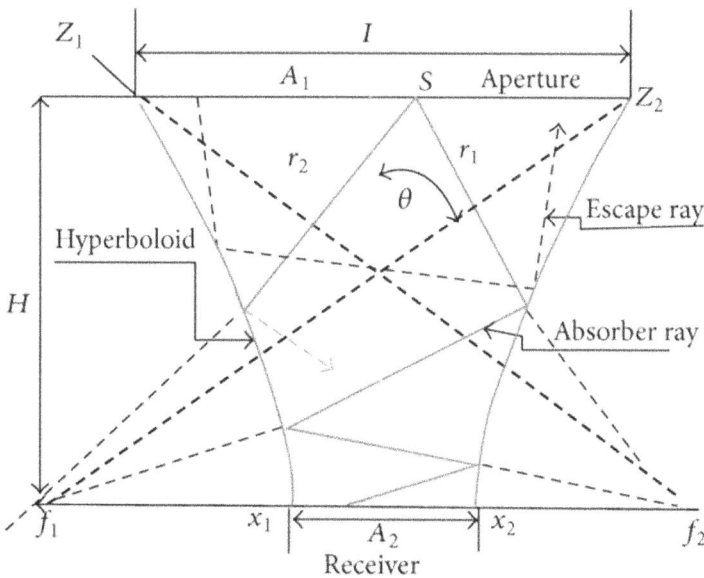

**FIGURE 6.24**  2-D Hyperboloid concentrator [17].

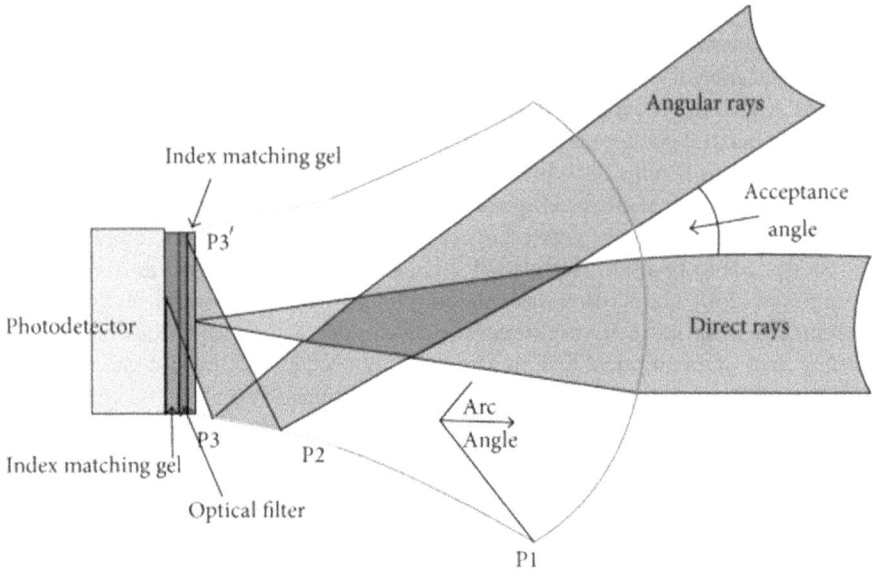

**FIGURE 6.25**    Side view of a DTIRC [17].

designing a static 3-D solar elliptical hyperboloid concentrator (EHC). They and Milano et al. [240–242] suggested RR, XX, XR, RX, and RXI concentrators. The advantages of RR, XX, XR, RX and RXI concentrators are :(a) they achieve the theoretical maximum acceptance angle concentration, (b) achieve high concentration, (c) lighter weight and (d) less expensive tracking system. On the other hand, the size of the cell must be kept minimum to reduce the shadowing effect.

Dielectric totally internally reflecting concentrator (DTIRC) is one of the most important non-imaging optical concentrators [243]. In addition to the solar application, these lenses were proposed for IR detection [244] and optical wireless communication systems [245,246]. As shown in Figure 6.25 [17], DITRCs consist of three main parts: a curved front surface, a totally internally reflecting profile, and an exit aperture [186]. For this type of concentrator, in order for rays to reach the exit aperture, it should be within the designed acceptance angle of the concentrator. Ning et al. [247] discussed two-stage photovoltaic concentrators with Fresnel lenses as primaries and dielectric totally internally reflecting non-imaging concentrators as secondaries. The results indicated that the two-stage concentrator provides higher concentration and more uniform flux distribution on the photovoltaic cell than the point focusing Fresnel lens alone. Muhammad-Sukki et al. [248] designed DTIRC using the maximum concentration method (MCM). DTIRC itself cannot efficiently pass all of the solar energy that it accepts into a lower index media [249]. Muhammad-Sukki et al. [250] presented a study about a mirror-symmetrical dielectric totally internally reflecting concentrator (MSDTIRC) which is a new type of DTIRC. While DTIRC has (a) higher efficiency and concentration ratio than CPC and (b) work without any need for cooling features, it can not efficiently pass all of the solar energy that it accepts into a lower index media.

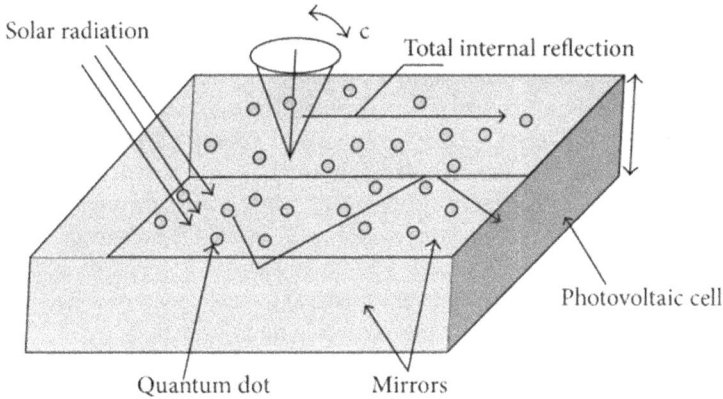

**FIGURE 6.26**   Principle of the QDC [17].

Quantum dot concentrator (QDC) is a non-tracking concentrator that includes three main parts; transparent flat sheet of glass or plastic doped with quantum dots (QDs), reflective mirrors placed on three different edges and the back surface, and a PV cell which is attached to the exit aperture. As it is shown in Figure 6.26 [213] when the sun radiation hits the surface of concentrator, a part of the radiation will be refracted by a fluorescent material and absorbed by quantum dots (QDs); photons are reemitted isotropically at a lower frequency and guided to the PV cell [219]. The size of quantum dots, which are made of nanostructures, typically varies from tens to hundreds of nanometers in size [251]. Mićić et al. [252] showed that QDs are capable of absorbing light over an extremely broad wavelength range and the absorption spectra also depict the spectral shift to higher energy as QD size decreases. The main advantages of QDC are (a) they are without any tracking system, they can concentrate both diffuse and direct radiations [253], (b) due to the geometries of these concentrators, they have fewer problems of heat dissipation [219], and (c) sheets are inexpensive and are suitable architectural components [254]. The development of QDCs is restricted by the stringent requirements of the luminescent dyes such as high quantum efficiency, suitable absorption spectra and red shifts, and illumination stability [255,256]. The problems of organic dyes can settle by replacing them with QDs which have the advantages of less degradation and high luminescence [257]. Schüler et al. [258] proposed that quantum dot containing nanocomposite coatings might be an alternative for the production of planar quantum dot solar concentrators. The concentration ratios of QDCs are extensively discussed by Gallagher et al. [198] who determined concentration ratios of different types by comparative analysis. A maximum comparative concentrating factor (MCCF) was determined at specific solar intensities using the power maximum for the test device and the power maximum for the reference devices [259]. The advantages of quantum dot concentrators are (a) they are non-tracking concentrators, (b) have fewer problems of heat dissipation and (c) sheets are inexpensive and are suitable architectural components. On the other hand, developing QDCs is restricted by stringent requirements of the luminescent dyes.

The use of flat-panel solar photovoltaics (FPV) is growing dramatically as its cost decreases. By contrast, more efficient concentrated PV systems (CPV), which focus direct sunlight onto a single point, have not been widely adopted because of their high cost, large size, and expensive tracking systems. However, CPV systems are cheaper than concentrated solar thermal energy systems. Moreover, CPV systems do not take up much land because they are installed onto tracking systems. It is also possible to use the place under the panels for growing crops, allowing for multi-purpose land utilization. Solar thermal power plants usually require a large amount of water and, in contrast, CPV stations use only a small amount of water. A new approach, micro-scale CPV systems (micro-CPV), may deliver the cost and size benefits of conventional FPV systems, but with an estimated 50% performance improvement. Micro-CPV modules would use cost-effective trackers and generate more electrical power in a given area. This allows installation on space-constrained residential rooftops and decreased costs for commercial and utility applications. Finally, the MOSAIC systems would have the ability to capture both direct and diffuse sunlight, which could make CPV economical in more geographical regions. These innovations could spur the expanded use of PV to generate clean, renewable energy.

The Massachusetts Institute of Technology (MIT) with partner Arizona State University is developing a new concept for PV power generation that achieves the 30% conversion efficiency associated with traditional concentrated PV systems while maintaining the low cost, low profile, and lightweight of conventional FPV modules. MIT aims to combine three technologies to achieve its goals: a dispersive lens system, laterally arrayed multiple bandgap (LAMB) solar cells, and a low-cost power management system. The dispersive lens concentrates and separates light that passes through it, providing 400-fold concentration for direct sunlight and threefold concentration for diffuse sunlight. The dispersive lens is a thin layer consisting of inexpensive, lightweight materials that can be manufactured at low cost using plastic molding, an improvement over traditional methods. The lens focuses the direct light onto the array of LAMB solar cells, while also focusing the diffuse light onto common PV cells integrated beneath the LAMB array. The power management system combines power from multiple cells into a single output so that the power from a panel of LAMB arrays can be processed with grid-interface power electronics, enabling as much as 20% additional energy capture in applications where the roof is partially shaded. If successful, innovations from MIT's project may lower the cost of solar systems by allowing economical, high-volume manufacturing of micro-CPV arrays. Improved systems could encourage greater adoption of solar power in all three primary markets – residential, commercial, and utility [260–263].

Swiss startup Insolight builds planar optical micro-tracking systems that keep each cell in focus regardless of the sun's position. The resulting flat panel combines higher efficiency with easier mounting on a standard rooftop or ground-mounted installation. The technology enables subtle sun-tracking by employing horizontal sliding movements, while the module stays at a fixed tilt. Insolight's modules are capable of enduring extreme radiation and temperature, especially useful in satellite applications. Swanson [195] performed a review study on the characteristics of CPV systems which approached the economical aspects of the systems. Table 6.10 summarizes

**TABLE 6.10**
**Different CPV Projects with Specifications [17,195]**

| Companies/Institutions | Type of Concentrator | Type of Focus | Concentration Ratio | Tracking System | Cooling System | Efficiency | Cost |
|---|---|---|---|---|---|---|---|
| Sun power corporation | Fresnel lens | Point | 25–400 | - | - | 27% | - |
| Solar research corporation | Parabolic dish | Point | 239 | Yes | Yes | 22% | - |
| Photovoltaics International | Fresnel lens | Linear | 10 | Yes | - | 12.7% | 4–6 cent kwh (110 MW/yr production rate) |
| Polytechnical University of Madrid | Flat concentration devices (RXI) | point | 1,000 | No | - | - | Low cost (need no tracking system due to high acceptance angle) |
| Fraunhofer-Institut fur Solare Energie systeme | Parabolic and trough | Linear and point | 214 | Yes | yes | 77.5% | - |
| Entech | Fresnel lenses | Linear | 20 | Yes | - | 15% | 7–15 cent kwh (30 MW/yr production rate) |
| BP Solar and the Polytechnical University of Madrid | Parabolic trough | Linear | 38 | Yes | Yes | 13% | 13 cent kwh (15 MW/yr production rate) |
| Australian National University | Parabolic trough | Linear | 30 | Yes | - | 15% | - |
| AMONIX and Arizona Public Service | Fresnel lens | Point | 250 | Yes | Yes | 24% | - |

**TABLE 6.11**

**Comparative Analysis of Different CPV Systems from Economic Aspects [17]**

| Primary Concentrator | Secondary Concentrator | Tracking System | Concentration Ratio | Cost ($/Wp) |
|---|---|---|---|---|
| Point focus Fresnel lens | No | Gimbals | 36 | 1.48 |
| Cylindrical paraboloid | Point-focus CPC | Polar | 65 | 1.78 |
| Linear Fresnel lens | Solid CPC | Gimbals | 37 | 2.02 |
| Curved TIR lens | No | Polar | 28 | 1.97 |
| Curved Fresnel lens | No | Polar | 15 | 2.18 |
| V-trough, screen printed | No | Polar | 2 | 4.31 |

*The costs given in the table are for cells, optical systems, mountings, and trackers only, including construction costs; balance of system costs are omitted as they are similar for all types of collector. The cost in $/Wp is for collectors at operating temperature and, for concentrators, is based on direct beam irradiance of 850 W/m²; the cost for the flat plate is based on a total irradiance of 1,000 W/m².*

Swanson's study which represents different CPV with their characteristics [17,195]. The cost comparison of different CPV systems is illustrated in Table 6.11 [17,208].

## 6.8 CHARACTERISTICS OF PVT AND CPVT SYSTEMS BASED ON LEVEL OF CONCENTRATION AND TEMPERATURE OF WORKING FLUID

As mentioned earlier, PV (PVT) and CPV (CPVT) can be classified based on the level of concentration and the temperature of the working fluid. Here we briefly examine the characteristics of PV and CPV with different levels of concentration.

### 6.8.1 Low Concentration (LCPVT) and Temperature

In the LCPVT, the flat-plate reflectors and the CPC (compound parabolic collectors) optical systems can provide geometrical CRs lower than 4 suns and optical CRs (concentration ratio) lower than 2.5 suns. The systems prefer to be stationary to avoid expensive tracking systems. At a higher CR of $4 < C_g < 10$, the linear focused system with 1-axis tracking is more widely accepted. The integration of the LCPVT system with the building and greenhouse highly depends on its CR. The 1-axis tracking systems can be employed with East-West oriented axis or North-South oriented axis. Diffuse radiation plays an important role in the LCPVT systems. At CR lower than ten suns, the commercially available silicon solar cells, including monocrystalline, polycrystalline and amorphous silicon (c-Si, pc-Si and a-Si) solar cells, are usually selected. The literature shows that the Heterojunction with Intrinsic Thin Layer (HIT) system has the highest: (a) module efficiency, (b) net annual electrical energy, (c) overall annual thermal energy, (d) exergy output.

Low concentration and low-temperature PVT and CPVT can use air, water, and air and water and nanofluids as heat transfer fluids. Active cooling is preferred over passive cooling. Since temperature is low, often no thermoelectric elements are used to generate additional electricity. Nazir et al. [264] however reported maximum thermal efficiency of 84% when the PVT/air system is accompanied by thermoelectric elements. In most cases, the maximum temperature of the working fluid (air) has ranged from around 30°C–55°C with thermal efficiencies ranging from 10% to 90%. Many systems show PV conversion efficiencies of around 6%–16% with high overall efficiencies (80% or higher). Mass flow rate considerably influences thermal-efficiency results [265]. A variety of configurations including Compound Parabolic Concentrators (CPCs), Fresnel lenses, micro-concentrators, and parabolic-trough concentrators can be used.

LCPVT is largely used for building integrated systems, agricultural applications and dryers (e.g. greenhouses, dryers of agricultural products). There are a few investigations about PVT systems with PCMs, integrated collector-storage, heat pipes, chillers, desiccants and polymer PVT collectors. Systems using PCM are appropriate for BI applications. Assoa and Ménézo [266] investigated a roof-integrated PVT/air system. It was found that forced ventilation offers advantages from a thermal production point of view. However, natural ventilation provides adequate cooling of the PV panels.

Alves et al. [267] investigated a PVT/water with a Solarus reflector (CR: 1.7X) and mono-Si PV cells. The system was appropriate for domestic water heating and showed an electrical efficiency of 19.1%. An increase in water flow rate reduced PV-cell temperature and increased PV efficiency. Khelifa et al. [268] studied a PVT/water system (without solar concentration) and found 94% thermal efficiency when collector was perfectly insulated and covered by a flat glass [268]. Wu et al. [269] investigated a CPVT system with nanofluids and thermoelectric elements. Glazed and unglazed configurations were examined. It was noted that nanofluid gave better heat removal rate than water. The set-up used by Wu et al. [269] is illustrated in Figure 6.27.

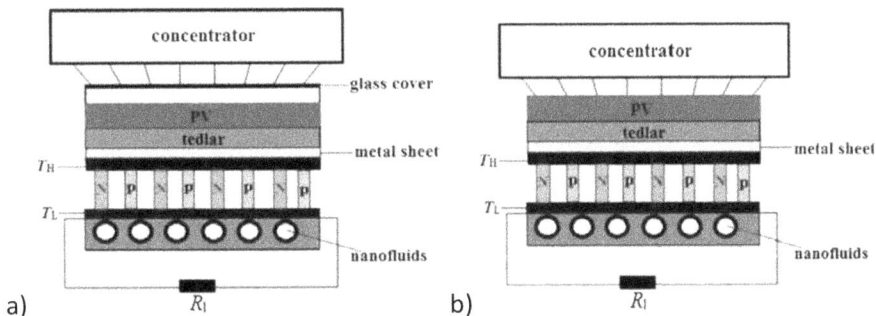

**FIGURE 6.27**   The low-temperature CPVT system: (a) system with a thermoelectric component, (b) nanofluid cooling tubes [10].

### 6.8.2 MEDIUM CONCENTRATION (MCPVT) AND TEMPERATURE

For MCPVT, the linear concentration optics, including the parabolic trough, linear Fresnel lens and reflector, can provide a CR of 10–100 suns. Since stationary ones cannot achieve such a high concentration, both 1-axis and 2-axis tracking systems are generally used. The cost of 2-axis tracking linear systems is generally higher than that for 1-axis tracking ones. For Fresnel lens concentrator, the domed Fresnel lens are preferred over the flat ones because of their advantages of reduced coma, minimum reflectance and shorter focal length among others. For greenhouses and buildings with large surfaces and illumination demand, a stationary concentrator with a moving receiver is an acceptable choice where tracking of the receiver is possible with the use of two motors; one for the distance to the Fresnel lens, and the other one for the translocation parallel to the lens. This type of optical system can provide a low or medium CR for CPVT systems. The transmissive linear Fresnel reflector provides another option for the building elements with illumination. At the CRs of over ten suns, the uniformity of solar radiation plays an important role in PV efficiency. The system performance can be improved by better mirror reflectivity, and by pursuing a suitable focal line with uniform illumination.

The studies for PVT systems with working-fluid temperatures between 60°C and 90°C, mostly used silicon-based PV cells with thermal efficiencies of around 55%–60% and electrical/PV efficiencies of approximately 13%–15%. A large majority of the investigations are about PVT/water and PVT/air configurations, for industrial and domestic applications (including BI systems). Some studies used configurations with PCMs, heat pumps, ethylene-tetrafluoroethylene cushions and polymeric PVT collectors. Since heat load on MCPVT systems are higher than that for LCPVT, passive cooling or thermosiphon methods are not suitable. Cooling systems generally include fins with water as heat transfer fluid. Ren et al. [270] investigated an amorphous silicon PVT/water configuration appropriate for medium-temperature applications. For CPVT, Haiping et al. [271] developed and tested a flash tank integrated CPC system demonstrating its feasibility for cogeneration of fresh water and electricity. CPVT systems are generally used for CRs up to 190X and the working-fluid temperatures around 62°C–90°C. The set-up used by Haiping et al. [271] is illustrated in Figure 6.28. In most of the reported studies, mono-Si and multi-/triple-junction PV cells are used with thermal efficiencies of approximately 50%–68% and PV efficiencies are around 10%–19% and 35%–38%, depending on the materials used for PV cells (single-junction vs. multi-junction, etc.). The applications of CPVT in this temperature range are largely restricted to domestic applications and desalination with few applied to greenhouses.

### 6.8.3 HIGH CONCENTRATION (HCPVT) AND TEMPERATURE

HCPVT systems generally employ concentrating optics such as dish reflectors or spot Fresnel lenses that concentrate sunlight to intensities of 100 suns or more. Two-axis tracking parabolic trough or tower-heliostat systems are not preferred. The design of DA-CPV modules becomes an important issue under non-uniform irradiation. To achieve the current matching, various PV connection methods have

a)

b)

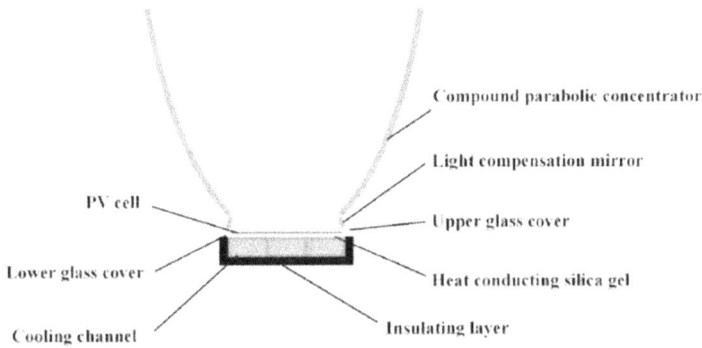

Compound parabolic concentrator

Light compensation mirror

PV cell

Upper glass cover

Lower glass cover

Heat conducting silica gel

Cooling channel

Insulating layer

**FIGURE 6.28**   The medium-temperature CPVT: (a) The Low Concentrating Photovoltaic/Thermal - Solar Thermal Collector (LCPVT-STC) experimental set-up, (b) details about the CPVT module [10].

been investigated, and the total cross tied (TCT) method achieved less mismatching loss [16]. Irregularly shaped PV cells would be another option to address this issue. The solar cells operating at over 100 suns require high-capacity heat sinks to prevent thermal destruction and to manage temperature-related performance losses. Microchannel heat sinks have the advantages of high heat transfer performance, easy integration with moving components, light weight, and high stability, and are therefore considered to be a viable method to cope with the extremely high heat load. Water-glycol mixture and pressurized water are preferred cooling fluids at temperatures above 100°C. A closed thermal cycle and a secondary heat exchanger are often necessary for HCPVT systems.

Most literature studies (a) do not include thermoelectric elements, (b) use CRs up to 1,000×, (c) use water (except some used oil, salts and nanofluids) as the working fluid, (d) use parabolic concentrators and (e) operate in the temperature range of around 100°C–250°C. Most of the systems examined used triple-junction or silicon-based PV cells systems with thermal efficiencies of around 50%–60% and electrical/PV efficiencies of around 20%–30%. CPVT systems in this temperature range were mostly investigated for domestic installations, desalination, polygeneration and large-scale systems. Riggs et al. [272] showed that concentrating solar systems offer multiple industrial and commercial applications with significant cost savings and with economics that are competitive with coal and propane. Crisostomo et al. [273] noted that spectral beam splitting is a solution that allows PV cells to work at low temperatures whereas thermal receiver operates at high temperatures. Unfortunately, in this case, the fraction of solar irradiation that without beam splitting would be converted into heat by the PV cell is lost. The experiments revealed that the PV cells, illuminated by the light reflected by the filters, show 9.2% (absolute) higher efficiency in comparison to the same PV cells without the filters [273]. Widyolar et al. [274] investigated a spectrum splitting hybrid concentrating solar power/CPV collector system with double-junction InGaP/GaAs solar cells. Experiments using a fluid with suspended solid (alumina-based) particles heated up to 600°C were conducted. The set-up used by Widyolar et al. [274] is illustrated in Figure 6.29.

### 6.8.4 Ultra High Concentration (UHCPVT) and Temperature

Muller et al. [275] examined ultra-high concentration photovoltaic-thermal systems based on microfluidic chip-coolers. In this study, the electrical efficiency of a photovoltaic-thermal system for coolant inlet temperatures ranging from 25°C to 75°C and concentrations from 500 to 1,500 suns was investigated experimentally and theoretically. A triple-junction solar cell was tested in two different configurations. At 1,500 suns the electric efficiency of a silicon microchannel cooler package exceeded the efficiency of a reference package with a copper cooler by 2% and it remained fully functional up to concentrations of 4,930 suns. Codd et al. [276] examined solar cogeneration of electricity with very high temperature process heat. In this study by using an all-in-one, spectrum-splitting hybrid receiver, electricity and high-temperature heat were generated with a single efficient system. Here, the performance of a transmissive concentrator photovoltaic/thermal (tCPV/T) system was demonstrated on the sun, with a total energy efficiency of 85.1% G 3.3%, 138 W electric power at 304 suns (with average cell temperatures <110°C), 903 W hot water output (average 34°C and 1.7 bar, peak temperatures to 56°C), and 1,139 W high-temperature steam output (average 201°C and 45 bar, peak temperatures up to 248°C). The spectrum-splitting hybrid receiver used a sparse array of III–V triple-junction solar cells on GaAs substrates contained within a transparent microchannel water cooling stack, followed by a structured flow path thermal receiver cooled with pressurized water. Based on the performance of a 2.72 m² prototype, economics of the system appear to be close to competitiveness to natural gas-produced process heat for a variety of locations. For an installation in San Diego, California, the levelized cost of electricity was 0.03 $/kWth [122]. The set-up used by Codd et al. [276] is illustrated in Figure 6.30.

a)

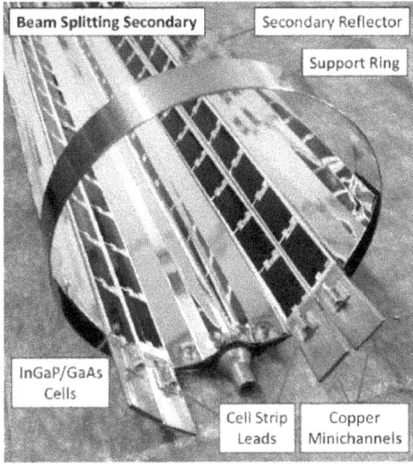

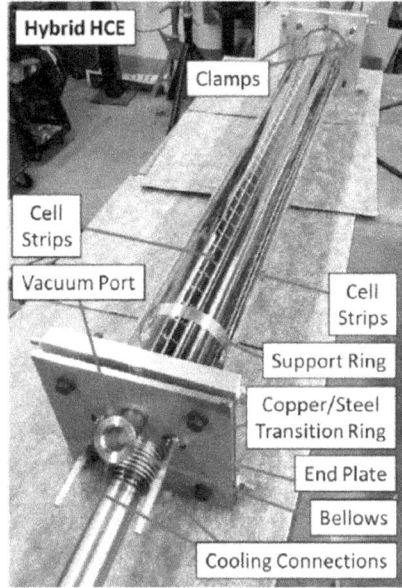

b)

c)

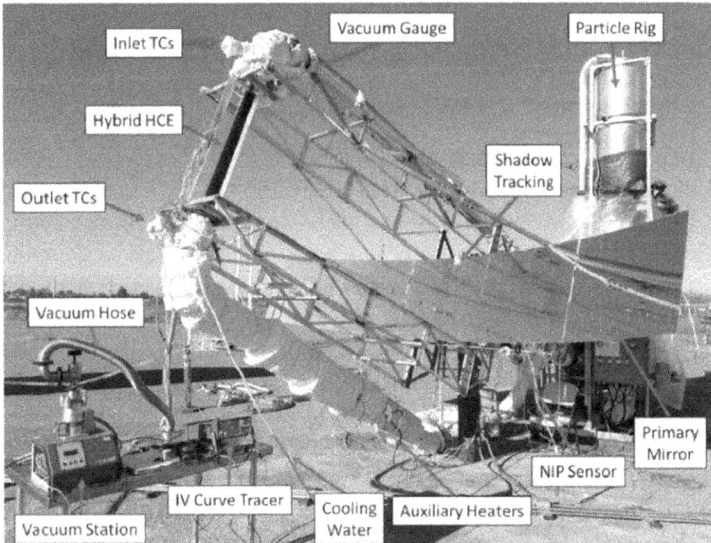

**FIGURE 6.29** The high-temperature CPVT system: (a) details about the secondary reflector, (b) hybrid Heat Collection Element (HCE), (c) primary mirror, Normal Incidence Pyranometer (NIP) sensor, shadow tracking, etc. [10].

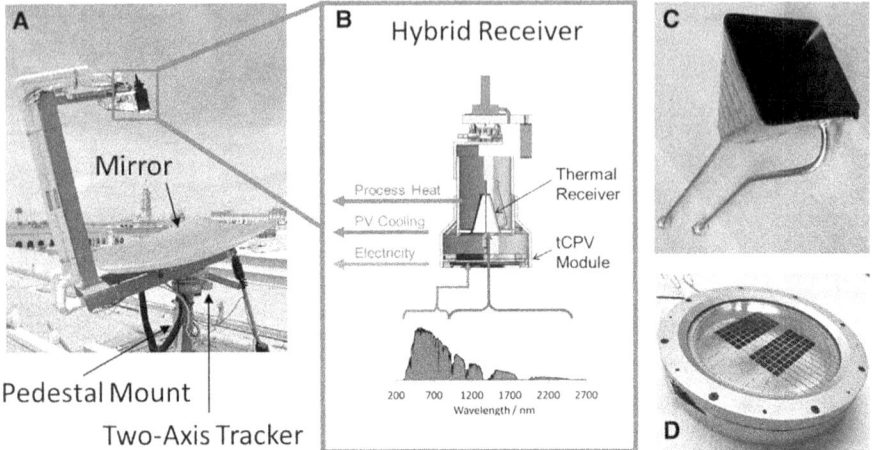

**FIGURE 6.30** Very High-Temperature Hybrid Solar System Overview (a–d) The proto-type mounted at an outdoor test facility (a); schematic of the power flow through the hybrid receiver cross-section (b); and photographs of (c) cavity thermal receiver and (d) tCPV module subassemblies. Sunlight is concentrated by the paraboloidal mirror on the 2-axis tracker and directed to the hybrid receiver. There, the tCPV module converts a portion of the high-energy photons to electricity and low-temperature heat, while the balance transmits through to the thermal receiver, where it is absorbed and converted to very high-temperature heat [276].

## 6.9   COOLING TECHNOLOGIES FOR PVT AND CPVT SYSTEMS

### 6.9.1   COOLING TECHNOLOGIES FOR PVT SYSTEMS

In a PV system, if heat is allowed to stay on the cell surface for a long period, it can cause structural damage to the cell. In a PVT system, the heat is recovered from the module and used for numerous applications such as crop drying, floor heating, hairdryer, etc. This is illustrated in Figure 6.31 where the air is used as a coolant medium. Packing factor, air mass flow rate, glazing (additional glass cover), wind velocity and temperature of the coolant, etc. are the parameters that have an impact on the PV/T air-collector performance. Modifications in the channel like the use of fins, TMS, hexagonal honeycomb heat exchanger, and v-grooved absorber have also been adopted by various analysts to improve its overall performance.

The structure of PVT when water is used as a coolant is very similar to the one for air collector except for the air channel. In the PVT water collector, water is forced to flow through the tubes beneath the PV module to cool down the PV module which in turn increases the $\eta_{ele}$ of PVT water collector. Zondag et al. [277] discussed various configurations of PV/T water collector based on water-flow patterns below the PV module, as shown in Figure 6.32. PVT combi system combines the use of air and water for cooling of PV cell. This improves the overall efficiency of the cell. The schematic diagram of the PVT combi collector is shown in Figure 6.33. The concept of the PVT combi system was first introduced by Tripanagnostopoulos [278]. By combining two types of heat collector media, the temperature of PV cells is greatly

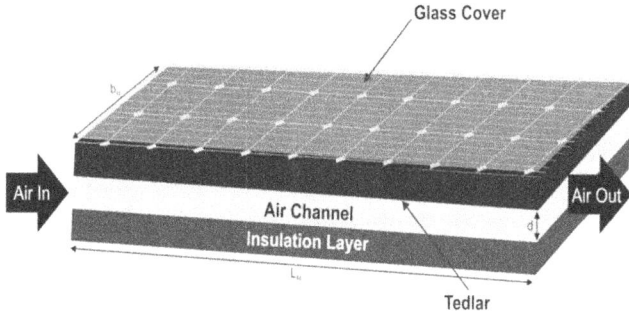

**FIGURE 6.31**    Layer diagram of PV/T air collector [1].

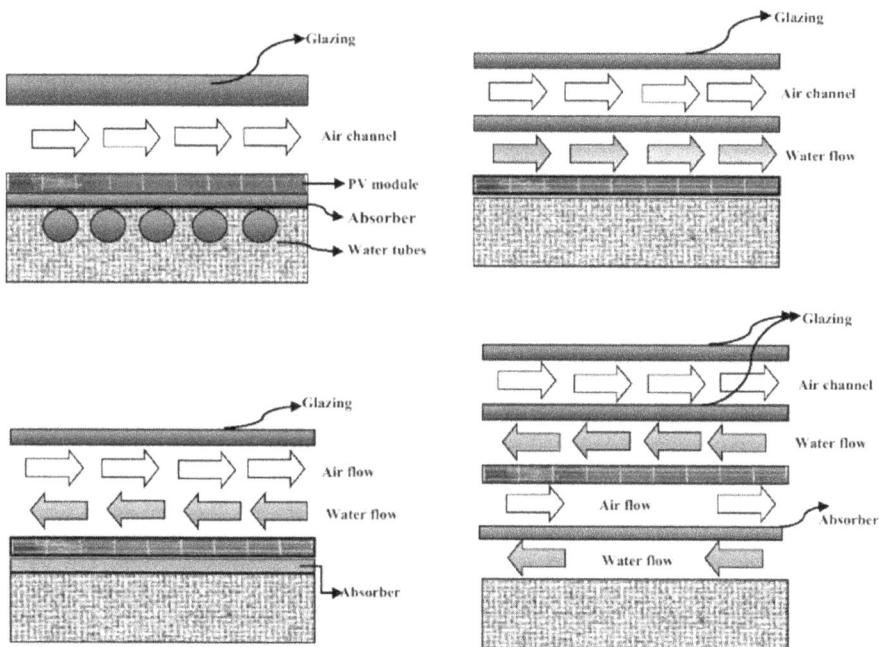

**FIGURE 6.32**    Classification of PVT water collector based on water-flow pattern [1].

reduced, and cell efficiency is improved as compared to PVT water or PVT air collector. Recently, phase-change materials (PCM) are used with PVT module to minimize the temperature of the solar cell. A PCM material swings between its solid and liquid phases when a change in temperature is observed [279]. A perfect PCM material must be non-corrosive and it must offer characteristics like high heats of freezing and dissolution. An experimental analysis has been carried out by Preet et al. [280] on the PVT-PCM system with different mass flow rates to observe the impact of PCM on its performance. The PVT setup with PCM used in this study is shown in Figure 6.34. Liang et al. [281] proposed a model in which the graphite layer

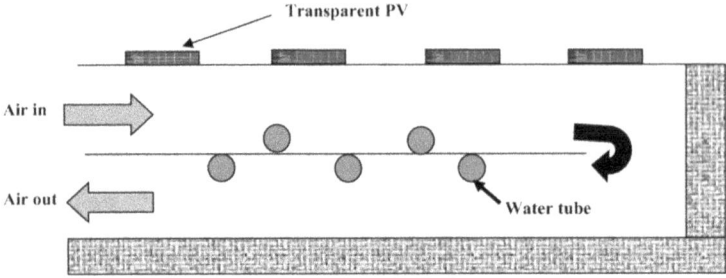

**FIGURE 6.33**   PV/T combi collector [1].

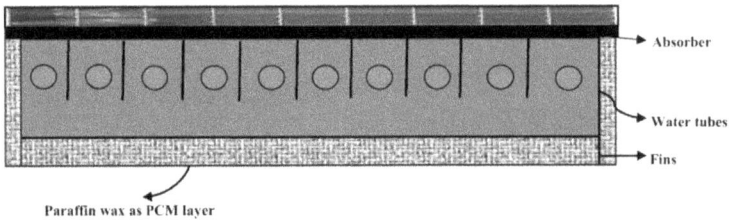

**FIGURE 6.34**   PV/T water collector with PCM [1].

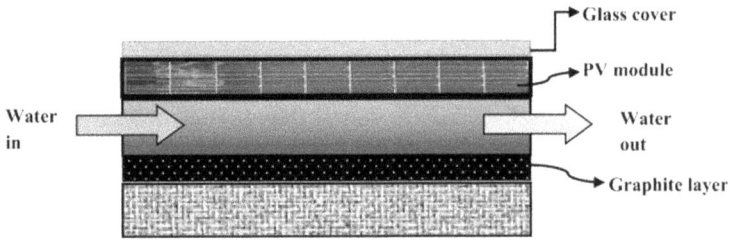

**FIGURE 6.35**   PV/T water collector with the graphite layer [1].

is used underneath the water channel as shown in Figure 6.35 and compared the performance with the conventional PV. The average *electrical efficiency* reported for the proposed PVT model and conventional PV system was 6.46% and 5.15%, respectively. Ong et al. [282] worked upon an experimental model in which a solar water heater is integrated with the TE module, which is used for the generation of electrical energy with hot water. The performance of the combined system was evaluated at a different fluid flow rate. From the observations, it has been observed that 0.16% of electrical efficiency ($\eta_{ele}$) was also achieved along with hot water.

The thermal absorber is the major component in the PVT system that greatly influences its overall performance. Over the last few years, various changes in design, material, connection methodology, and manufacturing techniques of thermal absorber have been observed. Sheet and tube design is the typical thermal absorber configuration used in PV/T system. The other configurations are roll-bond absorber and fully wetted type absorber. In the roll-bond absorber, channels are embedded between two rolled aluminum sheets, while in conventional sheet and tube design, the channels are welded under the plate. The performance of the flat-plate PV/T collector with roll-bond absorber was investigated by Haurant et al. [283]. The modified system was used for hot water production. The performance of PVT water collector has been compared based on different roll-bond absorber configurations, i.e., serpentine pipe and harp pipe [284].

## 6.9.2 COOLING TECHNOLOGIES FOR CPVT

The existing literature indicates that the cooling method and the nature of coolant used in CPVT systems should consider the temperature of solar battery, temperature uniformity, practicability and reliability of the cooling system, as well as its economic and environmental impact. The general aim should be to reduce cell temperature to the maximum extent taking account of cooling cost and temperature limit. Temperature decrease has been proven to enhance the solar-to-electricity conversion efficiency, extend the solar cells' service life and further reduce the cost of the CPVT system. The conversion efficiency is also somewhat affected by the temperature uniformity of the solar cell. The traditional solar concentrators including dish concentrators, parabolic trough concentrators (PTCs) and compound parabolic concentrators (CPCs) usually cannot produce a uniform concentration (thus temperature) on the absorber's surface [285]. Thus, "high-flux hot spot" usually occurs on the CPV module with a non-uniform concentration. These "hot spot" regions have a heat flux several times higher than the module average that causing exceedingly high temperatures and resulting in dramatic degradation in the system performance and reliability [286,287]. Baig et al. [288] pointed out the uneven concentration on solar cells increases the cell temperature and cell resistance resulting in power and efficiency loss.

Proper thermal management of CPVT systems is very important to maximize their power output. Solar cells are usually paralleled after being connected in series to enhance power generation efficiency. The output power of the entire assembled battery is limited by the battery which has the highest resistance, which generally occurs at the highest cell temperature. This battery, thus, limits the efficiency of the entire photovoltaic system. There are two common approaches to solve this problem. The first is adding bypass diodes to automatically cut off the least efficient battery power output when the cell temperature achieved a certain value. The second is adopting a heat-removal system to ensure the temperature uniformity of the battery pack. The second approach requires good thermal management system for CPVT which provides good heat dissipation without uneven temperature distribution.

The ultimate aims of good thermal dissipation system are to reduce costs and improve cell performance. The initial investment, maintenance cost, energy consumption, material and simple construction, etc. factors should be taken into account for the practicability of the cooling system. There is a trade-off between power required to provide cooling and power gained by the efficiency improvement. This trade-off should be optimized [289]. The reliability of the heat removal system determines the efficiency, maintenance cost and service life of the solar cell. Thus extreme conditions, such as maximum temperature limit and system failure, along with system weight should be considered in the final design.

The installation of solar power systems should consider cost-effectiveness and environmental issues [290]. For a solar power system, the major economic benefits arise from waste heat utilization and resulting cost savings during operations [291]. Such a combined heat and power system increases system reliability with the reduction in cost and the improvement of overall thermal efficiency. Environment pollution is also significantly reduced [291]. Both coolant technology and nature of coolants are important for cell efficiency and cell degradation. The coolant or working fluid should be compatible, which means that it should not attack or corrode the envelope or wick and there is no chemical reaction between the working fluid and the envelope or wick structure that liberates non-condensable gas (NCG).

Cooling technologies of the CPVT system can be divided into two types: passive cooling and active cooling. Passive cooling system has the advantages of zero energy consumption, low cost and easy installation. Back metal plate of high conductivity material with or without fins and heat pipe cooling system are the representative models. Metal plates were usually utilized as the heat sink for passive cooling of the photovoltaic module [292–294]. Cheknane et al. [295] proposed a heat sink integrated with a heat pipe for cooling CPV cells (see Figure 6.36). The passive cooling system has mainly been applied in low concentration ratio cell due to its limited cooling effect. Earlier studies by Royne et al. [296] indicate that while passive cooling by thermally conductive heat sink was insufficient for linear concentrator of CR above 20 suns, the active cooling using forced circulation caused additional power loss. Thermally conductive heat sink is also used in active cooling. Its characteristics of high heat dissipating capacity make it commonly adopted in high concentrated PV power generation system [297]. A schematic of passive and active cooling design option is illustrated in Figure 6.37. For the traditional cooling systems, the working mediums usually are water or air. Since forced convective heat transfer coefficient is generally higher than natural convective heat transfer, heat transfer in active cooling systems is much higher than that of passive cooling systems and water is a better cooling medium compared to air. Thus, the air is usually used in some low concentrating systems to reduce the temperature [298].

Air cooling either by forced or natural flow provides a low-cost and simple heat-dissipating way, but air's poor thermo-physical characteristic offers a relatively low cooling efficiency [298,299]. The optimization of heat transfer during air cooling requires consideration of parameters such as the air flow rate, flow length and height, etc. Brinkworth and Sandberg [300] indicated that for a certain length of a PV array, the best cooling performance would be achieved when the ratio of the PV array's length ($L$) to the air flow's hydraulic diameter ($D$) is 20 ($L/D = 20$), without consideration of the influence of other factors. This helps to choose the power rating of the blower to avoid unnecessary energy waste. Air cooling also depends on the flow pattern and insertion of internals in the air duct. Water cooling technology also can be divided into two types, natural and forced circulation cooling. The important factor in the design is to ensure good heat conduction and keep the electric insulation between the PV cell and the surface of heat exchanger. The leakage problem of the working medium should also be taken into consideration. A typical water-cooling system is composed of a heat exchanger, a water tank, valves, etc. Besides pure water cooling, often PCM materials with water are used to improve cooling efficiency.

Besides the conventional air and water-cooling methods, some advanced cooling technologies, such as microchannel cooling, impinging jet cooling, liquid immersion cooling, heat pipe cooling, hybrid cooling phase change material, promising ground-coupled cooling as well as earth water heat exchanger cooling, are applied and assessed for cooling the CPVT system. Two of these are graphically illustrated in Figures 6.38 and 6.39. The advantages, disadvantages and applications of these various cooling technologies are illustrated by Jakhar et al. [301].

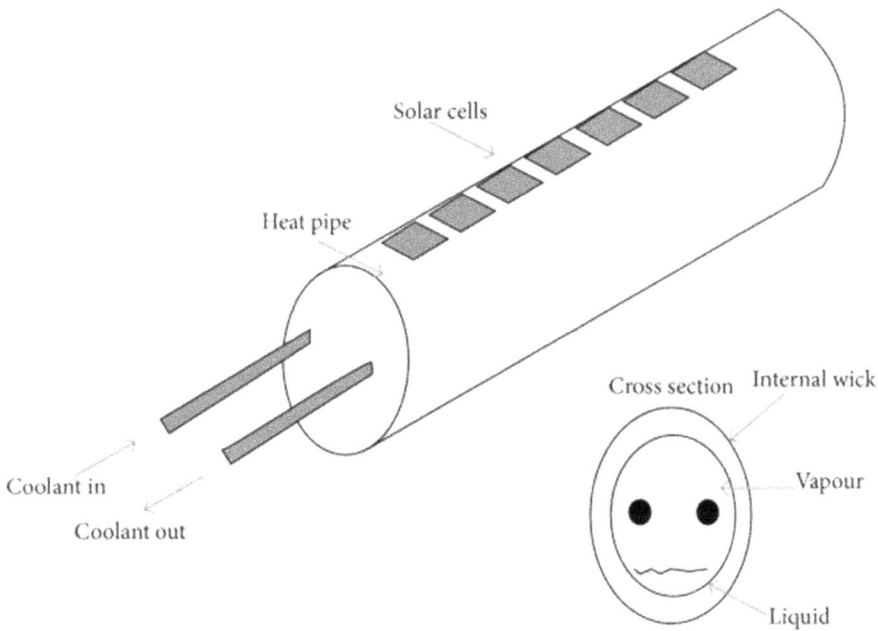

**FIGURE 6.36**   Heat pipe-based cooling system [17].

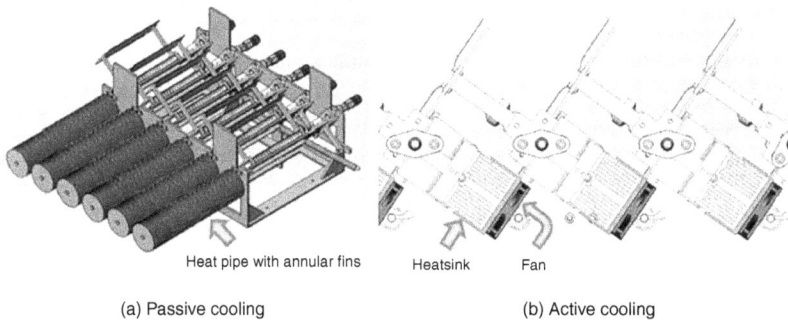

**FIGURE 6.37**   Schematic of two cooling design options: (a) passive heat pipe cooling; (b) active fan cooling [19].

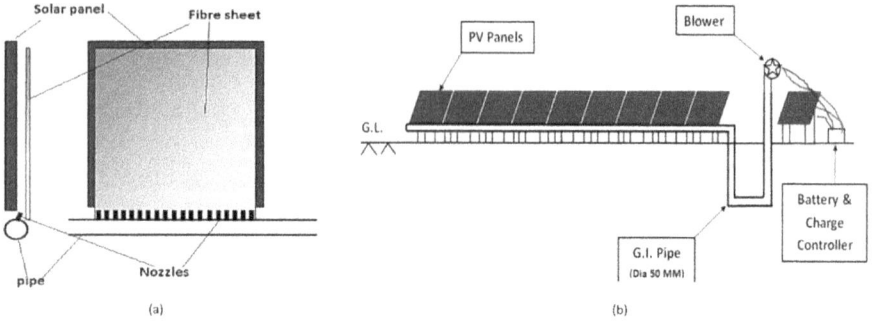

**FIGURE 6.38** Arrangement of solar panel, fiber sheet and nozzles (a) and schematic design of Ground-Coupled Central Panel Cooling System (GC-CPCS) (b) [19].

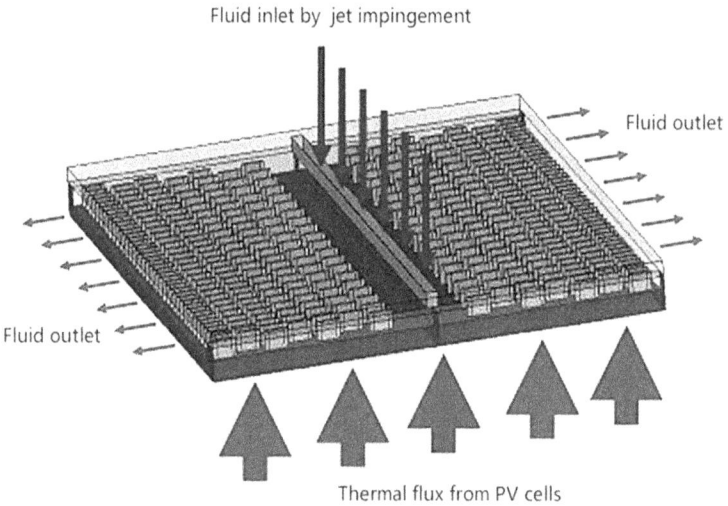

**FIGURE 6.39** Principle and operation of the hybrid cooling device [19].

One of the efficient collectors for the CPVT systems is point-focus Fresnel systems. In these units, the modules are cooled passively. Xie et al. [194] reported that for single cell geometries and high concentration ratio like 1,000 suns passive cooling provides a large surface area for heat sink. On the other hand, a large portion of solar radiation is dissipated as waste heat from the modules. In active cooling, the heat transfer from the central absorber to the working fluid can be recovered or recycled resulting in an enhancement of overall thermal efficiency. The recovered heat can be used for a domestic and industrial system through applications of solar cooling, air conditioning [302] and solar desalination [303] among others. There is significant literature on both passive and active cooling for CPVT systems. These are well reviewed by Khamooshi et al. [17], Xiao et al. [19], Bandaru et al. [18] and others [1,4,10–16,20].

## 6.9.3 ROLE OF NANOFLUIDS IN PVT AND CPVT SYSTEMS

There is extensive literature on the use of nanofluid as a coolant for PVT and CPVT systems. The application of nanofluids for heat removal in PVT system is extensively reviewed by Diwania et al. [1]. Abbas et al. [304] identified the necessity to study the effect of nanofluid usage on the lifetime enhancement of PVT systems. Yazdanifard et al. [305] indicated that due to economic and emission reduction reasons, a comprehensive review of the parameters affecting nanofluid-based PVT's performance is needed. Shah and Ali [306] investigated the stability of nanofluid for PVT applications. Said et al. [307] reviewed the environmental effects of conventional and nanofluid-based PVT systems. The performance of a copper-sheet-laminated PVT using copper oxide/water and nanofluid was studied by Michael and Iniyan [308]. They reported that the nanofluid-based system significantly improved the thermal efficiency compared to water-based PVT. Lari et al. [309] conducted a techno-economic performance analysis of a hybrid parallel-serpentine heat exchanger-based PVT system and reported that nanofluid gave 13.5% increase in electrical output compared to uncooled PVT system. Lu et al. [310] showed that heat transfer coefficient in an evacuated tubular solar collector increased with nanofluid (CuO/water). Kang et al. [311] conducted an economic analysis of flat-plate and U-tube solar collectors with nanofluid ($Al_2O_3$/water). Karami and Rahimi [312] examined the cooling performance of channels using Boehmite ($AlOOH \cdot xH_2O$) nanofluid for the PV module. Al-Waeili et al. [313] examined the PVT system using nanofluid (SiC/water). The results indicated that the overall efficiency of the PVT system was high compared to the PV system. Shmani et al. [314] showed that SiC-water nanofluid gave thermal and electrical efficiencies at 81.73% and 13.52%, respectively for a PVT system.

The study of Faizal et al. [315] indicated that higher density and lower specific heat of nanofluids offer higher thermal efficiency than water. Ladjevardi et al. [316] numerically examined the effects of nanofluid containing different diameters and volume fractions of graphite nanoparticles on the performance of a solar collector. Filho et al. [317] studied silver nanoparticles as direct sunlight absorbers for solar thermal applications. Saidur et al. [318] investigated the effects of nanofluid on the efficiency of a direct absorption solar collector (DASC) while Karami et al. [319] examined the stability of nanofluid for a DASC. Said et al. [320] found that nanofluids with single wall carbon nanotubes (SWCNTs) performed better than the ones with $Al_2O_3$, $TiO_2$ and $SiO_2$ nanoparticles in a flat plate solar collector. Tang et al. [321] showed that PEG/$SiO_2$/MWCNT composites can effectively improve the efficiency of solar energy applications. Sokhansefat et al. [322] numerically investigated the heat transfer enhancement by a nanofluid in a parabolic trough collector tube at different operational temperatures. Liu et al. [323] showed that CuO/water nanofluid improved the performance of a solar collector integrated with an open thermosyphon. The experiments by Manikandan and Rajan [324] showed a higher temperature rate in the case of nano fluid of sand-PG-water (0.5 vol%) than in that of using only PG-water. Al-Waeli et al. [313] found that 3 wt.% nanofluid gave promising enhancements in both the electrical and thermal efficiencies at the levels of 24.1 and 100.19%, respectively. Al-Waeli et al. [325] also built a novel design of the PVT system using nano particles in a PCM-enhanced process. Sardarabadi et al. [326]

examined the effects of $SiO_2$/water nano fluid on thermal and electrical efficiencies for a PVT system with a flat plate solar collector.

Karami and Rahimi [324] experimentally investigated the cooling performance of water-based Boehmite (AlOOH. $xH_2O$) nanofluid in a hybrid photovoltaic (PV) cell. Sardarabadi and Passandideh-Fard [326,327] presented a numerical and experimental study of a PVT system cooled by different types of nanofluids. Khanjari et al. [328] performed a CFD analysis of a PVT system using Ag-water and aluminum-water nanofluids. Hashim et al. [329] showed that for a nanofluid with a concentration of 0.3%, the temperature dropped significantly and the electrical efficiency rose to 12.1%. Elmir et al. [330] presented a simulation study for a one-way channel at the back side of the PV/T system with nanofluid as a working medium. Rejeb et al. [331] tested different types of nanoparticles (and Cu) at several concentrations with different base fluids on the electrical and thermal efficiencies of the system. Nada et al. [332] showed that by using the PCM and nanoparticles, both efficiency and temperature decline increased compared to the use of PCM only.

Michael and Iniyan [308] carried out an experimental study by adding a thin copper sheet instead of a Tedlar layer to the silicon cell and used it in the form of nanofluid as a cooling medium to enhance the performance of the system. They found that the thermal efficiency when using glazing and nanofluid was enhanced by about 45% in comparison with water only. Ghadiri et al. [333] experimentally examined cooling a PVT system using a ferrofluid under a magnetic field. The results showed that ferrofluid enhanced the overall efficiency by about 76% at 3 wt% concentration of nanofluid compared to distilled water. Various scientists have also examined the effects of nanofluids on the performance of the direct absorption solar collector (DASC), flat-plate and U-tube solar collectors (FP&UTC) [334–338]. In all cases, nanofluid improved the performance of solar collectors. The uses of nanofluid with different types of nanoparticles and concentrations have also shown improvement in the performance of various types of evacuated tube solar collector under different operating conditions [339–345]. All this literature clearly indicates that the use of nanofluids improves heat transfer and thermal efficiency of PVT and CPVT. More work is needed to demonstrate the overall economic impact of nanofluids and long-term stability of nanofluids in commercial PVT and CPVT operations.

### 6.9.4 APPLICATIONS OF THERMAL ENERGY OF COOLANT FLUIDS IN PVT AND CPVT SYSTEMS

Along with the generation of power, the use of thermal energy carried by coolant in PVT and CPVT systems improves the overall energy efficiency of the photovoltaic systems. The thermal energy in PVT and CPVT can be used for domestic hot water, local heating, heat pump, air conditioning, absorption cooling, vapor compression, thermal desalination, district heat, organic Rankine cycle, and so on. On the other hand, the electric energy can be used to drive the auxiliary devices of a thermal system or to drive the processes which can easily be combined with the thermal processes.

CPVT can be used in a number of different ways to provide heating and cooling. These include the use of CPVT for heat pumps, absorption cooling, vapor recompression, air conditioning, water heating, etc. The heat pump can be integrated with CPVT systems, such that the HTF temperature can be increased without sacrificing the PV efficiency. This kind of integration is more proper for LCPVTs. Xu et al. [346] fabricated a CPC-LCPVT integrating a heat pump system. Tsai [347] studied a refrigerant-based PVT system integrated with a heat pump water heating (HPWH) device to evaluate the electrical and thermal performance. Fang et al. [348] investigated the electrical and thermal performance of a PVT heat pump air-conditioning system. Different types of absorption solar cooling systems which include single effect, double effect and half effect absorption cycle have been also reported in numerous review papers [349–351]. The dominant driving power in solar absorption cooling systems is the thermal power from solar energy collectors. Solar radiation is absorbed by solar collectors then delivered to the storage tank through a hydraulic pump. This was analyzed by Alobaid et al. [352] and Vokas et.al [353]. Mittelman et.al [354] studied the performance and the economic viability of using triple junction cell concentrating photovoltaic thermal collectors (CPVT) for cooling and power generation. Bunomano et al. [355], Fumo et al. [356] and Eicker et al. [357] carried out theoretical and economic analysis of the use of CPVT for different types of heating and cooling systems at various locations.

Sanaye and Sarrafi [358] carried out a multi-objective optimization approach for combined solar cooling, heating and power generation CCHP system based on energy, exergy and economic evaluation. Al-Alili et al. [359] reviewed solar thermal air conditioning technologies and reported a number of research outcomes from the point of view of working fluid temperature, collector type, collector area, storage volume and COP values. Al-Alili et al. [360,361] also studied the application of the CPVT hybrid system for air conditioning of buildings in hot and humid climates. The system consisted of a solid desiccant wheel cycle (DWC) and a traditional vapor compression cycle (VCC). Guo et al. [362] reviewed the utilization of PVT for desiccant cooling and dehumidification that required a temperature in the range of 50°C⁻60°C. Lin et al. [363] investigated the use of PVT collectors and phase change materials (PCMs) that were integrated into the ceiling, in order to provide heating and cooling by utilizing solar radiation during winter daytime and radiative cooling during summer night-time in Sydney. Liang et al. [364] studied the dynamic performance of a PVT heating system which consisted of a PVT, to provide electricity and low-grade heat energy, hot water storage tank, heat exchanger to transfer the heat to under floor piping system and electric backup heater to maintain the temperature of the under floor system within the design points. Hartmann et al. [365] carried out a comparison between a solar electric compression refrigeration system and a solar adsorption refrigeration system to evaluate the primary energy savings and the cost to meet the demand for heating and cooling of a typical building in Germany and Spain.

In polygeneration CPVT systems, the PV part provides electricity for system parasitic consumption, local electricity supply or grid demands and the thermal part provides heat, used for space heating, domestic hot water and driving the absorption chiller to produce cooling energy. Kribus and Mittelman [366] theoretically compared several configurations for polygeneration based on the CPVT conception. Polygeneration led to increased conversion efficiency, and the PV-based polygenerations showed better performance than the heat engine-based ones. The feasibility of using a CPVT trigeneration system to satisfy the electricity and heating and cooling demands was examined by by Petrucci et al. [367], Garcia-Heller et al. [368] and Buonomano et al. [355,369,370]. Calise et al. [371] also showed that a similar type of polygeneration CPVT system was technically feasible for the university building. Calise et al. [372,373] also investigated The CPVT polygeneration system for energy demands of small isolated communities in European Mediterranean countries. Calise et al. [374] investigated a dynamic simulation system for cooling, heating and other building demand for electricity in order to find the optimal capacity of a solar collector field. Calise et al. [375] simulated the operation of a CPVT coupled air handling unit.

CPVT can be used for desalination in two different ways. The power generated from CPVT can be used for RO and MD processes. The waste heat can be used for MED and MEED processes. Thus, the combined CPVT-desalination plant can produce electricity and desalinated water. Ong et al. [376] developed a multi-effect membrane distillation (MEMD) system for isolated islands or coastal regions. Wisenfarth et al. [21] examined the application of CPVT mirror dish system for water desalination. The power generated from CPVT can also be used for the energy-efficient reverse osmosis process for desalting brackish water. The dual use of CPVT for both power and waste heat is most appropriate for this two-stage (RO and MD) desalination process for brackish water with high salinity. The hybrid process also reduces the quantity of brine that needs to be disposed when concentrating to a high salinityKelley et al. [377] used another novel way the LCPVT system for PV-powered RO desalination unit. Along with PV power, the study also exploited the complementary thermal energy of the solar panel for the reverse osmosis unit. Hughes et al. [378] examined an application of the CPVT system for MD. Seawater was used to cool III-V junction solar cells, and then further heated in an additional evacuated tube thermal collector. Mittleman et al. [303] combined a multi-effect evaporation desalination (MEED) system with a CPVT system.

PVT and CPVT can have a major impact on the reduction of $CO_2$ emission in the building sector. Integrating CPV/PV with building façades or windows not only can generate electricity but also can decrease direct sunlight and thus reduce air conditioning loads (up to 65% of heat gain reduction) [301,379]. A novel building façade integrated asymmetric Compound Parabolic Photovoltaic concentrator (BFI-ACP-PV) was designed and developed for saving materials and electricity costs by Lu et al. [380].

Compared to ground-mounted PV systems, rooftop CPV must balance low-cost and high-energy production, and it must be compatible with the rooftop environment for safety and wind loading. In order to meet these requirements, a system consisting of low-profile carousel-mounted array of Fresnel concentrators with triple-junction solar cell and a passive cooling scheme was proposed by an energy innovation company. The system was demonstrated to perform reliably for 15–20 years by Gleckman [381,382]. In CPV systems, increasing the concentration ratio monotonically yields higher cell temperature. The waste heat at the back of CPV cells can be recovered and used for space heating and residential water. The waste heat can also be injected into various industrial applications. The energy consumption in food, beverage, paper and textile industries can be supplied by low grade heat (<150°C) from CPVT. Also, CPV temperature frequently rises above 200°C as CPVs are mostly installed and operated under immense and stable solar radiation intensity [383]. Since PCM can store and deliver thermal energy on demand [394,395], temperatures up to 200°C can be used to serve various other process industries by using PCM as the heat storage material. Different PCMs with suitable melting points are available to store and supply heat demand at desired temperatures for various industrial requirements.

Finally, McConnell and Fthenakis [13] showed that in addition to generating clean, carbon-free solar electricity at low cost, concentrator PV systems potentially can produce hydrogen through the electrolysis of water. Concentrator photovoltaic systems potentially can generate cheaper electricity, primarily by utilizing high-efficiency multi-junction III-V solar cells. But it is the heat boost from concentrator PV systems that can dramatically improve and enhance the electrolysis efficiency of water in a high-temperature solid-oxide electrolyzer. This heat boost, ~40%, measured by Solar Systems in Australia above 1,100°C [386,387], was estimated in theoretical analyses [388]. McConnell and Fthenakis [13] suggested that this new pathway provides significant engineering- and economic benefits for generating electrolytic hydrogen from solar energy, thereby creating opportunities for PV directly to contribute to future transportation markets with low-cost hydrogen, or by producing liquid hydrogen-carrier fuels, such as methanol [389].

## 6.10   ADVANCES IN POWER GENERATION BY PV AND CPV

Large-scale PV and CPV power plants can be operated in a number of different ways. CPV has been used to generate utility-scale power generation. Table 6.12 lists all CPV power plants with a capacity of 1 MW or more using high concentration PV. Low concentration PV is used only in a few power plants such as SunPower, the USA in Arizona (7 MW), Solaria, the USA in Sardinia and Puglio (2 MW), SolFocus, the USA in Cerro Prieto (1 MW), Sungrow, China in Qinghai and Wuwei, Gansu (1 MW) and Abengoa Solar, Spain in Sanlucar La Mayor (1.2 MW) [162,164].

### 6.10.1   Economics of Power Generation by Micro-Tracked CPV

Tracking system is very important for CPV. Besides pedestal-mounted tracking systems, CPV power plants can also be operated as micro-tracked CPV. Wright et al. [263] evaluated the cost competitiveness of Micro-Tracked CPV with PV in Behind-The-Meter Applications with Demand Charges. To date, 86% of CPV deployments have been utility-scale projects using pedestal-mounted trackers. Research into micro tracking technology for CPV, in which the cell is moved in relation to fixed optics can result in a form factor for the modules that are similar to that of fixed PV, allowing CPV to compete with PV in behind-the-meter applications. In economic assessment, cost of land/roof area can be a significant factor, conferring an advantage on CPV over PV since CPV's greater efficiency allows it to occupy less land/roof area. Wright et al. [263] estimate target prices for micro-tracked CPV ($/W) at which it can compete with PV in behind-the-meter situations using microgrids with battery capacities optimized for each situation. This analysis took into account the price of batteries used in microgrids, the load profile of the customer, the demand and time-of-use charges, future trends in electricity prices and the hourly variation of

## TABLE 6.12
## Completed CPV Power Plants with a Capacity of 1 MW or More

| Company | Origin Company | Power in MW | Country | Location |
|---|---|---|---|---|
| Soitec | France/Germany | 5.8 | China | Hami |
| Soitec | France/Germany | 44.2 | South Africa | Touwsrivier |
| Soitec | France/Germany | 9.2 | USA | Borrego Springs |
| Soitec | France/Germany | 1.3 | Portugal | Alcoutim |
| Suncore Photovoltaic | China | 1.3 | Portugal | Evora |
| Soitec | France/Germany | 1.1 | Saudi Arabia | Tabuk |
| Solar Systems/Silex Systems | Australia | 1.0 | Saudi Arabia | Nofa |
| Suncore Photovoltaic | China | 79.8 | China | Goldmud |
| Soitec | France/Germany | 2.6 | China | Hami |
| Solaria | USA | 2.0 | Italy | Sardinia |
| Soitec | France/Germany | 1.7 | USA | Newberry Springs |
| Solar Systems/Silex Systems | Australia | 1.5 | Australia | Mildura |
| SolFocus | USA | 1.3 | Mexico | Guanajuato |
| Suncore Photovoltaic | China | 1.2 | USA | Albuquerque |
| Soitec | France/Germany | 1.2 | Italy | Saletti |
| Suncore Photovoltaic | China | 58.0 | China | Goldmud |
| Amonix | USA | 30.0 | USA | Alamosa |
| Magpower | Portugal | 3.0 | Portugal | Estoi |
| Solaria | USA | 2.0 | Italy | Puglia |
| Arima EcoEnergy Tech. Corp. | Taiwan | 1.7 | Taiwan | Linbian |
| Soitec | France/Germany | 1.2 | Italy | Santa Lucia |
| Soitec | France/Germany | 1.2 | Italy | Cerignola |
| Soitec | France/Germany | 1.1 | Italy | Bucci |
| Solaria | USA | 1.1 | USA | California |
| BEGI (Beijing General Industries) | China | 1.0 | China | Golmud |
| Solaria | USA | 1.0 | Italy | Sardinia |
| SolFocus | USA | 1.0 | Italy | Lucera |
| Amonix | USA | 5.0 | USA | Hatch |
| Amonix | USA | 2.0 | USA | Tucson |
| SolFocus | USA | 1.6 | USA | Yucaipa |
| SuncorePhotovoltaic | China | 1.5 | China | Xiamen |
| SolFocus | USA | 1.3 | USA | Hanford |
| SolFocus | USA | 1.3 | Greece | Crete |
| SolFocus | USA | 1.3 | USA | Yuma |
| Greenvolts | USA | 1.0 | USA | Yuma |
| SolFocus | USA | 1.0 | Chile | Santiago |
| Suncore Photovoltaic | China | 3.0 | China | Goldmud |
| Soitec | France/Germany | 1.4 | USA | Questa |
| SolFocus | USA | 1.3 | USA | Victorville |
| Amonix/Guascor Foton | Spain | 2.0 | Spain | Murcia |

*Note:* Power Plants Marked with an Asterisk Are Listed in the Project Library of the CPV Consortium (http://cpvconsortium.org/projects) [162,164].

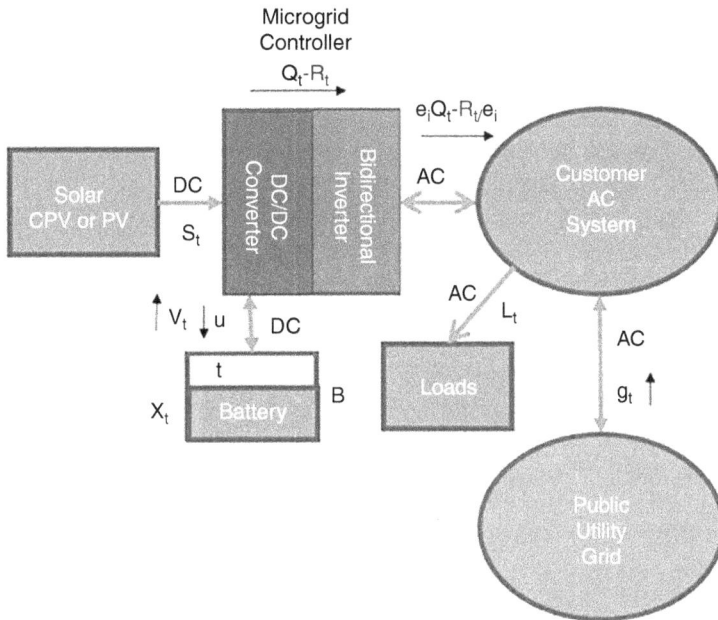

**FIGURE 6.40**   Utility-scale micro-tracked CPV plant with battery storage [263].

solar irradiance. The details are given by Wright et al. [263]. A typical micro-tracked CPV set-up with battery storage is illustrated in Figure 6.40.

An important finding of the study by Wright et al. [263] was that the target CPV prices are insensitive to (a) the prices of batteries used in the solar microgrids (b) the load profile of typical office buildings (c) the demand and time-of-use charges and (d) future trends in electricity tariffs. This robustness in the results was due to the fact that the analysis showed that these factors affect the profitability of PV and CPV in similar ways even though the optimal battery size in the microgrid was different for PV and CPV. The two factors that have a major influence on target CPV prices were (a) land price and (b) the ratio of CPV to PV annual energy yield. For the data chosen by Wright et al. [263], the energy yield ratio varied from 1.1 to 1.27. The results were obtained after optimization of PV and CPV microgrids, including optimal scheduling of power in/out of the battery and optimal battery sizing. The internal rate of return was then calculated over a 32-year life of the solar modules, replacing the battery and inverter as necessary within that time frame. The results allowed one to derive conclusions about optimized CPV microgrids in comparison with the optimized PV microgrids. The study showed that optimized CPV microgrids have the following advantages over optimized PV microgrids: (a) they achieve a higher reduction in peak time-of-use electricity charges (b) they achieve a greater reduction in monthly peak consumption from the grid, thus reducing demand charges (c) they require less battery cycles, thus reducing battery cell replacement costs.

## 6.10.2 Hybrid PV/TE System

The temperature of waste heat coming out of solar PV or CPV cell depends on the levels of concentration and temperature of the cell. Chávez-Urbiola et al. [390] demonstrated the feasibility of a PV-TE hybrid system based on a simulated model. PV-TE hybrid system can be created in two different ways; (a) using spectrum-splitter, (b) sticking the TE generator at the bottom of the PV cell [391]. Lee et al. [392] fabricated highly conductive PEDOT: PSS film as the TE generator with a PV cell to convert the heat generated in the organic solar cell to electrical energy. The literature indicates that the ambient temperature is very important for the effectiveness of the PV-TE hybrid system [393]. Teffah and Zhang [61] integrated the TEC-TEG module with the CPV-TE system. The study showed that under the concentration of 300 and 1,000 suns, the thermoelectric cooler cooled the triple junction solar cell from 368.2 K to about 322.6 K and 529 K to about 403 K respectively, with 1 W and 3 W increase in the corresponding total direct electrical power.

For a flat plate solar energy collector, Telkes [394] demonstrated efficiency of 0.63% and 3.35% at 1× and 50× suns (1 sun = 1,000 W/m²), respectively. ZnSb-type alloys in combination with a negative Bi-alloy were used. He et al. [395] carried out a theoretical and experimental study on the integration of thermoelectric modules into solar vacuum tube heaters (SHP-TE). Their experimental data showed an electrical efficiency of 1%. Miljkovic and Wang [396] and Fan et al. [397] carried out similar creative investigations using the same principle for the parabolic solar concentrator. Kraemer et al. [398] used nanostructured thermoelectric materials to develop the solar flat panel thermoelectric generators. These TEs achieved a maximum efficiency of 4.6% under an irradiation of 1 kW/m². The efficiency was seven to eight times higher than the best value previously reported for a flat panel display.

Amatya and Ram [399] combined a commercial $Bi_2Te_3$ module with a parabolic concentrator (solar concentration of 66× suns. System efficiency of 3% was measured and an output power of 1.8 W was achieved. Rehman et al. [400] proposed a novel collector design for a solar concentrated thermoelectric generator. The system had an electrical efficiency of 1.45% and a maximum optical efficiency of 93.61%. Li et al. [401] evaluated a prototype, consisting of a solar concentration thermoelectric generator (CTG) with a Fresnel lens. Their results showed that the highest possible CTG efficiency could reach 9.8%, 13.5% and 14.1%, for $Bi_2Te_3$, skutterudite and silver antimony lead telluride (LAST) alloys, respectively. Sahin et al. [402], Suleebka [403], Motiei et al. [404] and Mahmoudinezhad et al. [405] similarly examined hybrid solar PV and TEG systems for other applications. Sripadmanabhan Indira et al. [406] reported that the integrated CPV-TEG-based solar thermal systems achieve higher electrical and thermal performances than those of non-concentrated PV-TEG systems. Li et al. [407] compared a hybrid photovoltaic–thermoelectric (PV-TE) system employing a micro-channel heat pipe array together with PV electricity generation. The results showed that the electrical efficiencies of the hybrid PV-TE system were about 14% higher than those obtained with the PV system. Mizoshiri et al. [408] built a hybrid module composed of a thin-film thermoelectric module and a photovoltaic module. The total no-load voltage of the hybrid thermo-photovoltaic generator showed an increase of 1.3% when compared with the photovoltaic module alone.

According to Xiao et al. [19], the optimum operating temperature of the PV-TE hybrid system depends on the fluctuation of solar radiation. The use of a PCM cooling medium to reduce temperature variation has not been successful [409]. Zhang and Xuan [410] examined a novel PV-TE hybrid system with adjustable cooling blocks under different concentrated radiation (from 15.8 to 31.4 W) to reduce the temperature fluctuation caused by variation of the solar radiation. The results demonstrated this novel PV-TE hybrid system could achieve a lower temperature fluctuation than traditional hybrid system, and the output power can be increased in the PV-TE hybrid system [19,411]. Mohsenzadeh et al. [412], on the other hand, examined a novel PVT-TE hybrid system in which the entire system was cooled by water with a flow rate of 0.025 m/s. In this integrated system, a larger portion of solar radiation was directly converted to electricity thereby improving its electrical efficiency. Mohsenzadeh et al. [412] also showed that the usage of a glass cover tube on the receiver of the CPVT+TE system reduced the heat dissipation to the ambient air and increased the total efficiency, but it also increased the average PV cell temperature by 7.89%. More details on PV/TE hybrid systems are given by excellent reviews of Xiao et al. [19] and Zoui et al. [413].

## REFERENCES

1. Diwania S., Agrawal S., Siddiqui A.S., and Singh S. Photovoltaic–thermal (PV/T) technology: A comprehensive review on applications and its advancement. *Int J Energ Environ Eng* 2020;11:33–54 Doi: 10.1007/s40095-019-00327-y.
2. IRENA. *Renewable Power Generation Costs in 2018*. Abu Dhabi: International Renewable Energy Agency; 2019.
3. IRENA. Climate change and renewable energy – National policies and the role of communities, cities and regions. A report to the G20 Climate and Sustainability Working Group (CSWG). Abu Dhabi: International Renewable Energy Agency; 2019.
4. Ju X., Xu C., Liao Z., Du X., Wei G., Wang Z., and Yang Y. A review of concentrated photovoltaic-thermal (CPVT) hybrid solar systems with waste heat, recovery (WHR). *Sci Bull* 2017;62:1388–1426.
5. IRENA. *Innovation Landscape for a Renewable-Powered Future: Solutions to Integrate Renewables*. Abu Dhabi: International Renewable Energy Agency; 2019.
6. IRENA. *Renewable Capacity Statistics 2019*. Abu Dhabi: International Renewable Energy Agency; 2019.
7. IRENA. *Global Energy Transformation: A Roadmap to 2050*, 2019 ed. Abu Dhabi: International Renewable Energy Agency; 2019.
8. IEA. *Renewables 2018 – Analysis and Forecasts to 2023*. Paris: Organisation for Economic Co-operation and Development/International Energy Agency; 2018.
9. IEA et al. *Tracking SDG 7: The Energy Progress Report 2019*. International Energy Agency, International Renewable Energy Agency, United Nations Statistics Division, World Bank and World Health Organisation, Paris, France; 2019.
10. Lamnatou C., Vaillon R., Parola S., and Chemisana D. Photovoltaic/thermal systems based on concentrating and non-concentrating technologies: Working fluids at low, medium and high temperatures. *Renew Sustain Energ Rev* Elsevier, 2021;137:110625. Doi: 10.1016/j.rser.2020.110625.hal-03064762.
11. Dupeyrat P., Helmers H., Fortuin S., and Kramer K. Recent advances in the development and testing of hybrid pv-thermal collectors. A report by Fraunhofer Institute for Solar Energy Systems ISE, Freiburg, Germany; 2014.

12. Resch A., and Höller R. Design concepts for a spectral splitting CPVT receiver. *International Solar Energy Society EuroSun 2020 Proceedings*; 2020. The Authors. Published by International Solar Energy Society Selection and/or peer review under responsibility of Scientific Committee. Doi: 10.18086/eurosun.2020.05.07 Available from http://proceedings.ises.org.

13. McConnell R., and Fthenakis V. Concentrated photovoltaics, open access peer-reviewed chapter. In: Fthenakis V., editors. *Third Generation Photovoltaics*. An intech open access; 2012. Doi: 10.5772/39245.

14. Nishioka K., Takamoto T., Agui T., Kaneiwa M., Uraoka Y., and Fuyuki T. Annual output estimation of concentrator photovoltaic systems using high efficiency InGaP/InGaAs/Ge triple-junction solar cells based on experimental solar cell's characteristics and field-test meteorological data. *Sol Energ Sol Cells* 2006;90:57–67.

15. Babu C., and Ponnambalam P. The role of thermoelectric generators in the hybrid PV/T systems: A review. *Energ Convers Manage* 2017;151:368–385.

16. Azarian R.D., Cuce E., and Cuce P.M. An overview of Concentrating Photovoltaic Thermal (CPVT) Collectors, an open access science publication, distributed under a Creative Commons Attribution (CC-BY) 3.0 license. *Energ Res J* 2017;8(1):11–21. Doi: 10.3844/erjsp.2017.11.21.

17. Khamooshi M., Salati H., Egelioglu F., Faghiri A.H., Tarabishi J., and Babadi S. A review of solar photovoltaic concentrators. *Int J Photoenerg* 2014;2014:17 p, Article ID 958521. Doi: 10.1155/2014/958521.

18. Bandaru S.H., Becerra V., Khanna S., Radulovic J., Hutchinson D., and Khusainov R. A Review of Photovoltaic Thermal (PVT) technology for residential applications: Performance indicators, progress, and opportunities. *Energies* 2021;14:3853. Doi: 10.3390/en14133853.

19. Xiao M., Tang L., Zhang X., Lun I.Y.F., and Yuan Y. A review on recent development of cooling technologies for concentrated photovoltaics (CPV) systems. *Energies* 2018;11:3416. Doi: 10.3390/en11123416.

20. Ahmed A., Baig H., Sundaram S., and Mallick T.K. Use of nanofluids in solar PV/thermal systems. *Int J Photoenerg* 2019;2019:17 p, Article ID 8039129. Doi: 10.1155/2019/8039129.

21. Wiesenfarth M., Went J., Bösch A., Dilger A., Kec T., Kroll A., Koschikowski J., and Bett A.W. CPV-T mirror dish system combined with water desalination systems. *AIP Conf Proc* 2016;1766:020008, Published Online: 01 September 2016. Doi: 10.1063/1.4962076.

22. Bielecki A., Ernst S., Skrodzka W., and Wojnicki I. Concentrated solar power plants with molten salt storage: Economic aspects and perspectives in the European Union. *Int J Photoenerg* 2019;2019:10 p, Article ID 8796814. Doi: 10.1155/2019/8796814.

23. Bozorgan N., and Shafahi M. Performance evaluation of nanofluids in solar energy: A review of the recent literature. *Micro Nano Syst Lett* 2015;3:5. Doi: 10.1186/s40486-015-0014-2.

24. Imenes A.G., and Mills D.R. Spectral beam splitting technology for increased conversion efficiency in solar concentrating systems: A review. *Sol Energ Mat Sol Cells* 2004;84:19–69.

25. Chow T.T. A review on photovoltaic/thermal hybrid solar technology. *Appl Energ* 2010;87:365–379.

26. Sharaf O.Z., and Orhan M.F. Concentrated photovoltaic thermal (CPVT) solar collector systems: Part II – Implemented systems, performance assessment, and future directions. *Renew Sust Energ Rev* 2015;50:1566–633.

27. Sharaf O.Z., and Orhan M.F. Concentrated photovoltaic thermal (CPVT) solar collector systems: Part I – Fundamentals, design considerations and current technologies. *Renew Sust Energ Rev* 2015;50:1500–1565.

28. Zhang L., Jing D., Zhao L., Wei J., and Guo L. Concentrating PV/T hybrid system for simultaneous electricity and usable heat generation: A review. *Int J Photoenerg* 2012;2012:869753.
29. Chow T.T., Tiwari G.N., and Menezo C. Hybrid solar: A review on photovoltaic and thermal power integration. *Int J Photoenerg* 2012;2012:307287.
30. Zhang X., Zhao X., Smith S., Xu J., and Yu X. Review of R&D progress and practical application of the solar photovoltaic/thermal (PV/T) technologies. *Renew Sust Energ Rev* 2012;16:599–617.
31. Hasan M.A., and Sumathy K. Photovoltaic thermal module concepts and their performance analysis: A review. *Renew Sust Energ Rev* 2010;14:1845–1859.
32. Shan F., Tang F., Cao L., and Fang G. Performance evaluations and applications of photovoltaic–thermal collectors and systems. *Renew Sust Energ Rev* 2014;33:467–483.
33. Tyagi V.V., Kaushik S.C., and Tyagi S.K. Advancement in solar photovoltaic/thermal (PV/T) hybrid collector technology. *Renew Sust Energ Rev* 2012;16:1383–1398.
34. Pikk P., and Annuk A. Case study of increasing photovoltaic energy solar fraction in a conventional office building in northern lattitudes, *Agron Res* 2014;12(2):563–574.
35. Dupeyrat P., Helmers H., Fortuin S., and Kramer K. Recent advances in the development and testing of hybrid pv-thermal collectors. *Proceedings of ISES Solar World Congress*, Kassel, Germany; 2011, pp. 4548–4559.
36. Avezov R., Akhatov J., and Avezova N. A review on photovoltaic-thermal (PV-T) air and water collectors. *Appl Sol Energ* 2011;47:169–183.
37. Singh B., and Othman M.Y. A review on photovoltaic thermal collectors. *J Renew Sustain Energ* 2009;1:062702.
38. Charalambous P.G., Maidment G.G., Kalogirou S.A., and Yiakoumetti K. Photovoltaic thermal (PV/T) collectors: A review. *Appl Therm Eng* 2007;27:275–286.
39. Ju X., Xu C., Han X., Du X., Wei G., and Yang Y. A review of the concentrated photovoltaic/thermal (CPVT) hybrid solar systems based on the spectral beam splitting technology. *Appl Energ* 2017;187:534–563.
40. Meng X., Xia X., Sun C., Li Y., and Li X. A novel free-form Cassegrain concentrator for PV/T combining utilization. *Sol Energ* 2016;135:864–873.
41. Ju X., Xu C., Liao Z., Du, X., Wei, G., Wang, Z. and Yang, Y. A review of concentrated photovoltaic-thermal (CPVT) hybrid solar systems with waste heat recovery (WHR). *Sci Bull* 2017;62. Doi: 10.1016/j.scib.2017.10.002.
42. Han X., Zhao G., Xu C., Ju X., Du X., and Yang Y. Parametric analysis of a hybrid solar concentrating photovoltaic/concentrating solar power (CPV/CSP) system. *Appl Energ* 2017;189:520–533.
43. Han X., Xu C., Ju X., Du X., and Yang Y. Energy analysis of a hybrid solar concentrating photovoltaic/concentrating solar power (CPV/CSP) system. *Sci Bull* 2015;60:460–469.
44. Branz H.M., Regan W., Gerst K.J., Borak J.B., and Santori E.A. Hybrid solar converters for maximum exergy and inexpensive dispatchable electricity. *Energ Environ Sci* 2015;8 (11). Doi: 10.1039/C5EEQ1998B.
45. Branz H.M. Full-spectrum optimized conversion and utilization of sunlight (FOCUS) program. A paper presented at *20th Solar Power and Chemical Energy Systems Conference (SolarPACES 2014)* Beijing, China; 2014.
46. DeJarnette D., Otanicar T., Brekke N., Hari P., Roberts K., Saunders A.E., and Morad R. Plasmonic nanoparticle based spectral fluid filters for concentrating PV/T collectors. *Proceedings of SPIE 9175, High and Low Concentrator Systems for Solar Energy Applications IX*, San Diego, CA: International Society for Optics and Photonics; 2014, p. 917509.
47. Huang G., Curt S.R., Wang K., and Markides C.N. Challenges and opportunities for nanomaterials in spectral splitting for high-performance hybrid solar photovoltaic-thermal applications: A review. *Nano Mater Sci* 2020;2:183–203.

48. Mojiri A., Taylor R., Thomsen E., and Rosengarten G. Spectral beam splitting for efficient conversion of solar energy—A review. *Renew Sustain Energ Rev* 2013;28:654–663.
49. Stefancich M., Zayan A., Chiesa M., Rampino S., Roncati D., Kimerling L., and Michel J. Single element spectral splitting solar concentrator for multiple cells CPV system. *Opt Exp* 2012;20:9004–9018.
50. Vincenzi D., Busato A., Stefancich M., and Martinelli G. Concentrating PV system based on spectral separation of solar radiation. *Physica Status Solidi A* 2009;206:375–378.
51. Taudien J.Y., and Kern L.A. Concentrating and spectrum splitting optical device in high efficiency CPV module with five bandgaps. *Proc SPIE* 2013;8821:88210A.
52. Orosz M., Zweibaum N., Lance T., Ruiz M., and Morad R. Spectrum-splitting hybrid CSP-CPV solar energy system with standalone and parabolic trough plant retrofit applications. *AIP Conf Proc* 2016;1734:070023.
53. Hu F., Liu Y., Zhang Q., and Chen Z. *Solar Concentration and Frequency Division Photovoltaic Photo-Thermal Comprehensive Utilization Device*. CN103236463A; 7 August 2013.
54. Van Sark W.G.J.H.M., Barnham K.W.J., Slooff L.H., Chatten A.J., Büchtemann A., Meyer A., McCormack S.J., Koole R., Farrell D.J., Bose R., and Bende E.E. Luminescent solar concentrators—A review of recent results. *Opt Exp* 2008;16:21773–21792.
55. Zhu L., Yoshita M., Nakamura T., Imaizumi M., Kim C., Mochizuki T., Chen S., Kanemitsu Y., and Akiyama H. Characterization and modeling of radiation damages via internal radiative efficiency in multi-junction solar cells. *Phys Simul Photonic Eng Photovolt Devices V* 2016;9743:97430U.
56. Widyolar B., Jiang L., and Winston R. Spectral beam splitting in hybrid PV/T parabolic trough systems for power generation. *Appl Energ* 2018;209:236–250.
57. Yaping J., Ollanik A., Farrar-Foley N., Qi X., Madrone L., Lynn P., Romanin V., Codd D., and Escarra M. Transmissive spectrum splitting multi-junction solar module for hybrid CPV/CSP system. *Proceedings of the 2015 IEEE 42nd Photovoltaic Specialist Conference (PVSC)*, New Orleans, LA; 14–19 June, 2015, pp. 1–5.
58. Widyolar B.K., Abdelhamid M., Jiang L., Winston R., Yablonovitch E., Scranton G., Cygan D., Abbasi H., and Kozlov A. Design, simulation and experimental characterization of a novel parabolic trough hybrid solar photovoltaic/thermal (PV/T) collector. *Renew Energ* 2017;101:1379–1389.
59. Sun X., Silverman T.J., Zhou Z., Khan M.R., Bermel P., and Alam M.A. Optics-based approach to thermal management of photovoltaics: Selective-spectral and radiative cooling. *IEEE J Photovolt* 2017;7:566–574.
60. Rey-Stolle I., Olson J.M., and Algora C. Concentrator multijunction solar cells. In *Handbook of Concentrator Photovoltaic Technology*. John Wiley & Sons, Ltd.: Hoboken, NJ; 2016, pp. 59–136.
61. Teffah K., and Zhang Y. Modeling and experimental research of hybrid PV-thermoelectric system for high concentrated solar energy conversion. *Sol Energ* 2017;157:10–19.
62. Sadewasser S., Salomé P.M.P., and Rodriguez-Alvarez H. Materials efficient deposition and heat management of CuInSe2 micro-concentrator solar cells. *Sol Energ Mater Sol Cells* 2017;159:496–502.
63. Burhan M., Shahzad M.W., and Ng K.C. Development of performance model and optimization strategy for standalone operation of CPV-hydrogen system utilizing multi-junction solar cell. *Int J Hydrogen Energ* 2017;42:26789–26803.
64. Fraunhofer Institute for Solar Energy Systems, ISE with support of PSE Projects GmbH, Freiburg, Germany; 2019. www.ise.fraunhofer.de.
65. *Best Research-Cell Efficiency Chart – NREL*. Golden, CO: U.S. Department of Energy; 2021. https://www.nrel.gov›pv›cell-efficiency.
66. Dambhare M.V., Butey B., and Moharil S.V. Solar photovoltaic technology: A review of different types of solar cells and its future trends. *J Phys Conf Ser* 2021;1913:012053.

67. Goetzberger A., and Hebling C Photovoltaic materials, past, present, future. *Sol Energ Mater Sol Cells* 2000;62:1–19.
68. Green M.A., Zhao J., and Wang A. Progress and outlook for high-efficiency crystalline silicon solar cells. *Sol Energ Mater Sol Cells* 2001;65:9–16.
69. Zhao J., Wang A., and Greea M.A. High-efficiency PERL and PERT silicon solar cells on FZ and MCZ substrates. *Sol Energ Mater Sol Cells* 2011;65:429–435.
70. Zhao J., Wang A., and Green M.A. 19.8% efficient 'honeycomb' textured multicrystalline and 24.4% monocrystalline silicon solar cells. *Appl Phys Lett* 1998;73:1991–1993.
71. SunPower Corporation. 2015. Available from: http://www.sunpowercorp.co.uk/ (accessed: 18 December 2015).
72. Becker C., Sontheimer T., and Steffens S. Polycrystalline silicon thin films by high-rate electronbeam evaporation for photovoltaic applications – Influence of substrate texture and temperature. *Energ Procedia* 2011;10:61–65.
73. GlobalData. *Solar PV Module, Update 2018: Global Market Size, Competitive Landscape and Key Country Analysis to 2022.* London: GlobalData; 2018.
74. Parida B., Iniyan S., and Goic R. A review of solar photovoltaic technologies. *Renew Sustain Energ Rev* 2011;15:1625–1636.
75. Boutchich M., Alvarez J., and Diouf D. Amorphous silicon diamond based hetero junctions with high rectification ratio. *J Non-Crystal Solids* 2012;358:2110–2113.
76. Mah O. *Fundamentals of Photovoltaic Materials.* National Solar Power Research Institute, Inc.; 1998. Available at: http://userwww.sfsu.edu/ciotola/solar/pv.pdf (accessed: 18 December 2015).
77. Carlson D.E., and Wronski C.R. Amorphous silicon solar cell. *Appl Phys Lett* 1976;28:671–673.
78. Rech B., and Wagner H. Potential of amorphous silicon for solar cells. *Appl Phys Mater Sci Proces* 1999;69:155–167.
79. Extance A. The reality behind solar power's next star material. *Nat– Int J Sci* 2019. https://www.nature.com/articles/d41586-019-01985-y (accessed: 1 September 2019).
80. Boer K.W. Cadmium sulfide enhances solar cell efficiency. *Energ Conver Manage* 2011;52:426–430.
81. Compaan A.D. The status of and challenges in CdTe thin-film solar-cell technology. *MRS Sympos Proceed* 2004;808:545–555.
82. Schock H.W., and Pfisterer F. Thin-film solar cells: Past, present and future. *Renew Energ World* 2011;4:75–87.
83. Nowshad A., Takayunki I., and Akira Y. High efficient 1 μm thick CdTe solar cells with textured TCOs. *Sol Energ Mater Sol Cells* 2001;67:195–201.
84. Upadhayaya H.M., Razykov T.M., and Tiwari A. Photovoltaics fundamentals, technology and application. In: Goswami D.Y., and Kreith F., editors. *Handbook of Energy Efficiency and Renewable Energy.* New York: CRC Press; 2007, pp. 23-1–23-63.
85. Britt J., and Ferekides C. Thin film CdS/CdTe solar cell with 15.8% efficiency. *Appl Phys Lett* 1993;62:2851–2852.
86. Aramoto T., Kumaza S., and Higuchi H. 16.0% efficient thin-film CdS/CdTe solar cells. *J Appl Phys Part 1: Regul Papers Short Notes Rev Papers* 1997;36:6304–6305.
87. Wu X., Keane J.C., DeHart C., Albin D.S., Duda A., Gessert T.A., Asher S., Levi D.H., and Sheldon P. 16.5% efficient CdS/CdTe polycrystalline thin film solar cell. *Proceedings of the 17th European Photovoltaic Solar Energy Conference*, Munich; 14–17 October, 2001, pp. 995–999.
88. Cherradi N. *Solar PV Technologies What's Next?* Brussels: Becquerel Institute; 2019.
89. Schock H.W., and Shah A. *Proceedings of the 14th European PV Solar Energy Conference*, Barcelona, Spain; 1997.
90. Repins I., Conteras M., and Egaas B. 19.9%-efficient ZnO/CdS/CuInGaSe2 solar cell with 81.2% fill actor. *Progres Photovolt Res Appl* 2008;16:235–239.

91. Powalla M. The R&D potential of CIS thin-film solar modules. *Proceedings of the 21st European Photovoltaic Solar Energy Conference*, Dresden; September, 2006, pp. 1789–1795.

92. Iles P.A. Evolution of space solar cells. *Sol Energ Mater Sol Cells* 2001;68:1–13.

93. Saas F. A bright spot in a dark economy. 2009. Available from: http://www.pddnet.com/articles/2009/10/bright-spot-dark-economy (accessed: 18 December 2015).

94. Streetman B.G., and Banerjee S. *Solid State Electronic Devices*, 6th ed. Upper Saddle River, NJ: Prentice Hall; 2005.

95. Satyen K.D. Recent developments in high efficiency photovoltaic cells. *Renew Energ* 1998;15:467–472.

96. Yablonovitch E., Miller O.D., and Kurtz S.R. The opto-electronic physics that broke the efficiency limit in solar cells. *38th IEEE Photovoltaic Specialists Conference (PVSC)*, IEEE, Austin, TX; 3–8 June, 2012, pp. 001556–001559.

97. Sharp. Sharp develops solar cell with world's highest conversion efficiency of 36.9%. 2011. Available from: http://sharp-world.com/corporate/news/111104.html (accessed: 17 December 2015).

98. Osborne M. PV Tech. GaAs solar cell from Spire sets 42.3% conversion efficiency record. 2010. Available from: http://www.pv tech.org/news/gaas_solar_cell_from_spire_sets_42.3_conversion_efficiency_record (accessed: 16 December 2015).

99. Manna T.K., and Mahajan S.M. Nanotechnology in the development of photovoltaic cells. *IEEE Conference on Clean Electrical Power*, Capri, New York, IEEE; 21–23 May, 2007, pp. 79–86.

100. Meiller R. University of Wisconsin-Madison News. Future looks bright for carbon nanotube solar cells. 2013. Available from: http://www.news.wisc.edu/21890 (accessed: 19 December 2015).

101. Shi E., Zhang L., Li Z., Shang Y., Jia Y., Wei J., Wang K., Zhu H., Wu D., and Zhang S. $TiO_2$-coated carbon nanotube-silicon solar cells with efficiency of 15%. *PMC Scientific Report*, China; 2012, p. 884.

102. Multi junction solar cell, Wikipedia, the free encyclopedia, last edited 20 September, 2022. https://en.wikipedia.org/wiki/Multi-junction_solar_cell

103. Solar cell, Wikipedia, the free encyclopedia, last edited 23 September 2022. https://en.wikipedia.org/wiki/Solar_cell

104. Building integrated photovoltaics, Wikipedia, the free encyclopedia, last edited 12 March, https://en.wikipedia.org/wiki/Building-integrated_photovoltaics

105. Shravan K., and Chunduri K. PERC solar cell technology-2018 Edition. *TaiyangNews*. China; 2018.

106. Dullweber T., and Schmidt J. Industrial silicon solar cells applying the passivated emitter and rear cell (PERC) concept—A review. *IEEE J Photovolt* 2016, September; 6(5):1366–1381.

107. Roedern B. Status of amorphous and crystalline thin film silicon solar cell activities. *NCPV Sol Prog Rev Meet* 2003;5:552–555.

108. Hybrid solar cell. Wikipedia, the free encyclopedia, last edited 20 September, 2022. https://en.wikipedia.org/wiki/Hybrid_solar_cell

109. Thin film solar cell, Wikipedia, the free encyclopedia, last edited 23 September, 2022. https://en.wikipedia.org/wiki/Thin-film_solar_cell

110. Itoh M., Takahashi H., and Fujii T. Evaluation of electric energy performance by democratic module PV system field test. *Sol Energ Mater Sol Cells* 2001;67:435–440.

111. Wu L., Tian W., and Jiang X. Silicon based solar cell system with a hybrid PV module. *Sol Energ Mater Sol Cells* 2005;87:637–645.

112. Zipp K. Solar power world. Hybrid solar panel generates more power. 2011. Available from: http://www.solarpowerworldonline.com/2011/05/hybrid-solar-panel-generates-more-power/ (accessed: 10 December 2015).

113. Evoenergy. 2015. Available from: http://www.evoenergy.co.uk/ (accessed: 19 December 2015).
114. Panasonic. 2014. Panasonic HIT® solar cell achieves world's highest energy conversion efficiency of 25.6% at research level. Available from: http://panasonic.co.jp/corp/news/official.data/data.dir/2014/04/en140410-4/en140410-4.html (accessed: 15 December 2015).
115. Twidell J., and Weir T. *Renewable Energy Resources*, 2nd ed. London: Taylor & Francis; 1986.
116. Gerischer H., and Tributsch H. Elecrochemische Untersuchungen zur spectraleu sensibilisierung von ZnO-Einkristalien. *Berichte der Bunsengesellschaft für Physikalische Chemie* 1968;72:437–445.
117. Hauffe K. Danzmann H.J., and Pusch H. New experiments on the sensitization of zinc oxide by means of the electrochemical cell technique. *J Electrochem Soc* 1970;117:993–999.
118. Deb S.K., Chen S., and Witzke H. United States Patents US 4117510, 4080488, 4118246, and 4118247; 1978.
119. Ayre J. Clean technica: Dye-sensitized solar cells achieve record efficiency of 15%. 2013. Available from: http://cleantechnica.com/2013/07/15/dye-sensitized-solar-cells-achieve-record-efficiency-of-15/ (accessed: 19 December 2015).
120. Papageorgiou N. École Polytechnique Fédérale De Lausanne News. 2013. Available from: http://actu.epfl.ch/news/dye-sensitized-solar-cells-rival-conventional-ce-2/ (accessed: 15 December 2015).
121. Gratzel M. Dye-sensitized solar cells. *J Photochem Photobiol C: Photochem Rev* 2003;4:145–153.
122. Bifacial solar cell, Wikipedia, the free encyclopedia, last edited 24 September, 2022. https://en.wikipedia.org/wiki/Bifacial_solar_cells
123. Goetzberger A., and Hoffmann V.U. Photovoltaic solar energy generation. *Springer Series Opt Sci* 2005;112.
124. Yamaguchi M., Nishimura K., and Sasaki T. Novel materials for high-efficiency III–V multi-junction solar cells. *Sol Energ* 2008;82:173–180.
125. Yamaguchi M., Takamoto T., and Arak K. Multi-junction III–V solar cells: Current status and future potential. *Sol Energ* 2005;79:78–85.
126. Shockley W., and Queisser H.J. Detailed balance limit of efficiency of p–n junction solar cells. *J Appl Phys* 1961;32(3):510–519.
127. Guha S. Thin film silicon solar cells grown near the edge of amorphous to microcrystalline transition. *Sol Energ* 2004;77:887–892.
128. Diefenbach K.H. Wiped away. Photon International February, 2005, 48–67.
129. Yamamoto K., Nakajima A., and Yoshimi M. A high efficiency thin film silicon solar cell and module. *Sol Energ* 2004;77:939–949.
130. Liang T.S., Pravettoni M., Deline C., Stein J.S., Kopecek R., and Singh J.P. A review of crystalline silicon bifacial photovoltaic performance characterisation and simulation. *Energy Environ Sci* 2019;143:1285–1298.
131. Gu W., Ma T., Ahmed S., Zang Y., and Peng J. A comprehensive review and outlook of bifacial photovoltaic (bPV) technology. *Energ Convers Manage* 2020;223(223):113283. Doi: 10.1016/j.enconman.2020.113283. S2CID 224867963.
132. Guerrero-Lemus R., Vega R., Kim T., Kimm A., and Shephard L.E. Bifacial solar photovoltaics – A technology review. *Renew Sustain Energ Rev* 2016;60(60):1533–1549. Doi: 10.1016/j.rser.2016.03.041.
133. Shravan K., and Chunduri K. Advanced solar module technology – The time for a new generation of solar modules has come. *TaiyangNews*, China; 2019.

134. Roselund C. Bifacial module demand. *PV Magazine Photovoltaic Markets and Technology*, June 2019; 2019.

135. Half-Cell, Wikipedia, the free encyclopedia, last edited 18 July, 2022. https://en.wikipedia.org/wiki/Half-cell

136. Solomin E., Sirotkin E., Cuce E., Selvanathan S.P., and Kumarasamy S. Hybrid floating solar plant designs: A review, *Energies* 2021;14:2751. Doi: 10.3390/en14102751 https://www.mdpi.com/journal/energies.

137. World Bank. *Where Sun Meets Water: Floating Solar Market Report.* Washington, DC: World Bank Group; Energy Sector Management Assistance Program, Solar Energy Research Institute of Singapore; 2018.

138. Willuhn M. Floating PV's watershed moment. *PV Magazine Photovoltaic Markets and Technology*, June 2019; 2019.

139. Thoubborn K. Floating solar: What you need to know. *EnergySage*. 2018. https://news.energysage.com/floating-solar-what-you-need-to-know/ (accessed: 10 September 2019).

140. SolarPower Europe. *Solar Skins: An Opportunity for Greener Cities.* Brussels: SolarPower Europe; 2019.

141. Solar trees, Wikipedia, the free encyclopedia, last edited 31 July, 2021. https://en.wikipedia.org/wiki/Strawberry_Tree_(solar_energy_device)

142. Thurstom C.W. Solar carports thrive on dual-use concept. *PV Magazine Photovoltaic Markets and Technology*, March 2019; 2019.

143. Beck M., Bopp G., Goetzberger A., Obergfell T., Reise C., and Schindele S. *Combining PV and Food Crops to Agrophotovoltaic? Optimization of Orientation and Harvest.* Fraunhofer Institute for Solar Energy Systems (ISE), ResearchGate; 2019. Doi: 10.4229/27thEUPVSEC2012-5AV.2.25.

144. Hanley S. Combining solar and farming benefits both. *CleanTechnica*. 2019. https://cleantechnica.com/2019/09/09/combining-solar-farming-benefits-both/ (accessed 1 September 2019).

145. Dupraz C., Marrou H., Talbot G., Dufour L., Nogier A., and Ferard Y. Combining solar photovoltaic panels and food crops for optimising land use: Towards new agrivoltaic schemes. *Renew Energ* 2011;36(10):2725–2732. Doi: 10.1016/j.renene.2011.03.005.

146. Agrivoltaics, Wikipedia, the free encyclopedia, last edited 12 October, 2022. https://en.wikipedia.org/wiki/Agrivoltaics

147. Green M.A. Photovoltaic technology and visions for the future. *Progres Energ I* 2019;1(1), Doi: 10.1088/2516-1083/ab0fa8/pdf.

148. Svarc J. Most powerful solar panel 2022. A website report by clean energy reviews; March, 2022. https://www.cleanenergyreviews.info/blog/most-powerful-solar-panels

149. Svarc J. Most efficient solar panels 2022. A website report by Clean energy reviews; March, 2022. https://www.cleanenergyreviews.info/blog/most-efficient-solar-panels

150. Braun S., Hahn G., Nissler R., Pönisch C., and Habermann D. The multi-busbar design: An overview. *Energy Procedia* 2013;43:86–92. Doi: 10.1016/j.egypro.2013.11.092.

151. SolarPower Europe O&M Task Force. *Operation & Maintenance – Best Practices Guidelines* (Version 3.0). Brussels: SolarPower Europe O&M Task Force; 2018.

152. Kumar N.M., Sudhakar K., Samykano M., and Jayaseelan V. On the technologies empowering drones for intelligent monitoring of solar photovoltaic power plants. *Procedia Comput Sci* 2018;133:585–593. Doi: 10.1016/j.procs.2018.07.087.

153. KIC InnoEnergy. *Future Renewable Energy Costs: Solar Photovoltaics. How Technology Innovation is Anticipated to Reduce the Cost of Energy from European Photovoltaic Installations.* KIC InnoEnergy; 2015.

154. Filatof N. The importance of staying cool. *PV Magazine Photovoltaic Markets and Technology*; July 2019.

155. Mesbahi M. *White Paper on Soiling: The Science and Solutions.* Merida, Mexico: Solar Plaza; 2018.

156. Mansouri N., Lashab A., Sera D., Guerrero J.M., and Cherif A. Large photovoltaic power plants integration: A review of challenges and solutions, an open access MDPI article. *Energies* 2019;12:3798. Doi: www.mdpi.com/journal/energies.

157. Karimi M., Mokhlis H., Naidu K., Uddin S., and Bakar A.H.A. Photovoltaic penetration issues and impacts in distribution network—A review. *Renew Sustain Energ Rev* 2016;53:594–605.

158. Widén J., Wäckelgård E., Paatero J., and Lund P. Impacts of distributed photovoltaic on network voltages: Stochastic simulations of three Swedish low-voltage distribution grids. *Electr Power Syst Res* 2010;80:1562–1571.

159. Gallo D., Langella R., Testa A., Hernandez J.C., Papic I., Blažic B., and Meyer J. Case studies on large PV plants: Harmonic distortion, unbalance and their effects. *Proceedings of the IEEE Power and Energy Society General Meeting*, Vancouver, BC, Canada; 21–25 July, 2013.

160. Marinopoulos A., Papandrea F., Reza M., Norrga S., Spertino F., and Napoli R. Grid integration aspects of large solar PV installations: LVRT capability and reactive power/voltage support requirements. *Proceedings of the 2011 IEEE PES Trondheim PowerTech: The Power of Technology for a Sustainable Society*, Trondheim, Norway; 19–23 June, 2011.

161. Morjaria M., Anichkov D., Chadliev V., and Soni S. A grid-friendly plant: The role of utility-scale photovoltaic plants in grid stability and reliability. *IEEE Power Energ Mag* 2014;12:87–95.

162. Concentrator Photovoltaics, Wikipedia, the free encyclopedia, last edited 30 June, 2022. https://en.wikipedia.org/wiki/Concentrator_photovoltaics

163. Fraunhofer ISE. *Levelized Cost of Electricity - Renewable Energy Technologies*. 2013. www.ise.fraunhofer.de/de/veroeffentlichungen/veroeffentlichungen- pdf-dateien/stu-dien-und-konzeptpapiere/studie-stromgestehungskosten- erneuerbare-energien.pdf.

164. Philipps S.P. Bett, Dr. A.W. Horowitz K., and Kurtz S. Current status of concentrator photovoltaic (CPV) technology. A report by Fraunhofer Institute for solar energy systems ISE National Renewable Energy Laboratory NREL, NREL, Golden CO; 2015, December.

165. Gerstmaier T., Röttger M., Zech T., Moretta R., Braun C., and Gombert A. Five years of CPV field data: Results of a long-term outdoor performance study. *Proceedings of the 10th International Conference on Concentrating Photovoltaic Systems*, Albuquerque, NM; 2014.

166. Soitec. Four-junction solar cell developped using Soitec's expertise in semiconductor materials sets new efficiency record of 38.9% for CPV module. 2015. http://www.soitec.com/en/news/press-releases/article-1737.

167. Sasaki K., Agui T., Nakaido K., Takahashi N., Onitsuka R., and Takamoto T. Development of InGaP/GaAs/InGaAs inverted triple junction concentrator solar cells. *Proceedings of the 9th International Conference on Concentrator Photovoltaic Systems*, Miyazaki, Japan; 2013, pp. 22–25.

168. Fraunhofer ISE. New world record for solar cell efficiency at 46%. 2014. http://www.ise.fraunhofer.de/en/press-and-media/press-releases/press-releases-2014/new-world-record-for-solar-cell-efficiency-at-46-percent.

169. Philipps S.P., and Bett A.W. III–V multi-junction solar cells. In: Nozik A.J., Conibeer G., and Beard M.C., editors. *Advanced Concepts in Photovoltaics*. Cambridge: The Royal Society of Chemistry; 2014, pp. 87–117.

170. Friedman D.J. Olson J.M., and Kurtz S. High-efficiency III–V multijunction solar cells. In: Luque A., and Hegedus S., editors. *Handbook of Photovoltaic Science and Engineering*, 2nd ed. West Sussex: John Wiley & Sons; 2011, pp. 314–364.

171. Luque A. Will we exceed 50% efficiency in photovoltaics? *J Appl Phys* 2011;110(3):11.

172. Da Silva R.M., and Fernandes J.L.M. Hybrid photovoltaic/thermal (PV/T) solar systems simulation with Simulink/Matlab. *Sol Energ* 2010;84:1985–1996.

173. Li M., Ji X., Li G.L., Yang Z.M., Wei S.X., and Wang L.L. Performance investigation and optimization of the trough concentrating photovoltaic/thermal system. *Sol Energ* 2011;85:1028–1034.

174. Araújo G.L., and Martí A. Absolute limiting efficiencies for photovoltaic energy conversion. *Sol Energ Mater Sol Cells* 1994;33:213–240.

175. Rosell J.I., Vallverdú X., Lechón M.A., and Ibáñez M. Design and simulation of a low concentrating photovoltaic/thermal system. *Energ Convers Manage* 2005;46:3034–3046.

176. Dupeyrat P., Ménézo C. Rommel M., and Henning H.M. Efficient single glazed flat plate photovoltaic- thermal hybrid collector for domestic hot water system. *Sol Energ* 2011;85:1457–1468.

177. Makki A., Omer S., and Sabir H. Advancements in hybrid photovoltaic systems for enhancedsolar cells performance. *Renew Sustain Energ Rev* 2015;41:658–684.

178. Sharan S.N., Mathur S.S., and Kandpal T.C. Analytical performance evaluation of combined photovoltaic-thermal concentrator–receiver systems with linear absorbers. *Energ Convers Manage* 1987;27:361–365.

179. Algora C., Rey-Stollel I., Galiana B., Gonzalez J., Baudrit M., and Garcia I. Strategic options for a led-like approach in III-V concentrator photovoltaics. *Proceedings of the IEEE 4th World Conference on Photovoltaic Energy Conversion, Conference Record*; May 7–12, IEEE Xplore Press, Waikoloa, HI, 2006, pp. 741–744. Doi: 10.1109/WCPEC.2006.279562.

180. Verlinden P.J., Lewandowski A., Kendall H., Carter S., Cheah K., Varfolomeev I., Watts D., Volk M., Thomas I., Wakeman P., and Neumann A. Update on two-year performance of 120 kWp concentrator PV systems using multi-junction III-V solar cells and parabolic dish reflective optics. *Proceedings of the IEEE 33rd Photovoltaic Specialists Conference*; May 11–16, IEEE Xplore Press, San Diego, CA, 2008, pp. 1–6. Doi: 10.1109/PVSC.2008.4922734.

181. Lasich J.B., Verlinden P.J., Lewandowski A., Edwards D., Kendall H., Carter S., Thomas I., Wakeman P., Wright M., Hertaeg W., and Metzke R. World's first demonstration of a 140 kWp Heliostat Concentrator PV (HCPV) system. *Proceedings of the IEEE 34th Photovoltaic Specialists Conference*; June 7–12, IEEE Xplore Press, Philadelphia, PA, 2009, pp. 2275–2280. Doi: 10.1109/PVSC.2009.5411354.

182. Yadav P., Tripathi B., Rathod S., and Kumar M. Real-time analysis of low-concentration photovoltaic systems: A review towards development of sustainable energy technology. *Renew Sustain Energy Rev* 2013;28:812–823.

183. Luque A.L., and Andreev V.M. *Concentrator Photovoltaics*, 1st ed. Berlin: Springer; 2007.

184. Schuetz M.A., Shell K.A., Brown S.A., Reinbolt G.S., and French R.H. Design and construction of a ~7× low-concentration photovoltaic system based on compound parabolic concentrators. *IEEE J Photovolta* 2012;2:382–386.

185. Kandilli C. Performance analysis of a novel concentrating photovoltaic combined system. *Energ Convers Manage* 2013;67:186–196.

186. Zhao J., Song Y., Lam W.H., Liu W., Liu Y., Zhang Y., and Wang D. Solar radiation transfer and performance analysis of an optimum photovoltaic/thermal system. *Energ Convers Manage* 2011;52:1343–1353.

187. Kurtz S., and Geisz J. Multijunction solar cells for conversion of concentrated sunlight to electricity. *Opt Exp* 2010;18:A73–A78.

188. Yamaguchi M., Takamoto T., Araki K., and Ekins-Daukes N. Multi-junction III-V solar cells: Current status and future potential. *Sol Energ* 2005;79:78–85.

189. Dimroth F., Grave M., Beutel P., Fiedeler U., Karcher C., Tibbits T.N., Oliva E., Siefer G., Schachtner M., Wekkeli A., and Bett A.W. Wafer bonded four-junction GaInP/GaAs//GaInAsP/GaInAs concentrator solar cells with 44.7% efficiency. *Progres Photovolt Res Appl* 2014;22:277–282.

190. Araki K., Uozumi H., Egami T., Hiramatsu M., Miyazaki Y., Kemmoku Y., Akisawa A., Ekins-Daukes N.J., Lee H.S., and Yamaguchi M. Development of concentrator modules with dome-shaped fresnel lenses and triple-junction concentrator cells. *Progres Photovolt Res Appl* 2005;13:513–527.

191. Arak K., Yano T., and Kuroda Y. 30 kW concentrator photovoltaic system using dome-shaped Fresnel lenses. *Opt Exp* 2010;18:A53–A63.

192. Kinsey G.S., Nayak A., Liu M., and Garboushian V. Increasing power and energy in amonix CPV solar power plants. *IEEE J Photovolt* 2011;1:213–218.

193. Van Riesen S., Gombert A., Gerster E., Gerstmaier T., and Jaus J. Concentrix Solar's progress in developing highly efficient modules. *AIP Conf Proc* 2011;1407:235–238.

194. Xie W.T., Dai Y.J., Wang R.Z., and Sumathy K. Concentrated solar energy applications using Fresnel lenses: A review. *Renew Sustain Energ Rev* 2011;15:2588–2606.

195. Swanson R.M. The promise of concentrators. *Progres Photovolt Res Appl* 2000;8(1):93–111.

196. Andreev V., Rumyantsev V.D., and Grilikhes V.A. *Photovoltaic Conversion of Concentrated Sunlight*. Chichester: John Wiley & Sons; 1997.

197. Pérez-Higueras P., Muñoz E., Almonacid G., and Vidal P.G. High concentrator photovoltaics efficiencies: Present status and forecast. *Renew Sustain Energ Rev* 2011;15(4):1810–1815.

198. Gallagher S.J., Norton B., and Eames P.C. Quantum dot solar concentrators: Electrical conversion efficiencies and comparative concentrating factors of fabricated devices. *Sol Energ* 2007;81(6):813–821.

199. Lushetsky J. *Accelerating Innovation in Solar Technologies Overview of the DOE Solar Energy Technology Program*. Washington, D.C.: US Department of Energy: Solar Energy Technologies Program; 2008.

200. Butler B.A., van Dyk E.E., Vorster F.J., Okullo W., Munji M.K., and Booysen P. Characterization of a low concentrator photovoltaics module. *Physica B: Conden Mater* 2012;407(10):1501–1504.

201. Winston R., O'Gallagher J.J., and Gee R. Nonimaging solar concentrator with uniform irradiance. In: Winston R., and Koshel R.J., editors. *Nonimaging Optics and Efficient Illumination Systems*; August 2004, pp. 237–239.

202. Garcia-Botella A., Fernandez-Balbuena A.A., Vázquez D., and Bernabeu E. Ideal 3D asymmetric concentrator. *Sol Energ* 2009;83(1):113–117.

203. Sierra C., and Vázquez A.J. High solar energy concentration with a Fresnel lens. *J Mater Sci* 2005;40(6):1339–1343.

204. Leutz R., and Suzuki A. *Nonimaging Fresnel Lenses: Design and Performance of Solar Concentrators*. Berlin: Springer; 2001.

205. Leutz R., Suzuki A., Akisawa A., and Kashiwagi T. Developments and designs of solar engineering Fresnel lenses. *Proceedings of the Symposium on Energy Engineering*, Hong Kong; 2000.

206. Nakata Y., Shibuya N., Kobe T., Okamoto K., Suzuki A., and Tsuji T. Performance of circular Fresnel lens photovoltaic concentrator. *Jpn J Appl Phys* 1980;19:75–78.

207. Harmon S. Solar-optical analyses of a mass-produced plastic circular Fresnel lens. *Sol Energ* 1977;19(1):105–108.

208. Whitfield G.R., Bentley R.W., Weatherby C.K., Hunt A.C., Mohring H.D., Klotz F.H., Keuber P., Miñano J.C., and Alarte-Garvi E. The development and testing of small concentrating PV systems. *Sol Energ* 1999;67(1–3):23–34.

209. Franc F., Jirka V., Malý M., and Nábĕlek B. Concentrating collectors with flat linear fresnel lenses. *Sol. Wind Technol.* 1986;3(2):77–84.

210. Sierra C., Michie E., and Vázquez A.J. Production improvement of NiAl coatings achieved by self-propagating high-temperature synthesis with concentrated solar energy. *Revista de Metalurgia* 2005;469–474.

211. Sierra C., and Vázquez A.J. NiAl coatings on carbon steel by self-propagating high-temperature synthesis assisted with concentrated solar energy: Mass influence on adherence and porosity. *Sol Energ Mater Sol Cells* 2005;86(1):33–42.
212. Sierra C., and Vázquez A.J. NiAl coating on carbon steel with an intermediate Ni gradient layer. *Surf Coat Technol* 2006;200(14–15):4383–4388.
213. Clark J.A. An analysis of the technical and economic performance of a parabolic trough concentrator for solar industrial process heat application. *Int J Heat Mass Trans* 1982;25(9):1427–1438.
214. Kumar K.R., and Reddy K.S. Effect of porous disc receiver configurations on performance of solar parabolic trough concentrator. *Heat Mass Trans* 2012;48(3):555–571.
215. Riffelmann K.-J., Neumann A., and Ulmer S. Performance enhancement of parabolic trough collectors by solar flux measurement in the focal region. *Sol Energ* 2006;80(10):1303–1313.
216. Omer S.A., and Infield D.G. Design and thermal analysis of a two stage solar concentrator for combined heat and thermoelectric power generation. *Energ Convers Manage* 2000;41(7):737–756.
217. Al-Nimr M.A., and Alkam M.K. A modified tubeless solar collector partially filled with porous substrate. *Renew Energ* 1998;13(2):165–173.
218. Kumar K.R., and Reddy K.S. Thermal analysis of solar parabolic trough with porous disc receiver. *Appl Energ* 2009;86(9):1804–1812.
219. Odeh S.D., Morrison G.L., and Behnia M. Modelling of parabolic trough direct steam generation solar collectors. *Sol Energ* 1998;62(6):395–406.
220. Reddy K.S., Kumar K.R., and Satyanarayana G.V. Numerical investigation of energy-efficient receiver for solar parabolic trough concentrator. *Heat Trans Eng* 2008;29(11):961–972.
221. Reddy K.S., and Satyanarayana G.V. Numerical study of porous finned receiver for solar parabolic trough concentrator. *Eng Appl Comput Fluid Mech* 2008;2(2):172–184.
222. Zhang Q.-C., Zhao K., Zhang B.-C., Wang L.F., Shen Z.L., Zhou Z.J., Lu D.Q., Xie D.L., and Li B.F. New cermet solar coatings for solar thermal electricity applications. *Sol Energ* 1998;64(1–3):109–114.
223. Duffie J.A., and Beckman W.A. *Solar Engineering of Thermal Processes*. Hoboken, NJ: John Wiley & Sons; 2006.
224. Kreith F., and Kreider J.E. *Principles of Solar Engineering*. Washington, DC: Hemisphere Publishing Corporation; 1978.
225. Suzuki A., and Kobayashi S. Yearly distributed insolation model and optimum design of a two dimensional compound parabolic concentrator. *Sol Energ* 1995;54(5):327–331.
226. Senthilkumar S., Perumal K., and Srinivasan P.S.S. Construction and performance analysis of a three dimensional compound parabolic concentrator for a spherical absorber. *J Sci Indus Res* 2007;66(7):558–564.
227. Yehezkel N., Appelbaum J., Yogev A., and Oron M. Losses in a three-dimensional compound parabolic concentrator as a second stage of a solar concentrator. *Sol Energ* 1993;51(1):45–51.
228. Khalifa A.-J. N., and Al-Mutawalli S.S. Effect of two-axis sun tracking on the performance of compound parabolic concentrators. *Energ Convers Manage* 1998;39(10):1073–1079.
229. Mallick T.K., Eames P.C., Hyde T.J., and Norton B. The design and experimental characterization of an asymmetric compound parabolic photovoltaic concentrator for building façade integration in the UK. *Sol Energ* 2004;77(3):319–327.
230. Muhammad-Sukki F., Ramirez-Iniguez R., McMeekin S.G., Stewart B.G., and Clive B. Solar concentrators. *Int J Appl Sci* 2010;1(1):1–15.

231. Ali I.M.S., Mallick T.K., Kew P.A., O'Donovan T.S., and Reddy K.S. Optical performance evaluation of a 2-D and 3-D novel hyperboloid solar concentrator. *Proceedings of the 11th World Renewable Energy Congress*, Abu Dhabi, UAE; 2010.

232. Sellami N., Mallick T.K., and McNeil D.A. Optical characterisation of 3-D static solar concentrator. *Energ Convers Manage* 2012;64:579–586.

233. García-Botella A., Fernández-Balbuena A.Á., Vázquez D., Bernabeu E., and González-Cano A. Hyperparabolic concentrators. *Appl Opt* 2009;48(4):712–715.

234. Gordon J.M. Complementary construction of ideal nonimaging concentrators and its applications. *Appl Opt* 1996;35(28):5677–5682.

235. Chen C.-F., Lin C.-H., Jan H.-T., and Yang Y.-L. Design of a solar concentrator combining paraboloidal and hyperbolic mirrors using ray tracing method. *Opt Commun* 2009;282(3):360–366.

236. Saleh Ali I.M., Srihari Vikram T., O'Donovan T.S., Reddy K.S., and Mallick T.K. Design and experimental analysis of a static 3-D elliptical hyperboloid concentrator for process heat applications. *Sol Energ* 2014;102:257–266.

237. Minano J.C., Gonzalez J.C., and Zanesco I. Flat high concentration devices. *Proceedings of the 24th IEEE Photovoltaic Specialists Conference*, IEEE, New York, NY; December, 1994, Vol. 1–2, pp. 1123–1126.

238. Winston R., Minano J.C., and Benitez P. *Nonimaging Optics*. San Diego, CA: Elsevier Academic Press; 2005.

239. Minano J.C., Gonzalez J.C., and Benitez P. A high-gain, compact, nonimaging concentrator: RXI. *Appl Opt* 1995;34(34):7850–7856.

240. Minano J.C., Benitez P., and Gonzalez J.C. RX: A nonimaging concentrator. *Appl Opt* 1995;34(13):2226–2235.

241. Benitez P., and Minano J.C. Analysis of the image formation capability of RX concentrators. In: Winston R., editor. *Nonimaging Optics: Maximum Efficiency Light Transfer III*; 1995, Vol. 2538, pp. 73–84.

242. Minano J.C., Gonzalez J.C., and Benitez P. New nonimaging designs: The RX and the RXI concentrators. In: Winston R., and Holman R.L., editors. *Nonimaging Optics: Maximum-Efficiency Light Transfer II*. vol. 2016 of Proceedings of SPIE; 1993, pp. 120–127.

243. Ning X., Winston R., and O'Gallagher J. Dielectric totally internally reflecting concentrators. *Appl Opt* 1987;26(2):300–305.

244. Ning X.H. Application of nonimaging optical concentrators to infrared energy detection. *Nonimaging Optics: Maximum Efficiency Light Transfer*, vol. 1528 of Proceedings of SPIE; 1991, p. 88.

245. Ramirez-Iniguez R., and Green R. Elliptical and parabolic totally internally reflecting optical antennas for wireless infrared communications. *A paper presented at the IrDA/IEE/IEEE Conference on Optical Wireless*, Warwick University, Coventry, UK; 2003.

246. Ramirez-Iniguez R., and Green R.J. Optical antenna design for indoor optical wireless communication systems. *Int J Commun Syst* 2005;18(3):229–245.

247. Ning X.H., O'Gallagher J., and Winston R. Optics of two-stage photovoltaic concentrators with dielectric second stages. *Appl Opt* 1987;26(7):1207–1212.

248. Muhammad-Sukki F., Ramirez-Iniguez R., McMeekin S.G., Stewart B.G., and Clive B. Optimised dielectric totally internally reflecting concentrator for the solar photonic optoelectronic transformer system: Maximum concentration method. In: Setchi R., Jordanov I., Howlett R.J., and Jain L.C., editors. *Knowledge-Based and Intelligent Information and Engineering Systems*, vol. 6279 of Lecture Notes in Computer Science. Berlin: Springer; 2010, pp. 633–641.

249. Piszczor M.F., and Macosko R.P. *A High-Efficiency Refractive Secondary Solar ConcentratorforHighTemperatureSolarThermalApplications.*TechnicalMemorandum. Washington, D.C.: NASA; 2000.

250. Muhammad-Sukki F., Abu-Bakar S. H., Ramirez-Iniguez R., McMeekin S.G., Stewart B.G., Sarmah N., Mallick T.K., Munir A.B., Yasin S.H.M., and Rahim R.A. Mirror symmetrical dielectric totally internally reflecting concentrator for building integrated photovoltaic systems. *Appl Energ* 2014;113:32–40.

251. Gerion D., Pinaud F., Williams S.C., Parak W.J., Zanchet D., Weiss S., and Alivisatos A.P. Synthesis and properties of biocompatible water-soluble silica-coated CdSe/ZnS semiconductor quantum dots. *J Phys Chem B* 2001;105(37):8861–8871.

252. Mićić O.I., Cheong H.M., Fu H., Zunger A., Sprague J.R., Mascarenhas A., and Nozik A.J. Size-dependent spectroscopy of InP quantum dots. *J Phys Chem B* 1997;101(25):4904–4912.

253. Reisfeld R., and Jorgensen C.K. Luminescent solar concentrators for energy conversion. *Struct Bond* 1982;49:1–36.

254. Barnham K., Marques J.L., Hassard J., and O'Brien P. Quantum-dot concentrator and thermodynamic model for the global redshift. *Appl Phys Lett* 2000;76(9):1197–1199.

255. Wittwer V., Heidler K., Zastrow A., and Goetzberger A. Theory of fluorescent planar concentrators and experimental results. *J Luminescence* 1981;24–25(2):873–876.

256. Goetzberger A., Stahl W., and Wittwer V. Physical limitations of the concentration of direct and diffuse radiation. *Proceedings of the 6th European Photovoltaic Solar Energy Conference*, Reidel, Dordrecht, The Netherlands; 1985.

257. Alivisatos A.P. Perspectives on the physical chemistry of semiconductor nanocrystals. *J Phys Chem* 1996;100(31):13226–13239.

258. Schüler A., Python M., del Olmo M.V., and de Chambrier E. Quantum dot containing nanocomposite thin films for photoluminescent solar concentrators. *Sol Energ* 2007;81(9): 1159–1165.

259. Mousazadeh H., Keyhani A., Javadi A., Mobli H., Abrinia K., and Sharifi A. A review of principle and sun-tracking methods for maximizing solar systems output. *Renew Sustain Energ Rev* 2009;13(8):1800–1818.

260. Jared B.H., Saavedra M.P., Anderson B.J., Goeke R.S., Sweatt W.C., Nielson G.N., Okandan M., Elisberg B., Snively D., Duncan J., Gu T., Agrawal G., and Haney M.W. Micro-concentrators for a microsystems-enabled photovoltaic system. *Opt Exp* 2014;22:A521–A527.

261. Sato D., Masuda T., Araki K., Yamaguchi M., Okumura K., Sato A., Tomizawa R., and Yamada N. Stretchable micro-scale concentrator photovoltaic module with 15.4% efficiency for three-dimensional curved surfaces. *Commun Mater* 2021;2:7. Doi: 10.1038/s43246-020-00106-x.

262. *Integrated Micro-Optical Concentrator Photovoltaics.* Cambridge, MA: Arpa-e project by MIT; 2021. https://arpa-e.energy.gov/technologies/projects/integrated-micro-optical-concentrator-photovoltaics

263. Wright D.J., Badruddin S., and Robertson-Gillis C. Micro-tracked CPV can be cost competitive with PV in behind-the-meter applications with demand charges. *Front Energ Res* 25 September 2018. Doi: 10.3389/fenrg.2018.00097.

264. Nazri N.S., Fudholi A., Bakhtyar B., Yen C.H., Ibrahim A., Ruslan M.H., Mat S., and Sopian K. Energy economic analysis of photovoltaic–thermal-thermoelectric (PVT-TE) air collectors. *Renew Sustain Energy Rev* 2018a;92:187–197.

265. Kamthania D., and Tiwari G.N. Energy metrics analysis of semi-transparent hybrid PVT double pass facade considering various silicon and non-silicon based PV module Hyphen is accepted. *Sol Energ* 2014;100:124–140.

266. Assoa Y.B., and Ménézo C. Dynamic study of a new concept of photovoltaic–thermal hybrid collector. *Sol Energ* 2014;107:637–652.

267. Alves P., Fernandes J.F.P., Torres J.P.N., Branco P.J.C., Fernandes C., and Gomes J. From Sweden to Portugal: The effect of very distinct climate zones on energy efficiency of a concentrating photovoltaic/thermal system (CPV/T). *Sol Energ* 2019;188:96–110.

268. Khelifa A., Touafek K., Ben Moussa H., and Tabet I. Modeling and detailed study of hybrid photovoltaic thermal (PV/T) solar collector. *Sol Energ* 2016;135,169–176.

269. Wu Y.Y., Wu S.Y., and Xiao L. Performance analysis of photovoltaic–thermoelectric hybrid system with and without glass cover. *Energy Convers Manage* 2015;93:151–159.

270. Ren X., Li J., Hu M., Pei G., Jiao D., Zhao X., and Ji J. Feasibility of an innovative amorphous silicon photovoltaic/thermal system for medium temperature applications. *Appl Energ* 2019;252:113427.

271. Haiping C., Xinxin G., Heng Z., Yang L., Haowen L., and Yuegang B. Experimental study on a flash tank integrated with low concentrating PV/T (FT-LCPVT) hybrid system for desalination. *Appl Therm Eng* 2019;159:113874.

272. Riggs B.C., Biedenharn R., Dougher C., Ji Y.V., Xu Q., Romanin V, Codd D.S., Zahler J.M., and Escarra M.D. Techno- economic analysis of hybrid PV/T systems for process heat using electricity to subsidize the cost of heat. *Appl Energ* 2017;208:1370–1378.

273. Crisostomo F., Taylor R.A., Zhang T., Perez-Wurfl I., Rosengarten G., Everett V., and Hawkes E.R. Experimental testing of $SiN_x/SiO_2$ thin film filters for a concentrating solar hybrid PV/T collector. *Renew Energ* 2014;72:79–87.

274. Widyolar B., Jiang L., Ferry J., Winston R., Kirk A., Osowski M., Cygan D., and Abbasi H. Theoretical and experimental performance of a two-stage (50X) hybrid spectrum splitting solar collector tested to 600°C. *Appl Energ* 2019;239:514–525.

275. Müller M., Escher W., Ghannam R., Goicochea J., Michel B., Ong C.L., and Paredes S. Ultra-high-concentration photovoltaic-thermal systems based on microfluidic chip-coolers. *AIP Conf Proc* 2011;1407:231. Doi: 10.1063/1.3658333.

276. Codd D.S., Escarra M.D., Riggs B., Islam K., Ji Y.V., Robertson J., Spitler C., Platz J., Gupta N., and Miller F, Solar cogeneration of electricity with high-temperature process heat. *Cell Rep Phys Sci* 26 August 2020;1(8):100135.

277. Zondag H.A., De Vries D.W., Van Helden W.G.J., Van Zolingen R.J.C., and Van Steenhoven A.A. The yield of different combined PV-thermal collector designs. *Sol Energ* 2003;74:253–269.

278. Tripanagnostopolous Y., Nousia T., and Souliotis M. *A paper presented at the 17th European PV Solar Energy Conference*, Munich, Germany, 2001; 2002, pp. 2515–2518.

279. Gunther E., Hiebler S., Mehling H., and Redlich R. Enthalpy of phase change materials as a function of temperature: Required accuracy and suitable measurement methods. *Int J Thermophys* 2009;30:1257–1269.

280. Preet S., Bhusan B., and Mahagan T. Experimental investigation of water based photovoltaic/thermal PV/T system with and without phase change material. *Sol Energ* 2017;155:1104–1120.

281. Liang R., Zhang J., Ma L., and Li Y. Performance evaluation of new type hybrid photovoltaic/thermal solar collector by experimental study. *Appl Therm Eng* 2015;75:487–492.

282. Ong K.S., Naghavi M.S., and Lim C. Thermal and electrical performance of a hybrid design of a solar thermoelectric system. *Energ Convers Manage* 2017;133:31–40.

283. Haurant P., Menezo C., and Dupeyrat P. The Phototherm project: Full scale experimentation and modelling of a photovoltaic-thermal (PV-T) hybrid system for domestic hot water applications. *Energy Proced* 2014;48:581–587.

284. Aste N., Del Pero C., and Leonforte F. Thermal-electrical optimization of the configuration a liquid PVT collector. *Energy Proced* 2012;30:1–7.

285. Chen J., Yang L., Zhang Z., Wei J., and Yang J. Optimization of a uniform solar concentrator with absorbers of different shapes. *Sol Energ* 2017;158:396–406.

286. Anton I., Sala G., and Pachon D. Correction of the Voc vs. temperature dependence under non-uniform concentrated illumination. *Proceedings of the 17th European Photovoltaic Solar Energy Conference*, Munich, Germany; 22–26 October, 2001, pp. 156–159.

287. Shakouri A., and Yan Z. On-chip solid-state cooling for integrated circuits using thin-film microrefrigerators. *IEEE Trans Compon Packag Technol* 2005;28:65–69.

288. Baig H., Heasman K.C., and Mallick T.K. Non-uniform illumination in concentrating solar cells. *Renew Sustain Energ Rev* 2012;16:5890–5909.

289. Faiman D. Large-area concentrators. *Proceedings of the Second Workshop on the Path to Ultrahigh Efficiency Photovoltaics, JRC-Ispra*, Italy; 3–4 October, 2002.

290. Hosenuzzaman M., Rahim N.A., Selvaraj J., Hasanuzzaman M., Malek A.B.M.A., and Nahar A. Global prospects, progress, policies, and environmental impact of solar photovoltaic power generation. *Renew Sustain Energ Rev* 2015;41:284–297.

291. Cucchiella F., D'Adamo I., and Rosa P. End-of-Life of used photovoltaic modules: A financial analysis. *Renew Sustain Energ Rev* 2015;47:552–561.

292. Gray A., Boehm R., and Stone K.W. Modeling a passive cooling system for photovoltaic cells under concentration. *Proceedings of the ASME/JSME 2007 Thermal Engineering Heat Transfer Summer Conference*, Vancouver, BC, Canada; 8–12 July, 2007, pp. 447–454.

293. Cui M., Chen N., Yang X., Wang Y., Bai Y., and Zhang X. Thermal analysis and test for single concentrator solar cells. *J Semicond*. 2009;30:044011.

294. Sun J., Israeli T., Reddy T.A., Scoles K., Gordon J.M., and Feuermann D. Modeling and experimental evaluation of passive heat sinks for miniature high-flux photovoltaic concentrators. *J Sol Energ Eng* 2005;127:138–145.

295. Cheknane A., Benyoucef B., and Chaker A. Performance of concentrator solar cells with passive cooling. *Semicond Sci Technol* 2006;21:144.

296. Royne A., Dey C.J., and Mills D.R. Cooling of photovoltaic cells under concentrated illumination: A critical review. *Sol Energ Mater Sol Cells* 2005;86:451–483.

297. Yeom J., and Shannon M.A. 3.16—Micro-Coolers A2—Gianchandani, Yogesh B. In: Tabata O., and Zappe H., editors. *Comprehensive Microsystems*. Oxford: Elsevier; 2008, pp. 499–550.

298. Mallick T.K., Eames P.C., and Norton B. Using air flow to alleviate temperature elevation in solar cells within asymmetric compound parabolic concentrators. *Sol Energ* 2007;81:173–184.

299. Tonui J.K., and Tripanagnostopoulos Y. Air-cooled PV/T solar collectors with low cost performance improvements. *Sol Energ* 2007;81:498–511.

300. Brinkworth B.J., and Sandberg M. Design procedure for cooling ducts to minimize efficiency loss due to temperature rise in PV arrays. *Sol Energ* 2006;80:89–103.

301. Jakhar S., Soni M.S., and Gakkhar N. Historical and recent development of concentrating photovoltaic cooling technologies. *Renew Sustain Energ Rev* 2016;60:41–59.

302. Mittelman G., Dayan A., Dado-Turjeman K., and Ullmann A. Laminar free convection underneath a downward facing inclined hot fin array. *Int J Heat Mass Transf* 2007;50:2582–2589.

303. Mittelman G., Kribus A., Mouchtar O., and Dayan A. Water desalination with concentrating photovoltaic/thermal (CPVT) systems. *Sol Energ* 2009;83(8):1322–1334. ISSN 0038-092X, Doi: 10.1016/j.solener.2009.04.003.

304. Abbas N., Awan M.B., Amer M., Ammar S.M., Sajjad U., Ali H.M., Zahra N., Hussain M., Badshah M.A., and Jafry A.T. Applications of nanofluids in photovoltaic thermal systems: A review of recent advances. *Phys A Stat Mech Appl* 2019;536:122513.

305. Yazdanifard F., Ameri M., and Ebrahimnia-Bajestan E. Performance of nanofluid-based photovoltaic/thermal systems: A review. *Renew Sustain Energ Rev* 2017;76:323–352.

306. Shah T.R., and Ali H.M. Applications of hybrid nanofluids in solar energy, practical limitations and challenges: A critical review. *Sol Energ* 2019;183:173–203.

307. Said Z., Arora S., and Bellos E. A review on performance and environmental effects of conventional and nanofluid-based thermal photovoltaics. *Renew Sustain Energ Rev* 2018;94:302–316.

308. Michael J.J., and Iniyan S. Performance analysis of a copper sheet laminated photovoltaic thermal collector using copper oxide—water nanofluid. *Sol Energ* 2015;119:439–451.

309. Lari M.O., and Sahin A.Z. Design, performance and economic analysis of a nanofluid-based photovoltaic/thermal system for residential applications. *Energ Convers Manage* 2017;149:467–484.

310. Lu L., Liu Z.-H., and Xiao H.-S. Thermal performance of an open thermosyphon using nanofluids for high-temperature evacuated tubular solar collectors Part 1: Indoor experiment. *Sol Energ* 2011;85:379–387. Doi: 10.1016/j.solener.2010.11.008.

311. Kang W., Shin Y., and Cho H. Economic analysis of flat-plate and U-tube solar collectors using an $Al_2O_3$ nanofluid. *Energies* 2017;10:1911. Doi: 10.3390/en10111911 www.mdpi.com/journal/energies.

312. Karami N., and Rahimi M. Heat transfer enhancement in a PV cell using Boehmite nanofluid. *Energ Convers Manage* 2014;86:275–285. Doi: 10.1016/j.enconman.2014.05.037.

313. Al-Waeli A.H.A., Sopian K., Chaichan M.T., Kazem H.A., Hasan H.A., and Al-Shamani A.N. An experimental investigation of SiC nanofluid as a base-fluid for a photovoltaic thermal PV/T system. *Energ Convers Manage* 2017;142:547–558.

314. Al-Shamani A.N., Sopian K., Mat S., Hasan H.A., Abed A.M., and Ruslan M.H. Experimental studies of rectangular tube absorber photovoltaic thermal collector with various types of nanofluids under the tropical climate conditions. *Energ Convers Manage* 2016;124:528–542.

315. Faizal M., Saidur R., Mekhilef S., and Alim M.A. Energy, economic and environmental analysis of metal oxides nanofluid for flat-plate solar collector. *Energ Convers Manag* 2013;76:162–168.

316. Ladjevardi S.M., Asnaghi A., Izadkhast P.S., and Kashani A.H. Applicability of graphite nanofluids in direct solar energy absorption. *Sol Energ* 2013;94:327–334.

317. Filho E.P.B., Mendoza O.S.H., Beicker C.L.L., Menezes A., and Wen D. Experimental investigation of a silver nanoparticle-based direct absorption solar thermal system. *Energ Convers Manage* 2014;84:261–267.

318. Saidur R., Meng T.C., Said Z., Hasanuzzaman M., and Kamyar A. Evaluation of the effect of nanofluid-based absorbers on direct solar collector. *Int J Heat Mass Trans* 2012;55:5899–5907.

319. Karami M., AkhavanBahabadi M.A., Delfani S., and Ghozatloo A. A new application of carbon nanotubes nanofluid as working fluid of low-temperature direct absorption solar collector. *Sol Energ Mater Sol Cells* 2014;121:114–118.

320. Said Z., Saidur R., Rahim N.A., and Alim M.A. Analyses of exergy efficiency and pumping power for a conventional flat plate solar collector using SWCNTs based nanofluid. *Energ Build* 2014;78:1–9.

321. Tang B., Wang Y., Qiu M., and Zhang S. A full-band sunlight-driven carbon nanotube/ $PEG/SiO_2$ composites for solar energy storage. *Sol Energ Mater Sol Cells* 2014;123:7–12.

322. Sokhansefat T., Kasaeian A., and Kowsary F. Heat transfer enhancement in parabolic trough collector tube using Al2O3/synthetic oil nanofluid. *Renew Sustain Energ Rev* 2014;33:636–644. Doi: 10.1016/j.rser.2014.02.028.

323. Liu Z.H., Hu R.L., Lu L., Zhao F., and Xiao H.S. Thermal performance of an open thermosyphon using nanofluid for evacuated tubular high temperature air solar collector. *Energy Convers Manag* 2013;73:135–143.

324. Karami N., and Rahimi M. Heat transfer enhancement in a hybrid microchannel-photovoltaic cell using Boehmite nanofluid. *Int Commun Heat Mass Trans* 2014;55. Doi: 10.1016/j.icheatmasstransfer.2014.04.009.

325. Al-Waeli A.H., Sopian K., Kazem H.A., Yousif J.H., Chaichan M.T., Ibrahim A., Mat S., and Ruslan M.H. Comparison of prediction methods of PV/T nanofluid and nano-PCM system using a measured dataset and artificial neural network. *Sol Energ* 2018;162:378–396.
326. Sardarabadi M., Passandideh-Fard M., and Zeinali Heris S. Experimental investigation of the effects of silica/water nanofluid on PV/T (photovoltaic thermal units). *Energy.* 2014;66:264–272. Doi: 10.1016/j.energy.2014.01.102.
327. Sardarabadi M., and Passandideh-Fard M. Experimental and numerical study of metal-oxides/water nanofluids as coolant in photovoltaic thermal systems (PVT). *Sol Energ Mater Sol Cells* 2016;157:533–542.
328. Khanjari Y., Pourfayaz F., and Kasaeian A.B. Numerical investigation on using of nanofluid in a water-cooled photovoltaic thermal system. *Energ Convers Manage* 2016;122:263–278.
329. Hashim A.R.A., Hussien A., and Noman A.H. Indoor investigation for improving the hybrid photovoltaic/thermal system performance using nanofluid ($Al_2O_3$-water). *Eng Technol J* 2015;33(4):889–901.
330. Elmir M., Mehdaoui R., and Mojtabi A. Numerical simulation of cooling a solar cell by forced convection in the presence of a nanofluid. *Energ Procedia* 2012;18:594–603.
331. Rejeb O., Sardarabadi M., Ménézo C., Passandideh-Fard M., Dhaou M.H., and Jemni A. Numerical and model validation of uncovered nanofluid sheet and tube type photovoltaic thermal solar system. *Energ Convers Manage* 2016;110:367–377.
332. Nada S.A., El-Nagar D.H., and Hussein H.M.S. Improving the thermal regulation and efficiency enhancement of PCM-integrated PV modules using nano particles. *Energ Convers Manage* 2018;166:735–743.
333. Ghadiri M., Sardarabadi M., Pasandideh-Fard M., and Moghadam A.J. Experimental investigation of a PVT system performance using nano ferrofluids. *Energ Convers Manage* 2015;103:468–476.
334. Otanicar T.P., Phelan P.E., Prasher R.S., Rosengarten G., and Taylor R.A. Nanofluid-based direct absorption solar collector. *J Renew Sustain Energ* 2010;2(3):033102, Doi:10.1063/1.3429737.
335. Yousefi T., Veysi F., Shojaeizadeh E., and Zinadini S. An experimental investigation on the effect of $Al_2O_3$-$H_2O$ nanofluid on the efficiency of flat-plate solar collectors. *Renew Energ* 2012;39(1):293–298.
336. Kiliç F., Menlik T., and Sözen A. Effect of titanium dioxide/water nanofluid use on thermal performance of the flat plate solar collector. *Sol Energ* 2018;164:101–108.
337. Verma S.K., Tiwari A.K., Tiwari S., and Chauhan D.S. Performance analysis of hybrid nanofluids in flat plate solar collector as an advanced working fluid. *Sol Energ* 2018;167:231–241.
338. Chougule S.S., Pise A.T., and Madane P.A. Performance of nanofluid-charged solar water heater by solar tracking system. *IEEE-International Conference on Advances in Engineering, Science and Management*, Tamil Nadu, India; 2012, pp. 247–253.
339. Ghaderian J., and Sidik N.A.C. An experimental investigation on the effect of $Al_2O_3$ / distilled water nanofluid on the energy efficiency of evacuated tube solar collector. *Int J Heat Mass Trans* 2017;108:972–987.
340. Iranmanesh S., Ong H.C., Ang B.C., Sadeghinezhad E., Esmaeilzadeh A., and Mehrali M. Thermal performance enhancement of an evacuated tube solar collector using graphene nanoplatelets nanofluid. *J Clean Prod* 2017;162:121–129.
341. Mahendran M., Lee G.C., Sharma K.V., and Shahrani A. Performance of evacuated tube solar collector using water-based titanium oxide nanofluid. *J Mech Eng Sci* 2012;3:301–310.
342. Hussain A.H., Jawad Q., and Sultan K.F. Experimental analysis on thermal efficiency of evacuated tube solar collector by using nanofluids 2. Preparation of silver and zirconium. *Int J Sustain Green Energ* 2015;4:19–28.

343. Kaya H., Arslan K., and Eltugral N. Experimental investigation of thermal performance of an evacuated U-tube solar collector with ZnO/etylene glycol-pure water nanofluids. *Renew Energ* 2018;122:329–338.

344. Tong Y., Kim J., and Cho H. Effects of thermal performance of enclosed-type evacuated U-tube solar collector with multi-walled carbon nanotube/water nanofluid. *Renew Energ* 2015; 83:463–473.

345. Ozsoy A., and Corumlu V. Thermal performance of a thermosyphon heat pipe evacuated tube solar collector using silver-water nanofluid for commercial applications. *Renew Energ* 2018;122:26–34.

346. Xu G., Zhang X., and Deng S. Experimental study on the operating characteristics of a novel low-concentrating solar photovoltaic/thermal integrated heat pump water heating system. *Appl Therm Eng* 2011;31:3689–3695.

347. Tsai H.-L. Modeling and validation of refrigerant-based PVT-assisted heat pump water heating (PVTA–HPWH) system. *Sol Energ* 2015;122:36–47.

348. Fang G., Hu H., and Liu X. Experimental investigation on the photovoltaic–thermal solar heat pump air-conditioning system on water-heating mode. *Experim Thermal Fluid Sci* 2010;34(6):736–743.

349. Aliane A., Abboudi S., Seladji C., and Guendouz B. An illustrated review on solar absorption cooling experimental studies. *Renew Sustain Energ Rev* 2016;65:443–458.

350. Hassan H., and Mohamad A. A review on solar cold production through absorption technology. *Renew Sustain Energ Rev* 2012;16(7):5331–5348.

351. Siddiqui M.U., and Said S.A.M. A review of solar powered absorption systems. *Renew Sustain Energ Rev* 2015;42:93–115.

352. Alobaid M., Hughes B., Calautit J.K., O'Connor D., and Heyes A. A review of solar driven absorption cooling with photovoltaic thermal systems. *Renew Sustain Energ Rev* 2017;76:728–742. ISSN 1364-0321.

353. Vokas G., Christandonis N., and Skittides F. Hybrid photovoltaic–thermal systems for domestic heating and cooling—a theoretical approach. *Sol Energ* 2006;80(5):607–615.

354. Mittelman G., Kribus A., and Dayan A. Solar cooling with concentrating photovoltaic/thermal (CPVT) systems. *Energ Convers Manage* 2007;48(9):2481–2490.

355. Buonomano A., Calise F., and Palombo A. Solar heating and cooling systems by CPVT and ETsolar collectors: A novel transient simulation model. *Appl Energ* 2013;103:588–606.

356. Fumo N., Bortone V., and Zambrano J. Comparative analysis of solar thermal cooling and solar photovoltaic cooling systems. *J. Sol Energ Eng* 2013;135(2):021002.

357. Eicker U., Colmenar-Santos A., Teran L., Cotrado M., and Borge-Diez D. Economic evaluation of solar thermal and photovoltaic cooling systems through simulation in different climatic conditions: An analysis in three different cities in Europe. *Energy Build* 2014;70:207–223.

358. Sanaye S., and Sarrafi A. Optimization of combined cooling, heating and power generation by a solar system. *Renew Energ* 2015;80:699–712.

359. Al-Alili A., Hwang Y., and Radermacher R. Review of solar thermal air conditioning technologies. *Int J Refrigerat* 2014;39:4–22.

360. Al-Alili A., Hwang Y., Radermacher R., and Kubo I. A high efficiency solar air conditioner using concentrating photovoltaic/thermal collectors. *Appl Energ* 2012;93:138–147.

361. Al-Alili A., Hwang Y., Radermacher R., Kubo I., and Rodgers P. High efficiency solar cooling technique. *The Second International Energy 2030 Conference*, AbuDhabi, UAE; 2016 pp. 239–251.

362. Guo J., Lin S., Bilbao J.I., White S.D., and Sproul A.B. A review of photovoltaic thermal (PV/T) heat utilisation with low temperature desiccant cooling and dehumidification. *Renew Sustain Energ Rev* 2017;67:1–14.

363. Lin W., Ma Z., Sohel M.I., and Cooper P. Development and evaluation of a ceiling ventilation system enhanced by solar photovoltaic thermal collectors and phase change materials. *Energ Convers Manage* 2014;88:218–230.

364. Liang R., Zhang J., and Zhou C. Dynamic simulation of a novel solar heating system based on hybrid photovoltaic/thermal collectors (PVT). *Procedia Eng* 2015;121:675–683.

365. Hartmann N., Glueck C., and Schmidt F. Solar cooling for small office buildings: Comparison of solar thermal and photovoltaic options for two different European climates. *Renew Energ* 2011;36(5):1329–1338.

366. Kribus A., and Mittelman G. Potential of polygeneration with solar thermal and photovoltaic systems. *J Sol Energ-t Asme* 2008;130:011001.

367. Petrucci L., Fabbri G., Boccaletti C., and Cardoso A.J.M. Powering and cooling of a server room using a hybrid trigeneration system. *Int J Comp Theory Eng* 2013;5:263–267.

368. Garcia-Heller V., Paredes S., Ong C.L., Ruch P., and Michel B. Exergoeconomic analysis of high concentration photovoltaic thermal co-generation system for space cooling. *Renew Sust Energ Rev* 2014;34:8–19.

369. Buonomano A., Calise F., d'Accadia M.D., and Vanoli L. A novel solar trigeneration system based on concentrating photovoltaic/thermal collectors. Part 1: Design and simulation model. *Energy.* 2013;61:59–71.

370. Buonomano A., Calise F., Ferruzzi G., and Vanoli L. A novel renewable polygeneration system for hospital buildings: Design, simulation and thermo-economic optimization. *Appl Therm Eng* 2014;67:43–60.

371. Calise F., d'Accadia M.D., Palombo A., and Vanoli L. Dynamic simulation of a novel high-temperature solar trigeneration system based on concentrating photovoltaic/thermal collectors. *Energy* 2013;61:72–86.

372. Calise F., Dentice d'Accadia M., and Piacentino A. Exergetic and exergoeconomic analysis of a renewable polygeneration system and viability study for small isolated communities. *Energy* 2015;92 (3):290–307.

373. Calise F., Dentice d'Accadia M., and Piacentino A. A novel solar trigeneration system integrating PVT (photovoltaic/thermal collectors) and SW (seawater) desalination: Dynamic simulation and economic assessment. *Energy* 2014;67:129–148.

374. Calise F., d'Accadia M.D., Palombo A., and Vanoli L. Dynamic simulation of a novel high-temperature solar trigeneration system based on concentrating photovoltaic/thermal collectors. *Energy* 2013;61:72–86.

375. Calise F., Dentice d'Accadia M., Roselli C., Sasso M., and Tariello F. Desiccant-based AHU interacting with a CPVT collector: Simulation of energy and environmental performance. *Sol Energ* 2014;103:574–594.

376. Ong C.L., Escher W., Paredes S., Khalil A.S.G., and Michel B. A novel concept of energy reuse from high concentration photovoltaic thermal (HCPVT) system for desalination. *Desalination* 2012;295:70–81.

377. Kelley L., Bilton A.M., and Dubowsky S. ASME. Enhancing the performance of photovoltaic powered reverse osmosis desalination systems by active thermal management. *Proceedings of the ASME 2011 International Mechanical Engineering Congress and Exposition*, Denver, Colorado, 2012; p. 427–36.

378. Hughes A.J., O'Donovan T.S., and Mallick T.K. Experimental evaluation of a membrane distillation system for integration with concentrated photovoltaic/thermal (CPV/T) energy. *Energ Procedia* 2014;54:725–733.

379. Davidsson H., Perers B., and Karlsson B. Performance of a multifunctional PV/T hybrid solar window. *Sol Energ* 2010;84:365–372.

380. Lu W., Wu Y., and Eames P. Design and development of a building façade integrated asymmetric compound parabolic photovoltaic concentrator (BFI-ACP-PV). *Appl Energ* 2018;220:325–336.

381. Gleckman P. A high concentration rooftop photovoltaic system. *Proc. SPIE* 2007;6649:664903.
382. Gleckman P. Rooftop photovoltaic system using high concentration: An optical perspective. *A paper presented at the Workshop on Concentrating Photovoltaic Power Plants: Optical Design and Grid Connection*, Marburg, Germany; 11 October, 2007.
383. Lee D.I. Oh G.S., and Baek S.W. Development of a heating device using CPV and heat pipe. *a paper presented at the 10th International Conference on Heat Transfer, Fluid Mechanics and Thermodynamics*, Orlando, FL; 14–16 July, 2014.
384. Hasan A., McCormack S., Huang M., and Norton B. Energy and cost saving of a Photovoltaic-Phase Change Materials (PV-PCM) system through temperature regulation and performance enhancement of photovoltaics. *Energies* 2014;7:1318–1331.
385. Huang M.J., Eames P.C., and Norton B. Phase change materials for limiting temperature rise in building integrated photovoltaics. *Sol Energ* 2006;80:1121–1130.
386. Lasich J.Production of hydrogen from solar radiation at high efficiency, U.S. Patent 5658448, Solar system PTY Ltd. August 19; 1997.
387. Lasich J. Production of hydrogen from solar radiation at high efficiency U.S. Patent 5973825, Solar systems PTY Ltd. October 26, 1999.
388. Licht S. Solar water splitting to generate hydrogen. Fuel: Photothermal electrochemical analysis. *J Phys Chem* 2003;107(107):4253 4260.
389. Lewis N. 2006, August. http://www7.nationalacademies.org/bpa/SSSC_Presentations_Oct05_Lewis.pdf.
390. Chávez-Urbiola E., Vorobiev Y.V., and Bulat L. Solar hybrid systems with thermoelectric generators. *Sol Energ* 2012;86:369–378.
391. Huen P., and Daoud W.A. Advances in hybrid solar photovoltaic and thermoelectric generators. *Renew Sustain Energ Rev* 2017;72:1295–1302.
392. Lee J.J., Yoo D., Park C., Choi H.H., and Kim J.H. All organic-based solar cell and thermoelectric generator hybrid device system using highly conductive PEDOT: PSS film as organic thermoelectric generator. *Sol Energ* 2016;134:479–483.
393. Attivissimo F., Di Nisio A., Lanzolla A.M.L., and Paul M. Feasibility of a photovoltaic–thermoelectric generator: Performance analysis and simulation results. *IEEE Trans Instrum Meas* 2015;64:1158–1169.
394. Telkes M. Solar thermoelectric generators. *J Appl Phys* 1954;25:765.
395. He W., Su Y., Wang Y., Riffat S., and Ji J. Astudy on incorporation of thermoelectric modules with evacuated-tube heat-pipe solar collectors. *Renew Energ* 2012;37:142–149.
396. Miljkovic N., and Wang E.N. Modeling and optimization of hybrid solar thermoelectric systems with thermosyphons. *Sol Energ* 2011;85:2843–2855.
397. Fan H., Singh R., and Akbarzadeh A. Electric power generation from thermoelectric cells using a solar dish concentrator. *J Electron Mater* 2011;40:1311–1320.
398. Date A., Date A., Dixon C., and Akbarzadeh A. Progress of thermoelectric power generation systems: Prospect for small to medium scale power generation. *Renew Sustain Energ Rev* 2014;33:371–381.
399. Amatya R., and Ram R.J. Solar thermoelectric generator for micropower applications. *J Electron Mater* 2010;39:1735–1740.
400. Rehman N.U., Uzair M., and Siddiqui M.A. Optical analysis of a novel collector design for a solar concentrated thermoelectric generator. *Sol Energ* 2018;167:116–124.
401. Li P., Cai L., Zhai P., Tang X., Zhang Q., and Niino M. Design of a concentration solar thermoelectric generator. *J Electron Mater* 2010;39:1522–1530.
402. Sahin A.Z., Ismaila K.G. Yilbas B.S., and Al-Sharafi A. A review on the performance of photovoltaic/ thermoelectric hybrid generators. *Int J Energ Res* 2020;44:3365–3394.
403. Suleebka K. High temperature solar thermoelectric generator. *Appl Energ* 1979;5:53–59.

404. Motiei P., Yaghoubi M., GoshtashbiRad E., and Vadiee A. Two-dimensional unsteady state performance analysis of a hybrid photovoltaic-thermoelectric generator. *Renew Energ* 2018;119:551–565.

405. Mahmoudinezhad S., Rezania A., and Rosendahl L.A. Behavior of hybrid concentrated photovoltaic- thermoelectric generator under variable solar radiation. *Energ Convers Manage* 2018;164:443–452.

406. Indira S.S., Vaithilingam C.A., Chong K.-K., Saidur R., Faizal M., Abubakar S., and Paiman S. A review on various configurations of hybrid concentrator photovoltaic and thermoelectric generator system. *Sol Energ* 2020;201:122–148.

407. Li G., Shittu S., Zhou K., Zhao X., and Ma X. Preliminary experiment on a novel photovoltaic-thermoelectric system in summer. *Energy* 2019;188:116041.

408. Mizoshiri M., Mikami M., and Ozaki K. Thermal–photovoltaic hybrid solar generator using thin-film thermoelectric modules. *Jpn J Appl Phys* 2012;51:06FL07.

409. Cui T., Xuan Y., and Li Q. Design of a novel concentrating photovoltaic–thermoelectric system incorporated with phase change materials. *Energ Convers Manage* 2016;112:49–60.

410. Cui T., Xuan Y., Yin E., Li Q., and Li D. Experimental investigation on potential of a concentrated photovoltaic-thermoelectric system with phase change materials. *Energy* 2017;122:94–102.

411. Zhang J., and Xuan Y. Performance improvement of a photovoltaic—Thermoelectric hybrid system subjecting to fluctuant solar radiation. *Renew Energ* 2017;113:1551–1558.

412. Mohsenzadeh M., Shafii M.B., and Jafari mosleh H. A novel concentrating photo-voltaic/thermal solar system combined with thermoelectric module in an integrated design. *Renew Energ* 2017;113(C):822–834, Elsevier.

413. Zoui M.A., Bentouba S., Stocholm J.G., and Bourouis M. A review on thermoelectric generators: Progress and applications. *Energies* 2020;13:3606. Doi: 10.3390/en13143606 www.mdpi.com/journal/energies.

# 7 TPV Technology

## 7.1 INTRODUCTION

Thermophotovoltaics (TPVs) are a class of power generating systems that directly convert thermal energy into electrical energy. A TPV system uses light as an intermediary and consists of (at least) three components: a heat source, an emitter, and a photovoltaic (PV) cell with a low bandgap. The heat source brings the emitter to high temperature ($\geq$1,000 K), causing the emitter to emit thermal radiation, which is absorbed and converted to electricity by the PV cell. The hot side is made up of a heat source in thermal contact with an emitter and converts heat to light. On the cold side, the PV cell converts the thermal radiation from the emitter into electricity. Sometimes, the cold side includes a front side filter or back surface reflector. A near-field TPV device has a subwavelength gap between the emitter and the PV cell. A graphical illustration of TPV is shown in Figure 7.1 [1,2].

A TPV system uses the same principle as that of photovoltaic (PV) systems, but the main energy source is the infrared radiation of an emitter that can be heated by an external energy source such as combustion of a gas or liquid fuel, nuclear fission, the decay of a radioactive isotope or the waste heat of a furnace for glass production, etc. The main differences between TPV and PV systems are the temperature level and the distance of the source of the electromagnetic radiation. In PV systems, the sun generates an almost blackbody radiation at a temperature of about 6,000 K and a distance of about 150 Å~ 106 km, whereas in TPV systems the radiator emits at a sensibly lower temperature (1,500–1,800 K) and at a distance of the order of a few centimeters [3]. There are several advantages to using a TPV system instead of a PV system. First of all, there is the energy density; a TPV system works with an energy density of 5–30 W/cm$^2$ compared with 0.1 W/cm$^2$ received by a traditional PV system. This means that, for a given electric power, a TPV system is several times smaller than a PV system and thus more practical and usable for mobile devices [4–9]. TPV provides numerous advantages. TPV can operate with a variety of heat sources and fuels. The electric power generated by TPV systems is available on user demand and not dependent on the availability of sunlight; thus these systems are not limited to use during the day or linked to a battery pack or an energy buffer. Other advantages of the TPV are: low noise generation; possible use in co-generation energy system of heat and electricity; low pollution and dispatchable power delivery.

On the other hand, there are some disadvantages related to these devices: such as, the low efficiency of traditional TPV cells that can convert only a few per cent of the incident infrared radiation; the huge cost of the device due to the expensive materials that the system components are made of (capable of resisting very high temperatures); the fragility and instability of the system. Coutts [3,10] has given a complete overview of a wide range of TPV systems that have been designed so far. Nuclear fuel has been considered to be employed for aerospace applications by Schock

DOI: 10.1201/9781003328087-7

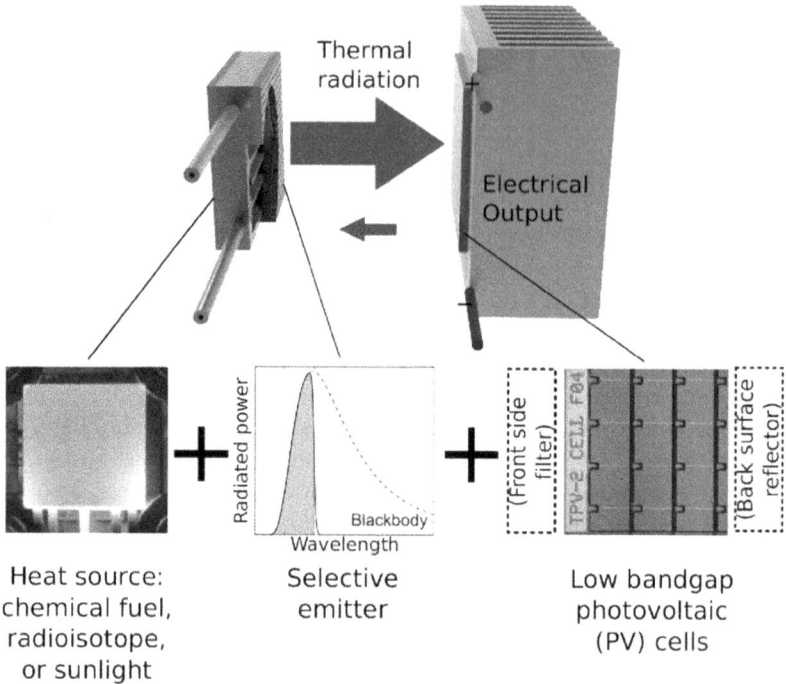

Heat source: chemical fuel, radioisotope, or sunlight

Selective emitter

Low bandgap photovoltaic (PV) cells

**FIGURE 7.1** Graphical illustration of TPV [1].

et al. [11] because it has no disadvantage in the environment where it has to be used. Potentially, nuclear-TPV systems are able to replace the radioisotope thermoelectric generators by reducing the consumption of nuclear fuel on the space probes or stations up to one-third compared to thermoelectric systems. In Schock's design, the RTPV device is able to produce up to 75 W of electric power using GaSb cells. The first application of the system was for military use, for portable devices, suitable for soldiers, to be employed in different missions and to work with any kind of fuel. Diesel fuel is mostly employed for military applications due to its extensive use as a fuel for vehicles and electric power generators. Scotto and DeBellis [6] have developed a prototype of a diesel-fuel-fired TPV system for portable devices in military applications. It is capable of 500 Wof dc power with 8% efficiency and it operates at about 1,600 K with a ceramic emitter (SiC) and GaSb cells with a power density of 2 W/cm². It is equipped with a heat regenerator from high-temperature exhaust gas able to boost the overall efficiency of the system.

Together with aerospace (12) and military applications, other possible application of TPV such as electric power generators or electric vehicles (4,5,13) have been investigated. Schubnell's [13] TPV application to residential heating systems gives an example of the potential everyday use of thermophotovoltaics in small co-generation systems that produce electricity and heat. Although this system, based on Si-PV cells, has only 5% efficiency, it is still convenient because the exhaust gas is employed to heat water for domestic use. The system developed at Western Washington University [4,5] to power the electric vehicle 'Viking 29' is an interesting device. 'Viking 29' uses a compressed natural gas fueled TPV electric generator, based on GaSb cells, to recharge the battery pack that feeds the electric motor of the car. The system is designed to generate 10 kW of dc power with a power density of 2 W/cm$^2$ and an efficiency of 7.5% and it is smaller, lighter and more silent than DC generators powered by internal combustion engines. The radiated power density from the TPV emitter is fundamentally limited only by Planck's law for blackbody emission [2]. However, high-performance TPV systems are particularly challenging to realize in part because of the need to coordinate multiple subsystems and the difficulties in designing a good emitter.

Most TPV systems include additional components such as concentrators, filters and reflectors. The basic principle is similar to that of traditional photovoltaics (PV) where a p-n junction is used to absorb optical energy, generate and separate electron/hole pairs, and in doing so convert that energy into electricity. The difference is that the optical energy is not directly generated by the Sun, but instead by a material at high temperature (termed the emitter), that causes it to emit light which is converted to electrical energy by TPV cell. The emitter can be heated by sunlight or other techniques. In this sense, TPVs provide a great deal of versatility in potential fuels. In the case of solar TPVs, large concentrators are needed to provide reasonable temperatures for efficient operation.

Improvements can take advantage of filters or selective emitters to create emissions in a wavelength range that is optimized for a specific photovoltaic (PV) converter. In this way TPVs can overcome a fundamental challenge for traditional PVs, making efficient use of the entire solar spectrum. For black body emitters, photons with energy less than the bandgap of the converter cannot be absorbed and are either reflected and lost or pass through the cell. Photons with energy above the bandgap can be absorbed, but the excess energy, is again lost, generating undesirable heating in the cell. In the case of TPVs, similar issues can exist, but the use of either selective emitters (emissivity over a specific wavelength range), or optical filters that only pass a narrow range of wavelengths and reflect all others, can generate emission spectra that can be optimally converted by the PV device. To maximize efficiency, all photons should be converted. A process often termed photon recycling can be used to approach this. Reflectors are placed behind the converter and anywhere else in the system that photons might not be efficiently directed to the collector. These photons are directed back to the concentrator where they can be converted, or back to the emitter, where they can be reabsorbed to generate heat and additional photons. An optimal TPV system would use photon recycling and selective emission to convert all photons into electricity.

The flexibility of converting various heat energy sources into high electrical power density broadens the TPV application ranging from micro-scale to large-scale TPV generators [2]. For instance, a worldwide potential of 3.1 GW electricity generation using TPV system in steel industry (>1,373 K) alone was estimated by Fraas et al. [4]. In comparison to a solar photovoltaic system, a TPV system works for a longer operation time at a lower radiator heating temperature [5]. A comprehensive analysis of the four elements of TPV system, namely generator, emitter, filter, and TPV cell, has been conducted in the literature to enhance the overall performance. TPV cell, which converts the photon radiation directly into electricity is the core component that contributes to the overall TPV system performance [10]. This technology is most applicable to narrow bandgap TPV cells made of gallium antimonide (GaSb), indium gallium arsenide (InGaAs) and a few other potential narrow bandgap materials such as germanium (Ge), indium arsenide (InAs), indium gallium arsenide antimonide (InGaAsSb), indium arsenide antimonide phosphide (InAsSbP), and indium gallium arsenide antimonide phosphide (InGaAsSbP). More discussion on the materials used for TPV cell is discussed later in Section 7.8.

## 7.2   TPV SYSTEM OVERVIEW

Ali Gamel et al. [14] illustrated a TPV system shown in Figure 7.2 which includes a generator, a radiator (emitter), a filter and an array of TPV cells [15]. This system can be accompanied by a reflector before or after TPV cells. The generator produces power from various heating sources ($P_{source}$) to the emitter with certain heat loss ($P_{source,loss}$). Next, the emitter generates the radiant power ($P_{radiant}$) to the PV cells via the filter. The filter then narrows the emission band from the emitter. The filtered radiated energy from the emitter should exceed the bandgap of PV cells with the bandgap power ($P_{gap}$), whereas losses ($P_{gap,loss}$) are induced by the photons with lower energy than the bandgap of PV cells. These photons are recuperated to the emitter ($P_{recuperate}$) to conserve heat and reduce $P_{source}$ at the required radiator temperature. The output power ($P_{out}$) at PV cells is measured through the optical-to-electrical signal conversion process [15].

A generator is a heat-driven source for TPV system with a typical working temperature range from 1,000 to 2,000 K [16,17]. This generator can be concentrated solar radiation, radioisotope thermal generator, combustion of hydrocarbon fuels or industrial waste heat [18,19]. Solar radiation produces the highest temperature among the generators. An emitter (radiator) emits electromagnetic energy by translating heat from generators into an emission spectrum to provide appropriate receiver cell sensitivity [20]. Selective emitters such as silicon carbide (SiC), tungsten (W), W-SiO$_2$ rare-earth oxide, and photonics crystal (PhC) provide narrow spectral range emission by enhancing in-band radiation and suppressing out-of-band radiation [21,22]. Bitnar et al. [23] reported that the maximum emissivity of ytterbia and erbia emitters is 0.82 at a photon energy of 0.80 eV, for temperature of 1,680 K. The bandgap of

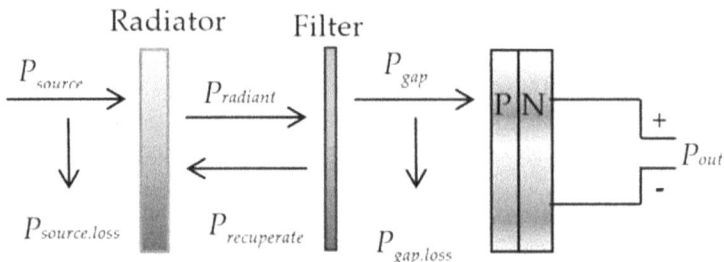

FIGURE 7.2   Schematic diagram of an overall TPV system [14].

the selective radiator must be higher than the bandgap of TPV cell to minimize the build-up of recombination for higher electrical energy conversion efficiency purpose. A promising emitter has been achieved by vacuum plasma spray coating of rare earth oxides on intermetallic alloy $MoSi_2$. The emitter can operate in an oxygen-containing atmosphere at a temperature of 1,873 K, which is highly thermal-shock stable and shows good selective-emitting properties [24].

Broadband radiator (emitter) establishes the emission across a wide range of wavelength for temperature range between 1,000 and 2,000 K [25,26]. Examples of broadband emitters are alumina, zirconia, magnesia, silica, yttria, and more, which possess a major challenge in low thermal shock resistance and low emissivity [27]. SiC with a minimum bandgap of 2 eV [28] has proven to be a suitable TPV emitter which can endure high melting point and high emissivity close to a 0.9 μm wavelength at an operating temperature up to 1,923 K [27,29]. A SiC porous super adiabatic radiant burner (emitter) was experimentally proposed for TPV system, achieving emitter efficiency up to 32% and a system output power of 5–10 W [30]. A broadband emitter shows advantages over a selective emitter due to the simplicity in fabrication, higher durability and less labor-intensive [31]. Nevertheless, these advantages are attained at the cost of lower TPV system efficiency and power density as compared to selective radiator [32]. Gentillon et al. [33,34] experimentally characterized and analyzed a design of a porous media combustion-based thermophotovoltaic reactor with controlled radiant emission using yttria-stabilized zirconia/alumina composite (YZA). It was found that the erbia coating on YZA foam increases the emissivity by ~10%.

A filter is located in between the emitter and the TPV cell to spectrally filter the emission from the emitter, it is matched to the bandgap of TPV cell to block the energy of photons that is lower than the energy bandgap of a TPV cell [13]. The TPV cell performance is optimized by selectively filtering the thermal radiation depending on the emitter temperature and bandgap of TPV cell. This is to promote photon recycling and to improve system conversion efficiency [23,35]. Catchpole et al. [36] demonstrated that 99% of photon energy sit above the TPV bandgap using a highly idealized filter with 0.7 V applied voltage. Tong et al. [37] proposed on the utilization of intermediate frequency filter and photon recycling back to the radiator. Interference filter or dielectric filter is realized as a multilayer stack which can be deposited over the cell or to be placed between the emitter and the cell. The interface creates a low pass filter that cuts at a specific wavelength [38,39]. TPV cell converts the photon radiation into electricity and share similar principal as PV cell.

TPV devices operate based on the principle of photovoltaic effect, the phenomenon in which electromagnetic radiation incident upon a $p–n$ junction induces an electromotive force. This requires an internal potential barrier with a built-in electric field to separate the photogenerated electron–hole pairs. The barrier is formed in a single crystal semiconductor material containing both $p$- and $n$-type regions. While the $p$-type region contains a large concentration of holes with few electrons, the opposite is true for the $n$-type region. The interface between the two regions is commonly referred to as the junction. Due to the charge concentration gradient, holes from the $p$-side diffuse into the $n$-side, whereas the electrons from the $n$-side diffuse to the $p$-side. Consequently, the region near the junction (known as the depletion layer), is depleted of majority carriers and contains only ionized dopants. The carrier concentration remains unperturbed from that in an isolated material beyond the depletion region on both $p$ and $n$ sides.

An antireflection coating of ZnS and $MgF_2$ can be formed on top of the cell, to reduce reflection of the incident radiation. For an ideal TPV cell, most of the incident energy should be absorbed within the depletion region corresponding to a length of $3\delta_\lambda$, where $\delta_\lambda$ is the penetration depth of the incident radiation. The InGaSb epitaxial layers are grown on GaSb, which is around $100\,\mu m$ thick to avoid absorption [4]. When incident radiation with photon energy greater than the band gap ($E_g$) of the cell strikes the $p$–$n$ junction, electron–hole pairs are generated. Those pairs generated in the depletion region are swept by the built-in electric field. These are then collected by the electrodes at the two ends of a TPV cell, thus resulting in drift current. On the other hand, the electron–hole pairs generated within the diffusion length outside the depletion region can diffuse as minority carriers to the edge of the depletion region, yielding diffusion current. Note that the diffusion current is the sum of the hole and electron diffusion currents. Due to charge accumulation at the two electrodes, an open circuit voltage develops across the terminals of the cell with the $p$-side positive and the $n$-side negative. If the load across the TPV cell is a short circuit, the current flowing in the circuit is a short-circuit current, or photo current. The photocurrent is the sum of diffusion and drift currents. When a load is added in the circuit, a positive voltage ($V$) develops across the $p$–$n$ junction and reduces the built-in potential of the TPV cell. This results in minority carrier diffusion, producing a small current called forward diode current.

Thermophotovoltaics (TPV) research focuses on developing semiconductor materials and devices that convert radiation from heat energy to electricity, with emphasis on alloys from gallium-indium arsenide and indium-phosphide arsenide, which achieve low-energy band gaps (0.40–0.70 eV). Solar cells made out of new material based on perovskite absorb light and screens emit light when electricity passes through it. It is a high-quality material, and very durable under light exposure, where it can capture light particles and convert them to electricity, or vice versa. What a third generation solar cell system does is to modify the spectral nature of the light which impacts the semiconductor solar cells. Finally, behind the solar cells, and next to the outermost wall of the thermophotovoltaic system, is a heat sink/heat pump layer. The function of this layer is to maintain the solar cells at their design operating temperature while, at the same time, delivering thermal energy to some external consumer.

No one, to date, has completely succeeded in fabricating a complete thermophotovoltaic system. There exist a number of interesting technical problems which require operational solutions before a working thermophotovoltaic system is established. First, the converter needs to be suspended approximately in the center of the thermophotovoltaic device by some mechanism which minimizes thermal losses from the converter. A second technical problem is the design and fabrication of the bandpass filter. This filter must have extremely sharp "corners" so that only the desired photons reach the solar cells. The reflectivity of the bandpass filter must be high, both to keep it cool and to maintain the temperature of the converter. Furthermore, the bandpass filter must not adversely affect the surface properties of the solar cells.

A third problem lies in the design of the solar cells themselves. They are exposed to a high density of incoming photons and will generate a significant photocurrent. The twin requirements of high electrical and thermal energy delivery place severe limitation on the resistance (both electrical and thermal) of the solar cells. The ideal optical configuration for the solar cells in a thermophotovoltaic system would be inverted. This allows for maximum photocurrent with the junctions located close to the heat sink. The cooler junctions permit a lower saturation current density and higher optical-to-electrical energy conversion efficiency. Finally, we must select a thermal energy extraction engine. This device is responsible for maintaining a proper solar cell junction temperature as-well-as delivering thermal energy to consumers.

The challenge of increased internal temperature of the TPV cells can be addressed by several methods. Apart from the incorporation of emitter and filter to protect the cell from thermalization effect, active cooling systems such as heat sinks, cycling coolant and forced-air coolant are commonly employed in numerous TPV prototypes [40]. Wu et al. [30] experimentally illustrated the integration of water-cooled mechanism in TPV system. A setup of low-iron soda-lime glass cooling system was studied with opto-electro-thermal coupled simulation by Zhou et al. [41] to investigate the effect of temperature under different condition for the TPV application. The integration of TPV with PCM such as paraffin wax, graphene, nano-PCM [42] will lead to an effective cooling and thermal energy storage TPV system. Furthermore, separating the cell from the emitter by transparent insulator layer or undoped semiconductor layer will help to protect the TPV cell from the hot emitter.

Ferraria et al. [15] presented and discussed a critical review of the TPV prototypes. Daneshvar et al. [38] reviewed the development of all main components, discussed the fundamental and technical challenges facing commercial adoption of TPV and prospects of TPV. Mustafa et al. [16] summarized the progress of combustion-driven thermoelectric (TE) and TPV power generation systems for the years 2000–2016. Datas and Martí [43] reviewed the state of the art and historical development of TPV for space application along with the main competing technologies. Tian et al. [44] reported the recent progress of near-field and far-field radiative heat transfer, various design structures of metamaterials and their properties, and focused on the exploration of tunable radiative wavelength selectivity of nano-metamaterials. More recently, in 2019, Sakakibara et al. [1] reviewed the state of the art of radiator (emitter) and presented a systematic approach for assessing radiators. A recent paper from Rashid et al. [45] has highlighted the recent development of TPV for waste heat harvesting application and investigated the potential implementation in coal-fired thermal power plant. Furthermore, Burger et al. [46] studied numerous decades of experimental TPV works and compared the energy-conversion of different systems with respect to experiment-specific thermodynamic limit.

## 7.3  TYPES OF TPV

As mentioned earlier, TPV can be operated with different types of heat sources. Here we briefly examine the workings of different types of TPV systems.

### 7.3.1  SOLAR THERMOPHOTOVOLTAIC

An STPV system is based on a principle of conversion of concentrated solar energy into radiation by heating an intermediate photon emitter with subsequent photovoltaic conversion of this radiation in low-band gap photo-converters. In the STPV system, concentrated solar radiation is thus absorbed and reemitted as a thermal radiation before illumination of the TPV cells. Essential advantage of such a system, compared to conventional ones, is the possibility to choose emitter material with selective emission spectrum and to use sub-band gap photons owing to recycling process. The theoretically estimated values of the STPV efficiency are in a wide interval from 30% to 70%, depending on the approach used. For ideal system elements, maximal theoretical efficiency was found to be about 85% which is identical to the efficiency of an unlimited stack of tandem cells. In practice, the expected efficiency of STPV converters is 30%–35%.

A STPV system consists of a sun tracker, concentrator, and an STPV module. One of the most important parts of the STPV system is the concentrator itself. Because of the use of a high-temperature emitter, the solar concentration ratio needs to be as high as possible to obtain high emitter temperature values. The maximum achievable concentration ratio on earth can be 46,164 suns, determined by the aperture angle of the sun, this value cannot be reached in STPV systems since expensive optics will be needed. Therefore, cost-effective Fresnel lenses are used in STPV systems. A Fresnel lens based on two-stage sunlight concentrator system was developed for TPV systems, consisting of a $0.36\,m^2$ Fresnel lens together with a secondary quartz concave–convex lens which gives a total concentration ratio of 4,600 suns.

STPV systems are energy conversion methods that are capable of overcoming the Shockley-Queisser efficiency limit of 32.1% for silicon photovoltaic (PV) cells [47]. In fact, the upper theoretical limit for the efficiency of STPV systems is 85.4% [48]. This is possible because in STPV systems the broadband solar spectrum is converted into a narrow spectrum tailored for use in a PV cell. This emission spectrum typically consists of photons with an energy just above the bandgap energy ($E_{bg}$) of the PV cell, resulting in greatly reduced thermalization and transmission losses in the cell [49,50].

In STPV systems, the sunlight is concentrated on an emitter, which is made of a refractory_metal like tungsten because the emissivity of tungsten matches the used TPV cells. The emitter needs to be protected from oxidation by placing it in a quartz tube, filled with an inert gas. The emitter can be made in a cylindrical shape with a sealed bottom such that the TPV cells can be placed around the emitter. In the STPV system developed at Ioffe Psysico-Technical Institute, TPV cells are fabricated from GaSb cells. The cells are mounted on a BeO ceramic plate which is water-cooled to prevent heating up of the cells because in a TPV generator the cells operate at high current densities and are placed close to the heat source. BeO is well suited for this application since it has an electrical resistivity of more than $1 \times 10^{14}\ \Omega$ cm with the best thermal conductivity of 250 W/(mK) and it has thermal expansion coefficient of $6 \times 10^{-6}\ K^{-1}$ which matches with the thermal expansion coefficient of GaSb. By means of ray-tracing it is also possible to calculate the optimal cavity structure to reach the highest possible energy conversion efficiency.

A module of three $1 \times 1\,cm^2$ GaSb cells has been measured in the STPV setup resulting in a total output power of 0.762 W by a total incident irradiation density of about $770\,W/cm^2$. Higher values can be expected for higher irradiation densities, the use of a larger emitter and the application of optical confinement or the use of

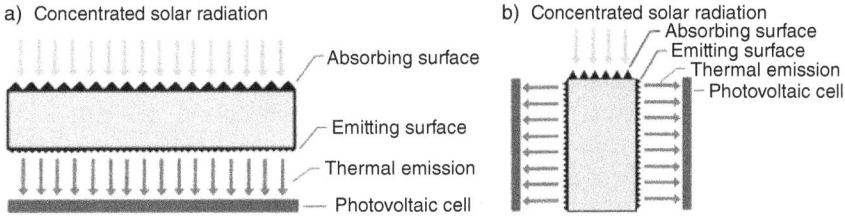

**FIGURE 7.3** Diagram of (a) a flat STPV system and (b) a cylindrical STPV system, both with solar absorber and thermal emitter [51].

tandem TPV cells based on InGaAsSb alloys. The use of a selective emitter, like 2D- or 3D-texturized tungsten may lead to the increase in emissivity in the desired wavelength region. All these approaches should ensure an STPV system efficiency increase of up to 30% and higher.

Figure 7.3a and b show two common system architectures in this field: planar and cylindrical systems. While the cylindrical architectures allow more control over the size of the emitting surface, they require large PV cell areas and are extremely difficult to achieve. Planar structures are much more simple and allow for a reduced PV cell area, making them ideal for this application. The key to realizing highly efficient STPV systems is precise control of the optical properties of the light absorbing and emitting surfaces. The absorbing surface must efficiently absorb solar energy, while simultaneously minimizing the emission of thermal energy, and the emitting surface must have high emission in a narrow band just above the $E_{bg}$ of the PV cell used. This spectral control is typically achieved through the use of nanostructures or thin film coatings; this paper will consider a combination of both methods.

The choice of PV cell in an STPV system is primarily determined by the operating temperature. The maximum efficiency for an STPV system occurs when the blackbody peak of the emitting surface is near the $E_{bg}$ of the PV cell used [48]. This allows the PV cell to absorb a large portion of emitted thermal energy, increasing the efficiency of the system. Narrowing the emission spectrum by controlling the optical properties of the emitting surface will further increase the portion of emitted energy that is usable by the PV cell. The wavelength of peak emission of a blackbody at a certain temperature is given by Wien's displacement law. Due to these restrictions, PV cells commonly used in STPV systems include germanium (Ge), gallium antimonide (GaSb), and indium gallium arsenic antimonide (InGaAsSb) cells. Silicon (Si) cells would require an operating temperature of about 2,600 K for efficient operation, making them a poor choice for most STPV systems. Ge and GaSb cells have high efficiencies at operating temperatures around 1,600 K, while InGaAsSb cells operate efficiently at temperatures around 1,250 K [52].

Despite the extremely high theoretical efficiency possible in STPV systems, experimental efficiencies remain low. Early STPV systems focused on cylindrical geometries, with large cavities for an absorbing surface and bulk tungsten or rare earth compounds as emitters [53–55]. Sunlight was concentrated onto these systems via Fresnel lens, raising their temperatures as high as 1,680 K, and emitted radiation was collected by Ge or GaSb PV cells. These early designs all had <1% efficiency, with PV cell heating, temperature gradients along the cylinders, and a lack of efficient absorbing and emitting surfaces, lowering the efficiency.

Lenert et al. [56] demonstrated an STPV system with 3.2% efficiency, which sparked a renewed interest in the field. This system used a planar geometry to simplify the architecture, reduced the required PV cell area, and removed the problem of a temperature gradient across the emitting surface. InGaAsSb PV cells with a 0.55 eV bandgap were used to allow the system to operate at lower temperatures (the 3.2% efficiency was recorded at 1,285 K), which afforded increased material stability and reduced the level of solar concentration required for operation. A multi-walled carbon nanotube blackbody absorber was used for the absorbing surface, and a $Si/SiO_2$ Bragg stack was used for the emitting surface. Experimental validation of this system was performed using a Xenon-arc light source and concentrating lens system to simulate the solar spectrum. Due to the fact that a blackbody absorber was used, spectral mismatch between solar energy and the Xenon-arc light source used in the experiment would have a negligible effect on the results; however, solar concentrator systems have optical losses up to 50% that would not be present in this simulated setup. The low efficiency of this system is primarily due to the low operating temperature and low efficiency PV cells.

Shimizu et al. [57] showed a ground-breaking experimental efficiency of 8% using a planar geometry and GaSb PV cells. Due to the 0.75 eV $E_{bg}$ of these cells, high operating temperatures were required for efficient operation, and the system was operated at a temperature of 1,640 K. Both the absorbing and emitting surfaces consisted of a stack of a yttria-stabilized zirconia (YSZ) layer followed by a tungsten (W) layer, followed by an additional YSZ layer and a W substrate. This resulted in reduced thermal emission from the absorbing surface; however, the absorption band was narrow and a 20% reflection loss from the absorbing surface was reported. Again, a solar simulator was used to illuminate the setup, and potential solar concentrator losses were not taken into account. Despite the large advances in STPV systems, further research is needed to improve system efficiencies to make them competitive with traditional PV systems. Shimizu et al. [57] also examined the losses in an experimental system in detail through modeling and experimental studies, in order to provide a path forward to a more efficient STPV system.

### 7.3.2 Combustion and Waste Heat Based Thermophotovoltaics

Apart from solar radiation and radioisotope, liquid and gas fuels such as oil, butane, propane, methane, and hydrogen have been employed to drive a generator [20,23,32,58]. Various TPV combustor–regenerator systems for electric vehicles have been studied both theoretically and experimentally [2]. The performance of the combustion system depends on chamber geometry, fuel injection, and mass flow rate. Furthermore, matrix material can be substituted with a ceramic material that is lighter than metal and with a greater heat capacity to store a high amount of energy. Additionally, ceramic matrix was able to obtain a greater porosity and thus a greater surface for the heat exchange with a reduction of volume and weight. Colangelo et al. [59] designed and tested various TPV combustors and heat recovery systems for different testing conditions. It was found that a rotary heat exchanger is an optimal design since it is very compact and has higher effectiveness in comparison with other types of regenerators with the same number of transfer units. Furthermore, the study developed a model which accurately predicts the performance of the heat exchanger, taking into account two different values for the physical properties (such as thermal conductivity, heat capacity) for the hot and cold sides of the regenerator.

Kim et al. [60,61] designed a novel combustor as a thermal heat source for a 10–30 W power- generating TPV system. The combustor consisted of an emitter (combustion chamber), injection nozzles, a mixing chamber and a quartz shield. To satisfy the primary requirements for designing the combustor (i.e., stable burning in the combustor chamber, maximized heat transfer through the emitting walls, but uniform distribution of temperature along the walls), the multiple injection configuration with annularly arranged nozzles and the cylindrical emitter with the quartz shield to apply a heat recirculation concept were adopted. Results showed that the heat recirculation substantially improved the performance of the combustor. Compared with conventional combustors with no heat-recirculation, the efficiency of the combustor was enhanced and the observed thermal radiation from the emitter walls indicated that heat generated in the emitter was uniformly emitted. Thus, the combustor configuration used in this study can be applied to the practical TPV systems without any moving parts (i.e., without frictional losses and clearance problems). The fuel nozzle length substantially affected flame behaviors. The study concluded that:

1. In order to satisfy the primary requirements for designing the combustor, i.e., stable burning in the combustor chamber, and maximized heat transfer through the emitting walls but uniform distribution of temperature along the walls, the combustor consisted of the multiple injection configuration with annularly arranged nozzles and the cylindrical emitter with the quartz shield to adopt a heat-recirculation concept should be designed.
2. For the optimized design condition, the heat recirculation substantially improved the performance of the combustor: the observed thermal radiation from the emitter wall indicated that heat generated in the emitter was uniformly emitted.
3. Three distinct flame stability behaviors were observed: flashback, stable flame and lift flame. The flashback was observed for a short fuel nozzle due to the intensified burning with the long residence time of the fuel-air mixture, while lift flame was observed for a long fuel nozzle due to the limited residence time of the mixture.

Molina et al. [62] examined porous media combustion based TPV energy conversion. Utlu et al. [63,17] used GaSb cell to convert waste heat to power by TPV. In these studies, the TPV systems were considered as an alternative energy source to generate power from waste heat in cost effective and efficient manner. The conversion of the high temperature applied to the cell to electrical energy was examined using the GaSb photovoltaic cell. Using cell temperature and source temperature as parameters, energy efficiency, fill factor, effect of open-circuit voltage and short-circuit current values were determined. The efficiency value of the GaSb TPV cell systems was calculated for the radiation source temperature between 1,300 and 3,100 K.

The study showed the optimum energy conversion efficiency values of GaSb solar cell structure to be 21.57%. The study also outlined opinions and recommendations about the feasibility, efficiency and development of thermophotovoltaic energy conversion systems for waste heat source. Finally, AliGamel et al. [14,64] carried out Multi-dimensional optimization of $In_{0.53}Ga_{0.47}As$ thermophotovoltaic cell using real coded genetic algorithm.

### 7.3.3   NEAR-FIELD THERMOPHOTOVOLTAICS

A simple solution to increase the heat transfer significantly is to reduce the distance between the emitter and the TPV receiver cell to a submicron range. This allows the energy to 'evanescently couple' or tunnel directly to the TPV receiver cell. The large increase of heat transfer creates a potentially large gain in electrical output power density. The transfer of heat between two surfaces is analogous to the transfer of light between two prisms with a spacing which is much less than the wavelength of the incident light. The enhancement is limited to $n^2$ of the lowest index of refraction material and will thus depend on the wavelength-dependent nature of the material.

Because of the very close spacing between the heat source and the cell, the requirements are different compared to far-field TPV. First of all, because of the large energy transfers, large currents will be generated in the cell which requires TPV cells with a high current carrying density. Because of the low spacing, the surface of the TPV cell must be flat, so no extending contacts can be present on the system. There are three possible cell designs to solve this issue: a gridless design, a design with recessed front contacts, and an inverted design with interdigitated contacts. There is an increase in photogenerated current visible because of the MTPV effect, but side effects as series and shunt resistance and diode temperature make it impossible to determine the actual enhancement of the photogenerated current.

To ensure a constant spacing between the heater and the cell, tubular spacers are used which are designed to minify the heat transfer along the spacers. Analysis shows that the use of spacers results in a final reduction of the parasitic conductive heat flow to less than 3% of the total heat transferred. The application of a back-surface reflector will also improve the cell results as sub-band gap photons will be reflected back to the emitter. The technology looks very promising because of the high amount of transferred energy and the potential high energy conversion efficiency, but the more complex cell technology together with the strict requirements for its mechanical stability will be a challenge for this technology. Progress toward functional near-field thermophotovoltaic devices has been limited by challenges in creating thermally robust planar emitters and photovoltaic cells designed for near-field thermal radiation.

The performance of a TPV system is characterized by two metrics: efficiency, which is defined as the ratio of electrical power output to the total radiative heat transfer from the hot emitter to the PV cell at room (or ambient) temperature, and the power density that is the electrical power output per unit area. Recently, efficiencies of up to 30% in the far field have been reported [65], where the emitter (at ~1,450 K) and the PV cell are separated by distances larger than the characteristic thermal wavelength. However, the power densities of far field TPV systems are constrained by the Stefan–Boltzmann limit, since only propagating modes contribute to energy transfer. This limit can be overcome by placing the hot emitter in close proximity (nanoscale gaps) to the PV cell, where, in addition to the propagating modes, evanescent modes also contribute and dominate the energy transfer. The enhancements in heat transfer via near-field (NF) effects have long been predicted and directly demonstrated in recent work paving the way for TPV applications [65]. In fact, several computational studies have suggested that it is possible to achieve high-power, high-efficiency TPV energy conversion via NF effects [65].

In spite of these predictions, few experiments have probed NFTPV energy conversion. This limited progress is due to multiple challenges associated with creating thermal emitters that are robust at high temperatures, creating high-quality PV cells for selectively absorbing above-band-gap NF thermal radiation and maintaining parallelization while precisely controlling the gap between the heated emitter and the PV cell. Most recently, Mittapally et al. [65] experimentally examined near field TPV for efficient heat to electricity conversion at high power density. Their experimental set-up is described in Figure 7.4. They demonstrated record power densities of ~5 kW/m² at an efficiency of 6.8%, where the efficiency of the system was defined as the ratio of the electrical power output of the PV cell to the radiative heat transfer from the emitter to the PV cell. This was accomplished by developing novel emitter devices that can sustain temperatures as high as 1,270 K and positioning them into the near-field (<100 nm) of custom-fabricated InGaAs-based thin film photovoltaic cells. In order to demonstrate efficient heat-to-electricity conversion at high power density, they reported the performance of thermophotovoltaic devices across a range of emitter temperatures (~800 K–1,270 K) and gap sizes (70 nm–7 µm). The methods and insights achieved in this study represent a critical step toward understanding the fundamental principles of harvesting thermal energy in the near-field.

### 7.3.4 RADIOISOTOPE THERMOPHOTOVOLTAICS

Thermal-based nuclear power generators can provide electrical energy for interstellar probes and remote terrestrial sensors for decades when solar energy is not available [66]. In land missions, radioisotope generators are mostly used in polar areas, floating buoys on water, and underwater in deep ocean floor to power devices like transmitters, sensors, and coast guard lights. In the 1960s, several modules were successfully launched in the polar regions and had demonstrated for as long as 11 years of continuous operation. In water applications, the power generators were tested and used in a wide range of projects, from sea surface to as deep as 2,200 feet on the ocean floor [67]. Table 7.1 shows some of the terrestrial thermal-based radioisotope generators developed and used before [67,68].

In comparison to the space generators, most of the terrestrial ones are smaller, have shorter lifetime, and lower output level. The most common output level ranges from several watts to tens of watts, with only 1–2 modules exceeding 100 W. In theory, the power generator can be designed and implemented into any output level. Practically, it is convenient to design the system as a standard power module with the output level around ~40 $W_e$ that can satisfy the power budget in most cases. If more power is needed, multiple modules can be used together to satisfy the output

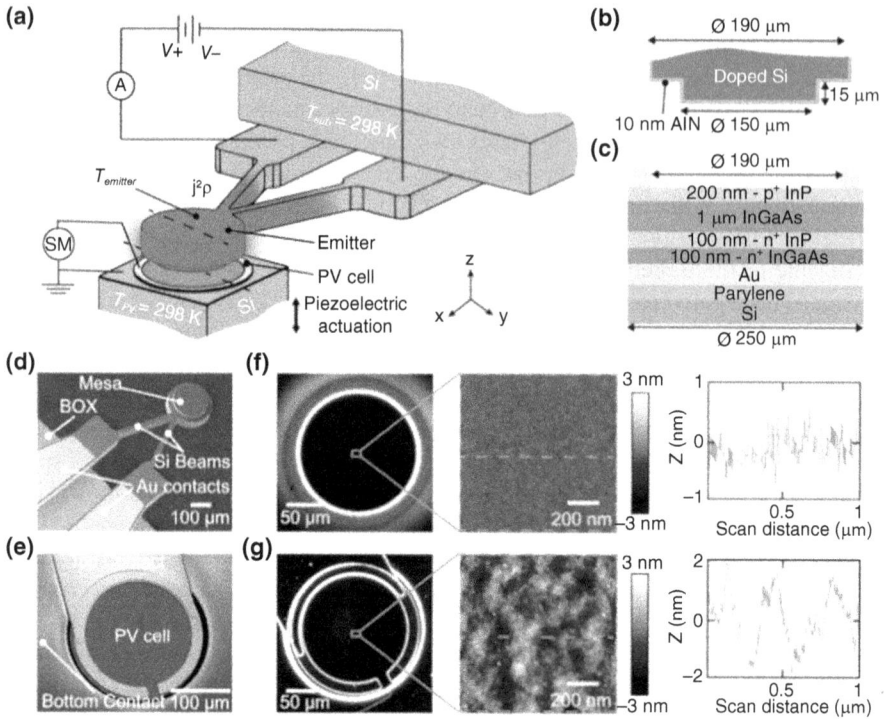

**FIGURE 7.4** (a) Schematic depiction of the experimental setup employed for near-field thermophotovoltaic measurements. The custom-fabricated Si emitter features a suspended mesa (see panel d) that is Joule heated (heat dissipation quantified with an ammeter 'A') up to1270 K by applying a bipolar voltage ($V+$, $V-$) to the two beams. The epitaxially grown InGaAs photovoltaic (PV) cell is moved toward the emitter via a piezoelectric actuator to systematically control the gap size while the electrical power generated is quantified with a source meter (SM). The emitter substrate and the PV cell are at a temperature of ~298 K. (b, c), Cross-sectional profiles of the emitter and the PV cell at the sections along the black dashed lines in (a). (d), False-colored scanning electron micrograph (SEM) of the emitter with mesa, showing the buried oxide layer (BOX) and the gold contacts on the Si beams. The Si beams featuring a temperature gradient are depicted by solid dark gray (e), False-colored SEM of the PV cell showing the central active layer of the PV cell (by lighter grey) as well as top (by white circle) and bottom (by very light grey) Au contacts. (f, g), Dark-field microscope (left panels), atomic-force microscopy (AFM) images (middle panels), and surface roughness profiles (corresponding to the blue dashed lines in the AFM images) of the mesa (f) and the PV cell (g) are shown in the right panels. The peak–peak roughness of the mesa is ~1 nm, while that of the PV cell's active surface is ~4 nm [65].

requirements. The system's critical figure of merit is the efficiency, which is the ratio of the electrical out- put power to the decay heat released by the fuels. Reducing the system weight becomes a lower priority in comparison to their space counterparts because of the lower cost of deployment. On the other hand, terrestrial generators require a higher degree of tamper-proof and shielding to prevent radiation

**TABLE 7.1**

**Radioisotope Generators for Terrestrial Surface and Underwater Applications [66]**

| Model | Year | Application | Output Power [W] | Location |
|---|---|---|---|---|
| NAP-100 | 1960 | Prototyping | 131 | Unfueled |
| Weather station generator | 1961 | Transmitter | 5 | Arctic region |
| SNAP-7A | 1961 | Coast flashing light | 11.6 | Buoy in Curtis Bay |
| SNAP-7B | 1963 | Coast guard light | 68 | Navy floating buoy |
| SNAP-7E | 1962 | Underwater acoustic beacon | 6.5 | Atlantic Ocean bottom |
| Sentinel 21A | 1966 | Oceanographic sensor | 25 | Island in Bering Strait |
| SNAP-21 | 1976 | Sensor and telemetry | 10 | Antarctica |
| URIPS-8 | 1976 | Transmitter | 8 | Antarctica |
| Sentinel-25C1 | 1977 | Transmitter | 25 | San Juan Seamount |
| Sentinel-100F | 1974 | Quartz clock timer | 125 | Eleuthera Island, Bahamas |

contamination and vandalism. Even though the power modules are used in remote places, they are actually accessible for installation and occasional maintenance.

In thermal-based radioisotope power generators, plutonium-238, which is an alpha emitter with a large decay heat, is used as the fuel source. The fuels are processed and pressed into a pellet containing 151 g radioactive materials and encapsulated in the iridium clads [4]. Four fuel pellets are further packaged into the rectangular shaped general purpose heat source (GPH- S) that releases ~250 W thermal power ($W_t$) at the beginning of the mission. A radioisotope generator's thermal source normally contains one or multiple GPHS units stacked together and operates at a temperature from 800°C to 1,200°C. The thermal energy is converted to electricity by various ways, such as thermal-electric materials, Stirling engines, alkali-metal thermal to electrical converters (AMTEC) [69], and thermophotovoltaics [70–73].

The radioisotope thermophotovoltaic system, abbreviated as (RTPV), uses the infrared emission from the high temperature emitter attached to the heat source to generate electricity by the low-bandgap thermophotovoltaic (TPV) cells. A high efficiency system depends on high performance TPV cells and good thermal management to drive the heat to the emitter to get converted. Spectral control is one approach to improve the system efficiency by shaping the emission spectrum of the high temperature emitter patterned with periodic micro-fabricated cavities [74]. The emission spectrum is shaped so that more convertible photons above the cell bandgap are emitted while the radiation in the far infrared is suppressed to reduce the waste heat as shown in Figure 7.5. With spectral control and high performance cells, an RTPV system is expected to reach higher efficiency than the currently deployed radioisotope thermoelectric generators (RTGs) using Seebeck materials. The modeling in this work is based on an experimentally tested prototype that has resolved the material compatibility issues and demonstrated the benefits of spectral control [75].

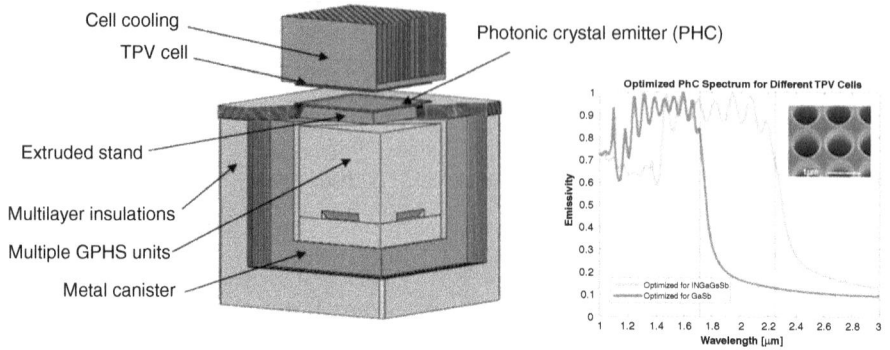

**FIGURE 7.5**   Radioisotope TPV [66].

## 7.3.5 OTHER ALTERNATIVES

TPV energy conversion stands out as the most efficient solid-state thermal-to-electric energy converter, with a record efficiency of nearly 30% at heat source temperatures higher than 1,000°C. As a matter of fact, it is the most efficient small-sized heat engine. Spectral control strategies implemented on the emitter and/or photovoltaic cell, such as back-surface reflectors or front-side filters, along with a proper selection of semiconductor bandgap energy, are the key design elements that enable reaching such high conversion efficiencies. High bandgap energies combined with excellent photon recycling leads to the highest theoretical conversion efficiencies at the expenses of very low output power density. When optical losses are considered the optimal semiconductor bandgap energy shifts to lower values. Moreover, low bandgap energies are required to maximize the output power density at the expenses of reducing the maximum attainable conversion efficiency. Low bandgap semiconductors also bring undesirable practical limitations, mostly regarding the large amount of nonradiative recombination losses. Thus a trade-off exists between power density and conversion efficiency that mostly drives selection of semiconductor bandgap energy. The fulfillment of such a trade-off strongly depends on particular applications. Closed-TPV systems (i.e., those in which the heat source is enclosed within the system, such as radioisotope-TPV) tend to require maximization of TPV efficiency, while open-TPV systems (i.e., those in which the heat source is external to the system, such as solar-TPV) typically require maximization of TPV power density, to compensate the heat losses through the heat inlet aperture. In any case all current conventional TPV devices require very high temperatures, near or beyond 1,000°C, to achieve decent output power densities of 1 W/cm$^2$ or more.

Novel TPV concepts are aimed at breaking trade-offs and limitations and relaxing some of the main design constrains of TPV system. NF-TPV (near field TPV described above), thermophotonics, and LTPV (light pipe TPV), all aim to increase output power density at lower emitter temperatures. On top of that, thermophotonics also allows the use of large TPV cell semiconductor bandgaps. Thermophotonics enable higher power density and lower ohmic and shadowing losses, which are particularly relevant in combination with NF-TPV devices. None of these concepts has reached high technology readiness levels yet. However, progress has been very fast in the past few years, especially for NF-TPV devices. These developments, combined with recent applications of TPV in ultrahigh temperature energy storage, are causing a renewed interest in TPV energy conversion.

TPV technology has also raised a strong interest among scientists in the past decades because it is able to capture sunlight in the entire solar spectrum and has the technical potential to beat the Shockley-Queisser limit of traditional photovoltaics. The efficiencies reported so far, however, are still too low to make it commercially mature, as STPV devices still suffer from a series of optical and thermal losses. With this in mind, a group of researchers from the University of Michigan and the U.S. Army Research Laboratory has proposed a new approach for STPV consisting of reducing the separation between the emitter and the photovoltaic cell to a nanoscale. The researchers call their approach 'near-field thermophotovoltaics' and claim it is able to achieve high power density and high power conversion efficiencies.

The scientists created an STPV device with an emitter that can reach temperatures as high as 1,270 K, and a thin-film photovoltaic cell based on indium gallium arsenide (InGaAs) which is said to be capable of absorbing above-band-gap (ABG) thermal radiation while minimizing absorption of sub-band-gap (SBG) photons. The photons above the band-gap of the cell are efficiently absorbed in the micron-thick semiconductor while those below the band-gap are reflected back to the silicon emitter and recycled. The solar cell was grown on thick semiconductor substrates and the thin semiconductor, active region of the cell was then peeled off and transferred to a silicon substrate. A nano positioning platform in a high-vacuum environment was used to parallelize, and gauge the distance between, the emitter and the PV cell. The emitter and PV cell were initially separated by only 7 μm and the cell was then placed closer to the emitter using a feedback-controlled piezoelectric actuator, which is a tool able to convert an electrical signal into a controlled physical displacement.

According to the researchers, the STPV device exhibited record power densities of around 5 kW/m$^2$ at an efficiency of 6.8%, which they stated is an order of magnitude larger than systems previously reported in the literature. The reported efficiency defines the ratio of electricity output of the PV cell to the radiative heat transfer from the emitter. This current demonstration meets theoretical predictions of radiative heat transfer at the nanoscale, and directly shows the potential for developing future near-field TPV devices for army applications in power and energy, communication and sensors.

Recently, researchers have also spurred research efforts in harvesting waste heat through a hybrid TPV power generation system. The advantage of a hybrid TPV system includes exposure to high-temperature waste heat in continuous operation with a steady condition. Thermoelectric generator (TEG), Brayton-Rankine combine cycle (TBRC), molten carbonate fuel cell (MCFC), solid oxide fuel cell (SOFC), direct carbon fuel cell (DCFC), and direct carbon SOFC (DC-SOFC) are among the reported systems which are paired with TPV system [45]. For example, Chubb and Good [76] investigated a TPV-TEG hybrid system and found that the system generates larger output power density as compared to stand-alone TEG or TPV system. Besides, the integration of TPV system at the exhaust waste heat of SOFC outperforms other SOFC-based coupling systems [77,78].

As regards TPV/TE hybrid systems, thermophotovoltaic cells (TPV) are capable of converting infrared radiation into electricity. They consist of a heat source, an emitter, a filter and photovoltaic (PV) cells [15]. Unlike photovoltaic solar panels, TPV cells are illuminated by radiant combustion sources. Given that the radiant power density of these sources can be much higher than that of the sun, the electrical power density of TPV cells is much higher than that of solar cells, with an efficiency of 24.5% [2]. As of yet, few studies have been conducted on integrated TPV/TE systems. Qiu and Hayden [79] reported that the efficiency of an integrated system with TPV GaSb cells and TEG was superior to that of individual TPV and TE. For this reason, the TPV/TE hybrid system is an interesting alternative system, and further research is required in the future [80]. The major concern with hybrid systems is to achieve optimal hybridization. This means ensuring that the sum of the maximum powers produced separately by the PV and TE systems equals the power produced by the hybrid system [81].

In general, the development of a high-performance hybrid system, however, comes with several challenges. For example, in SOFC-TPV hybrid system at 1,073 K, vacuum gap is limited to nanoscale due to the fluctuation dynamics at extreme near-field region [82]. The design of SOFC current density and heat leak ratio is important to obtain high performance hybrid system [78]. For TPV-DC-SOFC hybrid system, high cost of manufacturing and fabricating efficient catalysts are challenges [83]. For TPV-TEG hybrid systems, large temperature difference between TPV and TEG system decreases the system efficiency and performance depends on thermal characteristics (fuel-air equivalent ratio) of the burner [76,84]. For TPV-DCFC hybrid system, performance depends on the DCFC temperature and the number of slabs in DCFC [85]. For TPV-MCFC hybrid system, higher operating temperature is required to increase hybrid system performance [2]. Finally TPV-TBRC hybrid system requires expensive TPV emitter materials and manufacturing process [86].

## 7.4  EMITTER DESIGN

Since, only photons hitting the photodiode (PV cell) produce electric current, any non-radiative heat coupling between a warm body and the photodiode would be non-convertible and detrimental [87–89]. The emitter stage of a TPV system absorbs heat from an energy source and then emits radiation toward the TPV cell. Use of an emitter stage in the TPV system allows energy from the source transferred via convection, conduction, or radiation, to be turned into usable radiation, making the system energy agnostic in terms of the heat source. The emitter also serves to average the photon flux toward the TPV cell over time, leading to a more constant current.

Emitters are divided into two main categories: broadband and selective emitters. A broadband emitter has a high emissivity over a large range of wavelengths, resulting in a spectral response similar to that of a black body. They are often referred to as "grey bodies" and tend to be made of bulk materials, such as silicon carbide [87–89]. The use of broadband emitters results in more power being incident on to the TPV cell and more possible power output; however, the broad spectrum of frequencies leads to increased cell heating from thermalization, decreasing device performance, and lowering conversion efficiencies. Selective emitters are more popular as modern

TPV system components as they only emit a narrow range of wavelengths, resulting in higher conversion efficiencies. Ideally, a selective emitter has an emissivity of zero except for in the narrow desired band, where the emissivity is at or near unity. Regardless of the heat source, any selective emitter will only radiate in its characteristic spectrum. A narrow emission spectrum just above the band gap energy would result in a high conversion efficiency nearing the external quantum efficiency of the photocell in the limiting case. However, the narrow frequency band emission results in less power. Thus, a balance between efficiency and power must be achieved to maximize the benefit of a selective emitter for a given application [90].

A perfectly radiating body, a blackbody, has an emissivity of 1 at all wavelengths and the power output for the blackbody at any given temperature can be calculated using Planck's law [87–89]. Other objects emit a percentage of the blackbody curve depending on their emissivity at a given wavelength, but the power output scales as the object's temperature is increased. To make the best use of the available power, the peak of a selective emitter spectrum should also occur at the peak of the blackbody curve. The peak wavelength of a radiating blackbody can be calculated using Wien's law [87–89]. For the highest efficiency, the emitter spectrum peak should occur slightly above the band gap of the diode in use. For example, to match the band gap of presently available GaSb TPV cells and optimize the output based on the blackbody curve, a selective emitter needs to operate at 1,400°C. The designed emitter must be able to withstand these high temperatures. According to Kirchhoff's law of thermal radiation, at equilibrium the emissivity of an object is equal to the absorptivity of that object. This relationship is taken advantage of frequently in the development of emitters for any application. Heating a sample over 1,000 K to test emission brings in new safety concerns and the potential for damaging the sample. High temperature testing is particularly difficult for many nanostructured materials as the structures experience degradation below the melting point. Testing first for absorption and then for emission allows the separation of optical and mechanical properties of a given sample, providing more information for future designs.

Efficiency, temperature resistance and cost are the three major factors for choosing a TPV emitter. Efficiency is determined by energy absorbed relative to incoming radiation. High temperature operation is crucial because efficiency increases with operating temperature. As emitter temperature increases, black-body radiation shifts to shorter wavelengths, allowing for more efficient absorption by photovoltaic cells. Polycrystalline silicon carbide (SiC) is the most commonly used emitter for burner TPVs. SiC is thermally stable to ~1,700°C. However, SiC radiates much of its energy in the long wavelength regime, far lower in energy than even the narrowest bandgap photovoltaic. Such radiation is not converted into electrical energy. However, non-absorbing selective filters in front of the PV, or mirrors deposited on the back side of the PV can be used to reflect the long wavelengths back to the emitter, thereby recycling the unconverted energy. In addition, polycrystalline SiC is inexpensive. Tungsten is the most common refractory metal that can be used as a selective emitter. It has higher emissivity in the visible and near-IR range of 0.45–0.47 and a low emissivity of 0.1–0.2 in the IR region. The emitter is usually in the shape of a cylinder with a sealed bottom, which can be considered a cavity. The emitter is attached to the back of a thermal absorber such as SiC and maintains the same temperature.

Emission occurs in the visible and near IR range, which can be readily converted by the PV to electrical energy.

## 7.5 ADVANCED SELECTIVE EMITTER MATERIALS

As early as the 1930s, the Lathanides, or rare earth metals, were shown to have unique absorption and emission spectra [1,91,87–89]. It was not until 1972 that the potential for use as a selective emitter was recognized and the emission spectra specifically studied [46,87–89,93–95]. Since then, rare earth oxides have been shown to be a promising source of radiation for TPV and other applications [87–89]. Oxides are the most commonly used rare earth compound, particularly for high temperature applications, as they are the most thermodynamically stable form. Incorporation of rare earth ions into host materials has also garnered much interest as a way to combine the spectral performance of the lanthanides with high thermal performance of other ceramics and crystals [1,46,87–89,93–95]. These emitters are good candidates for TPV applications due to their highly selective peaks, but the peak wavelengths are linked to the atomic structure of the materials used and cannot be shifted. The addition of other compounds can broaden the overall peak increasing the power output of the emitter, but the spectral position in terms of wavelength remains the same.

Rare-earth oxides such as ytterbium oxide ($Yb_2O_3$) and erbium oxide ($Er_2O_3$) are the most commonly used selective emitters. These oxides emit a narrow band of wavelengths in the near-infrared region, allowing the emission spectra to be tailored to better fit the absorbance characteristics of a particular PV material. The peak of the emission spectrum occurs at 1.29 eV for $Yb_2O_3$ and 0.827 eV for $Er_2O_3$. As a result, $Yb_2O_3$ can be used a selective emitter for silicon cells and $Er_2O_3$, for GaSb or InGaAs. However, the slight mismatch between the emission peaks and band gap of the absorber costs significant efficiency. Selective emission only becomes significant at 1,100°C and increases with temperature. Below 1,700°C, selective emission of rare-earth oxides is fairly low, further decreasing efficiency. Currently, 13% efficiency has been achieved with $Yb_2O_3$ and silicon PV cells.

New techniques in nanofabrication and an increased understanding of the way electromagnetic waves interact with materials has led to the development of spectral engineering, or the controlled manipulation of a device's constituent materials or spatial design to generate a desired absorption/emission spectrum [92]. Though it was first mentioned off-hand in 1992 in reference to adjusting a potential well in order to change the quantum allowed and forbidden state spectra [1], it was not until 2001 that "spectral engineering" was used to correlate physical differences between nanostructures and the resulting changes in their optical spectra [1,46,93–95]. Today, plasmonic devices, generally in the form of photonic crystals and metamaterials, offer the best solution for selective emitters in terms of customizing their spectrum for specific applications.

Photonic crystals (PhCs) offer a more engineerable option of selective emitter than what is available naturally [1]. Photonic crystals allow precise control of electromagnetic wave properties [96]. These materials give rise to the photonic bandgap (PBG). In the spectral range of the PBG, electromagnetic waves cannot propagate. Engineering these materials allows some ability to tailor their emission and absorption properties, allowing for more effective emitter design. Selective emitters with peaks at higher energy than the black body peak (for practical TPV temperatures) allow for wider bandgap converters. These converters are traditionally cheaper to manufacture and less temperature sensitive. A photonic crystal is an array of multiple materials that repeats in one, two, or three dimensions. The dimensions of repetition determine how many planes of photonic response will be induced while the size of the repeated geometries will change what wavelength is affected [1]. PhC behavior was shown as early as 1987 [1,87–89], but application to high-temperature cases such as TPV did not begin until the turn of the millenium. Metallic photonic crystals utilizing tungsten or tantalum show promise as high temperature emitters [97], but in many cases require special packaging such as maintenance of an inert atmosphere. Researchers at Sandia Labs predicted a high-efficiency (34% of light emitted converted to electricity) based on TPV emitter demonstrated using tungsten photonic crystals. However, manufacturing of these devices is difficult and not commercially feasible. The sharp turn-on frequency induced by these devices reduces the number of low energy photons that hit the TPV cell [1,46,93–95]. However, many of them maintain a high emissivity of high energy photon at frequencies above the turn on frequency. The relaxation of these high energy photons can lead to parasitic device heating and reduced conversion efficiency. Also, difficulties in manufacturing nanostructures using tungsten can limit the potential applications [1,46,93–95].

In the past 2 decades, there has been a significant research focus on the development of metamaterials (MMs). In general, a metamaterial is an engineered material consisting of periodic patterns that are smaller than the wavelength of interest and can produce or behave with characteristics that are not found in nature. Using this definition, photonic crystals qualify as a subset of metamaterials. However, they are generally discussed separately as the PhC structure period is on the order of the size of the target wavelength, meaning interference and diffraction dominate the optical interactions. MM patterns are much smaller than the target wavelength, so more subtle electromagnetic field interactions dominate. MMs are designed to exhibit particular permittivity and permeability coefficients, thus impacting the impedance, or refractive index, of the material [89]. Natural materials can either have a negative permittivity or a negative permeability, but never both at the same time. This means a natural refractive index is always positive and may contain an imaginary component. Recently, negative refractive indices have been realized using man-made structures [98] and have led to research advances in the development of invisibility cloaks [99], perfect lenses [100], and other custom dielectrics. The advanced manipulation of electromagnetic radiation with MMs has also led to the development of perfect absorbers and emitters, which have additional applications in photodetector [101,102] and photovoltaic enhancement [103–105].

A standard metamaterial consists of a dielectric substrate with a subwavelength, periodic pattern made of a conducting material on top. They are usually fabricated using photo- or electron beam lithography, depending on the necessary resolution required for the pattern, on top of the dielectric substrate. Physical vapor deposition is used to lay down the metal on top of the resist pattern. A process called liftoff removes the extra metal so that only the area without resist remains metalized. The designs used for MMs are as varied as their many applications and creators. What

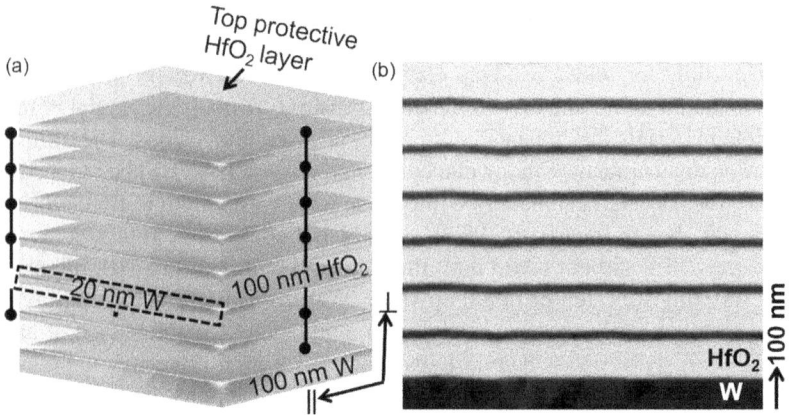

**FIGURE 7.6** Metamaterial emitter for thermophotovoltaics stable up to 1,400°C [91]. (a) two dimensional view of emitter, (b) front view showing bilayers

they have in common is the induction of fields that interact with light in customized, novel ways. For example, Chirumamilla et al. [91] realized a 1D structured emitter based on a sputtered W-HfO$_2$ layered metamaterial and demonstrated desired band edge spectral properties at 1,400°C. The spatial confinement and absence of edges stabilized the W-HfO$_2$multilayer system to temperatures unprecedented for other nano-scaled W-structures. Only when this confinement was broken W started to show the well-known self-diffusion behavior transforming to spherical shaped W-islands. The study further showed that the oxidation of W by atmospheric oxygen could be prevented by reducing the vacuum pressure below $10^{-5}$ mbar. When oxidation was mitigated the study observed that the 20 nm spatially confined W films survived temperatures up to 1,400°C. The demonstrated thermal stability was limited by grain growth in HfO$_2$, which led to a rupture of the W-layers, thus, to a degradation of the multilayer system at 1,450°C.

A schematic of the W and HfO$_2$-based layered metamaterial emitter is shown in Figure 7.6a. Six bilayers of W and HfO$_2$, with thicknesses of 20 and 100 nm, respectively, are sandwiched between a top protective HfO$_2$ layer and bottom thick W layer, each 100 nm thick. Cross-sectional view of the high-angle annular dark-field (HAADF) scanning transmission electron microscopy (STEM) image of the as-fabricated emitter structure is shown in Figure 7.6b. The number of bilayers was six in order to avoid residual transmission through the metamaterial. According to Kirchhoff's law of thermal radiation [91,87–89], the emissivity of a hot radiating body equals its absorptivity. Therefore, we can assess the TPV-relevant spectral emissivity by measuring the absorptivity of our metamaterial layer. At room temperature, the as-fabricated emitter structure showed a step function-like steep spectral cutoff around 1.7 μm and low absorptivities/emissivities above the wavelength corresponding to the bandgap of the PV cell, i.e. low emission of such photons. The metamaterial emitter structure after annealing at 1,400°C for 6 hours, measured at room temperature, showed similar band-edge characteristics with even a slight improvement of the spectral characteristics, e.g., a reduction of the absorptivity/

## Towards practical emitter implementation

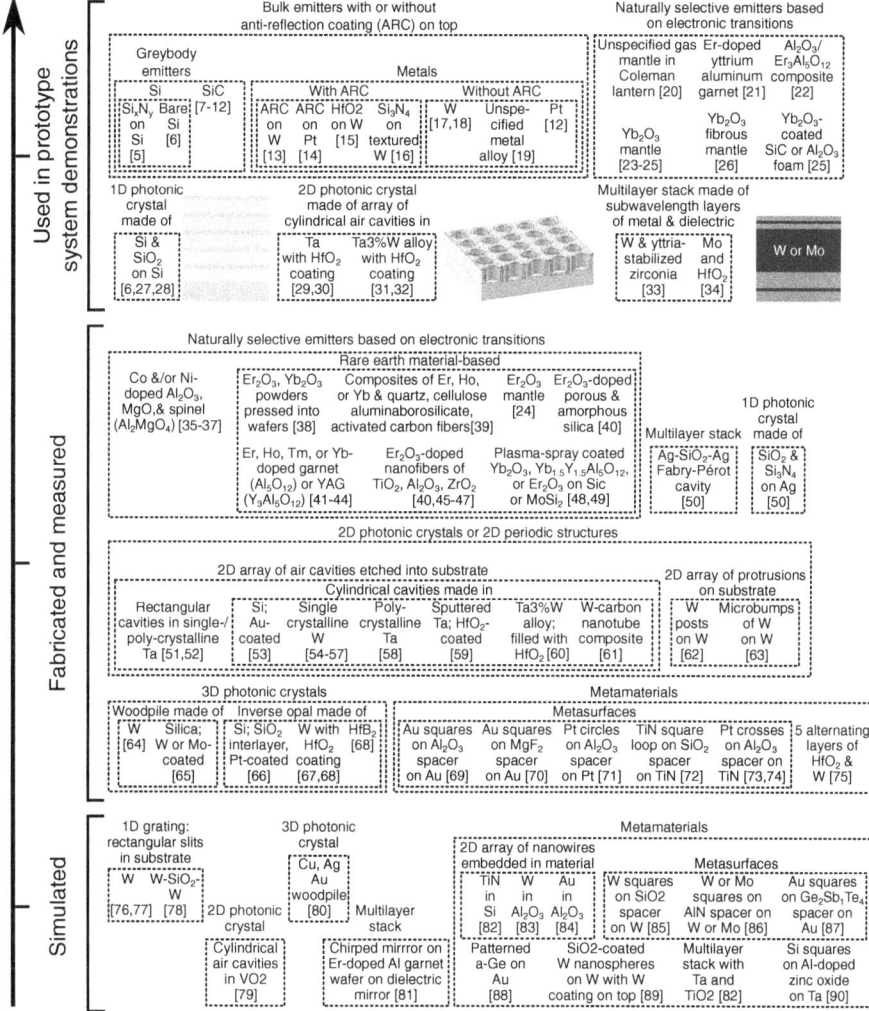

**Used in prototype system demonstrations**

Bulk emitters with or without anti-reflection coating (ARC) on top | Naturally selective emitters based on electronic transitions

Greybody emitters

| Si | SiC [7-12] |
| Si₃N₄, Bare on Si | |
| Si [6] | |
| [5] | |

Metals

With ARC

| ARC on W [13] | ARC on Pt [14] | HfO2 on W [15] | Si₃N₄ textured |

Without ARC

| W [17,18] | Unspe- cified metal [12] | Pt alloy [19] |

Unspecified gas mantle in Coleman lantern [20]

Er-doped yttrium aluminum garnet [21]

Al₂O₃/ Er₃Al₅O₁₂ composite [22]

| Yb₂O₃ mantle [23-25] | Yb₂O₃ fibrous mantle [26] | Yb₂O₃- coated SiC or Al₂O₃ foam [25] |

1D photonic crystal made of

| Si & SiO₂ on Si [6,27,28] |

2D photonic crystal made of array of cylindrical air cavities in

| Ta with HfO₂ coating [29,30] | Ta3%W alloy with HfO₂ coating [31,32] |

Multilayer stack made of subwavelength layers of metal & dielectric

| W & yttria- stabilized zirconia [33] | Mo and HfO₂ [34] | W or Mo |

**Fabricated and measured**

Naturally selective emitters based on electronic transitions

Rare earth material-based

Co &/or Ni- doped Al₂O₃, MgO,& spinel (Al₂MgO₄) [35-37]

| Er₂O₃, Yb₂O₃ powders pressed into wafers [38] | Composites of Er, Ho, or Yb & quartz, cellulose aluminaborosilicate, activated carbon fibers[39] | Er₂O₃ mantle [24] | Er₂O₃-doped porous & amorphous silica [40] |

1D photonic crystal made of

| Er, Ho, Tm, or Yb- doped garnet (Al₅O₁₂) or YAG (Y₃Al₅O₁₂) [41-44] | Er₂O₃-doped nanofibers of TiO₂, Al₂O₃, ZrO₂ [40,45-47] | Plasma-spray coated Yb₂O₃, Yb₁.₅Y₁.₅Al₅O₁₂, or Er₂O₃ on Sic or MoSi₂ [48,49] |

Multilayer stack

| Ag-SiO₂-Ag Fabry-Pérot cavity [50] | SiO₂ & Si₃N₄ on Ag [50] |

2D photonic crystals or 2D periodic structures

2D array of air cavities etched into substrate

Cylindrical cavities made in

| Rectangular cavities in single-/ poly-crystalline Ta [51,52] | Si; Au- coated W [53] | Single crystalline W [54-57] | Poly- crystalline Ta [58] | Sputtered Ta; HfO₂- coated [59] | Ta3%W alloy; filled with HfO₂[60] | W-carbon nanotube composite [61] |

2D array of protrusions on substrate

| W posts on W [62] | Microbumps of W on W [63] |

3D photonic crystals | Metamaterials

Woodpile made of | Inverse opal made of

| W [64] | W or Mo- coated [65] | Silica; Si; SiO₂ interlayer, Pt-coated [66] | W with HfB₂ HfO₂ coating [67,68] |

Metasurfaces

| Au squares on Al₂O₃ spacer on Au [69] | Au squares on MgF₂ spacer on Au [70] | Pt circles on Al₂O₃ spacer on Pt [71] | TiN square loop on SiO₂ spacer on TiN [72] | Pt crosses on Al₂O₃ spacer on TiN [73,74] | 5 alternating layers of HfO₂ & W [75] |

**Simulated**

1D grating: rectangular slits in substrate

| W [76,77] | W-SiO₂- W [78] |

3D photonic crystal

| Cu, Ag Au woodpile [80] |

2D photonic crystal

| Cylindrical air cavities in VO2 [79] |

Multilayer stack

| Chirped mirror on Er-doped Al garnet wafer on dielectric mirror [81] |

Metamaterials

2D array of nanowires embedded in material

| TiN in Si [82] | W in Al₂O₃ [83] | Au in Al₂O₃ [84] |

Metasurfaces

| W squares on SiO2 spacer on W [85] | W or Mo squares on AlN spacer on W or Mo [86] | Au squares on Ge₂Sb₂Te₄ spacer on Au [87] |

| Patterned a-Ge on Au [88] | SiO2-coated W nanospheres on W with W coating on top [89] | Multilayer stack with Ta and TiO2 [82] | Si squares on Al-doped zinc oxide on Ta [90] |

**FIGURE 7.7**  Three categories of TPV emitter [1].

emissivity at long wavelengths, which is attributed to a reduced electron collision frequency due to grain growth in the tungsten layer leading to an improved metallic reflection.

As shown by Sakakibara et al. [1], the literature on materials research for TPV emitters can be broken down in three categories, as shown in Figure 7.7. Three categories of TPV emitters include those that have been (a) used in published system demonstrations of TPV prototypes, (b) fabricated and measured, and (c) simulated. The emitters in this figure emit in ~1–3 μm range. Abbreviations and some terminology: atomic symbols are used, ARC is an antireflection coating, a photonic crystal is a periodic structure, a metamaterial is a manmade material that has optical properties not usually found in nature, and metasurfaces are a class of metamaterials that consist of a 2-D array of metal features on a dielectric spacer on a metal substrate [1].

### 7.5.1 EMITTERS USED IN PROTOTYPE SYSTEM DEMONSTRATIONS

Sakakibara et al. [1] indicate that broadly there are five types of emitters that have been implemented in prototype system demonstrations. These are:

1. **Bulk emitters**: There are two types: graybody emitters and metals with and without ARC. Both graybody emitters, such as silicon and silicon carbide, and metals with or without antireflection coating (ARC) can be easy and inexpensive to fabricate in large areas. Graybody emitters often can be fabricated onto or with the heat source. However, they have both high in-band emission and out-of-band emission, which is why they are often coupled with cold side filters. The emission in metals with and without ARC depends on the metal optical properties. The role of the ARC layer is typically to enhance emittance in a narrow band around the bandgap.

2. **Naturally, selective emitters**: These are made primarily from rare earth metals, especially erbium and ytterbium. They are easy to fabricate in large areas and with high-temperature stability, especially by doping high-temperature ceramics. However, the emission wavelength range of naturally selective emitters is not tunable and narrow band, which can lead to low in-band emitted power density.

3. **1-D photonic crystals**: This is often known as dielectric mirrors, and it consists of alternating layers of materials with a high contrast in indices of refraction. They are typically not made from high-temperature materials although they can be directly fabricated onto a heat source. They are easy and inexpensive to fabricate at large areas, but have multiple interfaces. Interference effects in this structure lead to a fairly broad reflection bandwidth, which can be used to suppress high natural emittance of a material for a wavelength region. They may have high out-of-band emission outside of the region of suppression.

4. **2-D photonic crystals**: These are laid out as 2-D array on top of a substrate, such as cylindrical posts or air cavities, with feature sizes on the order of the wavelength of interest. For photonic crystals with air cavities, each individual cavity acts as a waveguide to enhance emission of wavelengths below a cutoff (half a wavelength corresponds roughly to the cavity diameter). Generally high temperature material used for the substrate and whether an emitter can be fabricated inexpensively with large area and can be integrated with the heat source depends largely on the substrate.

5. **The multilayer stacks**: These differ from 1-D photonic crystals in that there is no periodicity in the thicknesses of the layers. These can be combined absorber/emitters for solar TPV that consist of alternating metal and dielectric layers of varying thicknesses. The layer thicknesses, which are sometimes subwavelength, are optimized to enable both broadband absorption and emission. While the optical performance of the stacks is good and

fabrication cost can be low, the ability to fabricate large areas depends on the available size of the metal. Furthermore, while high-temperature materials are used, there are many interfaces, and their long-term high-temperature stability is unclear.

Sakakibara et al. [1] pointed out that none of the emitters or types of emitters identified have yet satisfied all-needed criteria for practical TPV emitters. An emitter with good optical performance shows high, broadband, preferential in-band emittance. Some emitters with promising optical performance include the 1-D photonic crystal, 2-D photonic crystal, and multilayer stacks. For the latter two types, there have been some studies on the high-temperature stability and system integration. The multilayer stack is fabricated directly with the absorber (heat source), but its high-temperature stability has not been shown beyond 1 hour. On the other hand, a 2-D photonic crystal made in refractory metals that can be fabricated on the order of cm$^2$ has some slightly high out-of-band emittance but has been successfully integrated with micro combustor heat sources and has shown promising high-temperature stability of a few hundred hours at 900°C–1,000°C.

In future, studies on the practical aspects of each TPV emitter are critical for the maturation of TPV technology. So far an emitter that satisfies the following performance metrics: spectral selectivity exceeding 90%–95%, high in-band emittance of 0.9–0.95, and high-temperature stability on the order of thousands of hours and hundreds of cycles is not available. High-temperature stability and large-area fabrication can be addressed independently of the overall TPV system setup. One promising class of emitter is metamaterials, which show high optical performance; however, studies of high-temperature stability are at the moment limited. The ability of TPV systems to be commercialized hinges on the practical attributes of selective emitters.

## 7.6   REQUIREMENTS FOR EFFECTIVE EMITTERS

Although the primary purpose to develop an emitter is for its optical performance, the emitter with the best optical performance is not necessarily the best emitter for practical implementation. There are six requirements for effective emitters: (a) optical performance, (b) ability to scale to large areas, (c) long-term high-temperature stability, (d) ease of integration within the TPV system, (e) TPV subsystem efficiency, and (f) cost to evaluate overall suitability of TPV system. We evaluate these in detail.

### 7.6.1   OPTICAL PERFORMANCE

The optical performance depends on emittance at each angle, radiated in-band power density and spectral selectivity and efficiency. An emitter with good optical performance has, at all angles, preferential emission of in-band photons and suppression of out-of-band photons. Optical performance refers to the emission of photons as a function of both angle and photon energy, in particular, in two regimes for the latter, in-band photons that have energy higher than the PV cell bandgap and out-of-band photons that have energy lower. Spectral control refers to the methods that enable preferential in-band emission. Some designs of spectral control are designed for broadband emission while others for narrow-band emission. In the latter case, the emitted photons have energies slightly above the PV cell bandgap. Typically, broadband emitters yield higher output electrical power density while narrow-band emitters can increase the TPV conversion efficiency [82]. Prior reports have sought to describe emitter-specific spectral efficiency as the ratio of in-band power to total emitted power, [47] and therefore defined cell efficiency in terms of the conversion of in-band power. This formulation, however, neglects the cell's role in modifying the spectrum of $Q_{abs}$ and cannot be easily generalized to pairs with reflective cells. To provide a more general description of component-wise contributions to spectral management, one investigates the properties of a single component by considering its spectral efficiency when paired with a theoretical blackbody (non-selective) counterpart. This metric is termed "individual *SE*."

An emitter with good optical performance may have (a) either broadband emission, where any in-band photons (energy greater or wavelength shorter than the PV cell bandgap, where EPV and λPV are the bandgap energy and wavelength, respectively) are preferentially emitted, or narrow-band emission, where only photons with energy slightly above the bandgap are emitted. Here photons or radiation are referred both in terms of energy and wavelength. (b) The goal of angular control is to ensure good spectral control (preferential in-band emission) over a wide range of angles, as thermal radiation can be off-normal. (c) View factor loss, where photons are lost through the emitter-PV cell gap, is a significant source of loss. The purpose of angular control, which is often an implicit aspect of spectral control, is to ensure spectral control over all angles (polar and azimuthal, $\theta$ and $\phi$), because an emitter radiates photons over a wide range of angles. This is especially important because most thermal radiation is off-normal as according to Lambert's law. For TPV, the wavelength regions of interest are around 1–3 μm, approximately the regions of peak emission. For an emitter heated to realistic temperatures of 1,000–1,500 K, the peak emission wavelengths are 1.9–2.9 μm, as according to Wien's displacement law. As such, one of the main requirements of TPV is to have low-bandgap PV cells, with typical bandgaps in the range of about 0.50–0.74 eV or 1.7–2.3 μm.

Although a PV cell can generate electricity only from in-band photons, a real emitter emits both in-band and out-of-band photons at any given angle. This leads to the following problems: (a) if out-of-band photons reach the PV cell, they overheat the PV cell and reduce the PV cell efficiency and (b) when out-of-band photons are emitted and not recovered, this leads to both reduced heat-to-radiation efficiency and emitter temperature. On the other hand, selective emitters suppress out-of-band emission relative to in-band emission. This reduces view factor and absorption losses for out-of-band photons, although both losses, especially view factor losses, remain significant for in-band photons. The selective emitters should mostly emit in-bound photons and little or no out-of-band photons. These dilemma about in-bound versus out-bound photons, the effects of out-bound photons on thermalization of TPV cell and the desire to get high power intensity lead to a number of questions regarding the design criteria for emitter. These include (a) is it better to prioritized broadband or narrow band emission? While narrow band reduces the thermalization of TPV cell, it is accompanied by the reduction in power density. One way to go around this issue is to use TPV cells with multiple bandgaps [82]. (b) how to choose in an engineered emitter between its level of emission and capacity for in-band wavelength regions? In general, suppression of naturally high emission works only for a limited wavelength range [1]; ideally, the emission should be suppressed for wavelengths up to about 15 µm, which accounts for >96% of the energy emitted by a blackbody at 1,000 K. and (c) should one focus on the level of in-bound emission irrespective to the level of out-bound emission? It is clear that the strategy in design should be such that both increase in in-bound and decrease in out-bound emissions are considered and simultaneously optimized to get the best power density. An emitter is not better than others simply because it has reached high temperatures because temperatures beyond 1,500 K are hard to achieve, the amount of input power required to heat an emitter may vary widely, and it is unclear if a given emitter can sustain high optical performance at high temperature for prolonged periods. As shown later, the issue of stability at high temperature is important. One should focus on power density Mrad obtained by in-band emission because that affects the maximization of high-output electric power.

### 7.6.1.1 Spectral Control of Selective Emitters

In general, good optical performance is achieved by properly manipulating in-bound and out-bound emissions. There are two possible strategies to achieve this. The first is to enhance in-band and suppress out-of-band emission via selective emitters. This can be achieved by choosing emitters that have been heat treated at or above 1,023 K (750°C) for over 1 h and whose optical properties were characterized at or above 1,023 K (750°C). The second strategy is to reflect out-of-band photons back to the emitter, or photon recycling, via cold side filters or reflectors (CSFR) in front of (front side filter) or behind the PV cell (back surface reflector). These two strategies can be combined. One can use a selective emitter and a filter or reflector, or even all three together. Although an emitter with a CSFR performs better than a blackbody or graybody (relatively higher temperature and mitigated PV cell efficiency reduction), it suffers from view factor and absorption losses. In view factor loss, which is inherent to systems with diffuse emitters, photons are lost in the finite gap between the emitter and the filter/reflector, and in absorption loss, photons are absorbed at any interface (at the filter, reflector, or PV cell). While it is possible to reduce the view factor loss by reducing the emitter area relative to the PV cell area, keeping an emitter arbitrarily small decreases its absolute radiated power [46].

Often the first strategy is preferred because besides poor spectral performance, the stability of emitter is important. If an emitter is unstable, the material may evaporate and subsequently deposit onto the cell. The thermal stability of emitter affects the performance of overall TPS system and its maintenance cost. First strategy gives good indication of thermal stability of emitter around 1,000°C [106]. It is also important to calculate individual SE based on reported spectra measured at high temperatures because emissivity is temperature dependent [107]. Figure 7.8a shows the individual *SE (spectral efficiency)* of various emitters that meet the above strategies.

The results shown in Figure 7.8a show that the decreasing out-of-band emission and/or absorption has a greater effect on *SE* than increasing in-band emission and/or absorption due to the relative power in each band. The results are also shown for two different types of emitters; (a) intrinsically selective materials such as transition- metal and rare earth oxides (Figure 7.8b) and (b) structurally tunable thermal

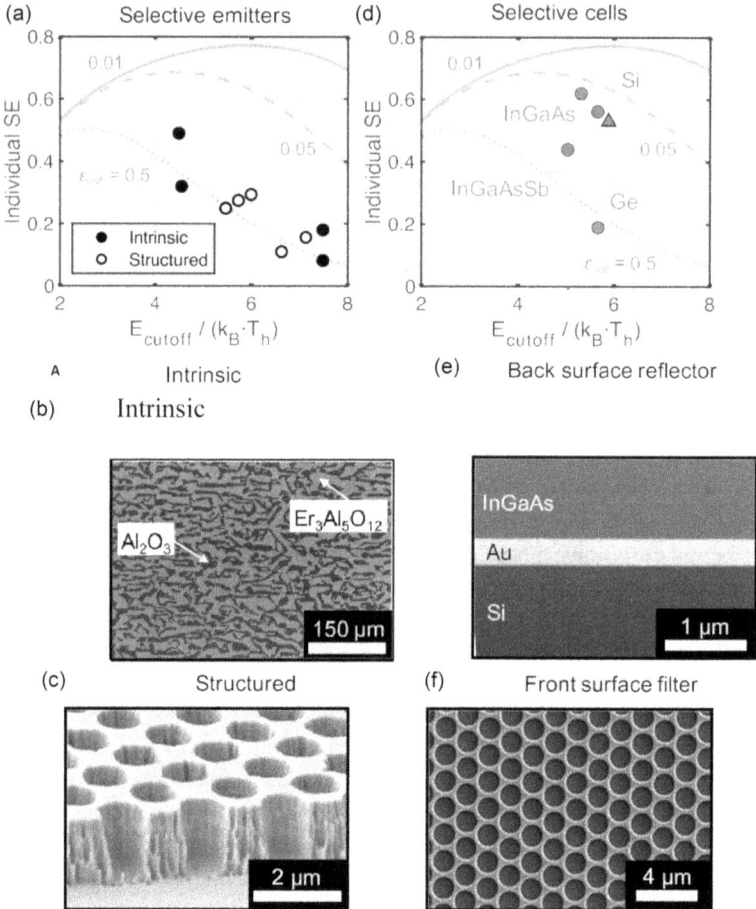

**FIGURE 7.8** Component-wise spectral control strategies. (a–e) Individual spectral efficiencies of (a) selective emitters and (d) selectively absorptive cells. Experimental values are compared to *SE* curves for various out-of-band emissivities with ideal in-band absorption ($\varepsilon_{in} = 1$). Examples of spectral control structures: (b) $Al_2O_3/Er_3Al_5O_{12}$ eutectic ceramic, [41] (c) 2D W cavity array, [42] (e) thin-film LM InGaAs with Au BSR, [14] and (f) 2D photonic crystal front-surface filter on a GaSb cell [51,46].

emitters exhibiting geometry-dependent spectral properties (Figure 7.8c). Tunable emitters typically leverage periodic architectures with one-dimensional (i.e., alternating stacks), two-dimensional (i.e., cavity or pillar arrays), or three-dimensional (i.e., inverse opal networks) periodicity at a length scale on the order of the wavelength of interest [108].

In terms of the best spectral efficiency exhibited among intrinsically selective materials, a MgO emitter with NiO loading [109] achieves a notable 49% spectral efficiency. Introducing transition-metal dopant ions within a low emissivity MgO host lattice leads to selective emission due to electronic transitions. Intrinsic emitters generally exhibit better thermal stability. Alternatively, structured emitters offer improved control over the emission cutoff energy. The leading structured thermal emitter is a W 2D photonic crystal emitter with a cavity array geometry that exhibits 29.4% individual *SE* [110]. Other structured emitters, characterized near room temperature, exhibit promising spectral properties. For example, a $HfO_2/Mo/HfO_2$ emitter leverages its ultrathin Mo absorber layer and a Fabry-Perot cavity created between the top interface and the bottom reflector to achieve coherent perfect absorption at a wavelength near the cell's bandgap [111]. The study by Burger et al. [46] indicate that the primary failure mechanisms of structured emitters appear to be oxidation of the metal layers and growth of dielectric grains, both of which are activated by high temperatures [91]. However, the kinetics of these mechanisms can be slowed by operating under moderate vacuum and below the grain growth temperature threshold. One promising way to stabilize emitters appears to be the use of transparent refractory coatings. Low-defect refractory coatings enhance thermal stability by impeding surface reactions and inhibiting diffusion [112,113]. A widely proposed plan to protect the cell from material deposition is the use of an intermediate glass cover or a gas purge [114]. However, out-of-band absorption in a deposition shield may limit the effectiveness of cell-side spectral control.

As introduced earlier, an alternative approach for spectral control is reflection of out-of-band radiation back to the emitter using a selectively absorptive cell. This is typically achieved through the use of a back-surface reflector (BSR) [115] (Figure 7.8e) and/or a front-surface filter (FSF) (Figure 7.8f). One practical advantage of this spectral control strategy is that pairs are not constrained by the requirement of material stability at high operating temperature. This may enable the design and use of a richer set of photonic architectures. Figure 7.8d shows the individual spectral efficiencies of representative TPV cells, given a blackbody emitter at the temperature indicated by the relevant publications. A 0.6 eV InGaAs cell developed by Wernsman et al. has the highest individual spectral efficiency to date (62%) [39]. Other leading designs have performed similarly well out-of-band. Beyond simple semiconductor-on-metal architectures, use of dielectric spacers at the back of the active layer has also been shown to improve out-of-band reflectance for Ge, LM InGaAs, and InGaAsSb cells [116–119]. There is also room for significant improvements to in-band absorption in many TPV cells. Deposition of an ARC or surface texturing can improve in-band absorption, *SE*, and output power. Spectral utilization may also be improved through the integration of additional absorbers. TPVs utilizing tandem cells may theoretically surpass the radiative limit *SE* defined above for a single-junction cell. This approach reduces thermalization and Ohmic losses. Table 7.2 describes the properties of some of the selective emitters. Blandre et al. showed that active cooling techniques may be necessary to meet cooling demands at high power densities [120]. As materials transition to prototypes, power consumed for circulating coolant may reduce overall efficiency, but this effect is expected to be small (<5%) with state-of-the-art thermal management systems [121–123].

## 7.6.2 SCALABILITY TO LARGE AREAS

Because the fundamental limit for emitters is on the power radiated per unit area, one way to increase the absolute radiated power is by increasing the emitter area (its macroscopic exterior dimensions). In terms of practical implementation, it is important to

## TABLE 7.2
### Selective Thermal Emitters with Emissivity Measurements and Heat Treatments Performed at Temperatures >1,023 K [46]

| Emitter Description | | $E_{cutoff}$ [eV] | Measurement Temperature $T_h{}^a$ [K] | $E_{cutoff}/k_B T_h{}^b$ | $\varepsilon_{out}{}^b$ | $\varepsilon_{in}{}^b$ | Ind. $SE^c$ | $E$ Range [eV]/ BB Fraction$^d$ | Heat Treatment Conditions |
|---|---|---|---|---|---|---|---|---|---|
| Intrinsic | MgO with 2 wt % NiO loading | 0.65 | 1,677 | 4.5 | 0.18 | 0.71 | 0.49 | 0.14–1.14/93% | 1,793 K, duration omitted |
| | Al₂O₃/Er₃Al₅O₁₂ eutectic | 0.73 | 1,850 | 4.58 | 0.27 | 0.43 | 0.32 | 0.62–1.4/40% | In air at 1,973 K for 1,000 hours |
| | Yb₂O₃ foam | 1.12 | 1,735 | 7.49 | 0.29 | 0.54 | 0.082 | 0.075–1.6/99% | 1,750 K for 200 cycles, duration omitted |
| | Yb₂O₃ mantle | 1.12 | 1,735 | 7.49 | 0.14 | 0.62 | 0.18 | 0.024–1.6/99% | 1,750 K for 200 cycles, duration omitted |
| Structured | Pt array on Al₂O₃/Pt | 0.6 | 1,273 | 5.47 | 0.47 | 0.93 | 0.25 | 0.25–1.4/76% | In Ar at 1,273 K for 2 hours |
| | W 2D cavity (D = 1.1 mm) array | 0.62 | 1,200 | 6.00 | 0.25 | 0.86 | 0.294 | 0.16–1.3/90% | Under vacuum at 1,200 K for 10 hours |
| | Ta 2D cavity array with HfO₂ coating | 0.62 | 1,255 | 5.73 | 0.34 | 0.89 | 0.275 | 0.41–0.89/40% | Under vacuum at 1,273 K for 1 hour/1,173 K for 144 hours |
| | HfO₂ coated W inverse colloidal 3D PhC | 0.67 | 1,173 | 6.63 | 0.62 | 0.91 | 0.11 | 0.25–0.98/71% | In Ar at 1,673 K for 1 hour |
| | W 2D cavity (D = 900 nm) array | 0.73 | 1,186 | 7.14 | 0.30 | 0.94 | 0.156 | 0.16–1.3/90% | Under vacuum at 1,200 K for 10 hours |

Emissivity data extracted from relevant publications is available for download (see Supplemental Information).

a $T_h$ refers to the measurement temperature at which spectral emissivity was characterized.

b Weighted average out-of-band emissivity ($\varepsilon_{out}$) and in-band emissivity ($\varepsilon_{in}$) have been calculated using spectral properties collected from graphical data. Error may have resulted from the data extraction process.

c Individual $SE$ calculations are based on spectral emissivity data. We extrapolate average emissivity values by band to the limits of integration to account for truncated data. Therefore, our calculated values may deviate from those values reported elsewhere. We note the reported spectral range as a measure of certainty for individual $SE$ calculations.

d The reported spectral range used to calculate $SE$ and the fraction of the emissive power at the given $T_h$ captured by this range are provided.

consider the following: (a) the substrates must be available in large sizes and (b) the fabrication methods should accommodate large-area samples relatively easily. For example, for 1, the single-crystalline substrates of tungsten and tantalum are typically available in small diameters 1–1.5 cm (area ~3–7 cm$^2$) [124–126], while naturally selective emitters made of rare earth metals can be on the order of tens of cm$^2$ [23,127,128]. An example for 2 is that electron beam lithography, which is typically used for features <500 nm, is both costly and time-consuming. The overall complexity of the fabrication process, including the number of steps and the complexity of each individual step, can impact the scalability as well as the cost. However, many of the fabrication techniques reported in the literature are best suited for proof-of-concept demonstrations, rather than mass production.

### 7.6.3 LONG-TERM HIGH-TEMPERATURE STABILITY

Sakakibara et al. [1] evaluated long-term high-temperature stability of emitters in detail. The TPV emitter must sustain its optical performance at high temperatures for extended periods of time, either continuously or over multiple thermal cycles. However, as pointed out by them at high temperatures, the kinetic energy of atoms increases and atoms diffuse more easily, leading to a number of potential thermodynamic effects such as: (a) Sharp edges and features can become more rounded. (b) A phase change may occur (e.g., the emitter might melt), accompanied also by changes in morphology and optical properties. This can happen also for crystalline phases. However, it is important to keep in mind that the melting point of a material at nanometer scale is lower than for bulk. (c) The sizes of grains can grow in polycrystalline materials. However, this can actually stabilize the material, so some substrates such as polycrystalline tantalum are annealed prior to use. It is also possible to use large-grain or single-crystal substrates. (d) Chemical degradation may occur, such as the formation of tungsten oxides and tantalum carbide. This can necessitate that the emitter operate in inert atmosphere or vacuum, which requires special packaging and complicates the TPV system integration. Chemical degradation of 2-D and 3-D tungsten and 2-D tantalum photonic crystals can be mitigated by capping the surface with a 20–40 nm protective coating of hafnium dioxide ($HfO_2$). One comparison of $HfO_2$ and $Al_2O_3$ in 3-D photonic crystals [129] has found $HfO_2$ to be more thermally robust than $Al_2O_3$, but $Al_2O_3$ is less expensive and has been used to protect a meta surface emitter [105]. (e)Thermal expansion could lead to the cracking of a material. Also, emitters with interfaces between different materials are at risk of delamination because different materials have different thermal expansion coefficients.

Some strategies for improving the high-temperature stability include selecting materials that are known to have good high temperature properties, alloying to promote a solute drag effect [1,130], and modifying the geometry of a structure to change diffusion rates [1,130]. There do not appear to be any published long-term (>1,000 hours) studies; in some cases, it appears the emitter is only heated to measure its high-temperature optical properties. One long study is 168 hours (7 days) at 1,000°C (1,273 K) for a 2-D structure made with tungsten and carbon nanotubes [131]. The literature studies on emitter stability are described in detail by Sakakibara et al. [1]. The longest reported studies are 300 hours each for an erbium-doped yttrium aluminum garnet

(Er-YAG) crystal used in an solar TPV system [132] and a 2-D photonic crystal made of tantalum-tungsten alloy and capped with 20–40 nm HfO2$\underline{1}$ used in a radioisotope TPV prototype [133]. The former cracked and darkened after 300 hours in the sun, although the authors attribute it potentially to a water leak. The 2-D photonic crystal showed little to no degradation in optical performance after annealing for 300 hours at 1,000°C (1,273 K) [134] and also 1 hour at 1,200°C (1,473 K) [135]. Other emitters used in both system demonstrations and some high-temperature stability experiments include a $Yb_2O_3$ foam ceramic [128], a 2-D photonic crystal made of polycrystalline tantalum and coated with 20–40 nm $HfO_2$ [112], and a multilayer stack made of tungsten and $HfO_2$ [136]. The foam ceramic was robust under 200 thermal cycles, the 2-D tantalum photonic crystal showed no visible degradation after 144 hours at 900°C (1,173 K) and 1 hour at 1,000°C (1,273 K) [112], and the multilayer stack showed little to no degradation after at least 1 hour at 1,423 K in vacuum $<5 \times 10^{-2}$ Pa and two rapid thermal cycles up to 1,250 K, but showed degradation after 1 hour at 1,473 K.

### 7.6.4 EASE OF INTEGRATION WITHIN THE TPV SYSTEM

The design choices for the emitter can present several challenges for its integration within the TPV system, in particular, when (a) putting the emitter and heat source physically together for thermal contact and (b) packaging the system for operation in vacuum or inert gas environment. In some cases, the emitter and heat source are made out of the same material, such as silicon carbide [137] or platinum, 12 or the emitter is directly fabricated onto the heat source, for example, through the deposition of emitter materials, such as silicon and silicon dioxide [56] or tantalum [138] onto a micro combustor, or the fabrication of a combined absorber/emitter for solar TPV [1,51].

In other cases, it may be required to cut the emitter to the correct size and to machine and weld it onto the heat source, such as a micro combustor. It is possible to use foil, sputtered coating, or a solid-state substrate. In the last case especially, the mechanical properties of the emitter material become significant. As an example, three different substrates have been explored in the development of 2-D photonic crystal emitters. These include single crystalline tungsten, polycrystalline tantalum, and tantalum-tungsten alloy [1,74]. Tungsten is brittle and difficult to machine and weld [74], while polycrystalline tantalum is more compliant and easier to weld and machine but is soft so needs to be thicker than tungsten to achieve the same mechanical stability. Tantalum-tungsten alloy combines the better thermomechanical properties of tungsten with tantalum's ability to be more easily machined and welded [134,135]. In addition, high-temperature stability concerns also apply: operating the heat source and emitter and high temperatures can lead to cracking and delamination of the emitter or heat source. Also, the emitter often must be in vacuum or inert gas environment in order to prevent chemical degradation processes, such as oxidation and heat losses, due to convective heat transfer processes.

### 7.6.5 TPV SUB-SYSTEM EFFICIENCY

Major improvements in efficiency measured under idealized test conditions are necessary, but not sufficient, for widespread adoption of TPV technology.

The performance of emitter-cell pairs needs to translate to prototypes and ultimately generator. In particular, significant advancements have been made toward practical implementation of TPV materials by improving emitter stability and developing quality narrow-bandgap cells that require lower heat source temperatures. However, more affordable manufacturing technologies are needed for large-scale cell production. TPV sub-system efficiency $h_{TPV}$ is *considered* as an intermediate performance metric for the transition from emitter-cell pairs to prototypes, which captures losses related to imperfect component integration, such as absorption by inactive regions of the cell, non-ideal cavity geometries, and convective loss from the emitter.

The literature indicates that for 0.6 eV InGaAs- and Si-based sub-systems, $h_{TPV}$ has been measured directly. For InGaAsSb-, GaSb-, and LM InGaAs-based sub-systems, it was deduced from simulated loss breakdowns. For all cases, a notable efficiency gap between leading TPV emitter-cell pairs and leading TPV sub-systems is observed, as shown in Figure 7.9a. The gaps between $h_{pairwise}$ and $h_{TPV}$ should not be fully attributed to CE. Overall, that sub-systems with wider-bandgap cells (e.g., Si) appear to be more susceptible to these issues. While it is possible that different testing conditions may be a variable, Figure 7.9b shows that the observed gap persists even when comparing thermodynamic efficiencies. This may be explained by considering the effects of imperfect component integration and scale up.

Each leading TPV sub-system utilizes a vacuum environment to ensure emitter stability and eliminate convective heat transfer losses. Therefore, the cavity efficiencies of these TPV sub-systems are primarily dependent on their geometrical design. In the case of 0.6 eV InGaAs. The highest reported $h_{TPV}$ for a 0.6 eV InGaAs sub-system is 20% [139], short of the 23.6% pairwise efficiency reported for the same cell under a lamp [39]. In a related report, Crowley et al. attribute this gap to imperfect cavity efficiency, non-uniform cell illumination, and inefficiencies related to cell interconnections [140]. This sub-system relies on a selective cell to achieve spectral

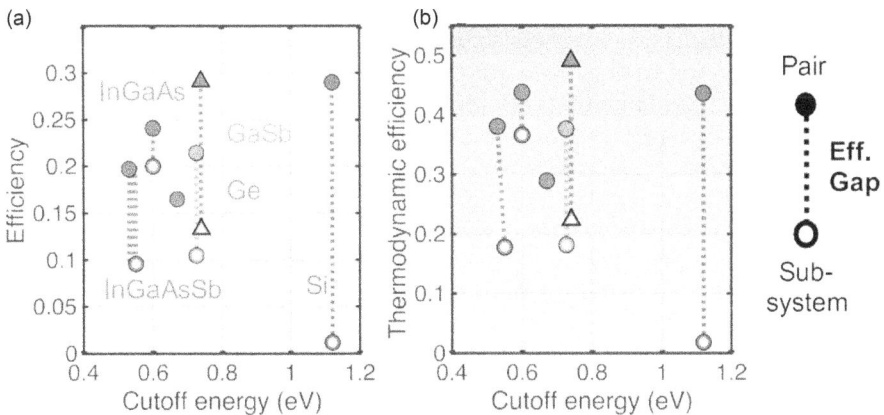

FIGURE 7.9 Gaps between Pairwise and TPV Sub-system efficiencies. (a and b) Leading pairwise efficiencies compared to record sub-system efficiencies: (a) absolute and (b) thermodynamic efficiency. Reporting literature: InGaAsSb; 0.6 eV InGaAs; Ge; GaSb; LM InGaAs (triangular marker); Si [46].

control and is likely more susceptible to cavity losses. These loss pathways are common among other sub-systems as well. In the case of GaSb, Bhatt et al. report a view factor of 0.85 due to the small area of the cell relative to the emitter [141]. It is important to note that many of the sub-system heat losses that severely restrict efficiency can be minimized as emitter-cell pairs reach the kW scale [140].

In addition to cavity imperfections, it appears that material supply and scaling issues may also be a factor in some of the observed performance gaps. Both the leading ImInGaAs- and InGaAsSb-based sub-systems have well-designed cavities with view factors of 0.97 and 0.96, respectively [140]. Thus, the drop in performance is likely because the ImInGaAs and InGaAsSb cells used in the sub-system did not perform as well as the champion cells. As TPVs transition toward commercialization, it will become increasingly necessary to address cavity inefficiencies and other scaling issues. One prior study has proposed several design strategies for reducing system sensitivity to cavity losses [142]. For example, improving the spectral overlap by increasing the ratio of the emitter temperature to the bandgap is expected to make TPV sub-systems less susceptible to such losses. Multi-junction designs offer a practical means of increasing spectral overlap, lowering Ohmic losses, and improving resistance to cavity inefficiencies. Further, utilization of a selective emitter, even in tandem with a selectively absorptive cell, can reduce sensitivity to parasitic optical loss.

## 7.6.6  COST

The overall cost of emitter production depends on the cost of the raw materials, fabrication, as well as system integration. In particular, many emitters make use of relatively scarce materials, such as hafnium and rare earth metals. The cost of fabrication may increase with increased number of processing steps or complexity of fabrication processes. Given current efficiencies and module costs, TPVs remain prohibitively expensive for widespread commercialization. One promising pathway to reduced cost ($/W) is to leverage high power densities, characteristic of local thermal emission. In theory, high power densities represent one of the technology's greatest assets. In practice, however, Ohmic losses at high current densities can inhibit efficient conversion under practical illumination levels. Utilization of a MIM architecture reduces individual cell area and, therefore, enables low current operation that can alleviate the stringent series resistance requirements of TPV configurations [143]. However, small cells may reduce cavity efficiency and complicate scale up. Accordingly, efficient operation at high power densities ($>3\,W/cm^2$) presents a challenging problem that remains to be addressed.

Whereas Ohmic losses in leading solar PV configurations are effectively negligible [144–146], TPV pairs are prone to Ohmic losses, which adversely affect fill factor. For example, the lmlnGaAs cell in the leading pair exhibits $R_s$ of $0.044\,U\,cm^2$, resulting in an 8% loss in power output [100]. Given that high power density could be a major advantage for TPV generators, in terms of cost per power ($/W), alleviating Ohmic losses for high power systems is essential for enabling practical implementation. Lowering series resistance to acceptable levels may require further development and optimization of transparent lateral conduction layers, low interfacial resistance contacts, and metal grids. These considerations are also important for near-field TPV configurations because of their enhanced power density. A complimentary approach to lower costs is to grow multiple cells from a single growth substrate. Crystalline III–V substrates persist as the largest single cost of TPV modules [147]. Recovery of a substrate after growth and subsequent reuse can therefore reduce module costs considerably. This approach requires non-destructive liftoff techniques to enable substrate reuse [148,149]. Despite the promise of substrate reuse, this technique has not been demonstrated for cells in TPV systems. Nevertheless, this manufacturing development appears to be important for commercialization and process sustainability.

TPV module costs stand to benefit from production at a higher volume than current lab-scale manufacturing. Notably, this may require utilization of TPVs in high-volume applications, such as grid-scale thermal energy storage or residential co-generation. Alternatively, it may be necessary for cell fabrication techniques to make use of more mature technologies already in use for production of solar PV or telecommunication components. As development of high-quality Si cells for solar PV application benefited from advances in integrated circuit technology, concurrent development of other materials for separate applications may expedite their deployment in TPV systems. For example, lmlnGaAs and Ge photodiodes are commonly used for optical detection in the near-IR. Further, Ge and various InGaAsSb and InGaAs alloys are common sub-cell materials in multi-junction solar PV technologies. Ongoing research efforts in these areas may benefit the performance and cost metrics of corresponding TPV cell materials. Whereas Ohmic losses in leading solar PV configurations are effectively negligible [144–146], TPV pairs are prone to Ohmic losses, which adversely affect fill factor. For example, the lmlnGaAs cell in the leading pair exhibits $R_s$ of 0.044 U cm$^2$, resulting in an 8% loss in power output. Given that high power density could be a major advantage for TPV generators, in terms of cost per power ($/W), alleviating Ohmic losses for high power systems is essential for enabling practical implementation. Lowering series resistance to acceptable levels may require further development and optimization of transparent lateral conduction layers, low interfacial resistance contacts, and metal grids. These considerations are also important for near-field TPV configurations because of their enhanced power density.

## 7.7 SPECTRAL CONTROL AND FILTER/REFLECTOR DESIGN

Spectral control is a key technology for thermophotovoltaic (TPV) direct energy conversion systems because only a fraction (typically less than 25%) of the incident thermal radiation has energy exceeding the diode bandgap energy, $E_g$, and can thus be converted to electricity. The goal for TPV spectral control in most applications is twofold; maximize TPV efficiency by minimizing transfer of low energy, below bandgap photons from the radiator to the TPV diode and maximize TPV surface power density by maximizing transfer of high energy, above bandgap photons from the radiator to the TPV diode.

TPV spectral control options include: front surface filters (e.g. interference filters, plasma filters, interference/plasma tandem filters, and frequency selective surfaces), back surface reflectors, and wavelength selective radiators. System analysis shows that spectral performance dominates diode performance in any practical TPV system, and that low bandgap diodes enable both higher efficiency and power density when spectral control limitations are considered. Lockheed Martin has focused its efforts on front surface tandem filters which have achieved spectral efficiencies of ~83% for $E_g = 0.52\,\text{eV}$ and ~76% for $E_g = 0.60\,\text{eV}$ for a 950°C radiator temperature [14].

The basic components of a TPV conversion system include: a photon radiator, a spectral control device, and a TPV diode with a bandgap ($E_g$). The TPV process can be conceptualized as the selective conversion of photons from the radiator which have energies greater than the diode bandgap ($E > E_g$). The conversion process incorporates a spectral control device to minimize parasitic absorption of photons of unusable energy ($E < E_g$). In essence, the spectral control device serves as a photon recuperator to minimize the heat required to keep the radiator at temperature. The goal for TPV spectral control in most applications is twofold:

1. Maximize TPV efficiency ($\eta_{\text{TPV}} = \eta_{\text{diode}} \cdot \eta_{\text{spectral}} \cdot \eta_{\text{module}}$) by minimizing transfer of low energy, below bandgap photons from the radiator to the TPV diode.
2. Maximize TPV surface power density by maximizing transmission ($T > E_g$) of high energy, above bandgap photons from the radiator to the TPV diode.

In practice, TPV spectral control may not be separable into independent components. For example, the radiator may provide spectral control by tailoring the emission spectrum; or the diode may provide spectral control if it is designed to be transparent to below-bandgap energy and a reflective surface is placed on the back side of the diode to reflect the below-bandgap energy back to the radiator [150].

TPV spectral control technologies can be divided into two categories according to the temperature at which they operate: cold side or hot side. Cold side TPV spectral control technologies are coupled to the TPV system heat sink, are kept at a low temperature (typically 20°C–50°C), and provide spectral control by reflecting low energy (long wavelength) photons back to the radiating surface. Cold side spectral control technologies include front surface filters and back surface reflectors (BSR). Examples of front surface filters that have been considered for use in TPV applications include: photonic bandgap filters with periodicity in one dimension (e.g. interference filters), two dimensions (e.g. frequency selective surfaces), and three dimensions; plasma filters; and various combinations (e.g. interference/plasma tandem filter). Hot side TPV spectral control technologies are coupled to the heat source (at approximately the same temperature as the heat source) and provide spectral control by suppressing emission of low energy photons from the radiating surface. Hot side spectral control can be accomplished by texturing the surface of the radiator, applying coatings or filters to the surface of the radiator, using bulk radiator materials with selective emission characteristics (e.g. rare earth oxides), or utilizing a three dimensional photonic bandgap (PBG) structure. TPV spectral control technologies are summarized in Table 7.3.

In general, selective emitters have had limited success. More often spectral control through filters and coatings are used with black body emitters to pass wavelengths matched to the bandgap of the PV cell and reflect mismatched wavelengths back to the emitter. A good filter should have high transmission of above band gap photons, high reflection of sub-band gap photons, a sharp transmission from reflection

**TABLE 7.3**
**TPV Spectral Control Technologies [150]**

| | |
|---|---|
| Radiator | • Selective Radiator |
| | • Textured Radiator |
| | • Filtered Radiator |
| | • 3D Photonic Bandgap Radiator |
| | ↑ **Hot Side** |
| Gap | |
| Front surface | ↓ **Cold Side** |
| | • Plasma Filter |
| | • 1-D Photonic Bandgap Filters (Interference Filter) |
| | • 2-D Photonic Bandgap Filters (Frequency Selective Surface (FSS) |
| | • 3-D Photonic Bandgap Filter |
| | • Tandem Filters (Combination of Interference Filter and Plasma Filter) |
| TPV cell | |
| Back surface | • Back Surface Reflector |

to transmission at the band gap energy, and minimal parasitic absorption of energy in the filter. A filter in addition to a selective emitter can aid in emitter temperature maintenance by reflecting unused photons back to the emitter to be absorbed. A good filter should have high transmission of above band gap photons, high reflection of sub-band gap photons, a sharp transmission from reflection to transmission at the band gap energy, and minimal parasitic absorption of energy in the filter. Ideally, any photons with energies far above the band gap, which would increase thermalization or be absorbed too quickly and generate carriers too far from the diode junction, would be reflected as well, but this behavior is difficult to achieve with good sub-band rejection. Filters rely on advanced wave interactions to operate; combined with their tunability based on material choice and design dimensions, many filter technologies are now considered a class of photonic crystal.

One type of filter that is used in TPV systems is a dielectric filter, or a distributed Bragg reflector (DBR), which utilize thin layers of materials with alternating indices of refraction to create interference between incident photons. The materials are selected with the goal of maximizing the transmission of the desired wavelengths and the reflection of all other wavelengths. For a small window of wavelengths, the DBR has high transmittance; however, very long wavelength sub-band gap photons have a low reflection rates. Increasing the number of layers in the filter can reduce this problem, but it can lead to more complicated film growth, increased production costs, and unintentional absorption.

Plasma filters are another 1D filter made of a single layer of heavily doped semiconductor film. The dopant concentration determines the plasma wavelength for the material which establishes the cutoff between reflected and transmitted photons. Common materials for plasma filters are silicon, indium, gallium-arsenide, and transparent conducting oxides (TCOs), such as indium tin oxide, zinc oxide, and cadmium stannate. Absorption around the plasma frequency can lead to a gradual transition between reflectance and transmission wavelength ranges, rather than the preferred sharp changeover. A balance must be struck to minimize above band gap photon absorption and maximize the sub-band gap reflection. Plasma filters can also be combined with interference filters to improve the cutoff sharpness as well as extend the transmission window into longer wavelengths.

Frequency selective surface (FSS) filters increase periodicity to two dimensions with an array of repeating conductive structures. Incident light induces a current in the metallic structures, creating a scatter field that interacts with the incident field to selectively reflect and transmit radiation. The size, shape, and spacing of the pattern influences the photon interactions. Similar structures can also be used to enhance absorption in the TPV cell by increasing light trapping or serving as an antireflection coating (ARC). As with plasma filters, the FSS generates responses that lack a sharp cutoff on either side of the spectrum]. Also, ohmic losses due to the conducting layer can increase absorption in the filter]. Spectral control does not only include filters; a back surface reflector (BSR) can reflect photons back toward the emitter, giving above band gap photons a second chance to get absorbed and allowing sub-band gap photons to contribute to emitter temperature maintenance. One way to reduce the resulting increased sub-band gap absorption is the use of a hybrid dielectric–metallic BSR. The use of a metallic layer reduces the number of dielectric layers required to achieve high reflectance, but it can cause some parasitic photon absorption. Combining a BSR with a front-side filter may still be beneficial enough to outweigh the additional sub-band gap absorption [28]. Application of other surface treatments such as photonic crystals and metamaterials can generate near-field effects that enhance thermal transfer when the gap between the emitter and diode is on the order of 100 nm.

## 7.8  CHARACTERISTICS OF MATERIALS FOR TPV CELL

Early TPV cell work focused on the use of silicon. Silicon's commercial availability, low cost, scalability and ease of manufacture makes this material an appealing candidate. However, the relatively wide bandgap of Si (1.1 eV) is not ideal for use with a black body emitter at lower operating temperatures. Calculations indicate that Si PVs are only feasible at temperatures much higher than 2,000 K. No emitter has been demonstrated that can operate at these temperatures. These engineering difficulties led to the pursuit of lower-bandgap semiconductor PVs. No efficient TPVs have been realized using Si PVs. Early investigations into low bandgap semiconductors focused on germanium (Ge). Ge has a bandgap of 0.66 eV, allowing for conversion of a much higher fraction of incoming radiation. However, poor performance was observed due to the high effective electron mass of Ge. Compared to III–V semiconductors, Ge's high electron effective mass leads to a high density of states in the conduction band and therefore a high intrinsic carrier concentration. As a result, Ge cells have fast decaying "dark" current and therefore, a low open-circuit voltage. In addition, surface passivation of germanium has proven difficult.

In principle, TPV cells operate at the optimum efficiency when the semiconductor energy bandgap is spectrally matched to the blackbody spectrum generated by the heat source [151]. Therefore, TPV cells material shall be chosen based on the bandgap, which correspond to the spectrum. This is to minimize optical loss cause by the spectral mismatch and poor absorption of photons by the TPV cell. Tan et al. [152] compared the performance of $In_{0.53}Ga_{0.47}As$ (0.74 eV) and $In_{0.68}Ga_{0.32}As$ (0.6 eV) under various range of blackbody temperatures. It was found that the efficiencies of both cells gradually increased from 800 to 1,323 K. The key reason for the efficiency increment is due to the positioning between cutoff wavelength of the materials and the peak emissivity ($\lambda_p$) of each temperature. The conversion of electricity effectively occurs at photon wavelengths near the $\lambda_c$ of a particular material.

Figure 7.10 shows that higher source of temperature is desirable for material with a higher bandgap energy. Two extremes are InSb and InGaAs. TPV cell can be matched with the emissivity of the practical radiator. Selective radiators can be engineered to emit at setting range of IR. In short, mismatching between the heat source spectrum and bandgap of the cell is one of the considerable issues that usually causes lower output energy and reduces the efficiency of TPV system. Therefore, the matching of blackbody spectrum to the suitable range of TPV cell bandgap energy is essential to generate the optimum amount of energy per unit area. The reflection and recycling of sub-bandgap photons in the radiator significantly enhances the conversation efficiency of the TPV system.

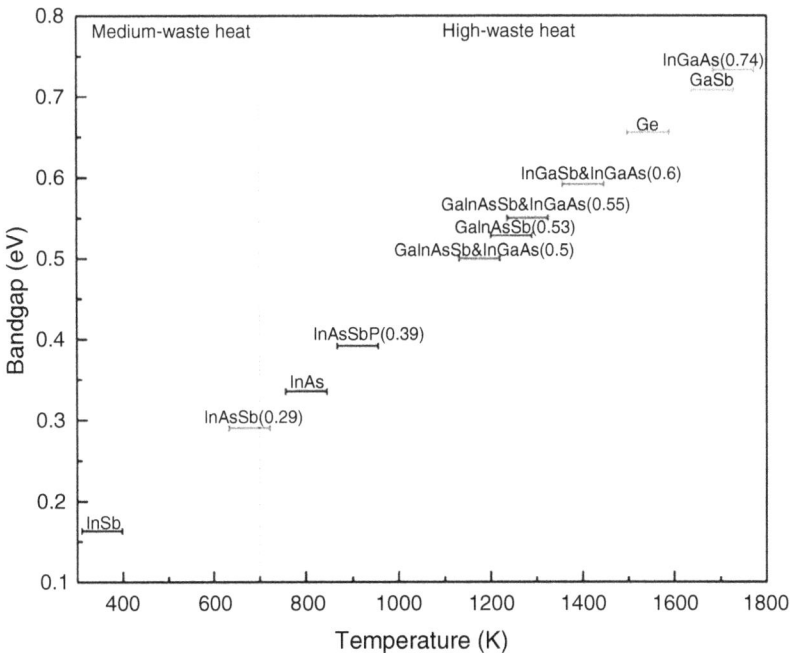

**FIGURE 7.10** TPV cells and the blackbody temperature range of their optimum performance [14].

## 7.8.1 GaSb-Based TPV Cell

The gallium antimonide (GaSb) PV cell is the basis of most PV cells in modern TPV systems. GaSb is a III–V semiconductor with the zinc blende crystal structure and the narrow bandgap of 0.72 eV. This allows GaSb to respond to light at longer wavelengths than silicon solar cell, enabling higher power densities in conjunction with manmade emission sources. A solar cell with 35% efficiency was demonstrated using a bilayer PV with GaAs and GaSb, setting the solar cell efficiency record. Manufacturing a GaSb PV cell is quite simple. Czochralski tellurium-doped n-type GaSb wafers are commercially available. GaSb is a III–V binary semiconductor compound. It is considered as one of the ideal semiconductor materials for TPV applications with temperature source ranging from 1,300 to 1,500 K. Due to its low bandgap energy ($E_g$), which spectrally matched with the medium blackbody temperature, it produces an excellent quantum efficiency of greater than 90%, especially at IR wavelength up to 1,800 nm [153]. GaSb TPV cell was used to generate electricity from waste heat in the steel industry which generates 3.1 GW of power under 1,673 K [154]. Recently, an evaluation study of GaSb TPV cell performance at low-, mid- and high-waste heat temperature was conducted by researchers in Turkey targeting the potential waste heat location in the Turkish Industrial Sector [14,155]. Significant research has been carried out to improve efficiency of GaSb TPV cells, and these are described in detail by Ali Gamel et al. [14].

Khvostikov et al. [156] compared the performance of Zn-diffused GaSb-based TPV arrays connected in both series and parallel configuration. The parallel connection in the cylindrical system was found to produce a better performance when compared to the series connection in a conical system. Tang et al. [157] performed a doping and depth optimization on a Zn-diffused GaSb cell for both emitter and base regions through simulation work. The doping of both regions was optimized to increase the $QE$ in a range of long wavelengths. The study showed that for an n-type GaSb substrate, increasing the doping concentration tends to reduce the power output of the cell while for the p-type GaSb emitter layer, a moderate Zn concentration is sufficient to achieve high QE. Tang et al. [158] also performed a numerical simulation to investigate the GaSb cell designed to be an inverted "n on p" configuration through Tellurium diffusion into unintentionally p-doped GaSb TPV cell. The results show that n on "p" configuration with a thin optimal diffusion n-emitter depth of 0.1 μm and a thick p-base shows better performance than the "p on n" GaSb TPV cell configuration.

According to Ali Gamel et al. [14], the literature has also examined the effects of spectral control [159], metal contact [14,160,161] and the layer thickness [158,162,153] on the cell performance. Good metal contact with low electrical resistance (often known as ohmic contact) is essential to minimize the loss of electrical power and maximize the overall efficiency [163]. The properties of an ohmic contact include good thermal stability, low resistance and good adhesion on the semiconductor surface [161]. Milnes et al. [164] found that the contact resistivity is inversely proportional to the doping concentration. From their finding, it was demonstrated that lower contact resistances are favorable to form ohmic contact on p-GaSb. On the other hand, for n-GaSb, Ni and Pd metals are found to be suitable candidates to form an ohmic contact with Ni performing better than Pd [161]. The performance of a GaSb TPV cell exhibits a strong dependence on the layer thickness and junction configuration and efforts are made to optimize these parameters. Sulima and Bett [165] showed that an optimization of Zn-diffused GaSb emitter can improve the cell performance. Most studies reported that good quantum efficiency can be produced by a thin layer of p-type emitter on the GaSb TPV cells. Rajagopalan et al. [166] found that the optimum junction depth that produces the maximum power output from the fabricated cell is around 0.4 μm. Abdallah et al. [162] emphasized the trade-off relation between the emitter thickness and the performance parameter. Bett et al. [167] demonstrated that a detailed optimization with the precise control of the emitter thickness is very crucial.

Ali Gamel et al. [14] also pointed out that the cell performance is also dependent on anti-reflective coating (ARC). Licht et al. [168] demonstrated that an incorporation of a metallic photonic crystal as the front-surface filters (MPhCs) results in 10% improvement in $IQE$ with optimized design of a GaSb TPV cell. Fraas et al. [169] discussed the design optimization incorporating an "n+" transparent conductive oxide (TCO) layer with a hydrogenated amorphous silicon interface (Si:H) passivation between the n+ and p-type GaSb base. They also reported higher $QEs$ at longer photon wavelengths and higher output power density in comparison to a traditional p-on-n GaSb TPV cell structure.

## 7.8.2 INGAAS-BASED TPV CELL

Indium gallium arsenide (InGaAs) is a compound III-V semiconductor and it can be applied in two ways for use in TPVs. When lattice-matched to an InP substrate, InGaAs has a bandgap of 0.74 eV, no better than GaSb. The bandgap energy of InGaAs can be engineered from 1.42 to 0.36 eV by having a variation of $x$ composition in $In_xGa_{1-x}As$, corresponding to cut-off wavelength from 0.87 to 3.34 μm [172]. Woolf [19] at GA Technologies (San Diego, CA, USA) was the first to propose InGaAs photoconverters for TPV applications. Devices of this configuration have been produced with a fill factor of 69% and an efficiency of 15%. However, in order to absorb higher wavelength photons, the bandgap may be engineered by changing the ratio of In to Ga. The range of bandgaps for this system is from about 0.4 to 1.4 eV. While these different structures cause strain with the InP substrate, it can be controlled with graded layers of InGaAs with different compositions. This was done to develop of device with a quantum efficiency of 68% and a fill factor of 68%, grown by MBE. This device had a bandgap of 0.55 eV, achieved in the compound $In_{0.68}Ga_{0.33}As$. InGaAs can be made to lattice match perfectly with Ge resulting in low defect densities. Ge as a substrate is a significant advantage over more expensive or harder-to-produce substrates.

According to Ali Gamel et al. [14], lower bandgap lattice-mismatched InGaAs cells have recently received tremendous attention due to the diversified applications. This can be achieved by increasing the indium ratio and decreasing the gallium ratio [170]. Reducing the bandgap of InGaAs results in higher output power [171]. Zhou et al. [172] investigated the utilization and optimization of TPV system cavity, which consists of emitter, PV cells, and mirrors to modify the spatial and spectral distribution within the system. The study showed that careful design of the cavity configuration can significantly enhance the performance of the TPV cell. Tan et al. [152] showed that the change in the radiation spectrum has a significant effect on current mechanisms of InGaAs cell. Zhang et al. [173] reported that the large lattice-mismatched between the epilayer and substrate cause defect, which affects the quality of the material in both electrical and optical properties. The lattice-mismatched problem was solved using two-step growth methods. More work is, however, needed [173,174]. For solar heat, n-p and p-n InGaAs have similar performance due to the long minority carrier diffusion length and good surface passivation [175,176]. However, larger contact resistance in the p-n configuration required a greater surface grid coverage to avoid $FF$ loss, leading to much higher shading loss and a significant reduction in efficiency per total area (7.2%) [176].

For p-InGaAs, higher contact resistance can be achieved as compared to n-InGaAs. Since the variation of layers' thickness and doping concentration directly affect the output performance of InGaAs TPV cell, they should be optimized to get maximum output performance [177,178]. Most of the literature reported a very thin layer of emitter layer from 0.1 to 0.3 µm [14]. Thinner highly doped emitter is able to produce lower resistance depending on the metal contact and able to generate a feasible electric field with the base layer. Tuley et al. [14,179] reported an optimum thickness (doping concentration) of 0.5 µm ($1 \times 10^{17}$ cm$^{-3}$) for emitter layer and 3.5 ($1 \times 10^{17}$ cm$^{-3}$) for a base layer under 1400 K blackbody radiation. Emrnzian et al. [180] optimized the thickness and doping concentration of the emitter and base layers of $In_{0.53}Ga_{0.47}As$ under 0.5 W/cm$^2$ at 3,300 K blackbody temperature. Optimum cell performance was reported at base and emitter thickness of 2 and ~0.2 µm. Gamel et al. [64] reported a multi-dimensional optimization of $In_{0.53}Ga_{0.47}As$ TPV structure using the real coded genetic algorithm at radiation temperatures between 800 and 2,000 K. Cell efficiency increased by an average percentage of 11.86% as compared to the non-optimized cell.

According to Ali Gamel et al. [14], the majority of InGaAs devices are epitaxial grown heterojunction. The active junction is combined with cap, window and buffer layers to improve device performance. Optimization of the ARC materials and thickness play a key role in decreasing the reflection at the front surface of the InGaAs cell. Other methods like the rear mirror reflector, Lambertian rear reflector and textured surface are used to improve the light absorption and light trapping in the cell [181]. Jurczak et al. [182] showed a 3%–5% improvement in the performance of InGaAs when the light trapping was implemented. Burger et al. [183] suggested the potential for a dramatic increase in conversion efficiency through improved spectral selectivity and recycling of longer IR, combined with the potential for reduced module costs through wafer reuse using thin-film TPV. Monolithic interconnected modules (MIM) were implemented for the TPV system to improve the system efficiency [184]. The series connected MIM cells generate high voltage and low current and the array size can be designed to minimize the loss due to non-uniform radiator temperature. MIM cell can also be directly connected to the substrate or heat sink without any electrical isolation limitation concern [185]. Utlu [155] showed that series connected MIMs generate high voltage and low current. The reliability of MIMs could be further improved by designing series-parallel string MIMs. Wehrer et al. [186] further explored the TPV tandem converter technology to improve TPV efficiency while sustaining high power densities. Performance of InGaAs device can also be improved by matching the radiation spectrum to the spectrum response by the device. Woolf et al. [105] reported 52% of spectral efficiency at 1,500 K for $In_{0.68}Ga_{0.32}As$ TPV cell with cell efficiency of 26%. Woolf et al. [14,19] presented the novel selective thermal radiator to improve the efficiency of TPV system with source temperature above 1,000°C and a $In_{0.68}Ga_{0.32}As$ TPV cell. Maremi et al. [14,187] numerically analyzed a multilayer optical radiator based on metamaterial design which exhibited polarization and azimuthal angle independent selective emission for $In_{0.68}Ga_{0.32}As$ TPV cell. Recent improvement have increased the feasibility and practicality of InGaAs TPV cell in various terrestrial and space applications.

### 7.8.3 NARROW BANDGAP MATERIALS FOR TPV CELLS

TPV cells utilize narrow bandgap materials which allow them to harvest the maximum amount of infrared radiation. Ali Gamel [14] pointed out that aside from the available GaSb and InGaAs TPV materials, there are several narrow bandgap materials that worth highlighting. Single element semiconductor like Ge with a bandgap of 0.66 eV, binaries like InAs with a bandgap of 0.35 eV and indium antimonide (InSb) with a bandgap of 0.17 eV and ternary like indium gallium antimonide (InGaSb) with a bandgap energy range of 0.17–0.72 eV and indium arsenide antimonide (InAsSb) with a bandgap energy range of 0.1–0.4 eV and quaternary like InGaAsSb with a bandgap energy range of 0.5–0.55 eV and InAsSbP with a bandgap energy range of 0.3–0.5 eV are accessible narrow bandgap materials [14]. These materials can be explored for the potential to harvest waste heat temperature below 1,200 K.

Theoretically, TPV cells made by narrower bandgap semiconductors are able to absorb photons up to longer wavelengths, resulting in better cell performance. Narrow band (<0.7 eV) TPV cells has several drawbacks such as low $V_{oc}$, high dark current, and immature growth technology, and nonoptimal structure design. Recently, Gamel et al. [188] compared the performance of various reported narrow bandgap cells under 1,000 K blackbody temperature. Based on their simulation result, InGaAsSb is more promising as compared to GaSb and Ge, due to the high cell efficiency and its ability to operate in the low range blackbody temperature. While extended InGaAs, InGaAsSb has the best potential among the narrow bandgap materials for radiation low-temperature (<1,273 K), further cell optimization is needed to improve the cell efficiency. As illustrated in Figure 7.11, InGaAs (0.74 eV) and GaSb TPV cells have recorded the highest cell efficiencies.

Indium gallium arsenide antimonide (InGaAsSb) is a compound III–V semiconductor ($In_xGa_{1-x}As_ySb_{1-y}$). The addition of GaAs allows for a narrower bandgap (0.5–0.6 eV), and therefore better absorption of long wavelengths. When the bandgap was engineered to 0.55 eV, the compound achieved a photon-weighted internal quantum efficiency of 79% with a fill factor of 65% for a black body at 1,100°C. This device was grown on a GaSb substrate by organometallic vapour phase epitaxy (OMVPE). Devices can also be grown by molecular beam epitaxy (MBE) and liquid phase epitaxy (LPE). The internal quantum efficiencies (IQE) of these devices approach 90%, while devices grown by the other two techniques exceed 95%. The largest problem

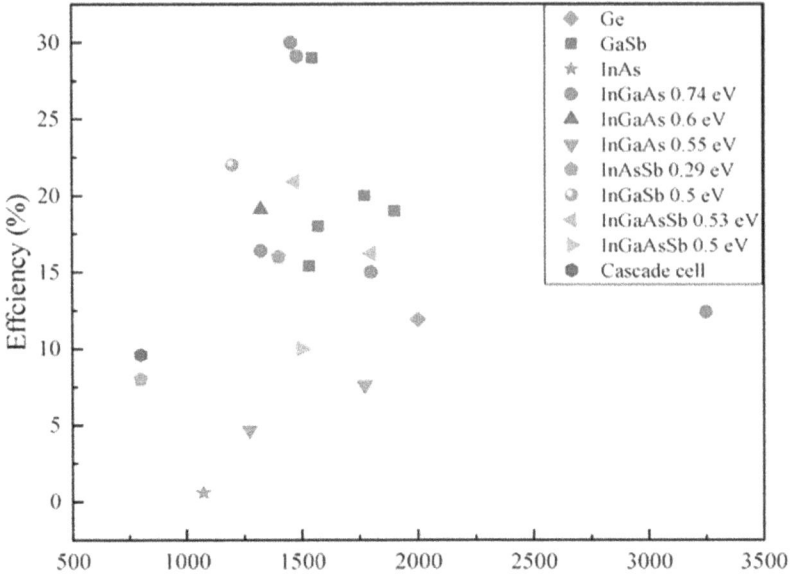

**FIGURE 7.11**   Record TPV cell efficiencies for various materials [14].

with InGaAsSb cells is phase separation. Compositional inconsistencies throughout the device degrade its performance. When phase separation can be avoided, the IQE and fill factor of InGaAsSb approach theoretical limits in wavelength ranges near the bandgap energy. However, the $V_{oc}/E_g$ ratio is far from the ideal. Current methods to manufacture InGaAsSb PVs are expensive and not commercially viable [14].

The InPAsSb quaternary alloy has been grown by both OMVPE and LPE. When lattice-matched to InAs, it has a bandgap in the range of 0.3–0.55 eV., but the benefits of such low bandgap are not studied. The performance of cells incorporating InPAsSb has as yet not been optimized. The longest spectral response from an InPAsSb cell studied was 4.3 μm with a maximum response at 3 μm. For this and other low-bandgap materials, high IQE for long wavelengths is hard to achieve due to an increase in Auger recombination. Ali Gamel et al. [14] pointed out that PbSnSe/PbSrSe quantum well materials, which can be grown by MBE on silicon substrates, have been proposed for low-cost TPV device fabrication. These IV–VI semiconductor materials can have bandgaps between 0.3 and 0.6 eV. Their symmetric band structure and lack of valence band degeneracy result in low Auger recombination rates, typically more than an order of magnitude smaller than those of comparable bandgap III–V semiconductor materials.

Finally, a tandem junction configuration has the ability to further improve performance. This corresponds to the optimization with an added bandgap parameter, subject to the constraint that the bandgap in front must have a higher energy bandgap than the one in back (otherwise, no useful photons would reach the junction in back). For an emitter at 2,360 K, a dual bandgap structure with bandgaps of 1.01 and 0.82 eV yield a power conversion efficiency of 66.3% (22.3% higher than a single junction configuration). Even for an emitter of only 1,000 K, the efficiencies can be maintained at a quite respectable level of 44.0% with a tandem-junction, thus representing a 45-fold improvement over the previously observed conversion efficiency of a plain silicon wafer with an InGaAsSb TPV cell. This substantially exceeds the Shockley–Quiesser limit for a single-junction PV cell of 31% without concentration ($C=1$) or 37% under full concentration ($C=46200$) [29].

## 7.9  TPV APPLICATIONS

TPV system received tremendous attention due to its promising contribution as economical, efficient, practicable power systems, and clean power generation. The application of TPV can be categorized based on the chemical reaction or nuclear fusion reaction types of thermal heat source. These heat sources are categorized into solar heat, combustion of fuels, nuclear sources and waste heat. Figure 7.12 illustrates the applications for TPV technology.

### 7.9.1  SOLAR TPV SYSTEMS

In principle, solar TPV (STPV) system utilizes solar energy to heat up the radiator to <3,273 K through a solar concentrator. These high-temperature radiators will then emit thermal radiation to the TPV cells that convert the infrared photons into electricity [189]. STPV work has been reported on Fresnel point-focus and dish concentrators, which recorded a temperature up to 1,623 K [151]. Xuan et al. [189]

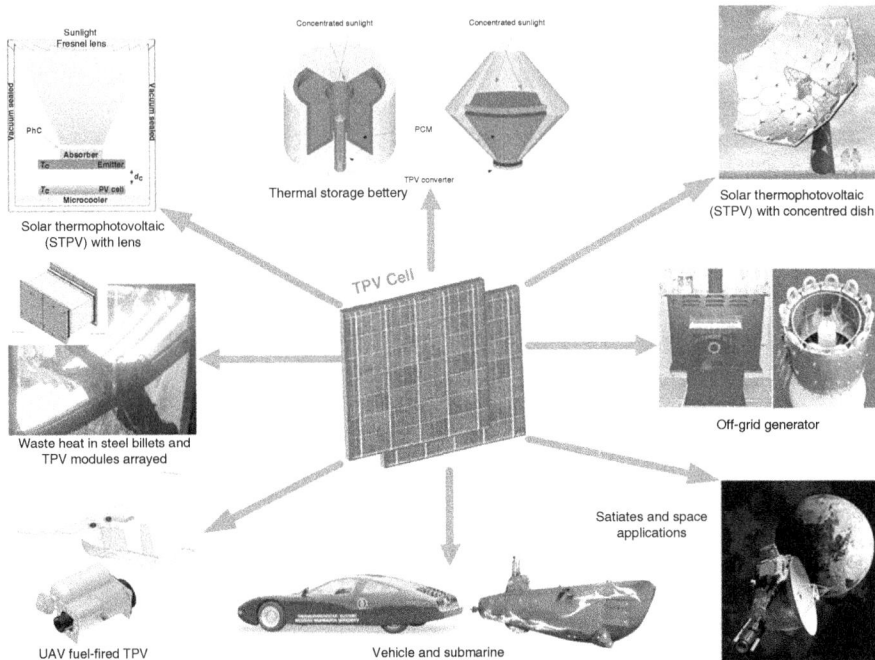

**FIGURE 7.12**   TPV applications [14].

emphasized that the configuration of radiators influences the distribution effect of radiator temperature. A cylindrical radiator was found to give the most stable performance for the STPV system. Zhou and associates [190] performed a comprehensive review of TPV cells material in the STPV system. In contrast to Si and Ge cell, GaSb and InGaAs TPV cells show better performance in the STPV system. Nevertheless, the cost of material and fabrication process remains a big challenge. Furthermore, a multi-bandgap cell such as GaInP/GaAs/Ge shows the potential of integration into a STPV system, which possesses higher conversion efficiency as compared to a single junction cell [151]. Nowadays, the STPV system in a hybrid system attracts considerable interest. Hussain et al. [191] conducted a performance analysis on different arrangements of hybrid STPV design using solar-biomass/gas power generation system. The hybrid STPV system can be operated at temperature lower than 1,273 K, which widens the opportunity for other applications such as portable battery charger and microgenerator for household applications. In addition, STPV hybrid with solar-natural gas has been demonstrated to give an electrical power output of around 500 W and a conversion efficiency of 22% [192]. This architecture offers a promising development in the near future.

Thermophotovoltaic devices use the same physics as the solar panels (photovoltaics) typically used on rooftops—they both capture light in a semi-conductor material that excites electrons, which then can flow out of the device as current. One major difference is that TPV devices can convert infrared light, which is the portion of the electromagnetic spectrum that is felt as heat directly into electricity. Currently, for TPVs to work efficiently, heat sources have to be in excess of 1,000°C. The glass and steel industries with their large, hot furnaces are low-hanging fruit best suited to TPV waste heat harvesting technology. For example, in the glass industry, furnaces must be maintained at not less than 1,575°C to keep the glass in a molten state. In steel production, iron ore must first be heated in blast furnaces up to 2,300°C. Studies have shown that the technology implemented to capture industrial waste heat can pay for itself in at least 3–5 years, after which any additional electricity produced is essentially a free source of energy.

In recent years, considerable research is being carried on to (a) extend the operating range of these thermophotovoltaic devices, giving them the ability to convert low-temperature and high-temperature heat sources more efficiently and (b) improve the efficiency of thermophotovoltaic devices. Traditionally, this is done using exotic alloys of different semiconductor materials; however, this approach can only go so far. With the use of nanomaterials, quantum effects, and some novel and clever design, the technology is being pushed to new boundaries. Initial results are promising, showing electricity generation from thermal sources at 500°C and below, with increased conversion efficiency. At present, TPV device development is very much focused on capturing and retaining solar energy to its direct conversion to electricity. TPV technology thus adds another dimension to traditional photovoltaics, which converts a portion of solar energy directly to electricity.

### 7.9.2 TPV for Waste Heat

TPV systems could potentially enable new methods for WHR. A small number of prototype systems have been built for small burner applications and in a helicopter gas turbine. TPV technology can be used for many chemical and industrial factories and processes that meet temperature requirements. Beyond industrial applications, if TPV devices were made efficient enough at lower temperature operation, they could be used to harvest heat in electronic devices, car engines, or battery packs. Even the human body could be used as a heat source, if the research is successful in allowing TPV devices to operate with temperatures as low as 37°C. If the TPV devices could harvest heat in these ways, batteries could last longer, farther distances could be traveled with fewer emissions, and huge advances could be made in bioengineering and medicine.

Effort has been made by researchers and private companies in exploring the possibility of utilizing TPVs for power generation [193–195]. In the UK industrial activities, the potential of the TPV system to recover waste heat was realized in high-temperature industry [193]. Turkish industrial sectors [155,195] estimated that the potential energy recovery using the TPV system is around 22.40–67.45 PJ/year. Moreover, a recent thermodynamic analysis of TPV system for industrial WHR application reported that up to 7.31% of annual waste energy can be recovered [17]. Despite the great promise of a TPV system in the WHR application, low power conversion efficiency at waste heat temperature of less than 1,273 K has been a major concern [194]. This can be solved by implementing TPV cells with material bandgap lower than 0.7 eV such as InGaAs and InAs. Furthermore, the potential of TPV heat recovery in glass and steel manufacturing industry has been an attractive research area nowadays. In glass manufacturing industry, the TPV system is estimated to produce 270 kW power output from high-temperature object with a surface area of 27 m² [196]. In addition, Fraas et al. [154] successfully demonstrated a TPV cell with power density output of 1.5 W/cm² per cell generated from a hot glowing radiant tube burner (1,548 K) at a steel industry. Besides, the potential of worldwide electricity production is estimated with the possibility to reach over 3.1 GW with TPV heat recovery system in the steel industry.

### 7.9.3 TPV GENERATORS FOR COMBUSTION HEAT

Combustion-driven TPV system can be widely considered for micro-, meso-, and macro-scale power supply applications [197]. Practical applications of combustion-driven TPV generator include portable electric generator, combined heat and power (CHP) as well as hybrid electrical vehicles. Portable TPV system was highly recognized in military application, aiming to replace heavy batteries and noisy diesel-electric generators [198]. Scalability allows TPVs to be smaller and lighter than conventional generators. Also, TPVs have few emissions and are silent. Multifuel operation is another potential benefit.

Collaboratively, JX Crystal Inc. developed the first hydrocarbon-fueled TPV generators using GaSb TPV cell known as "Midnight Sun," producing electrical power up to 100 W [199]. In early 2001, JX Crystals delivered a TPV-based battery charger to the US Army that produced 230 W fueled by propane. This prototype utilized an SiC emitter operating at 1,250°C and GaSb photocells and was approximately 0.5 m tall. The power source had an efficiency of 2.5%, calculated as the ratio of the power generated to the thermal energy of the fuel burned. Subsequently, McDermott Technology Inc (MTI) has then teamed up with JX Crystal and developed a 500 W diesel-fueled portable TPV power generator. In addition, Chan et al. [200] developed a portable microgenerator with the combination of propane-fueled TPV microgenerator, photonic crystal radiator, and low bandgap InGaAs-based TPV cells. After a thorough system analysis, Chan et al. [201] redesigned the micro burner and developed vacuum packaging to prevent convective loss to the photonic crystal. Results suggested that further development can be promoted for higher conversion efficiency. For home-scale operation, this technology is often called the off-grid generator where the system aims to provide heat and electricity to remote houses during the winter and nighttime.

The CHP-based TPV has not only been realized to provide good performance for home use, but the system is also applicable on a larger scale such as the central heating furnaces in large buildings and industrial furnaces [202]. The limitation of the TPV system in CHP is the low-grade heat generation due to heat dissipation during the cooling of TPV cells. The typical cell temperature in CHP mode is around 333 K, which has been suggested with an additional heat exchanger to upgrade the heat for the exhaust gas [151]. Exploitation of micro-CHP has been investigated on domestic boilers in residential area to provide central heating which converts waste heat into electricity [203]. Bianchi et al. [204] investigated CHP TPV for residential buildings, considering both the energetic and economical point of view. In addition, low bandgap tandem cell such as InGaAsP/InGaAs shows the potential of incorporation in CHP systems. In comparison to thermodynamic limit, a significant improvement was made for lower bandgap materials on the TPV cell conversion efficiency [205]. De Pascale et al. [206] introduced a thermodynamic analysis of CHP by integrating a TPV generator into an Organic Rankine Cycle. A 56% thermal efficiency and 24% electrical efficiency were achieved in optimum electrical load configuration.

TPVs can provide continuous power to off-grid homes. Traditional PVs do not provide power during winter months and nighttime, while TPVs can utilize alternative fuels to augment solar-only production. The greatest advantage for TPV generators is cogeneration of heat and power. In cold climates, it can function as both a heater/stove and a power generator. JX Crystals developed a prototype TPV heating stove/generator that burns natural gas and uses a SiC source emitter operating at 1,250°C and GaSb TPV cell to output 25,000 BTU/h (7.3 kW of heat) simultaneously generating 100 W (1.4% efficiency). However, costs render it impractical. Many TPV CHP scenarios have been theorized, but a study found that generator using boiling coolant was most cost-efficient. The proposed CHP would utilize a SiC IR emitter operating at 1,425°C and GaSb photocells cooled by boiling coolant. The TPV CHP would output 85,000 BTU/h (25 kW of heat) and generate 1.5 kW. The estimated efficiency would be 12.3% (1.5/25 kW = 0.06 = 6%) requiring investment or 0.08 €/kWh assuming a 20-year lifetime. The estimated cost of other non-TPV CHPs are 0.12 €/kWh for gas engine CHP and 0.16 €/kWh for fuel cell CHP. This furnace was not commercialized because the market was not thought to be large enough.

In hybrid electric vehicle (HEV), the exhaust heat from gas-powered engine is converted by a TPV device to charge the batteries with sufficient power to accelerate and maintain a cruise speed. TPV generators that produce a power range from 6 to 10 kW were investigated for HEV [4,207]. The major concern for TPV generators in HEV is the requirement of high efficiency. For instance, a 10 kW output capability is needed to maintain a steady-state cruising at 113 km/h without drawing any power from the battery [4]. The first TPV-powered automobile prototype was built in 1999 and named as "Viking 29" [207], which utilized GaSb- based TPV cells. The TPV generator consists of 20 GaSb-based TPV cells connected in series and $V_{DC}$ of 126 V [4]. The challenge of low conversion efficiency for the currently available TPV systems would be more reliable to support a small vehicle with smaller power range [198]. Research efforts have been devoted toward future development for better heat recovery and improvement of the system efficiency to achieve high-performance HEVs. The design system with selective emitter and quantum well TPV cell is theoretically capable of yielding 24.5% of efficiency and 6 kW of electric power to recharge the battery pack of an electric city car. However, further experimental investigations are needed for the emitter-cell system [2]. TPVs have been proposed for use in recreational vehicles. Their ability to use multiple fuel sources makes them interesting as more sustainable fuels emerge. TPVs silent operation allows them to replace noisy conventional generators (i.e. during "quiet hours" in national park campgrounds). However, the emitter temperatures required for practical efficiencies make TPVs on this scale unlikely.

### 7.9.4 SPACE APPLICATIONS

High-efficiency TPV cells are essential in space technology application [151]. There are two feasible power sources to power up a small spacecraft for long duration, which are solar and nuclear generators [190]. Recently, a critical review was conducted on space power generation to compare the main competing technologies [43]. TPV system promises up to 40% of efficiency and additional advantages of lightweight, mechanically static, and direct electricity production from radiant heat in space power generation. Also, the TPV system generates high power density per unit area as compared to other technologies, which is suitable for medium electrical system. The power density can be further increased with the use of radioisotope TPV generators with the use of nuclear fission sources [208,209]. For nuclear generators, the heat range is typically from 10 kW to MWs [210], and these generators are currently under consideration for power generation in future planetary settlement missions [43]. The most recent development of RTPV has been made in the Institute for Soldier Nanotechnologies [75,211] where simulation and measurement results have been experimentally reported on RTPV prototype system using photonic crystal spectral control in terrestrial application. Moreover, a RTPV system designed with an InGaAsSb cell was reported with 8.26% efficiency and output power of ~40 W [66,212]. A 0.6 eV InGaAs TPV cell is very useful in the space application with less than 1% of cell degradation in performance over the years caused by the damaging effect on the system. The Plutonium-238 radioisotope was investigated with up to 3 W power output and 10% system efficiency at the Los Alamos National Lab [213].

### 7.9.5 THERMAL ENERGY STORAGE SYSTEM

TPV devices are of particular interest for thermal energy storage application. Conceptually, thermal energy storage system utilizes ultra-high temperature phase change materials (PCM). In this system, the energy is stored in the form of latent heat and transformed to electricity upon demand via TPV cells. Thermal storage system enables an enormous thermal energy storage density of ~1 MWh/m$^3$, which is 10–20 times higher than that of lead-acid batteries, 2–6 times higher than Li-ion batteries as reported by Datas et al. [214]. Seyf and Henry [215] modelled a thermal energy storage system and identified the significant design parameters that affect the overall power cycle system efficiency. It was found that building systems at sufficiently large scales, integrating an effective BSR, increasing the *EQE* for photons above the bandgap will improve the system performance. Following that, the multi-junction or multiple TPV cells arranged optically in series can reduce the thermalization losses and may have the potential to exceed the efficiencies of combined cycles (~60%). In 2019, Amy et al. [216] employed ultra-high temperatures and multi-junction photovoltaics (2-junction cells) for thermal energy storage system. This new approach has several benefits including the ability to reach >50% roundtrip efficiency with a cost per unit power < $0.5 per W-e, and the potential to offer load following capabilities to grid operators.

## 7.10 CHALLENGES, RECOMMENDATIONS AND CLOSING PERSPECTIVES

At present, the market penetration of the TPV system remains inconclusive. There have been several challenges in the development of TPV which includes the high cost of TPV cells. Some of the TPV structures are fabricated using expensive growth technology such as MBE. Additionally, most TPV cells require front and back surface field layers to optimize cell performance and enhance the collection of long and short photogenerated carriers [217]. Lattice-matched TPV substrate is another issue that makes the TPV production expensive. The cost can be reduced by using commercially available and cost-effective substrates to grow TPV cells. Currently, the ELOG method is used to grow GaSb junction on GaAs substrate with low mismatch defect [218]. Furthermore, the combination of Zn diffusion and epitaxial growth method was reported for InAsSbP/InAs p/n junction on InAs substrate by Krier et al. [219]. The combination allows higher cell performance due to low surface recombination.

Ge with a bandgap of 0.66 eV is a cost-effective candidate for the fabrication of TPV devices. Ge structure can be easily deposited on either Ge or GaAs substrates. However, the cell efficiency of Ge is lower than those of GaSb and InGaAs TPV cells [220]. In addition, the quality of the Ge crystal with indirect bandgap tends to induce the recombination rate in the TPV cell, which decreases the overall efficiency. A number of studies have been conducted to improve the performance and to reduce the fabrication cost of Ge cell. For example, amorphous silicon (a-Si:H) was demonstrated to create a surface passivating material for Ge [221]. The investigation of P-i-N Ge cell and surface treatment on c-Ge using PH$_3$ exposure shows significant improvement of $V_{oc}$ with better temperature coefficient [222,223]. Further studies are needed to improve the light absorption of Ge cell at thin absorber layer integrated with light-trapping techniques. It is worth noting that the high output power per unit area of TPV cells may offset the drawbacks related to their high cost, especially if the radiation temperature is high. For example, GaSb TPV under 1,473 K radiation

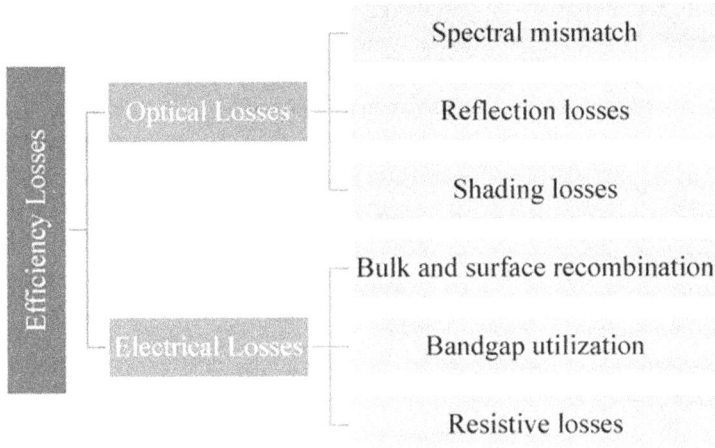

FIGURE 7.13 The TPV cells conversion efficiency losses [14].

temperature reported an output power density of 0.82 W/cm², which was almost 30 times higher than the output generated from the best GaAs solar cell [64].

The low conversion efficiency of TPV cells also remains the key challenge in realizing viable TPV system. As depicted in Figure 7.13, the TPV cells efficiency losses can be categorized into two main categories; optical losses and electrical losses. The prediction of optical-to-electrical conversion efficiencies via fundamental calculation in TPV is comparable to solar PV technology. However, as shown by Hitchcock et al. [224], lower optical-to-electrical conversion efficiency of the overall TPV system is observed in the real condition. This is mainly due to the poor spectral control from the generator up to the TPV cell, resulting in serious optical losses which are associated with the reflection and shading losses. The reflection problem can be mitigated by implementing an ARC as well as back surface reflector, while the shading loss can be reduced by optimizing the metal coverage area on the cell active surface.

Electrical losses caused by bulk and surface recombination affect the cell conversion efficiency. For instance, the majority of the reported works on InGaAs cell optimization focused on optimizing the electrical issue of the cell and improving the $V_{oc}$ (open circuit voltage) This was achieved by reducing the thickness of the absorber layer. However, the cell needs to be optically thick to absorb most of the incident illumination. Generally, TPV illumination flux is usually shifted to infrared wavelengths, and thicker absorber is needed to improve the absorption of infrared radiation and significantly increase $J_{sc}$. The use of intrinsic absorber layer in P-i-N or N-i-P TPV cells can increase $J_{sc}$, (short circuit current density) and increase the ratio of generation to recombination as the generated carriers in i-layer has high mobility and lifetime. Therefore, to maximize the cell efficiency, both optical and electrical losses must be considered to obtain an optimized designed TPV cell. On the other hand, resistive losses are related to the effectiveness of the metal contact, which allows smooth transportation of the photogenerated carriers to the external circuit. Although the importance of metal contact is highly recognized, the study of ohmic contact with semiconductor materials received less attention. For example, the process technology for the fabrication and metallization techniques needs to be further developed to facilitate a good metal contact formation onto the GaSb semiconductor material. Other than that, the optimization of TPV cell ohmic superconductive metal contact would ease the flow of high current density. This can be achieved by designing transparent metamaterial metal with ohmic work function.

Near-field TPV system is designed by separating the radiator from the TPV cell by only a few nanometers to centimeters gap distance [44]. The amount of power transferred significantly decreases with longer travelling distance, which can be presented by the inverse square law [176]. In nanometer gap distance, there will be near-field coupling, resulting in Super-Planckian characteristics. In recent years, rapid developments in the theoretical, computational, and experimental investigation of near-field TPV have taken place due to its ability to increase the amount of energy transfer into TPV converter. The near-field TPV is still in the early stages of modeling and simulation, where different materials and techniques are being investigated. For example, the characterization of multi-layer graphene on the top of InSb TPV cell [225] and the optimization for near-field TPV using a genetic algorithm method [226]. Near-field TPV system performs much better than far-field system for gap distance between 12 μm and 60 nm under low-radiation temperature (<655 K) [227,228]. Near-field was also implemented in hybrid thermophotonic-PV and thermionic-PV converters. While the former uses light-emitting diode at the hot side to produce an output power of 9.6 W/cm² at 600 K radiator and 10 nm gap distance, the latter depends on both electrons and photons emission at the radiator [227,229]. Optimum near-field thermionic-PV converter has the potential to generate an output power density > 100 W/cm² at 2,000 K radiation temperature [230].

Ali Gamel et al. [14] pointed out that the thermophotovoltaic system shows great benefits as an energy source with a full-time operating regime of blackbody temperature 500–2,000 K as compared to the solar photovoltaic system. The two dominant TPV technologies, which are the gallium antimonide and indium gallium arsenide, were extensively reviewed due to the vast integration in both industries and research areas. The fabrication of both cells was conducted based on non-epitaxial growth such as zinc diffusion and ion implantation methods, as well as epitaxial growth through liquid phase epitaxy, metal-organic vapor phase epitaxy and molecular beam epitaxy techniques. Non-epitaxial method is the preferable option as the manufacture can be done with low cost and less tedious procedures. On the other hand, an epitaxial of ternary of quaternary layer is essential to reduce the bandgap of the cell for a thermal photovoltaic system with heat source temperature of less than 1,000 K. The performances for both cells were summarized in terms of several parameters such as open-circuit voltage, short-circuit current, current density and cell efficiencies. In addition, the improvement for the cell performances was recapped with structural and functional optimization such as metal contact, doping concentration, thickness, window, cap, buffer, surface field layers, graphene, metamaterial selective radiator and monolithic interconnected modules.

Ali Gamel et al. [14] showed that the practical applications for a thermal photovoltaic system include a nuclear generator for space applications, hybrid electric vehicles, industrial and residential power supplies, waste heat recovery, solar thermophotovoltaic systems and portable electric generators. However, challenges in terms of spectral mismatching, internal cell temperature, and fabrication cost should be overcome to achieve a complete and more efficient thermophotovoltaic system. They [14] pointed out that future research and development should include (a) investigations to reduce the cost of the TPV cells. The use of ELOG fabrication which sacrifices layer between the junction and the substrate and allows the integration of cheap substrates such as Si, GaAs and InP should be further examined. In order to improve cheap Ge TPV cell, further studies on light trapping techniques are required to improve the light absorption of Ge cell at a thin absorber layer; (b) the development of the effective cooling system for TPV which consume the minimum energy. The integration of PCM can help to cool the TPV system without any energy consumption. Furthermore, separating the cell from the radiator using transparent insulator or undoped semiconductor layer can help to reduce the cell temperature; (c) selection of a suitable narrow bandgap material based on the heat source temperature. The research on efficient reflection and recycling of sub-bandgap photons in the radiator will significantly enhance the conversion efficiency of the TPV system; (d) investigation of TPV cell conversion efficiency. Research on structures optimization for near-field TPV spectrums would help to reduce the optical and electrical losses in the semiconductor layers; (e) development of effective metal contact to significantly ease the flow of high current density. This can be achieved by engineering the ohmic superconductive/transparent metamaterial metal with ohmic work function and (f) the standard by which TPV cells are characterized, and the performances are determined need to be developed. The existing characterization based on radiator temperature, cell temperatures, cavity geometry etc. are too complex and cumbersome.

## REFERENCES

1. Sakakibara R., Stelmakh V., Chan W.R., Ghebrebrhan M., Joannopoulos J.D., Soljacic M., and Čelanović I. Practical Emitters for thermophotovoltaics: A review. *J Photonics Energ* 26 February 2019;9:032713. Doi: 10.1117/1.jpe.9.032713.
2. Colangelo G., de Risi A., and Laforgia D. New approaches to the design of the combustion system for thermophotovoltaic applications. *Semicond Sci Technol* 2003;18:S262–S269 PII: S0268-1242(03)59245-8.
3. Coutts T.J. A review of progress in thermophotovoltaic generation of electricity. *Renew Sustain Energ Rev* 1999;3:77–184.
4. Seal M., Christ S., Campbell G., West E., and Fraas L. *Thermophotovoltaic Generation of Power for Use in a Series Hybrid Vehicle.* SAE-972648. Detroit: SAE; 1997.
5. Christ S., and Seal M. *Viking 29—A Thermophotovoltaic Hybrid Vehicle Designed and Built at Western Washington University.* SAE-972650. Detroit: SAE; 1997.
6. Scotto M.V., and DeBellis L. Diesel burner development for 500 Watt portable thermophotovoltaic generator. *American Flame Research 1999 Fall International Symposium*, San Francisco, CA; 3–6 October, 1999; 1999.
7. Ayr U., Barnham K., Cirillo E., de Vittorio M., de Risi A., Diso D., Laforgia D., Mazzer M., and Passaseo A. Generation of electric power by thermophotovoltaic conversion of heat. *54th ATI National Congress*, Italy; 1999, pp. 851–859.
8. Mazzer M., de Risi A., and Laforgia D. *High Efficiency Thermophotovoltaic for Automotive Applications SAE 2000 World Congress.* SAE 2000-01-0991. Detroit; 2000.

9. Congedo P.M., de Risi A., and Laforgia D. Optimization of a liquid fuel fired burner, for TPV system in electrical automotive applications. *ECOS 2002 Conference*, Berlin; 2002.

10. Coutts T.J. An overview of thermophotovoltaic generation of electricity. *Sol Energ Mater Sol Cells* 2001;66:443–452.

11. Schock A., Or C., and Kumar V. Design and integration of small RTPV generators with new millennium spacecraft for outer solar system Paper IAF 95.R1.01. *46th International Astronautical Congress*, Oslo, Norway; 1995.

12. Barnham K., Ballard I., Connoly J., Ekins-Daukes N., Kluftinger B., Nelson J., Rhor C., and Mazzer M. Recent results on quantum well solar cells. *J Mater Sci Mater Electron* 2000;11:531–536.

13. Schubnell M., Benz P., and Mayor J.C. Design of a thermophotovoltaic residential heating system. *Sol Energ Mater Sol Cells* 1998;52:1–9.

14. Gamel M.M.A., Lee H.J., Rashid W.E.S.W.A., Ker P.J., Yau L.K., Hannan M.A., and Jamaludin M.Z. A review on thermophotovoltaic cell and its applications in energy conversion: Issues and recommendations. *Materials (Basel)*. 30 August, 2021;14(17):4944. Doi: 10.3390/ma14174944. PMID: 34501032.

15. Ferrari C., Melino F., Pinelli M., Spina P.R., and Venturini M. Overview and status of thermophotovoltaic systems. *Energ Procedia* 2014;45:160–169. Doi: 10.1016/j.egypro.2014.01.018.

16. Mustafa K.F., Abdullah S., Abdullah M., and Sopian K. A review of combustion-driven thermoelectric (TE) and thermophotovoltaic (TPV) power systems. *Renew Sustain Energ Rev* 2017;71:572–584.

17. Utlu Z., and Önal B.S. Thermodynamic analysis of thermophotovoltaic systems used in waste heat recovery systems: An application. *Int J Low Carbon Technol* 2018;13:52–60. Doi: 10.1093/ijlct/ctx019.

18. Minotti A. Energy converter with inside two, three, and five connected $H_2$/air swirling combustor chambers: Solar and combustion mode investigations. *Energies* 2016;9:461. Doi: 10.3390/en9060461.

19. Woolf D.N., Kadlec E.A., Bethke D., Grine A.D., Nogan J.J., Cederberg J.G., Burckel D.B., Luk T.S., Shaner E.A., and Hensley J.M. High-efficiency thermophotovoltaic energy conversion enabled by a metamaterial selective emitter. *Optica* 2018;5:213–218. Doi: 10.1364/OPTICA.5.000213.

20. Chen Z., Adair P.L., and Rose M.F. Investigation of energy conversion in TPV power generation prototype using blackbody/selective emitters. *Proceedings of the IECEC-97 Thirty-Second Intersociety Energy Conversion Engineering Conference* (Cat. No.97CH6203), Honolulu, HI; 27 July–1 August, 1997, pp. 1097–1100.

21. Ghanekar A., Tian Y., Zhang S., Cui Y., and Zheng Y. Mie-metamaterials-based thermal emitter for near-field thermophotovoltaic systems. *Materials* 2017;10:885. Doi: 10.3390/ma10080885.

22. Nam Y., Yeng Y., Lenert A., Bermel P., Celanovic I., Soljačić M., and Wang E.N. Solar thermophotovoltaic energy conversion systems with two-dimensional tantalum photonic crystal absorbers and emitters. *Sol Energ Mater Sol Cells* 2014;122:287–296. Doi: 10.1016/j.solmat.2013.12.012.

23. Bitnar B., Durisch W., Mayor J.C., Sigg H., and Tschudi H.R. Characterisation of rare earth selective emitters for thermophotovoltaic applications. *Sol Energ Mater Sol Cells* 2002;73:221–234. Doi: 10.1016/S0927-0248(01)00127-1.

24. Tobler W., and Durisch W. High-performance selective Er-doped YAG emitters for thermophotovoltaics. *Appl Energy* 2008;85:483–493. Doi: 10.1016/j.apenergy.2007.10.006.

25. Iles P.A., and Chu C.L. Design and fabrication of thermophotovoltaic cells. *Proceedings of the 1994 IEEE 1st World Conference on Photovoltaic Energy Conversion—WCPEC*, A Joint Conference of PVSC, PVSEC and PSEC, Waikoloa, HI; 5–9 December, 1994, pp. 1750–1753.

26. Sai H., Yugami H., Nakamura K., Nakagawa N., Ohtsubo H., and Maruyama S. Selective emission of $Al_2O_3/Er_3Al_5O_{12}$ eutectic composite for thermophotovoltaic generation of electricity. *Jpn J Appl Phys* 2000;39:1957–1961. Doi: 10.1143/JJAP.39.1957.

27. Ferrari C., Melino F., Pinelli M., and Spina P.R. Thermophotovoltaic energy conversion: Analytical aspects, prototypes and experiences. *Appl Energy* 2014;113:1717–1730. Doi: 10.1016/j.apenergy.2013.08.064.

28. Casady J., and Johnson R. Status of silicon carbide (SiC) as a wide-bandgap semiconductor for high-temperature applications: A review. *Solid State Electron* 1996;39:1409–1422. Doi: 10.1016/0038-1101(96)00045-7.

29. Yang W., Chou S., Shu C., Li Z., and Xue H. Research on micro-thermophotovoltaic power generators. *Sol Energ Mater Sol Cells* 2003;80:95–104. Doi: 10.1016/S0927-0248(03)00135-1.

30. Wu H., Kaviany M., and Kwon O. Thermophotovoltaic power conversion using a superadiabatic radiant burner. *Appl Energy* 2018;209:392–399. Doi: 10.1016/j.apenergy.2017.08.168.

31. Coutts T.J., Wanlass M.W., Ward J.S., and Johnson S. A review of recent advances in thermophotovoltaics. *Proceedings of the Conference Record of the Twenty Fifth IEEE Photovoltaic Specialists Conference—1996*, Washington, DC; 13–17 May, 1996, pp. 25–30.

32. Yang W.M., Chou S.K., Shu C., Li Z.W., and Xue H. A prototype microthermophotovoltaic power generator. *Appl Phys Lett* 2004;84:3864–3866. Doi: 10.1063/1.1751614.

33. Gentillon P., Southcott J., Chan S., and Taylor R.A. Stable flame limits for optimal radiant performance of porous media reactors for thermophotovoltaic applications using packed beds of alumina. *Appl Energ* 2018;229:736–744. Doi: 10.1016/j.apenergy.2018.08.048.

34. Gentillon P., Singh S., Lakshman S., Zhang Z., Paduthol A., Ekins-Daukes N.J., Chan Q.N., and Taylor R.A. A comprehensive experimental characterisation of a novel porous media combustion-based thermophotovoltaic system with controlled emission. *Appl Energy* 2019;254:113721. Doi: 10.1016/j.apenergy.2019.113721.

35. Chubb L., Flood J., and Lowe A. *High Efficiency Thermal to Electric Energy Conversion Using Selective Emitters and Spectrally Tuned Solar Cells*, Vol. 105755 Washington, DC: National Aeronautics and Space Administration; 1992.

36. Catchpole K., Lin K., Green M., Aberle A., Corkish R., Zhao J., and Wang A. Thin semiconducting layers and nanostructures as active and passive emitters for thermophotonics and thermophotovoltaics. *Phys E: Low-dimensional Syst Nanostruct* 2002;14:91–95.

37. Tong J.K., Hsu W.C., Huang Y., Boriskina S.V., and Chen G. Thin-film 'thermal well' emitters and absorbers for high-efficiency thermophotovoltaics. *Sci Rep* 2015;5:10661. Doi: 10.1038/srep10661.

38. Daneshvar H., Prinja R., and Kherani N.P. Thermophotovoltaics: Fundamentals, challenges and prospects. *Appl Energ* 2015;159:560–575. Doi: 10.1016/j.apenergy.2015.08.064.

39. Wernsman B., Siergiej R., Link S., Mahorter R., Palmisiano M., Wehrer R., Schultz R., Schmuck G., Messham R., Murray S., and Murray C.S. Greater than 20% radiant heat conversion efficiency of a thermophotovoltaic radiator/module system using reflective spectral control. *IEEE Trans Electron Dev* 2004;51:512–515. Doi: 10.1109/TED.2003.823247.

40. Fraas L.M. *Thermophotovoltaics Using Infrared Sensitive Cells in Low-Cost Solar Electric Power*. Berlin/Heidelberg: Springer International Publishing; 2014.

41. Potter W.R. Radiative cooling for thermophotovoltaic systems. *Proc SPIE* 2016;997:997308.

42. Karthikeyan V., Prasannaa P., Sathishkumar N., Emsaeng K., Sukchai S., and Sirisamphanwong C. Selection and preparation of suitable composite phase change material for PV module cooling. *Int J Emerg Technol* 2019;10:385–394.

43. Datas A., and Marti A. Thermophotovoltaic energy in space applications: Review and future potential. *Sol Energ Mater Sol Cells* 2017;161:285–296. Doi: 10.1016/j. solmat.2016.12.007.

44. Tian Y., Ghanekar A., Ricci M., Hyde M., Gregory O., and Zheng Y. A review of tunable wavelength selectivity of metamaterials in near-field and far-field radiative thermal transport. *Materials* 2018;11:862. Doi: 10.3390/ma11050862.

45. Rashid W.E.S.W.A., Ker P.J., Bin Jamaludin Z., Gamel M.M.A., Lee H.J., and Rahman N.B.A. Recent development of thermophotovoltaic system for waste heat harvesting application and potential implementation in thermal power plant. *IEEE Access* 2020;8:105156–105168. Doi: 10.1109/ACCESS.2020.2999061.

46. Burger T., Sempere C., Roy-Layinde B., and Lenert A. Present efficiencies and future opportunities in thermophotovoltaics. *Joule* 2020;4:1660–1680. Doi: 10.1016/j. joule.2020.06.021.

47. Shockley W., and Queisser H.J. Detailed balance limit of efficiency of p-n junction solar cells. *J Appl Phys* 1961;32:510–519.

48. Chubb D.L. *Fundamentals of Thermophotovoltaic Energy Conversion.* Netherlands: Elsevier; 2007.

49. Wang Y., Liu H., and Zhu J. Solar thermophotovoltaics: Progress, challenges, and opportunities *APL Mater* 2019;7:080906. Doi: 10.1063/1.5114829.

50. Davies P.A., and Luque A. Solar photovoltaics: Brief review and a new look. *Sol. Energ Mater Sol Cells* 1994;33(1):11–22.

51. Ungaro C., Gray S.K., and Gupta M.C. Solar thermophotovoltaic system using nanostructures. *Opt Exp.* 2015;23:A1149–A1156.

52. Green M.A., Emery K., Hishikawa Y., Warta W., and Dunlop E.D. Solar cell efficiency tables (Version 45). *Progres Photovolt Res Appl* 2014;23:1–9.

53. Yugami H., Sai H., Nakamura K., Nakagawa H., and Ohtsubo H. Solar thermophotovoltaic using $Al_2O_3/Er_3Al_5O_{12}$ eutectic composite selective emitter. *IEEE Photovolt Special Conf* 2000;28:1214–1217.

54. Datas A., and Algora C. Development and experimental evaluation of a complete solar thermophotovoltaic system. *Progres Photovolt Res Appl* 2012;890:327–334.

55. Vlasov A.S., Khvostikov V.P., Khvostikova O.A., Gazaryan P.Y., Sorokina S.V., and Andreev, V.M. TPV systems with solar powered tungsten emitters. *AIP Conf Pro* 2007;890:327–334.

56. Lenert, A., Bierman D.M., Nam Y., Chan W.R., Celanovic I., Soljacic M., and Wang E.N. A nanophotonic solar thermophotovoltaic device. *Nat Nanotech* 2014;9:126–130.

57. Shimizu M., Kohiyama A., and Yugami H. High-efficiency solar thermophotovoltaic system equipped with a monolithic planar selective absorber/emitter. *J Photon Energ* 2015;5:053099.

58. Butcher T., Hammonds J., Horne E., Kamath B., Carpenter J., and Woods D. Heat transfer and thermophotovoltaic power generation in oil-fired heating systems. *Appl Energ* 2011;88:1543–1548. Doi: 10.1016/j.apenergy.2010.10.033.

59. Colangelo G., De Risi A., and Laforgia D. Experimental study of a burner with high temperature heat recovery system for TPV applications. *Energy Convers Manag* 2006;47:1192–1206. Doi: 10.1016/j.enconman.2005.07.001.

60. Kim T., Kim, H., Ku J., and Kwon O. A heat-recirculating combustor with multiple injectors for thermophotovoltaic power conversion. *Appl Energ.* 2017;193:174–181. Doi: 10.1016/j.apenergy.2017.02.040.

61. Kim T.Y., Kim H.K., Ku J.W., and Kwon O.C. Design of a heat-recirculating combustor for a thermophotovoltaic system. *Mater Energ Efficien Sustain TechConnect Briefs* 2016;60–63, TechConnect.org, ISBN 978-0-9975-1171-0.

62. Gentillon Molina P., Taylor R., Ekins-Daukes N.J, Chan S., and Paduthol A. Rshikesan (): Porous media combustion-based thermophotovoltaic (PMC-TPV) reactor experiment. *UNSW Dataset* Doi: 10.26190/5d7acb06e8488.

63. Utlu Z., and Önal B.S. Performance evaluation of thermophotovoltaic GaSb cell technology in high temperature waste heat. *IOP Conf Ser Mater Sci Eng* 2018;307:012075.

64. Gamel M.M.A., Ker P.J., Lee H.J., Rashid W.E.S.W.A., Hannan M.A., David J.P.R. and Jamaludin M.Z. Multi-dimensional optimization of $In_{0.53}Ga_{0.47}As$ thermophotovoltaic cell using realcoded genetic algorithm. *Nat Sci Rep* 2021; 11:7741, Springer Nature. Doi: 10.1038/s41598-021-86175-5.

65. Mittapally R., Lee B., Zhu L., Reihani A., Lim J.W., Fan D., Forrest S.R., Reddy P. and Meyhofer E. Near-field thermophotovoltaics for efficient heat to electricity conversion at high power density. *Nat Commun* 2021;12:4364. Doi: 10.1038/s41467-021-24587-7.

66. Wang X., Chan W., Stelmakh V., and Fisher P. Radioisotope thermophotovoltaic generator design and performance estimates for terrestrial applications. *2017 25th International Conference on Nuclear Engineering*, Volume 3: Nuclear Fuel and Material, Reactor Physics and Transport Theory; Innovative Nuclear Power Plant Design and New Technology Application, Shanghai, China; July 2–6, 2017, July 2, 2017. Doi:10.1115/ icone25-66607.

67. Navy, U. S. Radioisotope thermoelectric generators of the U.S. Navy. Final report. Hueneme, CA: Naval Nuclear Power Unit Port; July 1978. Available from: http://handle. dtic.mil/100.2/ADA057483.

68. Corliss W.R., and Harve D.G. *Radioisotopic Power Generation*. Englewood Cliffs, NJ: Prentice-Hall; 1964.

69. Crowley C.J., Elkouh N.A., Murray S., and Chubb D.L. Thermophotovoltaic converter performance for radioisotope power systems *AIP Conf Proc* 2005;746:601–614.

70. Rinehart G.H. Design characteristics and fabrication of radioisotope heat sources for space missions. *Prog Nucl Energ* 2001;39(3):305–319.

71. Chan J., Wood J., and Schreiber J. Development of advanced stirling radioisotope generator for space exploration. *AIP Conf Proc* 2007;800:615–623.

72. Shock, A., Noravian, H., Or, C., and Kumar V. Design, analyses, and fabrication procedure of amtec cell, test assembly, and radioisotope power system for outer planet missions. *Acta Astron* 2002;50(8):471–510.

73. Koudelka R., Murray C., Fleming J., Shaw M., Teofilo V., and Alexander C. Radioisotope micropower system using thermophotovoltaic energy conversion. *AIP Conf Proc* 2006;813:545–551.

74. Rinnerbauer V., Ndao S., Yeng Y., Senkevich J., Jensen K., and Joannopoulos J. Large-area fabrication of high aspect ratio tantalum photonic crystals for high-temperature selective emitters. *J Vac Sci Tecchnol B* 2013;31:011802.

75. Wang X., Chan W., Stelmakh V., Celanovic I., and Fisher P. Toward high performance radioisotope thermophotovoltaic systems using spectral control. *Nucl Instrum Methods Phys Res Sect A* 2016;838:28–32.

76. Chubb D.L. and Good B.S. A combined thermophotovoltaic-thermoelectric energy converter. *Sol Energ* January 2018;159:760–767.

77. *Waste Heat Recovery: Technology and Opportunities in U.S. Industry*. A report prepared by BCS Inc. for U.S. Department of Energy, Industrial Technology program, Washington, DC; March, 2008.

78. Bauer T., Penlington R., and Pearsall N. The potential of thermophotovoltaic heat recovery for the glass industry. *AIP Conf Proc*, 2003 653:101. Doi: 10.1063/1.1539368.

79. Qiu K., and Hayden A. Development of a novel cascading TPV and TE power generation system. *Appl Energ* 2012;91:304–308.

80. Huen P. and Daoud W.A. Advances in hybrid solar photovoltaic and thermoelectric generators. *Renew Sustain Energ Rev* 2017;72:1295–1302.

81. Park K.-T., Shin S.-M., Tazebay A.S., Um H.-D., Jung J.-Y., Jee S.-W., Oh M.-W., Park S.-D., Yoo B., Yu C., and Lee J.H. Lossless hybridization between photovoltaic and thermoelectric devices. *Sci Rep* 2013;3:2123.

82. Dong Q., Cai L., Liao T., Zhou Y., and Chen J. An efficient coupling system using a thermophotovoltaic cell to harvest the waste heat from a reforming solid oxide fuel cell. *Int J Hydrogen Energ* 2017;42(27):17221–17228.

83. Liao T., He Q., Xu Q., Dai Y., Cheng C., and Ni M. Harvesting waste heat produced in solid oxide fuel cell using near-field thermophotovoltaic cell. *J Power Sourc* 2020;452(November 2019):227831.

84. Shan S., Zhou Z., and Cen K. An innovative integrated system concept between oxy-fuel thermo-photovoltaic device and a Brayton-Rankine combined cycle and its preliminary thermodynamic analysis. *Energ Convers Manag* 2019;180(October 2018):1139–1152.

85. Yang Z., Xu H., Chen B., Tan P., Zhang H., and Ni M. Numerical modeling of a cogeneration system based on a direct carbon solid oxide fuel cell and a thermophotovoltaic cell. *Energ Convers Manag* 2018;171(April):279–286.

86. Liao T., Cai L., Zhao Y., and Chen J. Efficiently exploiting the waste heat in solid oxide fuel cell by means of thermophotovoltaic cell. *J Power Sourc* 2016;306:666–673.

87. Brace D. B. *The Laws of Radiation and Absorption: Memoirs by Pre´vost, Stewart, Kirchhoff, and Kirchhoff and Bunsen.* Knoxville, TN: American Book Company; 1901.

88. Siegel R., and Howell J. *Thermal Radiation Heat Transfer* New York: Hemisphere Publishing Corporation; 1981.

89. Mertens K. *Photovoltaics: Fundamentals, Technology and Practice.* Chichester: John Wiley & Sons; 2014.

90. Pfiester N.A., and Vandervelde T.E. Selective emitters for TPV applications. *Physics Status Solidi* 2017;A214:1600410.

91. Chirumamilla M., Krishnamurthy G.V., Knopp K., Krekeler T., Graf M., Jalas D., Ritter M., Stormer M., Yu Petrov A., and Eieh M. Metamaterial emitter for thermophotovoltaics stable up to 1400°C. *Sci Rep* May 2019;9(1):1–12. Doi: 10.1038/s41598-019-43640-6.

92. Sakurai A., and Matsuno Y. Design and fabrication of a wavelength-selective near-infrared metasurface emitter for a thermophotovoltaic system. *Micromachines.* 2019;10:157. Doi: 10.3390/mi10020157.

93. Wang Z., Kortge D., He Z., Song J., Zhu J., Lee C., Wang H. and Bermel P. Selective emitter materials and designs for high-temperature thermophotovoltaic applications. *Sol Energ Mater Sol Cells* May 2022;238:111554.

94. Chen B., Shan S., Liu J., and Zhou Z. An effective design of thermophotovoltaic metamaterial emitter for medium-temperature solar energy storage utilization. *Sol. Energ* 1 January 2022;231:194–202.

95. Pfiester N.A., and Vandervelde T.E. Selective emitters for thermophotovoltaic applications. *Physica Status Solidi (a)* 02 November 2016;214(1):1600410. Doi: 10.1002/pssa.201600410.

96. Rinnerbauer V., Lenert A., Bierman D.M., Yeng Y.X., Chan W.R., Geil R.D., Senkevich J.J., Joannopoulos J.D., Wang E.N., Soljačić M., and Celanovic I. Metallic photonic crystal absorber-emitter for efficient spectral control in high-temperature solar thermophotovoltaics. *Adv Energ Mater* 2014;4:1400334. Doi: 10.1002/aenm.201400334.

97. Sai H., and Yugami H. Thermophotovoltaic generation with selective radiators based on tungsten surface gratings. *Appl Phys Lett* 2004;85:3399–3401. Doi: 10.1063/1.1807031.

98. Jurczak P., Onno A., Sablon K., and Liu H. Efficiency of GaInAs thermophotovoltaic cells: The effects of incident radiation, light trapping and recombinations. *Opt Exp* 2015;23:A1208. Doi: 10.1364/OE.23.0A1208.

99. Fan D., Burger T., McSherry S., Lee B., Lenert A., and Forrest S.R. Near-perfect photon utilization in an air-bridge thermophotovoltaic cell. *Nature* 2020;586:237–241. Doi: 10.1038/s41586-020-2717-7.

100. Omair Z., Scranton G., Pazos-Outón L.M., Xiao T.P., Steiner M.A., Ganapati V., Peterson P.F., Holzrichter J., Atwater H., and Yablonovitch E. Ultraefficient thermophotovoltaic power conversion by band-edge spectral filtering. *Proc Natl Acad Sci USA* 2019;116:15356–15361. Doi: 10.1073/pnas.1903001116.

101. Wilt D.M., Fatemi N.S., Jenkins P.P., Hoffman R.W., Landis G.A., and Jain R.K. Monolithically interconnected InGaAs TPV module development. *Proceedings of the Conference Record of the Twenty Fifth IEEE Photovoltaic Specialists Conference—1996*, Washington, DC; 13–17 May, 1996, pp. 43–48.

102. Datas A., and Linares P. Monolithic interconnected modules (MIM) for high irradiance photovoltaic energy conversion: A comprehensive review. *Renew Sustain Energ Rev* 2017;73:477–495. Doi: 10.1016/j.rser.2017.01.071.

103. Charache G.W., Depoy D.M., Baldasaro P.F., and Campbell B.C. Thermophotovoltaic devices utilizing a back surface reflector for spectral control. *AIP Conf Proc* 1996;358:339–350.

104. Wehrer R., Wanlass M., Wilt D., Wernsman B., Siergiej R., and Carapella J. InGaAs series-connected, tandem, MIM TPV converters. *Proceedings of the 3rd World Conference on Photovoltaic Energy Conversion*, Osaka, Japan; 11–18 May, 2003, pp. 892–895.

105. Woolf D., Hensley J., Cederberg J.G., Bethke D.T., Grine A.D., and Shaner E.A. Heterogeneous metasurface for high temperature selective emission. *Appl Phys Lett* 2014;105:081110. Doi: 10.1063/1.4893742.

106. Arpin K.A., Losego M.D., Cloud A.N., Ning H., Mallek J., Sergeant N.P., Zhu L., Yu Z., Kalanyan B., Parsons G.N., and Girolami G.S. Three-dimensional self-assembled photonic crystals with high temperature stability for thermal emission modification. *Nat Commun* 2013;4:2630.

107. Cao F., Kraemer D., Tang L., Li Y., Litvinchuk A.P., Bao J., Chen G., and Ren Z. A high-performance spectrally selective solar absorber based on a yttria stabilized zirconia cermet with high temperature stability. *Energ Environ Sci* 2015;8:3040–3048.

108. Baranov D.G., Xiao Y., Nechepurenko I.A., Krasnok A., Alù A., and Kats M.A. Nanophotonic engineering of far-field thermal emitters. *Nat Mater* 2019;18:920–930.

109. Ferguson L.G., and Dogan F. A highly efficient NiO-Doped MgO matched emitter for thermophotovoltaic energy conversion. *Mater Sci Eng B* 2001;83:35–41.

110. Yeng Y.X., Ghebrebrhan M., Bermel P., Chan W.R., Joannopoulos J.D., Soljačić M., and Celanovic I. Enabling high temperature nanophotonics for energy applications. *Proc Natl Acad Sci USA* 2012;109:2280–2285.

111. Wang Y., Zhou L., Zhang Y., Yu J., Huang B., Wang Y., Lai Y., Zhu S., and Zhu J. Hybrid solar absorber-emitter by coherence enhanced absorption for improved solar thermophotovoltaic conversion. *Adv Opt Mater* 2018;6:1800813.

112. Rinnerbauer V., Yeng Y.X., Chan W.R., Senkevich J.J., Joannopoulos J.D., Soljačić M., and Celanovic I. High temperature stability and selective thermal emission of polycrystalline tantalum photonic crystals. *Opt Exp* 2013;21:11482–11491.

113. Nagpal P., Josephson D.P., Denny N.R., DeWilde J., Norris D.J., and Stein A. Fabrication of carbon/refractory metal nanocomposites as thermally stable metallic photonic crystals. *J Mater Chem* 2011;21:10836.

114. Abbas R., Muñoz J., and Martínez-Val J.M. Steady-state thermal analysis of an innovative receiver for linear Fresnel reflectors. *Appl Energ* 2012;92:503–515.

115. Charache G.W., Depoy D.M., Baldasaro P.F., and Campbell B.C. Thermophotovoltaic devices utilizing a back surface reflector for spectral control. *AIP Conf Proc* 1996;358:339. Doi: 10.1063/1.49697.

116. Fernández J., Dimroth F., Oliva E., Hermle M., and Bett A.W. Back-surface optimization of germanium TPV cells. *AIP Conf Proceed* 2007;890:190–197.

117. Burger T., Fan D., Lee K., Forrest S.R., and Lenert A. Thin-film architectures with high spectral selectivity for thermophotovoltaic cells. *ACS Photon* 2018;5:2748–2754. Doi: 10.1021/acsphotonics.8b00508.

118. Wang C.A., Shiau D.A., Murphy P.G., O'Brien P.W., Huang R.K., Connors M.K., Anderson A.C., Donetsky D., Anikeev S., Belenky G., and Depoy D.M. Wafer bonding and epitaxial transfer of GaSb-based epitaxy to GaAs for monolithic interconnection of thermophotovoltaic devices. *J Electron Mater* 2004;33:213–217.

119. Wu X., Duda A., Carapella J.J., Ward J.S., Webb J.D., and Wanlass M.W. A Study of Contacts and Back-Surface. Reflectors for 0.6 -eV Ga 0.32 $In_{0.68}$ As/$InAs_{0.32}P_{0.68}$ thermophotovoltaic monolithically interconnected modules. *AIP Conf Proc* 1999;460:517. Doi:10/1063/1.57836.

120. Blandre E., Vaillon R., and Drévillon J. New insights into the thermal behavior and management of thermophotovoltaic systems. *Opt Exp* 2019;27:36340–36349.

121. Stark A.K., and Klausner J.F. An R&D strategy to decouple energy from water. *Joule* 2017;1:416–420.

122. Wen R., Ma X., Lee Y.C., and Yang R. Liquid-vapor phase-change heat transfer on functionalized nanowired surfaces and beyond. *Joule* 2018;2:2307–2347.

123. Weinstein L.A., McEnaney K., Strobach E., Yang S., Bhatia B., Zhao L., Huang Y., Loomis J., Cao F., Boriskina S.V., and Ren Z. A hybrid electric and thermal solar receiver. *Joule* 2018;2:962–975.

124. Chan W., Stelmakh V., Ghebrebrhan M., Soljačić M., Joannopoulos J.D. and Čelanović I. Enabling efficient heat-to-electricity generation at the mesoscale. *Energ Environ Sci* 2017;10:1367.

125. Pralle M.U., Moelders N., McNeal M.P., Puscasu I., Greenwald A.C., Daly J.T., Johnson E.A., George T., Choi D.S., El-Kady I. and Biswas R. Photonic crystal enhanced narrow-band infrared emitters. *Appl Phys Lett* 2002;81:4685–4687.

126. Jovanic´ N., Čelanović I., and Kassakian J. Two-dimensional tungsten photonic crystals asthermophotovoltaic selective emitters. *AIP Conf Proc* 2008;890:47–55.

127. Chen K., Osborn D., Sarmiento P., Earath, S., and Prasad, A.J. *Small, Efficient Thermophotovoltaic Power Supply.* Tech. Rep. DAAG55-97-C-0003. Research Triangle Park, NC: U.S. Army Research Office; 1999.

128. Bitnar S., Durisch W., Palfinger G., Von Roth F., Vogt U., Brönstrup A., and Seiler D Practical thermophotovoltaic generators. *Semi-conductors* 2004;28:941–945.

129. Arpin K., Losego M., and Braun P. Electrodeposited 3D tungsten photonic crystals with enhanced thermal stability. *Chem Mater* 2011;23:4783–4788. Doi: 10.1021/cm2019789 CMATEX0897-4756.

130. Lee H.J., Smyth K., Bathurst S., Chou J., Ghebrebrhan M., Joannopoulos J., Saka N., and Kim S.G. Hafnia-plugged microcavities for thermal stability of selective emitters. *Appl Phys Lett* 2013;102(24):241904. Doi: 10.1063/1.4811703 APPLAB0003-6951.

131. Cui K., Lemaire P., Zhao H., Savas T., Parsons G. and Hart A.J. Tungsten-carbon nanotube composite photonic crystals as thermally stable spectral-selective absorbers and emitters for thermophotovoltaics. *Adv Energ Mater* 2018;8:1801471. Doi: 10.1002/aenm. v8.27ADEMBC1614-6840.

132. Stone K., Chubb D.L., Wilt D.M., and Wanlass M.W. Testing and modeling of a solar thermophotovoltaic power system. *AIP Conf Proc* 1996;199:199–209. Doi: 10.1063/1.49687APCPCS0094-243X.

133. Wang X., Chan W.R., Stelmakh V., Soljacic M., Joannopoulos J.D., Celanovic I., and Fisher P.H. Prototype of radioisotope thermophotovoltaic system using photonic crystal spectral control. *J Phys Conf Ser* 2015;660:012034. Doi: 10.1088/1742-6596/660/1/012034 JPCSDZ1742-6588.

134. Stelmakh V. A practical high temperature photonic crystal for high temperature thermophotovoltaics. 2017.

135. Stelmakh V., Rinnerbauer V., Geil R.D., Aimone P.R., Senkevich J.J., Joannopoulos J.D., Soljačić M., and Celanovic I. High-temperature tantalum tungsten alloy photonic crystals: stability, optical properties, and fabrication. *Appl Phys Lett* 2013;103:123903. Doi: 10.1063/1.4821586APPLAB0003-6951.

136. Kohiyama A., Shimizu M., and Yugami H. Unidirectional radiative heat transfer with a spectrally selective planar absorber/emitter for high-efficiency solar thermophotovoltaic systems. *Appl Phys Exp* 2016;9:112302. Doi: 10.7567/APEX.9.112302APEPC41882-0778.

137. Wenming Y., Siawkiang C., Chang S., Hong X., and Zhiwang L. Research on micro-thermophotovoltaic power generators with different emitting materials. *J Micromech Microeng* 2005;15:S239. Doi: 10.1088/0960-1317/15/9/S11JMMIEZ0960-1317.

138. Stelmakh V., Chan W.R., Ghebrebrhan M., Senkevich J., Joannopoulos J.D., Soljačić M., and Celanović, I. Sputtered tantalum photonic crystal coatings for high-temperature energy conversion applications. *IEEE Trans Nanotechnol* 2016;15(2):303–309. Doi: 10.1109/TNANO.2016.2522423ITNECU1536-125X.

139. Wilt D., Chubb D., Wolford D., Magari P., and Crowley C. Thermophotovoltaics for space power applications. *Seventh World Conference on Thermophotovoltaic Generation of Electricity AIP Conference Proceedings*; 2007, Vol. 890, pp. 335–345. https://aip.scitation.org/doi/abs/10.1063/1.2711751?casa_token=uadJ5AAjtFwAAAAA:9NpjIzUbcyVPQP2-pQIhsCnRdzUVgiptmElcUauWD8fMBZIfibgF7JIDMlZLK6gDZiF8_2lhplo.

140. Bierman D.M., Lenert A., Chan W.R., Bhatia B., Celanović I., Soljačić M., and Wang E.N. Enhanced photovoltaic energy conversion using thermally based spectral shaping *Nat Energ* 2016;1:16068.

141. Bhatt R., Kravchenko I., and Gupta M. High-efficiency solar thermophotovoltaic system using a nanostructure- based selective emitter. *Sol Energ* 2020;197:538–545.

142. Raman V.K., Burger T., and Lenert A. Design of thermophotovoltaics for tolerance of parasitic absorption. *Opt Exp* 2019;27:31757–31772.

143. Wilt D., Wehrer R., Palmisiano M., Wanlass M., and Murray C. Monolithic interconnected modules (MIMs) for thermophotovoltaic energy conversion *Semicond Sci Technol* 2003;18:S209–S215.

144. Haase F., Hollemann C., Schäfer S., Merkle A., Rienäcker M., Krügener J., Brendel R., and Peibst R. Laser contact openings for local poly-Si-metal contacts enabling 26.1%-efficient POLO-IBC solar cells. *Sol Energ Mater Sol Cells* 2018;186:184–193.

145. Yoshikawa K., Kawasaki H., Yoshida W., Irie T., Konishi K., Nakano K., Uto T., Adachi D., Kanematsu M., Uzu H., and Yamamoto K. Silicon heterojunction solar cell with interdigitated back contacts for a photoconversion efficiency over 26 *Nat Energ* 2017;2:17032.

146. Kayes B.M., Nie H., Twist R., Spruytte S.G., Reinhardt F., Kizilyalli I.C., and Higashi G.S. 27.6% Conversion efficiency, a new record for single-junction solar cells under 1 sun illumination. *37th IEEE Photovoltaic Specialists Conference (IEEE)*; 2011, pp. 000004–000008.

147. Horowitz K.A., Remo T.W., Smith B., and Ptak A.J. A techno-economic analysis and cost reduction roadmap for III-V solar cells office of scientific and technical information; 2018. Doi: 10.2172/1484349.

148. Lee K., Zimmerman J.D., Hughes T.W., and Forrest S.R. Non-destructive wafer recycling for low-cost thin-film flexible optoelectronics. *Adv Funct Mater* 2014;24:4284–4291.

149. Lee K., Shiu K.T., Zimmerman J.D, Renshaw C.K., and Forrest S.R. Multiple growths of epitaxial lift-off solar cells from a single InP substrate *Appl Phys Lett* 2010;97:10–12.

150. DePoy D.M., Fourspring P.M., Baldasaro P.F., Beausang J.F., Brown E.J., Dashiel M.W., Rahner K.D., Rahmlow T.D., Lazo-Wasem J.E., Gratrix E.J., and Wernsman B. *Thermophotovoltaic Spectral Control*. LM-04K053. Washington, DC: Department of Energy report; 2004, June 9.

151. Bauer T. *Thermophotovoltaics: Basic Principles and Critical Aspects of System Design*, Vol. 7. Berlin/Heidelberg: Springer; 2011.

152. Tan M., Ji L., Wu Y., Dai P., Wang Q., Li K., Yu T., Yu Y., Lu S., and Yang H. Investigation of InGaAs thermophotovoltaic cells under blackbody radiation. *Appl Phys Exp*. 2014;7:096601. Doi: 10.7567/APEX.7.096601.

153. Ni Q., Ye H., Shu Y., and Lin Q. A theoretical discussion on the internal quantum efficiencies of the epitaxial single crystal GaSb thin film cells with different p–n junctions. *Sol Energy Mater Sol Cells* 2016;149:88–96. Doi: 10.1016/j.solmat.2015.12.039.

154. Fraas L.M. Economic potential for thermophotovoltaic electric power generation in the steel industry. *Proceedings of the 2014 IEEE 40th Photovoltaic Specialist Conference, PVSC 2014;* Denver, CO; 8–13 June, 2014, pp. 766–770.

155. Utlu Z. Investigation of the potential for heat recovery at low, medium, and high stages in the Turkish industrial sector (TIS): An application. *Energy.* 2015;81:394–405. Doi: 10.1016/j.energy.2014.12.052.

156. Khvostikov V.P., Gazaryan P.Y., Khvostikova O.A., Potapovich N.S., Sorokina S.V., Malevskaya A.V., Shvarts M.Z., Shmidt N.M., and Andreev V.M. GaSb applications for solar thermophotovoltaic conversion. *Proceedings of the Thermophotovoltaic Generation of Electricity: Seventh World Conference on Thermophotovoltaic Generation of Electricity (AIP Conference Proceedings),* Madrid, Spain; 9 March, 2007, pp. 139–148.

157. Tang L., Fraas L.M., Liu Z., Duan H., and Xu C. Doping optimization in Zn-diffused GaSb thermophotovoltaic cells to increase the quantum efficiency in the long wave range. *IEEE Trans Electron Dev* 2017;64:5012–5018. Doi: 10.1109/TED.2017.2764528.

158. Tang L., Fraas L.M., Liu Z., Xu C., and Chen X. Performance improvement of the GaSb thermophotovoltaic cells with n-type emitters. *IEEE Trans Electron Dev* 2015;62:2809–2815. Doi: 10.1109/TED.2015.2455075.

159. Fraas L., Minkin L., Avery J., Ferguson L., and Samaras J. Spectral control development for thermophotovoltaics. *Proceedings of the 2016 IEEE 43rd Photovoltaic Specialists Conference (PVSC),* Portland, OR; 5–10 June, 2016, pp. 1–5.

160. Rahimi N., Aragon A.A., Romero O.S., Shima D.M., Rotter T.J., Balakrishnan G., Mukherjee S.D., and Lester L.F. Ultra-low resistance NiGeAu and PdGeAu ohmic contacts on N-GaSb grown on GaAs. *Proceedings of the 2013 IEEE 39th Photovoltaic Specialists Conference,* Tampa, FL; 16–21 June, 2013, pp. 2123–2126.

161. Rahimi N., Aragon A.A., Romero O.S., Kim D.M., Traynor N.B.J., Rotter T.J., Balakrishnan G., Mukherjee S.D., and Lester L.F. Ohmic contacts to n-type GaSb grown on GaAs by the interfacial misfit dislocation technique. *Phys Simul Photonic Eng Photovolt Dev II* 2013;8620:86201K.

162. Abdallah, S.A., Herrera, D.J., Conlon, B.P., Rahimi, N., and Lester, L.F. Emitter thickness optimization for GaSb thermophotovoltaic cells grown by molecular beam epitaxy. *Proceedings of the Next Generation Technologies for Solar Energy Conversion VI,* San Diego, CA; 9 August, 2015, Vol. 9562, p. 95620L.

163. Rahimi N., Aragon A.A., Romero O.S., Kim D.M., Traynor N.B.J., Rotter T.J., Balakrishnan G., Mukherjee S.D., and Lester L.F. Electrical and microstructure analysis of nickel-based low-resistance ohmic contacts to n-GaSb. *APL Mater* 2013;1:062105. Doi: 10.1063/1.4842355.

164. Milnes A.G., Ye M., and Stam M. Ohmic contacts of Au and Ag to p-GaSb. *Solid State Electron* 1994;37:37–44. Doi: 10.1016/0038-1101(94)90101-5.

165. Sulima O., and Bett A. Fabrication and simulation of GaSb thermophotovoltaic cells. *Sol Energ Mater Sol Cells* 2001;66:533–540. Doi: 10.1016/S0927-0248(00)00235-X.

166. Rajagopalan G., Reddy N., Ehsani E., Bhat I., Dutta P., Gutmann R., Nichols G., Charache G., and Sulima O. A simple single-step diffusion and emitter etching process for high-efficiency gallium-antimonide thermophotovoltaic devices. *J Electron Mater* 2003;32:1317–1321. Doi: 10.1007/s11664-003-0029-y.

167. Bett A.W., Keser S., Stollwerck G., Sulima O.V., and Wettling W. GaSb-based (thermo) photovoltaic cells with Zn diffused emitters. *Proceedings of the Conference Record of the Twenty Fifth IEEE Photovoltaic Specialists Conference—1996,* Washington, DC; 13–17 May, 1996.

168. Licht A.S., Shemelya C.S., DeMeo D.F., Carlson E.S., and Vandervelde T. Optimization of GaSb thermophotovoltaic diodes with metallic photonic crystal front-surface filters. *Proceedings of the 2017 IEEE 60th International Midwest Symposium on Circuits and Systems (MWSCAS);* Boston, MA; 6–9 August, 2017, pp. 843–846.

169. Fraas L.M., Tang L., and Zhang Y. Designing a heterojunction N+ on P GaSb thermophotovoltaic cell with hydrogenated amorphous silicon interface passivation. *Proceedings of the 2018 IEEE 7th World Conference on Photovoltaic Energy Conversion (WCPEC) (A Joint Conference of 45th IEEE PVSC, 28th PVSEC & 34th EU PVSEC)*, Waikoloa, HI; 10–15 June, 2018, 0887–0890.

170. Green M.A., Hishikawa Y., Dunlop E.D., Levi D.H., Hohl-Ebinger J., and Ho-Baillie A.W.Y. Solar cell efficiency tables (version 52). *Prog Photovolt Res Appl* 2018;26:427–436. Doi: 10.1002/pip.3040.

171. Wanlass M.W., Carapella J.J., Duda A., Emery K., Gedvilas L., Moriarty T., Ward S., Webb J.D., Wu X., and Murray C.S. High-performance, 0.6-eV, $Ga_{0.32}In_{0.68}As/InAs_{0.32}P_{0.68}$ thermophotovoltaic converters and monolithically interconnected modules. *AIP Conf Proc* 1999;460:132–141.

172. Zhou T., Sun Z., Li S., Liu H., and Yi D. Design and optimization of thermophotovoltaic system cavity with mirrors. *Energies*. 2016;9:722. Doi: 10.3390/en9090722.

173. Zhang Y., Gu Y., Zhu C., Hao G., Li A., and Liu T. Gas source MBE grown wavelength extended 2.2 and 2.5 µm InGaAs PIN photodetectors. *Infrared Phys Technol* 2006;47:257–262. Doi: 10.1016/j.infrared.2005.02.031.

174. Arslan Y., Oguz F., and Besikci C. Extended wavelength SWIR InGaAs focal plane array: Characteristics and limitations. *Infrared Phys Technol* 2015;70:134–137. Doi: 10.1016/j.infrared.2014.10.012.

175. Karlina L.B., Vlasov A.S., Kulagina M.M., and Timoshina N.K. Thermophotovoltaic cells based on $In_{0.53}Ga_{0.47}As/InP$ heterostructures. *Semiconductors* 2006;40:346–350. Doi: 10.1134/S1063782606030171.

176. Tuley R.S., Orr J.M.S., Nicholas R., Rogers D.C., Cannard P.J., and Dosanjh S. Lattice-matched InGaAs on InP thermophovoltaic cells. *Semicond Sci Technol*. 2012;28:015013. Doi: 10.1088/0268-1242/28/1/015013.

177. Sodabanlu H., Watanabe K., Sugiyama M., and Nakano Y. Growth of InGaAs(P) in planetary metalorganic vapor phase epitaxy reactor using tertiarybutylarsine and tertiarybutylphosphine for photovoltaic applications. *Jpn J Appl Phys* 2018;57:08RD09. Doi: 10.7567/JJAP.57.08RD09.

178. Kao Y.-C., Chou H.-M., Hsu S.-C., Lin A., Lin C.-C., Shih Z.-H., Chang C.-L., Hong H.-F., and Horng R.-H. Performance comparison of III–V//Si and III–V//InGaAs multijunction solar cells fabricated by the combination of mechanical stacking and wire bonding. *Sci Rep* 2019;9:4308. Doi: 10.1038/s41598-019-40727-y.

179. Tuley R.S., and Nicholas R.J. Material parameters and device optimization: Supplementary information for bandgap dependent thermophotovoltaic device performance using the InGaAs and InGaAsP material system. *J Appl Phys* 2010;108:156018. Doi: 10.1063/1.3488903.

180. Emziane M., Nicholas R.J., Rogers D.C., and Cannard P.J. Fabrication and assessment of optimized InGaAs single-junction TPV cells. *AIP Conf Proc* 2007;890:149–156.

181. Mertens K. *Photovoltaics: Fundamentals, Technology and Practice*. Chichester: John Wiley & Sons; 2014.

182. Jurczak P., Onno A., Sablon K., and Liu H. Efficiency of GaInAs thermophotovoltaic cells: The effects of incident radiation, light trapping and recombinations. *Opt Exp*. 2015;23:A1208. Doi: 10.1364/OE.23.0A1208.

183. Fan D., Burger T., McSherry S., Lee B., Lenert A., and Forrest S.R. Near-perfect photon utilization in an air-bridge thermophotovoltaic cell. *Nature*. 2020;586:237–241. Doi: 10.1038/s41586-020-2717-7.

184. Wilt D.M., Fatemi N.S., Jenkins P.P., Hoffman R.W., Landis G.A., and Jain R.K. Monolithically interconnected InGaAs TPV module development. *Proceedings of the Conference Record of the Twenty Fifth IEEE Photovoltaic Specialists Conference—1996*, Washington, DC; 13–17 May 1996, pp. 43–48.

185. Datas A., and Linares P. Monolithic interconnected modules (MIM) for high irradiance photovoltaic energy conversion: A comprehensive review. *Renew Sustain Energ Rev* 2017;73:477–495. Doi: 10.1016/j.rser.2017.01.071.

186. Wehrer R., Wanlass M., Wilt D., Wernsman B., Siergiej R., and Carapella J. InGaAs series-connected, tandem, MIM TPV converters. *Proceedings of the 3rd World Conference on Photovoltaic Energy Conversion*; Osaka, Japan; 11–18 May 2003, pp. 892–895.

187. Maremi H.H.C., Tolessa F., and Lee N., Choi G., and Kim T. Design of multilayer ring emitter based on metamaterial for thermophotovoltaic applications. *Energies* 2018;11:2299.

188. Gamel M.M.A., Ker P.J., Lee H.J., Rashid W.E.S.W.A., Jamaludin M.Z., and Mohammed A.I.A. Performance comparison of narrow bandgap semiconductor cells for photovoltaic and thermophotovoltaic application. *Proceedings of the 2020 IEEE 8th International Conference on Photonics (ICP)*, Kota Bharu, Malaysia; 12 May–30 June, 2020.

189. Xuan Y., Chen X., and Han Y. Design and analysis of solar thermophotovoltaic systems. *Renew Energy* 2011;36:374–387. Doi: 10.1016/j.renene.2010.06.050.

190. Zhou Z., Sakr E., Sun Y., and Bermel P. Solar thermophotovoltaics: Reshaping the solar spectrum. *Nanophotonics.* 2016;5:1–21. Doi: 10.1515/nanoph-2016-0011.

191. Cmih C.M.I.H., Norton B., Duffy A., and Oubaha M. Hybrid solar thermophotovoltaic-biomass/gas power generation system with a spectrally matched emitter for lower operating temperatures. *Proceedings of the 12th Conference on Sustainable Development of Energy, Water and Environment Systems*, Dubrovnik, Croatia; 4–8 October, 2017, p. 0563.

192. Davis G. Hybrid thermophotovoltaic power systems. California energy commission consultant report, 2002 US; P500-02-048F. (accessed: 17 August 2021). Available from: https://www.scribd.com/document/138324024/Hybrid-Thermophotovoltaic-Power.

193. Bauer T., Forbes I., and Pearsall N. The potential of thermophotovoltaic heat recovery for the UK industry. *Int J Ambient Energ* 2004;25:19–25. Doi: 10.1080/01430750.2004.9674933.

194. Yugami H. An overview of TPV research activities in Japan. *Proceedings of the AIP Conference Proceedings 738; TPV6: Sixth World Conference on Thermophotovoltaic Generation of Electricity*, Freiburg, Germany; 7 December, 2004, pp. 15–23.

195. Utlu Z., and Parali U. Investigation of the potential of thermophotovoltaic heat recovery for the Turkish industrial sector. *Energy Convers Manag* 2013;74:308–322. Doi: 10.1016/j.enconman.2013.05.030.

196. Bauer T., Forbes I., Penlington R., and Pearsall N. The potential of thermophotovoltaic heat recovery for the glass industry. *AIP Conf Proc* 2003;653:101–110.

197. Bitnar B., Durisch W., and Holzner R. Thermophotovoltaics on the move to applications. *Appl Energ* 2013;105:430–438. Doi: 10.1016/j.apenergy.2012.12.067.

198. Basu S., Chen Y.-B., and Zhang Z.M. Microscale radiation in thermophotovoltaic devices—A review. *Int J Energ Res* 2007;31:689–716. Doi: 10.1002/er.1286.

199. Fraas L., Ballantyne R., Samaras J., and Seal M. Electric power production using new GaSb photovoltaic cells with extended infrared response. *AIP Conf Proc* 1994;321:44–53.

200. Chan W.R., Stelmakh V., Waits C.M., Soljacic M., Joannopoulos J.D., and Celanovic I. Photonic crystal enabled thermophotovoltaics for a portable microgenerator. *J Phys Conf Ser* 2015;660:012069. Doi: 10.1088/1742-6596/660/1/012069.

201. Chan W.R., Stelmakh V., Karnani S., Waits C.M., Soljacic M., Joannopoulos J.D., and Celanovic I. Towards a portable mesoscale thermophotovoltaic generator. *J Phys Conf Ser* 2018;1052:012041. Doi: 10.1088/1742-6596/1052/1/012041.

202. Fraas L.M., Avery J.E., Daniels W.E., Huang H.X., Malfa E., Venturino M., Testi G., Mascalzi G., and Wuenning J.G. TPV tube generators for apartment building and industrial furnace applications. *AIP Conf Proc* 2003;653:38–48.

203. Qiu K., and Hayden A.C.S. Implementation of a TPV integrated boiler for micro-CHP in residential buildings. *Appl Energ* 2014;134:143–149. Doi: 10.1016/j.apenergy.2014.08.016.

204. Bianchi M., Ferrari C., Melino F., and Peretto A. Feasibility study of a thermo-photovoltaic system for CHP application in residential buildings. *Appl Energ* 2012;97:704–713. Doi: 10.1016/j.apenergy.2012.01.049.

205. Kilner J., Skinner S., Irvine S., and Edwards P. *Functional Materials for Sustainable Energy Applications*. Soston: Woodhead Publishing Limited; 2012.

206. De Pascale A., Ferrari C., Melino F., Morini M., and Pinelli M. Integration between a thermophotovoltaic generator and an organic Rankine cycle. *Appl Energ* 2012;97:695–703. Doi: 10.1016/j.apenergy.2011.12.043.

207. Morrison O., Seal M., West E., and Connelly W. Use of a thermophotovoltaic generator in a hybrid electric vehicle. *Proceedings of the Thermophotovoltaic Generation of Electricity, Fourth NREL Conference AIP Conference Proceeding*, Denver, CO; 11–14 October, 1998, pp. 488–496.

208. Schock A., and Kumar V. Radioisotope thermophotovoltaic system design and its application to an illustrative space mission. *AIP Conf Proc* 1995;321:139–152.

209. Wilt D., Chubb D., Wolford D., Magari P., and Crowley C. Thermophotovoltaics for space power applications. *AIP Conf Proc* 2007;890:335–345.

210. Teofilo V.L., Choong P., Chang J., Tseng Y.-L., and Ermer S. Thermophotovoltaic energy conversion for space. *J Phys Chem C* 2008;112:7841–7845. Doi: 10.1021/jp711315c.

211. Li Q., Shen K., Yang R., Zhao Y., Lu S., Wang R., Dong J., and Wang D. Comparative study of GaAs and CdTe solar cell performance under low-intensity light irradiance. *Sol Energ* 2017;157:216–226. Doi: 10.1016/j.solener.2017.08.023.

212. Wang X., Liang R., Fisher P., Chan W., and Xu J. Radioisotope thermophotovoltaic generator design methods and performance estimates for space missions. *J Propuls Power* 2020;36:593–603. Doi: 10.2514/1.B37623.

213. Strauch J.E., Klein A., Charles P., Murray C., and Du M. General atomics radioisotope fueled thermophotovoltaic power systems for space applications. *Proceedings of the 13th International Energy Conversion Engineering Conference*, Orlando, FL; 27–29 July, 2015, pp. 366–374.

214. Datas A., Ramos A., Marti A., del Cañizo C., and Luque A. Ultra high temperature latent heat energy storage and thermophotovoltaic energy conversion. *Energy*. 2016;107:542–549. Doi: 10.1016/j.energy.2016.04.048.

215. Seyf H.R., and Henry A. Thermophotovoltaics: A potential pathway to high efficiency concentrated solar power. *Energy Environ Sci* 2016;9:2654–2665. Doi: 10.1039/C6EE01372D.

216. Amy C., Seyf H.R., Steiner M.A., Friedman D.J., and Henry A. Thermal energy grid storage using multi-junction photovoltaics. *Energy Environ Sci* 2019;12:334–343. Doi: 10.1039/C8EE02341G.

217. Gamel M., Jern K.P., Rashid E., Jing L.H., Yao L.K., and Wong, B. Effect of front-surface-field and back-surface-field on the performance of GaAs based-photovoltaic cell. *Proceedings of the 2019 IEEE International Conference on Sensors and Nanotechnology*, Penang, Malaysia; 24–25 July, 2019, pp. 1–4.

218. Bumby C.W., Shields P.A., Nicholas R.J., Fan Q., Shmavonyan G., May L., and Haywood S.K. Improved efficiency of GaSb/GaAs TPV cells using an offset p-n junction and off-axis (100) substrates. *Proceedings of the Thermophotovoltaic Generation of Electricity: Sixth Conference on Thermophotovoltaic Generation of Electricity TPV6 (AIP Conference Proceedings)*, Freiberg, Germany; 15 December, 2004, pp. 353–359.

219. Krier A., Yin M., Marshall A., Kesaria M., Krier S., McDougall S., Meredith W., Johnson A., Inskip J., and Scholes A. Low bandgap mid-infrared thermophotovoltaic arrays based on InAs. *Infrared Phys Technol* 2015;73:126–129.

220. Sulima O.V., Bett A.W., Mauk M.G., Dimroth F., Dutta P.S., and Mueller R.L. GaSb-, InGaAsSb-,InGaSb-,InAsSbP-andGe-TPV cells for low-temperature TPV applications. *AIP Conf Proc* 2003;653:434.
221. Van Der Heide J., Posthuma N., Flamand G., Geens W., and Poortmans J. Cost-efficient thermophotovoltaic cells based on germanium substrates. *Sol Energ Mater Sol Cells* 2009;93:1810–1816.
222. Kaneko T., and Kondo, M. High Open circuit voltage and its low temperature coefficient in crystalline germanium solar cells using a heterojunction structure with a hydrogenated amorphous silicon thin layer. *Jpn J Appl Phys* 2011;50:23–26.
223. Nakano S., Takeuchi Y., Kaneko T., and Kondo M. Influence of surface treatments on crystalline germanium hetero junction solar cell characteristics. *J Non Cryst Solids* 2012;358:2249–2252.
224. Hitchcock C., Gutmann R., Borrego J., and Ehsani H. GaInSb and GaInAsSb thermophotovoltaic device fabrication and characterization. *Proceedings of the Thermophotovoltaic Generation Electricity Third NREL Conference*, Colorado Springs, CO; May 1997; Vol. 89.
225. Lim M., Lee S.S., and Lee B.J. Effects of multilayered graphene on the performance of near-field thermophotovoltaic system at longer vacuum gap distances. *J Quant Spectrosc Radiat Transf* 2017;197:84–94.
226. Lim M., Song J., Kim J., Lee S.S., Lee I., and Lee B.J. Optimization of an ear-field thermophotovoltaic system operating at low temperature and large vacuum gap. *J Quant Spectrosc Radiat Transf* 2018;210:35–43.
227. Zhao B., Santhanam P., Chen K., Buddhiraju S., and Fan S. Near-field thermophotonic systems for low-grade waste-heat recovery. *Nano Lett* 2018;18:5224–5230.
228. Fiorino A., Zhu L., Thompson D., Mittapally R., Reddy P., and Meyhofer E. Nanogap near-field thermophotovoltaics. *Nat Nanotechnol* 2018;13:806–811.
229. Datas A. Hybrid thermionic photovoltaic converter. *Appl Phys Lett* 2016;108:143503.
230. Bellucci A., Mastellone M., Serpente V., Girolami M., Kaciulis S., Mezzi A., Trucchi D.M., Antolin E., Villa J., García-Linares P., and Martí A. Photovoltaic anodes for enhanced thermionic energy conversion. *ACS Energ Lett* 2020;5:1364–1370.

# 8 Waste Heat to Power—Thermoelectricity

## 8.1 INTRODUCTION

Four major sectors for thermal energy consumption are power, industrial and manufacturing processes, heating and cooling residential and commercial building, and transportation. Although waste heat is generated in each case, the largest sources of waste heat are in industrial and manufacturing sector, power sector (Figure 8.1), and transportation sector. Waste heat generated in residential and commercial buildings is generally at low temperature and of low quantity to make them less economical to recover. Both quality and quantity of waste heat generated in industrial (manufacturing), power, and transportation sectors can be high, medium, or low but they can be recovered using a number of techniques. The applications of waste heat from these sources depend on the quality (temperature level, impurities involved, etc.) and quantity of waste heat (mass flow rate, specific heat, etc.).

The two most common approaches used to treat waste heat in power industry are combined cycles and cogeneration or trigeneration techniques. These techniques are applied for power generation from nuclear, solar, or combustion processes. The techniques and their modifications can significantly improve the thermal efficiency of power

**FIGURE 8.1** Coal-fired power station that transforms chemical energy into 36%–48% electricity and the remaining 52%–64% into waste heat [1, 3].

DOI: 10.1201/9781003328087-8

## TABLE 8.1
## Examples of Waste Heat Sources and End Uses [10]

| Waste Heat Sources | Uses for Waste Heat |
|---|---|
| • Combustion exhausts:<br>  Glass melting furnace<br>  Cement kiln<br>  Fume incinerator<br>  Aluminium reverberatory furnace boiler<br>• Nuclear reactor:<br>  Various applications of solar energy<br>  Various applications of geothermal energy<br>• Off-gases from internal combustion engine<br>• Process off gases:<br>  Steel electric arc furnace<br>  Aluminium reverberatory furnace<br>• Cooling water from: Furnaces<br>  Air compressors<br>  Internal combustion engines<br>• Conductive, convective, and radiative losses from equipment: Hall–Hèroult cells[a]<br>• Conductive, convective, and radiative losses from heated products:<br>• Hot cokes<br>• Blast furnace slags[a] | • Combustion air preheating<br>• Boiler feedwater preheating<br>• Load preheating<br>• Power generation<br>• Steam generation for use in:<br>  Power generation<br>  Mechanical power<br>  Process steam<br>• Space heating<br>• Water preheating<br>• Transfer to liquid or gaseous process streams<br>• Solar fuels<br>• Various applications of nuclear waste heat depending on temperature level<br>• Various applications of solar and geothermal energy depending on the temperature levels |

*Source:* Johnson and Choate [2].
[a] Not currently recoverable with existing technology.

generation processes. Cogeneration can have numerous applications, depending on the quality and quantity of heat. In all cogeneration applications, the proximity of the applied process is very important. The waste heat from the vehicle industry (using an internal combustion [IC] engine) can be used by numerous methods, which include combined cycles and co (or tri) generation methods. Lots of recovery technologies [1] and applications used for industrial waste heat are similar to the methods used in cogeneration processes.

The industrial sector accounts for approximately one-third of all energy used in the United States, consuming approximately 32 quadrillion Btu ($10^{15}$ Btu) of energy [2–5,1] annually and emitting about 1,680 million metric tons of carbon dioxide associated with this energy use. It is estimated that somewhere between 20% and 50% of industrial energy input is lost as waste heat in the form of hot exhaust gases, cooling water, and heat from hot equipment surfaces and heated products. As the industrial sector continues its efforts to improve its energy efficiency, recovering waste heat losses provides an attractive opportunity for an emission-free and less costly energy resource. Numerous technologies and variations/combinations of technologies are commercially available for waste heat recovery (WHR). Many industrial facilities [5–8] have upgraded or are improving their energy productivity by installing these technologies. However, heat recovery is not economical or even possible in many cases. Table 8.1 briefly summarizes examples of waste heat sources and end uses.

## TABLE 8.2
## Temperature Classification of Waste Heat Sources and Related Recovery Opportunity [10]

| Temp Range | Example Sources | Temp (°F) | Temp (°C) |
|---|---|---|---|
| Very high > 1,600°F | Electrical refractory furnace exhaust | 2,900 – 4,500 | 1,600 – 2,700 |
| High > 1,200°F | Nickel refining furnace | 2,500 – 3,000 | 1,370 – 1,650 |
| (> 650°C) | Steel electric arc furnace | 2,500 – 3,000 | 1,370 – 1,650 |
| | Basic oxygen furnace | 2,200 | 1,200 |
| | Aluminium reverberatory furnace | 2,000 – 2,200 | 1,100 – 1,200 |
| | Copper refining furnace | 1,400 – 1,500 | 760 – 820 |
| | Steel heating furnace | 1,700 – 1,900 | 930 – 1,040 |
| | Copper reverberatory furnace | 1,650 – 2,000 | 900 – 1,090 |
| | Hydrogen plants | 1,200 – 1,800 | 650 – 980 |
| | Fume incinerators | 1,200 – 2,600 | 650 – 1,430 |
| | Glass melting furnace | 2,400 – 2,800 | 1,300 – 1,540 |
| | Coke oven | 1,200 – 1,800 | 650 – 1,000 |
| | Iron cupola | 1,500 – 1,800 | 820 – 980 |
| Medium | Steam boiler exhaust | 450 – 900 | 230 – 480 |
| 450°F – 1,200°F | Gas turbine exhaust | 700 – 1,000 | 370 – 540 |
| (230°C – 650°C) | Reciprocating engine exhaust | 600 – 1,100 | 320 – 590 |
| | Heat treating furnace | 800 – 1,200 | 430 – 650 |
| | Drying and baking ovens | 450 – 1,100 | 230 – 590 |
| | Cement kiln | 840 – 1,150 | 450 – 620 |
| Low < 450°F | Exhaust gases exiting recovery devices in gas fired boilers, ethylene furnaces, etc. | 150 – 450 | 70 – 230 |
| (< 230°C) | Process steam condensate | 130 – 190 | 50 – 90 |
| | Cooling water from: Furnace doors | 90 – 130 | 30 – 50 |
| | Annealing furnaces | 150 – 450 | 70 – 230 |
| | Air compressors | 80 – 120 | 30 – 50 |
| | Internal combustion engines | 150 – 250 | 70 – 120 |
| | Air conditioning and refrigeration condensers | 90 – 110 | 30 – 40 |
| | Drying, baking, and curing ovens | 200 – 450 | 90 – 230 |
| Ultra low < 250°F | Hot processed liquids/ solids | 90 – 450 | 30 – 230 |
| | Kitchen, ventilation, fryer, condenser exhaust | 90 – 130 | 30 – 45 |
| | Cooling water from power plants | 60 – 140 | 15 – 50 |
| | Cooling water from air compressor | 75 – 140 | 25 – 50 |

*Source:* Johnson and Choate [2].

Sources of waste heat and their levels are also outlined in Table 8.2. The recovery of waste heat has some advantages and drawbacks. Very high-quality heat can be available for a diverse range of end uses with varying temperature requirements. However, they also create increased thermal stresses on heat exchanger materials.

High-temperature heat can generate power with significant thermal efficiency and can provide high heat transfer rate per unit area with greater compatibility with heat exchanger materials [9]. It, however, can increase chemical activity with materials and can induce higher corrosion, particularly, if the heat is in the so-called *harsh environment*. Medium- and low-temperature heats are practical for power generation, however, with lower efficiency. The uses of low-temperature heat are fewer and sometimes impractical. The quantity of low-temperature heat, however, can be very large.

The steps to improve industrial energy efficiency focus on two paths: (a) reducing the energy consumed by the equipment that is used in manufacturing (e.g., boilers, furnaces, dryers, reactors, separators, motors, and pumps) and (b) changing the processes or techniques to manufacture products. A valuable alternative approach to improving overall energy efficiency is to capture and reuse the lost heat or *waste heat* that is intrinsic to all industrial manufacturing. During these manufacturing processes, as much as 20%–50% of the energy consumed is ultimately lost via waste heat contained in streams of hot exhaust gases and liquids (and sometimes solids), as well as through heat conduction, convection, and radiation from hot equipment surfaces and from heated product streams. In some cases, such as industrial furnaces, efficiency improvements resulting from WHR can improve energy efficiency by 10% to as much as 50% [8,11].

Although some waste heat losses from industrial processes are inevitable, facilities can reduce these losses by improving technologies. WHR entails capturing and reusing the waste heat in industrial processes for heating or for generating mechanical or electrical work. Examples of waste heat use include generating electricity, preheating combustion air, preheating furnace loads, feedwater preheating, transfer to low- and medium-temperature processes, absorption cooling, domestic water heating, upgrading heat via a heat pump to increase the temperature for end use, and space heating, among others. Power generation at low temperature can be facilitated using organic Rankine cycle (ORC) or using newly developed thermoelectric or thermophotovoltaic generators for direct transfer of heat to electricity.

Captured and reused waste heat is an emission-free substitute for costly purchased fuels or electricity. Numerous technologies are available for transferring waste heat to a productive end use. Nonetheless, anywhere from 513 quadrillion Btu/year of waste heat energy remains unrecovered as a consequence of industrial manufacturing. Three essential components are required for WHR: (a) an accessible source of waste heat, (b) a recovery technology, and (c) use for the recovered energy. Large energy-consuming processes (totaling 8,400 trillion Btu/year or TBtu/year) identify unrecovered waste heat losses in exhaust gases totaling ~1,500 TBtu/year. Industrial manufacturing facilities will invest in WHR only when it results in savings that yield a reasonable payback period (less than 3 years) and when the perceived risks are negligible. Heat recovery technologies frequently reduce the operating costs for facilities by increasing their energy productivity [1,5–9,11].

## 8.2 FACTORS AFFECTING WASTE HEAT RECOVERY

Waste heat losses arise both from equipment inefficiencies and from thermodynamic limitations on equipment and processes. For example, consider a reverberatory furnace that is frequently used in aluminum melting operations. Exhaust gases leaving the furnace can have temperatures as high as 2,200°F–2,400°F (1,200°C–1,300°C) or in some specific cases even higher. Consequently, these gases have high heat content, carrying away as much as 60% of furnace energy inputs [2,12,13]. Efforts can be made to design more energy-efficient reverberatory furnaces with better heat transfer and lower exhaust temperatures; however, the laws of thermodynamics place a lower limit on the temperature of exhaust gases. As heat exchange involves energy transfer from a high-temperature source to a low-temperature sink, the combustion gas temperature must always exceed the molten aluminum temperature in order to facilitate aluminum melting. The gas temperature in the furnace will never decrease below the temperature of the molten aluminum because this would violate the second law of thermodynamics. Therefore, the minimum possible temperature of combustion gases immediately exiting an aluminum reverberatory furnace corresponds to the aluminum pouring point temperature of 1,200°F–1,380°F (650°C–750°C). In this scenario, at least 40% of the energy input to the furnace is still lost as waste heat.

Recovering industrial waste heat can be achieved via numerous methods. The heat can either be *reused* within the same process or transferred to another process. Ways of reusing heat locally include using combustion exhaust gases to preheat combustion air or feedwater in industrial boilers. By preheating the feedwater before it enters the boiler, the amount of energy required to heat the water to its final temperature is reduced.

Alternately, the heat can be transferred to another process; for example, a heat exchanger could be used to transfer heat from combustion exhaust gases to hot air that is needed for a drying oven. In this manner, the recovered heat can replace fossil energy that would have otherwise been used in the oven. Such methods for recovering waste heat can help facilities significantly reduce their fossil fuel consumption, as well as to reduce associated operating costs and pollutant emissions. Typical sources of waste heat and usage options are listed in Tables 8.1 and 8.2. Combustion air preheat can increase furnace efficiency by as much as 50% as shown in Table 8.3. The quality of waste heat is often dictated by its temperature. Five levels of temperature for waste heat normally used in practice are described in Table 8.2.

**TABLE 8.3**
**Furnace Efficiency Increases with Combustion Air Preheat [10]**

| Furnace Outlet Temperature | Combustion Air Preheat Temperature | | |
|---|---|---|---|
| | 400°F (204°C) (%) | 800°F (427°C) (%) | 1,200°F (649°C) (%) |
| 2,600°F (1,427°C) | 22 | 37 | 48 |
| 2,200°F (1,204°C) | 16 | 29 | 39 |
| 1,800°F (982°C) | 13 | 24 | 33 |
| 1,400°F (760°C) | 10% | 20% | 28% |

*Source:* EPA, Wise Rules for Energy Efficiency. Based on a natural gas furnace with 10% excess air, 2003; Johnson and Choate [2].

## 8.3  WASTE HEAT TO POWER BY THERMODYNAMIC CYCLES

The generation of power from waste heat follows two paths: indirect method that uses thermodynamic cycles to generate mechanical energy from heat, which in turn generates electrical energy through turbine and generator and direct method in which thermal energy is directly converted to electrical energy via photovoltaic, thermoelectric, thermophotovoltaic, electro-chemical, piezoelectric, thermo-galvanic, pyro-electric, thermionic and other processes. Most progresses are made in first three processes. Photovoltaic process was described in detail in Chapter 6. Thermophotovoltaic process was described in details in Chapter 7. Electrochemical process will be described in detail in my next book. This chapter describes in detail thermochemistry and its use in power generation in Sections 8.4–8.6.

For indirect methods, various implementation pathways can be adopted depending on the quality of waste heat sources. The main approach is to utilize an appropriate thermodynamic cycle to convert the waste heat into useful power. Two important indicators for choosing a proper cycle are the operation temperature and level of power generation. Figures 8.2a and b depict the possible thermodynamic cycles and the ranges of their relevant operation temperature and level of power. Figure 8.2b shows the expanded range of temperature for some of the cycles shown in Figure 8.2a.

The most suitable waste heat temperature range for the Steam Rankine Cycle (SRC) is medium-high temperature, at about more than 250°C. Steam Rankine cycles are usually preferred to ORC to exploit waste heat sources with higher thermal capacities (from 10 MW up to hundreds of MW) [16], because of its higher efficiency and the lower capital cost resulting from the use of more standard components [16]. The operating range of this technology can be further extended to power scales lower than 10 MW using micro steam turbines (Figure 8.2b) which are however characterized by lower performance than large machines due to high tip leakage losses [16]. The temperature range at which the Steam Rankine technology is usually employed goes from 250°C up to 700°C (Figure 8.2a and b) [16,17]. The lower limit is given by the low vapor pressure of water, while the upper one from material and technological constraints. More advanced units, such as the ultra-supercritical steam power systems, can also exploit heat sources beyond 620°C, but they require significant additional investment costs [16].

Systems for lower temperature heat sources are much less cost-effective and may lead to surface corrosion problem. For low-temperature waste heat, the Organic Rankine Cycle (ORC) which uses lower boiling point organic fluids, has been extensively investigated in the last few decades. Although Organic Rankine Cycle (ORC) systems proved to be a successful technical solution, especially for large-scale applications [18], the use of this technology is limited in a range of temperatures of the waste heat source that goes from 90°C–100°C up to 250°C–400°C [19]. The upper limit is imposed by the flammability and low chemical stability of the organic fluids at high temperatures, while the lower one is by their vapor pressure which, in turn, limits the efficiency and the output of ORC units at extremely low temperatures. For waste source capacities from 10 up to 200 kW, ORC systems equipped with positive displacement machines are preferred to the ones with axial or centrifugal turbines [20] (Figure 8.2b). In fact, in this power range, volumetric machines can achieve higher efficiencies compared to dynamic ones, whose reduced size leads to losses.

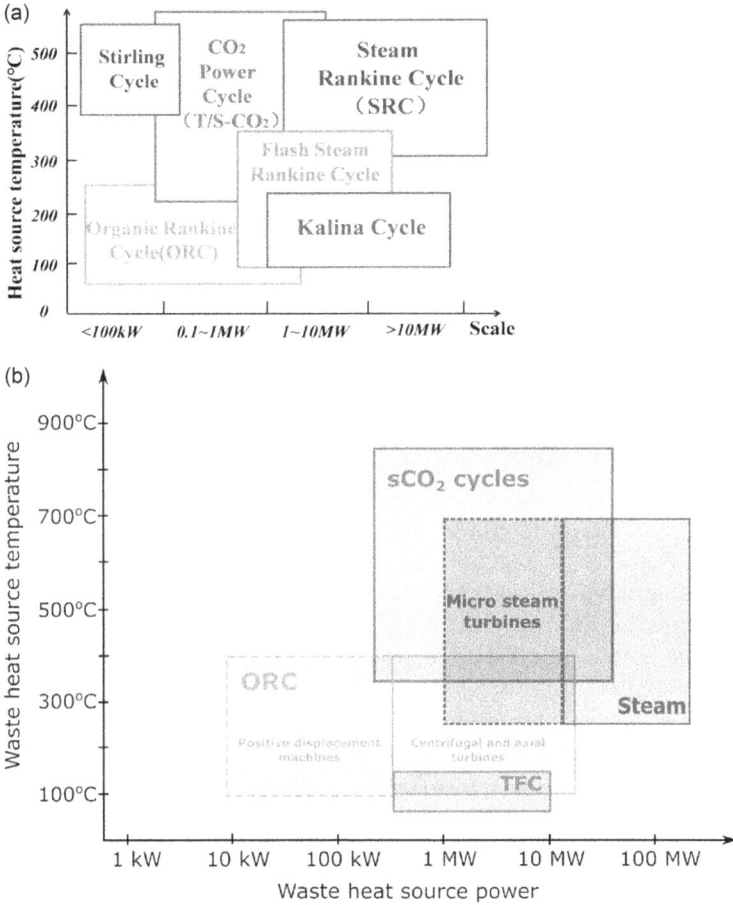

**FIGURE 8.2**   Ranges of various cycles [14,15]. (a) Low temperature range. (b) High temperature range.

Furthermore, positive displacement machines benefit from a reduced installation and maintenance complexity due to lower revolution speeds, reduced vibration levels and a wider range of optimal operating conditions [21]. For power scales between 200 kW and 15 MW (the larger ORC installation to date [22]), turbines are adopted since their size can be increased with consequent benefits in efficiency [21].

The Kalina Cycle uses a mixture of ammonia and water as working fluid to closely meet the temperature profile of waste heat sources during the phase change heat transfer process, between 100°C and 250°C. Waste heat source available at temperature levels lower than 100°C can be still exploited by adopting the Trilateral Flash Cycle (TFC) technology (Figure 8.2b). In these kinds of systems, the organic working fluid is heated until the saturated liquid conditions which then undergoes a two-phase expansion [23]. For these reasons, these units are suitable for ultra-low temperature

WHR applications, from 200°C down to 70°C. Furthermore, because of the two-phase expansion, volumetric machines are usually adopted since they guarantee a higher adiabatic efficiency. The size of these machines, however, limits the maximum thermal capacity of the waste heat source exploitable, which can go up to 5 MW [24,25]. For capacities lower than 1 MW, ORC systems are more competitive [24].

In recent years, the use of $CO_2$ power cycles for waste heat recovery has gained more and more attention. Such cycles are usually operated under trans-critical or supercritical conditions, due to the relatively low critical point of carbon dioxide. The $CO_2$ cycles are suitable for the recovery of waste heat sources with a wide range of temperatures [23], while the steam-based Rankine cycle is only efficient to recover waste heat at high temperatures. As shown in Figure 8.2a, the $CO_2$ cycle can replace SRC at high temperatures [17] particularly for smaller level of power generation. The special physical properties of $CO_2$ can also reduce the heat transfer loss between working fluid and heat source [24] and offer a wide range of temperature. Furthermore, the $CO_2$ is a non-combustible and nontoxic refrigerant which frees up space for the cycle operation temperature to rise [25,26]. In addition, the s-$CO_2$ power cycles occupy a smaller space, provide higher efficiency and show great potential for system downsizing and lower weight as compared with SRC, which has a large working fluid steam volume. This also makes the chemistry and condensing processes for s-$CO_2$ cycle more simpler [26,27]. Hence, s$CO_2$ systems fill an important gap in industrial WHR applications (Figure 8.2b). For waste heat source temperatures higher than 700°C, s$CO_2$ power cycles are the only option available and can thus constitute a breakthrough in the sector (Figure 8.2b). The high chemical stability of $CO_2$ allows to directly recover and convert heat at temperatures up to 850°C (Figure 8.2b), which is a limit posed by current materials [26]. The lower limit is instead set at 350°C considering a simple regenerated layout, which represents the most convenient option for WHR applications [27]. For such low cost systems, characterized by a low cycle pressure ratio, the achievement of higher temperatures at the turbine inlet to obtain a positive net electric output is required [27].

From a power scale perspective, several technological challenges and the high investment costs set the lowest feasible unit capacity at 50 kWe [28], which corresponds to a waste source thermal power of 300 kW assuming a 20% system thermal efficiency [27]. Among the technical limitations, the main ones arise from the reduced size of the turbomachines. Typical wheel diameters range from 30 to 50 mm, with consequent issues of leakage, high vibrations level and friction due to the elevated revolution speeds (over 60,000 RPM) [28,29]. On the other side, it is possible to scale up s$CO_2$ systems to tens of MW (Figure 8.2b), as has been proven by the SunShot program and Echogen units [30,31]. In this case, technological limitations arise in the scale-up of heat exchangers. The scale-up of turbomachinery can benefit from the knowledge acquired in the gas turbine and steam power plants sectors.

### 8.3.1 STEAM RANKIN CYCLE

The most frequently used system for power generation from waste heat involves using the heat to generate steam, which then drives a steam turbine. The traditional steam Rankine cycle [32] is the most efficient option for WHR from exhaust streams with temperatures about 650°F–700°F (340°C–370°C) [2,33]. It is used for waste heat from gas turbines, reciprocating engines, incinerators, and furnaces. At lower waste heat temperatures, steam cycles become less cost-effective because low-pressure steam will require bulkier equipment. Moreover, low-temperature waste heat may not provide sufficient energy to superheat the steam, which is a requirement for preventing steam condensation and erosion of the turbine blades. Therefore, low-temperature heat recovery applications are better suited for ORC [2,34–42] or Kalina cycle [43,44], which uses fluids with lower boiling point temperatures compared to steam. Evaporator waste heat from process turbine condenser generates electricity for driving pump for grid [2,34–37].

## 8.3.2 KALINA CYCLE

The Kalina cycle is a variation of the Rankine cycle, using an azeotropic mixture of ammonia and water as the working fluid. In Chapter 4, we examined the use of Kalina cycle to convert low temperature geothermal energy to power. Here we briefly assess its role for conversion of waste heat to power. As indicated earlier, Kalina cycles use binary fluids of water and ammonia. A key difference between single fluid cycles and cycles that use binary fluids is the temperature profile during boiling and condensation. For single fluid cycles (e.g., steam or organic Rankine), the temperature remains constant during boiling. As heat is transferred to the working medium (e.g., water), the water temperature slowly increases to boiling temperature at which point the temperature remains constant until all the water has evaporated. In contrast, a binary mixture of water and ammonia (each of which has a different boiling point) will increase its temperature during evaporation. This allows better thermal matching with the waste heat source and with the cooling medium in the condenser, resulting in significantly greater energy efficiency.

Kalina cycles can be used for the waste heat coming out of gas turbine exhaust, boiler exhaust, and cement kilns exhaust. The Kalina cycle is an innovative bottoming cycle in which the temperature of the cycle tracks the temperature of the turbine exit in the waste heat boiler. Nevertheless, during the condensing processes, the thermodynamic gain of the relatively small boiler temperature difference compared to a steam cycle would be gone. A basic Kalina cycle consists of a WHR vapor generator (in this case it is the HEX where the exhaust gas of the gas turbine is directed to supply heat to the bottoming Kalina cycle), a turbine which works with steam-ammonia, and the distillation condensation system. In the distillation condensation subsystem, first the flow coming from the turbine is cooled by the heater (recuperator), and then the stream is mixed with a lean solution of $NH_3$ to increase the condensation temperature of the working fluid. Finally, the basic solution is condensed in the absorber. The condensed solution is brought to the heater under pressure. A portion of the stream is directed to dilute the ammonia-rich stream coming from the separator. The primary flow passes the recuperator, then it is flashed in the separator. The vapor is mixed with the basic solution. The vapor is condensed, then pressurized by the pump before it flows to the vapor generator. The literature has shown [45,47,48] 10%–30% more energy can be generated by the Kalina cycle when it is compared to a Rankine cycle. Furthermore, Park and Sontag [46] presented a case study showing that the

Kalina cycle exergy efficiency is 15% higher as compared to the steam power cycle. The main reason for the advantage of Kalina cycle is the fact that the exhaust pressure of the Kalina cycle is above the atmospheric conditions. The starting time of a Kalina cycle is much less because sustaining of vacuumed medium is not a necessity for the condenser during the operation of the cycle. The working fluid mixture can be easily varied to acquire the best performance with respect to changes in load or ambient condition [45–48].

The cycle was invented in the 1980s, and the first power plant based on the Kalina cycle was constructed in Canoga Park, California in 1991. It has been installed in several other locations for power generation from geothermal energy or waste heat. Applications include 6 million metric tons/year steelworks in Japan and heat recovery from a municipal solid waste incinerator and from a hydrocarbon process tower [2,34–42]. The steel works application involved using a Kalina cycle [43,44] to generate power from cooling water at 208°F (98°C). In this case, with a water flow rate of 1,300 metric tons/h, the electric power output was about 4,500 kW. The total investment cost was about $4 million or about $1,100/kW [2].

### 8.3.3 Organic Rankine Cycle

Chapter 4 described how low-temperature geothermal heat can be converted to power using organic Rankine cycle. The same principle applies to low-temperature waste heat. The conversion of low-temperature waste heat to power increases the overall efficiency of the plant, reduces energy cost, increases plant revenue and decreases the use of fossil fuel and resulting $CO_2$ emission. The working principle of the ORC is the same as that of the Rankine cycle: the working fluid is pumped to a boiler where it is evaporated, passed through an expansion device (turbine or another expander), and then through a condenser heat exchanger where it is finally recondensed. In the ideal cycle described by the engine's theoretical model, the expansion is isentropic, and the evaporation and condensation processes are isobaric.

Since ORC requires heat recovering at lower temperature compared to the steam Rankine cycle, it uses lower boiling point organic compounds such as a refrigerant, a hydrocarbon such as pentane, butane, perfluorocarbon or silicon oil. While ORC was first used in 1970s and 1980s, *currently, more than 200 ORC* power plants are being operated generating more than 1,800 MWe power using biomass, geothermal energy and waste heat. The layout of ORC is much simpler compared to the steam cycle as there is no water vapor attached to the boiler, and a single heat exchanger could be utilized for the three processes of evaporation including preheating, vaporization and superheating. ORC is able to use low-grade heat sources than steam Rankine cycle [49]. Since it could be utilized at a lower temperature at the turbine inlet, the process reduces thermal stresses in the boiler. Since organic fluids have condensation pressure higher than the atmospheric pressure, the infiltration of non-condensable gases in the condenser is avoided.

ORC plant is simple and less costly compared to steam-based plant. Due to the small difference in liquid and vapor densities of organic working fluids, the system can use once through boilers and avoid the use of steam drums. The use of deaerator is not necessary. Unlike in steam-based cycle, in ORC the usage of deaerator is not necessary. Higher organic fluid density fluid allows the use of compact appliances, especially in marine application where the available space for recovery plant of waste heat is restricted [50]. Enthalpy drop in ORC is much lower compared to steam cycle. The process in ORC can be done in a single stage with much simpler turbine compared to steam cycle which requires turbine with some expansion stages. ORC is normally operated at much lower pressure levels and pressure rarely exceeds 30 bars. Overall, ORC is beneficial in low to medium power range due to its cycle simplicity, less cost, low-stress level at boiler, easy control and simpler usage of components [51].

The fluids used in ORCs have a higher molecular mass, enabling compact designs, higher mass flow, and higher turbine efficiencies (as high as 80%–85%. However, as the cycle functions at lower temperatures, the overall efficiency is only around 10%–20%, depending on the temperature of the condenser and evaporator compare to 30%–40% efficiency achieved in high-temperature steam cycle. As shown in Chapter 4, the selection of the working fluid is of key importance in low-temperature Rankine cycles [50,52], because heat transfer depends very strongly on the thermodynamic characteristics of the fluid and on the operating conditions. The recovery of low-grade heat also requires low boiling point fluids such as refrigerants and hydrocarbons. Optimal characteristics of working fluid should include: (a) isentropic saturation vapor curve, (b) low freezing point and high-stability temperature, (c) high heat of vaporization and density, (d) low environmental impact, (e) safety, (f) good availability and low cost, and (g) acceptable pressures.

Three possible working fluids are HFCs (e.g., R134a R245fa), HCs (which are flammable, common by-products of gas processing facilities; e.g., isobutane, pentane, and propane), and PFCs [53]. Some of these were evaluated in Chapter 4. Yamamoto et al. [54] experimentally demonstrated that organic substances used in Rankine cycle can give higher turbine power than water when the turbine inlet temperature is below 120°C. As more ORCs enter commercial stage [55], researchers have mainly focused on working fluid selection [56–58], cycle design optimization [59–61] and expander technologies [62–64]. Borsukiewicz-Gozdur et al. [56] recommended that several pure organic fluids such as R134a, R123, R245ca and R245fa are suitable for the ORC system. Heberle et al. [57] presented several azeotropic mixtures used in the ORC system. The study indicated that mixtures lead to an efficiency increase of up to 15% when heat source temperatures is below 120°C. Mago et al. [58] proposed seven kinds of organic fluids performance in the different temperature ranges: R113 is suitable for a system with heat source temperature higher than about 157°C; isobutane has better performance for low-temperature heat source lower than 107°C; for low-grade heat source between 107°C and 157°C, R123, R245ca and R245fa have better performance. T-s diagram of the water saturation curves of water and a few typical ORC organic fluid applications are illustrated in Figure 8.3 (50).

As illustrated in Chapter 4, Yekoladio et al. [59] experimentally compared the regenerative and non-regenerative organic Rankine cycles and demonstrated that the power output increased exponentially with the geothermal resource temperature. In order to optimize the regenerative ORCs, the lower vapor specific heat capacity organic fluids were needed. Branchini et al. [60] developed a performance calculation method based on cycle efficiency, specific work, recovery efficiency, ORC fluid-to-hot source mass flow ratio, turbine volumetric expansion ratio and heat exchangers size parameter and provided useful guidelines to select the most appropriate fluid,

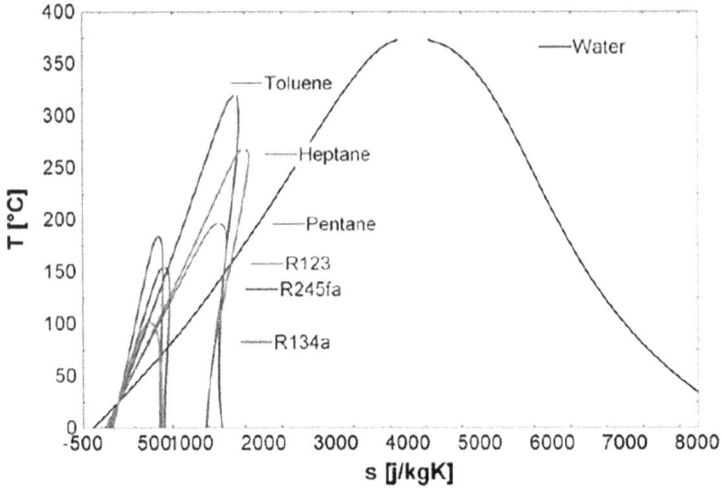

**FIGURE 8.3**  T-s diagram of the water saturation curves of water and a few typical ORC organic fluids applications [50].

the ORC configuration and operating parameters. Shengjun et al. [61] compared the subcritical ORC and transcritical power cycle systems for low-temperature geothermal power generation. The study showed that R123 yielded the maximum value of thermal efficiency and exergy efficiency in subcritical ORC system and R125 showed excellent economic performance in transcritical power cycle. Lemort et al. [62] experimentally studied the performance of the closed scroll expander using R245fa as the working fluid. The experimental results showed that the maximum output power was 2.2 kW and the expander efficiency was up to 71.03%. Hu et al. [63] showed that for a twin-screw expander an increase in rotational speed resulted in a greater loss of suction pressure and the associated efficiency. For the radial turbine as the expander and R245fa as the working fluid in the ORC system, Kang [64] showed that the output power of the system can reach 32.7 kW and the system efficiency and the turbine efficiency can reach 5.22% and 78.7%, respectively. Saadon and Islam [50] presented an excellent review of literature studies of ORC.

As illustrated in Chapter 4, ORC can be operated in a subcritical or supercritical cycle. In the supercritical cycle, the working fluid evaporation ends in the supercritical area and the heat rejection in the condenser occurs in the subcritical area. Many studies have been performed on the supercritical ORC. Yagli et al. [65] modeled subcritical and supercritical ORC to recuperate exhaust gas waste heat of biogas-fueled CHP engine. Comparing with subcritical condition, supercritical ORC showed greater performance. Guo et al. [66] studied the subcritical and transcritical ORC performance in regards to the evaporator pinch point locations. The study indicated that transcritical ORCs gave higher performance as the heat source outlet temperatures decreased. Ran et al. [67] showed that in transcritical ORCs, the thermophysical properties of the working fluid work at supercritical coefficient and logarithmic mean temperature difference (LMTD). Moloney et al. [68] indicated that the supercritical cycle is much more efficient than a subcritical cycle to optimize plant efficiency. The simulation of supercritical state from vehicle exhaust carried out by Chowdury et al. [69] showed that the key in transforming the operating temperature at the evaporator outlet was to modify the mass flow rate.

Regenerative ORC is formulated when ORCs and turbine bleeding are integrated to a heat exchanger. The cycle heats up the working fluid upon infiltrating the evaporator which is almost similar to the ORC with recuperator. Figure 8.4a and b provide the schematic cycle and T-s diagram of regenerative cycle. Le et al. [70] indicated that regenerative cycle gave greater efficiencies compared to simple cycle. Moloney et al. [68] studied the environmental fluids with critical temperature below 200°C in regenerative supercritical ORCs to upgrade the geothermal energy efficiency and noted that $CO_2$ operates the best. The same principle was applied by Muhammad et al. [71] to the basic ORC and single and double-stage regenerative ORC for recovering waste heat. Studies showed that the single and double-stage regenerative ORC has greater thermal efficiency with lower economic performance compared to the basic ORC.

Saadon and Islam [50] point out that in general, superheating of dry fluid negatively affects the ORC's efficiency while wet fluid positively affects the ORC's efficiency and isentropic fluid did not really affect ORC. The study by Guo et al. [66] indicates that even for dry working fluid, superheating is essential. Li et al. [72] found that for a small-scale ORC system with low-grade heat source to produce electricity, the fluid of ORC during superheat and pressure at the turbine inlet were two main variables they were able to manage with temperature of heat source and speed of the ORC pump. Roy et al. studied the consequences of level of superheat on the performance of ORC system [73]. Zhang et al. [74] indicated that the thermo-economic performance of internal heat exchanger ORC with dry condition surpasses the wet fluid as temperature of heat source load increases. Brizard et al. [51] suggested that for preventing condensation drops during the operation of superheating, the inlet of expander must exceed 20°C. Radulovic et al. [75] mentioned that superheat is important in cycle, especially in wet fluids. As the temperature of superheater rises, the cycle efficiency also rises and the chance of the working fluid condensation during pressure drop inside turbine, which results in lesser corrosion and efficiency drop.

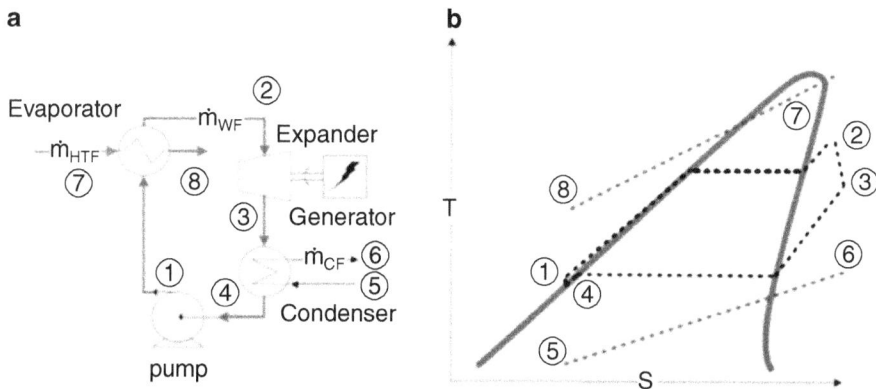

FIGURE 8.4    (a) ORC with regenerator and its (b) T-s diagram [50,70].

In order to get higher efficiencies and net power output, superheating is important. Feng et al. [76] found higher superheating reduces the overall heat transfer area of evaporator and thereby reduces the investment cost of the system. Li et al. [72] concluded that superheat is crucial in assuring an efficient and safe system operation of small ORC. Bianchi et al. [77] examined a micro-ORC setup for low-temperature application and concluded that efficiency increases as degree of superheating decreases. Ismail et al. [78] concluded that utilizing superheated vapor in the system with internal heat exchanger results in increasing of thermal efficiency of ORC. Superheating was essential to lower the mass flow rate and enhanced the performance of the system with the presence of internal heat exchanger.

Saadon and Islam [50] analyzed various applications of ORC for waste heat. Here we examine a few applications. The following summary follows the analysis of Saadon and Islam [50].

### 8.3.3.1 Applications of Organic Rankine Cycle for Waste Heat

Saadon and Islam [50] analyzed various applications of ORC for waste heat. The main emphasis in these applications was to generate cogeneration systems making use of waste heat. The applications include the use of waste heat from biomass combustion or gasification, recovery of geothermal energy, capturing solar radiation, waste heat from industrial processes and waste heat from aircraft engines. In each case, the use of ORC for waste heat improved the efficiency of the process.

The ORC technology has been utilized for heat to power in numerous applications. In recent years, biomass heating is actively used as a cogeneration system. Figures 8.5 and 8.6 illustrate the energy flows and the working principle respectively of such a cogeneration system as illustrated by Saadon and Islam [50]. In this illustration, in the range of temperature from 150°C to 320°C, the heat from flue gases is transmitted to the working fluid in two heat exchangers. When the temperature drops a little below 300°C, heat transfer fluid (thermal oil) is sent to ORC loop to evaporate the working fluid. Then, the evaporated fluid expands, to preheat the liquid using

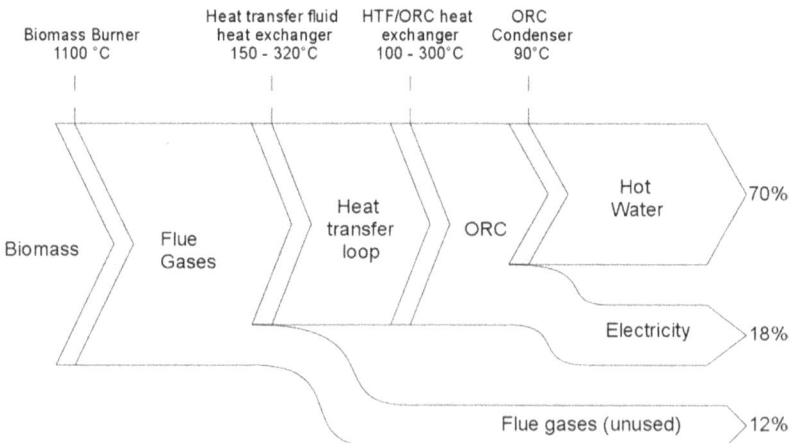

**FIGURE 8.5** Energy flows in a CHP system of biomass [40,50].

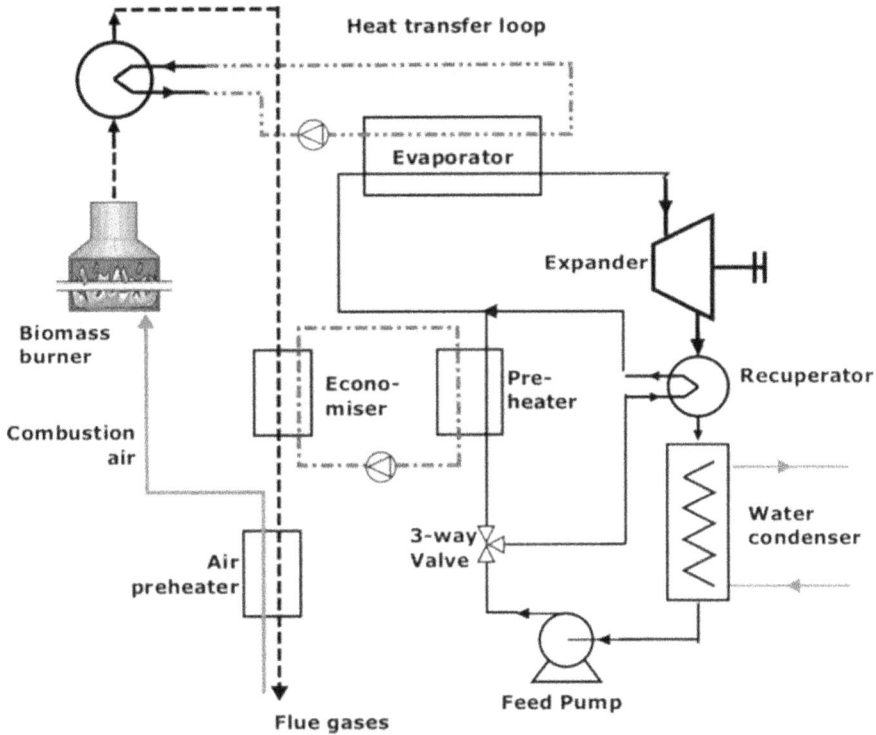

**FIGURE 8.6**   Biomass CHP ORC system working principle [40,50].

recuperator and when temperature reached 90°C, the fluid condensed to produce hot water. ORC efficiency is less compared to traditional steam cycles and gradually reduces for small-scale units. The overall conversion efficiency can be improved by using waste heat for space heating or industrial processes. Load of plant could be managed through on-site heat request or maximizing power generation. As shown in Figure 8.5, even though the CHP system's electrical efficiency is somewhat less (18%), the overall system efficiency is 88% greater than centralized power plants where most residual heat is lost. This technology can be compared with biomass gasification. The study showed that gasification results in higher investment costs (75%), higher maintenance costs (200%) and more power-to-thermal ratio. ORC is an established technology, while gasification plants are still being developed.

As shown in Chapter 4, ORC can also be used to recover low-temperature geo-thermal heat sources and convert them into power. Geothermal heat sources range from 10°C to 300°C. The actual lower technological limit to generate electricity is about 80°C and became less efficient with a temperature less than 80°C which results in uneconomical geothermal power plants. The high potential of geothermal energy in Europe is demonstrated in Table 8.4 [50]. For an efficient geothermal heat recovery, boreholes need to be drilled in the ground at a reasonable depth in order

TABLE 8.4
Geothermal Energy Potential in Europe for Different
Temperature Ranges of Heat Sources [50,79]

| Temperature (°C) | MWth | MWe |
|---|---|---|
| 65–90 | 147,736 | 10,462 |
| 90–120 | 75,421 | 7,503 |
| 120–150 | 22,819 | 1,268 |
| 150–225 | 42,703 | 4,745 |
| 225–350 | 66,897 | 11,150 |

to recover heat at an acceptable temperature. The degree and nature of Boreholes depend on the configuration of the geology and the cost of drilling [79]. At low temperature, pumps consume 30%–50% of gross power output [80]. Geothermal heat sources temperature (>150°C) allows for CHP, where the condensing temperature is limited to at least 60°C, which enables its district heating uses.

In a solar power plant electricity is generated when solar heat is transmitted to a power cycle. Point concentrating technologies consist of parabolic dishes and solar towers, which results in more concentration factor and greater temperatures. For solar towers, the Stirling engine (small-scale plants), the steam cycle or even the combined cycle is the best-suited power cycles. Parabolic troughs operate at a lower temperature (300°C–400°C). Up till now, they are combined to traditional steam Rankine cycles to generate electricity [81]. Sadoon and Islam [50] point out that the Organic Rankine cycle is a favorable technology that could lower the investment costs by working at lower temperatures and reducing total installed power to kW scale. Since Fresnel linear technology needs lower investment costs [82], they are suitable for ORCs operating at a lower temperature. So far only a few CSP plants with ORC are accessible on the market. In 2006, in Arizona, a 1 MWe solar concentration of ORC power plant was established. The ORC module utilized n-pentane as the working fluid with 20% efficiency. The overall solar energy efficiency was 12.1% [83]. Few small-scale operations for the applications of the remote-off grid have been studied. The only proof of concept obtained is that 1KWe system was installed for rural electrification in Lesotho by "STG International." The objective of this project was to integrate small-size solar thermal technology with medium-temperature collectors and an ORC to obtain the economics equivalent to big solar thermal installation.

Saadon and Islam [50] point out that at low temperature, most of the applications in the manufacturing industry reject waste heat because for large-scale plants, this heat cannot be used for on-site district heating. The release of waste heat causes pollution. The conversion of this waste heat to power by ORC can reduce pollution, provide on-site electricity or send it back to the grid. Normally, waste heat is recuperated through an intermediate heat transfer loop in such a system and used to evaporate the cycle's working fluid. In the USA, power generation from industrial waste heat sources is approximately about 750 MWe [84]. Some industries have greater potential in the recovery of waste heat. One of them, the cement industry [85], loses 40% of flue gas heat. These flue gases are in the temperature range of 215°C–315°C after the preheater of limestone or in the clinker cooler [24]. $CO_2$ released from the cement industry is 5% of the world's total $CO_2$ emissions, half of the results from fossil fuel combustion in kilns [86]. Other possible industries include iron and steel industries (for example, 10% of $CO_2$ emissions in China), refineries or chemical industries. Although their potential is higher and cost-effective (1,000–2,000 €/kWe), ORC recovery waste heat power cycles constitute only 9%–10% of the world's installed ORC plants. These are considerably smaller compared to biomass CHP and geothermal units [52].

Perullo et al. [87] integrated an ORC to an aircraft engine for power generation from waste heat. They indicated that with an aircraft engine, as bypass ratio keep on growing and the engine cores become effective, the diameter of engine fan increases and the core size decreases which causes pneumatic offset needing a greater percentage of the core flow and results in higher performance penalties. With the idea of no-bleed aircraft, performance penalties for shrinking cores and increased fan diameter are supposed to be eliminated. According to Saadon and Islam [50], ORC was used in the study due to the low temperature. The WHR system was placed in the core jet exhaust of a turbofan engine. The amount of heat extracted from the engine should be considered to avoid a reduction of thrust. The ORC WHR system was distributed in the nozzle, the nose cowl and the Pylon. It used R245fa as the working fluid having demonstrated the highest thermal efficiency in a wide range of operating pressure. The MathCAD 2001 software was used to model the design. Figure 8.7 describes the ORC schematics. Based on their analysis, Perullo et al. [87] concluded that an ORC WHR system could produce more power on the existing engine and can be utilized to supply sufficient power to a compressor driving air to the ECS (environment control

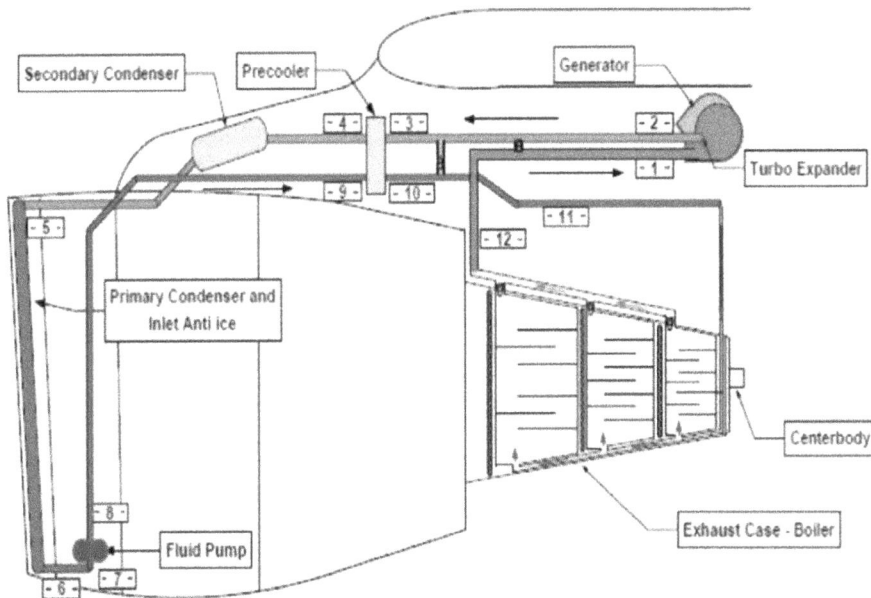

**FIGURE 8.7**   Aircraft engine ORC system working principle [87].

system). They suggested that the design system should be reconfigured to obtain the best results of fuel burn and take into account the need for an electric starting mechanism if the bleed system was removed in future research. The option of using the engine cowl or the anti-icing system in the wings as the condenser of the ORC system was also suggested.

### 8.3.4 ADVANCES IN $CO_2$ POWER CYCLES FOR WASTE HEAT RECOVERY

In general, a cogeneration system can consist of a topping- and a bottoming cycle based on the sequence of energy use. In the topping cycle, the input primary energy is used to first produce power and thermal energy, while in a bottoming cycle the waste heat rejected from the topping cycle is further used to generate power through a recovery heat exchanger and a turbine machine. The bottoming cycles are suitable to recover the low-grade waste heat produced by industrial processes. Sometimes waste heat is directly used for other heating purposes. In this section, studies on different applications of $CO_2$-based bottoming cycles for waste heat recovery are summarized and discussed. Liu et al. [14] have provided an excellent review on this subject. The following summary follows their analysis. A roadmap for the progress of research of the s-$CO_2$ cycle for different industrial waste heat recovery applications in the last 10 years is illustrated in Figure 8.8. Research has been mainly concentrated on three aspects, i.e., recovering waste heat from fuel cells, internal combustion engines (ICE), and gas turbine. Moreover, waste heat recovery from nuclear power plants and landfills has also been carried out. The use of s-$CO_2$ cycle for nuclear power plant was discussed earlier in Chapter 3.

The pioneering work for the employment of $CO_2$-based bottoming cycles for waste heat recovery in high temperature fuel cells was first reported in 2009 [88], and it applied a regenerative s-$CO_2$ cycle to recover flue gas waste heat from high-temperature

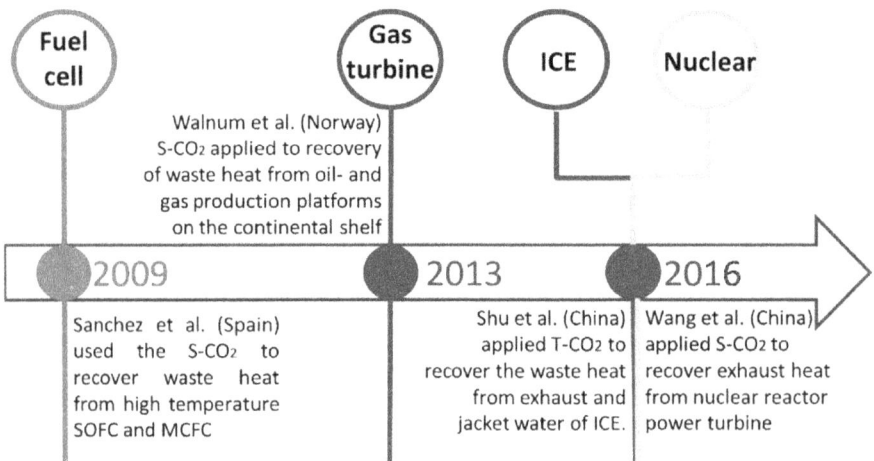

**FIGURE 8.8** Roadmap of s-$CO_2$ investigation in different industrial waste heat recovery applications [14].

solid oxide fuel cell (SOFC) and molten carbonate fuel cell (MCFC). With the use of the s-$CO_2$ cycle, the total system efficiency was increased by 4.4%, while the total net output power was increased by 583.6 kW. The study then compared the combinations of six different configurations of fuel cell and s-$CO_2$ cycles. The results indicated that the required power consumption of the compressor for the s-$CO_2$ bottoming cycle was far lower than the air bottoming cycle, and the operation performance of the bottoming cycle was less affected by the fuel cell operating temperature [89]. Baronci et al. [90] showed that the adoption of the s-$CO_2$ bottoming cycle under optimal conditions could enhance the total energy efficiency of the system by 8.15%. The study also found that the total energy efficiency of the system using s-$CO_2$ as the bottoming cycle could reach 55.3%, while the total energy efficiency of the system using ORC (with cyclohexane as the working fluid) as the bottoming cycle was only 53.3% [14].

Mahmoudi et al. [91] examined an MCFC/s-$CO_2$/ORC cascade system to create a combined supply of cooling, heating, and power, and optimized the system to obtain maximum exergic efficiency with a minimum initial investment of the system. The results showed that the largest exergy loss of the system came from the fuel cell cycle, and both the exergy efficiency and initial investment of the system were positively correlated to the operating temperature of the fuel cell. Ryu et al. [92], compared the thermodynamic performance of the MCFC cycle using three different configurations of the s-$CO_2$ bottoming cycle. The results showed that the total energy efficiency of the system could be increased by 3.41%–4.6% by adding the bottoming cycle, and the bottom cycle performance was mainly affected by the back pressure of the compressor. The study also showed that the combined cycle had obvious economic advantages over the traditional thermal and power cogeneration system, i.e., the heating cost was less than \$28/Gcal, and the cost of printed circuit board heat exchanger (PCHE) was lower than \$100/kW. Bae et al. [93] compared the thermodynamic performance of four different configurations of the s-$CO_2$ bottoming cycle to recover waste heat from the MCFC flue gas and compared it with the regenerative air Rankine cycle. The results showed that the total efficiency of the system could be improved by nearly 11% by using the cascade cycle, which was much higher than that of the system using the air Rankine cycle as the bottoming cycle [14].

Ahmadi et al. [94] proposed a combined cycle of proton exchange membrane fuel cell (PEMFC) and s-$CO_2$, which used s-$CO_2$ fluid to replace the cooling water of conventional fuel cells. It also reused the gasification cooling energy of liquefied natural gas (LNG) to reduce the condensation temperature of the combined cycle to improve the cycle efficiency. With parametric sensitivity analysis, the study demonstrated that the total energy efficiency of the system would decrease with the increase of operating temperature of the fuel cell and the increase of the pinch point temperature difference in the pre-heater of the bottoming cycle. The total energy efficiency of the system could be improved with the increase of the turbine inlet temperature of the bottoming cycle and the decrease of the pinch point temperature difference of the condenser. This study also showed that using s-$CO_2$ as the bottoming cycle could increase the net output power of the cycle by 39.56%, and the total energy efficiency could reach up to 72.36%.

Significant research is also done to use the $CO_2$ bottoming cycle for waste heat of internal combustion engines (ICE). Shu et al. [95], examined the system performance of different forms of $CO_2$ or ORC bottoming cycles for the cascade recovery of the waste heat of exhaust and jacket water, of a four-cylinder four-stroke water-cooled internal combustion engine. The results showed that the combined use of preheating and the regenerative $CO_2$ cycle could increase the total net output power of the system by 9.0 kW at the highest, and the corresponding thermal and exergic efficiencies of the system can be increased by 184% and 227%, respectively. Furthermore, pre-heater to the recovery of waste heat from jacket water and the use of regenerator significantly increased net output power of the system and exergic efficiency (at the level of 48%) of the cycle. Further economic studies by Shu et al. [96] indicated that while setting the regenerator is conducive to improving the system economy while adding the pre-heater is not as useful. The study also examined the performance of the $CO_2$ bottoming cycle under partial load condition of the ICE. In an additional study by Shu et al. [97], using the mass flow rate of the working fluid as the regulation target, system operation control strategy was proposed. Shu et al. [98] also developed an ICE-$CO_2$ cold-power cogeneration system. The results showed that compared with the traditional system, the proposed system could reduce fuel consumption of the ICE by 2.9% under the refrigeration mode, and increase the total net output power by 4.8%. Under the ice-making mode, the fuel consumption could be reduced by 3.4% and the total net output power of the system can be increased by 1.6% [98]. Shu et al. also adjusted the condensation temperature of the $CO_2$ bottoming cycle using mixed working fluid, and theoretically simulated the dynamics of the system with different mixed working fluids. The results showed that under the same operating conditions, with the increase of $CO_2$ concentration in the mixed working fluid, the dynamic response speed of the system became faster, while the thermal efficiency and net output power of the system decreased slightly. Moreover, the maximum value of net output power appeared at operation conditions with high working fluid pump speed [99]. Shu et al. [100] experimentally analyzed the effects of different concentration of $CO_2$/R134a mixture as the working fluid on the energy efficiency of the system. The results indicated that the energy efficiency and the net output work showed a trend of first rising and then falling as the mass fraction of R134a increased [100].

The $CO_2$ bottoming cycle performance under typical operating conditions of the ICE was experimentally tested with different cycle pressure ratios by Shi et al. [101]. The dynamic performance of the $CO_2$ bottoming cycle of three different configurations under given operating conditions was compared to examine the influence of different working fluid mass flow rates and cycle pressure ratios. Li et al. [102,103] obtained the time constant of the dynamic system performance. The operating performance of the $CO_2$ bottoming cycle with the pre-heater was examined under start, idling and emergency stop conditions. The results showed that the preheating effectively prevented the pressure surge at the inlet of the expander, so as to ensure the stable and safe operation of $CO_2$ bottoming cycle under special working conditions. It also improved the energy efficiency of the overall system under partial load conditions [104].

Liang et al. [105] showed that the s-$CO_2$/ORC combined cycle for waste heat recovery of the ICE increased the net output power of the overall system by 6.78%. Feng et al. [106] examined the influence of inlet temperature and pressure of the compressor and turbine on the performance of s-$CO_2$/Kalina combined cycle to recover waste heat of marine engines. The results showed that the annual fuel consumption of the engine was reduced by 16.62%. Liang et al. [107] investigated an engine waste heat-powered thermal-power cogeneration system which combined the s-$CO_2$ power cycle with a transcritical $CO_2$ refrigeration cycle. The results indicated that the proposed configuration reduces the size and weight of the system and is therefore proper on-board application. Pan et al. [108] proposed a cogeneration cycle which combined the s-$CO_2$ power cycle and ejector expansion refrigeration cycle as the bottoming cycle to recover the waste heat from ICE. The effects of the important operating parameters on system performance were investigated. Zhang et al. [109] developed a novel s-$CO_2$ power cycle based on a recompression cycle configuration to recover the waste heat from ICE. The results indicated that for the intermediate pressure the maximum system net output power can reach 39.49 kW. Song et al. [110] proposed a two-stage bottoming cycle for ICE waste-heat recovery. The s-$CO_2$ cycle was coupled with an ORC to further recover the heat rejected from the s-$CO_2$ cycle. The proposed cycle can contribute a maximum net power output of 215 kW, which accounted for ca. 18% of the rated power of ICE [14].

Choi [111] proposed to use the temperature difference between the jacket water of a marine engine and seawater to drive the two-stage reheat $CO_2$ power cycle. Thermodynamic analyses showed that the maximum net output power of the $CO_2$ bottom cycle was 383 kW, the highest thermal efficiency of the system was 7.87%, and the highest exergic efficiency was 5.96%. Sharma et al. [112] carried out thermodynamic analyses on the regenerative and recompressed s-$CO_2$ Brayton cycle used to recover waste heat from flue gas of marine engines. The influence of the inlet temperature of the turbine and compressor, as well as the equipment pressure drop on the overall performance of the combined cycle, was investigated. The results showed that the s-$CO_2$ bottoming cycle improved the overall cycle efficiency, and the net output power by 10%, and 25%, respectively. In addition, the exhaust composition and exhaust temperature of the gas turbine in the topping cycle had a significant effect on the performance of the s-$CO_2$ bottoming cycle. Hou et al. [113] proposed a trigeneration system by recovering the waste heat from the marine engine based on the recompression s-$CO_2$ cycle. The study showed that the high-temperature regenerator and evaporator of the refrigeration system were the key components that affected the thermal economy of the proposed system. Manjunath et al. [114] presented the energetic and exergetic performance analyses of a supercritical/transcritical $CO_2$-based bottoming cycle for a marine engine. The study showed that under the optimal operating conditions, the enhancement of the power output by the proposed system was nearly 18% and provided cooling of 892 TR having the COP (coefficient of performance) of 2.75 [14].

Besides the ICE, there are several studies that proposed to use the $CO_2$ power cycles for recovery waste heat from exhaust of gas turbines. Walnum et al. [115] conducted thermodynamic analyses on the applications of regenerative and two-stage $CO_2$ cycles for the recovery of waste heat from flue gas of offshore oil- and gas platforms. The operation performance of bottoming cycles under partial load conditions of the gas turbine was studied. The results showed that single-stage cycles could increase the total net output power and the overall system efficiency of the oil- and gas platform by about 27.6%, and 10.6%, respectively, while the improvement of the double-stage cycle was more significant. Moroz et al. [116] compared the thermo-dynamic performance of various combined s-$CO_2$ cycles for recovery of the waste heat from a 53 MW gas turbine. The results indicated that the simplest cascade cycle provided a power output of 16.13 MW, while the value of the cycle with the most complicated configuration was 17.05 MW, which represented 32% from the power output of the topping cycle. The output power of the single regenerative s-$CO_2$ bot-toming cycle or a recompression s-$CO_2$ bottoming cycle was 12.94, and 11.85 MW, respectively [14].

Khadse et al. [117] carried out an investigation of a simple construction of a s-$CO_2$ bottoming cycle to recover the waste heat recovery from a gas turbine. The results indicated that a maximum improvement of 22.9% can be gained by the use of recompression configuration. Cao et al. [118] propose a cascade configuration com-posed of an s-$CO_2$ Brayton cycle and a transcritical $CO_2$ Rankine cycle to recover the waste heat from a gas turbine. Both cycles were based on simple configurations, and the $CO_2$ was condensed by using the cold energy of LNG (liquid nature gas). The results indicated that the power output by cascade cycles contributed nearly 28.9%–39.1% to the power output of the whole system. Gao et al. [119] indicated that the partial heating cycle provided the highest power output compared to the single regenerative cycle due to its good waste heat absorption performance. Tozlu et al. [120] carried out a bi-objective optimization of a single regenerative s-$CO_2$ cycle for waste heat recovery from the exhaust gas of gas turbine. It was found that the s-$CO_2$ bottoming cycle showed a potential to increase the net power output of the turbines by 19.3%. Zhang et al. [121] proposed an improved cascade s-$CO_2$ bottoming cycle for recovering the waste heat from flue gas of the offshore oil- and gas platform. The results showed that the s-$CO_2$ bottoming cycle could improve the net output power of the overall system by 30% under rated conditions. Meanwhile, the high-temper-ature part of the cascade cycle had a greater impact on the thermal performance of the overall system, while the low-temperature part had a greater impact on the eco-nomic performance of the overall system. Sánchez et al. [122] used a partial heating s-$CO_2$ bottoming cycle to recover the waste heat from high-temperature exhaust of a gas turbine. The results showed that, compared to the conventional steam bottom-ing cycle, the proposed partial heating s-$CO_2$ bottoming cycle reached a high ther-mal efficiency and reduced the system's initial investment by a quarter. Zhou et al. [123] developed a novel supercritical-/transcritical-$CO_2$ combined cycle system for recovering waste heat from offshore gas turbines. Comprehensive parametric analy-sis was conducted to simultaneously optimize the net output work and net pres-ent value (NPV) under different conditions. Recently, Tao et al. [124] applied the two-stage reheat and recompression split s-$CO_2$ cycle to recover the waste heat of gas turbines in distributed energy system. The preliminary thermodynamic analysis results showed that the total thermal efficiency of the system could reach up to 48% under the optimal split ratio [14].

Cho et al. [125] investigated cascade systems which consisted of a recompres-sion (or pre-compression) cycles and a partial heating cycle for recovering waste heat from a 288 MW gas turbine. They found that the cascade systems were uncompetitive due to their complex configuration and lower power output. Huck et al. [126] carried out a thermodynamic performance comparison of different dual flow splitting s-$CO_2$ bottoming cycles and steam bottoming cycles to recover the waste heat from heavy-duty and aero-derivative gas turbine combined cycles. It was found that the efficiency improvement of s-$CO_2$ bottoming cycle was not significant when the system operated under more realistic assumptions. Wright et al. [127] compared three typical configurations of s-$CO_2$ cycles for recovering waste heat from a 25 MW gas turbine. The results showed that the total heat recovery efficiencies of the s-$CO_2$ cycles ranged from 20.3% to 21.2%, which was 4% higher than the baseline cycle. The single regenerative power cycle showed the best economy. Kim et al. [128] compared the thermodynamic performance of the waste heat recovery of gas turbines in a landfill plant with nine different configurations of the s-$CO_2$ bottoming cycle. The study showed that the recompressing cycle was not suitable for waste heat recovery, and the two-stage split-flow cycle had a significant effect on the improvement of the net output power of the overall system, but its structure was too complex [14].

Wang et al. [129] adopted a genetic algorithm to optimize the exergy-economy of an s-$CO_2$ bottoming cycle for recovering waste heat from combustion engines in nuclear reactors, which increased the total thermal efficiency and the net output power of the overall system by 7.92%, and 13.7 MW, respectively. Astolfi et al. [130] compared the performance of the dual regeneratives-$CO_2$ bottoming cycle against three traditional cycle layouts. The results indicated that the dual regenerative layout was found to be the best choice if the minimum heat source temperature has been not constrained. Olumayegun et al. [131] studied the dynamic performance of the recompression s-$CO_2$ cycle for recovering waste heat from cement plants. Those results indicated that the inlet temperature of the main compressor could be controlled by adjusting the mass flow of cooling water, and the inlet pressure of the compressor could be kept constant through the throttle valve to improve the dynamic performance of the whole cascade system. Luo et al. [132] proposed a multi-generation system which combined s-$CO_2$ cycle and transcritical $CO_2$ refrigeration cycles using waste heat as power source. The results indicated that the refrigeration cost was the highest, while the cost of the power was the lowest of total system operating cost. The literature also indicate that research on waste heat recovery from gas turbine accounts for half of the total listed literature, while the amount of investigation on waste heat recovery from the fuel cell and ICE are quite similar. Eight typical configurations of the s-$CO_2$ bottoming-cycle used for waste heat recovery from different sources are illustrated in Figure 8.9 [14].

As pointed out by Liu et al. [14], based on the available literature one can conclude that unless costs are significantly reduced, s$CO_2$ is primarily competitive for applications beyond 0.5 MWe and heat-source temperatures above 350°C. There is a need for high-temperature s$CO_2$ heat exchangers that are able to deal with high-temperature $CO_2$ corrosion issues; fouling, back-pressure and other thermo-structural challenges imposed by the flue gas and economic constraints imposed on materials selection and lifetime. Finally, s$CO_2$ power technology will face the same barriers as other waste-heat recovery technologies employed in industry as bottoming power cycles such as project financing challenges due to high capital cost; return on investment; disturbance of existing plant operations; space requirements and safety concerns [14].

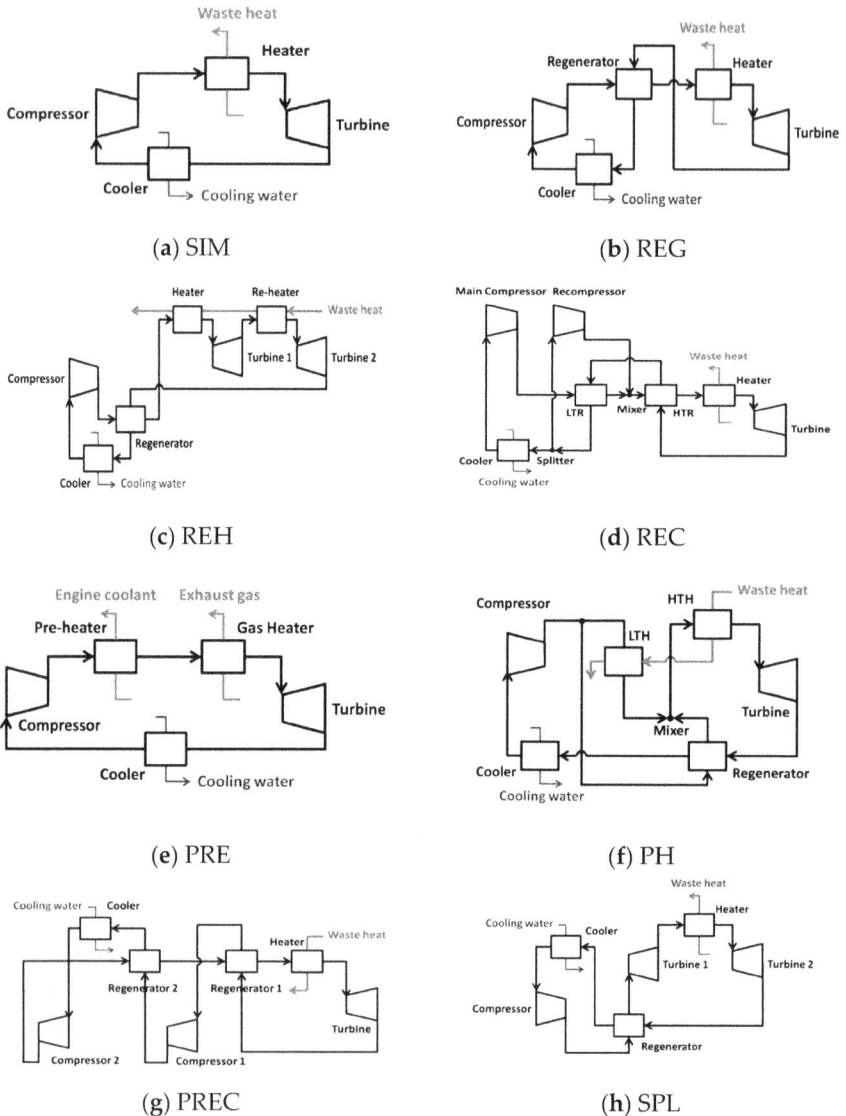

**FIGURE 8.9** Typical configurations of s-CO$_2$ bottoming cycle used for waste heat recovery [14]. (a) SIM, simple; (b) REG, regenerative; (c) REH, reheat; (d) REC, recompression; (e) PRE, preheat; REF, refrigeration; ORC, organic Rankine cycle; LNG, liquid nature gas; (f) PH, partial heating; (g) PREC, pre-compression; (h) SPL, flow split [14].

## 8.4 THERMOELECTRICITY

Thermoelectric materials have the specific capacity of converting a flow of heat into electrical energy (Seebeck effect) and vice versa (Peltier effect) [133]. Their use is becoming increasingly important because they convert waste heat from different sources (static and mobile) and convert into power and thereby increase system thermal efficiency, reduce operating costs and reduce environment pollution. The temperature of the waste heat can vary significantly depending on its source. For example, the temperature of exhaust gases emitted from vehicle engines, biomass combustion systems and matrix-stabilized porous mediated combustion can reach 500°C, while the operating temperature of micro-turbine power cycles can rise to 600°C, and even to 900°C in the case of a solar energy receiver [134]. Thermoelectric devices are particularly reliable, silent, and do not generate vibrations since their operation does not require the contribution of mechanical energy [135]. All these reasons have provided enormous momentum to increase research and development activities on new materials and new production procedures for thermoelectric systems [136].

Since the discovery of thermoelectricity (TE) by Seebeck [137], researchers have been trying to understand, improve and control its implementation. Peltier discovered the opposite effect [138], Lord Calvin formulated the laws that link these two phenomena [139] and Altenkirch [140] correctly calculated, for the first time, the energy efficiency of a thermoelectric generator now known as figure of merit (ZT), and developed thermoelectricity in cooling mode [141]. Altenkirch [142] also invented a thermoelectric heating and cooling apparatus, which was followed by several other prototypes developed by various scientists and companies. These attempts were, however, not successful due to lack of appropriate materials [143].

First breakthrough came when Ioffe [144] discovered the thermoelectric properties of semiconductors, which opened up new projections for thermoelectricity with a figure of merit ZT close to 1. This value was still low, but acceptable enough for some inventors and industrialists to design new applications to be commercialized. One such application was the thermoelectric refrigerator designed by Becket et al. [145]. At the same time, the idea of thermoelectric generators emerged, such as Ioffé's thermoelectric lamp. In 1993, Hicks and Dresselhaus [146] showed that quantum-well superlattice structures (small dimensions of matter) could affect thermoelectricity by reducing phonon thermal conductivity, and therefore improving the ZT by a factor of 13. This gave research on TE generators significant momentum [147].

### 8.4.1 Figure of Merit and Other Performance Parameters

The thermoelectric efficiency of a TE material is expressed by the dimensionless thermoelectric figure of merit ($ZT$), which is dependent on the transport properties of the material as shown in Equation (8.1).

$$ZT = S^2 \sigma \, T / K \tag{8.1}$$

where $S$ is the Seebeck coefficient (μV/K), $\sigma$ the electrical conductivity (1/Ωm) and K is the thermal conductivity (W/m K). This equation shows that to maximize the $ZT$ of a material, it must meet the following criteria: (a) low thermal conductivity to maintain a considerable temperature difference between the two ends of the material; (b) high electrical conductivity to reduce the internal resistance of the material and consequently the Joule effect; and (c) a high Seebeck coefficient, required to obtain a high

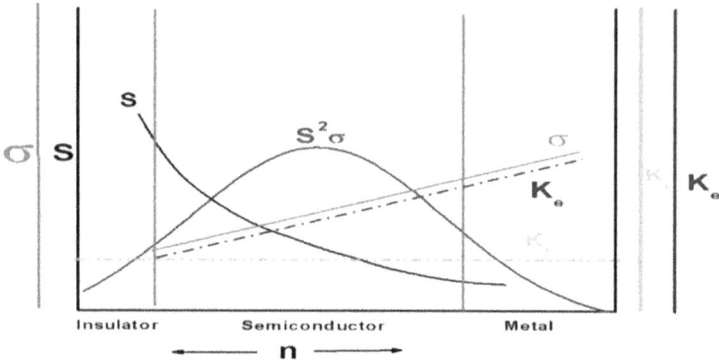

**FIGURE 8.10** Relationship between the figure of merit ZT and other parameters such as electrical conductivity $\sigma$, Seebeck coefficient $S$, power factor $S^2\sigma$, electronic thermal conductivity Ke, thermal conductivity of the network Kl and total thermal conductivity K [150,151].

voltage [148,149]. Unfortunately, according to the graph in Figure 8.10, these parameters are well correlated and it is very difficult to optimize them independently, especially for conventional metals. For example, for conventional materials, it is difficult to increase electrical conductivity and decrease thermal conductivity simultaneously.

In the literature, ZT is sometimes referred to as thermoelectric efficiency because it is related to the efficiency of a single element and the device. To determine the efficiency of a thermoelectric generator (TEG), it is necessary to calculate the ratio between the electrical power produced and the heat flow through the module. The basic standard equations for this purpose are typically built on four main hypotheses, namely (a) the electrical and thermal contact resistances are negligible, (b) the Thomson effect has a negligible effect on efficiency, (c) the convection and radiation heat transfer are negligible, and (d) the dependency on the thermoelectric transport properties of the TEG module with temperature, which makes the performance of TEGs change at different temperatures [152].

There are several ways to calculate the performance of a thermoelectric couple in terms of energy production, either averaging or using finite elements [153,154]. The averaging methods overestimate the efficiency but provide an immediate value based on the calculated properties of the TEG at the average junction temperature [155]. On the other hand, finite elements require a lot of iterations and therefore take more time to obtain results [156]. Although simple one-dimensional analytical models are frequently used to predict the performance of such devices [157], the diversity and complexity of thermoelectric applications generally require a complete three-dimensional (3D) numerical analysis [158], using simulation tools such as ANSYS [159], GT-SUITE [160], FLUENT [161] and GTPower [162]. Simple relationships among TE efficiency, temperature, ZT and Carnot efficiency are illustrated in the following Figure 8.11.

## 8.4.2 Thermoelectric Module

The promotion of TE generators required the manufacturing of TE modules that can be used by broader market at different scales. In 1959, the General Electric company

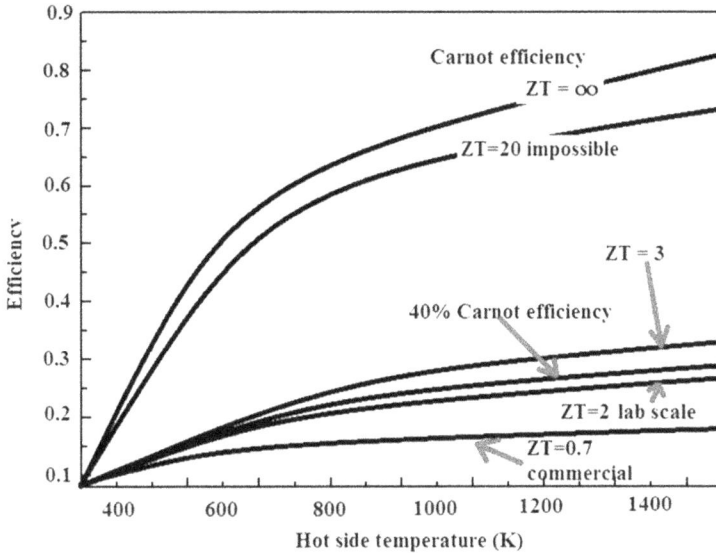

**FIGURE 8.11** Relationships among TE efficiency, temperature, ZT and Carnot efficiency [163].

commercialized [164] thermoelectric modules composed of 36 couples of bismuth telluride in flat bulk architecture. Nowadays, Champier [165] has shown that numerous companies all over the world manufacture TE modules. A typical thermoelectric generator (TEG) module consists of between 10 and 100 thermoelectric elements of type n and type p, electrically connected in series and thermally in parallel, and interposed between two ceramic layers, as shown in Figure 8.12. The p–n pairs are joined by conductive tabs connected to the elements via a low melting point solder (PbSn or BiSn). When a temperature gradient occurs between its two junctions, the TEG converts thermal energy into electrical energy according to the principle of the Seebeck effect. This flat bulk architecture is the most widely used and marketed. However, in some applications adapting heat source to thermoelectric device requires a shape different from a flat plate. While different shapes can make the device more costly, heavy and cumbersome, a flat shape is not always practical because of the difficulty in adapting the heat source to the thermoelectric device. As a result, other designs such as cylindrical shape [166–168], thick and thin films and flexible TE devices [169] are being developed.

In the flat plate design, the two ceramic plates serve as a support for the module and as electrical insulation, but their thermal resistance degrades the module's efficiency. From this, some investigations have proposed the concept of a direct contact thermoelectric generator (DCTEG), which is characterized by one of the surfaces of the module being directly exposed to the heat source and the other surface in direct contact with the coolant flow [171,172]. Several manufacturing technologies for TE modules are reported in the literature. Some examples include foil lithography [173], the lift-off process [174], flash evaporation [175], evaporation thin film [176], photolithography and etching [177], screen-printing [178], sputtering [179], dispenser

**Heat absorbed**

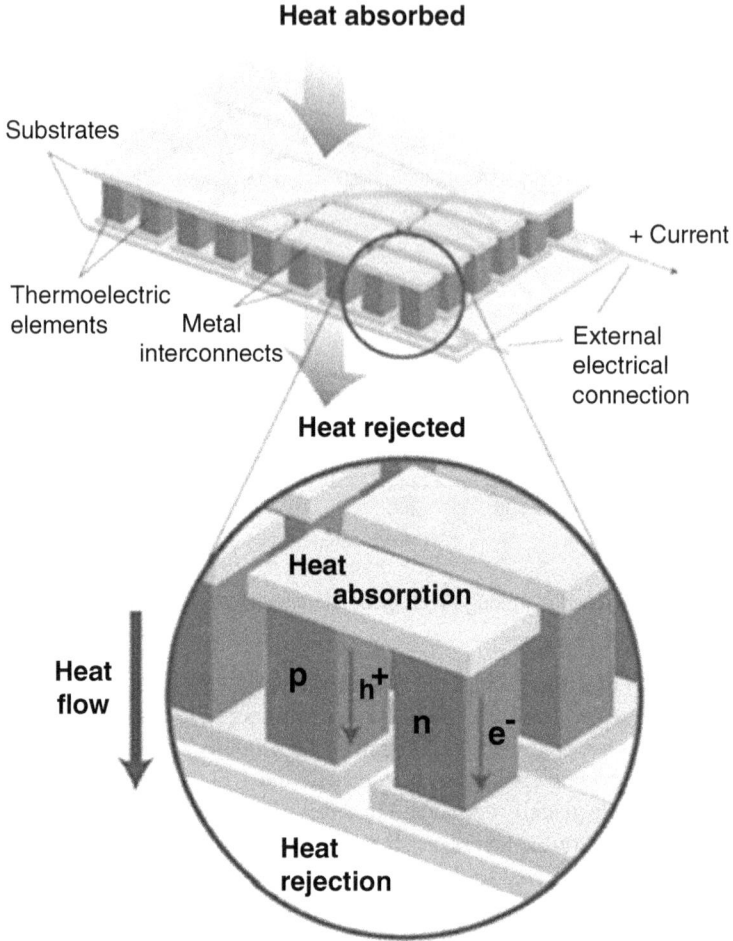

FIGURE 8.12   Diagram of a typical thermoelectric device [170,151].

printing [180], the spark plasma sintering technique [181], direct current (DC) magnetic sputtering [182] and the printing process [183].

Zoui et al. [151] point out that the critical challenge in the development of TEGs is the property degradation caused by thermal expansion and thermal shock [184] resulting in thermal fatigue. This degradation can result in a decrease in service life and efficiency. Thermal expansion of TE device caused by heating and cooling can be different from that of connected stunts. This difference causes stress at the interface which can ultimately result in the mechanism failure [185].

### 8.4.3  Thermoelectric Materials

Ever since the discovery of TE materials, their use has been limited to thermocouples for temperature measurements due to their very low efficiency [186]. The use of TEG for refrigeration and power generation started to develop in 1960 due to the discovery of semiconductor materials. From the discovery of thermoelectric semiconductors [144] until 1993, the figure of merit (ZT) experienced a modest improvement. After that date, theoretical predictions suggested that the efficiency of TE materials could be significantly improved by using nanostructure engineering [146]. The effectiveness of nano-structured materials for reducing the negative correlation between electrical and thermal conductivities was theoretically and experimentally explored by Rurali et al. [187]. For instance, 2D and 1D nanostructuring slows down the diffusion of phonons, leading to a decrease in thermal conductivity which in turn increases the efficiency of the TEG. Using modern synthesis and characterization techniques, conventional bulk materials containing nanostructured components were explored and developed with the aim of achieving higher yields [188]. A new research path showed that the ZT factor can be increased by using nanomaterials [189] or using bulk materials containing nanomaterial constitute. It should be noted that small (nanostructured) materials are difficult to manufacture with precision due to the high sample requirements at the nanoscale [190]. The research on thermoelectric materials bifurcated into two paths; the development of novel conventional materials using nanotechnology and new materials taking advantages of nanotechnology.

### 8.4.3.1 Conventional Thermoelectric Materials

Zoui et al. [151] pointed out that conventional thermoelectric materials, which are bulk doped semiconductor alloys or chalcogenide, can be grouped into three families according to the temperature range at which the performance is optimum: $Bi_2Te_3$-based materials for ambient temperature applications ($< 150°C$), TAGS $[(AgSbTe_2)_{1-x}(GeTe)_x]$ and PbTe-based materials for intermediate temperature range ($150°C–500°C$), and SiGe for use at temperatures over $500°C$ [191–193]. The temperature range can be widened by using a combination of materials characterized by different temperature ranges in a segmented structure [194]. The $Bi_2Te_3$ is well known and can have a ZT close to the unity at room temperature. However, as they are easily oxidized and vaporized, these materials cannot be used for high-temperature applications in the air [195]. Around 70% of the TE modules available on the market use Bismuth and Telluride as functional materials [196]. Recently, Mamur et al. [197] examined the growth of the $Bi_2Te_3$ nanostructure by various methods and concluded that the figure of merit (ZT) can be increased from 0.58 to 1.16 if materials are developed in a nanostructure form.

The Lead telluride (PbTe) is a good thermoelectric material for applications requiring mid-temperatures up to 900 K. PbTe has a high melting point of 1,190 K, good chemical stability, low vapor pressure and robust chemical strength [198]. Its high figure of merit, approaching 0.8, allowed for its successful use in several NASA space missions. Recent investigations have reported maximum ZT values of around 1.4 for single-phase PbTe-based materials, and 1.8 for homogeneous PbTe-PbSe materials [199]. Gayner and Kar [200] and Sootsman [201] presented an extensive survey of Research and Development on PbTe and its related compounds, alloys and composites, as well as PbTe-based nanostructured composites. Silicon–Germanium alloys ($Si_{1-x}Ge_x$) are among the best TE materials reported in the literature for high-temperature applications ($T_{hot\ side} > 500°C$). In addition, they are one of the cheapest and most nontoxic thermoelectric materials [191]. Delime-Codrin et al.

[202] reported a significant figure of merit, $ZT = 1.88$ at 873 K, for nanostructured $Si_{0.55}Ge_{0.35}(P_{0.10}Fe_{0.01})$ material.

### 8.4.3.2 New Thermoelectric Materials

Phonon-glass electron-crystal (PGEC) materials proposed by Slack [203] have a complex intermetallic cage structure, which gives the material good electronic characteristics like crystal, and at the same time, a low thermal conductivity, like glass [204]. Clathrates and Skutterudites are generally considered two categories of new PGEC materials [205]. Among other TE materials, half-Heusler alloys have attracted considerable attention with their attractive electrical transport properties, relatively high Seebeck coefficients and rich element combinations [206]. In addition, they boast robust mechanical strength, good thermal stability at high temperatures and multiform physical properties [207]. Gascoin et al. [208] proposed phase Zintl as an attractive candidate for thermoelectric materials. These are typically small-bandgap semiconductors with a complex structure. Several studies have been carried out on the use of this type of material, and the best values achieved for ZT ranged from 1 to a peak value of 1.5 [209]. Ferreira et al. [210] made significant efforts to obtain high-performance thermoelectric materials for energy conversion systems. TE oxides, such as $Ca_3Co_4O_9$ ($ZT \sim 1$), are good TE performers [148], and are environment friendly and essentially stable at high temperatures [211]. Other oxides that can be used at high temperatures without oxidation have also attracted significant attention [212].

The thermoelectric metal chalcogenide has high electrical properties and low thermal conductivity, so when advanced nanostructuring and band engineering were employed, they resulted in an improved figure of merit (ZT). Furthermore, chalcogenides are easy to process into different types of structures, thus offering huge potential for improvement in thermoelectric performance. The highest values of ZT recorded with lead selenide (PbSe) ranged from 1.4 to 1.7 at 800–900 K [213]. In the case of Tin chalcogenides Sn (Se, Te), numerical values of ZT above 2.3 were obtained at 723–973 K for single crystal SnSe [214], and about 1.6 at 923 K for SnTe-based materials [215]. While, these materials operate at both high and medium temperatures and are relatively inexpensive, their low mechanical properties and low thermal stability, and in some cases the presence of toxic elements (e.g., Pb), limit their use in real applications.

Since the discovery of conductive polymers, great interest has also been shown in organic TE materials [216]. These are lightweight, flexible and suitable for applications at room temperature, generally with relatively simple manufacturing processes compared to other semiconductor-based materials. Polymers are intrinsically poor thermal conductors, which makes them suitable for use in thermoelectricity, but their low electrical conductivity, Seebeck coefficient and stability have limited their use in thermoelectric applications [217]. However, compared to inorganic TE materials, organic or polymeric TE materials have low cost and simple synthesis processes. In addition, the physical and chemical properties of some polymers can be subjected to a fairly wide range of modifications in their molecular structures [218]. The properties of polymers and polymer-based TE composites have been significantly improved with ZT values up to 0.42 [219]. Yemata et al. [220] showed that hybridization of all the thermoelectric materials already mentioned (in particular combining organic and inorganic materials) can also achieve positive results [220].

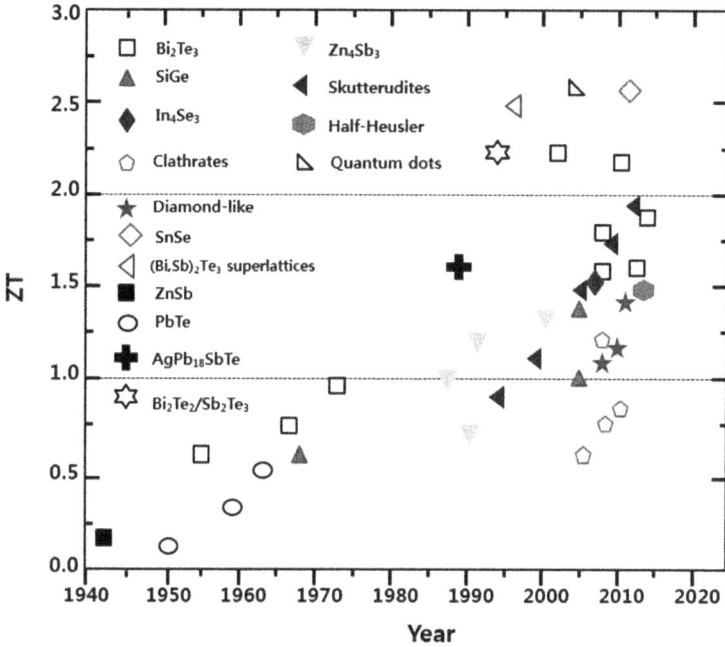

**FIGURE 8.13** Chronological evolution of TE materials [163].

Graphene (carbon atoms forming a crystalline two-dimensional material) also has many unusual thermoelectric and thermal transport properties [221]. A recent study has reported a thermoelectric figure of merit (ZT) up to 1.4 with graphene and $C_{60}$ clusters synthesized by chemical vapor deposition (CVD) [222]. Another theoretical investigation revealed three peak ZT values of 2.0, 2.7 and 6.1 at 300 K, with a twisted bilayer graphene nanoribbon junction [223]. Some of the materials examined over the years and their ZT are graphically illustrated in Figure 8.13. The figure also shows chronological evolution of TE materials [163].

### 8.4.3.3 Thermoelectric Materials: From Bulk to Nano

In recent years, many studies on improving the figure of merit (ZT) are moving toward the use of nanostructured materials [224]. Interestingly, although there have been several demonstrations of ZT > 1 in the past decade, no material has yet achieved the target goal of ZT ≥ 3 (Figures 8.13 and 8.14) [225,226]. In fact, the material systems that have achieved ZT > 1 till date, have all been based on systems using nanostructures [227]. For instance, Li et al. [228] demonstrated a microscopic picture connecting the electronic structure and phonon anharmonicity in a single crystal SnSe system. This offered new insights on how electron–phonon and phonon–phonon interactions may lead to the realization of ultralow thermal conductivity in nanostructured materials. Nanomaterials can be produced in the form of nanopowders, or nanotubes/ nanowires, nano-rods and nano coatings where the grain size falls in the nano-regime. Nanostructured materials such as alloye nanosystem [229], nanowires [230–235], superlattices [236], quantum wells [237], nanomesh [238], nanoribbons

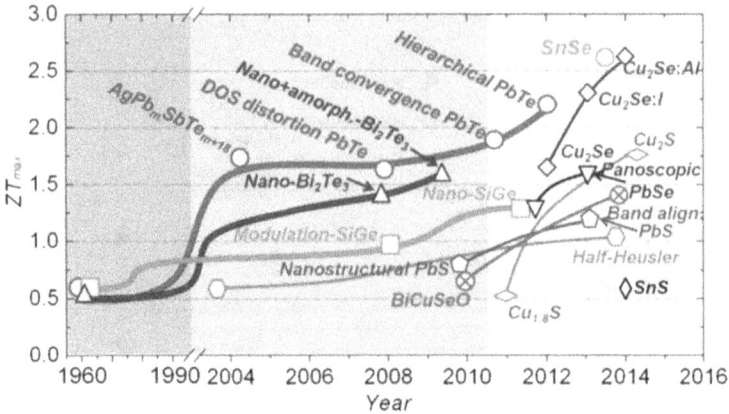

**FIGURE 8.14** ZT values of several nanostructured thermoelectric materials as a function of year [226,233].

[239] are often used in the manufacturing of new TE based systems. In fact, thermal conductivity can be significantly reduced owing to nanostructured materials [240]. The nanostructured thermo-electric materials have better figure of merit and conversion efficiencies as compared to their bulk counterparts [234,235]. The large decrease in the thermal conductivity as a consequence of the increased phonon scattering at the grain boundaries results in higher ZT value and higher power at higher temperatures [233,234].

Various kinds of nanostructured materials have been developed using nanotechnology (see Figure 8.14), such as, nanocomposites, ultrafine-grained materials and super lattices. In the case of nanocomposites, the selection of the reinforcement and the dispersion process must be optimized before actually being used in practice. In nanocomposites, at least one phase should be in the nano range. The incorporation of nanoparticles can solve the problem due to the formation of new interface in the matrix. However, the nanoparticles are very high surface active entities. Their selection as a reinforcement is critical for avoiding segregation and hence a depression in thermopower [234,241,243].

The interest in thermoelectric devices has gained momentum with the advances in nanostructural engineering, which led to overall efforts to demonstrate high-efficiency materials. At the same time, as mentioned earlier, complex bulk materials (such as skutterudites, clathrates, and Zintl phases) have also been explored, and it was found that high efficiencies could be obtained. These complex high-efficiency materials that manage to decouple the conflicting properties have led to many challenges in the field: a wide array of new approaches, from complexity within the unit cell to nano-structured bulk and materials combined in thin-film multi-layer structures, have all led to better and better solutions [242]. Skutterudites such as $CeFe_4Sb_{12}$, $Yb_{0.2}Co_4Sb_{12}$, thallium compounds: $Tl_2SnTe_5$, $Tl_9BiTe_6$, clathrates such as $Sr_8Ga_{16}Ge_{30}$ and $Zn_4Sb_3$ [6–8], rare earth chalcogenides ($La_{3-x}Te_4$), the half-Heusler alloys ($Hf_{0.75}Zr_{0.25}NiSb$), Zintls compounds ($Yb_{14}MnSb_{11}$), cobaltite oxides ($Na_xCoO_2$) and Si nanowires [242–244] have shown some positive features. However, all of these materials have a variety of drawbacks including incompatibility with NEMS technologies, toxicity and unsuitability for fabrication on flexible substrates. Thallium compounds have good ZT values of between 0.6 and 0.8 at room temperature, but the toxicity of thallium prohibits its use in many application areas. BiSbTe alloys are traditionally used in commercial devices and offer a significant advantage; unlike the thermoelectric materials previously mentioned, they can be deposited using electrochemical techniques under thin-film or nanostructures and are compatible with MEMS/NEMS processes.

Several novel approaches have also been extensively studied to enhance the thermoelectric performance such as enhancement of Seebeck coefficient through modifications of electronic band structures and band convergence [245–247], or reduction of lattice thermal conductivity via all-scale hierarchical rendering [248] and nanostructuring [249]. It has also been shown that in some cases the enhancement of ZT can be influenced many times more by the lattice thermal conductivity than that of the electric conductivity [250]. So far, several promising TE materials with intrinsically low thermal conductivities have been reported. Possible explanations to such low values are their liquid-like transports, high anharmonicity of chemical bonds, complex crystal structures or large molecular weights [233]. Additionally, it has been observed that doping of nanowires can be used to alter the scattering of phonons by modifying the surface roughness and size confinement, achieving the reduction of thermal conductivity while keeping the electrical conductivity unaffected, thus highly enhancing the overall ZT value. This technological approach could be encouraging and is compatible with the semiconductor industry [251].

Concerning polymeric-based TE materials, attempts have been reported to optimize the power factors of conducting polymer-based nanowires [252]. On the other hand, CNTs and graphene have also been explored to enhance the TE performance of conducting polymers due to their extremely high carrier mobility and unique electronic structures [253,254]. In this manner, nanocomposites have been used to channel the loss in effective efficiency by optimizing thermal conductivity. Moreover, further improvement in power factors of conducting polymers has been shown by the addition of inorganic semiconductors [255,256]. Finally, nanostructured binary metal sulfides have been recently identified as TE materials with great potential due to their great abundance and lower cost and toxicity than traditional TE materials [257].

It is of interest to note that the lowest thermal conductivity is always observed in an amorphous material since the average phonon mean free path is on the order of the lattice constant ($\sim$0.2–0.5 nm). The lowest thermal conductivity can be estimated to be $\kappa \sim 0.25$–1 W/m K, which is confirmed by more sophisticated theories [258]. Recent reports in superlattices of $WSe_2$/W layers, however, are quite intriguing as they suggested cross-plane lattice thermal conductivity values as low as 0.02 W/m K [259]. While the mechanism is not fully understood, it is likely that the layering creates large asymmetry in the directional phonon density of states and low coupling between phonons in different directions. Therefore, it is believed that a new physical understanding needs to be urgently developed to make much lower thermal conductivity become possible.

Up till now, achieving ZT values of 3 or greater has been difficult. It needs dramatic enhancements in the power factor, which depend on further reduction in the thermal conductivity and an increase in the electron conductivity. With regards to thermal conductivity, any further reduction below the amorphous limit can only occur if one can actively change the group velocity or reduce the number of phonon modes that propagate. This could result from coherent or correlated scattering effects, but so far this has remained elusive for phonons (although widely known for electrons) and thermal conductivity reduction through such mechanisms has never been conclusively demonstrated. These point to exciting scientific opportunities and create an open challenge to theorists and experimentalists alike to come up with new scattering mechanisms and concepts that will help achieve very large increases in the power factor and simultaneous decreases in thermal conductivity.

In summary, several different techniques and materials have been explored and are in constant evolution to improve efficiency and drive the transition toward a more stabilized commercial stage. Technically, high-performing thermoelectric properties appear to depend sensitively on the nanostructure, synthesis approach and device assembly. Various approaches will continue to examine thermoelectric nanomaterials with narrow bandgaps, heavy elements doping, point defects loading and nanostructuring. Especially, for practical thermoelectric applications, the synthetic approaches of thermoelectric nanomaterials, should be: (a) scalable, high-quality and low cost, with tuneable thermoelectric properties, (b) the nanostructured materials must be able to form dense compacts for machining/device integration (device-controllable), (c) the nanostructured material should demonstrate an enhanced ZT over the bulk material and finally (d) the compacted nanoscale features should be with high thermal stability for extended periods of time. Continued research to gain a more quantitative understanding is required to allow the rational design and preparation of optimized nanostructured thermoelectric materials and accelerate the wide adoption of thermoelectric technologies in power generation and cooling applications.

### 8.4.3.4   Thin Film Thermoelectric Technology

Due to the miniaturization of the electronic circuits and systems, the amount of heat dissipation is large enough to cause thermal breakdown of the device and eventually failure of the entire device. Traditional or bulk thermoelectric devices have been used for years to control the temperature of electronics where the cooling of the entire device is an efficient method for thermal management leading to the oversizing of the thermoelectric device. The modern thin film thermoelectric technology targets the heat flux source μm for traditional devices. This creates a huge heat flux around 20 times as compared to traditional ones. Tuyen et al. [260] nanostructured $Bi_2Te_3$ based thermoelectric thin films grown using pulsed laser deposition. However, there are various barriers to thin film technology which need to be sorted out to accomplish the full potential of this technology. For example, electrical contact resistance between metal electrodes and the semi-conductor layers could be comparable to that of the thermoelectric element itself, thus increasing the overall electric resistance of the thermoelectric device and reducing the cooling flux.

### 8.4.3.5   Magnesium-Based Thermoelectric Generators

Thermoelectric generator (TEG) that works in medium and high-temperature ranges has appeared as a necessity in combating with waste heat management from various industries. Therefore, the design of TEG with thermoelectric materials capable to exhibit a high capacity for thermoelectric power generation is of emerging technological interest. It is known that alloys based on $Mg_2(Si, Sn)$ and PbTe thermoelectric materials are best suited for the intermediate temperature scale (400–900 K). For constructing a TEG, higher manganese silicides (HMS) are often used for integration with n-type $Mg_2X$ (Si, Ge, Sn) alloys. The thermo-mechanical property-mismatch reasons out the option for such a combination in the fabrication of a sustainable thermoelectric generator. This makes one focus on evolving a generation system that comprises both n- and p-type thermoelements prepared from suitable $Mg_2X$ (Si, Ge, Sn) alloys. The study by Pandel et al. [261] presents a comprehensive take on the effect of thermoelectric leg dimensions and contact resistances on the output voltage, output power and efficiency, with alteration in the operating temperature span. The COMSOL modeling results obtained in this study indicate that the power output reduces considerably with increment in the thermoelectric leg length, while the conversion efficiency enhances. Contrarily, augmenting the cross-sectional area of a thermoelectric leg follows an opposite trend i.e., it increases the power output and decreases the conversion efficiency. The power output and the conversion efficiency values diminish when contact resistances are considered in the modeling study. This study also incorporates an efficient heat exchanger system including contact resistances and conductive and convective heat losses for accurate estimation of power output and efficiency of the optimized TEG module.

Magnesium based thermoelectric materials have also attracted the researchers in the past to overcome the issue of cost, toxicity and light weight compared to the bulk traditional thermoelectric materials ($Bi_2Te_3$, PbTe and $CoSb_3$). Mg is the light metal having density (1.73 g/cm$^3$) versus most popular aluminum (2.70 g/cm$^3$) and iron (7.86 g/cm$^3$). Mg-based alloys are high strength and low weight, moderate melting points and their intermetallic compounds or alloys form Zintl phase with a large electronegativity gap. Zintl phases are the potential candidates for obtaining high ZT value in thermoelectric materials, e.g. $(Eu_{1/2}Yb_{1/2})_{1-x}Ca_xMg_2Bi_2$. $Mg_2X$ (X = Si, Ge, Sn, Si-Sn, Si-Ge, etc) compounds have shown a higher figure of merit as compared to the Si-Ge and β-$FeSi_2$. It is also a noteworthy point that the band gap range of Mg-based thermoelectric materials are close band gaps of semiconductor materials.

However, the synthesis of these $Mg_2(Si, Sn)$ single phase is difficult due to the high vapor pressure and chemical to provide thermal control. Such types of thermoelectric device incorporate nanostructured p and n type materials in the form of a coating, typically 5–20 μm thick. The stability and performance of $Mg_2(Si, Ge)$ is better than $Mg_2(Si, Sn)$ but is highly expensive due to the cost of Ge. Application of nanotechnology has been also found to increase the figure of merit of these alloys via nanoparticles inclusion in the matrix.

## 8.5   APPLICATIONS OF THERMOELECTRIC GENERATORS FOR WASTE HEAT

Zoui et al. [151] presented an excellent review of the possible applications of TEG for waste heat. Here we briefly summarize important parts of their analysis. Thermoelectric applications are classified according to the two effects characterizing the process, namely the Seebeck effect, for any application that generates electricity with a temperature difference, and the Peltier effect for any cooling application powered by an electric current. In order to generate electricity from a TE module, it is necessary for there to be a temperature difference between its hot and cold surfaces. In other words, it is necessary that the heat recovered from the hot source scatters into the semiconductor elements p and n of the module, and then to the cold source, which is usually the environment. TEG applications can be classified into three categories, depending on the nature of the hot source: (a) radioisotope heat source, (b) natural heat source, and (c) waste heat source. The use of radioisotope heat source to generate power was described in Chapter 3. The use of heat from natural gas and biomass combustion to electricity by thermoelectricity was described in Chapter 2. The use of solar heat to generate power by thermoelectricity was described in Chapter 6. Here we mainly focus on the use of waste heat from various sources to generate power by thermoelectricity.

### 8.5.1  WASTE HEAT FROM HUMAN BODY

Besides significant advances made in the development of TEG materials, design and manufacturing of TEG for various applications have also made quantum progress. The scale, flexibility and range of applicability of TEG have been enormously expanded. For example, TEGs might play an important role in wearable technologies since they can convert the heat generated by the body into electrical energy useful to power up sensors, logic circuits and communication systems [226,262]. Acquiring the capability of stretching TEGs will enable an improved integration with the complex motions of the human body. Recently, the development of flexible TEG systems has attracted a lot of attention, including using thermoelectric (TE) films, thermoelectric bulks, printable thermoelectric inks, thermoelectric fibers, and organic thermoelectric materials. However, TEGs with good stretchability are critical to ensure conformal contact with complex geometries of the human body for optimal thermoelectric performance. Inspired by the self-healing capability of the human skin, self-healable electronics has also shown promising potential in wearable electronics for improved reliability and durability.

Smart clothing, fabrics and textiles, which can be flexed and stretched to some extent, might also be used for this purpose. Kim et al. [263] implemented a flexible and stretchy TEG by integrating a polymer-based fabric with inorganic TE materials to produce 224 nW from the human body at an ambient temperature of 5°C. Du et al. [264] developed power-generating clothing by coating a commercial fabric with a TE polymer, that generated 12.5 nW at a temperature difference of 75°C. While organic/polymeric materials can make stretchable devices, their performance is inferior to inorganic TE materials [265,266]. Another set of materials with potential stretchable capabilities are nano-engineered and 2D materials like silicine [267] and phosphorene [268], which can theoretically reach very high figures of merit (ZT), well above the unity. Oh et al. developed a stretchable and foldable TEG based on transition metal dichalcogenide (TMDC) nanosheets. While TEG generated maximum power of 38 nW at a temperature difference of 60°C while at rest, they observed a very strong power degradation while stretching [269].

Over the last several years, many different approaches have been explored to combine the high efficiency of inorganic thermoelectric materials with the mechanical flexibility and affordability provided by organic materials. More interestingly, composite materials of organics and inorganics with increased functionality and performance have been investigated. By preparing polymer-inorganic TE composites, there is a great potential to reduce the production cost while maintaining a good performance as well as exhibiting mechanical flexibility for various applications of human interest. More efficient and promising techniques have been developed to increase manufacturability and performance. Materials such as polyaniline (PANI), polythiophene (PTH) and numerous others have been explored as TEG materials [270–272]. Current research in polymer-based and composite-based TEGs is focusing on finding novel strategies to further enhance their figure of merit (ZT), such as including low-dimensional TE materials, or including additional treatment for encapsulation in order to mitigate the impact of humidity on the electrical conductivity [271,273].

The flexible thermoelectric generators can also provide a method to increase the energy production capabilities of wearable and implantable electronics. Madan et al. [274] demonstrated a novel method to integrate printing methods with thermoelectric generators fabrication. Madan's devices can produce output power in the range of μW at a temperature difference of 20°C. Several other devices have demonstrated the ability of simple and low-cost fabrication methods to create high-performance flexible TEGs that have advantages such as compact, silent, and high reliability [275–277]. For human comfort, TEGs are not likely to replace batteries in wearable devices, but instead, they can be used as a supplemental power source to extend the battery lifetime of state-of-the-art wearable sensor networks. An additional interesting approach involves the use of paper as an affordable substrate alternative. Paper is a widespread and inexpensive material, whose mechanical properties are of interest and, in fact, it has been the focus of several studies that demonstrate diverse applications ranging from electronics, displays, and sensors to energy harvesting among others [278]. Its use in the development of TEGs is also of great interest, as it has been shown to help improve the TE properties of PEDOT:PSS by reducing its thermal conductivity. Sun et al. demonstrated an interesting approach to fabricate paper-based TEGs by simply impregnating paper with n- or p-type colloidal semiconductor quantum dots, exhibiting power factors of $1 \times 10^{-5}$ W/m K$^2$ for the p-type material and $0.5 \times 10^{-5}$ W/m K$^2$ for the n-type material [279]. An important advantage of the use of paper is the possibility of employing out-of-the-box techniques, such origami and kirigami, to stack and fold the devices in an area-efficient way and thus improve their power density.

Sevilla et al. [280] developed a method to integrate bismuth and antimony telluride TEGs on flexible silicon platforms. These TEGs show power densities in the order of 140 μW/cm$^2$. Mature nanofabrication methods and materials allows high-performance TEGs to be integrated monolithically with silicon-based circuits. Additionally, due to the flexibility given by the thin silicon substrate, these devices can be used for wearable and implantable electronics where conformal integration of devices is required to take advantage of the waste heat generated by the human body. Francioso et al. developed a flexible TEG based on bismuth and antimony telluride with a power output of 4.18 nW at a temperature difference of 15°C [281]. Yang et al. showed a flexible thermoelectric nanogenerator (TENG) based on a Te-nanowire/poly(3-hexylthiophene) (P3HT) polymer composite, producing a current density of

32 nA/mm$^2$ under a temperature difference of 55°C, which can be also used as self-powered temperature sensor with a sensitivity of 0.15°C in ambient atmosphere [282]. Some successful previous attempts of stretchable electronic demonstrations include the introduction of novel, naturally stretchable materials such polymers/organics, composites, 2-dimensional (2D) and 1-dimensional (1D) materials [283,284] On the other hand, innovative ways to engineer the mechanical structure of devices can be used with out-of-the-box geometries, such serpentine, horseshoe, spiral, helical, fractals and others [285–288]. These materials and strategies have been used to demonstrate stretchable systems for diverse applications, such wearable systems, robotics, bio-integrated electronics and many others [289–292].

Zoui et al. [151] point out that the human body releases around 100 W of heat at rest, and 525 W during physical effort [293,294]. Several investigations have been conducted into wearable thermoelectric generators (WTEGs) since 2001 [293,295] (see Figure 8.15), with the aim of substituting lithium-ion batteries [296] as a power source for portable devices, given that the global market for portable technologies is growing rapidly and is expected to exceed about 80 billion dollars by 2022 [297]. WTEGs are classified according to their rigid or flexible [298] architectures in 2D or 3D configurations [299], or according to their TE component materials, which are inorganic, organic or hybrid [300]. Leonov and Vullers [301] concluded that the wearable thermoelectric generator style was mature, and that the major concern was to improve the efficiency of the generator and make it thinner and more flexible. These authors conducted extensive research on rigid substrate TEGs. They developed different WTEG products that used body heat, such as the wireless electrocardiography system integrated into an office shirt. This product was powered by 17

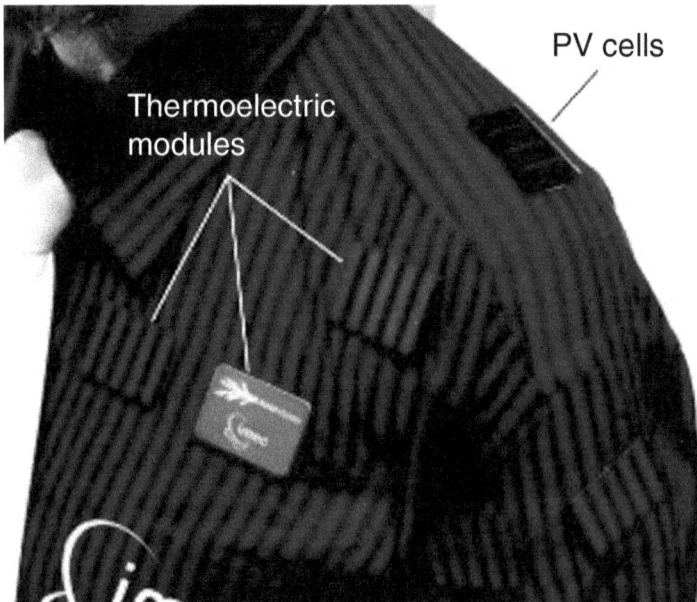

**FIGURE 8.15** Wearable Thermoelectric Generator (WTEG) integrated into a shirt [151,301].

small TE modules integrated into the front side of a shirt. They converted the body's natural heat flow into 0.8–3 mW of electrical energy, depending on the physical activity of the person [301].

The study of Ren et al. [302] report the TEG system which had the Lego-like reconfigurability, allowing users to customize the energy-harvesting device according to thermal and mechanical conditions. These properties are realized by integrating high-performance modular thermoelectric chips, dynamic covalent thermoset polyimine as substrate and encapsulation, and flowable liquid metal as electrical wiring through a novel mechanical architecture design of "soft motherboard-rigid plugin modules" (SOM-RIPs). A wavelength-selective metamaterial film was introduced to the cold side of the TEG to enhance the thermoelectric performance under solar irradiation, which is critically important for wearable energy harvesting during outdoor activities. It is worth pointing out that the flexibility and stretchability of this TEG are limited along the direction parallel to the thermoelectric chips. However, TEGs with ultrahigh flexibility and stretchability along one direction are well suited for cylindrical heat sources, such as arms, legs, and fingers for wearable applications and industrial pipelines for waste heat harvesting.

Often a wavelength-selective metamaterial film is integrated at the cold side of the TEG to simultaneously maximize the radiative cooling and minimize the absorption of solar irradiation. Therefore, the thermoelectric performance can be greatly enhanced under solar irradiation, which is critically important for wearable energy harvesting during outdoor activities. The overall design concept of this work is scalable and adaptable to other thermoelectric materials and fabrication methods, including roll-to-roll physical vapor deposition and printing techniques [303]. It is possible to further enhance the thermoelectric performance of the wearable TEG, by improving the fabrication process of thermoelectric films, adopting thermoelectric films with better thermoelectric properties [304–307], and using traditional thermoelectric legs with much smaller dimensions.

An additional advantage of developing a stretchable TEG is the ability to adjust its separation between the cold and hot junction. Naturally, the further away you move from a heat source, the higher would be the temperature difference, which can lead to higher power generation as long as the TE properties are not highly impacted by the device's deformation. Recently, a spiral-based architecture was used to demonstrate this effect on a TEG made from $Bi_2Te_3/Sb_2Te_3$ sputtered on top of polyimide (PI) and polyester paper. The maximum generated power increased by more than twice at 60% of elongation on the TEG fabricated on polyester- paper, a material with a lower thermal conductivity to ensure that the higher temperature difference is maintained [308]. In fact, this characteristic could be really useful in many practical scenarios, such as in moving engines, robotic parts, and others where a heat source and heat sink might be available in the system but their separation could be long and in constant movement. In such cases, a stretchable TEG could be used to leverage the situation and produce the highest possible power.

Zoui et al. [151] point out that one of the leading companies in this field is IMEC (Belgium), which has been working on the application of TEG for power electronic health care systems. IMEC and the Holst Centre have developed several wireless sensors, such as the Body-powered Electroencephalogram Acquisition System, which

(a)

(b)

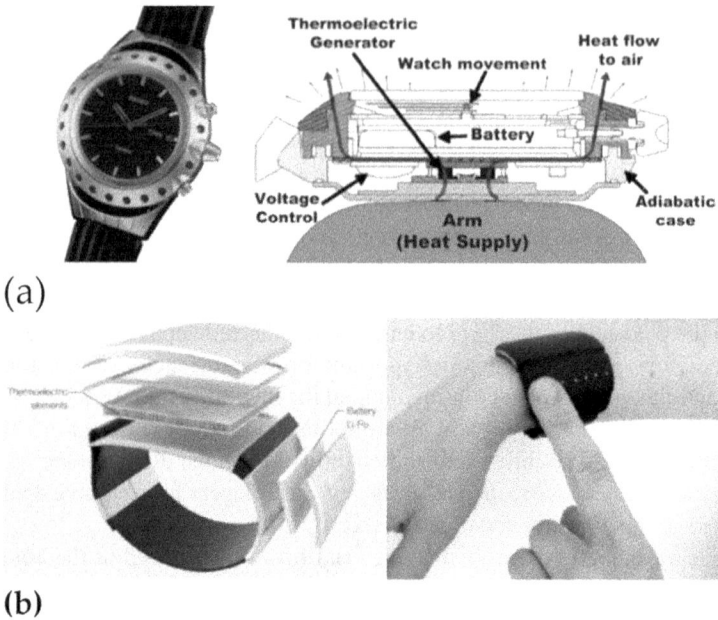

**FIGURE 8.16** (a) Seiko Thermic, wristwatch [136,317]; (b) Dyson bracelet [151,314].

produces 2–2.5 mW of power and is worn as a headband [309]. They have also developed a wireless pulse oximeter, powered entirely by a TEG-style watch using commercial $Bi_2Te_3$ thermopiles and in which the generator develops around 89 μW of power [310]. The first thermoelectric wristwatch powered by converting body heat into electrical power was marketed by Seiko and Citizen [311,312]. The Seiko watch (Figure 8.16a) produced 22 μW of electrical power and an open-circuit voltage of 300 mV with the efficiency of about 0.1% [313]. Another example is the Dyson TE bracelet, shown in Figure 8.16b, which, using body heat, charged a battery integrated into it for charging a mobile phone or any other mobile device.

Zoui et al. [151] point out that because of the high thermal resistance between skin and the TEG, flexible modules are more suitable for power generation from body heat, as they can be adapted to the shape of the body, thus increasing the useful surface area for heat capture and reducing thermal contact resistance [315]. Francioso et al. [316] developed a flexible and wearable micro thermoelectric generator, composed of an array of 100 thin film thermocouples of $Sb_2Te_3$ and $Bi_2Te_3$, designed to power very low-consumption electronical ambient assisted living (AAL) applications. The best result obtained was 430 mV in open circuit, and an electrical output power up to 32 nW at 40°C. Kim et al. [317] manufactured a flexible fabric-shaped TEG, with 3D printing technology, composed of 20 thermocouples and with a thickness of 0.5 mm. The TEG, when applied to a human body, generated an electrical power of 25 mV at an ambient temperature of 5°C. A new approach was presented by Suarez et al. [318], using standard bulk legs interconnected to a stretchable low-resistivity eutectic alloy of gallium and indium (EGaIn), all in a flexible elastomer package. Zadan et al. [319]

introduced a soft and stretchable thermoelectric generator (TEG) with the ability to expand, to explore the integration of this TEG into wearable technologies. There are, however, limitations to all approaches mentioned above. Environment conditions (particularly temperature) are important for their workability. They are mostly used for moderate indoor temperatures which limits their applications. TE modules are also expensive [320].

Soleimani et al. [321] reported that hybrid TE materials are the solution to the rigidity of inorganic TE materials and the low efficiency of organic TE materials. These hybrid TE materials are suitable for portable TEGs. Jiang et al. [322] concluded that flexible and wearable applications will eventually become a reality with the development of preparation technology for film or fiber legs, and the emergence of human thermoregulatory models for the designing of wearable devices and their integration with other wearable renewable energy conversion devices.

Yee and coworkers at Georgia Tech are developing flexible TEG using polymers. TEGs are typically made from inorganic semiconductors. Yet polymers are attractive materials due to their flexibility and low thermal conductivity. These qualities enable clever designs for high-performance devices that can operate without active cooling, which would dramatically reduce production costs. The researchers have developed P- and N-type semiconducting polymers with high performing ZT values (an efficiency metric for thermoelectric materials). Currently they have achieved ZT = 0.1 with goal of 0.5. In one project funded by the Air Force Office of Scientific Research, the team has developed a radial TEG that can be wrapped around any hot water pipe to generate electricity from waste heat. Such generators could be used to power light sources or wireless sensor networks that monitor environmental or physical conditions, including temperature and air quality. The researchers indicate that the use of polymer allow the development of paint or spray material that will generate electricity. It also opens up new opportunities in wearable devices, including clothing or jewelry that could act as a personal thermostat and send a hot or cold pulse to your body. While this can be done now with inorganic thermoelectrics, but this technology results in bulky ceramic shapes. Plastics and polymers would enable more comfortable, stylish options. While not suitable for grid-scale application, such devices could provide significant savings.

### 8.5.2  WASTE HEAT RECOVERY FROM INDUSTRY AND HOMES

Waste heat is a significant problem for homes and industry. A significant amount of waste heat is allowed to be released without recovery [148]. In recent years, the use of TEG for heat waste heat recovery from home and industry has increased and significant efforts are made to improve the efficiency of TEG systems [323]. TEG systems could be easily adopted to the pressure, temperature and heat transfer fluid required for these applications. Dai et al. [324] reported that, in United States, 33% of industrial manufacturing energy is released directly into the atmosphere or into cooling systems as waste heat, and this amount of heat could be used to produce 0.9–2.8 TWh of electricity per year if thermoelectric materials with average ZT values ranging from 1 to 2 were available. A large-scale application of TEG for industrial waste heat would make it a competitive source of clean energy [325]. Zou et al. [326] demonstrated that municipal wastewater can be used to produce electricity using a thermoelectric generator (TEG). Their theoretical study, performed for the Christiansburg Wastewater Treatment Plant, estimated energy generation of 1,094–70,986 kWh/year, with a saving of USD 163–6,076. Meng et al. [327] showed that for an air-cooling heat recovery device with 120-mm$^2$ pipes used for the wastewater, recovery efficiency was 1.28%, and the amortization period for the equipment extended to 8 years [327].

Araiz et al. [328] carried out a techno-economic study into the thermoelectric recovery of hot gases from a stone wool manufacturing plant. They reported a maximum net power production of 45 kW, and a Levelized Cost of Electricity at about 0.15 EUR/kWh. Mirhosseini et al. [329] performed a similar investigation, using an arc-shaped absorber designed for the thermoelectric heat recovery of waste heat from a rotary cement kiln. The economic evaluation showed that the dominant parameter in the system cost was the heat sink implying that in many cases heat recovery depends strongly on the ambient temperature. In hot regions, investigations were mainly focused on the recovery of heat released from air conditioning systems [330].

The investigations into heat recovery in cold regions were, however, more diversified. Killander and Bass [331] developed and tested a prototype of a thermoelectric generator designed to supply small amounts of electricity using heat from existing wood stoves in homes in the cold and isolated regions of northern Sweden. The device provided sufficient energy for electric lighting and watching television during the long winter nights. In the Netherlands, Gasunie Research, developed, built and tested 20 autonomous (self-powered) boilers that used the heater flame produced to co-generate enough electricity to operate its electrical components, using 6 Hz-20 thermoelectric modules [332]. They concluded that these thermoelectric generators supplied 60 W of electricity, which was enough to run their electrical components. Allen and Wonsowski [333] carried out tests for the use of TEG in residential-scale hydronic central heating units to demonstrate autonomous operation in a realistic environment. The thermoelectric stage was a set of 18 thermoelectric modules made of bismuth telluride alloy, which generates an electrical power of 109 W, sufficient to supply the blower, the gas control and the water pump of the hydronic central heating. Sornek et al. [334] pointed out that the commercial success of such an installation has to be focused on: (a) introduction of necessary modifications to the heating devices, and (b) the development of a dedicated structure of the TEG.

Bass and Farley [335] tested three thermoelectric generators designed to provide electrical energy in a natural gas field. These generators converted the waste heat produced by the equipment used in the gas field into a thermal energy source for the generators. Electricity was used for cathodic protection, telemetry power supply and lighting. US military [336] used thermoelectric technology to reduce the logistics of field feeding by integrating thermoelectric devices into the Assault Kitchen that was used to heat food rations out on the field. These devices eliminated the need for an external electrical generator to power the Assault Kitchen. In addition, they produced an excess of electricity that could be used for lighting, battery charging, radio power, communication equipment, etc.

### 8.5.3 Waste Heat Recovery from Transport Systems

#### 8.5.3.1 Automobiles and Motorcycles

Zoui et al. [151] point out that road transport in Europe contributes about 20% of the total carbon dioxide emissions, 75% of which come from private cars, and similar rates are observed in America and Asia [337]. European regulations aim to achieve a $CO_2$ emissions target of 68 g/km by 2025, for passenger cars and light commercial vehicles [338]. It is worth taking into account that two-thirds of a vehicle's combustion energy is lost as waste heat, 40% of which is in the form of hot exhaust gas [339,340]. If about 6% of exhaust heat could be converted to electrical energy, it would be possible to reduce fuel consumption by about 10% [341]. To this end, Agudelo et al. [342] tested a diesel passenger car in a climatic chamber in order to determine the potential for energy recovery from exhaust gases. They concluded that the potential fuel savings ranged from 8% to 19%, and the silencer showed the highest energy losses, so the installation of a TEG needed to be located prior to it. Moreover, the study showed that there are three main possible locations for the TEG [343], namely: (a) The TEG is placed at the end of the exhaust system (see Figure 8.17); (b) The TEG is located between the catalytic converter and the silencer—the best option; and (c) The TEG is located upstream of the catalytic converter and silencer. If the weight of the installed TEG and the additional pressure drops in the exhaust system are not optimized, the vehicle will consume more fuel than it needs to save, and consequently the system becomes totally inefficient [343].

Birkholt [344], in collaboration with Porsche, proposed a thermoelectric generator with a rectangular cross-section, which was able to produce up to 58 W under peak conditions with $FeSi_2$ elements. The Nissan Research Center from Japan [345] developed a TE generator with a rectangular cross-section of 72 modules. Later, they tested a thermoelectric generator composed of 16 $Bi_2Te_3$ modules operating at low temperature; the electrical power generated by the generator was 193 W [346]. Hi-Z Technology started the development of a 1-kW thermoelectric generator for diesel truck engines with funding from the U.S. Department of Energy and the California Energy Commission [347]. Amerigon (now Gentherm) developed thermoelectric generators for passenger vehicles in five phases [348–352]. In the final phase, power output improvement was achieved with a BMW X6 (Figure 8.17) and a Lincoln MKT (Ford), with more than 600 W produced in vehicle tests and more than 700 W in bench tests [348,351,352].

According to Zoui et al. [151], the success by Gentherm led Gentherm, BMW and Tenneco to launch a new program, projected over 7 years, using a new thermoelectric cartridge device. The electricity generated by the TEG could save fuel consumption by 2%, which was far from the program's aims. General Motors developed a prototype using Bi-Te and Skutterudite modules, which was installed on a Chevrolet suburban [354]. Average power developed by the TEG was expected to be 350 W for city driving cycles, and 600 W on motorways [355]. In 2013, Fiat and Chrysler announced the manufacture of the first light commercial vehicle equipped with a TEG [356] with a fuel economy of 4%. The TEG used cross-flow architecture, with segmented TE elements of TAGS, $Bi_2Te_3$-PTe and Skutterudites. RENOTER [357], an association between Renault and Volvo also developed TEG for automobiles. Bou Nader [358]

**FIGURE 8.17** TEG integration into the exhaust line of the BMW X6 prototype vehicle [151,353].

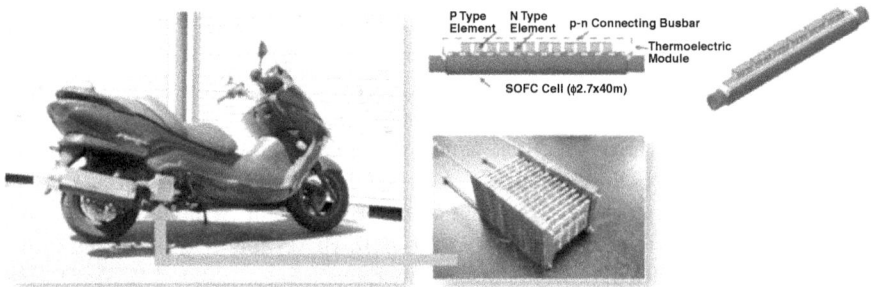

**FIGURE 8.18** TEG installed in the exhaust gases of a motorcycle [151,360].

proposed an innovative thermodynamic configuration, and investigated the fuel-saving potential of hybrid electric vehicles using a thermoelectric generator system as an energy converter, instead of the conventional internal combustion engine. Recently, Shen et al. [353] presented the current status, challenges and future prospects of automotive exhaust thermoelectric generators. The authors cited 11 challenges to overcome before the possibility of commercial applications could be realized.

The main mode of transportation in Indonesia and other Asian countries is motocycles. Septiadi et al. [359] investigated the usefulness of installing a thermoelectric generator on the exhaust of a motorcycle. Their results showed that the output voltage was 15.7 and 7.7 V, for TEGs of four modules and two modules, respectively. ATSUMITEC, in cooperation with the Nagoya Institute of Technology, applied the Heusler module to the underpower generator of a motorcycle [360], by integrating the thermoelectric device with a fuel cell, as illustrated in Figure 8.18. The total output power of the fuel cell plus the thermoelectric power was 400 W [361]. Schlichting et al. [362] showed that the potential of replacing the alternator with TEG units was quite low.

### 8.5.3.2   Aircraft

Modern aircrafts are increasingly equipped with sensors and transmitters for better control and monitoring, and greater safety. The supply of power to these sensors via power lines would result in additional heavy wiring and requiring additional fuel consumption. According to Zoui et al. [151], Boeing Research & Technology estimated that a 0.5% reduction in fuel consumption would be translated into a USD 12.075 million reduction in monthly operating costs for U.S. commercial aircrafts, and approximately a 0.03% reduction in carbon emissions for U.S. passenger aircrafts [363]. German Aeronautical Research Program (LuFo-5) showed that when TEG is integrated between the hot part of a propeller and the bypass flow of the cooler, the efficiency of the TEG ranged from 3% to 7%, with a power of 1–9 kW/m$^2$ depending on its location in the various hot parts of the propeller [364]. Lyras et al. [365] designed a TEG to be installed between the inside and outside walls of the aircraft which are at substantially different levels of temperature [366]. Allmen et al. [367] indicated that some sensors, such as the stress sensor which controls the state of health of the hull, must be installed on different parts of the aircraft. Therefore, it would be very useful to be able to employ a TEG attached directly to the fuselage and combined with a phase-change material (PCM) heat storage unit. This would create a temperature gradient during takeoff and landing, which could generate electricity to power a node of autonomous low-power wireless sensors [368,369]. The system was successfully integrated and functionally tested by Allmen et al. [367], qualifying it for use in a flight test facility. The results of the study for helicopters showed that the electrical energy produced under real operating conditions was significant but insufficient when one takes into account the weight/power ratio [370].

### 8.5.3.3   Ships

Maritime shipping alone represents about 2.8% of greenhouse gases in the world [371]. Moreover, the integration of thermoelectric power generation into ships is more advantageous than integration into other transportation systems, because cooling water is fully available. The integration of thermoelectricity in this sector is almost non-existent because of the absence of strict international regulations imposing permissible pollution rates for ships. The European Union is currently planning measures to reduce emissions from international maritime transport [165]. A project called ECOMARINE was carried out to implement a thermoelectric energy recovery unit, with the aim of maximizing electricity production by waste heat recovery. In this context, Loupis et al. [372] developed a tubular TEG with a diameter of 500 mm, which ensured a very low pressure drop of the exhaust gas flows during their passage through the RTG. The authors reported a conversion efficiency of 6.4%, a waste heat recovery of 1.2%, and an electricity supply of 20.3 kW [151].

### 8.5.4   Thermoelectricity for Regenerative Computing

Due to difficulty in advancing process technology as well as fundamental physics bottlenecks, the computing industry is now shifting away from a tick-tock process-architecture model to a process-architecture-optimization model. What this means is that future integrated-circuit technology nodes (down to 7 nm) will be around for a longer time and integrated device manufacturers may determine their own roadmap based on the requirements of system integrators. The focus of major system integrators is largely in mobile and distributed computing, given the growing consumer demands for faster and efficient electronics in an attractive form factor. In order to bridge the performance-power gap, new innovation is required at the transistor level. The paradigm shift from a planar topology to a three-dimensional architecture (FinFETs), is an excellent example [373]. Compared to past planar transistors, current generation of FinFETs [373] are faster while consuming less power and also run cooler temperatures due to lower heat dissipation. But due to the rising cost and complexity in manufacturing FinFETs on bulk silicon, there is a wide consensus in migrating to silicon on insulator (SOI) platforms for future technology nodes. Heat dissipation and transport in SOI-based nanoscale transistors is more challenging due to the presence of an insulating buried silicon dioxide layer [374,375]. This is even more profound in stacked 3-D integrated circuits [376] that are being considered for unprecedented computing parallelism and speed. The bottom line is that heat dissipation from such nanoscale devices will persist as transistor density continues to increase. The way this heat flux is managed in electronics will ultimately determine their performance, power consumption, operation lifetime and reliability.

One way to address the problems due to heat dissipation is by effectively removing this heat flux using thermoelectric coolers. Chowdhury et al. demonstrated chip-level site-specific cooling by making use of nanostructured thin-film super-lattices of $Bi_2Te_3$ to remove dissipated heat from a silicon test chip containing micro-heaters [377]. Using this, they were able to demonstrate nearly 15°C cooling on the test chip with a heat flux of ~1,300 W/cm$^2$. Another way to aid in tackling the heat dissipation problem as well as boost battery lifetime in mobile electronics, is by harvesting and reusing this heat at the device, circuit or die level. Using commercial off-the-shelf thermoelectric generators (TEG), Zhou et al. carried out experiments to measure and harvest the heat energy dissipated by microprocessors at different workloads showing the un-tapped potential of thermoelectricity in computing [378]. Building on this concept, in another work, Fahad et al. introduced the concept of quasi-self-powered integrated circuits by reusing the harvested waste heat from microprocessors [379], to drive an array of light emitting diodes (LED). However, these are all examples of off-chip/die implementations of standalone thermoelectric modules. A more fundamental and impactful approach would be at the device and circuit level, by integration of thermoelectric elements alongside transistor blocks as well as metal interconnects. A great example of this is demonstrated by Aktakka et al., where low-temperature high-quality $Bi_2Te_3$ and $Sb_2Te_3$ thin film thermoelectric devices were integrated with SOI FinFETs to harvest ~0.7 μW of power from a ~21°C thermal gradient [380]. Similar innovations can enable the development of quasi-self-powered electronics for the ever growing mobile consumer market.

### 8.5.5 SOLAR WASTE HEAT TO GENERATE THERMOELECTRIC POWER

The promise of an un-interrupted supply of energy from photovoltaic solar panels is severely impeded by the intermittent nature of available sunlight. Inayat et al. [381,382] implemented novel thermoelectric glass windows for mass scale energy generation using the temperature difference between the rich supplies of solar heat and the ubiquitous ambient environment around us. These novel energy generators scavenge electricity from direct solar irradiance during daytime, while during off sun hours are still capable of utilizing the hot outdoor temperature to generate ample amount of electricity. Thermoelectric systems were either fabricated to produce milli-watts of energy using a small temperature gradient of only a few Kelvins to power micro devices. On the other hand, thermoelectric generators in industrial applications or engines use temperature gradients of hundreds of Kelvins to produce kilowatts of energy. These thermoelectric windows extend the concept of TE systems at the lower end of the spectrum (used for micro-devices), to a large coverage area, using abundant and unlimited source of temperature gradient between outside solar and ambient inside, for mass scale energy generation.

Expanding the concept of micro TEGs onto large coverage areas simultaneously eliminate the two primary problems associated with the micro TEGs. First, the low-temperature gradient produces a very small output voltage in a micro TEG. This low output requires voltage boosters that are not only miniaturized but should also be highly efficient. In addition, designing and integrating power conditioning circuitry at micro-scale becomes a hard challenge. By expanding thermoelectric system over a large coverage multiplies the output voltage manifolds, thus eliminating the need for voltage booster altogether or at least relaxing the efficiency specs for these boosters. Additionally, the macro-sized mass scale energy harvesters also allow the use of large-sized maximum power point trackers (MPPT) with high efficiency.

Design limitation of commonly available TEGs has served as a major impediment toward the employment of thermoelectric systems for mass scale energy generation utilizing the temperature gradients existing between outdoor (environment) and indoor (inside room) [381,382]. Lateral thermoelectric generators demonstrated by H. Glosch [383] and D.M. Rowe [384] require the heat to flow through the thermocouples in the lateral direction. Alternatively, the vertical TEGs with vertically erected thermocouples are similar to the devices implemented by Bottner for Micropelt [385], Snyder for JPL [386], Kishi for Seiko [387], required the flow of heat from top to bottom or vice versa through the device. Both these design versions strictly required the two counter temperature environments to be simultaneously accessible to the hot and cold side of the generator, thus eliminating the presence of any interface between the two temperature environments. Given these conventional design limitations, interfaces like window glasses tend to make it impossible to generate thermoelectricity using the rich supply of hot outdoor environment and indoor ambient. Placing the thermoelectric generator on the outer side of a window (or vice versa) allows the outdoor temperature to influence the hot side of the generator but prevents the thermoelectric generator from accessing the cold indoor because of the intermediate blocking window. Inayat et al. [381,382] have, however, been able to generate mass scale thermoelectricity by innovatively using the interface, between outdoor solar heat and indoor ambient, as generator of thermoelectricity by placing the thermoelectric materials through the interface.

The validity of the concept was verified with the finite element behavioral analysis of a thermoelectric window system. The model was designed for a single pair of complimentary thermoelectric legs (pillars) integrated inside a glass substrate. The two pillar device model used can be further extrapolated to give the outcome for a full sized glass window with thermoelectric materials integrated across the entire coverage of the glass window. The presence of an output voltage of 0.12 mV on the p type pillar while −0.04 mV on the n type pillar of the device for a 20°C temperature gradient, confirms the validity of the model for predicting the behavior of a thermo-electric system, with thermoelectric materials embedded inside a glass window [388].

Based on trade-offs between thermoelectric properties and the processing delay involved in mechanical alloying of large amounts of powders, as purchased powders were selected for batch nano-manufacturing of thermoelectric pillars. The as pur-chased thermoelectric powders were hot pressed into vertical pellets. The tubular design of the dies facilitates the thermoelectric characterization of the pressed pil-lars by eliminating the unnecessary delay involved in cutting smaller pellets out of a large tablet obtained using conventional large sized dies. Prototype implementation of a thermoelectric window has been implemented with (132.25 cm$^2$) Plexiglas panel laser drilled with 144 holes to accommodate hot pressed thermoelectric pillars.

Inayat et al. [381,382], leveraged the mechanical alloying of commercially available powders of $Bi_{1.75}Te_{3.25}$ and $Sb_2Te_3$ to nano-manufacture pellets through ball milling for high performance thermoelectric windows. Effect of mechanical alloying and hot press-ing on thermoelectric properties of the as-purchased powders was investigated through the thermoelectric characterization of the hot pressed pillars of the as purchased pow-der samples. Nano-manufacturing through mechanical alloying showed a significant enhancement in electrical and thermal properties of the as purchased powder samples. Although the output power levels from the prototype panel were low, the presence of highly repeatable thermoelectric output from the prototype thermoelectric Plexiglas vali-dated the effectiveness of a batch fabrication process enabled efficient nano-manufactur-ing of large batches of thermoelectric pillars for large coverage thermoelectric window. Additionally, the property enhancement of thermoelectric powders with mechanical alloying also opened up an opportunity for further investigation into the ball milling of thermoelectric powders to bring a significant increase in the output power levels.

Maximum attainable limit of the hot press used and the contact resistance between the thermoelectric pillars and the metal interconnects have been the major perfor-mance impeding factors for the prototype thermoelectric panel. By using techniques like Spark Plasma Sintering (SPS) for performance enhancement of thermoelectric pillars and inkjet printing for low resistance interconnects, the work of Inayat et al. offers an opportunity to potentially generate 304 W of usable power from 9 m$^2$ win-dow at a 20°C temperature gradient [381]. If a natural temperature gradient exists, this can serve as a sustainable energy source for green building technology.

### 8.5.6 WASTE HEAT FROM NATURAL GAS IN REMOTE LOCATIONS

In recent years, thermopiles or thermoelectric generators are designed to supply energy to autonomous sensors, installed in remote locations subject to severe envi-ronmental conditions, i.e., very low-temperature and difficult-to-access locations, where conventional renewable energy sources, such as solar and wind energy, are not regularly available. Heat is usually supplied by a flameless catalytic burner [389]. A few manufactures of thermoelectric generators powered by natural gas have installed them in more than 55 countries. According to Zoui et al. [151], Gentherm manufactures TEGs with powers ranging from 15 to 550 W. These generators are mainly used on offshore platforms, along pipelines, at high altitudes or near gas wells (Figure 8.19) [390]. Farwest Corrosion Control, a company that manufactures and installs TEGs for cathodic protection against pipe corrosion, and they have installed more than 15,000 generators in 51 countries [391].

Several products designed for public use using natural gas have also been marketed. Horie [392] developed a thermoelectric candle radio, which used the heat from candles to power a radio via a $FeSi_2$ TE module. More recently, as illustrated in Figure 8.20, the CampStove designed for camping burns wood to produce 2 W of 0.4 A and 5 V power using a thermoelectric generator to which the connection of the electrical devices is made via a USB port [393].

FIGURE 8.19 Gentherm Gas TEG [116].

FIGURE 8.20 Picture of a CampStove [127].

## 8.6   PERSPECTIVES ON COMMERCIALIZATION AND FUTURE OUTLOOK

Demonstrations of high-power thermoelectric WHR are necessary to prove that such systems can take advantage of economies of scale. Demonstrated power has typically been less than 1 kW, with notable demonstrations of 169 W generation from a cement kiln [394], 250 W generation from glass furnace exhaust [395], and 240 W generation from a steel carburizing furnace afterburner [396]. A 1 kW generator was mounted successfully into the Bradley fighting vehicle by a Department of Defense contractor based on a module from Hi-Z Technology, Inc.

Two recent thermoelectric energy harvesting efforts have demonstrated power generation in excess of 1 kW. The first of these involved a high-power TEG installed over a continuous casting line. This system contained 896 Komatsu modules, 16 of which were used in the carburizing furnace demonstration, and it produced power on the order of 9 kW when exposed to the radiant heat of a 915°C slab [397]. The other is the first plug-and-play TEG available for purchase, the 25 kW Alphabet Energy™ E1, which was announced in October 2014 [398]. Efficient and cost-effective thermoelectric energy generation systems depend on a number of module- and system-level design factors in addition to the selection of materials with good thermoelectric properties. At the module level, the choice of materials can be complicated by the fact that p- and n-type materials are both typically required. In addition, assembly of thermoelectric materials together with conductive leads and dielectric plates to form modules at an industrial scale is challenging. The different materials and their interfaces must be robust enough to tolerate thermal expansion and contraction, and modules must be designed to properly conduct heat through thermoelectric materials, maximizing temperature gradients and thus power generation. Module assembly quality requirements are high: a single bad connection makes a module useless due to serial electrical connections between the thermoelectric legs.

At the system level, the challenge of maximizing the temperature difference over the thermoelectric modules to maximize efficiency means that heat exchangers on the hot and cold sides must be well designed to maximize heat flux to the modules. Hot-side heat exchangers are particularly challenging to design in many cases in which they must tolerate high heat flux fluids that can be corrosive or contaminated with particulate matter (PM). Cost optimization of heat exchangers that transfer source heat to the cold side and also include heat transfer within the module is important. In high-temperature applications, heat exchanger materials and protective coatings impact costs as the material thermal properties do. Electrical design challenges include proper matching of resistance between thermoelectric modules and loads as well as efficiently inverting the DC power for use on the grid. Costs for electrically insulating plates (usually ceramic) must be reduced while maintaining good thermal conductivity at elevated temperatures. Electrical interconnects and other interfaces must be engineered to minimize electrical and thermal losses and to provide oxidation protection in order to maximize device and system efficiency and reliability. Using more corrosion-resistant heat exchangers would allow thermoelectric heat recovery from a more diverse range of industrial exhaust streams. Studies to co-optimize the thermal and electrical properties of the whole TEG system while maintaining its mechanical integrity are also important [399]. Heat exchangers can present significant design challenges for applications with modest temperature gradients. Enabling high heat flux in these regimes requires more complex and highly engineered systems whose cost can significantly impact overall system cost.

Progress must also be made in the area of automated assembly so that thermoelectric devices can be made in a reliable and cost-effective manner. Traditionally, metal interconnects are attached to ceramic insulating plates by using one of several available processes, such as soldering, thin film sputtering, and plating [400]. The thermoelectric legs are then soldered to these interconnects. More than 90% of the thermoelectric modules assembled today require some manual operations, such as attaching leads and visual inspections [401]. Some manufacturers have implemented automation systems for the assembly process, but in most cases, price points do not justify straying away from the standard pick and place machines used to assemble electronic components [401]. Thin film thermoelectric modules offer an alternative to the manufacturing methods of conventional bulk materials because the p- and n-type materials can be sputtered onto separate wafers (using techniques from silicon microelectronics fabrication) that are then fused together. An overarching technology need is to reduce the cost of power generated by thermoelectric WHR systems. This need can be met by using lower-cost materials and automated methods of thermoelectric assembly. A commonly discussed cost target in the thermoelectric field is $1/W for an installed system. This, along with a system life of 5 years, a discount rate of 7%, a capacity factor of 75%, and an annual cost (for maintenance and operating costs) of $0.20/W would lead to a levelized cost of electricity (LCOE) of $0.067/kWh. This is comparable to the 2013 average U.S. industrial electricity price of $0.068/kWh [402]. A recent DOE workshop on manufacturing opportunities for low-cost thermoelectric modules indicated that labor is responsible for a significant portion of the cost of thermoelectric modules [403]. New manufacturing approaches, such as automation and wafer-based manufacturing, have the potential to reduce thermoelectric module costs. Additive manufacturing of thermoelectric modules and wafers should be considered. The Fraunhofer Institute for Manufacturing Technology and Advanced Materials has demonstrated additive manufacturing of thermoelectrics with embedded sensors [404]. More demonstrations with medium- and high-temperature thermoelectric materials are needed to prove that this approach is viable.

The lack of moving parts in TEGs holds the promise of reduced operation and maintenance costs and longer intervals between failures. These potential benefits make TEGs important to consider for industrial WHR applications because they have good reliability. Large-scale TEGs (greater than 1 kW) that generate general purpose power (i.e., power not meant for a particular dedicated application) from industrial waste heat is not in general use at the present time. The largest commercially available TEG systems are remote generators manufactured by Global Thermoelectric, Inc., a Calgary-based company recently acquired by Gentherm, Inc., which has installed over 20,000 TEGs worldwide. These systems are generally fuel burning with a maximum power of around 500 W, and they provide electricity along gas pipelines and on offshore oil platforms, among other similar locations. However, there are a few larger waste heat systems that have been installed and discussed in the literature [10]. An extensive study of a TEG in a working glass plant [395] discussed the difficulty of delivering heat from the exhaust stream to the generator via a heat pipe. The power generation in this study was not cost-effective, and the researchers faced great difficulties with the degradation of heat flux through the heat pipe, owing to corrosion in the exhaust flue [10]. The iron and steel industry is a common target for WHR technology based on its high-quality waste heat.

As such, the iron and steel industry has been the subject of the most extensive discussion of industrial thermoelectric waste heat applications in the literature. One notable study involved a TEG with 16 bismuth telluride modules placed above an afterburner flame in the exhaust system of a carburizing furnace at the Komatsu Ltd., Awazu steel plant in Japan [10]. The afterburner flame was estimated to produce up to 20 kW of heat and to induce temperatures between 120°C and 250°C on the heat collection place of the water-cooled TEG. The modules used were developed by a Komatsu subsidiary, had a potential power density of 1 W/cm², and had the highest conversion efficiency of any commercial thermoelectric module when they were announced in 2009 [10]. The modules were very expensive, however, with a price of roughly $30/W when they were released. A later study at this plant demonstrated power outputs on the order of 240 W for single generators and discussed the installation of power hardware to effectively manage the output of multiple generators [10]. Results for a modified version of a cost model from LeBlanc et al. [10,405] for thermoelectric generation system costs and the resulting LCOE values have been used to evaluate the economics of thermoelectric generation in the steel industry, based on a detailed waste heat breakdown. In these calculations, casting was assumed to have no exhaust losses (as all losses would be from cast products cooling to room temperature); therefore, estimates have been limited to non-exhaust sources in this case. The sources such as castings, BOFs, and blast furnaces/stoves are deemed as the most promising for thermoelectric WHR and are the non-exhaust waste heat sources with the lowest LCOE values. Non-exhaust waste heat sources are desirable because the corrosive chemicals in steel industry exhaust gases are a great obstacle to thermoelectric WHR [10].

Although thermoelectric systems could contribute to efficiency gains at manufacturing facilities, demonstration at the levels of prototype and full-scale production is needed, particularly to understand and address the 10%–20% drop in ZT that often occurs for large production volumes due to process defects. There are only a few large-volume thermoelectric material producers in the world today. Collaboration with manufacturers to perform cost-effective, system-level TEG demonstrations in near-term potential applications such as those performed in Japanese steel plants [397] would also be useful for establishing the feasibility of TEG WHR for commercial applications. Finally, encouragement of more and better wireless sensor network energy data collection without large infrastructure investments would allow manufacturers to develop more efficient processes by identifying which of their systems produce the most sensible and latent heat for WHR. Encouraging the use of self-power TEG sensor nodes would have the added benefit of increasing the exposure of TEG technology [10]. The possibilities for installing a similar system to the one at the JFE Steel Corporation plant in Japan [397] at the Nucor mini mill in Jewett, Texas were explored. At 13–30 cm wide, the five continuous casting lines at the Nucor plant are not as wide as the 1.3–1.7 m slabs at the JFE plant, but assuming that the slab temperatures are the same, a similar amount of heat flux (on the order of 17 kW thermal/m$^2$) could be intercepted by placing 50 cm wide generators roughly 20 cm from the cast slabs. Assuming that the same level of thermoelectric generation per unit area that Kuroki et al. [397] discussed (1.13 kW electric/m$^2$) could be achieved over 14 m for each strand, 39 kW electric of thermoelectric power could be produced by 35 m$^2$ of generators (based on a 14 m × 0.5 m generator above each of five strands) at the Nucor plant. Under these conditions, using a nanobulk $Bi_{0.52}Sb_{1.48}Te_3$ TEG of a higher ZT value of 1 with a 15-year life span and a capacity factor of 60%, a modified version of LeBlanc's cost model [10] predicts an LCOE of $0.31/kWh.

Besides the steel industry, any industry in which high-quality waste heat goes unused should be considered a possible target for thermoelectric WHR. These potential targets include glass, aluminum, cement, and ethylene manufacturing, all discussed in a 2006 report by Hendricks and Choate [400]. This report also considered industrial and commercial boilers for their large aggregate waste heat, but these were found to lack high enough temperatures to make thermoelectric energy harvesting with current technology feasible. The most promising source the report found in terms of potential annual electricity generation was aluminium melting, which the authors determined could produce 1.4 TBtu/year by using thermoelectric materials with a ZT of 1. Current technology can be cost-effective when the thermoelectric system adds value beyond electricity production, as is the case in generation for wireless sensor networks. TEGs combined with wireless sensor network nodes can add value by allowing sensing and automation without the need for wire runs or batteries that require checking and replacement. In addition, remote industrial facilities with abundant waste heat and expensive electricity—in the oil and gas industry, for example—might benefit from current thermoelectric technology. The recent developments on the use of TEG for self-powered human devices also have a bright future. The use of TEG for solar energy-operated devices is gaining ground.

Examples of these value additions outside of the manufacturing sector include the use of thermoelectrics to drive fans that increase the efficiency of woodstoves [406] and to power wirelessly controlled radiator valves that do not need batteries. Several companies have demonstrated woodstoves with TEG-powered fans to show the potential of TEGs for self- powered appliances, including BioLite, Hi-Z Technology, Research Triangle Institute, Greenway Grameen, and others [407]. High-efficiency, low-emission biomass stoves have the potential to significantly reduce the 4 million deaths per year caused by pollution from indoor cooking with biomass in developing countries, as estimated by the World Health Organization; thermoelectric-driven fans offer an important pathway to help achieve this [408]. The overall effort is being led by the U.S. Department of State and the Global Alliance for Clean Cooking [409]. In October 2014, Alphabet Energy, Inc. announced a TEG product called the E1 that fits in a standard shipping container, connects to the exhaust pipe of a generator and has a modular design so that thermoelectric components can be swapped out as materials improve. The thermoelectric materials used in the E1 are p-type tetrahedrites and n-type magnesium silicide ($Mg_2Si$), which will provide an average ZT of around 1, similar to that obtained by skutterudite (GM, Gentherm) and halfHeusler (Evident Thermoelectrics) materials used by the DOE Waste Heat 2 (WH2) projects [401]. Alphabet Energy states that the E1 can produce 25 kW from the exhaust of a

1,000 kW generator, 66 implying an efficiency of about 2.5% based on the exhaust heat of such a generator [10]. If running constantly, the generator would produce roughly 219,000 kWh and would save 50,000 L of diesel fuel per year [10].

To obtain an estimate of the amounts of waste heat in each industrial sector, the DOE Advanced Manufacturing Office (AMO) *Manufacturing Energy and Carbon Footprints* data [410] can be used. The potential of thermoelectric energy from manufacturing plant waste heat can be estimated with the knowledge of the fraction of the heat that can be recovered by the generation system and the efficiency value for those systems. The choice for the low end of the recoverable heat range was 10%, based on an estimate from Polcyn and Khaleel [395], and the high end of 25% was based on heat recovery calculations for boiler exhaust from a study by Hill [411]. These estimates assume the thermoelectric generation efficiency to be 2.5%. This efficiency figure was chosen because module efficiencies of around 5% are seen in the sales literature of market modules for temperature differences around 200°C–250°C [412], and generally only half of the temperature gradient across the TEG system is available for power conversion across the TEG material (whereas the other half is dissipated across the heat exchangers) [413]. In addition, this 2.5% figure matches the efficiency implied in early press related to the first published large-scale, off-the-shelf, exhaust-based thermoelectric generation system [414].

Based on this estimate, the thermoelectric recovery potential for U.S. manufacturing is about 1,880–4,701 GWh (6–16 TBtu). This is a conservative estimate based on the thermoelectric generation efficiencies of existing systems on the market and does not take future technology development into account. DOE estimates [401] of waste heat that can be recovered from major process industries by TEG technologies are illustrated in Table 8.5. The energy savings opportunity could be considerably enhanced

---

**TABLE 8.5**

**Estimate of Waste Heat That Could Be Recovered with Thermoelectric Generation Technologies for Major Process Industries [10]**

| Manufacturing Process Industry [97] | Process Heating Energy Use (TBtu/yr) [101] | Process Heating Energy Losses (TBtu/yr) [102] | Estimated Recoverable Heat Range (TBtu/yr) [103] | Estimated Thermoelectric Potential (TBtu/yr) | Estimated Thermoelectric Potential (GWh/yr) [104] |
|---|---|---|---|---|---|
| Petroleum refining | 2,250 | 397 | 40–99 | 1–2 | 291–727 |
| Chemicals | 1,460 | 328 | 33–82 | 1–2 | 240–601 |
| Forest products | 980 | 701 | 70–175 | 2–4 | 513–1,280 |
| Iron and steel | 729 | 334 | 33–84 | 1–2 | 245–612 |
| Food and beverage | 518 | 293 | 29–73 | 1–2 | 215–537 |
| Glass | 161 | 88 | 9–22 | 0–1 | 64–161 |
| Other manufacturing | 1,110 | 426 | 43–107 | 1–3 | 312–780 |
| All manufacturing | 7,200 | 2,570 | 257–642 | 6–16 | 1,880–4,700 |

*Source:* Department of Energy [401].

with advanced materials, better coupling through improved heat exchangers and other technology improvements [401]. Risks involved in encouraging R&D in thermoelectric WHR include the following three categories: (a) those related to the effectiveness (efficiency, power, and durability) of TEGs; (b) those related to competing technologies; and (c) those related to the amount of waste heat available. The practical potential for WHR using thermoelectric generation could be limited if TEGs never reach a low enough price point for wide adoption. TEGs might also be found to degrade when exposed to variable temperature environments over multiyear life spans.

There are some other barriers to the growth potential for TEG devices in the United States. Other countries could have advantages in thermoelectric production due to their extant research base (Japan) or manufacturing infrastructure (China, Vietnam). Other technologies for WHR (such as low-temperature Rankine cycle variants, load preheating, or exotic solutions, such as PCM generators) could see breakthroughs that would cause them to outcompete TEGs. Preheating and Rankine cycle variations are more commercially established than TEGs as industrial WHR solutions, but thermos electrics have advantages of low maintenance requirements as well as the option to be installed with minimal downtime and minimal effects to existing systems. Finally, if higher efficiency industrial processes are adopted or value chains change to lower waste heat options (integrated steel mills to minimills, for example), then the amount for waste heat available for thermoelectric recovery would decrease, potentially leading to a smaller return on thermoelectric R&D investment [10].

Earlier we suggested the need for seamless ultra-lightweight micro to milli-scale power supplies for emerging electronic applications. Efficient energy harvesters with ultra-large capacity (as well as ultra-lightweight) energy storage can be significantly useful to meet that demand. However, there is still a long way to go to achieve such goal. In that regard, thermoelectric generators have been identified as an alternative energy source with great potential. However, its efficiency needs to be improved significantly. In that pursuit, the constant development of nanotechnology and nanomaterials has been greatly benefiting TEG technology. Furthermore, the engineering challenges that are coming with emerging technologies, such as wearable electronics, bio-integrated systems, robotics, cybernetics and others, lead the way to novel engineering approaches and innovative devices that are the focus of this review, such as the introduction of mechanically reconfigurable and adaptable devices. Looking forward to the continuing development of thermoelectric, some applications have been discussed, such as the integration of TE materials in windows for mass-scale electricity production and in current- and next-generation computing for efficient cooling and energy harvesting. Previously, thermoelectric materials were used primarily in niche applications. With the advent of broader automotive applications and the effort to utilize waste-heat-recovery technologies, thermoelectric devices are becoming more prominent. This has helped spawn a program between the Energy Efficiency and Renewable Energy office of the US Department of Energy and automotive manufacturers to incorporate thermoelectric waste-heat-recovery technology into the design of heavy trucks. However, current available thermoelectric devices are not in common use, partly due to their low efficiency relative to mechanical cycles and engineer challenges related to using thermoelectric devices for general applications. Therefore, the need of developing high-efficiency thermoelectric materials for waste-heat-recovery systems is urgent and will bring vast economic and environmental benefits.

## 8.7    OTHER METHODS FOR WASTE HEAT TO POWER

DOE report outlines numerous other methods [415–420] for energy conversion, which are at different stages of maturity. Phase-change material (PCM) engine generators use the volume changes caused by the melting and solidification of a PCM material such as paraffin wax to drive a hydraulic system. That hydraulic power is used to drive a generator and to produce electric power. A study of these systems found that they had the potential for higher net present value than ORC systems at very low temperatures (60°C–75°C) over a 20-year life [417]. Magnetocaloric generators, which harvest energy from temperature-induced changes in a material's magnetic characteristics over time, have high-exergy efficiencies and would be useful in industrial WHR [418]. However, few magnetocaloric materials with a sufficiently large magnetocaloric effect have been discovered; they are generally expensive and can be difficult to manufacture; they are susceptible to cycling stability problems; and many materials lack the high Curie temperatures (temperature at represent an active research area, and progress is being made to develop magnetocaloric systems for practical applications [418,420]. A similar effect with time-dependent, temperature-induced electrical polarization changes that lead to voltage differences across a material is called the pyroelectric effect. Lee et al. [415] discussed the use of pyroelectric materials to generate electricity with the Olsen cycle. Waste heat can also be converted to power by Thermoionic, Piezo electricity and tribo electricity and Thermogalvanic devices. These devices have low efficiencies and they are all at the preliminary development stages. They are described in details in my previous book [10].

Garimella and coworkers at Georgia Tech have developed a novel textbook-sized cooling system that operates on waste heat rather than electricity. The underlying technology has been used in very large-scale installations, such as hospitals and university campuses. Yet his team takes the science to a new level by working at the micro scale and creating a self-contained unit. In this unit extremely small passages are etched into thin sheets of metal with different areas representing different components. Working fluids flow in the same order as they would in a larger system, albeit in one space. The minimization of plumbing inlets and outlets translates into greater compactness—and lower price tags. No synthetic refrigerants are used, and less fluid is required, which further lowers costs and increases safety. No compressor is needed and there are few moving parts, decreasing noise and increasing reliability. Modular design allows units to be configured to generate anywhere from a few watts to tens of kilowatts of cooling or heating.

Since unveiling a proof-of-concept unit in 2009, the researchers have developed heat pumps with cooling capacities of one and two refrigerant tons. (Capacity of current residential units ranges from one to four refrigerant tons.) Efficiency has been substantially improved, and fabrication techniques have also been improved to enable mass production. Although the initial cost to consumers might be higher than traditional heat pumps, lifecycle costs should be comparable because of lower operating costs. The researchers have also adapted the technology to provide cooling using waste heat from diesel-driven generators at military bases, where ambient temperatures are extremely high. Since diesel fuel is very expensive and risky to transport, using the energy in the diesel fuel to the fullest extent by providing power as well as cooling through these units, without consuming additional prime energy, will lower overall costs and increase personnel safety.

## REFERENCES

1. Reddy B.C.S., Naidu S.V., and Rangaiah G.P. Waste heat recovery methods and technologies. *Chem Eng* 2013;28–38.
2. Johnson I., and Choate W. *Waste Heat Recovery: Technology and Opportunities in U.S. Industry*, a report by U.S. department of energy–Industrial technologies program, Laurel, MD: prepared by BCS; 2008. Available from http://www1.eere.energy.gov/manufacturing/intensiveprocesses/pdfs/ waste_heat_recovery.pdf.
3. Waste Heat. Wikipedia, last edited 13 October, 2022. https://en.wikipedia.org/wiki/Waste_heat
4. US DOE EIA. *Annual Energy Review*; 2006, DOE, Washington, D.C.
5. *Energy Use, Loss, and Opportunities Analysis: U.S Manufacturing & Mining*, a report prepared by Energetics Incorporated and E3M for DOE, Washington, DC, December 2004, 168 p.
6. Viswanathan V.V., Davies R.W., and Holbery J.D. *Opportunity Analysis for Recovering Energy from Industrial Waste Heat and Emissions*. Richland, WA: PNNL, Pacific Northwest National Laboratory; 2006.
7. US EPA. *Industrial Waste Heat Recovery and the Potential for Emissions Reduction*, Vol. 1, Main Report, Cincinnati, OH: Center for Environmental Research Information; 1984.
8. Goldstick R. *Principles of Waste Heat Recovery*. Atlanta, GA: The Fairmont Press; 1986.
9. Thekdi A. Waste heat recovery for process heating equipment—Review of current status. *Presented at IHEADOE Conference*, Indianapolis, IN; 1 November, 2005.
10. Shah Y.T. Chapter 14, waste heat. In *Thermal Energy-Sources, Recovery, Applications*. New York: CRC Press; 2016.
11. Thekdi A., Nimbalkar S., Sundaramoorthy S., Armstrong K., Taylor A., Gritton J.E., Wenning T. and Cresko J. Technology assessment on low-temperature waste heat recovery in industry, a report by Oak Ridge national Laboiratory, Oak Ridge, TN, September, 2021.
12. Bruce A.H. Waste energy recovery opportunities for interstate natural gas pipelines. *Prepared for: Interstate Natural Gas Association of America (INGAA), Prepared by: Energy and Environmental Analysis, an ICF International Company*; February 2008.
13. Jonathan O., Morrell G., Besseling J., and Slater S. *Barriers and Enablers to Recovering Surplus Heat in Industry, a Qualitative Study of the Experiences of Heat Recovery in the UK Energy Intensive Industries*. URN 15D/541. London: Department of Business, Energy & Industrial Strategy; November 2016.
14. Liu L., Yang Q., and Cui G. Supercritical carbon dioxide(s-$CO_2$) power cycle for waste heat recovery: A review from thermodynamic perspective. *Processes* 2020;8:1461. Doi: 10.3390/pr8111461 www.mdpi.com/journal/processes.
15. Marchionni M., Bianchi G., and Tassou S.A. Review of supercritical carbon dioxide (s$CO_2$) technologies for high-grade waste heat to power conversion. *SN Appl Sci* 2020;2:611. Doi: 10.1007/s42452-020-2116-6.
16. Tanuma T. *Advances in Steam Turbines for Modern Power Plants*. Amsterdam: Elsevier Inc; 2016.

17. Li C., and Wang H. Power cycles for waste heat recovery from medium to high temperature flue gas sources —from a view of thermodynamic optimization. *Appl Energ* 2016;180:707–721. Doi: 10.1016/j.apenergy.2016.08.007.

18. Hung T. Waste heat recovery of organic Rankine cycle using dry fluids. *Energy Convers Manag.* 2001. Doi: 10.1016/S0196-8904(00)00081-9.

19. Dai X., Shi L., and Qian W. Review of the working fluid thermal stability for organic Rankine cycles. *J Therm Sci* 2019;28:597–607. Doi: 10.1007/s11630-019-1119-3.

20. Zywica G., Kaczmarczyk T.Z., and Ihnatowicz E. A review of expanders for power generation in small-scale organic Rankine cycle systems: Performance and operational aspects. *Proc Inst Mech Eng Part A J Power Energy* 2016;230:669–684. Doi: 10.1177/0957650916661465.

21. Weiß A.P. Volumetric expander versus turbine–which is the better choice for small Orc plants. *3rd International Seminar on ORC Power Systems*; October, 2015, pp. 12–14.

22. Tartière T., and Astolfi M. A world overview of the organic Rankine cycle market. *Energ Procedia* 2017;129:2–9. Doi: 10.1016/j.egypro.2017.09.159.

23. Smith I.K. Development of the trilateral flash cycle system: Part 1: Fundamental considerations. *Proc Inst Mech Eng Part A J Power Energ* 1993;207:179–194. Doi: 10.1243/PIME_PROC_1993_207_032_02.

24. Bianchi G., McGinty R., Oliver D., Brightman D., Zaher O., Tassou S.A., Miller J., and Jouhara H. Development and analysis of a packaged trilateral flash cycle system for low grade heat to power conversion applications. *Therm Sci Eng Prog* 2017;4:113–121. Doi: 10.1016/J.TSEP.2017.09.009.

25. Marchionni M., Zaher O., and Miller J. Numerical investigations of a trilateral flash cycle under system off-design operating conditions. *Energ Procedia* 2019;161:464–471 Doi: 10.1016/J.EGYPRO.2019.02.070.

26. Kung S.C., Shingledecker J.P., Thimsen D., Wright I.G., Tossey B.M., and Sabau A.S. Oxidation/corrosion in materials for supercritical $CO_2$ power cycles. *A paper Presented at 5th International Symposium - Supercritical $CO_2$ Power Cycles,* San Antonio, TX; March 28–31, 2016.

27. Marchionni M., Bianchi G., and Tassou S.A. Techno-economic assessment of Joule–Brayton cycle architectures for heat to power conversion from high-grade heat sources using $CO_2$ in the supercritical state. *Energy* 2018;148:1140–1152. Doi: 10.1016/J.ENERGY.2018.02.005.

28. Bianchi G., Saravi S.S., Loeb R., Tsamos K.M., Marchionni M., and Leroux A. Design of a high-temperature heat to power conversion facility for testing supercritical $CO_2$ equipment and packaged power units. *Energ Procedia* 2019;161:421–428. Doi: 10.1016/J.EGYPRO.2019.02.109.

29. De Miol M., Bianchi G., Henry G., Holaind N., Tassou S.A., and Leroux A. Design of a single-shaft compressor, generator, turbine for small-scale supercritical $CO_2$ systems for waste heat to power conversion applications. 2018. Doi: 10.17185/duepublico/46086.

30. Mehos M., Turchi C.S., Jorgenson J., Denholm P., Ho C., and Armijo K. On the path to sunshot-advancing concentrating solar power technology. *Perform Dispatc* 2016. Doi: 10.2172/1344199.

31. Held T.J. Initial test results of a megawatt-class supercritical $CO_2$ heat engine. *4th International Symposium $CO_2$ Power Cycles*, Pittsburgh, PA; 2014.

32. Rankine Cycle. Wikipedia, the free encyclopedia, last edited 17 August 2022. Available from: https://en.wikipedia.org/wiki/RankineCycle.

33. Erickson D.C., Anand G., and Kyung I. *Heat Activated Dual Function Absorption Cycle.* A paper presented at ASHRAE SYMP 00138. New Orleans, LA; 2004.

34. STOWA. Organic Rankine cycle for electricity generation, selected technologies, 2007. Available from: http://www.stowaselectedtechnologies.nl/Sheets/index.html.

35. Duffy D. *Better Cogeneration Through Chemistry: The Organic Rankine Cycle.* Distributed Energy, SOWA and Distributed Energy; November/December 2005.
36. Heidelberg Cement. Organic Rankine cycle method, 2007. Available from: http://www. heidelbergcement.com/global/en/company/products_innovations/innovations/orc.htm.
37. IEA. 2002. *A Power Generating System for Low Temperature Heat Recovery*, a report prepared by IEA, Paris, France.
38. Vanslambrouck B. The organic Rankine cycle (ORC). CHP: Technology Update, 2010. Available from: 29 April 2015 http://www.ibgebim.be/uploadedFiles/Contenu_du_site/ Professionnels/Formations_et_s%C3%A9minaires/S%C3%A9minaire_ URE_%28En ergie%29_2010_%28Actes%29/02-ORC_VANSLAMBROUCK.pdf.
39. Tchanche B.F., Lambrinos G., Frangoudakis A., and Papadakis G. Low-grade heat conversion into power using organic Rankine cycles—A review of various applications. *Renew Sustain Energ Rev* 2011;15(8):3963–3979. Doi: 10.1016/j. rser.2011.07.024.
40. Quoilin S., Vandenbroek M., Declaye S., Dewaller P., and Lemort V. Techno-economic survey of organic Rankine cycle (ORC) systems. *Renew Sustain Energ Rev* 2013;22:168–186.
41. Arvay P., Muller M.R., Ramdeen V., and Cunningham G. Economic implementation of the organic Rankine cycle in industry, a report by *ACEEE Summer Study on Energy Efficiency in Industry*; 2011, pp. 12–22.
42. Daccord R., Melis J., Kientz T., Darmedru A., Pireyre R., Brisseau N., and Fonteneau E. Exhaust heat recovery with Rankine piston expander. *Proceedings of ICE Powertrain Electrification & Energy Recovery*, France, Rueil-Malmaison; 28 May, 2013.
43. Nguyen T.-V., Knudsen T., Larsen U., and Haglind F. Thermodynamic evaluation of the Kalina split-cycle concepts for waste heat recovery applications. *Energy* 2014;71:277–288. Doi: 10.1016/j.energy.2014.04.060.
44. Kalina cycle. Wikipedia, the free encyclopedia, last edited 2 September, 2022, last modified 19 January 2017; 2017. Available from: https://en.wikipedia.org/wiki/Kalina cycle.
45. Hettiarachchi M., Golubovic M., Worek W., and Ikegami Y. The performance of the Kalina Cycle system 11(KCS11) with low-temperature heat sources. *J Energ Res Technol-Trans ASME* 2007;129. Doi: 10.1115/1.2748815.
46. Park Y.M., and Sonntag R.E. A preliminary study of the Kalina power cycle in connection with a combined cycle system. *Int J Energ Res* 1990;14:153–162.
47. Jonsson M. *Advanced Power Cycles with Mixtures as the Working Fluid*, doctoral thesis. Sweden: KTH; 2003.
48. Thorin E. *Power Cycles with Ammonia-Water Mixtures as Working Fluid Analysis of Different Applications and the Influence of Thermophysical Properties*, doctoral thesis. Sweden: KTH; 2000.
49. Sung T., and Kim K.C. An organic Rankine cycle for two different heat sources: Steam and hot water. *Energ Procedia* 2017;129:883–890.
50. Saadon S., and Islam S.M.S. A recent review in performance of organic Rankine cycle (ORC). An Intech open access paper; 2019. Doi: 10.5772/intechopen.89763.
51. Brizard A., Dolz V., Galindo J., and Royo-Pascaul L. Dynamic modeling of an organic Rankine cycle to recover waste heat for transportation vehicles. *Energy* 2017;129:192–199.
52. Enertime. The organic Rankine cycle and its applications, 2011. Available from: http:// www.cycleorganique-rankine.com.
53. Li L., Tao L., Li Q., and Hu Y. Experimentally economic analysis of ORC power plant with low-temperature waste heat recovery. *Int J Low-Carbon Technol* March 2021;16(1):35–44. Doi: 10.1093/ijlct/ctaa032.
54. Yamamoto T., Furuhata T., Arai N., and Mori K. Design and testing of organic Rankine cycle. *Energy* 2001:26:239–251.

55. Noroozian A, Naeimi A, Bidi M., and Ahmadi M.H. Exergoeconomic comparison and optimization of organic Rankine cycle, trilateral Rankine cycle and transcritical carbon dioxide cycle for heat recovery of low-temperature geothermal water. *Proc Inst Mech Eng A* 2019;233:1068–1084.

56. Borsukiewicz-Gozdur A, and Nowak W. Comparative analysis of natural and synthetic refrigerants in application to low temperature Clausius–Rankine cycle. *Energy* 2007;32:344–352.

57. Heberle F., Preißinger M., and Brüggemann D. Zeotropic mixtures as working fluids in organic Rankine cycles for low-enthalpy geothermal resources. *Renew Energ* 2012;37:364–370.

58. Mago P.J., Chamra L.M., and Somayaji C. Performance analysis of different working fluids for use in organic Rankine cycles. *Proc Inst Mech Eng A* 2007;221:255–263.

59. Yekoladio P.J., Bello-Ochende T., and Meyer J.P. Thermodynamic analysis and performance optimization of organic rankine cycles for the conversion of low-to-moderate grade geothermal heat. *Int J Energy Res* 2015;39:1256–1271.

60. Branchini L., Pascale A.D., and Peretto A. Systematic comparison of ORC configurations by means of comprehensive performance indexes. *Appl Therm Eng* 2013;61:129–140.

61. Shengjun Z., Wang H., and Tao G. Performance comparison and parametric optimization of subcritical organic Rankine cycle (ORC) and transcritical power cycle system for low-temperature geothermal power generation. *Appl Energ* 2011;88:2740–2754.

62. Lemort V., Declaye S., and Quoilin S. Experimental characterization of a hermetic scroll expander for use in a micro-scale rankine cycle. *Proc Instit Mech Eng Part J Power Energ* 2012; 226.

63. Hu F., Zhang Z., Chen W., He Z., Wang X., and Xing Z. Experimental investigation on the performance of a twin-screw expander used in an ORC system. *Energ Procedia* 2017;110:210–215.

64. Kang S.H. Design and experimental study of ORC (organic Rankine cycle) and radial turbine using R245fa working fluid. *Energy* 2012;41:514–524.

65. Yagli H., Koc Y., Koç A., Adnan G., and Tandiroglu A. Parametric optimization and exergetic analysis comparison of subcritical and supercritical organic Rankine cycle(ORC) for biogas fuelled combined heat and power (CHP) engine exhaust gas waste heat. *Energy* 2016;111:923–932.

66. Guo C., Du X., Yang L., and Yang Y. Performance analysis of organic Rankine cycle based on location of heat transfer pinch point in evaporator. *Appl Thermal Eng* 2014;62(1):176–186.

67. Ran T., Qingsong A.N., Lin S.H.I., Huixing Z., and Xiaoye D. Performance analyses of supercritical organic Rankine cycles (ORCs) with large variations of the thermophysical properties in the pseudocritical region. *3rd International Seminar on ORC Power Systems*, Brussels, Belgium; 12–14 October, 2015, pp. 1–10. Paper ID: 53.

68. Moloney F., Almatrafi E., and Goswami D.Y. Working fluid parametric analysis for regenerative supercritical organic Rankine cycles medium geothermal reservoir organic Rankine cycles for medium geothermal reservoir temperatures. *Energ Procedia.* 2017;129:599–606.

69. Chowdhury J.I., Nguyen K.B., Thornhill D., Douglas R., and Glover S. Modelling of organic Rankine cycle for waste heat recovery process in supercritical condition. *Int J Mech, Aerospace, Indus Mechatron Eng* 2015;9(3):477–482.

70. Le V.L., Feidt M., Kheiri A., and Pelloux-Prayer S. Performance optimization of low-temperature power generation by supercritical ORCs (organic Rankine cycles) using low GWP (global warmingpotential) working fluids. *Energy* 2014;67:513–526.

71. Muhammad I., Park B.S., Kim H.J., Lee D.H., Muhammad U., and Heo M. Thermoeconomic optimization of regenerative organic Rankine cycle for waste heat recovery applications. *Int J Energ Convers Manage* 2014;87(November):107–118.

72. Li L., Ge Y.T., and Tassou S.A. Experimental study on a small-scaleR245fa organic Rankine cycle system for low-grade thermal energy recovery. *Energ Procedia* 2017;105:1827–1832.

73. Roy J., and Misra A. Parametric optimization and performance analysis of a regenerative organic Rankine cycleusing R-123 for waste heat recovery. *Energy* 2012;39(1):227–235.

74. Zhang C., Liu C., Xu X., Li Q., Wang S., and Chen X. Effects of superheat and internal heat exchanger on thermoeconomic performance of organic Rankine cycle based on fluid type and heat sources. *Energy* 2018;159:482–495.

75. Radulovic J., and Benedikt K. Utilisation of diesel engine waste heat by organic Rankine cycle. *Appl Thermal Eng* 2015;78:437–448.

76. Feng Y., Zhang Y., Li B., Yang J., and Shi Y. Comparison between regenerative organic Rankine cycle (RORC) and basic organic Rankine cycle (BORC) based on thermoeconomic multiobjective optimization considering energy efficiency and levelized energy cost (LEC). *Energ Convers Manage* 2015;96:58–71.

77. Bianchi M., Branchini L., De Pascale A., Orlandini V., Ottaviano S., Pinelli M., Spina P.R., and Suman A. Experimental performance of a micro-ORC energy system for low grade heat recovery. *Energ Procedia.* 2017;129:899–906.

78. Ismail H., Aziz A.A., Rasih R.A., Jenal N., Michael Z., and Roslan A. Performance of organic Rankine cycle using biomass as source of fuel. *J Adv Res Appl Sci Eng Technol* 2016;4(1):29–46.

79. Rentizelas A., Karellas S., Kakaras E., and Tatsiopoulos I. Comparative technoeconomic analysis of ORC and gasification for bioenergy applications. *Energ Convers Manage* 2009;50(3):674–681.

80. Karytsas C., Mendrinos D., and Radoglou G. The current geothermal exploration and development of the geothermal field of Milos Island in Greece. *GHC Bull* 2004;25:17–21.

81. Frick S., Kranz S. and Saadat A. Holistic design approach for geothermal binary power plants with optimized net electricity provision. *A Paper Presented at World Geothermal Congress*, Bali, Indonesia; 25–29 April, 2010.

82. Müller-steinhagen H., and Trieb F. *Concentrating Solar Power—A Review of the Technology*, Vol. 18. Stuttgart: Institute of Technical Thermodynamics, German Aerospace Centre; 2004, pp. 43–50.

83. Ford G. CSP: Bright future for linear Fresnel technology? *Renewable Energy Focus* 2008;9(5):48–49.

84. Bundela P.S., and Chawla V. Sustainable development through waste heat recovery. *Am J Environ Sci* 2010;6(1):83–89.

85. Engin T., and Ari V. Energy auditing and recovery for dry type cement rotary kiln systems—A case study. *Energ Convers Manage.* 2005;46(4):551–562.

86. Canada S., Cohen G., Cable R., Brosseau D., and Price H. *Parabolic Trough Organic Rankine Cycle Power Plant*; 2005, p. 5.

87. Perullo C.A., Mavris D.N., and Fonseca E. An integrated assessment of an organic Rankine cycle concept for use in on board aircraft power generation. *ASME Turbo Expo: Turbine Technical Conference and Exposition American Society of Mechanical Engineers*, San Antonio, TX; 2013.

88. Sánchez D., Chacartegui R., Jiménez-Espadafor F., and Sánchez T. A new concept for high temperature fuel cell hybrid systems using supercritical carbon dioxide. *J Fuel Cell Sci Technol* 2009;6:021306.

89. Sanchez D., de Escalona J., Chacartegui R., Munoz A., and Sanchez T. A comparison between molten carbonate fuel cells based hybrid systems using air and supercritical carbon dioxide Brayton cycles with state of the art technology. *J Power Sourc* 2011;196:4347–4354.

90. Baronci A., Messina G., McPhail S., and Moreno A. Numerical investigation of a MCFC (molten carbonate fuel cell) system hybridized with a supercritical $CO_2$ Brayton cycle and compared with a bottoming organic Rankine cycle. *Energy* 2015;93:1063–1073.

91. Mahmoudi S., and Ghavimi A. Thermoeconomic analysis and multi objective optimization of a molten carbonate fuel cell–Supercritical carbon dioxide-organic Rankin cycle integrated power system using liquefied natural gas as heat sink. *Appl Therm Eng* 2016;107:1219–1232.

92. Ryu J., Ko A., and Park S. Thermo-economic assessment of molten carbonate fuel cell hybrid system combined between individual $sCO_2$ power cycle and district heating. *Appl Therm Eng* 2020;169:114911.

93. Bae S., Ahn Y., Lee J., and Lee J. Various supercritical carbon dioxide cycle layouts study for molten carbonate fuel cell application. *J Power Sourc* 2014;270:608–618.

94. Ahmadi M., Mohammadi A., Pourfayaz F., Mehrpooya M., Bidi M., and Valero A. Thermodynamic analysis and optimization of a waste heat recovery system for proton exchange membrane fuel cell using transcritical carbon dioxide cycle and cold energy of liquefied natural gas. *J Nat Gas Sci Eng* 2016;34:428–438.

95. Shu G., Shi L., Tian H., Li X., Huang G., and Chang L. An improved $CO_2$-based transcritical Rankine cycle (CTRC) used for engine waste heat recovery. *Appl Energ* 2016;176:171–182.

96. Shu G., Shi L., Tian H., Deng S., Li X., and Chang L. Configurations selection maps of $CO_2$-based transcritical Rankine cycle (CTRC) for thermal energy management of engine waste heat. *Appl Energ* 2017;186:423–435.

97. Shu G., Li X., Tian H., Shi L., Wang X., and Yu G. Design condition and operating strategy analysis of $CO_2$ transcritical waste heat recovery system for engine with variable operating conditions. *Energ Convers Manage* 2017;142:188–199.

98. Shi L., Tian H., and Shu G. Multi-mode analysis of a $CO_2$-based combined refrigeration and power cycle for engine waste heat recovery. *Appl Energ* 2020;264:114670.

99. Shu G., Wang R., Tian H., Wang X., Li X., Cai J., and Xu Z. Dynamic performance of the transcritical power cycle using $CO_2$-based binary zeotropic mixtures for truck engine waste heat recovery. *Energy* 2020;194:116825.

100. Liu P., Shu G., Tian H., Feng W., Shi L., and Wang X. Experimental study on transcritical Rankine cycle (TRC) using $CO_2$/R134a mixtures with various composition ratios for waste heat recovery from diesel engines. *Energ Convers Manage* 2020;208:112574.

101. Shi L., Shu G., Tian H., Chang L., Huang G., and Chen T. Experimental investigations on a $CO_2$-based transcritical power cycle (CTPC) for waste heat recovery of diesel engine. *Energ Procedia* 2017;129:955–962.

102. Li X., Shu G., Tian H., Shi L., Huang G., Chen T., and Liu P. Preliminary tests on dynamic characteristics of a $CO_2$ transcritical power cycle using an expansion valve in engine waste heat recovery. *Energy* 2017;140:696–707.

103. Li X., Shu G., Tian H., Huang G., Liu P., Wang X., and Shi L. Experimental comparison of dynamic responses of $CO_2$ transcritical power cycle systems used for engine waste heat recovery. *Energy Convers Manage* 2018;161:254–265.

104. Shi L., Shu G., Tian H., Chen T., Liu P., and Li L. Dynamic tests of $CO_2$-Based waste heat recovery system with preheating process. *Energy* 2019;171:270–283.

105. Liang Y., Bian X., Qian W., Pan M., Ban Z., and Yu Z. Theoretical analysis of a regenerative supercritical carbon dioxide Brayton cycle/organic Rankine cycle dual loop for waste heat recovery of a diesel/natural gas dual-fuel engine. *Energy Convers Manage* 2019;197:111845.

106. Feng Y., Du Z., and Shreka M. Thermodynamic analysis and performance optimization of the supercritical carbon dioxide Brayton cycle combined with the Kalina cycle for waste heat recovery from a marine low-speed diesel engine. *Energy Convers Manage* 2020;206:112483.

107. Liang Y., Sun Z., Dong M., Lu J., and Yu Z. Investigation of a refrigeration system based on combined supercritical $CO_2$ power and transcritical $CO_2$ refrigeration cycles by waste heat recovery of engine. *Int J Refrig* 2020;118:470–482.

108. Pan M., Bian X., Zhu Y., Liang Y., Lu F., and Xiao G. Thermodynamic analysis of a combined supercritical $CO_2$ and ejector expansion refrigeration cycle for engine waste heat recovery. *Energy Convers Manage* 2020;224:113373.

109. Zhang R., Su W., Lin X., Zhou N., and Zhao L. Thermodynamic analysis and parametric optimization of a novel s–$CO_2$ power cycle for the waste heat recovery of internal combustion engines. *Energy* 2020;209:118484.

110. Song J., Li X., Wang K., and Markides C. Parametric optimisation of a combined supercritical $CO_2$ (S-$CO_2$) cycle and organic Rankine cycle (ORC) system for internal combustion engine (ICE) waste-heat recovery. *Energy Convers Manage* 2020;218:112999.

111. Choi B. Thermodynamic analysis of a transcritical $CO_2$ heat recovery system with 2-stage reheat applied to cooling water of internal combustion engine for propulsion of the 6800 TEU container ship. *Energy* 2016;107:532–541.

112. Sharma O., Kaushik S., and Manjunatti K. Thermodynamic analysis and optimization of a supercritical $CO_2$ regenerative recompression Brayton cycle coupled with a marine gas turbine for shipboard waste heat recovery. *Therm Sci Eng Prog* 2017;3:62–74.

113. Hou S., Zhang F., and Yu L. Optimization of a combined cooling, heating and power system using $CO_2$ as main working fluid driven by gas turbine waste heat. *Energy Convers Manage* 2018;178:235–249.

114. Manjunath K., Sharma O., Tyagi S., and Kaushik S. Thermodynamic analysis of a supercritical/transcritical $CO_2$ based waste heat recovery cycle for shipboard power and cooling applications. *Energy Convers Manag* 2018;155:262–275.

115. Walnum H., Neksa P., Nord L., and Andresen T. Modelling and simulation of $CO_2$ (carbon dioxide) bottoming cycles for offshore oil and gas installations at design and off-design conditions. *Energy* 2013;59:513–520.

116. Moroz L., Burlaka M., Rudenko O., and Joly C. Evaluation of gas turbine exhaust heat recovery utilizing composite supercritical $CO_2$ cycle. *Proceedings of the International Gas Turbine Congress*, Tokyo, Japan; 15–20 November, 2015.

117. Khadse A., Blanchette L., Kapat J., Vasu S., and Ahmed K. Optimization of supercritical $CO_2$ Brayton cycle for simple cycle gas turbines exhaust heat recovery using genetic algorithm. *Proceedings of the ASME Turbo Expo 2017: Turbomachinery Technical Conference and Exposition*, Charlotte, NC; 26–30 June, 2017. Article No. 63696.

118. Cao Y., Ren J., Sang Y., and Dai Y. Thermodynamic analysis and optimization of a gas turbine and cascade $CO_2$ combined cycle. *Energy Convers Manage* 2017;144:193–204.

119. Gao W., Li H., Nie P., Yang Y., and Zhang C. A novel s-$CO_2$ and ORC combined system for exhaust gases. *Proceedings of the ASME Turbo Expo 2017: Turbomachinery Technical Conference and Exposition*, Charlotte, NC; 26–30 June, 2017. Article No. 65214.

120. Tozlu A., Abusoglu A., and Ozahi E. Thermoeconomic analysis and optimization of a recompression supercritical $CO_2$ cycle using waste heat of Gaziantep municipal solid waste power plant. *Energy* 2018;143:168–180.

121. Zhang Q., Ogren R., and Kong S. Thermo-economic analysis and multi-objective optimization of a novel waste heat recovery system with a t-$CO_2$ cycle for offshore gas turbine application. *Energy Convers Manag* 2018;172:212–227.

122. Sánchez V., and Vargas M. Thermoeconomic and environmental analysis and optimization of the supercritical $CO_2$ cycle integration in a simple cycle power plant. *Appl Therm Eng* 2019;152:1–12.

123. Zhou A., Li X., Ren X., and Gu C. Improvement design and analysis of a supercritical $CO_2$/transcritical $CO_2$ combined cycle for offshore gas turbine waste heat recovery. *Energy* 2020;210:118562.

124. Tao Z., Zhao Q., Tang H., and Wu J. Thermodynamic and exergetic analysis of supercritical carbon dioxide brayton cycle system applied to industrial waste heat recovery. *Proc CSEE* 2019;39:6944–6952. (In Chinese).

125. Cho S., Kim M., Baik S., Ahn Y., and Lee J. Investigation of the bottoming cycle for high efficiency combined cycle gas turbine system with supercritical carbon dioxide power cycle. *Proceedings of the ASME Turbo Expo 2015: Turbine Technical Conference and Exposition*, Montréal, QC, Canada; 15–19 June, 2015. Article No. 43077.

126. Huck P., Freund S., Lehar M., and Peter M. Performance comparison of supercritical $CO_2$ versus steam bottoming cycles for gas turbine combined cycle applications. *Proceedings of the 5th International Symposium on Supercritical $CO_2$ Power Cycles*, San Antonio, TX; 29–31 March, 2016.

127. Wright S., Davidson C., and Scammell W. Thermo-economic analysis of four s$CO_2$ waste heat recovery power systems. *Proceedings of the 5th International Symposium on Supercritical $CO_2$ Power Cycles*, San Antonio, TX; 29–31 March, 2016.

128. Kim M., Ahn Y., Kim B., and Lee J. Study on the supercritical $CO_2$ power cycles for landfill gas firing gas turbine bottoming cycle. *Energy* 2016;111:893–909.

129. Wang X., and Dai Y. An exergoeconomic assessment of waste heat recovery from a gas turbine-modular helium reactor using two transcritical $CO_2$ cycles. *Energy Convers Manage* 2016;126:561–572.

130. Astolfi M., Alfani D., Lasala S., and Macchi E. Comparison between ORC and $CO_2$ power systems for the exploitation of low-medium temperature heat sources. *Energy* 2018;161:1250–1261.

131. Olumayegun O., and Wang M. Dynamic modelling and control of supercritical $CO_2$ power cycle using waste heat from industrial processes. *Fuel* 2019;249:89–102.

132. Luo J., Morosuk T., and Tsatsaronis G. Exergoeconomic investigation of a multi-generation system with $CO_2$ as the working fluid using waste heat. *Energy Convers Manage* 2019;197:111882.

133. Schierning G., Chavez R., Schmechel R., Balke B., Rogl G., and Rogl P. Concepts for medium-high to high temperature thermoelectric heat-to-electricity conversion: A review of selected materials and basic considerations of module design. *Transl Mater Res* 2015;2:025001.

134. Ma T., Qu Z., Yu X., Lu X., and Wang Q. A review on thermoelectric-hydraulic performance and heat transfer enhancement technologies of thermoelectric power generator system. *Therm Sci* 2018;22:1885–1903.

135. Radousky H.B., and Liang H. Energy harvesting: An integrated view of materials, devices and applications. *Nanotechnol* 2012;23:502001.

136. Stockholm J.G. Génération thermoélectrique. Actes Des Journées Electrotechniques Du Club EEA-Energie PorTable 2002.: Autonomie et intégration dans l'environnement Humain, Cachan, France; 21–22 March, 2002. Available from: https://www.academia.edu/16362069/DISPOSITIFS_%C3%89LECTROM%C3%89CANIQUES_PERMETTANT_L_EXPLOITATION_DE_L_%C3%89NERGIE_DES_MOUVEMENTS_HUMAINS?auto=download (accessed: 9 June 2020).

137. Seebeck T.J. *Magnetische Polarisation der Metalle und Erze durch Temperatur-Di_erenz; Abhandlungen der Königlichen Preußischen Akademie der Wissenschaften zu.* Berlin: von Wilhelm Engelman; 1895. Available from: https://onlinelibrary.wiley.com/doi/abs/10.1002/andp.18260820302 (accessed: 9 June 2020).

138. Peltier J.C. Nouvelles expériences sur la caloricité des courants électriques. *Annales de Chimie et de Physique* 1834;56:371–386.

139. Thomson W. On the dynamical theory of heat transfer. *Trans R Soc Edinb* 1851;3:91–98.

140. Altenkirch E. Über den nutzeffekt der thermosäule. *Physikalische Zeitschrift* 1909;10:560.

141. Altenkirch E. Elektrothermische Kälteerzeugung und reversible elektrische Heizung. *Physikalische Zeitschrift* 1911;12:920–924.

142. Altenkirch W.W.E., and Gehlhoff G.R. *Thermo-Electric Heating and Cooling Body.* U.S. Patent 1120781A; 15 December, 1914. Available from: https://patents.google.com/patent/US1120781A/en (accessed: 9 June 2018).

143. Goldsmid H.J., and Douglas R.W. The use of semiconductors in thermoelectric refrigeration. *Br J Appl Phys* 1954;5:386–390.

144. Ioffe A.F. *Semiconductor Thermoelkements and Thermoelectric Colling.* London: Infosearch Ltd, 1956.

145. BJohn F., Harry B., and Reggie S. *Thermoelectric Cooling Units.* U.S. Patent 2932953A; 19 April 1960. Available from: https://patents.google.com/patent/US2932953A/en (accessed: 9 June 2018).

146. Hicks L.D., and Dresselhaus M.S. Effect of quantum-well structures on the thermoelectric figure of merit. *Phys Rev B* 1993;47:12727–12731.

147. Vining C.B. ZT~ 3.5: Fifteen years of progress and things to come. *5th European Conference on Thermoelectrics*, ECT2007, Odessa, Odessa House of Scientists; 2007.

148. Ohta H., Sugiura K., and Koumoto K. Recent progress in oxide thermoelectric materials: P-type $Ca_3Co_4O_9$ and n-Type $SrTiO_3$. *Inorg Chem* 2008;47:8429–8436.

149. Zide J.M., Vashaee D., Bian Z.X., Zeng G., Bowers J.E., Shakouri A., and Gossard A.C. Demonstration of electron filtering to increase the Seebeck coeffcient in $In_{0.53}Ga_{0.47}As$? $In_{0.53}Ga_{0.28}Al_{0.19}As$ superlattices. *Phys Rev B* 2006;74:205335.

150. Bin Masood K., Kumar P., Singh R.A., and Singh J. Odyssey of thermoelectric materials: Foundation of the complex structure. *J Phys Commun* 2018;2:062001.

151. Zoui M.A., Bentouba S., Stocholm J.G., and Bourouis M. A review on thermoelectric generators: Progress and applications. *Energies* 2020;13:3606. Doi: 10.3390/en13143606 www.mdpi.com/journal/energies.

152. Elarusi A., Fagehi H., Attar A., and Lee H. Theoretical approach to predict the performance of thermoelectric generator modules. *J Electron Mater* 2016;46:872–881.

153. Buist R.J. Design and engineering of thermoelectric cooling devices. *Proceedings of the 10th International Conference on Thermoelectrics*, Cardiff; 10 September, 1991.

154. Buist R. Calculation of peltier device performance. *CRC Handb Thermoelectr* 1995;143–155.

155. Kumar S., Heister S.D., Xu X., and Salvador J.R. Optimization of thermoelectric components for automobile waste heat recovery systems. *J Electron Mater* 2015;44:3627–3636.

156. Lau P., and Buist R. *Calculation of Thermoelectric Power Generation Performance Using Finite Element Analysis.* Piscataway, NJ: Institute of Electrical and Electronics Engineers (IEEE); 2002, pp. 563–566.

157. Angrist S.W. *Direct Energy Conversion.* Boston, MA: Allyn and Bacon; 1976.

158. Antonova E., and Looman D. *Finite Elements for Thermoelectric Device Analysis in ANSYS* Piscataway, NJ: Institute of Electrical and Electronics Engineers (IEEE); 2005, pp. 215–218.

159. Release A. *9.0 Documentation.* Canonsburg, PA: ANSYS, Inc.; 2004.

160. Kim T.Y., and Kim J. Assessment of the energy recovery potential of a thermoelectric generator system for passenger vehicles under various drive cycles. *Energy* 2018;143:363–371.

161. Chen M., Rosendahl L.A., and Condra T. A three-dimensional numerical model of thermoelectric generators in fluid power systems. *Int J Heat Mass Transf* 2011;54:345–355.

162. Hussain Q.E., Brigham D.R., and Maranville C.W. Thermoelectric exhaust heat recovery for hybrid vehicles. *SAE Int J Eng* 2009;2:1132–1142.

163. Sharma A., Lee J.H., Kim K.H., and Jung J.P. Recent advances in thermoelectric power generation technology. *J Microelectron Packag Soc* 2017;24(1):9–16. Doi 10.6117/kmeps.2017.24.1.009 Print ISSN 1226-9360 Online ISSN 2287-7525.

164. Roy C. *Manufacture of Thermoelectric Devices.* U.S. Patent 2980746A; 18 April 1961. Available from: https://patents.google.com/patent/US2980746A/en#patentCitations (accessed: 9 June 2018).

165. Champier D. Thermoelectric generators: A review of present and future applications. *Springer Proceedings in Energy*, Cham, Springer Science and Business Media LLC; 2016, pp. 203–212.

166. Merkulov O.V., Politov B.V., Chesnokov K.Y., Markov A.A., Leonidov I.A., and Patrakeev M.V. Fabrication and testing of a tubular thermoelectric module based on oxide elements. *J Electron Mater* 2018;47:2808–2816.

167. Jang H., Kim J.B., Stanley A., Lee S., Kim Y., Park S.H., and Oh M.-W. Fabrication of Skutterudite-based tubular thermoelectric generator. *Energies* 2020;13:1106.

168. Min G., and Rowe D.M. Ring-structured thermoelectric module. *Semicond Sci Technol* 2007;22:880–883.

169. He R., Schierning G., and Nielsch K. Thermoelectric devices: A review of devices, architectures, and contact optimization. *Adv Mater Technol* 2017;3:1700256.

170. Saidur R., Rezaei M., Muzammil W., Hassan M., Paria S., and Hasan M.H. Technologies to recover exhaust heat from internal combustion engines. *Renew Sustain Energ Rev* 2012;16:5649–5659.

171. Kim T.Y., Negash A., and Cho G. Direct contact thermoelectric generator (DCTEG): A concept for removing the contact resistance between thermoelectric modules and heat source. *Energy Convers Manage* 2017;142:20–27.

172. Kim T.Y., Negash A., and Cho G. Experimental and numerical study of waste heat recovery characteristics of direct contact thermoelectric generator. *Energy Convers Manage* 2017;140:273–280.

173. Qu W., Plötner M., and Fischer W.-J. Microfabrication of thermoelectric generators on flexible foil substrates as a power source for autonomous microsystems. *J Micromech Microeng* 2001;11:146–152.

174. Itoigawa K., Ueno H., Shiozaki M., Toriyama T., and Sugiyama S. Fabrication of flexible thermopile generator. *J Micromech Microeng* 2005;15:S233–S238.

175. Takashiri M., Shirakawa T., Miyazaki K., and Tsukamoto H. Fabrication and characterization of bismuth–telluride-based alloy thin film thermoelectric generators by flash evaporation method. *Sens Actuat A* 2007;138:329–334.

176. Yadav A., Pipe K.P., and Shtein M. Fiber-based flexible thermoelectric power generator. *J Power Sourc* 2008;175:909–913.

177. Khan S., Dahiya R., Lorenzelli L., and Khan S. *Flexible Thermoelectric Generator Based on Transfer Printed Si Microwires*. Piscataway, NJ: Institute of Electrical and Electronics Engineers (IEEE); 2014, pp. 86–89.

178. Lee H.B., We J.H., Yang H.J., Kim K., Choi K.C., and Cho B.J. Thermoelectric properties of screen-printed ZnSb film. *Thin Solid Films* 2011;519:5441–5443.

179. Sevilla G.A.T., Bin Inayat S., Rojas J.P., Hussain M.M., and Hussain M.M. Flexible and semi-transparent thermoelectric energy harvesters from low cost bulk silicon (100). *Small* 2013;9:3916–3921.

180. Madan D., Wang Z., Chen A., Wright P.K., and Evans J.W. High-performance dispenser printed MA p-Type $Bi_{0.5}Sb_{1.5}Te_3$ flexible thermoelectric generators for powering wireless sensor networks. *ACS Appl Mater Interf* 2013;5:11872–11876.

181. Delaizir G., Monnier J., Soulier M., Grodzki R., Villeroy B., Testard J., Simon J., Navone C., and Godart C. A new generation of high performance large-scale and flexible thermo-generators based on $(Bi, Sb)_2 (Te, Se)_3$ nano-powders using the spark plasma sintering technique. *Sens Actuat A* 2012;174:115–122.

182. Fan P., Zheng Z., Li Y.-Z., Lin Q.-Y., Luo J.-T., Liang G.-X., Cai X.-M., Zhang D.-P., and Ye F. Low-cost flexible thin film thermoelectric generator on zinc based thermoelectric materials. *Appl Phys Lett* 2015;106:73901.

183. Suemori K., Watanabe Y., and Hoshino S. Carbon nanotube bundles/polystyrene composites as high-performance flexible thermoelectric materials. *Appl Phys Lett* 2015;106:113902.

184. Cramer C.L., Wang H., and Ma K. Performance of functionally graded thermoelectric materials and devices: A review. *J Electron Mater* 2018;47:5122–5132.

185. Jovovic V., Kossakovski D., and Heian E.M. *Thermoelectric Devices with Interface Materials and Methods of Manufacturing the Same*. U.S. Patent 9865794B2; 9 January 2018. Available from: https://patents.google.com/patent/US9865794B2/en (accessed: 13 October 2018).

186. Zheng J.-C. Recent advances on thermoelectric materials. *Front Phys China* 2008;3:269–279.

187. Rurali R., Yu C., and Zardo I. Special issue on thermoelectric properties of nanostructured materials. *J Phys D* 2018;51:430301.

188. Beretta D., Neophytou N., Hodges J.M., Kanatzidis M.G., Narducci D., Martin-Gonzalez M., Beekman M., Balke B., Cerretti G., Tremel W., and Zevalkink A. Thermoelectrics: From history, a window to the future. *Mater Sci Eng R* 2019;138:100501.

189. Sarbu I., and Dorca A. A comprehensive review of solar thermoelectric cooling systems. *Int J Energ Res* 2017;42:395–415.

190. Chen K.-X., Li M.-S., Mo D.-C., and Lyu S.-S. Nanostructural thermoelectric materials and their performance. *Front Energ* 2018;12:97–108.

191. Romanjek K., Vešín S., Aixala L., Baffe T., Bernard-Granger G., and Dufourcq J. High-Performance silicon–germanium-based thermoelectric modules for gas exhaust energy scavenging. *J Electron Mater* 2015;44:2192–2202.

192. Vining C.B. The thermoelectric limit ZT 1: Factor artifact. *Proceedings of the XIth International Conference on Thermoelectrics*, Melville, NY, University of Texas; 1992, pp. 223–231.

193. Crane D.T., and Bell L. *Progress Towards Maximizing the Performance of a Thermoelectric Power Generator*. Piscataway, NJ: Institute of Electrical and Electronics Engineers (IEEE); 2006, pp. 11–16.

194. Yatim N.M., Sallehin N.Z.I.M., Suhaimi S., and Hashim M.A. *A Review of ZT Measurement for Bulk Thermoelectric Material*. Melville, NY: AIP Publishing; 2018, Vol. 1972, p. 030002.

195. Liu S., Wang J., Jia J., Hu X., and Liu S. Synthesis and thermoelectric performance of Li-doped NiO ceramics. *Ceram Int* 2012;38:5023–5026.

196. Rad M.K., Rezania A., Omid M., Rajabipour A., and Rosendahl L.A. Study on material properties effect for maximization of thermoelectric power generation. *Renew Energ.* 2019;138:236–242.

197. Mamur H., Bhuiyan M.R.A., Korkmaz F., and Nil M. A review on bismuth telluride ($Bi_2Te_3$) nanostructure for thermoelectric applications. *Renew Sustain Energ Rev* 2018;82:4159–4169.

198. Dughaish Z. Lead telluride as a thermoelectric material for thermoelectric power generation. *Phys B* 2002;322:205–223.

199. LaLonde A.D., Pei Y., Wang H., and Snyder G.J. Lead telluride alloy thermoelectrics. *Mater Today* 2011;14:526–532.

200. Gayner C., and Kar K.K. Recent advances in thermoelectric materials. *Prog Mater Sci* 2016;83:330–382.

201. Sootsman J.R., Chung D.Y., and Kanatzidis M.G. New and old concepts in thermoelectric materials. *Angew Chem Int Ed* 2009;48:8616–8639.

202. Delime-Codrin K., Omprakash M., Ghodke S., Sobota R., Adachi M., Kiyama M., Matsuura T., Yamamoto Y., Matsunami M., and Takeuchi T. Large figure of merit $ZT = 1.88$ at 873 K achieved with nanostructured $Si_{0.55}Ge_{0.35}(P_{0.10}Fe_{0.01})$. *Appl Phys Exp* 2019;12:045507.

203. Anno H., Yamada H., Nakabayashi T., Hokazono M., and Shirataki R. Gallium composition dependence of crystallographic and thermoelectric properties in polycrystalline type-I $Ba_8Ga_xSi_{46-x}$ (nominal $x = 14$–18) clathrates prepared by combining arc melting and spark plasma sintering methods. *J Solid State Chem* 2012;193:94–104.

204. Wan C., Wang Y., Wang N., and Koumoto K. Low-thermal-conductivity $(MS)_{1+x}(TiS_2)_2$ (M = Pb, Bi, Sn) misfit layer compounds for bulk thermoelectric materials. *Materials* 2010;3:2606–2617.

205. Iversen B.B., Palmqvist A., Cox D.E., Nolas G.S., Stucky G.D., Blake N.P., and Metiu H. Why are Clathrates good candidates for thermoelectric materials? *J Solid State Chem*. 2000;149:455–458.

206. Zou M., Li J., and Kita T. Thermoelectric properties of fine-grained FeVSb half-Heusler alloys tuned to p-type by substituting vanadium with titanium. *J Solid State Chem* 2013;198:125–130.

207. Yu J., Xia K., Zhao X.B., and Zhu T. High performance p-type half-Heusler thermoelectric materials. *J Phys D* 2018;51:113001.

208. Gascoin F., Ottensmann S., Stark D., Haile S.M., and Snyder G.J. Zintl phases as thermoelectric materials: Tuned transport properties of the compounds $Ca_xYb_{1-x}Zn_2Sb_2$. *Adv Funct Mater* 2005;15:1860–1864.

209. Shuai J., Mao J., Song S., Zhang Q., Chen G., and Ren Z. Recent progress and future challenges on thermoelectric Zintl materials. *Mater Today Phys* 2017;1:74–95.

210. Ferreira N., Rasekh S., Costa F.M., Madre M., Sotelo A., Diez J., and Torres M. New method to improve the grain alignment and performance of thermoelectric ceramics. *Mater Lett* 2012;83:144–147.

211. Wang Y.F., Lee K.H., Ohta H., and Koumoto K. Fabrication and thermoelectric properties of heavily rare-earth metal-doped $SrO(SrTiO_3)_n$ (n = 1, 2) ceramics. *Ceram Int* 2008;34:849–852.

212. Lu D., Chen G., Pei J., Yang X., and Xian H. Effect of erbium substitution on thermoelectric properties of complex oxide $Ca_3Co_2O_6$ at high temperatures. *J Rare Earths* 2008;26:168–172.

213. Gayner C., Kar K.K., and Wang H. Recent progress and futuristic development of PbSe thermoelectric materials and devices. *Mater Today Energ* 2018;9:359–376.

214. Zhao L.-D., Chang C., Tan G., and Kanatzidis M.G. SnSe: A remarkable new thermoelectric material. *Energ Environ Sci* 2016;9:3044–3060.

215. Li S., Li X., Ren Z., and Zhang Q. Recent progress towards high performance of tin chalcogenide thermoelectric materials. *J Mater Chem A* 2018;6:2432–2448.

216. Tsai T.-C., Chang H.-C., Chen C.-H., and Whang W.-T. Widely variable Seebeck coefficient and enhanced thermoelectric power of PEDOT: PSS films by blending thermal decomposable ammonium formate. *Org Electron* 2011;12:2159–2164.

217. Choi Y., Kim Y., Park S.-G., Kim Y.-G., Sung B.J., Jang S.-Y., and Wang H. Effect of the carbon nanotube type on the thermoelectric properties of CNT/Nafion nanocomposites. *Org Electron* 2011;12:2120–2125.

218. Yue R., and Xu J. Poly(3,4-ethylenedioxythiophene) as promising organic thermoelectric materials: A mini-review. *Synth Met* 2012;162:912–917.

219. McGrail B.T., Sehirlioglu A., and Pentzer E. Polymer composites for thermoelectric applications. *Angew Chem Int Ed* 2014;54:1710–1723.

220. Yemata T.A., Ye Q., Zhou H., Kyaw A.K., Chin W.S., and Xu J. Conducting polymer-based thermoelectric composites. *Hybrid Polym Compos Mater* 2017;169–195.

221. Duan W., Liu J., Zhang C., and Ma Z. The magneto-thermoelectric effect of graphene with intra-valley scattering. *Chin Phys B* 2018;27:097204.

222. Olaya D., Tseng C.-C., Chang W.-H., Hsieh W., Li L.-J., Juang Z.-Y., and Hernández Y. Cross-plane thermoelectric figure of merit in graphene-C60 heterostructures at room temperature. *Flat Chem* 2019;14:100089.

223. Deng S., Cai X., Zhang Y., and Li L. Enhanced thermoelectric performance of twisted bilayer graphene nanoribbons junction. *Carbon* 2019;145:622–628.

224. Rowe DM., editor. *Handbook of Thermoelectrics: Macro to Nano*. Boca Raton, FL: CRC Press; 2005.

225. Zulkepli N., Yunas J., Mohamed M.A., and Hamzah A.A. Review of thermoelectric generators at low operating temperatures: Working principles and materials. *Micromachines* 2021;12:734. Doi: 10.3390/mi12070734 https://www.mdpi.com/journal/micromachines.

226. Rojas J.P., Singh D., Inayat S.B., Sevilla G.A.T., Fahad H.M., and Hussain M.M. Review—micro and nano-engineering enabled new generation of thermoelectric generator devices and applications. *ECS J Solid State Sci Technol* 2017;6(3):N3036–N3044.

227. Vineis C.J., Shakouri A., Majumdar A., and Kanatzidis M.G. Nanostructured thermoelectrics: Big efficiency gains from small features. *Adv Mater* 2010;22:3970.

228. Li C.W., Hong J., May A.F., Bansal D., Chi S., Hong T., Ehlers G., and Delaire O. Orbitally driven giant phonon anharmonicity in SnSe. *Nat Phys* 2015;11:1.

229. Pan Y., and Li J.-F. Thermoelectric performance enhancement in n-type $Bi_2$ (TeSe) 3 alloys owing to nanoscale inhomogeneity combined with a spark plasma textured microstructure. *NPG Asia Mater* 2016;8:e275.

230. Boukai A.I., Bunimovich Y., Tahir-Kheli J., Yu J.-K., Goddard W.A., and Heath J.R. Silicon nanowires nasvefficient thermoelectric materials. *Nature* 2008;451:168.

231. Lee S.H., Jang S.Y., Roh J.W., Park J., and Lee W. *INEC 2010-2010 3rd International Nanoelectronics Conference Proceedings*, Hong Kong, China; 2010, p. 1203.

232. Ramayya E.B., Maurer L.N., Davoody A.H., and Knezevic I. Thermoelectric properties of ultrathin silicon nanowires. *Phys Rev B* 2012;86:115328.

233. Zhang X., and Zhao L.-D. Thermoelectric materials: Energy conversion between heat and electricity. *J Mater* 2015;1:92.

234. Xu E.Z., Li Z., Martinez J.A., Sinitsyn N., Htoon H., Li N., Swartzentruber B., Hollingsworth J.A., Wang J., and Zhang S.X. Diameter dependent thermoelectric properties of individual SnTe nanowires. *Nanoscale* 2015;7:2869.

235. Yang H., Bahk J.H., Day T., Mohammed A.M.S., Snyder G.J., Shakouri A., and Wu Y. Composition modulation of $Ag_2Te$ nanowires for tunable electrical and thermal properties. *Nano Lett* 2015;15:1349.

236. Venkatasubramanian R., Siivola E., Colpitts T., and O'Quinn B. Thin-film thermoelectric devices with high room temperature figures of merit. *Nature* 2001;413:597.

237. Harman T.C., Taylor P.J., Spears D.L., and Walsh M.P. Thermoelectric quantum-dot superlattices with high ZT. *J Electron Mater* 2000;29:L1.

238. Yu J.-K., Mitrovic S., Tham D., Varghese J., and Heath J.R. Reduction of thermal conductivity in phononic nanomesh structures. *Nat Nanotechnol* 2010;5:718.

239. Sadeghi H., Sangtarash S., and Lambert C.J. Enhancing the thermoelectric figure of merit in engineered graphene nanoribbons. *Beilstein J Nanotechnol* 2015;6:1176.

240. Yamasaka S., Watanabe K., Sakane S., Takeuchi S., Sakai A., Sawano K., and Nakamura Y. Independent contro of electrical and heat conduction by nanostructure designing for Si- based thermoelectric materials. *Sci Rep* 2016;6:22838.

241. Sootsman J.R., Chung D.Y., and Kanatzidis M.G. New and old concepts in thermoelectric materials. *Angew Chemie Int Ed* 2009;48:8616.

242. Snyder G.J., and Toberer E.S. Complex thermoelectric materials. *Nat Mater* 2008;7:105.

243. Alam H., and Ramakrishna S. A review on the enhancement of figure of merit from bulk to nano-thermoelectric materials. *Nano Energ* 2013;2:190.

244. Choi S., Lee H., Ghaffari R., Hyeon T., and Kim D.-H. Recent advances ion flexible and stretchable bio-electronic devices integrated with nanomaterials. *Adv Mater* 2016;28:4203.

245. Pei Y., Wang H., and Snyder G.J. Band engineering of thermoelectric materials. *Adv Mater* 2012;24:6125.

246. Zhang Q., Wang H., Liu W., Wang H., Yu B., Zhang Q., Tian Z., Ni G., Lee S., Esfarjani K., Chen G., and Ren Z. Enhancement of thermoelectric figure of merit by resonant states of aluminium doping in lead selenide. *Energy Environ Sci* 2012;5:5246.

247. Zhang Q., Liao B., Lan Y., Lukas K., Liu W., Esfarjani K., Opeil C., Broido D., Chen G., and Ren Z. High thermoelectric performance by resonant dopant indium in nanostructured SnTe. *Proc Natl Acad Sci* 2013;110:13261.
248. Zhao L.D., Wu H.J., Hao S.Q., Wu C.I., Zhou X.Y., Biswas K., He J.Q., Hogan T.P., Uher C., Wolverton C., Dravid V.P., and Kanatzidis M.G. All scale hierarchical thermoelectrics: MgTe in PbTe facilitates valence band convergence and suppresses bipolar thermal transport for high performance. *Energ Environ Sci* 2013;6:3346.
249. Minnich A.J., Dresselhaus M.S., Ren Z.F., and Chen G. Bulk nanostructured thermoelectric materials: current research and future prospoects. *Energ Environ Sci* 2009;2:466.
250. Demchenko D.O., Heinz P.D., and Lee B. Determining factors of thermoelectric properties of semiconductor nanowires. *Nanoscale Res Lett* 2011;6:502.
251. Hochbaum A.I., Chen R., Delgado R.D., Liang W., Garnett E.C., Najarian M., Majumdar A., and Yang P. Enhanced thermoelectric performance of rough silicon nanowires. *Nature* 2008;451:163.
252. Zhang K., Qiu J., and Wang S. Thermoelectric properties of PEDOT nanowire/PEDOT hybrids. *Nanoscale* 2016;8:8033.
253. Toshima N., Oshima K., Anno H., Nishinaka T., Ichikawa S., Iwata A., and Shiraishi Y. Novel hybrid organic thermoelectric matewrials: three component hybrid films consisting of a nano particle polymer complex, carbon nanotubes and vinyl polymer. *Adv Mater* 2015;27:2246.
254. Cho C., Stevens B., Hsu J.H., Bureau R., Hagen D.A., Regev O., Yu C., and Grunlan J.C. Completely organic multi layer thin film with thermoelectric power factor rivaling inorganic tellurides. *Adv Mater* 2015;27:2996.
255. Zhang K., Davis M., Qiu J., Hope-Weeks L., and Wang S. Thermoelectric properties of porous multi-walled carbon nanotube/polyaniline core/shell nanocomposites. *Nanotechnology* 2012;23:385701.
256. Meng C., Liu C., and Fan S. A promising approach to enhanced thermoelectric properties using carbon nanotube networks. *Adv Mater* 2010;22:535.
257. Ge Z.H., Zhao L.D., Wu D., Liu X., Zhang B.P., Li J.F., and He J. Low-cost abundant binary sulfides as promising thermoelectric materials. *Mater Today* 2016;19:227.
258. Pop E., Sinha S., and Goodson K.E. Heat generation and transport in nanoscale transistors. *Proc IEEE* 2006;94:1587.
259. Tavakkoli F., Ebrahimi S., Wang S., and Vafai K. Analysis of critical thermal issues in 3D integrated circuits. *Int J Heat Mass Transf* 2016;97:337.
260. Tuyen L.T.C., Le P.H., and Jian S-R Nanostructuring $Bi_2Te_3$-based thermoelectric thin-films grown using pulsed laser deposition, an Intech review paper. 2021. Doi: 10.5772/intechopen.99469.
261. Pandel D., Singh A.K., Banerjee M.K., and Gupta R. Optimizing thermoelectric generators based on $Mg_2(Si, Sn)$ alloys through numerical simulations. *Energ Convers Manage* September 2021;11:100097.
262. Leonov V., and Vullers R.JM. Wearable electronics self-powered by using human body heat: The state of the art and the perspective. *J Renew Sustain Energ* 2009;1:062701.
263. Kim M.-K., Kim M.-S., Lee S., Kim C., and Kim Y.-J. Wearabke thermoelectric generator for harvesting human body heat energy. *Smart Mater Struct* 2014;23:105002.
264. Du Y., Cai K., Chen S., Wang H., Shen S.Z., Donelson R., and Lin T. Thermoelectric fabrics: toward power generating clothing. *Sci Rep* 2015;5:6411.
265. Zhang Q., Sun Y., Xu W., and Zhu D. Organic thermoelectric materials: emerging green energy bmaterials converting heat to electricity directly and efficiently. *Adv Mater* 2014;26:6829.
266. Aranguren P., Roch A., Stepien L., Abt M., von Lukowicz M., Dani I., and Astrain D. Progress in polymer thermoelectrics. *Appl Therm Eng* 2016;102:402.

267. Sadeghi H., Sangtarash S., and Lambert C.J. Enhanced thermoelectric efficiency of porous Silcene nanoribbons. *Sci Rep* 2015;5:9514.
268. Zhang J., Liu H.J., Cheng L., Wei J., Liang J.H., Fan D.D., Shi J., Tang X.F., and Zhang Q.J. *Sci Rep* 2014;4:3621.
269. Oh J.Y., Lee J.H., Han S.W., Chae S.S., Bae E.J., Kang Y.H., Choi W.J., Cho S.Y., Lee J.-O., Baik H. K., and Il Lee T. Chemically exfoliated transition metal dichalcogenide nanosheet-based wearable thermoelectric generators. *Energ Environ Sci* 2016;9:1696.
270. Bahk J.-H., Fang H., Yazawa K., and Shakouri A. Flexible thermoelectric materials and device optimization for wearable energy harvesting. *J Mater Chem C* 2015;3:10362.
271. Du Y., Shen S.Z., Cai K., and Casey P.S. Research progress on polymer-inorganic thermoelectric nanocomposite materials. *Prog Polym Sci* 2012;37:820.
272. McGrail B.T., Sehirlioglu A., and Pentzer E. Polymer composites for thermoelectric applications. *Angew Chem Int Ed* 2015;54:1710.
273. Kamarudin M.A., Sahamir S.R., Datta R.S., Long B.D., Mohd Sabri M.F., and Mohd Said S. A review on the fabrication of polymer based thermoelectric materials and fabrication methods. *Sci World J.* 2013;2013:713640.
274. Madan D., Wang Z., Wright P.K., and Evans J.W. Printed flexible thermoelectric generators for use on low levels of waste heat. *Appl Energ* 2015;156:587.
275. Kim S.J., We J.H., and Cho B.J. A wearable thermoelectric generator fabricated on a glass fabric. *Energ Environ Sci* 1959;7.
276. Sheng P., Sun Y., Jiao F., Di C., Xu W., and Zhu D. A novel cuprous ethylenetetrathiolate coordination polymer: structure characterization, thermoelectric property optimizationand a bulk thermogenerator demonstration. *Synth Met* 2014;193:1.
277. Kato K., Hatasako Y., Kashiwagi M., Hagino H., Adachi C., and Miyazaki K. Fabrication of a flexible Bismuth Telluride power generation module using microporous polyimide films as substrates. *J Electron Mater* 2014;43:1733.
278. Tobjörk D., and Österbacka R. Paper electronics, *Adv Mater* 2011;23:1935.
279. Sun C., Goharpey A. H., Rai A., Zhang T., and Ko D.-K. *ACS Appl Mater Interf* 2016;8:22182.
280. Sevilla G.A.T., Bin Inayat S., Rojas J.P., Hussain A.M., and Hussain M.M. Flexible and semi transparent thermoelectric energy harvesters from low cost bulk silicon (100). *Small* 2013;9:3916.
281. Francioso L., De Pascali C., Farella I., Martucci C., Cretì P., Siciliano P., and Perrone A. Flexible thermoelectric generator for ambient assiste living wearable biometric sensors. *J Power Sourc* 2011;196:3239.
282. Yang Y., Lin Z.-H., Hou T., Zhang F., and Wang Z.L. Nanowire composite based flexible thermoelectric nanogenerators and self powered temperature sensors. *Nano Res* 2012;5:888.
283. Choi S., Lee H., Ghaffari R., Hyeon T., and Kim D.-H. Recent advances in flexible and stretchable bio-electronic devices integrated with nanomaterials. *Adv Mater* 2016;28:4203.
284. Hussain A.M., and Hussain M.M. CMOS-technology enabled flexible and stretchable electronics for internet of everything applications. *Adv Mater* 2016;28:4219.
285. Wang S., Huang Y., and Rogers J.A. Review of flexible and transparent thin film transistors based on zinc oxide and related materials. *IEEE Trans Compon, Packag, Manuf Technol* 2015;5:1201.
286. Xu S., Yan Z., Jang K.-I., Huang W., Fu H., Kim J., Wei Z., Flavin M., McCracken J., Wang R., and Badea A. Material science. Assembly of micro-nanomaterials into complex, three -dimensional architectures by compressive buckling. *Science* 2015;347:154.
287. Rojas J.P., Arevalo A., Foulds I.G., and Hussain M.M. Review-Micro and Nano engineering enabled new generation of thermoelectric generator devices and applications. *Appl Phys Lett* 2014;105:154.

288. Fan J., Yeo W.-H., Su Y., Hattori Y., Lee W., Jung S.-Y., Zhang Y., Liu Z., Cheng H., Falgout L., Bajema M., Coleman T., Gregoire D., Larsen R.J., Huang Y., and Rogers J.A. Factual design concepts for stretchable electronics. *Nat Commun* 2014;5:3266.

289. Kim J., Lee J., Son D., Choi M.K., and Kim D.-H. Deformable devices with integrated functional nanomaterials for wearable electronics. *Nano Converg* 2016;3:1.

290. Jung I., Xiao J., Malyarchuk V., Lu C., Li M., Liu Z., Yoon J., Huang Y., and Rogers J.A. Dynamically tunable hemispherical electronic eye camera system with adjustable zoom capability. *Proc Natl Acad Sci* 2011;108:1788.

291. Rus D., and Tolley M.T. Design, fabrication and control of soft robots. *Nature* 2015;521:467.

292. Xu L., Gutbrod S.R., Ma Y., Petrossians A., Liu Y., Webb R.C., Fan J.A., Yang Z., Xu R., Whalen J. J., Weiland J.D., Huang Y., Efimov I.R., and Rogers J.A. *Adv Mater* 2015;27:173.

293. Leonov V., and Vullers R. Wearable electronics self-powered by using human body heat: The state of the art and the perspective. *J Renew Sustain Energ* 2009;1:62701.

294. Bhatnagar V., and Owende P. Energy harvesting for assistive and mobile applications. *Energy Sci Eng* 2015;3:153–173.

295. Siddique A.R.M., Mahmud S., and Van Heyst B. A review of the state of the science on wearable thermoelectric power generators (TEGs) and their existing challenges. *Renew Sustain Energ Rev* 2017;73:730–744.

296. Karthikeyan V., Surjadi J.U., Wong J.C., Kannan V., Lam K.-H., Chen X., Lu Y., and Roy V.A.L. Wearable and flexible thin film thermoelectric module for multi-scale energy harvesting. *J Power Sourc* 2020;455:227983.

297. Nozariasbmarz A., Collins H., Dsouza K., Polash M.H., Hosseini M., Hyland M., Liu J., Malhotra A., Ortiz F.M., Mohaddes F., and Ramesh V.P. Review of wearable thermoelectric energy harvesting: Frombody temperature to electronic systems. *Appl Energ* 2020;258:114069.

298. Han S. Wearable thermoelectric devices. In *Thermoelectric Thin Films*. Cham: Springer Science and Business Media LLC; 2019, pp. 31–42.

299. Sun T., Zhou B., Zheng Q., Wang L., Jiang W., and Snyder G.J. Stretchable fabric generates electric power from woven thermoelectric fibers. *Nat Commun* 2020;11:572.

300. Xu Q., Qu S., Ming C., Qiu P., Yao Q., Zhu C., Wei T.-R., He J., Shi X., and Chen L. Conformal organic–inorganic semiconductor composites for flexible thermoelectrics. *Energ Environ Sci* 2020;13:511–518.

301. Leonov V., Torfs T., van Hoof C., and Vullers R.J. Smart wireless sensors integrated in clothing: An electrocardiography system in a shirt powered using human body heat. *Sens. Trans* 2009;107:165.

302. Ren W., Sun Y., Zhao D., Aili A., Zhang S., Shi C., Zhang J., Geng H., Zhang J., Zhang L., and Xiao J. High-performance wearable thermoelectric generator with self-healing, recycling, and Lego-like reconfiguring capabilities. *Sci Adv* 10 February 2021;7(7). Doi: 10.1126/sciadv.abe0586.

303. Jo S., Choo S., Kim F., Heo S.H., and Son J.S. Ink processing for thermoelectric materials and power-generating devices. *Adv Mater* 2019;31:1804930.

304. Jin Q., Jiang S., Zhao Y., Wang D., Qiu J., Tang D.-M., Tan J., Sun D.-M., Hou P.-X., Chen X.-Q., Tai K., Gao N., Liu C., Cheng H.-M., and Jiang X. Flexible layer-structured $Bi_2Te_3$ thermoelectric on a carbon nanotube scaffold. *Nat Mater* 2019;18:62–68.

305. Venkatasubramanian R., Siivola E., Colpitts T., and O'Quinn B. Thin-film thermoelectric devices with high room-temperature figures of merit. *Nature* 2001;413:597–602.

306. Goncalves L.M., Couto C., Correia J.H., Alpuim P., Min G., and Rowe D.M. Optimization of thermoelectric thin-films deposited by co-evaporation on plastic substrates. *Proceedings of 4th European Conference on Thermoelectrics*, Cardiff; 9–11 April, 2006.

307. Giani A., Boulouz A., Pascal-Delannoy F., Foucaran A., Charles E., and Boyer A. Growth of $Bi_2Te_3$ and $Sb_2Te_3$ thin films by MOCVD. *Mater Sci Eng B Solid State Mater Adv Technol* 1999;64:19–24.
308. Rojas J.P., Singh D., Conchouso D., Arevalo A., Foulds I.G., and Hussain M.M. Stretchable helical architecture inorganic-organic hetero thermoelectric generator. *Nano Energ* 2016;30:691.
309. van Bavel M., Leonov V., Yazicioglu R.F., Torfs T., van Hoof C., Posthuma N., and Vullers R. Wearable battery-free wireless 2-channel EEG systems powered by energy scavengers. *Sens Trans J* 2008;94:103–115.
310. Torfs T. Pulse oximeter fully powered by human body heat. *Sens Trans J* 2007;80: 1230–1238.
311. Yuan Z. *Etude et Réalisation de Microgénérateurs Thermoélectriques Planaires en Technologie Silicium.* Ph.D. Thesis. Lille: Université Lille 1; 2012.
312. Kishi M., Nemoto H., Hamao T., Yamamoto M., Sudou S., Mandai M., and Yamamoto S. Micro thermoelectric modules and their application to wristwatches as an energy source. *Proceedings of the Eighteenth International Conference on Thermoelectrics. Proceedings, ICT'99 (Cat. No.99TH8407)*, Baltimore, MD, Piscataway, NJ, Institute of Electrical and Electronics Engineers (IEEE); 29 August–2 September, 1999; 2003, pp. 301–307.
313. Fernandes A.E.S.D.S. *Conversão de Energia com células de Peltier.* Ph.D. Thesis. Caparica, Portugal: Faculdade de Ciências eTecnologia; 2012.
314. Dyson Energy Bracelet a Good Call. Available from: https://newatlas.com/dyson-energy-bracelet/12040/ (accessed: 30 September 2018).
315. Suarez F., Nozariasbmarz A., Vashaee D., and ztürk M.C. Designing thermoelectric generators for self-powered wearable electronics. *Energ Environ Sci* 2016;9:2099–2113.
316. Francioso L., De Pascali C., Farella I., Martucci C., Creti P., Siciliano P.A., and Perrone A. Flexible thermoelectric generator for ambient assisted living wearable biometric sensors. *J Power Sourc* 2011;196:3239–3243.
317. Kim M.K., Kim M.S., Jo S.E., Kim H.L., Lee S.M., and Kim Y.J. *Wearable Thermoelectric Generator for Human Clothing Applications.* Piscataway, NJ: Institute of Electrical and Electronics Engineers (IEEE); 2013, pp. 1376–1379.
318. Suarez F., Parekh D.P., Ladd C., Vashaee D., Dickey M.D., and ztürk M.C. Flexible thermoelectric generator using bulk legs and liquid metal interconnects for wearable electronics. *Appl Energ* 2017;202:736–745.
319. Zadan M., Malakooti M.H., and Majidi C. Soft and stretchable thermoelectric generators enabled by liquid metal elastomer composites. *ACS Appl Mater Interf* 2020;12:17921–17928.
320. Geisler M. *Récupération d'énergie Mécanique pour Vêtements Connectés Autonomes.* Ph.D. Thesis. Saint-Martin-d'Hères, France: Université Grenoble Alpes; 2017.
321. Soleimani Z., Zoras S., Ceranic B., Shahzad S., and Cui Y. A review on recent developments of thermoelectric materials for room-temperature applications. *Sustain Energ Technol Assess* 2020;37:100604.
322. Li C., Jiang F., Liu C., Liu P., and Xu J. Present and future thermoelectric materials toward wearable energy harvesting. *Appl Mater Today* 2019;15:543–557.
323. Barma M., Riaz M., Saidur R., and Long B. Estimation of thermoelectric power generation by recovering waste heat from Biomass fired thermal oil heater. *Energy Convers Manage* 2015;98:303–313.
324. Dai D., Zhou Y., and Liu J. Liquid metal based thermoelectric generation system for waste heat recovery. *Renew Energ* 2011;36:3530–3536.
325. Junior O.H.A., Calderon N.H., and De Souza S.S. Characterization of a thermoelectric generator (TEG) system for waste heat recovery. *Energies* 2018;11:1555.

326. Zou S., Kanimba E., Diller T.E., Tian Z., and He Z. Modeling assisted evaluation of direct electricity generation from waste heat of waste water via a thermoelectric generator. *Sci Total Environ* 2018;635:1215–1224.

327. Meng F., Chen L., Xie Z., and Ge Y. Thermoelectric generator with air-cooling heat recovery device from wastewater. *Therm Sci Eng Prog* 2017;4:106–112.

328. Araiz M., Casi Á., Catalan L., Martínez Á., and Astrain D. Prospects of waste-heat recovery from a real industry using thermoelectric generators: Economic and power output analysis. *Energ Convers Manage* 2020;205:112376.

329. Mirhosseini M., Rezania A., and Rosendahl L.A. Power optimization and economic evaluation of thermoelectric waste heat recovery system around a rotary cement kiln. *J Clean Prod* 2019;232:1321–1334.

330. Agrawal R. Development and analysis of waste heat recovery system by air conditioning application. *Int J Emerg Trends Eng Dev* 2017;1.

331. Killander A., and Bass J.C. *A Stove-Top Generator for Cold Areas.* Piscataway, NJ: Institute of Electrical and Electronics Engineers(IEEE); 2002, pp. 390–393.

332. Alien D., and Mallon W. *Further Development of "Self-Powered Boilers"* Piscataway, NJ: Institute of Electrical and Electronics Engineers (IEEE); 2003, pp. 80–83.

333. Allen D.T., and Wonsowski J. *Thermoelectric Self-Powered Hydronic Heating Demonstration.* Piscataway, NJ: Institute of Electrical and Electronics Engineers (IEEE); 2002, pp. 571–574.

334. Sornek K., Filipowicz M., Żołądek M., Kot R., and Mikrut M. Comparative analysis of selected thermoelectric generators operating with wood-fired stove. *Energy* 2019;166:1303–1313.

335. Bass J., and Farley R. *Examples of Power from Waste Heat for Gas Fields.* Piscataway, NJ: Institute of Electrical and Electronics Engineers (IEEE); 2002, pp. 547–550.

336. Pickard D., DiLeo F., Kushch A., Hauerbach M., and LeVine L. *A Self-Powered Field Feeding System.* Natick, MA: Army Natick Soldier Center; 2006.

337. Fontaras G., Zacharof N., and Ciuffo B. Fuel consumption and $CO_2$ emissions from passenger cars in Europe–Laboratory versus real-world emissions. *Prog Energ Combust Sci* 2017;60:97–131.

338. Di Battista D., Mauriello M., and Cipollone R. Waste heat recovery of an ORC-based power unit in a turbo charged diesel engine propelling a light duty vehicle. *Appl Energ* 2015;152:109–120.

339. Yu C., and Chau K.T. Thermoelectric automotive waste heat energy recovery using maximum power point tracking. *Energ Convers Manage* 2009;50:1506–1512.

340. Stabler F. Automotive applications of high efficiency thermoelectrics. *Proceedings of the DARPA/ONR Program Review and DOE High Efficiency Thermoelectric Workshop*, San Diego, CA; 24–27 March, 2002.

341. Vázquez J., Sanz-Bobi M.A., Palacios R., and Arenas A. State of the art of thermoelectric generators based on heat recovered from the exhaust gases of automobiles. *Proceedings of the 7th European Workshop on Thermoelectrics*, Pamplona, Spain; 3–4 October, 2002.

342. Agudelo A., García-Contreras R., Agudelo J.R., and Armas O. Potential for exhaust gas energy recovery in a diesel passenger car under European driving cycle. *Appl Energ* 2016;174:201–212.

343. Mori M., Yamagami T., Sorazawa M., Miyabe T., Takahashi S., and Haraguchi T. Simulation of fuel economy effectiveness of exhaust heat recovery system using thermoelectric generator in a series hybrid. *SAE Int J Mater Manuf* 2011;4:1268–1276.

344. Birkholt U. Conversion of waste exhaust heat in automobiles using $FeSi_2$ thermoelements. *Proceedings of the 7th International Conference on Thermoelectric Energy Conversion*, Arlington, TX; 16–18 March, 1988.

345. Ikoma K., Munekiyo M., Furuya K., Kobayashi M., Izumi T., and Shinohara K. *Thermoelectric Module and Generator for Gasoline Engine Vehicles.* Piscataway, NJ: Institute of Electrical and Electronics Engineers (IEEE); 2002, pp. 464–467.

346. Ikoma K., Munekiyo M., Furuya K., Kobayashi M., and Komatsu H. Thermoelectric generator for gasoline engine vehicles using $Bi_2Te_3$ modules. *J Jpn Inst Met* 1999;63:1475–1478.

347. Bass J.C., Elsner N.B., and Leavitt F.A. Performance of the 1 kW thermoelectric generator for diesel engines. *Proceedings of the AIP Conference Proceedings*, Melville, NY, AIP Publishing; 1994, Vol. 316, pp. 295–298.

348. Crane D., LaGrandeur J., Jovovic V., Ranalli M., Adldinger M., Poliquin E., Dean J., Kossakovski D., Mazar B., and Maranville C. TEG on-vehicle performance and model validation and what it means for further TEG development. *J Electron Mater* 2012;42:1582–1591.

349. Crane D.T., LaGrandeur J.W., Harris F., and Bell L.E. Performance results of a high-power-density thermoelectric generator: Beyond the couple. *J Electron Mater* 2009;38:1375–1381.

350. Crane D.T., and LaGrandeur J.W. Progress report on BSST-led US department of energy automotive waste heat recovery program. *J Electron Mater* 2009;39:2142–2148.

351. Bell L.E., LaGrandeur J.W., and Crane D.T. Progress report on vehicular waste heat recovery using a cylindrical thermoelectric generator. *Thermoelectr Goes Automot* 2010;2:83–91.

352. Ranalli M., Adldinger M., Kossakovski D., and Womann M. Thermoelectric generators from aerospace to automotive. *ATZ Worldw* 2013;115:60–65.

353. Shen Z.-G., Tian L.-L., and Liu X. Automotive exhaust thermoelectric generators: Current status, challenges and future prospects. *Energy Convers Manage* 2019;195:1138–1173.

354. Meisner G.P. Advanced thermoelectric materials and generator technology for automotive waste heat at GM. *Proceedings of the 2nd Thermoelectrics Applications Workshop*, San Diego, CA; 3–6 January, 2011.

355. Kumar S., Heister S.D., Xu X., Salvador J.R., and Meisner G.P. Thermal design of thermoelectric generators for automobile waste heat recovery. *Volume 1: Heat Transfer in Energy Systems; Theory and Fundamental Research; Aerospace Heat Transfer; Gas Turbine Heat Transfer; Transport Phenomena in Materials Processing and Manufacturing; Heat and*, New York, ASME International; 2012, Vol. 1, pp. 67–77.

356. Magnetto D. HeatReCar: First light commercial vehicle equipped with a TEG. *Proceedings of the 3rd International Conference Thermal Management for EV/HEV*, Darmstadt, Germany; 23–26 June, 2013.

357. Aixala L. RENOTER project. *3rd Thermoelectrics Applications Workshop*, Baltimore, MD, US Department of Energy; 2012.

358. Nader W.B. Thermoelectric generator optimization for hybrid electric vehicles. *Appl Therm Eng* 2020;167:114761.

359. Septiadi W.N., Iswari G.A., Rofiq M.A., Gitawan B., Gugundo J.M., and Purba C.A.D. Output voltage characteristic of heat pipe sink thermoelectric generator with exhaust heat utilization of motorcycles. *IOP Conf Ser* 2018;105:12129.

360. Fundamental Research-Atsumitec (Japan). Available from: http://www.atsumitec.co.jp/en/technology/basis (accessed: 13 June 2018).

361. Shinohara Y. The state of the art on thermoelectric devices in Japan. *Mater Today* 2015;2:877–885.

362. Schlichting A.D., Anton S.R., and Inman D.J. Motorcycle waste heat energy harvesting. *Proceedings of the 15th International Symposium on: Smart Structures and Materials & Nondestructive Evaluation and Health Monitoring*, San Diego, CA; 9–13 March, 2008, Vol. 6930, p. 69300.

363. Huang J. Aerospace and aircraft thermoelectric applications. *Proceedings of the DoE Thermoelectric Applications Workshop*, San Diego, CA; 29 September–1 October, 2009.

364. Bode C., Friedrichs J., Somdalen R., Köhler J., Büchter K.-D., Falter C., Kling U., Ziolkowski P., Zabrocki K., Müller E., and Kožulović D. Potential of future thermoelectric energy recuperation for aviation. *J Eng Gas Turbines Power* 2017;139:101201.

365. Lyras M., Zymaride L., Kyratsi T., Louca L., and Becker T. Simulation based design of a thermoelectric energy harvesting device for aircraft applications. *Proceedings of the Dynamic Systems and Control Conference. American Society of Mechanical Engineers*, TysonsCorner, VA; 11–13 October, 2017, p. V003T41A003.
366. Pidwirny M. Causes for Climate change, The layered atmosphere. *Fundam Phys Geogr* 2006.
367. Allmen L.V., Bailleul G., Becker T., Decotignie J.-D., Kiziroglou M.E., Leroux C., Mitcheson P.D., Muller J., Piguet D., Toh T.T., and Weisser A. Aircraft strain WSN powered by heat storage harvesting. *IEEE Trans Ind Electron* 2017;64:7284–7292.
368. Kiziroglou M.E., Wright S., Toh T.T., Mitcheson P.D., Becker T., and Yeatman E.M. Design and fabrication of heat storage thermoelectric harvesting devices. *IEEE Trans Ind Electron* 2014;61:302–309.
369. Samson D., Otterpohl T., Kluge M., Schmid U., and Becker T. Aircraft-specific thermoelectric generator module. *J Electron Mater* 2009;39:2092–2095.
370. Kousksou T., Bedecarrats J.-P., Champier D., Pignolet P., and Brillet C. Numerical study of thermoelectric power generation for an helicopter conical nozzle. *J Power Sourc* 2011;196:4026–4032.
371. Smith T.W.P., Jalkanen J.P., Anderson B.A., Corbett J.J., Faber J., Hanayama S., O'Keeffe E., Parker S., Johanasson L., and Aldous L. *Third IMO GHG Study*. London: International Maritime Organization; 2015.
372. Loupis M., Papanikolaou N., and Prousalidis J. Fuel consumption reduction in marine power systems through thermoelectric energy recovery. *Proceedings of the 2nd International MARINELIVE Conference on All Electric Ship*, Athens, Greece; 3–5 June, 2013, pp. 1–7.
373. Hisamoto D., Lee W.-C., Kedzierski J., Takeuchi H., Asano K., Kuo C., Anderson E., King T.-J., Bokor J., and Hu C. FinFET-a self aligned double gate MOSFET scalable to 20 nm. *IEEE Trans Electron Dev* 2000;47:2320.
374. Asheghi M., Touzelbaev M.N., Goodson K.E., Leung Y.K., and Wong S.S. Temperature dependent thermal conductivityof single crystal silicon layers in SOI substrates. *J Heat Transf* 1998;120:30.
375. Pop E., Sinha S., and Goodson K.E. Heat generation and transport in nanometer-scale transistors. *Proc IEEE* 2006;94:1587.
376. Tavakkoli F., Ebrahimi S., Wang S., and Vafai K. Analysis of critical thermal issues in 3D integrated circuits. *Int J Heat Mass Transf* 2016;97:337.
377. Chowdhury I., Prasher R., Lofgreen K., Chrysler G., Narasimhan S., Mahajan R., Koester D., Alley R., and Venkatasubramanian R. On chip cooling by superlattice-based thin film thermoelectrics. *Nat Nano* 2009;4:235.
378. Zhou Y., Paul S., and Bhunia S. *A Paper Presented at Design, Automation and Test in Europe conference*, Munich, Germany; March 10–14, 2008.
379. Fahad H., Hasan M., Li G., and Hussain M. Thermoelectricity from wasted heat of integrated circuit. *Appl Nanosci* 2013;3:175.
380. Aktakka E.E., Ghafouri N., Smith C.E., Peterson R.L., Hussain M.M., and Najafi K. Post-CMOS FinFET integration of Bismuth Telluride and Antimony Telluride thin film based thermoelectric devices on SoI substrate. *IEEE Electron Dev Lett* 2013;34:1334.
381. Inayat S.B., Rader K.R., and Hussain M.M. Nano-materials enabled thermoelectricity from window glasses. *Sci Rep* 2012;2:841.
382. Inayat S.B., Rader K.R., and Hussain M.M. Manufacturing of thermoelectric nano-materials (Bi 0,4 Sb 1.6 Te 3/Bi 1.75 Te 3.25) and integration into window glasses for thermoelectricity generation. *Energ Technol* 2014;2:292.
383. Glosch H., Ashauer M., Pfeiffer U., and Lang W. A thermoelectric converter for energy supply. *Sens Actuat A Phys* 1999;74:246.

384. Rowe D.M., Morgan D.V, and Kiely J.H. Miniature low power/ high voltage thermoelectric generator. *Electron Lett* 1989;25:166.

385. Bottner H., Nurnus J., Gavrikov A., Kuhner G., Jagle M., Kunzel C., Eberhard D., Plescher G., Schubert A., and Schlereth K.H. New thermoelectric components using microsystem technologies. *J Microelectromech Syst* 2004;13:414.

386. Snyder G.J., Lim J.R., Huang C.-K., and Fleurial J.-P. Thermoelectric microdevice fabricated by a MEMS-like electrochemical process. *Nat Mater* 2003;2:528.

387. Kishi M., Nemoto H., Hamao T., Yamamoto M., Sudou S., Mandai M., and Yamamoto S. Micro thermoelectric modules and their application to wristwatches as an energy source. *18th International Conference Thermoelectricity*, Baltimore, MD; 1999, p. 301.

388. Inayat S.B. *Nano-Micro Materials Enabled Thermoelectricity From Window Glasses.* Thesis. King Abdullah University of Science and Technology, Thuwal, Saudi Arabia; 2012.

389. Bonin R., Boero D., Chiaberge M., and Tonoli A. Design and characterization of small thermoelectric generators for environmental monitoring devices. *Energy Convers Manage* 2013;73:340–349.

390. Thermoelectric Generators (TEGs). Gentherm Global Power Technologies. Available from: http://www.genthermglobalpower.com/products/thermoelectric-generators-tegs (accessed: 22 June 2018).

391. Thermoelectric Generators for Cathodic Protection by Global Thermoelectric Inc. Farwest Corrosion Control. Available from: https://www.farwestcorrosion.com/thermoelectric-generators-for-cathodic-protection-byglobal-thermoelectric.html (accessed: 22 June 2018).

392. Horie S. *Thermoelectric Energy Conversion Systems.* Tokyo: REALIZE Science & Engineering; 1995, pp. 112–115.

393. BioLite-ROW. BioLite Outdoor & Off-Grid Energy|Rest-Of-World, BioLite-ROW. Available from: https://row.bioliteenergy.com/ (accessed: 25 June 2018).

394. Hsu C.-T., Won C.-C., Chu H.-S., and Hwang J.-D. A case study of thermoelectric generator application on rotary cement furnace. *2013 8th International Microsystems, Packaging, Assembly and Circuits Technology Conference (IMPACT)*; 2013, pp. 78–81. IEEE. Doi:10.1109/IMPACT.2013.6706644.

395. Polcyn A., and Khaleel M. Advanced thermoelectric materials for efficient waste heat recovery in process industries, 2009. Available from: https://www1.eere.energy.gov/manufacturing/industries_technologies/imf/pdfs/16947_advanced_thermoelectric_materials1.pdf.

396. Kaibe H., Kajihara T., Fujimoto S., Makino K., and Hachiuma H. Recovery of plant waste heat by a thermoelectric generating system. *Komatsu Technical Report* 2011;57(164):26–30. Available from: http://dcnwis78.komatsu.co.jp/CompanyInfo/profile/report/pdf/164-E-05.pdf.

397. Kuroki T., Kabeya K., Makino K., Kajihara T., Kaibe H., Hachiuma H., Matsuno H., and Fujibayashi A. Thermoelectric generation using waste heat in steel works. *J Electron Mater* 2014. Doi: 10.1007/s11664-014-3094-5.

398. Industrial-size generator makes waste heat valuable. Global Sources, a report in EE Times, India, 22 October 2014. Available from: http://www.eetindia.co.in/ART_8800705257_1800008_NT_3a933cb6.HTM.

399. Yazawa K., and Shakouri A. Energy payback optimization of thermoelectric power generator systems. In: *Energy Systems Analysis, Thermodynamics and Sustainability. Nano Engineering for Energy, Engineering to Address Climate Change*, Vol. 5, pp. 569–576, Parts A and B. ASME. Doi: 10.1115/IMECE2010-37957.

400. Hendricks T., and Choate W. *Engineering Scoping Study of Thermoelectric Generator Systems for Industrial Waste Heat Recovery.* Washington, DC: U.S. Department of Energy; 2006.

401. Department of Energy, a report on Innovating clean energy technologies in advanced manufacturing (Chapter 6). In: *Quadrennial Technology Review: An Assessment of Energy Technologies and Research Opportunities*. Washington, DC: Department of Energy; 2015.

402. U.S. Energy Information Administration. Electricity Data Browser. EIA.gov. 24 September 2014. Available from: http://www.eia.gov/ electricity/data/browser.

403. *Workshop: Manufacturing Opportunities for Low-Cost Thermoelectric Modules*, held February 2014, organized by U.S. Department of Energy, Washington, D.C.

404. Fraunhofer. Thermogenerator from the printer. Press Releases, 2012, 2 June 2015. Available from: https://www. fraunhofer.de/en/press/research-news/2012/october/thermogenerator-from-the-printer.html.

405. LeBlanc S., Yee S.K., Scullin M.L., Dames C., and Goodson K.E. Material and manufacturing cost considerations for thermoelectrics. *Renew Sustain Energ Rev* 2014;32:313–327.

406. Patyk A. Thermoelectrics: Impacts on the environment and sustainability. *J Electron Mater* 2009;39(9):2023–2028. Doi: 10.1007/s11664-009-1013-y.

407. Bass J.C., and Thelin J. Development of a self-powered pellet stove. Hi-Z Technical Papers, 2001, 6 January 2015. Available from: http://www.hi-z.com/uploads/2/3/0/9/23090410/7_dev_of_self-powered_pellet_stove.pdf.

408. World Health Organization. Household air pollution and health, fact sheet #292, 2014, March 2014. Available from http://www.who.int/mediacentre/factsheets/fs292/en/.

409. Global Alliance for Clean Cookstoves. Available from http://cleancookstoves.org/.

410. U.S. DOE Office of Energy Efficiency and Renewable Energy. Manufacturing energy and carbon footprints (2010 MECS), 15 April 2014. Available from: http://energy.gov/eere/amo/manufacturing-energy-and-carbon-footprints-2010-mecs.

411. Hill J.M. *Study of Low-Grade Waste Heat Recovery and Energy Transportation Systems in Industrial Applications*. The University of Alabama; 2011. Available from http://acumen.lib.ua.edu/content/u0015/0000001/0000628/u0015_0000001_0000628.pdf.

412. Hi-Z Technology. HZ-14 thermoelectric module. hi-z.com, 2014, 30 April 2015. Available from http://www.hi-z.com/uploads/2/3/0/9/23090410/hz-14.pdf.

413. Stevens J.W. Optimal design of small $\Delta T$ thermoelectric generation systems. *Energ Convers Manage* 2001;42(6):709–720.

414. Lamonica M. A thermoelectric generator that runs on exhaust fumes. *IEEE Spect*. 2014. Available from: http://spectrum.ieee.org/energywise/green-tech/conservation/a-thermoelectric-generator-that-runs-on-exhaust-fumes.

415. Lee F.Y., Navid A., and Pilon L. Pyro electric waste heat energy harvesting using heat conduction. *Appl Thermal Eng* 2012;37:30–37. Doi:10.1016/j. applthermaleng.2011.12.034.

416. Aladayleh W., and Alahmer A. Recovery of exhaust waste heat for ICE using the beta type stirling engine. *J Energ* 2015. Available from: http://downloads.hindawi.com/journals/jen/2015/495418.pdf.

417. Johansson M.T., and Söderström M. Electricity generation from low-temperature industrial excess heat—An opportunity for the steel industry. *Energ Effic* 2013;7(2):203–215. Doi: 10.1007/s12053-013-9218-6.

418. Vuarnoz D., Kitanovski A., and Gonin C. Quantitative feasibility study of magneto caloric energy conversion utilizing industrial waste heat. *Appl Energ* 2012;100:229–237. Doi: 10.1016/j.apenergy.2012.04.051.

419. Kitanovski A., Plaznik U., Tomc U., and Poredos A. Present and future caloric refrigeration and heat-pump technologies. *Int J Refrigerat* 2015;57:288–298. Available from http://www.sciencedirect.com/science/article/pii/S0140700715001759.

420. Bruck E., Dung N.H., Ou Z.Q., Caron L., Zhang L., Buschow H.J., de Wijs G.A., and de Groot R.A. Magneto caloric materials: Not only for cooling applications. *Presented at Delft Days on Magneto Calorics*, The Netherlands, Delft University; 2011. Available from: http://www.tnw.tudelft.nl/fileadmin/Faculteit/TNW/Images/Monday_Orall-Bruck.pdf.

# Index

For Product Safety Concerns and Information please contact our EU
representative  GPSR@taylorandfrancis.com
Taylor & Francis Verlag GmbH, Kaufingerstraße 24, 80331 München, Germany